WATER MANAGEMENT IN THE '90S
A TIME FOR INNOVATION

Proceedings of the 20th Anniversary Conference

Sponsored by the Water Resources
Planning and Management Division
of the American Society of Civil Engineers

Co-sponsored by the
American Consulting Engineers Council
American Water Resources Association
American Water Works Association
Bureau of Reclamation, Pacific Northwest Region
Canadian Society of Civil Engineers
Canadian Water Resources Association
Federal Emergency Management Agency
King County
Northwest Power Planning Council
St. Martin's College
Seattle City Light
Seattle University
Seattle Water Department
Society of American Military Engineers, Seattle and Portland
University of Washington, Department of Civil Engineering
U.S. Army Corps of Engineers, Seattle, Portland,
    Walla Walla, and North Pacific Division
U.S. Geological Survey
Washington Hydrologic Society
Washington State Department of Ecology

Hosted by the Tacoma-Olympia Section, ASCE

Seattle, Washington
May 1-5, 1993

Edited by Katherine Hon

Published by the
American Society of Civil Engineers
345 East 47th Street
New York, New York 10017-2398

ABSTRACT

This proceedings, *Water Management in the '90s: A Time for Innovation*, contains papers presented at the 20th Anniversary conference of ASCE's Water Resources Planning and Management Division held in Seattle, Washington, May 1-5, 1993. The conference theme is an acknowledgement of the need for water resources professionals to face major challenges in managing diminishing supplies for ever-increasing demands, in protecting valuable watersheds from urban and agricultural pollution, and in building and maintaining critical infrastructure with limited financial resources. The papers in this proceedings reflect a practical, problem-solving focus with emphasis on novel solutions for current and near future challenges. Included are papers from three symposia held as part of the conference: 1) Urban Runoff and the Environment, 2) Water Supply and Conservation, and 3) the National Drought Study. Other topics covered in this book are: 1) Computer-aided decision support systems; 2) the Endangered Species Act impact on major water systems; 3) international disasters; 4) geographic information systems; 5) global warming; and 6) hydropower planning.

Library of Congress Cataloging-in-Publication Data

Water management in the '90s : a time for innovation : proceedings of the 20th anniversary conference / sponsored by the Water Resources Planning and Management Division of the American Society of Civil Engineers ; co-sponsored by the American Consulting Engineers Council ... [et al.] ; hosted by the Tacoma-Olympia Section, ASCE, Seattle, Washington, May 1-5, 1993 ; edited by Katherine Hon.
    p.cm.
    Includes bibliographical references and index.
    ISBN 0-87262-912-0
    1. Water resources development — Congresses. 2. Water conservation — Congresses. 3. Urban runoff — Congresses. 4. Water — supply — Management — Congresses. I. Hon, Katherine. II. American Society of Civil Engineers. Water Resources Planning and Management Division. III. American Consulting Engineers Council.
TC401.W3626 1993
333.91—dc20   93-3571
                                            CIP

The Society is not responsible for any statements made or opinions expressed in its publications.

Authorization to photocopy material for internal or personal use under circumstances not falling within the fair use provisions of the Copyright Act is granted by ASCE to libraries and other users registered with the Copyright Clearance Center (CCC) Transactional Reporting Service, provided that the base fee of $1.00 per article plus $.15 per page is paid directly to CCC, 27 Congress Street, Salem, MA 01970. The identification for ASCE Books is 0-87262/93.$1+.15. Requests for special permission or bulk copying should be addressed to Reprinting/Permissions Department.

Copyright © 1993 by the American Society of Civil Engineers,
All Rights Reserved.
Library of Congress Catalog Card No: 93-3571
ISBN 0-87262-912-0
Manufactured in the United States of America.

FOREWORD

This volume contains papers presented at the 20th Anniversary Conference of ASCE's Water Resources Planning and Management Division, held in Seattle, Washington May 1 to 5, 1993. The conference theme, Water Management in the '90s: A Time for Innovation, reflects the crossroads where water resources professionals are finding themselves these days. With climates and populations all over the world in upheaval, we stand at the threshold of the 21st century carrying baggage from the Industrial Revolution. We face major challenges in managing diminishing supplies for ever-increasing demands, in protecting valuable watersheds from urban and agricultural pollution, and in building and maintaining critical infrastructure with limited financial resources. The technical sessions in this conference address our biggest problems in water resources planning and management, and the papers have a practical, problem-solving focus. The intent of the conference is to provide a forum where knowledge, experiences, applications, and innovative ideas can be shared.

The conference is organized into multiple daily tracks emphasizing specific topics, such as operations management, computer applications, Pacific Northwest projects, wetlands, water policy, groundwater modeling and management, urban infrastructure, international issues, and computer-aided decision support systems. Also, in recognition of the growing need to communicate critical issues to water management professionals, a 3-day track on issues for water resources executives is included. Papers address the Endangered Species Act impacts on major water systems; customer care and business practices, large-scale project planning, design, and construction; the Northwest power situation, and public involvement in water resources planning.

The conference features three symposia. The Symposium on Urban Runoff and the Environment is jointly sponsored by the Urban Water Resources Committee and the Urban Water Resources Research Council. It is organized under the direction of Dr. John J. Warwick, with the University of Nevada, Reno. This symposium encompasses nine sessions plus several poster presentations, and includes papers on runoff models; environmental impacts; best management practices; combined sewer overflow; and regulatory issues.

The Symposium on Water Supply and Conservation is sponsored by the Task Committee on Water Supply and Conservation, initiated by the Executive Committee. It is organized under the direction of Mr. Mark D. Rothenberg, with the Spring Valley Water Company, New York. This symposium encompasses eight technical sessions and two panel discussions. Paper topics include water availability, demand forecasting, recycling, new ideas, and pricing. The symposium opens with a panel discussion on competing needs for water, and closes with a panel of water suppliers and regulators discussing conservation's role and their responsibilities.

The Symposium on the National Drought Study is organized under the direction of Dr. Richard N. Palmer, with the University of Washington in Seattle. This symposium encompasses four sessions which represent the four cornerstones of the study: the national search for answers, regional demonstration studies, new uses for computers in water governance, and new ways to involve water users in water managment.

The conference is sponsored by the Water Resources Planning and Management Division and hosted by the Tacoma-Olympia Section of ASCE. The co-sponsors are:

American Consulting Engineers Council
American Water Resources Association
American Water Works Association
Bureau of Reclamation, Pacific Northwest Region
Canadian Society of Civil Engineers
Canadian Water Resources Association

Federal Emergency Management Agency
King County
Northwest Power Planning Council
St. Martin's College
Seattle City Light
Seattle University
Seattle Water Department
Society of American Military Engineers, Seattle and Portland

University of Washington, Department of Civil Engineering
U.S. Army Corps of Engineers, Seattle, Portland, Walla Walla, and North Pacific Division
U.S. Geological Survey
Washington Hydrologic Society
Washington State Department of Ecology

The members of the conference steering committee include:

Jack Mowreader, National Conference Chair
Katherine Hon, Technical Program Chair
Mark Killgore, Local Arrangements Chair
Phil Cohen, Local Arrangements Co-Chair
Mike Stansbury, Local Arrangements Co-Chair
John Abdalkhani
Walt Anton
Cass Brotherton
Jack Bjork
Benjamin Carr

Dan Clark
Stephen Hart
Karin Hilding
Diane Hilmo
Jill Marilley
Alan Murray
Cheryl Nakata
Ron Rossmiller
Rick Schaefer

Members of the Executive Committee of the Division are:
William Cox, Chair
Jack Mowreader, Vice Chair
Neil Grigg

Jerry Anderson
Don Phelps, Secretary
Darell Zimbelman, MGD Contact

Each of the papers included in the Proceedings has been accepted for publication by the Proceedings Editor. All papers are eligible for discussion in the Journal of the Water Resources Planning and Management Division. All papers are eligible for ASCE awards. The papers in each track are presented together so that topical continuity is maintained.

The photograph for the cover of these proceedings is the winner of the third annual Water Resources Planning and Management Division student photography contest. Members of all ASCE student chapters and clubs were invited to submit original work photographs appropriate to the theme of the conference, with judging by the sponsoring Membership Development committee, the conference chairs, and the ASCE publications office.

The winning photograph is of the Bonneville Hydroelectric Project. The photograph was taken and submitted by Keith G. Kaase, of University of Houston, who received the grand prize of a free conference registration. Five other entries received Honorable Mention: Amy Behrman (first runner-up, Cooper Union), Allen Deur (Ohio Northern University), Matthew W. Shiley (Cadet, Virginia Military Institute), Chris Smith (Virginia Tech) and Rebecca L. Boyden (Virginia Tech).

An effort such as this does not happen in a vacuum. Many people contribute to a successful conference. The Technical Program Chair would like to thank the local organizers, committee participants, session moderators, all of the presenters, and the ASCE staff, especially Julie Taylor, for their excellent work. Special debts are owed to Jack Mowreader for his innovative ideas, and the Benjamin Carr for his contributions to an outstanding conference brochure. I would like to thank my husband, Steve, for his calming influence and patient support. Also, the support of management at Ogden Environmental and Energy Services Company, especially David Potter, is gratefully acknowledged. The technical program could not have been compiled without the word processing and database management of LaTasca Morgan.

This volume is dedicated to William K. Johnson, mentor, teacher, role model, and friend, for bringing me into ASCE ten years ago as News Correspondent. Participation in this Division is the most rewarding aspect of my engineering career.

Katherine Hon
Technical Program Chair
Ogden Environmental and Energy Services Company
San Diego, CA

CONTENTS

Plenary Session
Moderator: William Cox

Reflections on the Beginnings of the Water Resources Planning and Management Division
Victor A. Koelzer .... *
The Water Resources Planning and Management Division: Now and 20 Years from Now
Kyle Schilling .... *
Senior Policy Advisor for Senator Slade Gorton
Don Moos .... *

Session A1
ENDANGERED SPECIES ACT IMPACTS ON MAJOR WATER SYSTEMS
Moderator: John W. Keys, III

Resource Implications of Listing Columbia River Basin Salmon Stocks Under the
Endangered Species Act
John E. Velehradsky 1
Economic Impacts of Columbia River Salmon Recovery: An Industry Prespective
Glen Vanselow .... *

Session A2
NATIONAL WATER POLICY ISSUES
Moderators: Greg Ten Eyck and Jerry R. Rogers

Panel Discussion
Jerry R. Rogers, James R. Hanchey, Darrell Zimbleman, Neil Grigg, Conrad G. Keyes, and
Walter Anton .... *

Session A3
INTERNATIONAL DISASTERS AFFECTING WATER RESOURCES
Moderator: Captain John Callahan

Mount St. Helens Recovery Project
John Sager and Chris Budai 4
Mt. Pinatubo Eruption Recovery Project
Michael P. Roll .... *
Landers Earthquake
Clarence Allen .... *

Session A4
CUSTOMER CARE AND BUSINESS PRACTICES
Moderator: John J. Healey

Total Quality Management and Partnering in the Public Sector
Tom Warne .... *
A/E Partnership
Stan Moon .... *

---

*Paper not available at the time of printing.

Session A5
CASE STUDIES IN WATER RESOURCES PLANNING
Moderator: Glenn Johnson

New Perspectives in Large Water Supply System Planning
  William Whipple, Jr. .... *
Basin Scale Management of Surface and Ground Water
  John C. Tracy and Munjed Al-Sharif 8
Fresno/Clovis Water Management Plan
  William E. Burmeister and Dave Peterson 12
Urban Water Planning—A Western Nevada Case Study
  John W. Fordham 16
Water Management Plan for Lake Okeechobee
  James W. Vearil 20

Session A6
LARGE-SCALE PROJECT PLANNING
Moderator: G. Edward Dickey

Los Angeles County Drainage Area (LACDA) Review
  Col. Robert L. Van Antwerp .... *
Planning of the Eastside Reservoir Project
  Dennis Majors, Glen Rockwell, and Bruce Ainsworth .... *
Panama Canal: Commodity and Traffic Projections Study
  Paul Sorenson .... *
A Computer Model for Multiplant Hydroelectric System Simulation
  S. O. Russell and D. A. Fayegh 24

Session A7
WATER RESOURCES PLANNING AND PROJECT MANAGEMENT CHALLENGES
Moderator: A. Wayne Pietz

Licensing and Permitting the Devil's Nose Project Under the 1990's Regulatory Conditions
  James E. Alverson and Roderick E. Schuler 30
FERC Third-Party EIS Contracting for Hydroelectric Project Licensing—Initial Indicators
  Robert A. Mohn and Wayne Dyok .... *
Managing Government Contracts: Tips to Keeping Your Sanity
  Lisa T.M. Vomero 34
Developing Power-Presentation Skills
  Ralph Pehrson 38

Session A8
DESIGN CHALLENGES IN WATER RESOURCES
Moderator: Steve Stockton

Ice Harbor Juvenile Fish Bypass Facilities
  Mark Lindgren and Brian Grant .... *
Seven Oaks Dam and Outlet Works
  Dave Illias and Glenn Ellis .... *
Recent Hydro Power Developments in the Dominican Republic
  Gerry Millar .... *

---

*Paper not available at the time of printing.

Session A9
CONSTRUCTION LESSONS LEARNED
Moderator: Carl F. Enson

A New Bonneville Navigation Lock—Results of a Partnered Project
   Col. Charles A. W. Hines and Howard Jones .... *
Loss and Rebuilding of the Lake Washington Floating Bridge
   Robert A. Josephson and Chuck Ruth .... *
New Waddell Dam Construction
   Joseph L. Ehasz .... *

Session A10
NORTHWEST POWER SITUATION
Moderator: Jack M. Mowreader

Northwest Power Situation
   Robert Myers .... *
Fish and Power: The Electric Energy Future of the Northwest
   Randall W. Hardy .... *
A Regional Planning Perspective
   Tom Trulove .... *

Session A11
PUBLIC INVOLVEMENT IN WATER RESOURCES PLANNING
Moderator: James Vearil

The Rewards of Public Involvement in Water System Planning
   Ikuno Masterson .... *
Partnering in Water Resources Planning in Florida
   Adil J. Salem 39
Public Involvement in the Design for Improving Spectacle Island
   Sallye E. Perrin .... *
Public process in Fresno/Clovis Water Supply Plan
   Paula Farris 43

Session B1
VALUING ENVIRONMENTAL SYSTEMS
Moderator: Claire Welty

Valuing Resources in New England's Merrimack River Basin
   Drew Parkin .... *
Evaluating Environmental Projects
   William Hansen 47
Finding the True Cost of Power: Incorporating Environmental Costs into Electric Utility Decision Making
   Shepard C. Buchanan .... *
Nonmarket Economic Values for Salmon and Wildlife Resources
   Darryll Olsen 51

Session B2
ENVIRONMENTAL ECONOMICS IN WATER PANNING
Moderator: Robert Hamilton

Operating the Columbia River: A Balancing Act
   David B. Smith 55
Economics of Endangered Salmon in the Pacific Northwest
   Joel R. Hamilton 59

---

*Paper not available at the time of printing.

Glen Canyon: The Economic Costs
  David A. Harpman, Timothy J. Randle, and S. Clayton Palmer 63
The Economic Value of Trout Fishery Management Programs
  Donn Michael Johnson 67

Session B3
INSTREAM FLOW ISSUES
Moderator: Marshall Flug

Meeting Instream Flow Needs of Lower Colorado River in Texas
  Quentin W. Martin 71
Managing Instream Flows for Salmonid Spawning Habitat
  Jeffrey B. Bradley 75
Managing Reservoir Storage for Instream Flow
  Terry Waddle 79
Natural, Cultural, and Environmental Resource Values Influencing Colorado River Basin Management
  Marshall Flug and William L. Jackson 83

Session B4
INNOVATIVE WETLANDS ENGINEERING AND DESIGN
Moderator: Glenn G. Gregory, Jr.

A Knowledge Based Expert System Engineering and Construction in Wetlands
  J. Spigolon .... *
Wetland Engineering, Design, and Construction: State-of-the-Science and Research Needs
  Donald F. Hayes, Timothy A. Crockett, and Michael T. Arends 88
Juanita Creek Channel Restoration, King County, Washington
  Jack C. Bjork and A. William Way 92
Double Rock Park Stream Restoration Project
  James W. Gracie, Candace Szabad, and Raymond I. Green 96
Use of Flood Control Features for Environmental Mitigation
  J. Craig Fischenich, E. A. Dardeau, Jr. and Kenneth D. Parrish 101

Session B5
WETLANDS INNOVATION AND THE ENVIRONMENT
Moderator: Mary C. Landin

Evaluation of the Weaver Bottoms Restoration Project on the Upper Mississippi River
  Mary M. Davis .... *
An Isolated Wetland Used for Stormwater Treatment
  Betty Rushton and David Carr 105
Innovative Wetlands in Urban Settings Using Dredged Material
  Mary C. Landin and T. R. Patin 109
Innovative Wetland Re-creation in Urban Settings
  C. Lynn Miller and George G. Feher 113

Session B6
WATER TEMPERATURE MANAGEMENT FOR THE 90's
Moderator: Steve Railsback

Water Temperature Issues in the 90's and Beyond
  S. F. Railsback 117
Field Comparison of Temperature Models and Study Design Considerations
  Kent Doughty .... *
Recent Developments in Determining Thermal Requirements of Fresh Water Fish
  Kate Sullivan .... *

---

*Paper not available at the time of printing.

Sacramento and Trinity River Systems: A Demonstration of River Temperature Management
for the 90's
    Perry Johnson and Chet Bowling 122

Session B7
BIOLOGICAL TREATMENT FOR WATER QUALITY IMPROVEMENT
Moderator: Chuck Findley

Development of Bioretention Practices for Stormwater Management
    Larry Coffman, Raymond Green, Michael Clar, and Susan Bitter 126
Design Considerations Associated with Bioretention Practices
    Larry Coffman, Raymond Green, Michael Clar, and Susan Bitter 130
Pollutant Removal Efficiency of Detention Basins
    A. Osman Akan 134
Long-term Performance of Wet Detention Ponds
    Agustin E. Maristany 138

Session B8
TOOLS AND MODELS USED IN WATER MANAGEMENT DECISIONS ON
THE COLUMBIA-SNAKE RIVER SYSTEM
Moderator: Lt. Col. Robert D.G. Volz

Biological Modeling in the Columbia Basin: An Organized Approach to Dealing
with Uncertainty
    Willis E. McConnaha 142
Managing Snake River Operations for Juvenile Salmon Migration
    Charles M. Brendecke 147
Application of the HEC Prescriptive Reservoir Model in the Columbia River System
    Richard Hayes, Michael Burnham, and David T. Ford .... *
Columbia River System Operations-Water Quality Assessment
    R. G. Willey, Bolyvong S. Tanovan, and Donald J. Smith 151
Hydro Models and Salmon Recovery in the Northwest
    Ken Dragoon 156

Session B9
ENDANGERED SALMON AND PACIFIC NW WATER IMPACTS
Moderator: Peter Klingeman

Snake River Salmon Recovery Plan Development
    Peter C. Klingeman and James Litchfield 160
Columbia River Basin Fish and Wildlife Program Strategy for Salmon
    James Ruff and John Fazio 164
BPA Efforts to Protect Salmon and Supply Energy
    Jim Geiselman and Roger Schiewe 168
Snake Reservoir Drawdown: A Brief Progress Report
    John J. Pizzimenti, Kevin Malone, Paul Tappel and Brain Sadden 169
Role of Economics in Endangered Species Act Activities Related to Snake River Salmon
    Edwin J. Woodruff and Daniel D. Huppert 173

Session B10
WORKSHOP ON WATER MANAGEMENT FOR
HYDROELECTRIC OPTIMIZATION
Moderator: Charles D. D. Howard

---

*Paper not available at the time of printing.

Session B11
WATER QUALITY MODELING
Moderator: Peter Schwartzman

Water Quality Impacts from Development of Port Ludlow, Washington
  Thomas Smayda, Clain Jones, Martin E. Harper and David Cuningham 177
Multiobjective Models for Determining Fresh Water Inflows to Bays and Estuaries
  Ning Mao and Larry W. Mays 181
City of Federal Way Panther Lake Surface Water Study: A Case Study of HSPF Modeling
in Urban Detention Basins
  Andrew B. Lukas and Cary M. Roe 185
Water Quality Model of the Chicago Waterway
  Bill Macaitis and Carl Johnson 189
Simulation of Water Quality Processes in the Chicago Waterway and Upper Illinois
River Systems
  Bill Macaitis, John Variakojis, and Gary Mercer 193

Session C1
PRACTICAL APPLICATIONS OF OPTIMIZATION IN WATER RESOURCE PROJECT OPERATION
Moderator: Aris P. Georgakakos

Experience with Optimization for Power Operations
  Charles Allen, Gregory Ott, and Richard Hartsock .... *
Optimization of Operations—The Manitoba Hydro Experience
  A. D. Cormie and P. E. Barritt-Flatt 197
HEC Prescriptive Reservoir Model
  Robert D. Carl .... *
Impacts of Hydropower Operation on Water Supply from Lower Colorado River in Texas
  Quentin W. Martin 201
Optimization of Operations of a Multi-Reservoir Water Distribution System
  Ali Diba, W. F. Louie, Manouchehr Mahjoub, and Williams W. G. Yeh 205

Session C2
ENVIRONMENTAL AND SOCIAL ISSUES IN WATER RESOURCES PROJECT MANAGEMENT
Moderator: Russell George

Columbia River System Operation Review
  Ray Jaren .... *
Updating Missouri River Master Water Control Manual to Meet Changing Meeds
  Duane J. Sveum .... *
Endangered Species Act Impact on Operation of Snake River Projects
  Bruce Glabau .... *
Reliability Curves for Reservoir Performance
  Loren Jangaard 212

Session C3
OPTIMIZATION VS. SIMULATION IN MULTIPURPOSE RESERVOIR MANAGEMENT
Moderator: George McMahon

Panel Discussion
  Charles D. D. Howard, Bill Eichert, Aris Georgakakos, and George McMahon .... *

---

*Paper not available at the time of printing.

Session C4
PRACTICAL APPLICATIONS OF GEOGRAPHICAL
INFORMATION SYSTEMS
Moderator: Walter Grayman

Geographic Information System Application: Water Right Places of Use as Polygon Attributes
  Tim Reierson and Hank Riddle 216
GIS Reconstruction of Exposure to Water Supply Contaminants
  Margrit von Braun, Ian von Lindern and Jim C. P. Liou 220
GIS Application for Urban Nonpoint Source Pollution Assessments
  Eric D. Loucks and Hans J. Peterson .... *
Computerized Data Processing and Geographic Information Systems Applications for
Development of a 3-Dimensional Groundwater Flow Model
  Shih-Huang Chieh, Mark V. Cromer, and William R. Swanson 224
Using Automated Mapping/Facilities Management to Manage Change
  Benjamin I. Carr .... *

Session C5
GEOGRAPHICAL INFORMATION SYSTEMS IN WATER RESOURCES:
EDUCATION AND CURRICULUM
Moderator: Daene C. McKinney

Panel Discussion
  James P. Heaney, Hugo Loaiciga, Walter Grayman, Daniel P. Loucks, and A. J. Fredrich .... *

Session C6
STREAMFLOW FORECASTING FOR RESERVOIR OPERATIONS:
CASE STUDIES
Moderator: Lindell E. Ormsbee

Stochastic Forecasts in Reservoir Planning and Operations
  Charles D. D. Howard 228
Characterization of Uncertainty in Reservoir Operations
  Lindell E. Ormsbee 231
Seasonal and Weekly Streamflow Forecasts with Modest Data Requirements
  Gary Freeman, Jan Grygier, Jery Stedinger, and Bin Zhang .... *
Operating the Savannah River System Using Multiple ESP Forecast
  Brian McCrodden, James B. Atkins, Dan Scheer, and Dean Randall .... *

Session C7
STREAMFLOW FORECASTING FOR RESERVOIR OPERATIONS: ISSUES
Moderator: Juan Valdés

Hydrologic Forecasting: What are the Issues?
  L. E. Brazil, D. P. Laurine, G. N. Day, and J. B. Valdés 235
Using Forecasts to Improve Reservoir Operations
  Jery R. Stedinger, Mohammed Karamouz, Brian McCrodden, George McMahon, Richard Palmer,
  and Juan Valdés 240

Session C8
DEVELOPING AND IMPLEMENTING DECISION SUPPORT SYSTEMS I
Moderator: Steven C. Chapra

The Development and Use of Policy Statements in River Simulation Models
  Jon S. Behrens, Daniel P. Loucks and Hussam Fahmy 244
The Law of the River: Policy Expression
  Hussam Fahmy, Jon Behrens, and Kenneth Strzepek 248

---

*Paper not available at the time of printing.

Optimal Model-predictive Real-time Control of Canals and Rivers
  Edith A. Zagona and David E. Clough .... *
Advanced Decision Support Systems for Environmental Simulation Modeling
  Steven C. Chapra, Jean M. Boyer and Robert L. Runkel 252
Can We Get More Benefits From Our Data
  David R. Sieh and Rene Reitsma .... *

Session C9
DEVELOPING AND IMPLEMENTING DECISION SUPPORT SYSTEMS II
Moderator: Daniel P. Loucks

The Modular Hydrologic Modeling System
  Pedro J. Restrepo, George H. Leavesley, Mike Dixon, Linda G. Stannard, and Tom Ryan .... *
Solute Transport Modeling Under Unsteady Flow Regimes: An Application of the Modular Modeling System
  Robert L. Runkel and Pedro J. Restrepo 256
Object Oriented River System Simulation
  Jon S. Behrens and Daniel P. Loucks 260
RSS: A Construction Kit for Visual Programming of River Basin Models
  Rene Reitsma, Andy Sautins, and Steve Wehrend 264
Integration of Real-time Hydrologic, Hydraulic and Reservoir Operation Models with a Geographic Information System: The Han River Case Study
  Pedro J. Restrepo, Edith A. Zagona, Mark Jourdan, and Juan Valdés .... *

Session C10
DEALING WITH UNCERTAINTY AND RISK
Moderator: George W. Ploudre

Managing Large-scale Water Systems under Conditions of Uncertainty
  Marvin Waterstone, Lucien Duckstein, and Donald Davis .... *
Stochastic Reservoir Operation under Drought with Fuzzy Objectives
  E. Parent and L. Duckstein 268
Seasonal Water Quality Management Given Sparse Data
  Andrews K. Takyi and Barbara J. Lence 272
Imprecise Probability and Water Resources Decisions
  W. F. Caselton and W. Luo 276
Risk Reduction Approaches for Hydrologic Forecasting System Development
  Lynn E. Johnson 280

Session C11
MODELING WATER QUALITY AND QUANTITY WITH COMPUTER-AIDED DECISION SUPPORT SYSTEMS
Moderator: Dan Clark

Development of a Raw Water Master Plan Model to Assist Long Range Drought and Water Conservation Planning for the City of Louisville, Colorado
  Noel Hobbs, Thomas Charles, Jeffrey Carpenter, and Brad Eaton 284
Application of Computer Support for Water Quality Management
  John F. DeGeorge, Stephen A. Breithaupt, Robert R. Klamt, and Theresa Wistrom 288
A Federated Water Database Management System for the South Platte River Basin of Colorado
  R. Craig Woodring and Timothy K. Gates .... *
CASS for Evaluating Hg Contamination in Clear Lake, CA
  A. E. Bale, P. L. Shrestha and G. T. Orlob 292
Graphics User Interface for Water Quality Model Calibration
  Stephen A. Breithaupt, John F. DeGeorge, and Gerald T. Orlob 296

---

*Paper not available at the time of printing.

Session D1
THE ROLE OF RESEARCH IN WATER RESOURCES PLANNING AND MANAGEMENT
Moderator: A. J. Fredrich

Research in Retrospect
  Randy Hanchey .... *
The Role of Research in Addressing Social and Environmental Objectives
  Roger T. Kilgore .... *
Research Needs in Water Resources Planning and Management
  Neil S. Grigg 300

Session D2
EVOLUTION OF COMPUTERS IN WATER RESOURCES PLANNING AND MANAGEMENT
Moderator: Richard M. Males

Remarks on Computer-Use History in Water Resources
  Leo R. Beard 304
A Look Ahead
  Richard N. Palmer .... *
The Path Not Taken: Implications of, and Alternatives to, Computer Usage
  David T. Ford and William Johnson .... *

Session D3
WATER RESOURCES EDUCATION
Moderator: James R. Groves

An Historical Perspective on Water Resources Education
  L. Douglas James 308
A Practitioner's Perspective on Water Resources Education
  Walter M. Grayman .... *
Future Directions in Water Resources Education
  Stuart G. Walesh 312

Session D4
INSTITUTIONAL MECHANISMS FOR MANAGING WATER RESOURCES
Moderator: Randall Parsons

Evolving Institutions for Managing the ACF Basin
  Andrew Dzurik, Wayne Hall, and Steve Leitman 316
Section 404 Under Seige; The Courts May Assure a More Balanced Approach
  William R. Walker and Elizabeth Crumbly .... *
Water Management by Endangered Species
  George R. Baumli 320
Innovation and Diversification—The Key to Our Future Water Supply
  Albert Muniz, R. David G. Pyne and Sharon M. Trost 324
Local Agency Regulation: The Policies and Process of Surface Water Protection
  Richard L. Schaefer and Jeffrey H. Stern 328

Session D5
PRIVATIZATION: INNOVATIVE WATER FINANCING FOR THE 90'S
Moderator: Ben Frerichs

Panel Discussion
  Ben Frerichs, Paul Eisenhardt, and Roger Wagner, Esq. .... *

---

*Paper not available at the time of printing.

Session D6
WATER TRANSFERS AND WATER MARKETING
Moderator: John T. Scott

Ag-to-Urban Water Transfer in California: Win-Win Solutions
   Lee A. Jacobi and Robert L. Carley 332
Opportunity and Proposals for Water Marketing within the Central Valley Project, California
   Michael L. Delamore .... *
Water Supply Systems Planning with Water Transfers
   Jay R. Lund and Morris Israel 336
Water Allocation: Matching Supply With Demand
   John F. Scott 340

Session D7
INNOVATION IN THE ASCE MODEL STATE WATER CODE
Moderator: Michael C. Sciacca

The Once and Future ASCE Model Water Allocation Law
   Ray Jay Davis 344
From Riparianism to Water-Use Permitting: Issues Confronting State Government
   William E. Cox 348
Instream Flows According to the ASCE Model State Water Code
   George William Sherk and Berton L. Lamb 352
Water Marketing in Washington: The Next Step?
   Donald Phelps 356

Session D8
URBAN DRAINAGE REHABILITATION
Moderator: Bill Macaitis

Optimal Planning of Combined Sewer Rehabilitation
   Santiago M. Reyna and Jorge A. Vanegas 360
City of Austin, Texas Drainage Management Cost-of-Service Study
   George E. Oswald, Nilo Priede, Joe Pantalion, David Strychalski, and Frank Houston .... *
Collection System Inspection and Rehabilitation Program
   Bill Macaitis and Amreek Paintal 364
Estimating Maintenance Cost of Existing Stromwater Retention Ponds
   Robert E. Molzahn 368

Session D9
RESERVOIR SYSTEMS AND FLOOD CONTROL
Moderator: Richard J. DiBuono

Wynoochee Lake and Dam Flood Storage Reevaluation Study
   Christopher J. Lynch 372
Real-Time Water Control System for the Trinity River, Texas
   David T. Ford and J. Russell Killen 376
Legal Basis for Reservoir Water Control Decisions: Assessing Conflicts and Opportunities
   William K. Johnson and Richard J. DiBuono .... *
Hydrologic Analysis of Leveed Interior Areas
   Harry W. Dotson and Michael W. Burnham 380
A Derived Flood Frequency Distribution for Ungaged Catchments
   Rafael S. Seoane and Juan B. Valdés 384

---

*Paper not available at the time of printing.

Session D10
PIPELINES-THEORY AND PRACTICE
Moderator: J. R. Fotheringham

Global Approaches for the Nonconvex Optimization of Pipe Networks
G. V. Loganathan and J. J. Greene 388
Optimal Operation of Looped Water Distribution Networks for Public Health and Hydraulic Objectives
James Uber, Cheri Bush, Mao Fang, and Ken Hickey .... *
Pipe Network Optimisation Using Genetic Algorithms
Angus R. Simpson, Laurie J. Murphy, and Graeme C. Dandy 392
Emergency Repair of Point Loma Ocean Outfall
Gregory McBain .... *
Alternative Designs for Permanent Protection of the San Luis Rey River Aqueduct Crossings
Chenchayya T. Bathala, E. Morris McClung, Ergun Bakall, and W. Jeffery Moncrief 396

Session D11
WATER DISTRIBUTION SYSTEM PERFORMANCE ASSESSMENT
Moderator: Arun K. Deb

Water Distribution System Performance Assessment Methodologies
Arun K. Deb .... *
Optimal Rehabilitation Model for Water Distribution Systems
Joong Hoon Kim and Larry W. Mays 400
Water Distribution System Research Needs
Gregory J. Kirmeyer and William G. Richards .... *
Non Contact, Remote Sensing of Buried Water Pipeline Leaks Using Infrared Thermography
Gary J. Weil 404

Session E1
TAKING A HOLISTIC VIEW
Moderator: Warren Viessman, Jr.

Watershed Management at Three Governmental Levels
Peter E. Black 408
Multiple Objective Planning: Emerging Trends and Pressures for Change
Kyle E. Schilling .... *
Urban Water Resources Planning: Balancing Demand and Supply
Duane D. Baumann .... *
Water Resources Decision Making in the 90's: A New Game Plan
Nancy C. Lopez and Holly E. Stoerker .... *
A Watershed Approach to Water Quality Criteria
Scott J. Kenner 412

Session E2
GLOBAL WARMING AND HYDROLOGIC VARIABILITY
Moderator: Hugo A. Loaiciga

The Impact of Global Warming on Water Resources: Overview and Implications for Coastal California
Jeffrey Garvey 416
Impacts of El Niño in Western Washington
Thomas D. Murphy and V. Bruce Sandoval 421
A Methodology for the Evaluation of Global Warming Impact on Soil Moisture and Runoff
Juan B. Valdés, Rafael S. Seoane, and Gerald R. North 425

---

*Paper not available at the time of printing.

A Stochastic Approach for Assessing the Effect of Changes in Regional Circulation Patterns
on Flood Frequency
  James P. Hughes, Dennis P. Lettenmaier, and Peter Guttorp .... *
Global Ground Water Fluxes and Greenhouse Warming
  Hugo Loaiciga and Igor S. Zekster .... *

Session E3
WATER SUPPLY PLANNING AND SOURCE PROTECTION: CONSERVING AND PRESERVING OUR PRECIOUS RESOURCES
Moderator: Teri Liberator

Integrated Water Resource Planning—A Northwest Case Study
  Jane Evancho and Kelly Lange 429
Source Protection for Surface Water Supplies
  Richard W. Robbins .... *
Ground Water Management Standards and Protection Tools
  Jon D. Witten 433
A Planning Strategy for Safe Drinking Water Act Compliance
  Scott Trusler 440

Session E4
MAJOR ACCOMPLISHMENTS OF THE NATIONAL DROUGHT STUDY
Moderator: Richard N. Palmer

National Study of Water Management During Drought: Results Oriented Water
Resources Management
  William J. Werick 445
The Rediscovery of Planning Principles
  Gene Stakhiv and Myron Fiering .... *
Empowering Stakeholders Through Simulation in Water Resources Planning
  Richard N. Palmer, Allison M. Keyes and Selene Fisher 451
The National Drought Atlas
  James R. Wallis 455

Session E5
REGIONAL CASE STUDIES: TESTING THE PHILOSOPHY AND METHODS
Moderator: William Werick

The Potomac Experience: A Forerunner of DPS
  Roland C. Steiner 463
Drought in the Emerald City
  Steven D. Babcock 467
The Challenges of Interstate Water Planning and Management
  Michael J. Bart and Christopher R. Erickson 471
The James River Case Study
  Thomas J. Lochen 475

Session E6
DEMONSTRATION OF OBJECT ORIENTED SIMULATION MODELS
Moderator: Allison M. Keyes

The Role of Object Oriented Simulation Models in the Drought Preparedness Studies
  Allison M. Keyes and Richard N. Palmer 479
Demonstrating Competition for a Limited Resource in the Cedar/Green Basins
  Christopher J. Lynch 483

---

*Paper not available at the time of printing.

Use of Interactive Simulation Environment to Model the Marais des Cygnes-Osage River Basin
  Kevin Low and Chris Erickson .... *
Including Expert System Decisions in a Numerical Model of a Multi-Lake System Using STELLA
  James M. Stiles and Richard E. Punnett 487
Multiparty Model Development Using Object-Oriented Programming
  Daniel N. Nvule 491

Session E7
BROADENING THE PLANNING PROCESS
Moderator: William Werick

Bringing People, Policies, and Computers to the Water (Bargaining) Table
  Richard E. Punnett and James M. Stiles 495
Circles of Influence
  Robert Waldman .... *
Politics and Drought Planning: Friends or Foes?
  Bruce D. McDowell and William Blomquist 498
Citizen Participation in Water Supply System Planning and Management
  William G. Elliott, Eileen R. Simonson, Alexandra D. Dawson, and Robie O. Hubley 502
Living with Public Involvement Ideals in the Real World
  Merle Lefkoff .... *

Session E8
WORKSHOP ON DAM SAFETY
Moderator: Mike Pavone

Session E9
CANADA AND U.S. ENVIRONMENTAL WATER POLICY AGREEMENTS: HOW THEY WORK, WHEN THEY WORK
Moderator: Barbara Lence

Water Resources Policy and Management in a Multi-jurisdictional Setting: A Great Lakes Case Study
  M. J. Donahue .... *
Managing Boundary and Trans-boundary Waters from an IJC Perspective
  A. L. Hamilton .... *
Analysis of Operating Criteria: Multiple Lakes at Voyageurs National Park
  Marshall Flug and Larry W. Kallemeyn 506
Rebuilding Salmon Runs: Lessons from the Canadian Experience
  R. Hilborn .... *

Session E10
WATER RESOURCES ISSUES AROUND THE WORLD
Moderator: Harold J. Day

Water Resources Management Issues Confronting Developing Countries
  Yin Au-Yeung 510
Technical and Organizational Constraints on Surface and Groundwater Irrigation in Small-Holder Areas of South Asia
  Donald E. Campbell 514
Monthly Water Balance for Blue Nile River Basin in Ethiopia
  Peggy A. Johnson and P. Douglas Curtis 518
Hydrologic and Structural Considerations for the Initial Filling of Jiguey Dam, Dominican Republic
  Guy S. Lund and Ed A. Toms 522
Geothermal and Hydropower Production in Iceland
  Duane J. Rosa 526

---

*Paper not available at the time of printing.

Session E11
RECENT U.S. AND CANADIAN WATER QUALITY INITIATIVES IN THE GREAT LAKES AND PACIFIC NORTHWEST
Moderator: Robert J. Montgomery

The North East Wisconsin (NEW) Waters for Tomorrow Initiative
 Paul E. Thormodsgard and Harold J. Day 530
The Rouge River National Wet Weather Demonstration Project
 John Bona and James E. Murray .... *
Developing an Ecosystem Approach for Restoration of the Bay of Quinte
 Murray German 531
The Puget Sound Water Quality Plan
 Nancy McKay .... *
The Fraser Basin Management Program: Water Quality Management and Sustainable Development
 Tony Dorsey .... *

Session F1
HYDROPOWER PLANNING, LICENSING, RELICENSING AND DESIGN IN AN ERA OF POLICY TRANSITION
Moderator: Neil Macdonald

Dealing with Public Perceptions of Flooding and Other Hydrologic Issues During the Relicensing Process
 Virginia Howell and Robert S. Barnes .... *
Plant Modernization and Its Role in the Efficient Use of Water Resources
 Michael J. Haynes 534
Developing a Workable Public Input Process for Aesthetics and Recreational Needs During Hydropower Licensing
 Deborah Howe and Mike Stimac 538
Identifying the True Issues of Hydropower Resource Development in an Era of Public and Regulatory Policy Transition
 Neil Macdonald and Mike Stimac 543
Monroe Street Hydroelectric Project
 Steve Silkworth and Michael Finn .... *

Session F2
IMPLEMENTATION OF COORDINATED WATER SYSTEM PLANNING IN NORTH SNOHOMISH COUNTY AND ASSOCIATED ADJACENT AREAS: AN OVERVIEW
Moderator: Jacqueline E. Hightower

"SNO WATER" Led Coordinated Water System Planning
 Pat Burnaroos 548
3 Counties Coordinated Water System Plans
 Jacqueline Hightower 551
A Satellite Water System Program in Snohomish County, Washington
 Mark D. Spahr 555
The Department of Health's Role in Implementing the Coordination Act in North Snohomish County
 Nancy Feagin 559

Session F3
CSO CONTROL IMPLEMENTATION IN SEATTLE METROPOLITAN AREA
Moderator: Joe Talbot

Environmental Citizen's Group Perspective
 Kathy Fletcher .... *

---

*Paper not available at the time of printing.

City of Seattle Perspective
Bob Chandler .... *
Municipality of Metropolitan Seattle Perspective
Laura Wharton .... *
Consultant Perspective
Steve Merrill .... *

Session F4
GROUNDWATER MODELING AND MANAGEMENT
Moderator: Miguel A. Mariño

Ground-water Modeling as an Integrating Factor: A Case Study
James O. Rumbaugh, III and Seth E. Matters .... *
Ecological Considerations in Groundwater Management
Eugene B. Yates 563
Coupled Simulation-Optimization Approach to Wellhead Protection Area Delineation to Minimize Contamination of Public Ground-Water Supplies
John M. Shafer and Mark D. Varljen 567
Effects of Canal Lining on the Ground-water Quality in the Imperial Valley
S. Alireza Taghavi and Young S. Yoon .... *

Session F5
GROUNDWATER CONTAMINATION AND CLEANUP
Moderator: Steve Browning

Assessment of TCE Concentration in South Tucson Water Network
Ian H. von Lindem, Jim C. P. Liou, and Margrit von Braun 571
Groundwater Contaminant Transport at a Hazardous Waste Disposal Site, Pullman, Washington
Bill Saur, M. Yavuz Corapcioglu and Kent Keller 575
Investigation and Rehabilitation of the Moses Lake Larson Wellfield
David Banton and Robert Anderson 581
Optimal Strategy for Aquifer Remediation
Fethi Ben-Jemaa and Miguel A. Mariño 585
Application of Oil Field Technology to Clean-up of Petroleum Spills
George E. Maddox .... *

Session F6
GROUNDWATER QUALITY PROTECTION
Moderator: C. T. Bathala

A Groundwater Flow Model Based on the Boundary Element Method
Mark A. Liebe .... *
Wellhead Protection Area Delineation for the Weyerhaeuser Wellfield, Springfield, Oregon
Mark Cunnane, David Banton, and Jonathan Snell 589
Computer Applications in the Development of a Wellhead Protection Plan for the Lower Issaquah Valley, Washington
Robert Anderson, Phil Beilin, and Ken Brettmann .... *
Optimum Operation of Recharge Basins
Hasan Mushtaq, Larry W. Mays, and Kevin E. Lansey 593
Unsaturated Flow Characteristics of Clay Soils
Lakshmi N. Reddi and Wong Y. Lee .... *

Session F7
INNOVATIVE CONCEPTS FOR GROUNDWATER MANAGEMENT
Moderator: Timotheus Hampton

Artificial Recharge to Manage Groundwater Quality in a Connected Surface Water Groundwater System
Seshadri Suryanarayana and A. Osman Akan 597

---

*Paper not available at the time of printing.

The Strategic Development of a Sole Source Aquifer to Improve Water Quality While
Minimizing Environmental Impact
  Marc V. Cromer, Mark J. Abbott,and Shih-Huang Chieh 601
Developing, Managing, and Protecting Urban Aquifers in the Pacific Northwest
  Michael R. Warfel 605
Large-Scale Conjunctive Use of the San Gabriel Basin: An Environmentally Beneficial
Water Supply Project
  Jonathan Harris and Timotheus Hampton 609

Session F8
STRATEGIES FOR DEVELOPING MAJOR CAPITAL FACILITIES I
Moderator: Walt Anton

Session F9
STRATEGIES FOR DEVELOPING MAJOR CAPITAL FACILITIES II
Moderator: Walt Anton

Strategies for Developing Major Capital Facilities
  Walt Anton, Kristie Langlow, David Every, Rosemary Menard, Dave Hilmoe, Jay B. Laughlin,
  Robert Ellis, Chips Barry, Tim J. Block, and Jim Goetz 613

Session F10
WEST POINT TREATMENT PLANT UPGRADE: INNOVATION IN PLANNING, DESIGN AND CONSTRUCTION
Moderator: John Lesniak

The West Point Story Chapter 1: Overview and Planning
  John Lesniak 619
The West Point Story Chapter 2: Project Management Approach
  Timothy J. Block 623
The West Point Story Chapter 3: An Integrated Park and Treatment Plant
  Linda Sullivan 627
The West Point Story Chapter 4: Design
  James G. Goetz 631
The West Point Story Chapter 5: Construction
  Jim Benedict 635

Session F11
INTEGRATED WATER RESOURCE PLANNING: AN ALTERNATIVE TO WATER RIGHTS LITIGATION
Moderator: Bob Wubbena

Introduction
  Bob Wubbena .... *
Utility Perspective
  Richard Cyr .... *
Hydrologist Perspectives
  Don Matlock .... *
Regulatory Agency Perspective
  Gary Hanson .... *

Session G1
WATER SUPPLY-COMPETING INTERESTS AND NEEDS
Moderator: Mark Rothenberg

Panel Discussion .... *

---

*Paper not available at the time of printing.

Session G2
WATER AVAILABILITY-A NORTH AMERICAN, NATIONAL, AND REGIONAL PERSPECTIVE
Moderator: Kyle E. Schilling

Water Conservation and Consumption—Future Trends for North America
  Harold J. Day .... *
Water Supply—A National Perspective
  Kyle E. Schilling .... *
Planning for Water Availability in the West
  Carroll Hamon .... *
Upper Guadalupe River Authority, Texas: Innovative Solution to Long-Term Water Needs
  Paul D. Thornhill, Robert Adams, John McLeod, and B. W. Bruns 639

Session G3
WATER DEMAND FORECASTING
Moderator: Thomas G. Sands

Projecting Customers Using a GIS Land Use Forecasting Model
  Jacqueline Borrego 866
Small Area Water Demand Forecasting at the Salt River Project
  Robert S. Nichols 644
A Comparison of Short Term Forecast Methods for Municipal Water Use
  Ashu Jain and Lindell Ormsbee 649
Water Use Pattern of Residential and Commercial Customers
  Pen C. Tao 653

Session G4
CONSERVATION'S IMPACT ON WASTEWATER FLOWS
Moderators: William Maddaus and Donald D. Gray

Low Flush Plumbing Fixtures and Wastewater Systems
  Thomas P. Konen, Srinivasan Pongavanam and R. Bruce Martin 657
Waste Transport and Treatment Effects from Complete System Conversion to Pressurized
Low Consumption (1.6 GPF) WC's
  R. B. Martin .... *
Use of a Statistical Model to Forecast Future Wastewater Flows
  Erick Health, John Calmer, William Maddaus, and Jack Weber 666
The Water and Wastewater Savings Achieved by Ultra Low Flush Toilets in Santa
Monica, California
  Craig Perkins and Susan Munves 670

Session G5
GRAY WATER AND REUSE OF WATER
Moderator: Claire Welty

Challenges in Implementing the Use of Reclaimed Water
  David B. Parkinson and James V. Wodrich 674
The City of Los Angeles Grey Water Pilot Project Shows Safe Use of Gray Water Is Possible
  Bahman Sheikh 678
Bellingham Frozen Foods Spray Irrigation System Operations
  Martin Harper, Clain Jones, and David B. Green 682
Water Reclamation in the Southwest: Where It Is Now and Where It Is Headed
  Wally Ambrose .... *

---

*Paper not available at the time of printing.

Session G6
SUPPLIER INITIATIVES
Moderator: Pen C. Tao

Water Management and Conservation in the Delaware River Basin: An Interstate Perspective
   Gerald M. Hansler and Jeffrey Featherstone .... *
Water Conservation—An Operator's Perspective
   Michael J. Barnes .... *
Residential and Commercial Audits—A Plan and a Program
   Kim Drury .... *
EBMUD's Approach to Demand Reduction
   John B. Lampe and Jacqueline A. Millet 686

Session G7
NEW IDEAS AND CONCEPTS
Moderator: Richard S. Siegel

Regional Benefits for MWDSC's Seasonal Storage Programs
   Tim Hampton and Nina Topjian .... *
Water Conservation Plan Development for Icicle—Peshastin Irrigation District
   Edwin T. Zapel and Robert A. Montgomery .... *
Subsea Dam for Freshwater Reservoir
   Mark Capron .... *
Global Distribution of Water Through the Oceans
   E. Robert Winter 690
Residential Water Conservation and Reuse Demonstration: Casa del Agua and Desert House
   Martin M. Karpiscak, Richard G. Brittain, and Mark A. Emelity 694

Session G8
AGRICULTURAL NEEDS AND IMPACTS
Moderator: Bert Clemmens

Agricultural Water Conservation Programs to Improve Water Use Efficiency
   Baryohay Davidoff 698
Impact of Groundwater Management Act and CAP Water Supply on Agricultural Water Conservation Programs
   Thomas Carr 702
Agricultural Water Conservation Programs in the Lower Colorado River Authority
   Jobaid Kabir 705
Agricultural Water Conservation Technology Transfer
   Gerald W. Buchleiter 709

Session G9
PRICING WATER—IMPACT ON CONSERVATION
Moderator: Michael Bennett

Price Elasticity and Conservation Potential
   David S. Hasson 713
Historical Impacts of Conservation Measures on Rates and Revenues in the City of Los Angeles
   Jerry Gewe .... *
Seasonal Rates—The Pros and Cons: A Case Study
   Frank Gradilone III and Mark D. Rothenberg 717
The Evolution of Conservation Rates in Phoenix, Arizona
   Edward G. Blundon and Jeffrey S. DeWitt 721

---

*Paper not available at the time of printing.

Session G10
CONSERVATION FROM THE OPERATORS' AND REGULATORS' PERCEPTIVE
Moderator: Mark D. Rothenberg

Panel Discussion .... *

Session G11
MORE GROUNDWATER MODELING
Moderator: Edmund A. Prych

Optimal State Feedback Estimating in Groundwater: Application to Leaky Aquifers
   Mohamed M. Hantush and Miguel A. Mariño 725
Estimation of Groundwater Recharge by Coupling Hydrological Balance and
One-dimensional Unsaturated Flow Equations
   Nien-Sheng Hsu .... *
Recharge of Stormwater to Groundwater Aquifers
   Erez Sela .... *
Flow Investigation for Landfill Leachate
   Reza M. Khanbilvardi and Shabbir Ahmed 729

Session H1
URBAN RUNOFF ASSESSMENT
Moderator: John Warwick

Urban Runnoff and the Environment
   J. J. Warwick 861
Application of HSPF to Evaluate Hydrologic and Water Quality Changes in the Tualatin
River, Oregon
   Roy W. Koch, J. Patrick Moore, Fei Tang, and Wayne C. Huber .... *
Impact of Spatial and Temporal Data Limitations on the Modeling of Runoff Quantity
and Quality
   J. J. Warwick and J. Litchfield 862
Methodology for Evaluating Urban Storm Water Management Strategies
   James Struck and Jim Heaney .... *
The PULSEQUAL Model: Using Combinations of Simple Mathematical Equations to
Evaluate Complex Storm Effects on Water Quality
   John K. Marr and Robert Eimstad 734
Vendor-developed Urban Stormwater Software, in Time for the 90's?
   Gary L. Lewis .... *

Session H2
URBAN RUNOFF ENVIRONMENTAL IMPACT
Moderator: Ed Herricks

Impacts of Urbanization on the Physical and Chemical Structures of Freshwater Wetlands in
the Puget Sound Region
   Richard R. Horner, Lorin E. Reinelt, and Brian Taylor .... *
Impacts of Urbanization on the Plant and Animal Communities of Freshwater Wetlands in
the Puget Sound Region
   Sarah S. Cooke, Klaus O. Richter, and Amanda Azous .... *
Ecologically Relevant Constructed Wetland Design to Minimize Effluent Water Quality Impacts
   Edwin E. Herricks .... *
Reducing Environmental Impact through Urban Pollution Management—Current Projects in
the United Kingdom
   John M. Tyson .... *

---

*Paper not available at the time of printing.

Session H3
URBAN RUNOFF BEST MANAGEMENT PRACTICES
Moderator: Joan Lee

Implementing a Watershed Plan for Lake Stevens
  Gene N. Williams 738
Design of Aquatic Treatment Systems
  Robert B. Aldrich 742
Selection of Optimal Best Management Practices (BMPs)
  Thomas R. Sear and Ronald L. Wycoff 747
A Discussion of the Effectiveness of Stormwater BMP's
  Daniel P. Clark and Molly Adolfson .... *

Session H4
URBAN RUNOFF CASE STUDIES I: CSO
Moderator: Jonathan Buckley

Urban Runoff Considerations in the Design of Wastewater Management Programs for Coastal Areas
  Larry A. Roesner 751
Urban Stormwater Runoff Control System, Village of Skokie, Illinois
  Eddy H. Nakai and Robert W. Carr 758
Storage/Treatment Isoquants for CSO Control Planning
  Ronald L. Wycoff and Lester E. Lee 762
Combined Sewer Overflow Abatement Planning Based on SWMM EXTRAN Modeling
  David M. Wood and David Crawford 766
Receiving Water Quality Bases for Evaluating CSO Control Alteratives
  Ken C. Hall and William A. Kreutzberger 770

Session H5
URBAN RUNOFF CASE STUDIES II: EASTERN
Moderator: Raymond Wright

Application of the Storm Water Management Model in Evaluating Combined Sewer Overflows
  Rajat Roy Chaudhury, Raymond M. Wright, Igor Runge and Daniel W. Urish 774
The Kettering Community Demonstration Project Non-point Pollution Control and Environmental Enhancement Program
  David B. Ennis, Michael L. Clar, and Larry S. Coffman 778
Application of Stormwater Monitoring Programs to Meet Regulatory Requirements
  Thomas F. Quasebarth, Kelly A. Cave, and Eric M. Harold .... *
Storm Water Management in the Greater New Orleans Area
  Kent B. Dussom, Gordon C. Austin, and Marnie Winter 782
Utilization of Roadway Crossings as BMP's in Urban Areas
  G. V. Loganathan, E. W. Watkins, A. B. Small and D. F. Kibler 785

Session H6
URBAN RUNOFF CASE STUDIES III: CENTRAL
Moderator: Jill C. Bicknell

Analysis of Historical Wet Weather Water Quality for Regional NPDES Stormwater Permitting Efforts in North Central Texas
  Alan H. Plummer, Jr., Jonathan Young, George Oswald and Robert W. Brashear .. *
A Need for Synthesized Federal, State, Regional, and Local Efforts in Drafting Non-point Source Regulations for an Urban Karst Water Supply in the Austin, Texas Area
  Robert D. Conti .... *
Evaluation of CSO Impacts on the Rouge River, Michigan
  James E. Murray, John M. Bona, and Larry A. Roesner .... *

---

*Paper not available at the time of printing.

Characterizing Urban Runoff Quantity and Quality
  G. Padmanabhan and Louis P. Erdrich 789
Storage of Combined Sewer Overflow: How Effective Is It Anyway?
  Khamis A. Al-Omari 793

Session H7
URBAN RUNOFF CASE STUDIES IV: WESTERN
Moderator: Steven J. Haness

Fairfield-Suisun Urban Runoff Management Program: The Approach of Small Communities
to NPDES Permitting
  Jill C. Bicknell and Michael J. Barnes 797
Modeling and Managing Storm Water Quantity and Quality—A Case Study for the City
of Yakima
  Kent K. Mao .... *
Seattle's Storm Water Application
  Neil F. Thibert 801
NPDES Municipal Storm Water Permit: A Utility Approach
  Jeff Niermeyer 805
Sources of Stormwater Contamination Originating from Urban/Agricultural Communities
  Louis A. Courtois and Stephen A. Grieg .... *

Session H8
URBAN RUNOFF MANAGEMENT
Moderator: Charles H. Call

Development and Implementation of Stormwater Utilities in Texas Cities
  C. Diane Palmer 809
When the Well Runs Dry: Paying for Storm Water
  Shaun Pigott 813
Storm Water Utility Experience in Bellevue, Washington
  Damon Diessner 817
Storm Water Utility Experience in Salt Lake City, Utah
  Charles H. Call .... *
Cross Section 90's—A Profile of User Fee Funded Stormwater Utility Practices in the U.S.
  Robert B. Benson 821

Session H9
URBAN RUNOFF REGULATORS PANEL
Moderator: Charles H. Call

Panel Discussion
  Rhonda Harris, Mimi Dannel, Ugene Bromley, David W. Smith, Steve Bubnick,
  and Bruce Cleland .... *

Session H10
EMERGING TECHNOLOGIES: MULTIMEDIA IN WATER RESOURCES
PLANNING AND MANAGEMENT
Moderator: Mark Houck

Panel Discussion .... *

---

*Paper not available at the time of printing.

Session H11
COST SCHEDULING WORKSHOP
Moderator: Roy S. Prakash

Poster Sessions

Continuous Simulation Modeling for Sewer Systems
   Eric C. M. Bergstrom 823
Case Study: Storm Water Analysis of Manatee Pocket in Martin County, Florida
   William C. H. Wang, Robert A. Laura, and E. Scott Webber 827
Nonpoint Source Surface Water Management Planning for Rural Areas Within the Tualatin River Basin
   Morton D. McMillen and Donna Hempstead .... *
Managing Stormwater with a Microcontroller Operated System
   Edward McCarthy 831
Development of a Regional Framework for Stormwater Quality Management Programs for NPDES Permitting Efforts in North Central Texas
   Robert W. Brashear, George Oswald, Alan Plummer, Jr., and Jonathan Young .... *
The Urban Runoff Tsunami: How a Small Community Manages Large Scale Development
   Jenna L. Getz .... *
Lake Washington Ship Canal Water Quality Monitoring
   Glen F. Singleton .... *
Trout Spawning Habitat Mitigation: A Constructed Example
   Gregory Koonce 835
Assessment and Mitigation for Endangered Vernal Pool Invertebrates
   E. J. Koford 839
Interactive Decision Support for Hydrologic, Hydraulic, and Instream Flow Criteria
   Marshall Flug and Darrell G. Fontane 842
Reservoir Management and Thermal Power Generation
   Yulianti and Barbara J. Lence 846
A Raft System for Large River Hydraulic Measurements
   Scott D. Wilcox and Ted M. Frink 850
Alternatives Analysis Using Two-Dimensional Modeling for the Owensboro Bridge and Approaches
   M. A. Ports, T. G. Turner, and D. C. Froehlich 854
The Winthrop Fish Hatchery
   Ken Pflueger .... *
Getting the Most from Our Existing Dams—Twin Falls Hydroelectric Project
   J. R. Fotheringham .... *
Getting the Most from Our Existing Dams—North Umpqua Hydroelectric Project
   Paul Carson .... *
Effects of Reservoir Management on Establishment of Riparian and Wetland Vegetation in Northeastern Washington
   Clayton J. Antieau 858
Hazel Creek Tunnel: Drought Planning
   Joseph H. Barnes .... *
Roller Compacted Concrete Projects in the Northwest
   Stephen Benson .... *
Aquatic Resources and Basin Planning
   Domini Glass .... *
Icing in Rivers
   Tom Cannon .... *
Mission River Basin Studies
   Lon Hachmeister .... *

---

*Paper not available at the time of printing.

Rehabilitating Water Supply Facilities
   George Kanakaris, Joe Falbo, and Ed O'Connor .... *
Streamflow Measurement & Communications and Data Transmission
   Jeff Paine and Don Christofferson .... *
Koma Kulshan Project
   William Shaffer .... *
Finite Element Analysis Methods for Complex Structures
   Marc Van Patten .... *
AMP Flow Augmentation Project: Expanding Pipeline Capacity from 366 to 556 cfs
   Lee A. Jacobi and Robert L. Carley .... *
Twentieth Anniversary of WRPMD
   A. J. Fredrich, TC on Observance of 20th Anniversary .... *
Energy Management of Houston's Ground and Surface Water System
   Theodore G. Cleveland, Jerry R. Rogers, and Kathlie Sheu .... *
Dynamic Evaluation of Sustainable Development Paradigms
   M. J. Bender and G. V. Johnson .... *

Subject Index 871

Author Index 877

---

*Paper not available at the time of printing.

RESOURCE IMPLICATIONS OF LISTING
COLUMBIA RIVER BASIN SALMON STOCKS
UNDER THE ENDANGERED SPECIES ACT

JOHN E. VELEHRADSKY

Abstract

The Columbia River and Snake River dams and reservoirs provide substantial benefits in the Northwest through their operation for hydropower, flood control, irrigation, navigation, and fish and wildlife. The listing of certain Snake River salmon stocks as endangered and threatened, under provisions of the Endangered Species Act, has surfaced major public policy issues. Protection and enhancement of these salmon stocks has resulted in proposals to significantly modify the operation of the reservoir projects. Implementation of these proposals could have significant economic, environmental and social impacts in the region.

Introduction

The Corps of Engineers operates twelve reservoir projects in the Columbia, Snake River system. These projects are operated primarily as part of the hydropower system; however, they also provide navigation, flood control, irrigation, water supply, recreation and fish and wildlife benefits. Since 1938, the Corps has expended over a billion dollars on fish ladders, barges, bypass systems, fish screens and hatcheries. Notwithstanding these efforts, certain Snake River sockeye and spring and fall chinook salmon stocks continue to decline resulting in the listing of these stocks as threatened and endangered under provisions of the Endangered Species Act.

The continuing decline of these stocks can be attributed to turbine mortality, predation, destruction of habitat, and harvest management practices. The existence of the Federal reservoirs and the operation of the hydropower system has been suggested by some as the primary cause of the declining stocks.

Director, Programs and Project Management, US Army Corps of Engineers, PO Box 2870, Portland, OR 97208-2870

Modification of the four lower Snake River reservoirs and modification of the John Day Dam operations on the Columbia River has been proposed. Major modifications of the magnitude suggested by the proponents would have significant impacts on other river uses.

Short-Term Operational Changes

In the spring of 1991, the Corps of Engineers, Bonneville Power Administration, and the Bureau of Reclamation initiated the preparation of an Options Analysis Environmental Impact Statement (EIS), to cover 1992 river operations in support of the declining salmon stocks. The Options Analysis (EIS) addressed operational changes that could be accomplished within existing legislative authorities. The combined Options Analysis EIS for the 1992 operations was completed in January 1992.

In addition, the three agencies entered into consultation with the National Marine Fisheries Service following the December 1991 listing of the Snake River sockeye and spring and summer chinook. This action resulted in changes in the operation at Dworshak Dam in Idaho and Grand Coulee Dam in Washington. In addition, the four lower Snake reservoirs were operated at minimum operating levels during the migration periods. John Day reservoir was operating at lower levels. These actions were taken to increase flow velocities through the system to reduce the travel time for smolts.

For 1993 a river operations supplement to the 1992 Options Analysis EIS was prepared. Consultation with the National Marine Fisheries Service was conducted concerning operational strategies for 1993.

Structural Modifications

Structural modifications to the lower Snake River dams and reservoirs has been proposed to increase flow velocities. The proponents would like to see these reservoirs operating below design operating levels during migration periods for spring, summer, and fall juvenile salmon. As currently designed, project operation at these lower levels would adversely impact hydropower production, navigation, recreation, and other users within the Snake River system. In addition, flows over the spillways would be increased causing high levels of dissolved nitrogen, which causes gas bubble disease. Upstream migration of adult salmon and steelhead would also be adversely affected.

In response to a request of the Northwest Power Planning Council, the Corps of Engineers initiated the System Configuration Study. The study identified alternative modifications to the dam structures along the lower Snake River to permit operation at less than minimum operating levels. These potential modifications were reported to the Northwest Power Planning Council in December 1992. The costs would range from $1.3 billion dollars to more than five billion dollars. Design and construction would take from fourteen to seventeen years to complete.

The Corps of Engineers is proceeding with an analysis of selected structural modifications and other structural measures to

---

Director, Programs and Project Management, US Army Corps of Engineers, PO Box 2870, Portland, OR 97208-2870

improve fish passage. These studies will include estimates of biological, economic, environmental, and survival effects of alternative measures. Results are scheduled to be reported to the region during the late summer of 1993.

Public Policy Issues

Biological monitoring and research activities have been conducted along the Snake and Columbia Rivers for decades. Yet there remains disagreement among the experts concerning assumptions, the validity of models, and relationships between flow and survival. Current studies will determine within reasonable certainty the cost and other impacts of an engineering measure to improve fish passage. The larger question is the risk and uncertainty associated with the biological effects of the proposed major changes in the reservoir system.

Additional research is needed to verify assumptions concerning flow-survival relationships. Such data is difficult to obtain without adversely affecting the threatened and endangered stocks. Discussions have been on-going concerning test protocols that would yield potential data with minimum impact on fish. Thus far, no consensus has developed concerning acceptable test protocols.

Without better flow-survival relationship information, decision makers are faced with making choices among alternatives which have significant financial and economic costs and environmental and social effects. Enlightened public policy requires innovation and the development of consensus among the engineers and scientists associated with the Columbia River System.

Director, Programs and Project Management, US Army Corps of Engineers, PO Box 2870, Portland, OR 97208-2870

Mount St. Helens Recovery Project

John Sager and Chris Budai[1]

Abstract

Work began immediately after the eruption of Mount St. Helens to restore Columbia River navigation, provide 100 year flood protection for the downstream communities located along the Cowlitz River, prevent deposition of sand and gravel into the Cowlitz/Columbia River Basin as far upstream as possible to minimize dredging, prevent Spirit Lake and others from overtopping debris barriers, and to minimize environmental impacts. More than 50 agencies plus approximately 50 contractors were involved with work that spanned ten years and cost a total of $507 million.

Unlike other engineering projects, which are original actions--ingenious, beneficial, but not strictly necessary--the Mount St. Helens recovery project represents a human response to catastrophic events beyond human control. No choice existed about whether or not to respond, rather the choices lay entirely in how to respond.

Introduction

The purpose of this project was to provide initial recovery, interim and long-term flood control on the Cowlitz and Toutle rivers, navigation on the Columbia River, and to restore adverse environmental impacts, such as diminished fish runs. Long-term protection was accomplished by constructing a permanent outlet tunnel for Spirit Lake, a Sediment Retention Structure on the North Fork Toutle River, and dredging sediment from the Cowlitz and Toutle rivers. Funding was initially provided under Public Law 84-99 for emergency response and Public Law 98-63 in FY-83 for interim flood control measures. Permanent solution features were authorized by the Supplemental

---

[1]Geologists, Geotechnical Engineering Branch, Army Corps of Engineers, Portland District, P.O. Box 2946, Portland, OR 97208-2946

Appropriations Act of 1985 (Public Law 88).

Initial recovery (emergency response) actions included the reestablishment of the navigation channel on the Columbia River and flood control on the Cowlitz River. Extensive dredging and construction of temporary levee raises and two temporary debris retaining structures were accomplished immediately after the eruption during 1980 and 1981. The eruption produced a 17-mile-long debris avalanche that blocked the natural outlets for Spirit, Coldwater, and Castle lakes, and caused significant sedimentation in the Cowlitz and Toutle rivers. The interim response goal was to stabilize the blocked lakes and rivers so further impact on downstream communities would be minimized until long-term solutions could be implemented.

Permanent outlets were constructed for Coldwater and Castle lakes, and an interim barge-mounted pumping system was installed on Spirit Lake. From November 1980 to April 1985, when permanent measures went into effect, this unprecedented pumping system on Spirit Lake had moved 263,000 acre-feet of water across the avalanche debris dam and held the lake to a safe elevation of 3,452 feet, averting flood danger to communities downstream while the long-term Spirit Lake solution was evaluated and designed. Dredging was used as an interim measure to control sedimentation. The Cowlitz River was dredged as needed, and the lower Toutle River was dredged every year, mostly from two large sediment retention basins. Dredging in combination with initial emergency levee reinforcement and later levee modification provided 100-year flood protection for Washington communities of Castle Rock, Lexington, Longview, and Kelso.

Permanent, long-term solutions to breaching of the avalanche debris dam at Spirit Lake and downstream sedimentation included construction of an 11-foot-diameter outlet tunnel at Spirit Lake and a 184-foot-high Sediment Retention Structure (SRS) about 20 miles downstream of Mount St. Helens on the North Fork Toutle River. The Spirit Lake Outlet Tunnel was designed to lower the lake level about 20 feet to elevation 3,440, and to maintain that level indefinitely with minimum fluctuation. The contractor selected the tunnel-boring machine (TBM) option of the contract and began excavation from the downstream portal during September 1984. The intake structure, designed as a concrete bulkhead with a single gated, four-foot-square opening allowing a maximum flow of 500 cfs into the tunnel through a vertical shaft, was built behind a natural rock cofferdam left in place in the intake channel. The upstream tunnel floor elevation is 40 feet below the lake level intake. If future geologic events require the lake to be lowered below its current design elevation, this can be done with minimal changes to the intake structure only, and with no change to the tunnel. Tunnel support consisted of fiber-reinforced shotcrete applied behind the TBM as necessary to prevent future erosion, and shotcrete-encased rib sets for support in nine shear/fault zones encountered in one long reach near the tunnel center.

Authorization to construct the SRS was received in 1985 and diversion and excavation for the 1,800-foot-long embankment began at the end of 1986. In November 1987, two years before final project completion, a lake formed behind a 100-foot-high upstream cofferdam and began retaining sediment. The problem of sedimentation into the Cowlitz River was thus addressed as efficiently as possible through resourceful planning and innovative design and construction. Given its proximity to the volcano, the SRS was designed to withstand major earthquake damage and to accommodate the largest projected floods and mudflows. The use of mudflows as design criteria was a particularly innovative aspect of the SRS design. The dam was constructed using rockfill with a tapered impervious clay core. All outlet works and spillway rock slopes were treated and stabilized with fiber-reinforced shotcrete and rock bolts. Unique and innovative Japanese design concepts were used in the concrete gravity monolith design of the SRS outlet works. This feature contains six rows of five outlet pipes or conduits through which water and fish can pass into the plunge pool and exit channel below without abrasion damage to the structure or injury to fish. Gates or valves installed inside the pipes to control water flow can be closed off permanently as the sediment level rises behind the structure. As each tier of pipes is closed off, the lake behind the structure will be newly enlarged and the sediment level will gradually rise to the next tier. The stilling action that occurs when the river current hits the lake causes waterborne sand and gravel to drop out of suspension. The resulting deposit will build gradually toward and away from the embankment as the 3200-acre storage area fills with about 258 mcy of sediment over the 50-year life of the project. By about the year 2035, when the sediment retention area has reached capacity, all the conduits in the monolith will be closed and the river will flow continuously over the unlined, ungated spillway.

Environmental awareness affected all design decisions from the start. An environmental task force was convened a few days after the initial volcanic eruption and played an active role in project planning and design. Sediment retention and excavation programs included measures to protect and restore fish runs, such as the trap and haul Fish Collection Facility built just downstream of the SRS on the North Fork Toutle River. Construction impacts were carefully controlled in the area immediately around the mountain and all physical disruptions (i.e., road work areas, etc.) were restored to their original condition as much as possible. While working to serve and protect the public, the Corps took care to allow natural recovery to proceed undisturbed.

Of the estimated nearly $1 billion costs associated with eruption of Mount. St. Helens, the Corps spent about $507 million to restore navigation and provide flood and sediment control to the three affected rivers. Benefits to navigation, transportation, and flood protection of the approximately 50,000 downstream residents have been estimated at several billion dollars. This project received the American Society of Civil Engineers (ASCE) Outstanding Civil Engineering Achievement Award for 1991. It also has been selected to receive a Federal Design Achievement Award from the National Endowment for the Arts.

Basin Scale Management of Surface and Ground Water

John C. Tracy[1], Assoc. Member ASCE and
Munjed Al-Sharif[2], Student Member ASCE

Introduction

An important element in the economic development of many regions of the Great Plains is the availability of a reliable water supply. Due to the highly variable nature of the climate through out much of the Great Plains region, non-controlled stream flow rates tend to be highly variable from year to year. Thus, the primary water supply has tended towards developing ground water aquifers. However, in regions where shallow ground water is extracted for use, there exists the potential for over drafting aquifers to the point of depleting hydraulically connected stream flows, which could adversely affect the water supply of downstream users (who in many situations have senior water rights). To prevent the potential conflict that can arise when a basin's water supply is being developed or to control the water extractions within a developed basin requires the ability to predict the effect that water extractions in one region will have on water extractions from either surface or ground water supplies else where in the basin. This requires the ability to simulate ground water levels and stream flows on a basin scale as affected by changes in water use, land use practices and climatic changes within the basin. In addition, since the control of ground water extractions within a basin typically resides in the hands of local or regional water development agencies the results of a basin scale simulation model must be easily understood and interpreted by management

---

[1]Assistant Professor, Department of Civil Engineering, Box 2219, South Dakota State University, Brookings, SD 57007.

[2]Research Assistant, Department of Civil Engineering, 119 Seaton Hall, Kansas State University, Manhattan, KS 66506.

personnel working for the agencies.

The outline for such a basin scale surface water-ground water model has been presented in Tracy (1991) and Tracy and Koelliker (1992), and the outline for the mathematical programming statement to aid in determining the optimal allocation of water on a basin scale has been presented in Tracy and Al-Sharif (1992). This previous work has been combined into a computer based model with graphical output referred to as the LINOSA model and was developed as a decision support system for basin managers. This paper will present the application of the LINOSA surface-ground water management model to the Rattlesnake watershed basin that resides within Ground Water Management District Number 5 in south central Kansas.

Basin Description:

The Rattlesnake watershed basin is located within the Great Bend Prairie of south central Kansas. The basin has an area of approximately 1,445 mi^2 (3,783 km^2) with a length of approximately 95 miles (152 km) and an average width of 18 miles (29 km) where the long axis is oriented in the southwest to northeast direction. The extreme southwest edge has a latitude of 37.26° and a longitude of 97.21°. Of the total area of the watershed, about one half is non-contributing to the surface runoff. The watershed is drained by the Rattlesnake Creek that meanders from the High Plains area northeasterly into the Great Bend lowlands area where it empties into the Arkansas River. The drainage area of the Rattlesnake Creek mainly consist of unconsolidated deposits of clay, silt, sand, and gravel of the Pleistocene age which overly the eroded surface of Cretaceous and Permian age rocks. The state geological map describes the Rattlesnake watershed as dune sand with a small area of loess in the head water area and a thick strip of alluvium adjacent to the Rattlesnake Creek. Currently, the water table depth is classified as shallow, with depths running from 10 to 20 feet (3 to 6 meters) deep in the northeastern part to 100 feet (30 meters) in the southeastern section of the watershed.

Model Inputs:

Use of the LINOSA model requires that the basin be discretized into nodes (representing the storage characteristics of each area in the basin) that are connected by a series of links (representing the ease at which water can move from one node to the next). In addition a simulation time step must be selected that is representative of the time scale at which management decisions are made within the basin and temporally averaged basin inputs must be developed for each time step of the simulation.

The properties associated with each link and node within the basin can be referred to as basin characteristics. These characteristics can change from one location to the next within the basin, but are assumed to remain relatively constant throughout the time of simulation unless changed due to a management decision. These basin characteristics can be classified in one of five categories, which are: (1) the land uses; (2) the soil types; (3) the aquifer parameters; (4) the stream flow properties; and (5) the water rights status.

The Rattlesnake Basin was simulated using the LINOSA model by dividing the ground water basin into square nodes that where approximately 1 mi^2 (2.6 km^2) and triangular nodes that were approximately one-half mi^2 (1.3 km^2) and the stream sections where divided into nodes of approximately 1 mi (1.6 km) length. This resulted in approximately 1,500 ground water and 100 stream nodes being required for simulating the surface and ground water flow system in the Rattlesnake Basin. The characteristics for each node in the basin were then assigned using a variety of information available through previous geologic, hydrologic and land use studies.

The time step that was selected for use in the Rattlesnake Basin simulations was one month. Many decisions on the use of water within the basin can only be implemented on monthly or seasonal periods and thus the use of a monthly time period in the model simulations should prove sufficient. Thus the data required for model inputs are total monthly precipitation and evapotranspiration for the Rattlesnake Basin. For the calibration and validation of the LINOSA model to the Rattlesnake Basin, historical climatological data can be used. However, for the model to be used as an aid in water management decisions, some type of climatological forecasting procedure must be developed.

Due to the highly variable nature of the climate in the Great Plains region of the country, it has proven very difficult to forecast precipitation and evapotranspiration amounts from one year to the next. Thus, instead of actually forecasting future values, data sets of extreme and average monthly evapotranspiration and precipitation totals for the entire management simulation period were developed. This allows the water manager within the basin to view the effects that decisions on water uses will have based on average, wet, or dry conditions to further refine the basin's water management plan.

The climatological data sets used in the simulation of the Rattlesnake Basin where developed by selecting monthly precipitation and evapotranspiration totals over consecutive months that represent the driest, wettest and average conditions over the management simulation. The monthly precipitation values were obtained relatively simply by inspecting the records produced by the National Oceanic and Atmospheric Administration (NOAA) publication of climatological data for Kansas. The monthly evapotranspiration values are not as readily available as that of precipitation, and for this study where developed using an energy budget and mass balance approach that utilizes daily meteorological inputs for each land use and soil type in the basin.

The results of the calibration and validation of the LINOSA model to the Rattlesnake Basin show that the monthly averaged basin conditions are simulated relatively well over the period of record. In addition, evaluations of the effects of hypothetical water management decisions within the basin using the model simulations demonstrate the usefulness of a modeling procedure of this type.

References:

Tracy, J. C., 1991. "A model for the management of groundwater and surface water rights during droughts," In <u>Water Resources Planning and Management and Urban Water Resources, Proceedings of the 18th Annual Water Resources Planning and Management Division Specialty Conference and Urban Water Resources Symposium</u>, May 20-22, 1991, New Orleans, Louisiana, pgs. 517-521.

Tracy, J. C. and Al-Sharif, M., 1992. "Scheduling of ground water pumpage in alluvial aquifers to minimize the impact on surface water diversions" <u>In the proceedings of the 1992 Annual Water Resources Planning and Management Division Specialty Conference</u>, August 3-5, 1992, Baltimore, MY, pgs 79-83.

Tracy, J. C., and Koelliker, J. K. (1992). "A model for the management of water rights in basins with interconnected surface and ground water supplies," <u>Contribution Number 294, Kansas Water Resources Research Institute</u>, Report No. G2020-04, 41 pgs.

FRESNO/CLOVIS WATER MANAGEMENT PLAN

William E. Burmeister[1], M. ASCE, and Dave Peterson[2], M.ASCE

INTRODUCTION

The Fresno/Clovis Metropolitan Area (FCMA) has historically relied solely on untreated undisinfected groundwater as a source of potable water to serve its 500,000 people. Contamination was discovered in some wells in the late 1970s, and cones of depression in areas of heavy pumping have caused contaminants to spread within the basin. Recent data indicate that at least 44 of the 352 public water agency wells in the FCMA have already been deactivated because of groundwater quality degradation. Major plumes of groundwater contamination occur throughout the study area. Most of the agricultural contaminants in the FCMA groundwater are the consequence of routine pesticide application over thousands of acres of surrounding farmland. Commercial and industrial contaminants are primarily due to poor storage and handling practices, careless or improper disposal, and leaking underground tanks.

Recent and anticipated water quality regulations will probably require some form of wellhead treatment at every public water agency well in the FCMA. Such treatment may consist of disinfection, corrosion control, and the removal of radionuclides and organic chemicals. Many of the well sites are not sized, located, or configured to accommodate wellhead treatment.

Potable water distribution systems in the FCMA were constructed based on dispersed wells and a local distribution network of relatively small water mains. The loss of wells has created local areas of low pressure during peak demand periods and inhibited the ability of the systems to provide fire protection.

---

[1]Water System Manager, City of Fresno, 1910 E. University Ave., Fresno, CA 93703

[2]Manager, Municipal Services Department, CH2M HILL, 4910 E. Clinton Wy., Fresno, CA 93727

These problems have manifested themselves in a short period of time, triggered by a rapid and drastic change in water quality regulations. The funding and implementation of remedial measures is slow, resulting in virtual building moratoriums in some portions of the FCMA.

A technical advisory committee (TAC) was formed in 1984 to promote a cooperative water planning effort. The TAC is composed of the five major water agencies in the metropolitan area: the Cities of Fresno and Clovis, Fresno County, the Fresno Metropolitan Flood Control District (FMFCD), and the Fresno Irrigation District (FID). The TAC prepared the work plan and, in August of 1991, selected CH2M HILL to develop the Fresno/Clovis Water Resources Management Plan.

STUDY DESCRIPTION

The principal objectives of the study were to:

- Provide safe, adequate, and dependable water supplies to economically meet the future needs of the metropolitan area

- Protect groundwater quality from further degradation

- Provide a plan that could be implemented

The Plan was developed in three phases spread over a 1-year period:

Phase I--Existing Water Supply System Assessment. Existing system information on facilities, data bases, hydrologic setting, institutional framework, and operational and management practices were assessed.

Phase II--Water Supply Alternatives. Alternative configurations of water supply elements were identified and screened. Based on the screening process, a preferred alternative was recommended.

Phase III--Implementation Plan. The process, schedule, and responsibility for implementing the preferred plan was formulated, along with a facilities plan, financing plan, and programmatic Environmental Impact Report (EIR).

An integral part of the study was a four-part public involvement program, which was designed to:

- Apprise the public of the Plan schedule, goals, and objectives

- Provide the public with informational updates at appropriate milestones
- Solicit input from the community during plan development
- Disseminate information regarding water supply issues, the range of resource alternatives considered, and the recommended alternative for providing an adequate water supply to the area

Three planning horizons were used in the study:

- 1992-1997—Project Horizon
- 1998-2010—Program Horizon
- 2011-2050—Policy Horizon

RECOMMENDED PLAN

The recommended plan consists of a facilities plan, financing plan, institutional plan, and implementation schedule. The recommended water supply makeup is given in Table 1. Major facilities are shown on Figure 1.

The total cost of facilities need through the year 2010 is estimated to be $343 million (1992 dollars). The recommended funding plan includes revenue bond issue every 3 years through 2010. The bonds will be repaid through user charges and connection fees based on capital cost allocation between existing and growth water demands, respectively.

The recommended institutional body for implementation of the Plan is a Joint Powers Authority of the Cities of Fresno and Clovis. The JPA will provide full retail water service to the 2 cities, assuming all the duties, responsibilities, and contractual obligations of the current water divisions, including operation, expansion, repair, maintenance, billing, and collection. The 2 adjacent water systems will be tied together, providing more reliable service. The most important benefit of a JPA will be effective management of groundwater quality and quantity within the boundaries of the FCMA.

To facilitate Plan implementation, a task force will be empowered to resolve details of JPA formation and Plan adoption. During the first 6 months, the task force must determine the composition, authority, and voting rights of the JPA Board of Directors, develop legal documents necessary to create and empower the JPA, secure adoption of the enabling agreement by the 2 City Councils, and secure Board appointments by the 2 Councils. During the next 12 months, the task force must assist the Board in securing an office and staff, interim rate setting, interagency agreements, and initial construction scheduling.

Table 1
Water Supply Summary
(1,000 acre-feet)

	1997	2010	2050
Demand			
FCMA Demand (without conservation)	180	228	365
Supply			
Conservation	22	35	48
Untreated Canal Water for Landscaping	3	7	13
Treated Surface Water			
Northeast Treatment Plant	15	15	30
Southeast Treatment Plant	0	15	30
Total Treated Surface Water	15	30	60
Groundwater			
Plume Management	47	47	47
Existing Wells without Wellhead Treatment	58	60	67
Existing Wells with Wellhead Treatment	10	15	20
New Wells without Wellhead Treatment	20	27	89
New Wells with Wellhead Treatment	5	7	21
Total Groundwater	140	156	244
Urban Intentional Recharge			
Existing Single-Purpose Basins	25	25	25
Flood Control Basins	15	30	90
New Single-Purpose Basins	0	0	0
Groundwater Level Restoration	10	10	10
Total Intentional Recharge	50	65	125

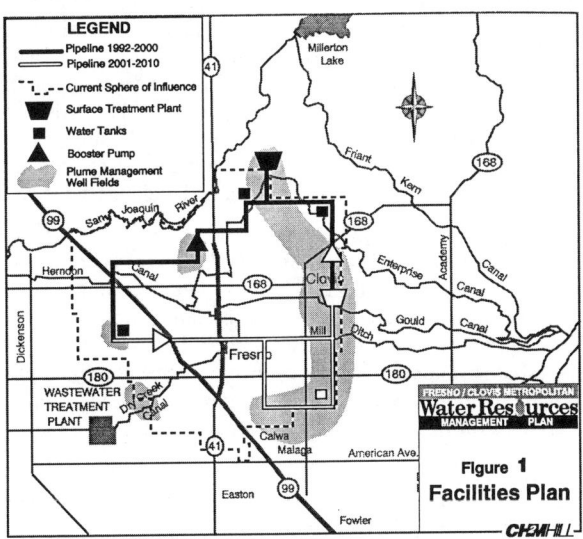

Figure 1
Facilities Plan

URBAN WATER PLANNING – A WESTERN NEVADA CASE STUDY

John W. Fordham, Member ASCE

Abstract

The result of the increased demands due to population growth in the Reno/Sparks urban area and new uses i.e., environmental and Pyramid Lake related, has created increased competition for limited available regional water resources. This competition has generated considerable water resource planning activity at all levels of government and the private sector to cope with the evolving situation. This has resulted in numerous plans often conflicting or overlapping, and considerable political maneuvering. The paper will trace this activity from before 1970 when most water planning for this area was performed by Federal agencies in a rather benign atmosphere to today where local and private entities dominate the process in a highly competitive atmosphere.

Background

The Reno/Sparks urban area is situated in a valley, the Truckee Meadows, on the lee side of the Sierra Nevadas. The climate is semi-arid with an annual precipitation on the valley floor of approximately 8 inches. However, visual appearances belie the area's aridity due to diversions of agricultural irrigation and M & I water from the Truckee River which courses through the valley on its way from alpine Lake Tahoe to desert Pyramid Lake. Population has grown rapidly in the urban area and adjacent outlying suburban areas over the past two decades placing severe stress on the area's water resources in terms of both water supply and water quality. These difficulties are compounded by physical limitations of available supplies and competition among users both in the urban area and downstream where significant water demands exist for irrigation, recreation and protection of threatened and endangered species.

Serious difficulties in development and management of water resources for the area arise from institutional/legal constraints at all levels of government (city, county, state and federal). A further complicating factor is that most water allocation is administered through numerous court decrees. Under Nevada water law all waters are under the doctrine of prior appropriation. Added to the problems related to the physical and institutional/legal constraints has been an evolving struggle among several entities both public and private to control the resource, thereby determining future directions for growth and economic development of the urban area.

Water Resources Center, Desert Research Institute, University and Community College System of Nevada, P.O. Box 60220, Reno, Nevada 89506

Prior to 1970

The demands for water in the Reno/Sparks urban area was easily met through diversion of surface waters together with a small amount of groundwater until about 1970. The primary water source, the Truckee River, rises in the Sierra Nevadas and is controlled through a series of structures on both the main stem and on several larger tributaries. These structures and resultant storage had been provided to primarily benefit agricultural interests both in the immediate vicinity of the urban area and further downstream.

The major diversions from the Truckee River at this time were for irrigation in the Truckee Meadows, for M & I use in the cities of Reno and Sparks, and downstream for irrigation of the Newlands Project. Water for the urban population was for the most part supplied by a private company from water rights on the river and from local groundwater. Water supplies for small outlying developments were also privately developed from underground sources. The river system operation was administered by a Federal watermaster serving the federal court system. At this time, the cities and county were little involved in any aspect other than wastewater management. The river system water planning was agriculturally driven and took place on a Federal level with input from the local interests. The facilities which were built to help satisfy and firm supplies were Federally financed. Local planning for urban water use was left to the major water purveyor, a part of the Sierra Pacific Power Company now known as Westpac Utilities. Since there was "enough" water to meet demands except in the most critical of droughts, that planning primarily consisted of infrastructure within the urban area with little thought of developing additional river or other supplies. The primary controls on the areas water resources and their allocation were the 1935 Truckee River Agreement (U.S., 1935) which set a maximum M&I diversion from the river and set operating criteria for the upstream reservoirs, and the Orr Ditch Decree (U.S., 1944) which finalized the adjudication of 744 claims to the waters of the Truckee River in Nevada.

1970 – 1983

Beginning in the early 1970's, there was increase in urbanization and increased interest in the water resources of the Truckee River basin. The increased urbanization created a dynamic situation with respect to water requirements as urban uses began to displace irrigated agriculture. In addition, there was a growing awareness of some of the past resource management practices and their consequences, primarily those resulting from diversion of the lower river to the Newlands Project. This awareness manifested itself in concern for the fate of Pyramid Lake, the natural terminus of the Truckee River. These concerns brought about the Pyramid Lake Task Force which was created to identify alternative solutions to alleviate the recession of Pyramid Lake. The Task Force was composed of representatives of the U.S. Department of Interior with interested parties from California and Nevada. This group attempted to find compromises and alternative operations for the river system to provide additional water to stop or slow the decline of the Lake. This task force concluded its work in 1971 without truly identifying a viable solution. At the same time the State Division of Water Resources was also active prepared a set of water resource plans which examined alternate futures for the major water resource systems in the State, including the Truckee

River System (DWR, 1973). This planning effort did not make recommendations, rather it provided an examination of the consequences of the four alternatives examined. A local group, the Blue Ribbon Task Force, was also formed to investigate growth and development problems in Reno, Sparks and Washoe County. This group considered ten growth parameters of which water was one. The recommendations (Washoe County, 1974) made by that group have been or currently are being used in some shape or form by the two cities and the County. With these activities water resource planning was being pursued at all levels of government with a considerable amount of interchange of ideas and information. The major private water utility was a key player in each of the efforts.

With the inability of any of these efforts to satisfactorily address the Pyramid Lake problem, a number of lawsuits were filed by the United States Justice Department on behalf of the Pyramid Lake Indian Tribe. These suits asserted claims to significant amounts of river waters to sustain both Pyramid Lake and its fisheries which include threatened and endangered species. These suits, which at times have exceeded 30 in number, covered both water quantity and quality issues and have impacted all aspects of water management for the area. They detracted significantly from formal planning efforts during the mid to late 70's when efforts at all levels concentrated on resolution of the major lawsuits which would have reopened the adjudication of the River and perhaps significantly changed the operation of the reservoir system. As these lawsuits worked their way through the court system, some of the uncertainties were laid to rest and planning efforts picked up once more. In 1977 the State created a separate Division of Water Planning in response to a severe drought and increasing water demands from urban areas. The Division put out yet another study (DWP, 1978) which presented a number of possible options which might meet future water demands. The effort did not recommend a specific supply option, but concluded that water supply goals for the urban area could be met through the year 2025 by utilizing a combination of demand management, efficiency improvements, and additional in−basin sources. The State water planning effort continued looking at several specific topics such as water reuse and land application of effluent until 1983 when State financial conditions forced elimination of the Division.

1983 − Present

By 1983, local entities had begun to put significant efforts into water related planning. A Regional Water Planning and Advisory Board of Washoe County was created by an act of the state legislature in 1983 to address regional issues. The board was composed of one general public member and three members from each of the three urban political jurisdictions. The act also created a technical advisory committee to give advice and technical assistance to the board. The Board's initial activity was to contract for a study (Beck, 1987) of possible groundwater importation to the urban area as a "new" source of water for not only the Truckee Meadows but for the outlying urbanizing areas which could not be served with Truckee River water. At the same time the major urban water purveyor, Westpac, was planning resource acquisition for their system looking at numerous alternative options (SPPCo, 1986 and 1989). These efforts of the Water Board and the utility were brought together in 1990 yielding a regional water resource plan (RWPAB, 1990). At this same time, Washoe County as a single entity began pursuing an active groundwater importation project in partnership with a private individual. Both the import project and those projects and options being pursued by Westpac are running parallel to the overall planning effort. In addition, the 1991 legislature directed the local governmental agencies to end their infighting and to coordinate their planning efforts. The three government agencies have now formed the Regional Water Management Agency, successor to the Water Board, which is directing a Regional Water Study identifying specific facilities to meet both water quantity and quality demands to the year 2012 for the entire urban/suburban area. This effort has brought to forefront several areas of competition among the cities and the county related to control of the water supply and the associated power that control has with respect to economic development. The major urban water purveyor is still a private company, and has considerable influence on decisions. The county has taken over a number of troubled private water companies in outlying areas and has embarked on an effort company to import groundwater to the area from a rural area 30 miles away. The conversion of agricultural lands to urban has resulted in conflict over just who and how those areas are to be served, by the county or by the major private water purveyor. The end result is that there is now serious division among local entities, both public and private, and infighting to determine who will ultimately be the water czar.

References

U.S. of America, Truckee River Agreement 1935. Appendix to the Decree of the Truckee River Adjudication Suit.

U.S. of America, Final Decree of September 8, 1944 in the Case of United States vs. Orr Water Ditch Company et al.

Division of Water Resources, "Alternative Plans for Water Resource Use, Carson-Truckee River Basins Area II," State Engineer's Office, 1973.

Washoe County, "Blue Ribbon Task Force Program – Growth and Development in Reno, Sparks, and Washoe County, Nevada," 1974.

Division of Water Planning, "Water Supply Report 1, Truckee River Summary and Recommendations," State Department of Conservation and Natural Resources, 1978.

R.W. Beck and Associates, "Washoe County Ground Water Importation Project," report to the Regional Water Planning and Advisory Board, 1987.

Sierra Pacific Power Company, "Water Resources Plan," 1986 and 1989.

Regional Water Planning and Advisory Board, "Regional Water Resources Plan," 1990.

Water Management Plan for Lake Okeechobee

James W. Vearil[1] M.ASCE

Introduction

Lake Okeechobee is a large (approximately 730 square mile), natural, freshwater lake in Southern Florida with a drainage area of about 5,650 square miles. Lake Okeechobee functions as a multipurpose reservoir and is considered the "heart" of the Central and Southern Florida (C&SF) Project (U.S. Congress, 1948). The Corps of Engineers operates and maintains project works on Lake Okeechobee and its outlets, and the main outlets for the Water Conservation Areas. The South Florida Water Management District (SFWMD) operates the remainder of the project based on Corps specified criteria. The SFWMD is responsible for allocating water supply from project storage, except where specified by Federal Law. Lake Okeechobee is regulated to provide flood control; navigation; water supply for agricultural irrigation, municipalities and industry, and Everglades National Park; regional groundwater control, and salinity control; enhancement of fish and wildlife; and recreation (U.S. Army Corps of Engineers, 1991). Water control plans for multi-purpose projects must blend all the varied, and often conflicting, purposes.

Background

Prior to 1900 it has been estimated that water levels in Lake Okeechobee were normally about 20.5 feet, with a fluctuation of about 2.5 feet. During wet years Lake Okeechobee would overflow its southern rim into the Everglades (U.S. Army Corps of Engineers, 1960). Between 1905-1931 the State of Florida's Everglades Drainage District constructed a number of features around Lake Okeechobee for the purpose of draining and reclaiming the Everglades. The hurricanes of 1926 and 1928 killed between 2000 to 3000 people and caused much property damage due to flooding from Lake Okeechobee.

---

[1] Hydraulic Engineer, Jacksonville District, U.S. Army Corps of Engineers, P.O. Box 4970, Jacksonville, FL 32232-0019

Following the hurricanes of 1926 and 1928, Congress provided disaster relief in the form of flood control and navigation in the 1930 Rivers and Harbor Act that authorized the Corps of Engineers to construct improvements to the Caloosahatchee River and St. Lucie Canal, construction of a levee and navigation canal along the south shore of the lake, and a levee around the town of Okeechobee. The Okeechobee Waterway is a Federal Navigation Project across Southern Florida via the St. Lucie Canal, Lake Okeechobee, and Caloosahatchee River. After the 1947 flood, the Flood Control Act of 1948 authorized a comprehensive water resource project known as the Central and Southern Florida Project to address the complex set of water related problems in that area.

The 1948 authorization did not specify a particular Lake Okeechobee regulation schedule. The Corps of Engineers analyzed a series of alternatives and a flat schedule of 16.4 feet, NGVD was approved for Lake Okeechobee (U.S. Corps of Engineers, 1951). Subsequent studies lead to the approval of a 15.5-16.5 foot seasonally variable regulation schedule (U. S. Army Corps of Engineers, 1959). The Corps of Engineers (1959) concluded that this schedule provided about the same level of agricultural water supply benefits as the flat 16.4 foot schedule, provide seasonal flood storage to help reduce damaging regulatory releases to the estuaries, and that large additional benefits could be obtained from sufficient conservation storage in the lake to supply the needs of the urban areas along the east coast during droughts. Considerable flood storage or very large outlets are necessary for Lake Okeechobee to function as a multiple-purpose reservoir. The top of the flood control pool was considered the maximum lake level reached during the Standard Project Flood (SPF). A variety of regulation schedules were used during construction of C&SF Project facilities necessary to permit implementation of the 15.5-16.5 foot schedule. In 1978 a 15.5-17.5 foot regulation schedule was approved, which retained the design parameters of the 15.5-16.5 foot schedule, but provided some additional water supply storage. A 15.65-16.75 foot regulation schedule is currently approved for May 1992 to May 1994. The Lake Okeechobee regulation schedules used since the C&SF Project was authorized have essentially been based on the 1948 Authorization. The Flood Control Act of 1968 authorized an additional 4 foot rise in the lake schedule for water supply, but that lake regulation schedule has not been implemented. The recommended plan approved in 1968 included the "pump first via agricultural canals" concept first introduced in 1966 which reduces releases to the estuaries and increases water available to Everglades National Park.

Water Control Plan and Regulation Schedule

The regulation schedule varies from high stages in the late fall and winter to low stages at the beginning of the wet season. Runoff during the wet season is stored for use during the dry season. The regulation schedule provides for about two feet of flood control storage being evacuated prior to the beginning of the wet season (U. S. Army Corps of Engineers, 1991). Water levels in Lake Okeechobee are normally below the regulation schedule. When lake levels exceed the regulation schedule, flood control releases are made from the lake. When the lake level is below schedule, releases are made from the lake as needed for water supply, navigation, prevention of saltwater intrusion, and environmental enhancement. The Everglades Agricultural Area, south of the lake, is irrigated with water from the lake during dry periods. Water is transferred to the Water Conservation Areas (WCA's) and the lower east coast, from Lake Okeechobee, when inadequate storage is available in the WCA's to meet demands. The SFWMD uses a Lake Okeechobee Supply-Side Management Plan to manage water supply releases during droughts (U. S. Army Corps of Engineers, 1992).

Levees completely encircle the huge lake except where they tie to high ground on either side of Fisheating Creek. Structures through the levee are closed completely in advance of a hurricane or tropical storm to maintain the integrity of the levee during a wind tide. Outlet capacity from the lake is small compared to the immense storage capacity of the lake. The total lake storage at specific lake stages is shown below:

	Stage (Ft. NGVD)	Storage (Acre-ft)
Top of Flood Control Pool (SPF)	25.1	8,617,000
Max Top of Conservation Pool	17.5	5,062,000
Min Top of Conservation Pool	15.5	4,165,000
Min Conservation Pool	10.5	2,203,000

Regulatory outlets from the lake are the St. Lucie Canal, Caloosahatchee River, and the Agricultural Canals. It is important to note that lake regulatory releases through these outlets are on a secondary basis, when sufficient capacity is available after removal of local runoff. The Caloosahatchee River and St. Lucie Canal provide for navigation, serve as outlets for Lake Okeechobee, provide flood protection for lands along the canals, and provide a source of water supply. The estimated maximum lake regulation capacity (U.S. Army Corps of Engineers, 1991) is:

West Palm Beach Canal	900 cfs
Hillsboro Canal	800 cfs
North New River Canal	1,600 cfs
Miami Canal	2,000 cfs
Caloosahatchee River	9,300 cfs
St. Lucie Canal	14,800 cfs

The shallow bottom topography of Lake Okeechobee combined with the long fetch, can result in significant wind tides on the lake. The principle factors that determined the required levee grades around Lake Okeechobee were lake level prior to the hurricane, and the wind tide and wave-runup expected during the hurricane. The three hydraulic conditions analyzed were Maximum Probable Hurricane (MPH) with the lake at the top of the conservation pool, Standard Project Hurricane (SPH) with the lake at the highest 30-day average 100-year flood stage, and Moderate Hurricane at highest 30-day average Standard Project Flood (SPF) stage (U.S. Army Corps of Engineers, 1959). According to the Corps of Engineers (1959) the characteristics and the area of formation of the MPH and SPH virtually preclude their passing over Lake Okeechobee after the middle of September. Routings indicated that the lake can be regulated so the stage very seldom rises appreciably above the top of the conservation pool before September 15.

Environmental Issues

There are number of environmental issues relative to Lake Okeechobee (SFWMD, 1989). They include water quality in the lake, algal blooms, sport and commercial fisheries, waterfowl and wading birds, the lake littoral zone, endangered species, and water requirements for Everglades restoration. Lake Okeechobee regulatory releases have adverse impacts on estuarine systems. Concerns have been expressed that the current lake regulation schedule could adversely impact the littoral zone, and waterfowl and wading bird communities in the lake. The SFWMD (1989) pointed out that Lake Okeechobee supports a highly productive, diverse ecosystem.

Acknowledgements

The author appreciates the support of the Jacksonville District, U. S. Army Corps of Engineers in preparing this paper. The views expressed in this paper are those of the author, and do not necessarily represent those of the Army Corps of Engineers.

Appendix I. References

South Florida Water Management District. (1989). Interim Surface Water Improvement and Management Plan For Lake Okeechobee. West Palm Beach, FL.

U.S. Army Corps of Engineers. (1951). Part I, Agricultural and Conservation Areas (With Preliminary Information on Lake Okeechobee and Principal Outlets). Jacksonville, FL.

U.S. Army Corps of Engineers. (1959). Part IV, Supplement 2, Section 7 - General Design Memorandum, Combinations of Hydrologic and Hydraulic Factors Affecting Height of Levees. Jacksonville, FL.

U.S. Army Corps of Engineers. (1960). Part I, Supplement 33 - General Design Memorandum, Conservation Area No. 3. Jacksonville, FL.

U.S. Army Corps of Engineers. (1991). Water Control Plan for Lake Okeechobee and Everglades Agricultural Area. Jacksonville, FL.

U.S. Army Corps of Engineers. (1992). Drought Contingency Plan for Lake Okeechobee. Jacksonville, FL.

U. S. Congress. (1948). House Document No. 80-643 Comprehensive Report on Central and Southern Florida For Flood Control and Other Purposes. Washington, DC.

A Computer Model for Multiplant Hydroelectric System Simulation

S. O. Russell M. ASCE and D. A. Fayegh

Professor and Research Assistant,
Civil Engineering Department, University of British Columbia,
Vancouver, B.C., Canada, V6T-1Z4

Abstract

An interactive computer model of a multi-plant hydrothermal electric power generating system has been developed that roughly captures the main features of the British Columbia Hydro system. The objective of the model was to help familiarize its users with the problems associated with the monthly operation of multi-reservoir systems when complete knowledge of future inflows, power demands, and market prices does not exist.

To encourage the use of the model, it has been designed as a "game" that is played on the computer by a hypothetical "operator" for 24 months. At the beginning of each month, the operator is required to make decisions on how to allocate the total forecast system load, given the reservoir elevations, forecast of inflows, and energy prices for the month. Once the operator enters his decisions at the beginning of the month, the system is simulated for that month and a number of graphs, bar charts, and tables are updated with the new values of the various variables, such as reservoir levels, operating costs, revenue, etc.. At the end of the simulation, the operator may repeatedly request the model to improve on her/his game score (net revenue) by running a deterministic differential dynamic programming procedure until an optimal decision sequence is found.

This study revealed that the players, who were not generally reservoir operation experts, initially found the model to be quite complex. However, with the aid of an appropriate graphical user interface and sound feedback, they consistently improved their scores and understanding of the complexities of the operation problem.

1 Introduction

In order to meet the demand for electrical energy, B.C. Hydro operates more than thirty hydroelectric power plants with storage reservoirs of various sizes. However, the bulk of the total supply is generated by the Peace and Columbia hydro subsystems. In this paper we report on the design and implementation of an interactive computer model of a simplified version of the B.C. Hydro's power generating system with the hypothetical Big Peace and Big Columbia subsystems as the main sources of supply. The Big Peace plant represents the main features of two plants on the Peace river (Peace + Shrum) with a combined capacity of 1700 GWh per month while the hypothetical Big Columbia plant represents the main features of two plants on the Columbia river (Mica + Revelstoke) with a combined capacity of 2000 GWh per month. The model also incorporates the Burrard thermal plant which is used by B.C. Hydro to supplement its electric power generation capability during periods of low inflow and high demand.

The computer model was developed as a result of a collaborative research project involving B.C. Hydro technical staff and a group of researchers with expertise in systems engineering and computer graphics at the University of British Columbia, Canada. The motives of this study were to demonstrate the complexities of operating a multi-plant hydro-thermal system in today's freewheeling commercial environment by means of a computer "game" that is both educational and entertaining. The major phases of the project involved implementation and augmentation of existing simulation and optimization models to incorporate a more complete set of operational strategies and the development of an interactive graphical user interface that incorporates colour graphs, charts, tables, and sound to add to the computer game's interest and appeal.

This paper outlines the overall operation and functional components of this computer model and discusses how they were designed and interfaced together to provide the player with an easy-to-use and interesting environment to explore the complexities of hydro-electric system management. The details of how the simulation and optimization components of the model were designed, implemented, and interfaced is the subject of a companion paper and will not be considered here.

2 The Problem

A typical hydro-electric power generating system consists of a number of reservoirs and power plants. The amount of electricity generated by any given plant depends on the flow through the various turbines and the total energy "head". The total energy head is the difference in elevation between the water level in the reservoir and the tailwater level - the level in the body of water into which the turbine discharges. Demand for electricity (or loads) change both seasonally and instantaneously as household electrical appliances are turned on and/or off during the day with a changing pattern of use throughout the year. The inflows into the various reservoirs also vary continuously. Standards in North America are rather high and hence, despite the varying energy demand and supply patterns, an electrical utility company, such as B.C. Hydro, is expected to meet its load at all times, except perhaps in times of real emergencies.

With a hydro plant, the lower the reservoir level, the more water is needed to generate a unit of electricity. However, if the reservoir is full and there is a large inflow, water over and above the maximum storage capacity of the reservoir must be spilled – and hence wasted as far as power generation is concerned. Operating a hydro plant may thus be likened to a juggling act where the operator must try to keep the reservoir full to assure sufficient head for running the turbines; try not to spill water during periods of high inflow and low demand; and try not to have to make up for energy deficits by purchasing energy from other more expensive sources.

In modern hydroelectric systems there are many generating stations interconnected to a number of load centers in a grid. B.C. Hydro operates over thirty hydro plants

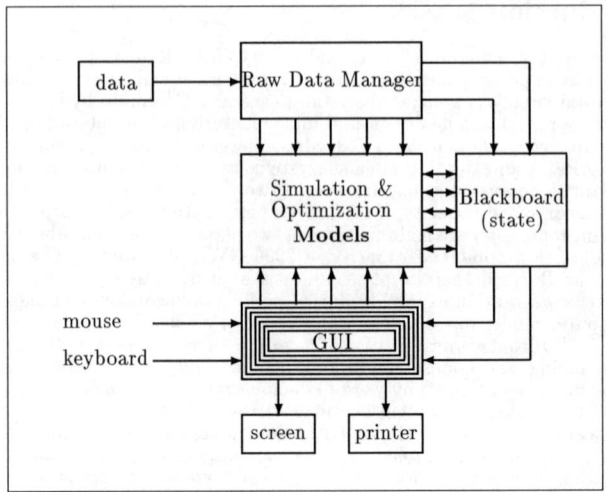

Figure 1: The Functional Components of the Hydro Operations Game

and a relatively large thermal plant at Burrard, near Vancouver. Depending on the location of the plant, the cost of generating electricity varies both with time and local conditions. Until recently and while B.C. Hydro was fairly self contained, their aim was to meet the total demand at minimum cost, no mean task in itself given the number of load centres and generating plants. In today's freewheeling commercial environment, there are now additional complexities that must be taken into consideration. B.C. Hydro can now buy or sell power from or to independent agencies within its service area as well as import or export power to other utilities through interconnecting transmission lines. The operating aim is to maximize net revenues, an extremely complex task given the varying prices for import and export power, as well as all the uncertainties associated with the prediction of future energy demand and supply patterns.

B.C. Hydro continually updates the operating plans for its system of hydroelectric plants, thermal plant, and storage reservoirs in light of the current situation and forecasts of future conditions. The system is operated on several levels. On the management level, overall operating policy decisions are made, usually by committees, at intervals of about a months, on how to allocate the load among the various subsystems, whether to buy or sell energy on contract, how much to buy or sell, for what length of contract period, and so on. Within the policy guidelines, more detailed operating plans are drawn up for shorter periods – such as daily and hourly. Finally, highly trained human operators must make real-time decisions in response to market demands and opportunities available.

3 The Model

The model simulates a simplified version of B.C. Hydro's system operation at the policy level by using a monthly time step for 24 consecutive months. This period was chosen since it is long enough to allow an operator to gain insight into the decision problem, but short enough to keep the game interesting.

Figure 1 illustrates the interaction between the main components of the model. Although the flow of data and/or instructions between the various functional components may have been specified differently, this configuration was chosen both for its conceptual simplicity and open-ended design philosophy. The Raw Data Manager (RDM), the Simulation and Optimization Models (SOM), and the Graphic User Interface (GUI) may be best thought of as independent *processes* that indirectly communicate by reading/writing data from/to the *blackboard*. This modular design philosophy allowed implementation of the GUI, each of the simulation and optimization models, and the RDM to proceed in parallel and almost independently.

The application program was intended for use on the Apple Macintosh computer. The development machine was a Mac IIsi equipped with 8 megabytes (MB) of RAM, math co-processor, an 40 megabyte (MB) internal drive, and a 120 MB external hard drive running Apple Computer's "System 7" operating system.

The code at the heart of the simulation and optimization modules were written in the C programming language and are hence architecture-independent and have in fact been compiled and run on a number of computers equipped with an ANSI C compiler without any modifications. The GUI component of the application program, is however machine-dependent.

Each simulation has a "control panel" window that is opened automatically when a new simulation is started and contains an icon for each of the remaining windows. Clicking on one of these icons brings the corresponding window to the foreground. Each window may contain a number of graphs, bar charts, and/or tables. The operator inputs his decision for the month by manipulating "sliders" on the appropriate window with the mouse.

When a new simulation is started, the RDM processes any available raw data (such as historic streamflow records) and produces statistical summaries of the data which are then used by the simulation model to randomly generate both an "actual" and a "forecast" value for the total domestic load and monthly inflows to each reservoir for each of the 24 months in the simulation from appropriate probability distributions. In addition, the simulation model generates "actual" short and long term contract energy prices for each of the 24 months. Note that although the actual prices are available for inspection by the player throughout the game, only forecast values of demand and inflows are displayed at the beginning of each month in the simulation.

At each step of the simulation, the operator or game player must make the following decisions by manipulating the appropriate sliders with the mouse:

- Amount of energy to be supplied by the thermal plant (up to a maximum of 700 GWh per month),

- Amount of energy to be bought/sold using either or both short-term (1 month) or long-term (3 months) contracts (up to a maximum of 1250 GWh per month for each),

- Percentage of the remaining (and as yet unknown) demand (load) to be allocated to each of the two reservoirs.

- The number of consecutive months (up to six) to run the simulation for using the input values.

Once the operator commits to his/her decision by clicking on the "go button" with the mouse, the outcome of the operator's decisions are simulated and the values of a number of variables are then updated on the blackboard. Using these values, the GUI then updates the appropriate screens and displays the results on the appropriate graphs, bar charts, etc. providing almost instant visual feedback of the outcomes. Examples of the updated values include, the reservoir elevation at the end of the month

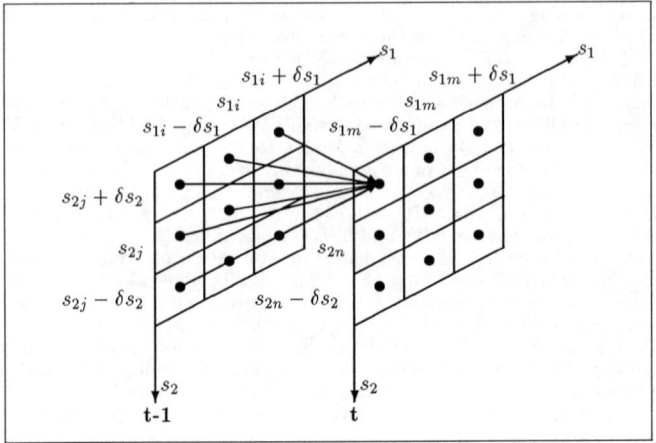

Figure 2: 3 X 3 Corridor used during a Forward Step of the Differential DP

being simulated, net and monthly costs and revenues, amount of energy generated by each hydro plant, and so on. Revenues are obtained from domestic sales at the fixed provincial rate, plus any export sales. Costs are calculated by summing the fixed and variable operating costs, the cost of operating the thermal plant, and any contract purchases. To ensure that the operator does not run the reservoirs empty, long term expected values are assigned to the reservoir levels at the end of the 24 months of simulation and added to the net revenues as a "bonus" or a "penalty".

Once the 24 month operation cycle is complete, the player may choose to run an optimization routine which provides feedback on how well s/he could have done with complete knowledge of the actual demands, inflows, and prices for the previous run. A two dimensional deterministic differential dynamic programming algorithm was implemented that only uses a limited 3 X 3 corridor (2 state increments above and 1 below) the reservoir levels that resulted from the operator's decisions during the simulation in order to keep the memory burden and computing time within reasonable bounds (see Figure 2). However, the optimization can be repeated as many times as desired, each time starting from the optimal policy found in the previous iteration. If repeated enough times, the model eventually stabilizes on the optimal decision sequence.

To achieve the game's objective of user friendliness in an educational context, the player may operate the model at four levels of difficulty:

- Run the two hydro-electric plants only

- Run the two hydro-electric plants plus the thermal plant

- Run the two hydro-electric plants plus the thermal plant plus the option to buy/sell energy using short term contracts

- Run the two hydro-electric plants plus the thermal plant plus the option to buy/sell energy using both short *and* long term contracts.

- Run the two hydro-electric plants plus the thermal plant plus the option to buy/sell energy using both short *and* long term contracts.

In addition, at any month during the simulation, the player can "undo" her/his decisions for any desired number of months up to the begining of the simulation after having been exposed to the "actual" values of the generated loads and inflows.

4 Some Insights

The authors found that "operating" and "playing with" and inspecting the results obtained from the optimization model provided a number of interesting insights and an intuitive understanding of the system behavior that had not been available beforehand – eventhough we fully understood and indeed specified and implemented all the details and equations used in the model. It was noticed that even experienced hydro operators became quite intrigued with the model and learned from it.

When the model is operated with only short-term contracts, the optimization routines consistently improve on the player's score. This is not surprising since the optimization model is deterministic and finds the optimal decision policy with complete knowledge of the actual demands, inflows, and prices. The player, however, makes monthly decisions under a probabilistic environment where only forecasts (in contrast to actual) values of the month's inflows and demands are available.

When 3-month as well as 1-month contracts were used, it was observed that a skilled player could occasionally score higher than the optimization routines, which is limited to 1-month contracts only. In this case, the player has one more degree of freedom (i.e., the three month contracts) than the optimization routine and hence is able to do better despite the lack of foresight with enough practice sessions. This observation could have important practical implications, since in real situations, there are always many more complexities than can be explicitly taken into account and modeled by optimization routines, which of necessity work on simplified abstractions of real systems. However, these complexities could possibly be modeled by elucidation of a set of fuzzy rules that captures the implicit knowledge used by skilled players suggesting that fuzzy logic in conjunction with standard optimization techniques could do better than either alone. The authors are presently pursuing this line of research.

5 Conclusions

The interactive computer model of a simplified multi-plant hydro-thermal electrical power generating system provided an improved understanding of the system behaviour and some interesting but unexpected insights.

6 Acknowledgements

Financial support from B.C. Hydro and help from their technical staff, namely, Eric Joa, Klaus Wensauer, and Ralph Legge are gratefully acknowledged. In addition, we would like to thank John Hogg and Helen Salter of the University of B.C. Computing Services and Dr. Kellogg S. Booth of the U.B.C Computer Science Department for their assistance with the design and implementation of the graphical user interface.

Licensing and Permitting the Devil's Nose Project
Under the 1990's Regulatory Conditions

James E. Alverson[1] and
Roderick E. Schuler[2]

Introduction and Objective

Amador County is one of California's mountain counties of the western slope of the Sierra Nevada range in the north central portion of the state. The present population in Amador County is estimated to be 31,000. Although this figure represents a rather low population density, growth has picked up significantly in recent years, and since 1960 the population has increased at an average annual rate of 3.4 percent. Because the western portion of Amador County is located within commuting distance of the metropolitan centers of Sacramento and Stockton, and because of the attractive characteristics of the countryside, growth in Amador County is projected to continue increasing into the 21st century.

The focus of this paper is on the difficulties encountered by Amador County officials in planning for and attempting to develop a reliable water supply to provide for projected future growth, in the current regulatory and economic environment. In normal runoff years, significant water resources, originating largely from the annual melting of the Sierra snowpack, flow in rivers along County borders. However, this relative abundance attracted out-of-County interests decades ago that claimed rights to water and developed major projects. Thus, cities on the east shore of San Francisco Bay divert a large portion of the Mokelumne River

---

[1] Member, ASCE, Associate, R. W. Beck and Associates,
   Seattle, Washington.

[2] Member, ASCE, Director of Water Resources, Amador
   County, Jackson, California.

streamflow for their municipal and industrial supplies, as do downstream agricultural areas for irrigation. A major hydroelectric project has been developed in the upper basin of the Mokelumne. These earlier developments by entities holding much greater financial resources than Amador County have in the past developed the most ideal sites, and have obtained water rights to the most reliable portion of the Mokelumne River streamflow. Development of the less desirable sites now remaining is impeded by both greater financial costs and greater costs of environmental mitigation, as well as by the much greater regulatory hurdles existing today.

Purpose and Need for Project

Studies show that Amador County by 2050 will need an additional water supply of about 22,000 acre-feet, based on projections of population and per capita water demand, and assuming that no additional water for irrigated agriculture will be provided after 2020. This forecast of demand also considers that conservation and water reclamation measures are implemented to reduce per capita demand, and to capture and use canal seepage. The increase in net water demand by 2050 represents a doubling of the County's present reliable dry year water supply.

Because of the prevalent mineralized hard rock geology, affecting both quantity and quality, ground water is not a potential source for additional supply for the County in the future. Surface water in Amador County is relatively abundant in normal water years during winter and early spring. The Mokelumne River produces an average annual runoff of about 724,000 acre-feet. Other smaller streams that may flow bank to bank at times in the winter and spring are typically dry in summer and fall. This hydrologic regime dictates the necessity for constructing reservoirs to store the winter and spring runoff. For the above reasons, neither local water projects on the County's small streams nor increased ground water development have the potential to satisfy the County's increasing future water needs. Instead, the County must turn to new water supply projects on the Mokelumne or Cosumnes rivers.

Of four major projects identified, development of the Devil's Nose/Cross County Water-Power Project on the North Fork Mokelumne River is the preferred project. The principal project features include a reservoir of 145,000 acre-foot capacity and three powerhouses with total installed capacity of 121.5 MW. The water supply yield of the project would be up to about 20,000 acre-feet per year of firm supply, depending on various assumptions regarding the water rights that are finally granted. The proposed Devil's Nose reservoir would be largely on federal lands administered by the U.S. Forest Service.

Project History

Since 1981 Amador County has undertaken a continuous series of activities directed, first, toward investigation and later toward development of the Devil's Nose project. Major activities and milestone events include the following:

- December 1981: FERC preliminary permit issued.

- December 1985: Amador County filed for water rights with the State Water Resources Control Board.

- July 1987: The County filed an Application for License with the FERC.

- July 1988: The FERC accepted the license application for processing and requested significant additional information on various fishery, wildlife and environmental topics, recreation and cultural resources.

- January 1989: The U.S. Forest Service announced it would undertake a "suitability study" of the potential for inclusion of the Project reach of the river in a "Wild and Scenic River" designation.

- 1989-91: A detailed operations and yield model was developed of Devil's Nose reservoir, the two East Bay Municipal Utility District reservoirs, and other consumptive water rights on the river. The model calculates reliable estimates of reservoir yield under various operating scenarios.

- 1987-92: Meetings were held with various water purveyors outside of Amador County that might have an interest in purchase of an interim water supply.

- 1989 and 1990: The County submitted proposals to two electric utilities offering to supply capacity and energy from the project.

- 1992: The County offered, in principal, to consider joint Project development with East Bay Municipal Utility District.

- 1992: The County participated in a proceeding before the State Water Resources Control Board considering reallocation of waters of the Mokelumne River to greatly increase minimum instream flow in the lower reach for fishery habitat purposes.

Permitting and Regulatory Aspects

In addition to the FERC Licensing requirements for consultation with resource agencies and mitigation plan development, Amador County is faced with other difficult permitting requirements. These include a potential designation of the Devil's Nose reservoir site as a Wild and Scenic river. On U.S. Forest Service issues alone, the County has logged more than 100 activities, such as meetings, correspondence, filing of appeals, etc., and more than 60 such activities with the California Department of Fish and Game. A final Forest Service decision on its recommendation to Congress regarding wild and scenic river designation remains to be issued.

The Forest Service also will place conditions on the FERC license pursuant to Section 4(e) of the federal power act, because much of the Project area is on National Forest land. Water rights are a difficult area to resolve, which will require costly negotiations to settle protests and a hearing process before the State Water Resources Control Board (SWRCB). Water rights are difficult because of the highly appropriated condition of the Mokelumne River basin, and because of two proceedings underway in 1992, before the FERC and before the SWRCB, which are considering greatly increasing minimum water releases for fish from existing reservoirs in the lower basin, under public trust doctrine. If greatly increased fish releases are mandated, the result for Amador County would be an adverse impact on the Project's firm water yield. Other permitting requirements include compliance with the California Environmental Quality Act (CEQA), a Section 401 water quality certification and a Corps Section 404 permit, both under the Clean Water Act. Other causes for difficulty include a sort of gridlock in review and decision making by some resource agencies, and the sometimes conflicting agency objectives and regulations faced by a water project proponent.

Conclusions

This paper has described briefly the difficulties surrounding development of a reliable future water supply for a relatively small mountain county in California. Even though located in a water-rich area, the County's water resources have been largely developed by outside entities having much greater financial and political strength. With significant growth anticipated in the future, the County now finds that competition for the remaining water is intense, and that the remaining sites for storage project development are much less economical than earlier projects, and are difficult to license.

MANAGING GOVERNMENT CONTRACTS: TIPS TO KEEPING YOUR SANITY

Lisa T.M. Vomero, ASCE Affiliate[1]

Abstract

The key to managing large government projects is to divide it into smaller more manageable pieces and organize it accordingly. Government projects are also unique in that they require much more detailed documentation and financial record keeping than their private client counterparts. Therefore, the purpose of this paper is to summarize some management and organizational tips for making a large government project more manageable and enjoyable so that you can keep your sanity!

Introduction

So now that you have finally landed that big government contract, now what? No, don't panic -

ORGANIZE !!!!!!!!!!

It is imperative that immediately upon receipt of a notice to proceed that all team members and subconsultants are notified and an action plan established. Assemble an organizational chart for use as a blueprint for the tasks and more importantly, to serve as a flow chart for communication of information regarding the project through its entirety. Also before the project begins, an internal organization and filing system should be established so that all pieces of the job can be readily accessible for review by all team members.

---

1 Project Manager, WEST Consultants, Inc.; 2111 Palomar Airport Road Suite 180, Carlsbad, CA 92009-1419; (619) 431-8113; FAX (619) 431-8220.

Divide and Conquer

The best approach to both cost effectiveness and time management is to divide the entire project into smaller tasks. This will help you focus and make you organize. It is suggested that the project be divided into approximately ten (10) basic categories, they are:

1. project goal(s);
2. background information;
3. establish design criteria and priorities;
4. acquire design specifications and standards, if any;
5. determine computer requirements;
6. type and quantity of production drawings, reports, etc.;
7. results and recommendations;
8. quality control;
9. assembly of all final products for the client; and finally,
10. follow-up with the client regarding performance and overall comments for use in future jobs.

Take the necessary time to establish a project time schedule and to assign each smaller task to an individual team member. Assuming of course, that the original team you put together on the proposal still exists! If not, rearrange your new team "ASAP". The process of delegation will make it necessary to assess the skills and shortcomings of each staff member so that you can effectively delegate the work assignments ahead. Your company structure should have a system that rewards individuals for work on time and within budget as well as penalizes individuals who are not preforming up to par and/or not working as a part of the overall team. As part of the management of the project, both short and long term individual and team goals should be developed, written down and distributed. This will help assess both the status of the job as well as team and member performance.

Delegate

It is necessary at this point to identify one person as the Point Of Contact (POC). This will centralize all communication within the management of the project as well as make the client very happy.

The most important decision to be made in this phase of the job is who will be the Project Leader. This should be one of the most important people in your organization whom you trust and can depend on; however, do not choose a Project Leader who has all the capabilities but IS OVERWHELMED with many other assignments. The end result will be burnout, frustration and none of the projects will be completed to anyone's satisfaction.

Another tip for success that can be taken is to assign an Assistant Project Leader. This will insure that the Leader will have someone readily accessible for help when many different things are happening in regards to the work effort. This will also allow for continuity on the project should something happen to the original team. This process is doubly important for the upward mobility of your staff. Each and every Leader and engineer should be a mentor for someone else in the firm. This will produce a united team as well as build good will and morale in the employees. The arrangement should be friendly, open and honest with a focus on teaching others all the pertinent aspects of what it means to be a professional. This will reflect positively on your firm and employees for this, as well as future endeavors.

Communication

Without a doubt, communication is the number one ingredient for the successful completion of any project large or small; however, it is especially important on large government contracts because of the amount of work that has to be completed as well as the interconnection between tasks. This concept has more recently been referred to as "partnering". Employees that have both poor and/or lacking communication and people skills will only bring down the efforts of the team in the overall scope of the project. Should difficulties arise and communication breakdown, it should be the duty of the Principle-In-Charge to meet with all parties involved to air disagreements and resolve issues that may later threaten the successful completion of the contract. Only if absolutely necessary should the original Project Leader be removed from the job.

It can not be stressed enough how very important communication is. You will do well to insure that the Project Leader in charge of the day to day activities has excellent communication and negotiation skills. Because the scope of project, regardless of size, will change and evolve as time goes on.

Budget

Assign a dollar value to each task along with corresponding "people-hours". Let your staff know exactly how much time and money they have to perform each item. Establish a system to reward your successful personnel. Monitor the progress of the job each month, reconciling the money and time spent with your previous estimate. Adjust the old budget to reflect what is actually happening with the billing. You may need to go to bimonthly review if the project involves: a tight time frame, a large staff and/or numerous billing hours.

Another tip is that, if any changes in the scope of work occur, as they usually do, or disagreements arise between yourself and the client over project scope, they should be addressed immediately. It should also be determined at that time, if the work will require a change in compensation from the original contract.

Paperwork...

The most overwhelming thing about any size contract is the amount of paper work that needs to be completed; therefore, force yourself to do it first thing in the morning, when you are fresh and most relaxed. As hard as it *seems*, stay on top of it - this will pay off many times over.

Time should be taken to create some simple forms using a spreadsheet or word processing software to track: telephone conversations, change orders, unanticipated problems, delays, schedules, deadlines and project milestones. There is also software available specifically designed for project management, no matter which one you choose its procurement is highly recommended.

Finally - Document Everything!

It is imperative to create a cohesive "paper trail" by keeping copies of all correspondence and documenting telephone conversations as well as all other pertinent data. Follow up verbal communication with a brief letter to the client or at least an internal memo to the project file. The use of a three ring binder/notebook is an ideal way to organize this data for convenient, future reference. In addition, add books as the project grows and keep them all in centrally located place that all staff members can freely access.

Summary

The most important steps, to the successful management and completion of any large government contract, are:

>organization,
>division,
>delegation,
>paperwork documentation and most importantly,
>communication (partnering).

Take extra time in the beginning to divide and setup the entire project. Before delegating tasks to each individual, remember that companies do not complete projects - PEOPLE DO! Keep the lines of communication open at all times and listen to the comments and ideas of your staff. Discuss and resolve any problems as soon as they are realized. Finally, document everything relating to the day to day activities of the contract. As a last resort, take a day off and relax so you can

KEEP YOUR SANITY!

DEVELOPING POWER-
PRESENTATION SKILLS
by Ralph W. Pehrson

1. Structure

2. Opening: Audience attention; How To Get It.

3. Body: "Says Who"

4. Conclusion: "What Did You Say?

5. You Can't Tell Them Everything.

6. Visual Aids, "listening with the eyes"

7. Lighten Up.

8. Voice: How Do You Sound?

9. Goal of Presentation

Ralph Pehrson · Pehrson & Associates Seminars
P.O. Box 1216, 19707 · 64th West, Suite 209, Lynnwood, WA 98046,USA • (206) 774-1695

Partering in Water Resources Planning
in Florida

Adil J. Salem[1] P.E.

Abstract

This paper discusses successful partnering efforts by the Corps of Engineers in water resources development in Florida. The partnering involves the Corps, local project sponsors and other local governments, the public, State and Federal agencies, as well as regional and national interest groups. Two specific examples of partnering are presented in this paper.

Introduction

The Central and Southern Florida (C&SF) Project was developed by the Corps of Engineers in 1947 and was authorized by Congress in 1948. It includes all or parts of 18 counties and encompasses about 16,000 square miles stretching from Orlando to the southern tip of the state. The C&SF Project is a multi-purpose project that provides flood control, water conservation, water control, water supply, navigation, wildlife preservation, and recreation. Due to the tremendous population growth in central and south Florida, conditions and water resources priorities in the area are vastly different today than they were in 1947. Understanding and appreciation of the complex and fragile ecosystem of this area have also increased in the years since the project was originally developed. As a result, the Corps has made a number of revisions to the project to meet current needs.

Upper St. Johns River Basin Project Partnering

The history of the Upper St. Johns River Project shows how partnering has evolved the project to meet changing needs. This project, which is located in east central Florida, is designed to

---

[1]Chief, Planning Division, U.S. Army Corps of Engineers, Jacksonville District, P.O. Box 4970, Jacksonville, Florida 32232-0019

provide not only flood protection, but to improve water quality and wildlife habitat, and to develop recreation facilities.

The plan for the Upper St. Johns River Basin presented in the 1947 report consisted of spillways, levees, and other control structures. As design of this plan began in the mid-1950's, a number of groups objected to the purchase of several hundred thousand acres of valley lands for water storage. Consequently, a new plan was developed in 1957 for flood control and irrigation purposes. This plan was also unacceptable to local interests because the plan required the creation of water conservation areas which encompassed over 200,000 acres of flood plain land. In 1962, another plan was developed which replaced valley storage with upland reservoirs. Construction of this plan began in 1968.

Construction of the Upper St. Johns River Project was stopped in 1972 in order to prepare an Environmental Impact Statement. The EIS was not well received, and the Corps began to modify the plan. In 1977, the newly created St. Johns River Water Management District took over responsibility for local sponsorship for the project. A technical advisory committee (TAC) was created in March 1980 to assist in plan development. The Water Management District staff, with TAC assistance, formally submitted a basic concept plan to the Corps in November 1980. Subsequent to the initiation of design studies, the Corps conducted numerous technical discussions with the sponsor between 1980 and 1984 which resulted in modifications to the basic concept plan. During the design phase, the Corps continued to work closely with the Water Management District, Federal and state agencies, and affected landowners. As a result of these activities, in 1985 a new plan for the Upper St. Johns River Basin emerged. This plan, which was approved for implementation in 1986, is the product of the extensive studies and public involvement activities conducted over a number of years.

The TAC, which was the key to the successful development of the Upper St. Johns Project, was composed of nine members and included the Corps; the Water Management District; state and local agencies; and agricultural, environmental, and homeowner interests. This group was formed to develop a hydrologically and environmentally balanced concept plan for surface water management in the area. The TAC, which met on a regular basis, reviewed and discussed Water Management District and Corps project proposals. As a result, plans were developed, discussed, and refined until a consensus on the project was developed.

The project plan consists of providing segregated areas intended for the restoration and enhancement of the historic floodplain wetlands and areas for the detention of agricultural runoff and storage of water supply. While maintaining flood control objectives, the project will also provide other major benefits which include: improved water quality throughout the river basin; improvement and recovery of river marsh and wildlife habitat; and quality of life benefits through recreational features. With implementation of the project, approximately 110,000 acres of marsh will be restored and/or enhanced in the historic floodplain.

The idea that the St. Johns River Water Management District and the Corps work as partners in the development and delivery of the project has been integral to our relationship. Advisory committees and other groups have been established to expand the partnership to others with a stake in the project. On the Federal level, the Fish and Wildlife Service is a member of the group established to review, advise, and support the environmental aspects of the project.

Kissimmee River Restoration Partnering

The Kissimmee River Restoration study represented a superb partnering effort with our study partner, the South Florida Water Management District (SFWMD), and other state and Federal agencies. This partnership among the Corps, SFWMD, U.S. Fish and Wildlife Service, Florida Game and Fresh Water Fish Commission, Florida Department of Natural Resources, and Florida Department of Environmental Regulation was critical to the successful completion of the study effort.

The Kissimmee River Basin lies in central Florida. Comprised of some 3,000 square miles, it extends from Orlando southward to Lake Okeechobee. C&SF Project works in the Kissimmee River basin were constructed between 1962 and 1971. The Upper Basin portion of the project provides structures, canals and regulation of the upper basin lakes. In the Lower Basin, the project provides for the channelization of Kissimmee River from Lake Kissimmee to Lake Okeechobee through a canal called C-38. Channelization reduced the river's length from 103 miles to 56 miles and includes 6 water control structures which form pools to step down the 36-foot drop between Lake Kissimmee and Lake Okeechobee.

While averting catastrophic floods, channelization of the Kissimmee River caused a number of ecological problems particularly due to the alteration of the hydrologic regime and drainage of wetlands. The elimination of the river-flood plain interaction has affected the functional integrity of both the river and the floodplain. The once complex natural ecosystem was significantly simplified, resulting in a substantial reduction in species diversity. About 20,000 acres of the original 35,000 acres of wetlands were either drained, covered with material dredged during construction, or converted to canal. Since channelization, there has been a 94 percent reduction in wintering waterfowl due to the elimination of plant species and community diversity that is required to support large waterfowl populations. Loss of wetland habitat diversity resulted in an 80 percent reduction of wading birds.

In response to the concern about the effects of the Kissimmee River flood control project, three major planning studies have been undertaken by the Corps or the SFWMD. Following Congressional resolutions requesting the Corps to review the completed Kissimmee River Project, the first study by the Corps began in 1978 and was completed in 1985.

Using the Corps' data as a starting point, in 1984 the SFWMD initiated a series of activities designed to test and evaluate the State's preferred alternative of dechannelization. Their efforts culminated in the release of a report in June 1990 which recommended that a dechannelization plan, called the Level II Backfilling Plan, be adopted to restore the ecological integrity of the Kissimmee River.

The Corps' most recent feasibility study was authorized by the Water Resources Development Act of 1990. The Act gave the Corps very specific direction and time frames for completing the study. It directed that the study be based on implementing the Level II Backfilling Plan developed by the SFWMD. The Act also directed that the feasibility report be transmitted to Congress not later than April 1, 1992. This date precipitated the need to initiate special partnering efforts with Federal, State, and local agencies in an effort to expedite the study process. Daily communication between the Corps and SFWMD concerning status and issues surrounding the project was instrumental in meeting study milestones. Four interagency conferences were held during the preparation of the draft report. These conferences provided a forum to present and resolve issues while facilitating communication for the benefit of each team member. These partnering efforts resulted in the completion of the study and submittal of the report to Congress on schedule.

The plan authorized by Congress will restore the essential physical and hydrologic characteristics of the Lower Kissimmee Basin, including a natural river channel and flood plain, with flows, depths, and hydroperiods like that of the historic condition. Restoration of these physical and hydrologic characteristics will provide the conditions necessary for reestablishment of an ecosystem similar to that which existed and functioned prior to construction of the basin's flood control project. The restored ecosystem will include 56 miles of restored river, about 29,000 acres of restored wetlands, improved water quality, and restored conditions for over 300 fish and wildlife species, including waterfowl, wading birds, and three endangered species - the Bald Eagle, the Wood Stork, and the Snail Kite.

Conclusion

The Corps' accomplishments in Florida depend to a large extent on successful partnering with Federal, state, and local agencies. Communications between interested parties is essential. Special partnering groups, like the Upper St. Johns TAC, can assist in solving controversial issues. In addition to information flow, these special groups provide a forum for surfacing views and ultimately help build consensus. Finally, there must be active participation and input by all.

Public Process in Fresno/Clovis Water Supply Plan

Paula Farris[1]

Groundwater is the principal reservoir for the Fresno/Clovis Metropolitan Area (FCMA) of Central California. However, significant portions of the aquifer are threatened by pesticides and other contaminants.

To provide water supplies that are safe, dependable, economical, and meet the future needs of the metropolitan area, five water agencies participated in a three phase water resource planning process.

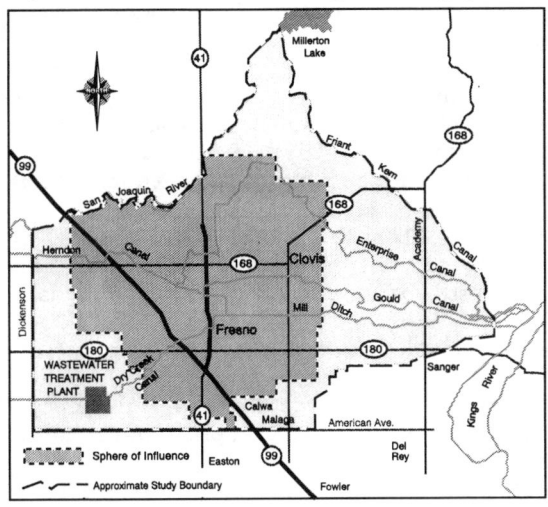

FCMA study area.

---

[1]Executive Vice President, Panagraph Marketing Communications, 2445 Capitol St. #105, Fresno, CA 93721

Participating agencies are the City of Fresno, the City of Clovis, the County of Fresno, the Fresno Irrigation District and the Fresno Metropolitan Flood Control District. The principal objective of the process is to prepare a plan that is implemented by the participating agencies.

Water supply planners were to assess the capability of the existing water supply systems to meet service demand and water quality standards within the FCMA study area.

The engineering firm awarded the contract, CH2M Hill, recognized that a technical approach to the plan, while necessary, was not adequate to assure implementation. With the help of Panagraph Marketing Communications, a Fresno-based sub-contractor with expertise in public information/involvement and water issues, a program was designed to integrate input from both the public sector and the governing boards and councils of the participating agencies.

The three-phase planning process included the following:

<u>Phase I</u>: The planning team assessed all elements of the existing water systems including facilities, hydrologic setting, institutional framework and operation and management practices in the FCMA.

<u>Phase II</u>: Planners analyzed those alternative configurations of water supply elements necessary to meet anticipated water demands over a 60 year period.

<u>Phase III</u>: An implementation plan was prepared. Water supply elements, process, schedule and identification of the Joint Powers Authority (JPA) for management of the FCMA water supply were developed.

The preferred water supply plan was developed through a public involvement and a technical screening process that involved the following:

Focused interviews with key community leaders representing urban, agrarian, and environmental groups were conducted early and late in the planning process. Key issues were identified and incorporated into the technical planning process and the refinement of the preferred plan.

Public Involvement Session II, Fresno, California.

Public involvement sessions were scheduled and held. The sessions, scheduled during each phase of the study process, were held to assure planners understood and responded to citizen concerns.

Community presentations were provided to targeted groups in a program designed to "take the information to the public."

An initial technical workshop brought water resource experts from public and private sectors together to help screen many methods of configuring water supply elements to three. Each method was required to (1) meet projected demand; (2) include conservation as a source of supply; (3) meet drinking water standards; (4) involve recharge sufficient to recover lost water levels; and (5) manage plumes to stop contaminant movement.

A second technical workshop presented a refinement and preliminary cost analysis of the alternatives. After careful evaluation, the workshop participants recommended the preferred plan.

After careful evaluation, the workshop participants recommended the preferred plan.

During interviews held early in the planning process, it became evident that agencies were interested in participating in a joint workshop for governing boards and councils. This event was held to provide participating agencies with the opportunity to discuss the preferred plan and associated costs in a common setting. Through this workshop, differences regarding growth policies, institutional arrangements, financing and rate impacts were identified.

Focused interviews held late in the planning process facilitated the refinement of the JPA for management of the FCMA. In addition, it became evident that slight modifications in the staging of facilities could enhance final adoption of the Plan. An implementation task force was recommended to determine the composition, authority and voting rights of the JPA Board of Directors. Other responsibilities of the task force include: (1) develop the legal documents necessary to create and empower the JPA; (2) secure adoption of the JPA enabling agreement(s); and, (3) recommend members of the JPA board and secure their appointments.

Providing the public with information regarding water quality, water distribution system limitations and the fiscal impact of increasingly stringent water quality regulations can help build a foundation for facing the challenges of water supply planning. These challenges are all the more critical when viewed in light of the deteriorating fiscal condition of state and county governments and the growing burden of fiscal responsibility that local citizens and elected officials are being asked to bear.

EVALUATING ENVIRONMENTAL PROJECTS

William Hansen[1]

Throughout the Nation, there is increased awareness of and concern for, environmental protection and restoration. Within the U.S. Army Corps of Engineers, new authorities are providing increased opportunities to pursue environmentally oriented projects. Currently, however, there is a lack of accepted methods for assessing the efficiency and effectiveness of investments for mitigating, protecting, or restoring environmental resources.

Objective

The objective of this paper is to identify, through the use of a simplified, but holistic evaluation framework, the need for an interdisciplinary approach to the evaluation of environmental projects. Three concerns are highlighted: 1) the need for better understanding of cause and effect relationships; 2) the need for better communication between disciplines, and 3) the need to clearly delineate evaluation criteria (study objectives) against which the effectiveness and efficiency of specific environmental plans or programs can be measured.

Evaluation Framework

Critical to the evaluation process is a clear understanding of the linkages (cause and effect relationships) between management actions, environmental impacts, and human behavior. Consider the simplified evaluation framework presented in Figure 1. For the purpose of this discussion, a **Management Measure** is defined in the broad sense to be any action taken that ultimately impacts on the environment. It could include such actions as a change in reservoir operations, the construction of a flood control project, aeration of an existing lake to improve water quality, or the implementation of an environmental mitigation or restoration project. Management Measures typically result in some **Physical Change,** such as an increase in downstream flows during certain times of the year, a reduction of pollutants in the water column, or alteration in vegetation through the planting of

---

[1] Economist, US Army Corps of Engineers Institute for Water Resources, Fort Belvoir, Virginia. (The views expressed are those of the author and not of the Department of the Army or the U.S. Army Corps of Engineers.)

food plots or other wildlife habitat. Physical changes can further be described in terms of a **Change in Ecological Resources**. For example, an increase in water releases during the spring season might be measured in terms of an increase in the quantity of spawning habitat for select fish species.

Some may want to stop here and say that measuring, for example, the change in habitat is sufficient. But if alternative plans and programs are to be compared, further evaluation of the effectiveness (accomplishment of study objectives) and efficiency (costs) is needed. The changes in Ecological Resources need to be further described in terms of **Environmental Outputs** that can be used in the evaluation process.

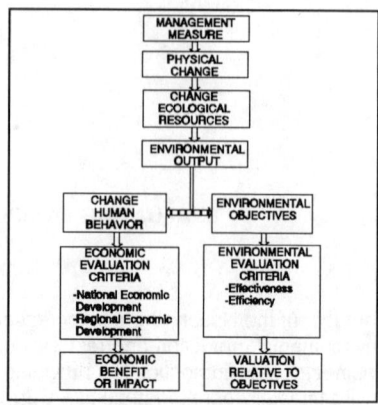

Figure 1: Components of Evaluation Framework

Environmental outputs need to be described in a manner so that their contribution to the study objectives can be measured. As described in more detail below, the specific measure used might vary depending on whether economic or environmental evaluation criteria are to be used.

Economic Evaluation

The economic evaluation criterion most frequently used in project evaluation is maximizing net benefits, under the objective of National Economic Development (NED). Regional Economic Development (RED), measured in terms of jobs or revenues that might be impacted by environmental projects or programs, can also be of concern, especially to state and local interests.

Measuring both NED and RED benefits requires an understanding of how **Human Behavior** and/or attitudes might be impacted by changes in Environmental Outputs. The interrelationships depicted in Figure 1 must be described not only in engineering, economic, and biological terms, but also in a meaningful way that can be perceived or understood by impacted individuals. For example, most individuals would find it very difficult to determine how a specific increase in habitat units would change their behavior or attitudinal values. Most are not technically capable of relating such a measure to their experiences. However, when the resultant impact is further described, for example, in terms of an increased chance to observe wildlife or in a change in fish catch, many individuals are able to express probable changes in behavior and attitude.

Non-market valuation techniques (e.g., Travel Cost, Contingent Value, and Hedonic Pricing Methods) and economic impact procedures (e.g., Input-Output Models) can then be used to derive the **Economic Benefit or Impact** for the associated Environmental Outputs. Significant improvements have been, and are continuing to be, made in the application of non-market valuation techniques to environmental outputs. This is not to imply that economic evaluation criteria can always be used. However, in those instances where economic evaluation is applicable, close cooperation and coordination between all disciplines is needed to fully identify potential applications and to insure that environmental outputs can be described in meaningful terms for the needed economic studies.

Environmental Evaluation

There are many applications where environmental outputs cannot be evaluated in monetary terms. Given limited resources, however, it is still necessary to evaluate the effectiveness and efficiency of alternative management measures. In these instances, tools such as cost-effectiveness frontiers and incremental (marginal) cost analysis can assist in the comparison of alternative plans and programs. Such techniques can help decision makers allocate limited resources and avoid the selection of economically irrational plans and programs. To use such techniques, **Environmental Objectives** need to be delineated and Environmental Outputs need to be identified and described such that a **Valuation** of their contribution **Relative to** these **Objectives** (effectiveness) can be made. To illustrate the use of cost effectiveness analysis, assume that five alternative restoration plans have been formulated, with the costs and levels of output included in Table I.

Alternative	Units of Output	Total Costs ($)
A	80	2,000
B	100	2,600
C	100	3,600
D	120	3,600
E	140	7,000

Table I

A cost-effectiveness frontier is developed for these alternatives in Figure 2 by plotting their total cost (vertical axis) against the output provided (horizontal axis). The most cost effective measures delineate the cost-effectiveness frontier. Any plan lying inside the frontier, such as C, should not be imple-

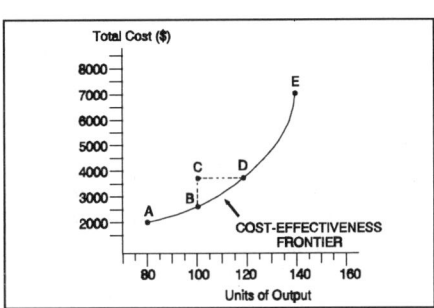

Figure 2: Cost-Effectiveness Frontier

mented, since the same level of output could be provided by an alternative plan (B, in this example) for less cost; or for the same cost, an alternative plan (D) could provide more output. The cost-effectiveness frontier can assist in screening "irrational" alternatives from the analysis. It can also be used to identify the single most cost-effective alternative that will provide a specified level of output. For example, if the objective is to provide 120 units of output, then alternative D would be the most cost effective alternative.

Often, alternative restoration plans are formulated without specific levels of output in mind. In these instances, information used to derive a cost-effectiveness frontier can also be used in an incremental (marginal) analysis. Although this analysis will not, in and of itself, identify the "most desirable" alternative, it can help illustrate the trade-offs between outputs and resource requirements of alternative plans. Continuing with the previous example, the incremental outputs and costs associated with each plan are shown in Table II.

Alt.	Output:		Cost:		
	Total	Inc.	Total	Inc.	Inc./Unit
A	80	80	2000	2000	25
B	100	20	2600	600	30
C	120	20	3600	100	50
D	140	20	7000	340	170

Table II

Alternative A would provide 80 units of output at a cost of $2,000 or $25 per unit. Alternative B would provide an additional 20 units of output (100-80) at an additional cost of $600 ($2,600-$2,000). The incremental cost per unit for the additional 20 units is, therefore, $30. Similar computations can be made for alternatives D and E. Alternative C has been deleted from the analysis, since it was previously identified as an irrational alternative. The

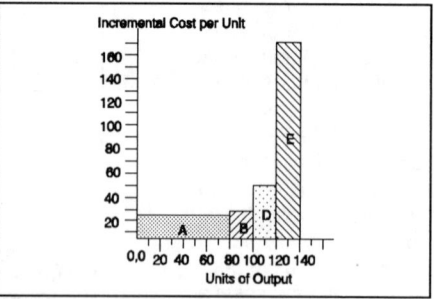

Figure 3: Incremental Analysis

results of the above calculations are illustrated in Figure 3. The incremental cost per unit is measured on the vertical axis, while the amount of output is measured on the horizontal axis. The width of the bar associated with each alternative identifies the incremental amount of output that would be provided over the next smaller alternative. The height of the bar illustrates the relative cost per unit of adding the additional output associated with that alternative.

Summary

Whether or not outputs are measured in monetary terms, economic analysis can assist in the evaluation of the efficiency and effectiveness of environmental projects and programs. Measuring the effectiveness of alternative management measures requires, not only a clear delineation of the objectives to be addressed, but also an understanding of the interrelationships between the measures being considered, and their impacts on ecological resources and human behavior and attitudes. Communication and cooperation between all disciplines can help insure all impacts are considered and the description and measurement of environmental outputs supports the evaluation criteria being used.

NONMARKET ECONOMIC VALUES FOR SALMON AND WILDLIFE RESOURCES

Changing Perspectives for Resource Decision Making

Darryll Olsen
Regional Planner/Resource Economist
Northwest Irrigation Utilities

Abstract

Until recently, resource economists primarily focused on measuring the direct net value of fish and wildlife resources, when analyzing management options. And the technical literature abounds with examples of how to estimate these nonmarket economic values (Olsen 1991). But for making important management decisions affecting fish and wildlife resources, direct net values are becoming less significant--the pendulum for resource management decision criteria is swinging toward a new direction. In the Pacific Northwest, the desire for local economic benefits derived from fish and wildlife resources and the concern about securing cost-effective resource mitigation and enhancement actions is outweighing more "conventional" economic benefit assessments.

Nonmarket Values, Some Examples

Nonmarket values can be described under two categories: user value, and nonuser value. An example of user value would be the value held by a sport fisherman to participate in a specific fishery. Resource users are assumed to hold expected consumer surplus (direct use) value, option value, and existence value for the resource. Combined, these three basic value components form the "total value" for the resource user.

To illustrate further, during the spring of 1992, many purchasers of Oregon State fishing licenses participated in a large-scale research effort to estimate economic values for sport fishing on the Rogue River (Olsen and Richards 1992). The study relied on a contingent valuation method (CVM) survey to solicit from Oregon residents, and others, their level of value for the Rogue's salmon and steelhead sport fisheries. The contingent valuation method seeks to identify respondents' willingness-to-pay (WTP) for their sport fishing experience. This value level defines the direct net value of the sport fisheries, an economic value that can be compared to other types of economic activities (through benefit-cost analysis).

Approximately 3,500 anglers who purchased a 1991 Oregon annual, seasonal, or daily fishing license or a salmon or steelhead tag were successfully contacted during the CVM survey. This large number of angler contacts formed two major survey groups: those who purchased fishing licenses and tags within the local, tri-county area; and those who purchased licenses and tags outside the local area.

By contacting license purchasers from the tri-county and non-tri-county areas, researchers were able to make estimates of the distribution of anglers' residences, the sport fisheries' direct net value, and the amount of expenditures anglers made during their Rogue River fishing trips. Once collected, the survey respondents' sample data were statistically "weighted" to conform to the known distribution of license purchasers, based on the Oregon Department of Fish and Wildlife's (ODFW) license data base. Using this approach, it was possible to make a comprehensive economic estimate of the fall chinook and summer steelhead sport fisheries (see Figure 1).

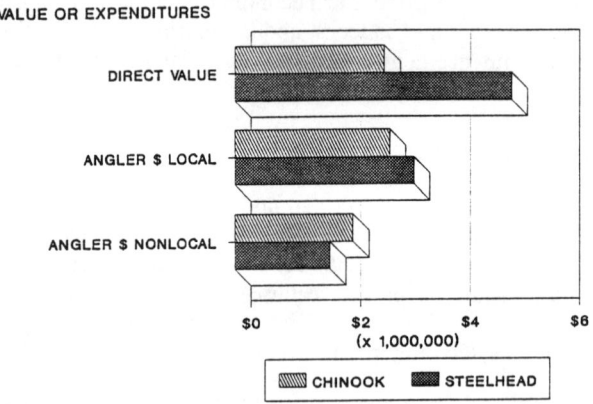

FIGURE 1. ANNUAL DIRECT NET VALUE AND SPORT ANGLER EXPENDITURES FOR ROGUE RIVER FISHERIES

In contrast to user value, nonuser value is a relatively pure form of existence value. Existence value essentially consists of an intrinsic or inherent value that individuals attach to the resource; it can be the value an individual is willing-to-pay for the satisfaction that a recreational or natural resource is protected.

For example, an existence value study was recently conducted to determine

Pacific Northwest residents' willingness-to-pay to protect and enhance Columbia River Basin salmon and steelhead runs (Olsen and Richards 1991). In this study, existence value represented the benefit that individuals gained from the knowledge that doubling the size of the fish runs would provide the runs with greater ecological stability and diversity. The runs are perceived as being in a more favorable state, even though an individual does not intend to use directly the fishery resource in the future.

From this study, relying on a CVM methodology, it was determined that Northwest households were willing-to-pay about $170 million (1989 $) annually for additional fisheries protection and enhancement, in the Columbia River Basin. Resource users contributed 65% to the total value; nonusers with zero probability of future participation in the sport fishery, 25%; and nonusers with some probability of sport fishery participation, 10%.

Focusing Away from Direct Net Value

While many resource economists prefer measures of direct net value to assess social welfare gains or losses from different resource management options, most resource decision makers and those attempting to influence management decisions are more concerned with other types of economic measures.

In the Rogue River sport fishery valuation study discussed above, both direct net values (WTP) and secondary values (angler expenditures) were estimated (see Figure 1). Although the initial purpose for the study was based on the need to acquire direct net values, this need substantially declined, as planners and resource managers began to shift their attention to local economic development actions. The primary questions for resource managers became: how much economic activity is created by the sport fisheries within the local Rogue River area, and how much of this "wealth" comes from non-local residents?

A second reason for the movement away from direct net value assessments is the emphasis on cost-effectiveness analysis for resource mitigation and restoration actions. Under a cost-effectiveness analytical framework, the decision to protect or enhance a resource has already been made--independent of the resource's economic value--and the economic issue concerns the direct net costs of resource protection or restoration. Following a cost-effectiveness path, the highest level of biological benefit (fish, wildlife) is sought for each mitigation or enhancement dollar spent.

For example, Figure 2 displays the results of a cost-effectiveness "quadrant" methodology for salmon recovery under the Endangered Species Act, for Snake River wild salmon stocks (Olsen 1992). Restoration measures that fall in quadrant 1 possess relatively high fish benefits and low economic costs, while measures entering quadrant 4 hold relatively low fish benefits but high economic costs. The point here is that the economic benefits of the fish runs do not have a bearing on the most cost-effective restoration decision.

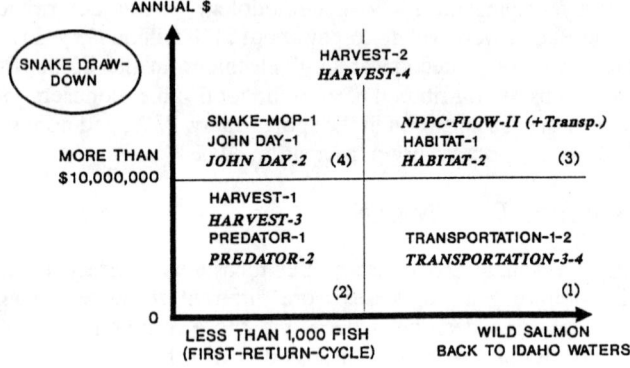

FIGURE 2. COST-EFFECTIVENESS ANALYSIS
RECOVERY MEASURE MARGINAL BENEFITS/COSTS
(QUADRANT RANKS 1,2,3,4)

Conclusion

In the Future, it can be anticipated that more emphasis will likely be placed on secondary expenditure values and cost-effectiveness measures for fish and wildlife resources. Direct net economic values appear to be less meaningful to decision makers and resource planners, at this time.

References Cited

Olsen, D. 1991. Using the contingent valuation method to estimate nonmarket values for energy and natural resource management decisions, an annotated bibliography. The Pacific Northwest Project, Lake Oswego, Oregon.

Olsen, D and J. Richards. 1991. Existence and sport values for doubling the size of Columbia River Basin salmon and steelhead runs. RIVERS 2(1):44-56.

Olsen, D. and J. Richards. 1992. Summary report: Rogue River summer steelhead and fall chinook sport fisheries economic valuation study. Prepared for ODFW by the Pacific Northwest Project, Lake Oswego, Oregon.

Olsen, D. 1992. The path toward cost-effectiveness for salmon recovery measures, working paper. Northwest Irrigation Utilities, Portland, Oregon.

Operating the Columbia River: A Balancing Act

David B. Smith[1]

Abstract

Agreements to market the Canadian share of the hydroelectric power generated at U.S. facilities as a result of the Columbia River Treaty will begin terminating in 1998. Renewal of the power marketing agreements is complicated by the increasing competition for the limited water resources in the Columbia River Basin and the lack of public and institutional forums for addressing and resolving those issues on a system-wide basis. This paper discusses how the three Federal agencies involved in renewing the marketing agreements developed a strategy for future system-wide operation.

Introduction

The Columbia River and its major water projects provide valuable products and services which are critical to the regional economy and way of life. Ever since the first dam was completed on the mainstem Columbia River in 1933, Northwest residents have sought to harness this mighty river for economic growth through providing abundant, low cost electrical energy and transportation of goods; taming the region's devastating spring floods; and storing water to irrigate hundreds of thousands of acres of rich farmland.

To a large extent, these goals have been met, but changing societal values are increasingly putting more emphasis on environmental preservation. Consequently, the challenge has been to find a way to operate the river system to preserve environmental values which contribute to the Northwest's quality of

---

[1]Resource Economist, U.S. Bureau of Reclamation, PO Box 25007, Denver, CO 80225-0007.

life without unduly impacting that part of the region's economic infrastructure which is river dependent.

Ratification of the Columbia River Treaty between the United States and Canada in 1964 made possible the construction of three large water storage projects in Canada and one in the United States. Under the treaty, Canada is entitled to one-half of the increased hydropower production at 11 U.S. dams downstream from the treaty projects. Canada has chosen to sell its half of the power during the first 30 years of operation under the treaty to a group of 41 U.S. utilities.

The Army Corps of Engineers (Corps), Bureau of Reclamation (Reclamation), and Bonneville Power Administration (BPA) each share portions of a complex set of responsibilities and legal authorities for the management of the Columbia River. Because the power marketing agreements with Canada will soon expire, this presented a historic opportunity to review the present operation of the river system to develop a strategy to reduce conflict by providing public input into the decisionmaking process. As a result, a study known as the Columbia River System Operation Review (SOR) was begun to help the agencies make decisions regarding power contracts and future operation of the Columbia River System.

The SOR focused on 14 large Federal dams. Responsibility for the operation of 12 of these dams lies with the Corps, while Reclamation is responsible for the other 2. The hydroelectric power generated at each of these dams is marketed by BPA.

Six dams are on the mainstem Columbia River: Bonneville, The Dalles, John Dam, McNary, Chief Joseph, and Grand Coulee; four are on the lower Snake: Ice Harbor, Lower Monumental, Little Goose, and Lower Granite; and two, Hungry Horse and Albeni Falls, are on the Flathead and Pend Oreille rivers, respectively.

The Process

During the summer of 1990, more than 800 people in 14 cities around the region attended meetings organized by BPA, Reclamation, and the Corps to express their opinions on managing the Columbia River. As a result, over 500 comments were received on how the SOR should proceed. Most comments revolved around how to strike a balance among hydropower, fish, and other uses.

After the public meetings, the interagency study team summarized and distributed the comments received during the public meetings in the Scoping Document. The public was also kept informed on study progress and other pertinent issues (e.g., how the river system is operated) through a newsletter entitled Streamline.

Nine work groups were formed representing the major uses of the river. They were: power, navigation, flood control, anadromous fish, resident fish and wildlife, recreation, irrigation, water quality, and cultural resources. In addition to members from each of the three agencies, other Federal and State agencies provided staff to the work groups. Individuals representing nongovernment groups and Indian Tribes were also invited to serve on the work groups.

Each group was tasked to develop a river system operation which maximized benefits for that use. In addition, the groups were to formulate alternatives which were less than optimum but which improved conditions for their use over present conditions. It soon became apparent that if the present way of operating the system is changed to improve conditions for one use, it usually made conditions worse for some other use(s). Consequently, each work group also evaluated the alternatives from the other work groups as to the impacts on their particular function, i.e., the recreation work group evaluated the impacts on recreation of the flood control proposals.

Several "nonuse" groups, were also formed. One of these groups was tasked to evaluate the economic and social impacts of alternatives. Another was to develop and carry out the public involvement aspects of the entire study. Another group took each alternative and ran it through the hydrologic/power generation models to determine the physical changes in water levels of the river/reservoirs, power production, etc.

Over 90 alternative river system operations were developed in the work groups. Time and budget constraints made it impossible to do a "full-scale" analysis, i.e., economic and social impacts of each alternative; consequently, representatives from each work group and management from each agency had a 2-day retreat to screen them to make a preliminary decision as to which alternatives should move forward for detailed analysis. During the retreat, the more than 90 alternative were blended into 10 system operating strategies. Most of the strategies were a blend of several alternatives (up to seven), but several were made up of only one alternative each.

Meetings were held in September of 1992 in the 14 cities in which the previous meetings had been held in order to present the 10 system operating strategies to the public. Prior to these meetings, the results of the 2-day screening were summarized in the Screening Analysis: A Summary which was sent to people who had indicated some interest in the study in order to help the public understand what the 10 strategies were and how the various river uses would be impacted by them.

Comments received from the public on the 10 strategies are now being evaluated to further refine them. In the near future, an economic and social impacts analysis will be made of the refined strategies.

Conclusion

The SOR study participants realize that there is no perfect operating strategy which will fully accommodate all of the competing needs of the Columbia River users. However, it is hoped that at the end of the study it will be possible to select an operating strategy which minimizes conflict and better addresses the changing needs of the region than is currently the case. It is also hoped that a process will have been developed through which the public users of the river and the Federal agencies which operate it can work together to better balance the river operation to meet both economic and environmental needs.

Literature Cited

The Columbia River System Operation Review Interagency Team. Scoping Document, May 1991. Portland, Oregon.
The Columbia River System Operation Review Interagency Team. Screening Analysis: A Summary, 1992. Portland, Oregon.
The Columbia River System Operation Review Interagency Team. Streamline, various dates. Portland, Oregon.

Economics of Endangered Salmon in the Pacific Northwest

Joel R. Hamilton[1]

INTRODUCTION

Under mandate of the Endangered Species Act (ESA), Idaho, Oregon and Washington are developing recovery plans to preserve several threatened and endangered stocks of salmon. These fish, some migrating as much as 900 miles inland to spawn, have been decimated by man's modification of their habitat. Downstream migrating juveniles must navigate through or around eight slackwater pools and hydropower dams. Adult upstream migrants must negotiate fish ladders, and endure modified temperature regimes and severely reduced spawning area and habitat quality.

While the ESA says little about economics, economic factors do play important roles in designating endangered species and the design of recovery plans. For species in severe trouble ESA pushes us to drastic strategies such as captive breeding, or cryogenic sperm and egg preservation without paying much attention to whether the cost is "worth it". For species in less dire straights, we often face a menu of possible recovery actions from which we can select the least cost set. Economics is important irrespective of the role it plays in decisions to classify salmon as endangered, or decisions about recovery plans. Economic analysis can provide those affected with information about their future. For individuals, communities, and businesses faced with adjusting to ESA actions, more good information is better than less, and economic impact estimates can be a valuable part of this information.

COSTS OF SALMON RECOVERY

There are a lot of players in this economic analysis business. Many of the contending interest groups have come

---

[1] Professor of Agricultural Economics and Director of Martin Institute for Peace Studies & Conflict Resolution, University of Idaho, Moscow, ID 83843.

up with their own competing estimates of the impacts of protecting salmon. Much of the economics being done is not being done well, and many of the estimates are designed to serve constituency interests. I want to showcase some estimates which are an exception to that rule. Cost estimates for salmon recovery were prepared in conjunction with the NMFS Economics Committee by Huppert, Fluharty, and Kenney. Their report serves as an example of the range of recovery scenarios which need to be analyzed, and the range of impacts that can result.

Economic Costs of Salmon Recovery Actions

	Flow Augmentation		Reservoir Drawdown	
	NPPC	NMFS	4 Dams	1 Dam
Hydropower	66 - 112	151 - 1,159	41 - 97	(4) - 23
Irrigation				
L. Snake & John Day	2.5 - 3.5	0	6.1 - 8.1	2.5 - 3.5
Upper Snake	(1.7) - 2.6	(1.7) - 2.6	0	0
River Navigation				
Snake & Columbia	0	0	4.9 - 6.5	2.5 - 3.3
Dworshak	0	0.3 - 0.3	0	0
Recreation	2.7 - 5.4	3.5 - 7.0	4.3 - 8.6	3.3 - 6.3
Dam Modification	0	0	47 - 91	12 - 23
Total	70 - 124	153 - 1,169	103 - 211	16 - 59

Huppert et.al. focused on four alternative sets of recovery actions:
1. The NPPC phase 2 flow augmentation proposal
2. A NMFS proposal which would allocate a larger water budget to flow enhancement.
3. The proposal to draw down the levels of the four lower Snake dams.
4. A proposal to draw down Lower Granite reservoir only.

They also identified five major types of costs:
1. Hydropower costs resulting from reduced head at dams and from shifting flows to periods of power surplus and away from times when the power would have been more valuable.
2. Costs related to irrigation. This includes pumps in the lower Snake and John Day reaches that would need to be relocated if reservoir pools are dropped to MOP or below. Also included is the cost of acquiring flow augmentation water from the upper Snake basin via water markets, net of power values associated with this additional water flow.
3. Drawdown would impose costs on those who use the river for transport of grain and other products. On the

other hand if one relies on massive use of stored water to augment flows, this will disrupt log transportation on Dworshak pool.
4. The various proposals could disrupt recreation on the lower Snake and John Day pools or on Dworshak.
5. Finally, the drawdown proposals would require significant modifications to the structures of the dams themselves.

Note that the estimates are still broad cost ranges; there were wide differences among NMFS Economics Committee members on both methodology and cost magnitude. While some of these estimated costs are large, even if they were all passed on to the region's electricity users the implied rate increases would be only a few percent. Costs to save salmon should not devastate the economy of the northwest.

CONCEPTUAL AND METHODOLOGICAL ISSUES

I will turn to some conceptual problems of doing impact analysis of ESA-type actions. One of the most important considerations is the choice of a base case to compare the ESA action to. Quite apart from the ESA process there are presently a large number of regional programs to preserve salmon runs. The Northwest Power Planning Council has a mandate to give fish and electricity equal priority in planning the power supply future of the region. The states, the Forest Service, BPA, the Bureau of Reclamation all have programs related to salmon survival. My point is that not all of the costs of saving salmon are attributable to ESA. Only to the extent that ESA pushes us to actions and costs beyond what would have been done without ESA, can these impacts be attributed to ESA.

Benefits present other conceptual problems. If salmon runs return to fishable levels, regional economic benefits would be substantial. Even short of full recovery, the struggling runs may be a significant tourist resource. Such benefits are valid offsets to the costs of salmon recovery.

Another complication results when economic activity is displaced but not really lost. Drawdowns will certainly reduce recreation use of affected reservoirs. However the money not spent for recreation on these reservoirs will not stay in the wallets of recreationists. Most will spend it on at another site, or for another activity. Much of the recreation lost to endangered species actions will be offset by increased activity elsewhere in the region.

Even where salmon recovery causes clear direct loss of economic activity and jobs, one must be careful in calling this a cost. Most workers, capital and other displaced resources won't remain unemployed forever. In time they will be reemployed in other communities or industries. This reemployment may offset much of the original loss.

Finally, there are the impacts of recovery spending itself. We may spend millions on new fish screens, fish barges, dam modifications, and legal, economic and biology studies. Some of this work will be supported by new money flowing into the region. The employment and income created will give a nontrivial boost to the regional economy.

I have asserted that costs of salmon recovery will not devastate the region. However I recognize that these costs and benefits will be distributed unevenly, and may be very important to the impacted individuals, communities and industries. Farmers who lose grain revenue to the drawdown will feel the hurt. A construction worker hired to install fish screens, or a guide whose business flourishes because of salmon recovery will feel the benefit.

There are some fairly clear regional interests. There is an upstream-downstream split; downstream interests defending their interest in low electric rates and commercial fishing, and upstream interests defending their interests in sport fishing and their water supply. There are subregional divisions like the situation in Idaho which pits northern Idaho interests in barge transportation against southern Idaho irrigation water users.

There seems to be a commitment to compensate those hurt most seriously by salmon recovery efforts. It will be interesting to watch this play out, especially the efforts to find a pocket deep enough to fund the compensation.

CONFLICT RESOLUTION

The diversity of parties interested in salmon recovery make it vital to try to manage and resolve the conflict. We all know that courts are a blunt and clumsy instrument for resolving such issues. Some view last year's "salmon summit" as a failure because it didn't produce a consensus recovery plan. I regard it as a success because it opened communication among interest groups, and helped everyone toward a more common information base from which to view the problem. I credit the salmon summit for the fact that the salmon dialogue has so far been conducted on a higher plane than the dialogue about spotted owls. Salmon issues will eventually reach the courts, but this stage of the process will be less acrimonious because of the legacy of the salmon summit. We will be better of if we mediate more and litigate less in managing endangered species conflicts. Good economic information is essential to such mediation.

Glen Canyon: The Economic Costs

David A. Harpman[1], Timothy J. Randle[2] ASCE, and S. Clayton Palmer[3]

Abstract

Revenues from power produced at Glen Canyon Dam are used to support Colorado River Storage Project (CRSP) purposes, to pay O&M costs, and to repay construction costs. Generation of peaking power causes downstream releases and river stage to fluctuate on an hourly basis. This has been shown to impact the downstream physical and biological environment. A number of alternative management regimes are being considered to reduce these impacts. This paper discusses the potential impacts of these regimes on power production.

Introduction

Generation of peaking power at Glen Canyon Dam typically results in hourly fluctuations in release and river stage. These fluctuations significantly affect the quality of white-water boating and angling (Bishop, et al. 1987), and the maintenance of the downstream trout fishery. Fluctuations are also thought to affect the reproduction, recruitment, and survival of native fish. Further, historic operations have been implicated in the depletion of pre-dam alluvial deposits with associated impacts on cultural and riparian resources. The Glen Canyon Dam Environmental Impact Statement (GCDEIS) was initiated in 1990 to examine options which "... minimize-- consistent with law-- adverse impacts on downstream environmental and cultural resources and Native American interests..." (U.S. Department of the Interior 1993).

---

[1]Resource Economist (D-5810), [2]NEPA Manager (D-117), U.S. Bureau of Reclamation, P.O. Box 25007, Denver, CO 80225.
[3]Resource Economist, Western Area Power Administration, P.O. Box 11606, Salt Lake City, UT 84147.

Analysis

Nine operational alternatives and their impacts are described in detail in the forthcoming GCDEIS. These range from operations which are largely unrestricted to baseloading of the powerplant. The parameters affecting operation of Glen Canyon Dam are summarized by alternative in Table 1.

Table 1. Summary of Parameters Affecting Generation

Alternative	Upramp rate (cfs/hr)	Downramp rate (cfs/hr)	Minimum Flow (cfs)	Allowable Daily Change (cfs)
No Action	unlimited	unlimited	1,000 winter 3,000 summer	30,500
Maximum Powerplant Capacity	unlimited	unlimited	1,000 winter 3,000 summer	32,200
High Fluctuating Flow	unlimited	5,000	3,000-8,000 depending on month	15,000 - 22,000
Moderate Fluctuating Flow	4,000	2,500	5,000	45% monthly flow
Low Fluctuating Flow	2,500	1,500	5,000 night 8,000 day	5,000-8,000
Seasonally Adjusted Fluctuating Flow	8,000 Oct-Apr 2,500 May-Sep	1,500	5,000 night 8,000 day	45% Oct-May 2,000 June 2,500 Jul-Sep
Existing Monthly Volume	2,000/day between months	2,000/day between months	8,000	1,000
Seasonally Adjusted Steady Flow	2,000/day between months	2,000/day between months	> 8,000 varies by month	1,000
Year Round Steady Flow	2,000/day between months	2,000/day between months	prorated annual volume	1,000

For each alternative, the economic and financial impacts on seven large utilities and over 100 small utilities were estimated for both the existing contract rate of delivery (CROD) institution and an optimal institutional arrangement (HYDROLOGY). Under both institutions examined, the total energy produced is the same but there are losses in summer and winter capacity for most alternatives. The loss of capacity has considerable financial and economic impact on the power system. The estimated economic impacts (Stone and Webster 1992, Moulton 1992) are illustrated in Figure 1. These impacts are calculated following the Principles and Guidelines (U.S. Water Resources Council 1983).

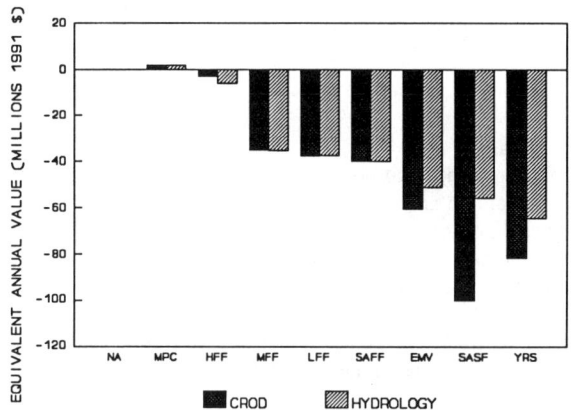

Figure 1. Estimated Economic Impact on Power System.

Restrictions, the specified ramping rates, allowable daily changes in flow, and minimum flows, largely determine the potential generation capacity for any given alternative. As shown in Figure 1, the system-wide economic impact of the alternatives increases with the degree of these operational restrictions. The Seasonally Adjusted Steady Flow alternative has the largest cost since the powerplant is baseloaded and monthly release volumes during peak load months are reduced compared to the No Action Alternative.

As noted in Figure 1, estimates of impact differ between the two institutions examined. Estimates made under the CROD institution represent potential impacts under the existing contractual framework. To the extent that the contractual framework changes, impacts are likely to be less than those portrayed under the CROD institution. Estimates made under the HYDROLOGY institution represent potential impacts if the contractual framework changes and optimal adjustments to the power system are made. To the extent that the contractual framework does not change and optimal adjustments are not made, impacts are likely to be greater than portrayed under the HYDROLOGY institution. Conceptually, these two institutions represent the extremes of likely economic impact.

The impacts displayed in Figure 1 are currently being revised to address several shortcomings. The estimates of economic impact presented here apparently reflect some degree of inter-system transfer payments thereby overstating the economic effect by an unknown amount. These payments will be eliminated in subsequent analyses. In addition, forthcoming analyses will be based on a framework more conducive to the assessment of existing facilities, will model energy conservation measures in a more appropriate manner, and will utilize improved hydrology series. These refinements will allow for a more accurate appraisal of national economic impact. However, the relative economic ranking illustrated in Figure 1 is unlikely to change.

Conclusion

Generation of peaking power at Glen Canyon Dam causes downstream releases to fluctuate on an hourly basis. The resulting variations in flow and river stage may have significant impacts on native and non-native fish, recreational, cultural, and riparian resources. Constraints on hydropower operations may well be imposed to reduce the impacts on these resources. These constraints will degrade demand following capability and decrease the capacity of this facility to generate power on peak. Large and significant economic effects on the power system will result. The magnitude of the estimates presented here argues for a careful assessment of the tradeoff.

Literature Cited

Bishop, R.C., K.J. Boyle, M.P. Welsh, R.M Baumgartner, and P.R. Rathbun. Glen Canyon Dam Releases and Downstream Recreation: An Analysis of User Preferences and Economic Values. Madison, WI: HBRS, Inc. January 1987.

Moulton, Ronald E. Division of Technical Analysis, Western Area Power Administration. Salt Lake City, UT. Personal Communication. November 1992.

Stone and Webster Management Consultants, Inc. Power System Impacts of Potential Changes in Glen Canyon Powerplant Operations. Draft Report to the Power Resources Committee. May 1992.

U.S. Department of the Interior. Operation of Glen Canyon Dam: Draft Environmental Impact Statement. Denver, CO: U.S. Bureau of Reclamation, 1993.

U.S. Water Resources Council. Economic and Environmental Principles and Guidelines for Water and Related Land Resources Implementation Studies. Washington, D.C.: U.S. Government Printing Office, 1983.

The Economic Value of Trout Fishery Management Programs

Donn Michael Johnson[1]

ABSTRACT

The contingent valuation method is used to estimate the economic value of a trout fishing day at the Cache la Poudre River in Colorado. In addition, changes in daily economic value and yearly participation are estimated for changes in the quality of fishing such as: (1) change in catch or size of catch; (2) wild trout fishing versus hatchery trout fishing; (3) change in catch or size under catch and release management.

Anglers are grouped by a distinguishing characteristic, skill level, and the economic value of trout fishing and changes in the quality of fishing are compared at different skill levels.

Introduction

The purpose of this study was to show how the contingent valuation method could be used to improve the cost effectiveness of fishery management programs by estimating the economic value of potential changes in the quality of the fishing experience. It may also be important to show managers that the value of an angling day or changes in quality are not the same for all anglers. Bryan (1977) suggested that anglers are heterogeneous but that within group types, preferences could be homogeneous. In this paper anglers will be grouped as low, medium, and high skill, which fit Bryan's grouping of "occasional," generalist," and "specialist" anglers. Skill level may be a variable that can help explain demand for possible fishing opportunities produced by different management programs.
----
[1]Assistant Professor, Economics Department, University of Northern Iowa, Cedar Falls, Iowa 50614-0401

The study site was the Cache la Poudre River located near Fort Collins, Colorado. The Poudre River was selected to represent a range of management practices and angler activity. Some sections of the river are heavily stocked with catchable rainbow trout while others are set aside for wild trout (rainbow and brown) fishing, where artificial lures must be used and catch and size is restricted.

From the Summer of 1985 to the fall of 1987 a total of 150 interviews were conducted on the Poudre River. The iterative bidding approach to the contingent valuation method was used to obtain anglers economic values for an angler day and changes in the quality of an angler day. Less than 2% of the anglers approached refused to participate in the survey. Protest responses were less than 2% of those interviewed and were removed from the sample.

Results

The following regression equation tests those variables that are hypothesized to influence angler net willingness to pay per day (consumer surplus). A weighted least squares approach was used (A Goldfield - Quant test suggested homoscedasticity) to estimate the equation.

Consumer surplus = 6.6502 - 3.3592 Low Skill (1yes,0no)
(1.32)   (-3.65)

+ 4.0469 High Skill (1yes,0no) - 0.9545 days (per year)
(3.75)   (-3.65)

+ 0.0422 days^2 - 0.0006 days^3 + 0.3073 income ($1,000)
(2.88)   (-2.56)   (1.78)

- 0.0077 Income^2 + 0.00008 Income^3 + 0.7251 Education yrs
(-1.77)   (2.45)   (1.16)

- 0.0370 Education^2 - 0.1126 Time Driving (% trip time)
(-1.77)   (-3.06)

+ 1.4714 Environmental Quality (1-5 Discrete Scale)
(3.07)

Adjusted R^2 = .57, F = 7.18
T - Statistics in Parenthesis
below coefficients.

The regression results suggest that the value of fishing may be influenced by angler skill (Mean value for all anglers was $13.10 per day). Table 1 groups mean economic values by skill group for different potential changes in the fishing experiences. Generally these results support the hypothesis that values vary with skill level. One notable result is that catch and release fishing would not seem to be an acceptable management tool to improve fishing quality for low or middle skill anglers.

Table 1
Differences in Value (Per Day) and Participation by Skill Group for Catch, Size, Wild Trout, Catch and Release Management, Cache la Poudre River, Colorado

	Low Skill	Medium Skill	High Skill
Change in Catch	60	58	32
Change in Dollars	$1.54* (1.46)	$0.52 (0.78)	$0.68* (0.90)
Change in Days	0.74* (0.67)	0.37 (0.54)	0.41* (0.51)
Change in Size			
Change in Dollars	$1.20 (1.55)	$1.31* (1.49)	$2.33* (1.43)
Change in Days	0.49* (0.49)	0.96** (1.00)	1.42* (1.18)
Wild Trout			
Change in Dollars	$0.30* (1.25)	$1.55** (3.11)	$2.97* (4.21)
Change in Days	0.08* (0.46)	1.22 (2.70)	2.06* (4.13)

	Low Skill	Medium Skill	High Skill
Change in Catch (catch and release)	43	34	26
Change in Dollars	$-0.58 (3.51)	$-0.57* (1.87)	$0.70* (1.40)
Change in Days	-0.19 (1.25)	-0.13 (1.44)	0.30* (0.58)

Table 1 Continued

	Low Skill	Medium Skill	High Skill
Change in Size (catch and release)	38	45	21
Change in Dollars	$-1.36* (2.73)	$0.28* (3.15)	$2.03* (1.49)
Change in Days	-0.12** (1.04)	0.48** (2.10)	1.21* (0.89)

Standard Deviation in parenthesis

* Means for skill levels are statistically different at the .05 level or better.
** Means for skill levels are statistically different at the .10 level or better.
Mean values (total sample) for changes in catch, $0.96 per fish and 0.53 days per fish, size, $1.48 per inch and 0.87 days per inch, wild trout, $1.35 per day and 0.95 days per year, catch (catch and release), $-0.26 per fish, -0.05 days per fish, size (catch and release), $0.03 per inch and 0.41 days per inch.

Appendix 1 - Skill Level

The skill rating was determined by total daily catch, whether browns or rainbows, and the observed skill in handling fishing tackle. The differences shown on table 1 would indicate that skill level can be used as a proxy for taste and preferences of anglers on the Poudre River. This suggests skill differences may be a useful way to group anglers at other sites to determine if differences exist in willingness to pay for fishing or changes in its quality.

Appendix 2 - References

Bryan H. (1977). "Leisure value systems and recreation specialization: The Case of Trout Fishermen." *Journal of Leisure Research*, 9(3), 174-187.

MEETING INSTREAM FLOW NEEDS OF
LOWER COLORADO RIVER IN TEXAS

Quentin W. Martin[1], M. ASCE

Abstract

The Lower Colorado River Authority (LCRA), an agency of the State of Texas, manages the surface waters of the lower Colorado River in Texas. The major water supply source in the lower basin is the Highland Lakes chain of reservoirs in Central Texas. The use of water from these lakes for environmental protection and enhancement has received increasing attention in recent years. The LCRA recently completed major revisions to its comprehensive Water Management Plan (WMP) for the Highland Lakes. These revisions included changes to incorporate the results of a three year study of instream flow needs in the lower Colorado River. The instream flow needs were determined to consist of two flow regimes: critical and target. The critical flows are considered to be the daily minimum flows needed to maintain minimum viable aquatic conditions for important fish species. The target flow needs are those daily flows which maximize the available habitat for a variety of fish. After evaluating numerous policy options, LCRA revised to WMP to allow the release of water from the Highland Lakes to maintain the daily river flows at no less than the critical flows in all years. Further, in those years when drought-induced irrigation water supply curtailments do not occur, LCRA will release water from the lakes, to the extent of daily inflows, to maintain daily river flows at no less than the target levels. To fully honor this pledge, LCRA committed an average of 28,700 acre-feet annually, during any ten consecutive years, from the dependable supply of the Highland Lakes.

---
[1] Manager, Water and Wastewater Engineering Program, Lower Colorado River Authority, P.O. Box 220, Austin, Texas 78767, 1-512-473-4064, Fax 1-512-469-6873.

Introduction

The Lower Colorado River Authority (LCRA) is a water conservation and reclamation district created by the State of Texas in 1934. It has a statutory service district of ten counties in Central Texas, covering approximately 10,000 mi^2 (Figure 1). LCRA operates a major reservoir system, called the Highland Lakes, on the lower Colorado River, and provides approximately 650,000 acre-feet of surface water annually for municipal, manufacturing and irrigation purposes. Approximately 80% of that supply is used for agriculture, specifically the irrigation of rice in Colorado, Wharton, and Matagorda Counties.

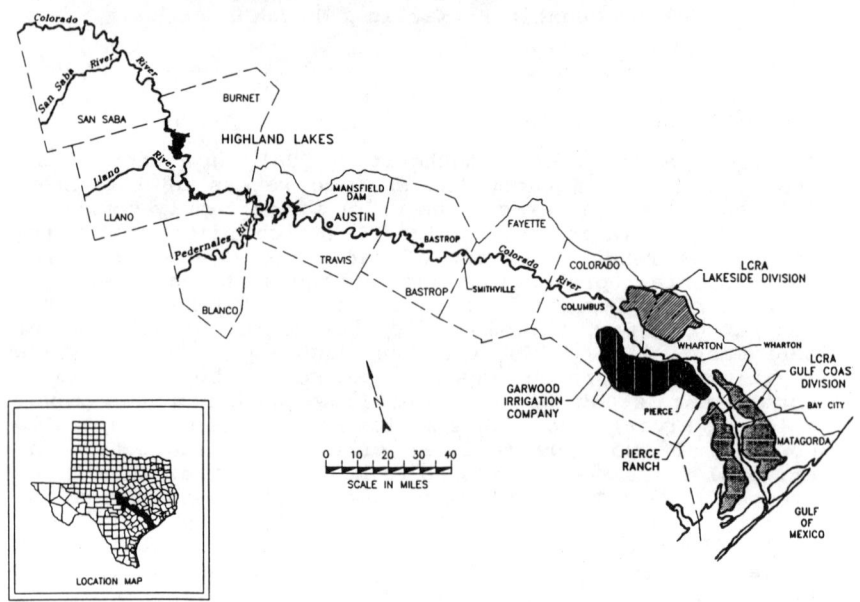

Figure 1. Lower Colorado River Authority District

In 1989, the Texas Water Commission (TWC) approved a water management plan for the LCRA's stored water rights in the Highland Lakes. The plan specifies an interim minimum instream flow requirement of 200 cubic feet per second at or downstream of Bastrop. The TWC directed LCRA to study the instream flow needs of the aquatic communities in the lower Colorado River and return to the TWC no later than 1992 with amendments to WMP reflecting the results of that study. In the plan, LCRA committed 25,000 acre-feet annual from the combined firm yield of the Highland Lakes for instream flow maintenance and freshwater inflows for the bays and estuaries.

Instream Flow Needs

After a three year study, LCRA adopted two sets of instream flow needs: critical flows and target flows (1). The critical flows are considered to be the daily minimum flows needed to maintain a viable aquatic habitat for fish species. The critical flows are:

1. **Minimum daily flow at the United States Geological Survey (USGS) Bastrop streamgage of 120 cubic feet per second (cfs),**

2. **Minimum daily flow at the USGS Austin streamgage of 46 cfs, and**

3. **Minimum daily flow at the USGS Bastrop streamgage of 500 cfs from the period April 15 through May 31.**

The first critical flow is based on the need to provide adequate dissolved oxygen levels in the river. The minimum flow at Austin represents a flow necessary to preserve the river flow at no less than the lowest seven day average flow that would be expected once every ten years. The final flow critical flow is that needed to provide adequate spawning conditions for the Blue Sucker fish species. This species is considered an important indicator species and potentially a threatened species.

A second set of instream flow needs are those which maximizes the available habitat for the fish species in the lower Colorado River. The target instream needs are not considered as biologically critical since the native species in the river are adapted to highly variable river flow conditions. The target instream needs are indicated in Table 1.

Adopted Policy for Meeting Instream Flow Needs

After evaluating numerous policy options, LCRA adopted a policy (2) of maintaining instream flows by releasing water from the Highland Lakes to:

1. **Maintain the daily river flows at no less than the critical instream flow needs in all years, and**

2. **Maintain daily river flows at the target instream flow needs in those years when the four major irrigation districts are not curtailed, to the extent of inflows each day to the Highland Lakes as measured at the upstream streamgages.**

This policy fully meets the most important instream flow

needs at all times and meets the desirable (target) flows during periods of normal or above normal streamflow conditions. To fully honor this commitment, LCRA increased its the present guarantee of dependable water supplies from the Highland Lakes for instream flow and bay and estuary inflows from 25,000 acre-feet annually to an average of 28,700 acre-feet annually during any ten consecutive years.

Table 1. Target Instream Flow Needs For the Lower Colorado River

MONTH	TARGET DAILY FLOWS (CFS)		
	BASTROP	EAGLE LAKE (NEAR COLUMBUS)	EGYPT (NEAR WHARTON)
JANUARY	370	300	240
FEBRUARY	430	340	280
MARCH	560	500	360
APRIL	600	500	390
MAY	1030	820	670
JUNE	830	660	540
JULY	370	300	240
AUGUST	240	200	160
SEPTEMBER	400	320	260
OCTOBER	470	380	310
NOVEMBER	370	290	240
DECEMBER	340	270	220

The additional water committed to meeting the instream flow needs reduces the water available for the four major irrigation districts under the existing LCRA Drought Management Plan (DMP). To mitigate that impact, LCRA revised the DMP to achieve as much cultivated acreage during the critical drought as would be expected under the current DMP. This was made possible by reduced from 400,000 to 325,000 acre-feet the minimum beginning of year storage required for annual irrigation water sale contacts.

Appendix: References

1. Mosier, D. T. (1992). "Instream Flows for the Lower Colorado River," Open File Report, Lower Colorado River Authority, Austin, Tx.

2. Lower Colorado River Authority (1992), <u>Water Management Plan for the Lower Colorado River Basin</u>, Austin, Texas.

Managing Instream Flows for Salmonid Spawning Habitat

Jeffrey B. Bradley[1], Member ASCE

Abstract

Detrimental sediment related effects on aquatic habitat for anadromous fisheries include (1) deposition of excessive amounts of fine sediments in spawning gravels, and (2) mobilization of the streambed during high stream flows thus destroying spawning redds or alevin. These effects can be managed by utilization of upstream reservoirs to modify instream flows to minimize negative impacts on the fishery. That management in some cases may be a tradeoff with other alternate uses of water such as water supply or power. In other cases, however, it can be a win-win alternative for all parties. This paper presents methods for assessing detrimental impacts on salmonid habitat which are directly related to instream flows.

Fine Sediment Intrusion

The effects of fine sediment on stream biota and their habitats include reduction of primary production, damage to respiratory organs, entombment of organisms, increased disease, reduction of intragravel dissolved oxygen and flow, and alteration of water chemistry. Studies have shown that fine sediment intrusion into streambed gravels can reduce permeability and intragravel water velocity, thereby limiting the supply of oxygen to developing embryos and the removal of metabolic wastes from them. Excessive fine sediment deposition can effectively smother incubating eggs and entomb alevin and fry. Most studies of egg incubation in sediments have related survival and emergence of young fish to amounts of one size class of sediment. The larger sediments most often studied include those with diameters of 0.84, 2.0, 3.3, 4.6, and 6.4 mm. In tests designed to evaluate intrusion of sediment into streambeds, Beschta and Jackson (1979) found that fine sands (less than 2 mm) moved through clean gravel (15 mm average diameter) and filled the gravel voids from the bottom up. In contrast the intrusion of larger material (5 mm) formed a sand seal in the upper layers of the gravel, the larger

---

[1] Chief Executive Officer, WEST Consultants, Inc., 2101 Fourth Avenue, Suite 1310, Seattle, WA 98121-2307

particles bridging the voids of adjacent gravels. In this situation, eggs deposited in the gravel should at least be able to continue embryo-genesis and hatch, if intragravel water velocity and dissolved oxygen are adequate, whereas fine sands will cause significantly greater mortality. Reiser and White (1988) looked at sixteen mixtures of two distinct size classes. Fine sediments were defined as less than 0.84 mm and coarse sediments from 0.84-4.6 mm. Their studies showed that egg survival decreased as the proportion of fine sediment increased. For low ambient levels of sediment (less than or equal to 10%), egg survival could be markedly increased if the fraction of material less than 0.84 mm were minimized or eliminated. Other investigators have suggested that productive spawning areas should contain less than 5% of material smaller than 0.84 mm. Reiser and White (1988) showed in their tests that the mixture containing 5% fine sediment and 5% coarse sediment resulted in egg survival exceeded only by survival in mixtures containing no fine sediment. In their studies fine sediment alone in the gravel caused greater egg mortality than the same amounts of fines mixed with coarse sediment.

It is clear that different sizes of fine sediment and the mixture of fine and coarse sediment has an important role in determining egg survival. It is insufficient to speak in terms of percentage of fines in connection with egg survival without defining a specific size of the "fine" sediment. Certainly, the smaller sediments (less than 0.84 mm) are the most detrimental to incubating eggs. A great deal more research will be needed to refine our understanding of the impact of fine sediment percentage and gradation on salmonid habitat. Standardization of data collection methods and definition of the sediment size constituting "fine" sediment is also needed.

Streambed Mobilization at High Flows

Conceptually, when the drag force (shear stress) is less than some critical value the bed material of a channel remains motionless. But when the shear stress over the bed attains or exceeds its critical value, particle motion begins. The actual beginning of motion is more difficult to define. That difficulty is a consequence of a phenomenon which is random in time and space. Many researchers have attempted to better identify incipient motion. Still the exact solution defies theoretical analysis. The complexity of the problem explains the diversity of experimental results. In reality, there is no truly critical condition for initiation of motion for which motion begins suddenly as that condition is reached. Data available on critical shear stress are based on more or less arbitrary definitions of critical conditions.

The beginning of motion, or incipient motion, has ramifications to fisheries in that once the "critical threshold" has been exceeded the armor of a gravel bed stream will be in general motion. The substrate beneath the armor will "blow out" once the armor is in general motion. Salmonid redds will at least be partially destroyed during such an occurrence. Salmonids will generally spawn in coarser sediments or in lower velocity areas mitigating this concern except at higher flows. Consequently, large floods may negatively effect fish habitat.

Incipient Motion Criteria for Use in Identifying Instream Flow Requirements

A movement parameter (a modified Shields parameter) can be defined (Milhous and Bradley, 1986) as

$$\beta = \gamma RS/\gamma(G_s - 1)D_{50}$$

where R is the hydraulic radius (depth in a wide channel), S is the energy gradient, γ the specific weight of water, G_s the specific gravity of bed sediment and D_{50} the median size of the bed surface layer. γRS is the bed shear stress. The movement parameter as defined above is the dimensionless shear stress. Work in Oak Creek, Oregon (Milhous, 1973) indicated that if the shear stress applied to the streambed is just large enough to move a small portion of the larger particles in the surface layer, fines deposited in and among the armor material will be moved at a higher shear stress (surface flushing). The movement parameter required to cause surface flushing of fine sediment from the bed was determined to be 0.02. The movement parameter must be sufficiently high to move the bed material below the armor in order to remove fines from within the voids and cause general motion of the bed at depth (depth flushing). Review of the literature shows that the movement parameter can vary from 0.017 to 0.076 for the beginning of motion. The higher values are probably for general movement and the lower for absolute stability. Gessler (1970) indicated that 50% of the particles will move when the movement parameter is 0.047. Milhous and Bradley (1986) indicated that for depth flushing, defined as the point where 30% of the armor is moved, allowing for depth flushing of fine sediments from the bed, the movement parameter is 0.035. These values of movement parameter for surface and depth flushing are supported some data from the Cedar River, Washington (Bradley, et al 1991).

Hydraulic parameters and bed sediment gradation can be used to compute surface and depth flushing movement parameters. Bradley, et al (1991) computed movement parameters in the Cedar River for each cell of the PHABSIM hydraulic model of the U.S. Fish and Wildlife Service. That model, though one-dimensional, is divided into cells or stream tubes with individual hydraulic parameters computed for each cell (based on one-dimensional concepts). The velocity distribution for turbulent flow may be approximated using a logarithmic or one-sixth power law distribution. The mean velocity and depth for specific flows at each cell were computed in PHABSIM. The equations assume that the mean vertical velocity is defined as that velocity measured at 0.6 of the depth. The shear velocity, the bed shear stress and the movement parameter can then be computed for each cell of the hydraulic model. Those values on a cell by cell basis can then be compared with the movement parameters for surface flushing, 0.02, and for depth flushing, 0.035. In order to obtain more accurate hydraulic parameters it may be advisable in some cases to use a two-dimensional hydrodynamic model.

The concept of surface and depth flushing tied to incipient motion criteria provides a method to determine instream flow requirements to provide flushing flows for removal of fine sediments from spawning gravel, and also for use in determining flows at which the entire streambed is mobilized. Upstream reservoir regulation can then be optimized in order to minimize negative impacts to the spawning gravels. If fine sediment intrusion is identified as a negative impact at a given site or river reach, the flow required to cause surface and depth flushing can be identified. Flushing flow criteria can then be developed to enhance the fishery. Conversely, by using the depth flushing movement parameter, an upper boundary can be established and reservoir regulation can be used to minimize the possibility of general streambed movement and destruction of spawning redds. Once flow requirements have been identified to enhance the fishery it will then be possible to develop a reservoir management scheme that will optimize all purposes of a multi-purpose dam.

Conclusions

Incipient motion criteria and computed movement parameters (dimensionless shear stress) can be used to identify instream flows to minimize fine sediment intrusion, and flows that will cause general mobilization of the streambed causing destruction of salmonid redds. That information can then be used to optimize reservoir regulation of a multipurpose dam to provide the most benefit to all purposes.

Appendix - References

Beschta, R.L., and Jackson, W.L., "The Intrusion of Fine Sediments into a Stable Gravel Bed," Journal of the Fisheries Research Board of Canada, 36:204-210, 1979.

Bradley, J.B., Williams, D.T., and Barclay, M., "Incipient Motion Criteria Defining "Safe" Zones for Salmon Spawning Habitat," Proceedings, Hydraulic Specialty Conference, ASCE, Nashville, TN, 1991.

Gessler, J., "Self Stabilizing Tendencies of Alluvial Channels," Journal of the Waterways and Harbors Division, ASCE, Vol. 96, No. WW2, May 1970.

Milhous, R.T., Sediment Transport in a Gravel-Bottomed Stream, Ph.D. Dissertation, Oregon State University, Corvallis, OR, 1973.

Milhous, R.T. and Bradley, J.B., "Physical Habitat Simulation and the Moveable Bed," Proceedings, Water Forum '86, ASCE, Long Beach, CA, 1986.

Reiser, D.W., and White, R.G., "Effects of Two Sediment Size Classes on Survival of Steelhead and Chinook Salmon Eggs," North American Journal of Fisheries Management, 8, 1988.

Managing Reservoir Storage for Instream Flow

Terry Waddle, AM ASCE[1]

Abstract: Two possible approaches to using a portion of reservoir storage to supply instream flows are 1) to determine a fixed amount to be released each year as minimum instream flows and 2) to set aside a fraction of storage and inflow for instream management. An example is presented showing operations of these two alternatives to provide instream habitat below a reservoir. A simplified fish population index model, the effective habitat time series, is used to determine when water budget releases will produce habitat benefits. The effective habitat time series acts as a surrogate for fish population and reflects the mid to long term influence of water management decisions on the life cycle of a fish species. An operation rule for the storage account that considers habitat events is developed. The paper contrasts the fish habitat benefits of storage account operation with the fixed minimum flow approach.

Introduction

Currently in the United States many water storage projects must satisfy tailwater instream flow (IF) constraints. When stored water is used for IF needs water managers need to know how to use it most efficiently. Traditional approaches assign a minimum flow as a project license condition. Minimum flows give certainty of timing and volume of instream flow water use. A new concept, dedicating a volume of storage to instream use, gives certainty of storage available for other purposes and causes instream flow uses to be subject to water availability. This paper contrasts the fish population implications of an algorithm to manage an instream flow storage account with those of a constant flow regime.

An Approximate Fish Population Model

An extension to the U.S. Fish and Wildlife Service's Physical

---

[1]Hydrologist, Nat. Ecology Research Cntr., U.S. Fish and Wildl. Serv., 4512 McMurry Ave., Ft. Collins, CO 80525

Habitat Simulation System (PHABSIM) (Milhous et al., 1989) was proposed by Bovee (1982) to approximately represent a fish population in terms of physical habitat. This model is called the effective habitat time series or effective habitat model. It is based on two assumptions. First, the fish population is assumed limited by physical habitat and second, the limiting habitat is assumed to be occupied at equal density from year to year. In river systems where these assumptions hold, the fish population can be expressed in terms of habitat actually occupied during limiting events.

Waddle (1992) formalized the structure of this model as the recursive relations given in equations 1 and 2.

$$EH_{\ell,y} = MIN(\ LAH_{\ell,y},\ HD_{\ell,y}) \qquad (1$$

$$HD_{\ell,y} = EH_{\ell-1,y-1} * M_{\ell-1,\ell} \qquad (2$$

where: EH = the approximate population equivalent called effective habitat, HD = habitat demand, M = a multiplier relating effective habitat for successive life stages between years, LAH = limiting available habitat; the lowest habitat value for lifestage ℓ in year y. This sequence describes the complete fish life cycle because EH for adults is coupled to HD for spawning (lifestage ℓ-1 for spawning is the adult life stage). In the water allocation algorithm that follows, HD is used to forecast the next year's seasonal pattern of target flows that would not limit the population.

Storage Account Management Algorithm

The instream storage account is operated as a discrete portion of a reservoir. A volume or fraction of reservoir inflow is allocated to the instream storage account by water right priority or other administrative procedure. Target instream flow deliveries for each time period in a year are derived from habitat demand by solving for $Q_t = q_t(HD_t)$ from the habitat-discharge relation given by the PHABSIM procedure and subject to a mass balance constraint on the storage account.

In dry years, reduced instream habitat targets are calculated by imposing uniform habitat reductions across all life stages ℓ for all time periods t within a year as shown in equation 3.

$$HD'_{\ell,t} = HD_{\ell,t} * \frac{\sum_t q_t(HD_t)}{Q\ available} \qquad (3$$

Then the revised instream flow releases are recursively calculated from the reduced habitat demand until the instream flows match the amount of water available in the storage account as shown in equations 4 and 5.

$$Q'_t = q_t(HD'_{l,t}) \qquad (4$$

$$\sum\nolimits_t (Q'_t) \le Q \text{ available} \qquad (5$$

Using this management logic, instream flows are varied so the fish population faces equally severe habitat constraints for all life stages throughout the year. In contrast, with a fixed minimum flow regime one life stage may face a more severe constraint than the others and thus become a more severe limit to the population. This approach attempts to maintain long term population robustness by avoiding severe restrictions to any single life stage during dry years.

Dolores River Example

The Dolores River is located in the southwest corner of Colorado. Its waters are impounded by McPhee Dam and diverted directly from the reservoir for out-of-the-basin irrigation near Cortez, Colorado. Releases to the river include instream flows and controlled releases to avoid spills. All other water is exported.

The system was modeled using a 54 year deterministic simulation in which water demands were represented using the appropriation doctrine. Water delivery volumes, patterns and priorities were assigned to similar groups of water users. An instream storage account of varying size was implemented with the instream flow delivery algorithm described above. The model was designed to recognize storage rights in the reservoir so the instream account was treated as one of several storage rights.

The physical habitat conditions in the river were derived from PHABSIM data collected by the Colorado Division of Wildlife (Nehring, 1990). Empirical water temperature data supplied by the Bureau of Reclamation (Lashmet, 1991) was used to seasonally adjust PHABSIM habitat representations. In particular, the lowest summer flows were associated with high water temperatures and reduced habitat suitability (LAH in this model).

Results and Conclusions

Figure 1 contrasts the effective habitat produced by a constant discharge flow regime and by flow regimes using a storage account and the habitat shortage spreading algorithm described above. Note that on average the same effective habitat can be supplied with a 24,500 acre foot storage account as with a constant 50 cfs minimum flow. Similarly, using approximately the same amount of water as is

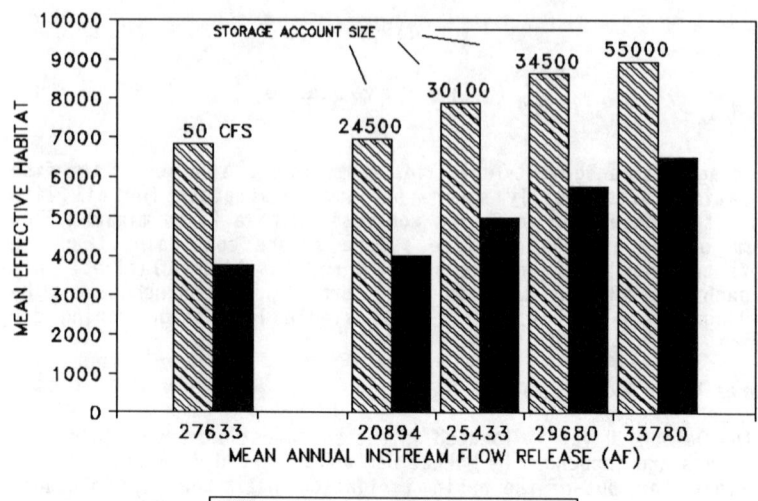

Figure 1. Effective Habitat Produced by Different Management Alternatives

required to maintain a 50 cfs constant flow the effective habitat for different species can be increased between 12 and 25 percent.

Appendix 1. References

Bovee, K.D., A Guide to Stream Habitat Analysis Using the Instream Flow Incremental Methodology, Instream Flow Information Paper 12, U.S. Fish Wildl. Serv., 1982, 247 pp.

Milhous, R.T., M.A. Updike, and D.M. Schneider, Physical Habitat Simulation System Reference Manual - Version II, Instream Flow Information Paper 26, U.S. Fish and Wildl. Serv. Biol. Rep. 89(16), 1989, v.p.

Lashmet, K., U.S. Bureau of Reclamation, Durango Projects Office, Personal communication transmitting data files, August, 1991.

Nehring, R. B., Colorado Div. of Wildlife, Personal communication transmitting data files, Jan. 13, 1990.

Waddle, T.J., A Method for Instream Flow Water Management, Ph.D. Dissertation, Colorado State U., Summer, 1992, 278 pp.

NATURAL, CULTURAL, & ENVIRONMENTAL RESOURCE VALUES INFLUENCING COLORADO RIVER BASIN MANAGEMENT

By Marshall Flug[1], M.ASCE; and William L. Jackson[2]

Abstract

The Colorado River Basin supplies water to the seven western states of Wyoming, Colorado, Utah, Arizona, New Mexico, Nevada, and California, and to Mexico. Originally the Colorado River was developed to help settle the semiarid lands in the west by means of irrigation. Other uses soon evolved for power generation, flood control, domestic and industrial supply, recreation, and fish and wildlife protection. Several large dams and reservoirs can store four times the natural flow of 18.5 Gm^3 per year. The Federal Government owns over half the lands within the Basin. Management agencies include the National Park, and Fish & Wildlife Services, and the Bureaus of Indian Affairs, Land Management, and Reclamation. Since 1991, agencies are cooperating to understand and manage multiple purposes that include protection of natural, cultural, recreational, and environmental resources, and endangered species for the enjoyment and use by future generations.

Introduction

The purpose of this paper is to provide a perspective to changing water use priorities in the Colorado River Basin. These changes have inspired an integration of natural resource values into the Colorado River Annual Operating Plan.

The Colorado River Basin, which drains an area of 629 Mm^2 (US Comptroller General 1979) in seven western states, comprises the Yampa, Green, San Juan, Gila, Colorado, and Little Colorado Rivers, as well as many others. Each tributary has unique characteristics owing to the diversity of geologic features, and large variations in elevation and associated climate. The Colorado River traverses some 2,240 km, originating as snowmelt at high altitudes in the Rocky Mountains with peaks reaching over 4,200 m elevation and eventually emptying into the Gulf of California at sea level. The average annual precipitation is 0.15 m in parts of the arid desert southwest, but as much as 1.5 m at higher mountain elevations. Since the majority of streamflow is

---

[1]Research Hydrologist, Water Resources Division, National Park Service, 1201 Oakridge Dr., Suite 250, Ft. Collins, CO 80525.
[2]Hydrologist and Chief Water Operations Branch, Water Resources Division, National Park Service, 1201 Oakridge Dr., Suite 250, Ft. Collins, CO 80525.

derived from melting snow, it is not surprising that large dams were constructed to tame highly variable seasonal and annual runoff. The regulated waters sustain large agricultural and urban economies in the states of Wyoming, Colorado, Utah, Arizona, New Mexico, Nevada, and California, and in Mexico.

Historical Basin Development

The Colorado River Basin has been actively developed and managed, and it's water fought over since the early 1900's. When President Theodore Roosevelt wanted to encourage settlement of arid lands in the west, the US Congress passed the Reclamation Act of 1902 creating the present day Bureau of Reclamation (BOR), to develop water projects that provide irrigation water and storage. President Roosevelt also began a legacy of conservation by expressing the duty of people to protect ourselves and our children against the wasteful development of natural resources. Natural aquatic and riparian ecosystems were poorly understood at that time; social and environmental objectives were undefined; and the needs of National Parks, Monuments, Recreation Areas, Wildlife Refuges, and Indian Reservations were unquantified. Thus, the Colorado River was developed to meet the demands and economic growth in agriculture, urban water supply, and electric power.

Major Water Projects. The Boulder Canyon Project Act of 1928 authorized construction of hydro-powerplants, Hoover Dam which formed Lake Mead, and the All American Canal for delivery of water to California. The Colorado River Storage Project of 1956 authorized construction of Glen Canyon Dam which formed Lake Powell and three other major storage units within the Upper Colorado River Basin. The Colorado River Basin was divided into Upper and Lower Basins at Lee Ferry, Arizona, by a 1922 Colorado River Compact that was agreed to by the seven Basin states.

More than a half century of water compacts, treaties, laws, and water rights cases (i.e., legal decisions), commonly known as the "law of the river", have clouded authorized consumptive and nonconsumptive uses of Colorado River water. Numerous studies and reports were prepared about the Colorado River Basin (US Comptroller General 1979), its water availability (Iorns, et al. 1965), quality (US Environmental Protection Agency 1971), and management strategies (Hyatt, et al. 1970). The energy crisis of the 1970's inspired a new wave of reports concerning the availability of water for uranium, tar sands, oil shale, and synthetic fuels industries (US Department Of Interior 1974; Hansen 1975; Flug 1979; Flug, et al. 1979). The Colorado River is considered over appropriated by many water users. This belief is attributable to the estimate of annual flow used to apportion water to beneficial consumptive uses among the seven basin states plus Mexico. At the time of the division of water between the Upper and Lower Basins, the Colorado River was experiencing the wettest ten-year period of record, about 23.2 Gm^3. Currently the BOR uses an annual natural flow of 18.5 Gm^3 which is the long term average computed for the period 1906-1977. The over appropriation exists even before consideration of unresolved federal reserved water rights, threatened and endangered species, and environmental water needs.

The Environmental Era

Federal Reserved Rights. The 1908 Winters Doctrine (207 US 564) first enunciated federal reserved water rights where the US Supreme Court held that water was set aside to fulfill the purposes of an Indian Reservation. This principle was expanded with the landmark 1963 US Supreme Court case of Arizona vs. California (373 US 546) which stated that "the United States intended to reserve water sufficient for the future requirements of Lake Mead National Recreation Area ... " and included national monuments, wildlife refuges, forests, parks, and other federal reserved lands. Federal Reservations and federally owned lands managed by the National Park Service, Fish & Wildlife Service, Bureaus of Indian Affairs, Land Management, and Reclamation comprise over half of the lands in the Colorado River Basin.

Changing societal values and greater environmental awareness in the 1960's have placed increasing emphasis on water use for preservation of fish and wildlife and for recreation. The original Organic Act (United States Code 1916) creating the National Park Service states its mission as "... to conserve the scenery and the natural and historic objects and the wildlife therein and to provide for the enjoyment ... as will leave them unimpaired for the enjoyment of future generations." The Organic Act was amended in 1970 and more notably by the 1978 Redwood Amendment which stresses that the Secretary of the Interior has an absolute duty, which is not to be compromised, to fulfill the mandate of the 1916 Act and take actions and seek relief to safeguard park units (Mantell 1990). In 1992, the Grand Canyon Restoration Act was signed into law, providing for protection of downstream resources in the Grand Canyon by including them as primary purposes of the Colorado River Storage Project Act of 1956. Other notable legislation influencing federal land management agencies include the Endangered Species Act of 1973, Wilderness Act of 1964, Wild and Scenic Rivers Act of 1968, National Environmental Policy Act (NEPA) of 1969 which required the writing of environmental impact statements (EIS), Executive Order 11988 in 1977 for Floodplain Management, and Executive Order 11990 for Wetlands Protection with section 404 of the Clean Water Act. Cultural preservation is covered by the Archaeological Resources Protection Act of 1979, National Historic Preservation Act of 1966, American Indian Religious Freedom Act of 1978, and Native American Graves and Repatriation Act of 1990.

The National Park Service manages Dinosaur and Black Canyon of the Gunnison National Monuments; Grand Canyon, Canyonlands, and Arches National Parks; Curecanti, Glen Canyon, and Lake Mead National Recreation Areas all located along the Colorado River or major tributaries. Values of concern to the National Park Service include providing favorable conditions for a myriad of resource attributes briefly identified in the following subsections.

Vegetation. Native riparian vegetation such as willow, cottonwood, and box elder provide habitat for native bird and animal species. Conversely, management strategies are sought to prevent the introduction and spreading of exotic species. Most notably, the prolific tamarisk has choked portions of the river within the Grand Canyon.

Endangered Species. Endangered species protection includes the humpback and bonytail chubs, razorback sucker, and Colorado squawfish, which require very limited remaining sand or gravel bars and water conditions (e.g., temperature) for spawning and other life stages. The restoration of backwaters, riparian wetlands, fish spawning areas, natural sediment deposition patterns, protection of shorelines by reduced wave erosion, and the seasonal timing of flow patterns all contribute to fish habitat and reproduction needs.

Recreation. Both whitewater and recreational boating demands are increasing at a rapid rate and are directly dependent on Colorado River flow release to determine the quality of enjoyment experienced by participants. In addition, fishing conditions, access to shorelines and boat docks on reservoirs, navigability in side canyons and hideaway spots, and fish feeding areas are dependent on flow levels.

Cultural Resources. Protection of archeological sites, cultural resources, and traditional cultural properties are important to Native Americans. Included are springs, marshes, willow gathering areas, places of religious worship, and geologic features that provide minerals, pigments, and other substances for traditional lifeways.

Ironically, many conflicts can exist between natural and environmental resource issues that compete for protection. Most notable is the situation that arose by introducing the cold water trout, a non-native, to the Grand Canyon. The release of clear and cold water has created a Blue Ribbon Trout Fishery in the waters below Glen Canyon Dam. The protected eagle has returned to theses canyons to feed on the widely available trout that are visible in the clear water discharge. This has created a dillema for management of how to balance the recreation demands of the trout fishery with habitat conditions required for some of the native and endangered fish species.

The latest Bureau of Reclamation's 1992 Strategic Plan sets its mission "To manage, develop, and protect water and related resources in an environmentally and economically sound manner in the interest of the American public." The plan also identifies several specific elements for protecting the environment. This change in philosophy is experienced in the Colorado River Basin where, in 1991, agencies with purposes and mandates to protect the natural resources of the Colorado River were invited to participate in the Management Work Group, which prepares an annual operating plan that is eventually implemented by the Secretary of Interior, and to provide input to the Long-Range Operating Criteria for Colorado River reservoirs. A cooperative interagency effort to identify natural resource issues and to develop acceptable methodologies for use in arriving at consensus recommendations for river operations is being pursued. The interagency efforts include identifying and gathering technical information for each of the impacted aquatic and riparian resources; quantifying Colorado River flow and reservoir water level preferences for protection of natural, cultural, recreational, and environmental resources, and endangered species for enjoyment and use by future generations; and development and application of appropriate methodologies and modeling for evaluating tradeoffs and priorities associated with protecting the resource values within the context of maintaining legislated water management values. The intent of this program is to develop solid technical information and data to assist with resource management decisions along the Colorado River Basin for the purpose of enhancing natural resources and ensuring preservation into succeeding generations.

References

Flug, M. 1979. "Impacts of Water Use Efficiency on Energy Development." *Water Resources Bulletin*, AWRA, 15(6), 1743-1752.

Flug, M., Walker, W., and Skogerboe, G. V. 1979. "Energy-Water-Salinity: Upper Colorado River Basin." *J. Water Resources Planning & Management*, ASCE, 105(WR2), 305-315.

Hansen, D. C. 1975. "Water Available for Energy-Upper Colorado River Basin." *Paper Preprint 2564*, ASCE, NY.

Hyatt, M. L., et al. 1970. *Computer Simulation of the Hydroilogic-Salinity Flow System Within the Upper Colorado River Basin*. Water Research Lab, Utah State University, Logan, Utah.

Iorns, W. V., Hembree, C. H., and Oakland, G. L. 1965. "Water Resources of the Upper Colorado River Basin-Technical Report." *Professional Paper 441*. US Geological Survey, US Government Printing Office, Washington, DC.

Mantell, M. A. (Ed). 1990. *Managing National Park System Resources: A Handbook @ Legal Duties, Opportunities, & Tools*. The Conservation Foundation, Washington, DC.

United States Code. Act of August 25, 1916 (39 Stat. 535), 1916.

US Comptroller General. 1979. *Colorado River Basin Water Problems: How To Reduce Their Impact*. CED-79-11, US General Accounting Office, Washington, DC.

United States Department of the Interior, Water for Energy Management Team. 1974. *Report on Water for Energy in the Upper Colorado River Basin*. US Government Printing Office, Washington, DC.

United States Environmental Protection Agency. 1971. *The Mineral Quality Problem n the Colorado River Basin*. US EPA Regions VIII and IX, Denver, CO.

WETLAND ENGINEERING, DESIGN, AND CONSTRUCTION: STATE-OF-THE-SCIENCE AND RESEARCH NEEDS

Donald F. Hayes, Member ASCE, Timothy A. Crockett, Student Member ASCE, and Michael T. Arends, Student Member ASCE[1]

ABSTRACT

Local, state, and federal agencies, special interest groups, and individual landowners are undertaking wetland construction projects at a feverish pace to mitigate past and anticipated wetland losses. Considerable resources are being allocated to these projects. Information on engineering aspects of wetland design and construction is available; however, it is located in diverse sources and usually pertains to a specific project or wetlands. This paper summarizes the results of a project to identify and assimilate previous efforts related to wetlands engineering including efforts from these diverse fields. The paper identifies areas where the state-of-the-science is adequate and areas in which additional research is needed.

INTRODUCTION

Wetlands continue to receive tremendous interest as society learns of their important ecological functions. An explosion of wetland enhancement, restoration, creation, and protection projects have resulted from intense public interest. Wetlands projects, whether aimed at enhancement, restoration, creation, or protection, usually focus on providing specific functions such as groundwater recharge, water quality improvement, fish and wildlife habitat, or recreation. In some cases, these functions have been determined to be particularly important to a local and regional ecology. Wetlands replacement may also be required to mitigate for wetlands loss due to construction activity; in these cases, complete functional replacement may be required.

---

[1]Asst. Professor, Graduate Student, and Undergraduate Student respectively, Department of Civil Engineering, University of Nebraska-Lincoln, 60th and Dodge Streets, Omaha, Nebraska 68182-0178.

The focus on wetland construction projects has brought engineers in the center of the wetlands issue. Although engineering involvement in wetlands projects is not new, there has not been a long history of continuously evolving engineering practices such as found in many other areas. Conversely, a diverse collection of professional fields outside of engineering have devoted considerable resources to wetlands projects. As a result, a number of successful techniques for wetland enhancement, restoration, creation, and protection have been developed; many of these techniques have engineering applications as well. The challenge is developing a set of engineering guidelines which ensures these wetland projects result in successful and productive wetlands.

This paper summarizes the results of an intensive literature search in which over 350 references related to wetlands engineering and design considerations were reviewed. An effective categorization of the literature for discussion and use has been difficult to establish. It would be convenient to organize the literature according to design requirements and construction techniques for specific wetland functions. The attraction of this organization is apparent; however, most wetland functions are not separable and in fact are closely interrelated. A second alternative would be to group multiple wetland functions into more encompassing categories based upon natural processes. This categorization would group wetland functions together by the natural process from which they result. This grouping also proves to be too simplistic. The categorization in this paper is not intended to be perfect, but allows some disaggregation of the existing literature based upon a combination of functions, processes, and application.

WETLAND HYDROLOGY

Proper hydrology is the common thread among successful wetlands. Hydrology controls important wetland characteristics such as water and nutrient availability, soil conditions, water depths, water chemistry, flow conditions, and vegetation. Each characteristic, in turn, affects the wetlands functions. Hydrologic characteristics also vary considerably depending upon the local topography and geography. For example, consider the hydrologic diversity among the southern swamps, prairie potholes of the mid-west, coastal marshes, and mountain pools.

The numerous studies of wetland hydrology attest to its importance; however, many studies are very site specific and consequently have limited general application. However, some authors have attempted generalized descriptions of wetland hydrology; a study of which is essential to understanding wetland functions, processes or policies. A few of the more fundamental discussions are found in O'Brien (1988), Kusler and Kentula (1990), Marble (1992), and Hammer (1992). If one consideration in the engineering design process can be identified as most important in the design process, it is the necessity of an accurate and reliable water budget. Carter, et al (1978) provides a concise summation of conducting a water budget for a wetland.

Hydrograph Attenuation and Water Quality Improvement

Physical, chemical, and biological processes occur naturally in wetlands and improve the quality of waters passing through them. These natural processes encourage the use of wetlands to help control stormwater flows and improve the quality of the stormwater before discharging into a receiving waterbody. Many communities now combine the utility of a stormwater retention basins to temporarily store runoff or provide hydrograph attenuation with the ecological functionality of wetlands in what many refer to as wet detention basins. The EPA Nationwide Urban Runoff Program (NURP) studies (EPA 1988) demonstrated that wet detention basins exhibit some of the highest pollutant removal efficiencies of any Best Management Practice (BMP). The integration of wetlands within a detention pond allows the efficient use of public funds to provide aesthetic, ecological, and public protection functions in an effective urban runoff treatment project.

Riparian wetlands provide important ecological and flood control functions. Large wetland areas in the flood plan reduce development and provide extensive flood storage hydrograph attenuation during peak flows. These areas may also improve water quality in a riparian system, if sufficient amount of the streamflow is allowed to pass through these complex wetland systems as it moves downstream. Unfortunately, man's desire for well defined and controlled stream channels has eliminated flow into these wetlands except during flood periods.

Wastewater Treatment

Constructed wetlands for wastewater treatment also utilize the natural water quality improvement characteristics of wetlands for the direct benefit of society. Few suggest the use of wetlands for primary treatment of wastewater; however, wetlands are currently being used in many areas for tertiary treatment and in some small systems for secondary treatment. Wetlands are not only being used for municipal and household wastes, but also for the treatment of industrial wastes.

Despite numerous successes and extensive publicity, the author would still classify the science of designing wetlands to treat wastewaters to be in the immature stage. Several publications are available which describe currently recommended design methodologies. Hammer (1989) provides a large array of papers related to the use of wetlands for wastewater treatment. Design procedures for treating municipal and industrial wastewaters are described in a number of articles found in Hammer (1989). The use of subsurface flow wetlands to treat wastewaters has also received considerable attention. Reed (1990) provides extensive design procedures and criteria for the application of subsurface flow wetlands for treating various wastewaters.

ACKNOWLEDGEMENTS

This study was funded by the Environmental Laboratory of the USAE Waterways Experiment Station and the Center for Infrastructure Research at the University of Nebraska-Lincoln.

REFERENCES:

Carter, Virginia, Bedinger, M.S., Novitzki, Richard P., Wilen, W.O., "Water Resources and Wetlands," Wetland Functions and Values: The State of our Understanding, November, 1978, pp. 344-376.

Hammer, Donald A. (ed.), Constructed Wetlands for Wastewater Treatment - Municipal, Industrial, and Agricultural, Lewis Publishers, Inc., Chelsea, MI, 1989.

Hammer, D.A., Creating Freshwater Wetlands, Lewis Publishers, Inc., 1992.

Kusler, Jon A. and Mary E. Kentula (ed.), Wetland Creation and Restoration - The Status of the Science, Island Press, Washington, DC, 1990.

Marble, Ann D., A Guide to Wetland Functional Design, Lewis Publishers, Inc., 1992.

O'Brien, Arnold L., "Evaluating the cumulative effects of alteration on New England wetlands", Environmental Management, Vol. 12, No. 5, 1988, pp. 627-636.

Reed, Sherwood C. (ed.), Natural Systems for Wastewater Treatment, Manual of Practice FD-16, Water Environment Federation (formerly Water Pollution Control Federation, Alexandria, VA, 1990.

U.S. Environmental Protection Agency, "Results of the National Urban Runoff Program (NURP), Final Report: Volume I," U.S. Environmental Protection Agency, Washington, DC, 1988.

Juanita Creek Channel Restoration
King County, Washington

Jack C. Bjork, P.E., Member ASCE[1] and A. William Way[2]

Abstract

Juanita Creek, a small basin near Seattle that contains anadromous fish, was experiencing degradation due to urbanization. A detention pond and channel restoration improvements were constructed to rehabilitate the stream. One year after construction, some success is evident. Some application criteria and design improvements are proposed for similar projects.

Introduction

Juanita Creek is a stream that drains 7.1 square miles (1,840 hectares) in the northeast portion of the Seattle metropolitan area. In the past 40 years, the land use in the basin has changed from rural, forest, and agricultural, to predominantly suburban residential, although the basin has not yet reached full build-out. The stream, which hosts stocks of both resident and anadromous fish, is a tributary of Lake Washington. The fish species of primary importance are cutthroat trout and coho salmon. In May 1989 observations revealed that coho and trout fry were abundant.

The surface material in the basin is soil derived from glacial recessional outwash and is predominantly sand and silt. Channel banks in some reaches are primarily poorly-graded sand, creating banks very susceptible to erosion. The

---

[1] Project Manager, R. W. Beck and Associates, 2101 4th Avenue, Seattle, Washington 98121.

[2] President, The Watershed Company, 10827 NE 68th Street, Kirkland, Washington 98033.

project area is on a major tributary, which has a drainage area of 1.3 square miles (340 hectacres) and is bisected by the I-405 freeway. The current 2-year and 25-year peak flows at the downstream end of the problem area were estimated to be 71 cfs and 176 cfs (2.0 and 5.0 m^3/s), respectively.

Problem Description

Due to the removal of natural vegetation, the increase in impervious area, and the installation of pipe systems, flows in Juanita Creek have become more volatile. Peak flows have already increased significantly above natural conditions, and peak runoff under future land use conditions is expected to increase by an additional 9 to 20 percent. In addition, reduction in vegetation and other urban impacts have increased the volume of sediment being introduced into the stream. The result has been channel erosion, property damage, sedimentation, and loss of fish habitat. Four stream reaches totaling 475 linear feet (145 m) had experienced serious bank erosion.

Problem Solution

The Surface Water Management Division of the King County Department of Public Works authorized an investigation, as well as subsequent design and construction, to help solve these problems. Improvements included the following major elements:

1. **Use of the large I-405 embankment to create a 35 acre-foot (43,000 m^3) detention pond to attenuate peak flows.** With this improvement in place, peak flows for the 2-year and 25-year events at the lower end of the problem reach under full build-out conditions will be reduced by 18 percent and 27 percent, respectively. That means peak flows will be lower than they were with undeveloped conditions. The work also included wetland relocation and enhancement, as well as a pedestrian bridge.

2. **Rehabilitation of four stream reaches using bioengineered techniques.** In-stream structures of natural materials were installed to stabilize the channel and create fish habitat in the channel. This included 7 deflector logs, 4 root wads, and 11 log check dams. Restoration of the bank included several other features, including bank restoration structures with a foundation of rock spalls that extended from below the channel up to the approximate level of the 2-year flood flow. Figure 1 illustrates a typical cross section.

    Above the rock spalls, the bank restoration structure for the next 3.5 feet (1.1 m) in height consisted of three layers of coconut fiber matting wrapped around earthen fill material, functioning in a manner similar to reinforced earth. This permits intensive bare-root planting of the soil-filled space between each layer of coconut matting. More than 1,800 plants were installed in 500 linear feet (150 m) of bank restoration. Growing root mass of plants, such as red twig dogwood, willow, snowberry, and salmonberry, provides soil-holding properties. Besides binding the soil, vegetation growing on the bank contributes to fish habitat by providing shade from the hot sun, cover from predators and food in the form of insects.

Above the zone of coconut fiber wrapping is a fill slope graded to no more than 1.5 horizontal to 1 vertical. In addition to hydroseeding, this slope was stabilized with bundles of willow wattles of live willow cuttings, laid horizontally along the slope contours.

A majority of this project was constructed by October 1991, although one the four bank restoration sites was completed in September 1992. The final construction cost was $251,000 for the detention pond and $202,000 for the channel restoration. The approximate cost for the channel work alone was about $360 per linear foot.

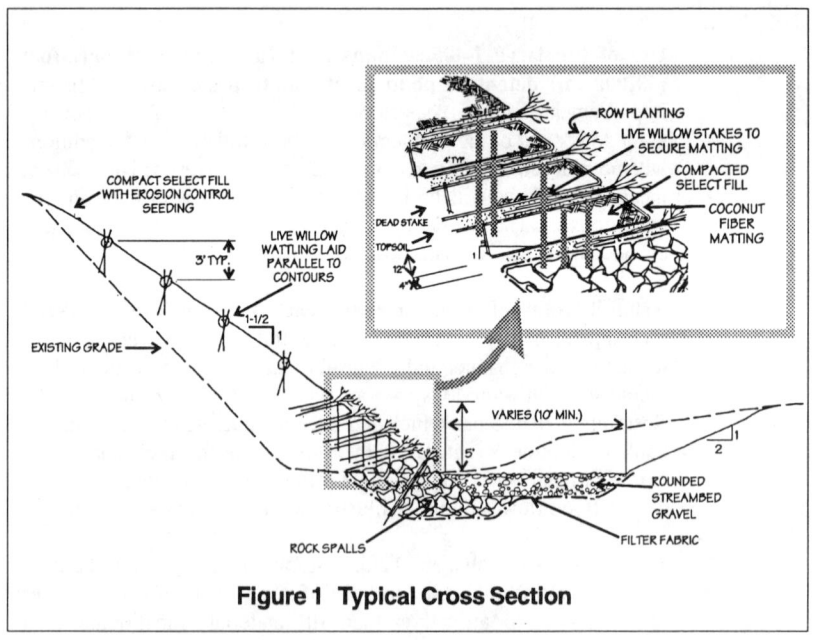

Figure 1 Typical Cross Section

Postconstruction Conditions

Most of the construction work has been in place for more than a year. The detention pond is working as expected, although a larger trashrack needed to be installed. Deposition the of sand and silt in the pond will tend to improve fish habitat downstream, but the trapping of gravel may be detrimental to habitat quality in the future because salmonids need a gravel stream bottom for spawning. Additional time will be needed to determine the long-term impacts.

The success of the restoration work has varied from reach to reach. Adjacent to the coconut fiber wrapping, a sufficient number of the plants have become established to hold stream bank soils, and further root binding is progressing. This portion has survived the initial critical period and will become stronger with time. The willow wattle zone is growing well and is successfully rooted in most areas.

Considerations and Recommendations

Use of bioengineered methods, which use plants as a structural element, should be carefully considered where public access cannot be restricted. Urban streams can also be playgrounds, and children's activities can be particularly damaging. Bare-root plants, which look like sticks, are easy to pull out. The fill slope can damaged by foot traffic. Similarly, riprap should be too heavy for children to lift.

The success of bioengineered systems relies on plant and root growth. The material used in the coconut fabric wrappings and the fill slope should be a good growth medium. Supplements should be added if necessary. Rock spalls need to be carefully placed and compacted so they do not slough into the channel and interfere with rounded spawning gravel. If the stream is considered an amenity, it is likely that people will want bankside access. Including access as a design element can prevent traveling across the fill slope and coconut fiber wrapping.

Bioengineered systems are relatively sensitive, compared to typical municipal projects. In addition, many contractors are not familiar with this type of work. Thorough inspection and monitoring must occur during both the construction and the plant establishment period.

DOUBLE ROCK PARK STREAM RESTORATION PROJECT

James W. Gracie, Candace Szabad, and Raymond I. Green, P.E.

Abstract: The Double Rock Park stream restoration is a pilot project for the Baltimore County, Maryland, Stream Restoration Program. The study assessed watershed conditions and the stream channel stability of Stemmer's Run and its tributaries from the downstream boundary of Double Rock Park to its headwaters. Double Rock Park is a highly utilized public recreation facility with unique physiographic features that include waterfalls which are uncommon in eastern Maryland. The objectives of the study were to diagnose the causes of stream bank erosion, and to recommend stream and watershed restoration strategies within the 754 acre watershed draining to the park boundary.
  The paper presents a methodology for diagnosing causes of the stream bank erosion through the use of the Rosgen Stream Classification System, a system of classifying stream channels by delineative criteria which integrate the eight variables which shape and maintain channels. The methodology includes an assessment of how the channel handles stress during frequent high flows based upon its width-depth ratio, slope, confinement, the size and cohesiveness of the materials in the active channel, vegetation on the banks and its depositional state. A method of evaluating the equilibrium between channel size and flow regime is also presented. The results of the analysis are used in the design of a stable channel configuration.
  Stream restoration concepts are specified for reaches of Stemmer's Run. The goal at each site is a stable stream type having the capacity to convey frequent high flows, the competence to transport sediment delivered from the watershed, and a dramatic reduction of sediment from bank erosion.

INTRODUCTION

When development occurs in the watershed, increased impervious area leads to increases in peak flows which result in channel enlargement. This enlargement can take one or a combination of two forms. Channels either incise or adjust laterally. When they incise streams no longer have access to a flood plain to relieve stress of high flows and adjacent wetlands become drier.
  Stemmer's Run is suffering from bank erosion which was initiated as a

result of development in the watershed. This development, by increasing the impervious area and speeding up the delivery of runoff to the receiving channels, increased the quantity of runoff delivered to the streams. Subsequent to the initial channel enlargement which followed this development, the stream channels adjusted in various ways. In most cases sediment from bank erosion exceeded the streams ability to transport it. This led to patterns of deposition which increased erosion downstream. The channel instability which is now present in Stemmer's Run is a result of those depositional features. In most cases they will persist for many years if not corrected.

BASIC PRINCIPLES

Stream channels develop their shape, size, slope and other morphological features as a result of the interaction of flowing water on the materials in the stream's valley. A stream or river is the manifestation of a process of energy transformation in which potential energy of elevation is transformed into the kinetic energy of flowing water. In this process of energy use and transformation there are eight variables. They are discharge, width, depth, velocity, slope, roughness, particle size and sediment quantity. The independent variables are discharge and sediment load. Discharge can be affected by climate which affects the amount of precipitation and land use which affects the fate of precipitation (i.e., changes in infiltration, evapotranspiration, storage or runoff). Sediment quantity is affected by the type and distribution of rocks and the effect of weathering and transport. The interactions of these variables follow the laws of physics including conservation of mass and energy, the relationship of friction to velocity, depth and slope, and the relationship of sediment load to power. Over geologic time the continual interaction of these parameters tends toward minimum work and minimum variance among the parameters (Leopold,L.B., Wolman, and Miller, 1964).

Stream channel stability is not a static state, but stable streams are those in which a dynamic equilibrium exists. Sediment supply is in equilibrium with sediment transport. Low rates of erosion on the outside of meander bends are matched by similar rates of deposition on point bars. Disequilibrium can come about as a result of a change in any one of the variables that govern stream morphology. One of the disturbances that can result in disequilibrium is an increase in frequency, magnitude and duration of bankfull flows that result from extensive land development.

Dramatic alterations in the hydrologic cycle occur as a result of extensive land development. Land development activities result in an increase in the quantity of runoff, and a decrease in infiltration, storage and evapotranspiration. Rooftops, sidewalks, roads, and driveways do not permit water to infiltrate into the ground and therefore generate increased runoff in any given storm event. In addition to generating more runoff, impervious area which is much smoother than natural features, causes faster concentration and delivery to receiving streams.

There is a close relationship between the magnitude of runoff from frequent storm events and stream channel dimensions that holds throughout regions of similar climate (Leopold et al, 1964). It has been established that the peak discharge from storms which occur on a recurrence interval of from 1 to 3 years produce the flows which shape, size and maintain stream channels (Leopold et al, 1964). This peak flow is called the bankfull flow.

It follows that a substantial increase in frequency and duration of the peak discharge which generates the bankfull flow will result in more stress on stream channels. How a channel handles this stress at frequent high flows depends on its width depth ratio, slope, confinement, the size and cohesiveness of the materials in the active channel, vegetation on the banks and its depositional state. Some of these factors such as width depth ratio, confinement, slope, and depositional state affect how the stress is distributed. For example, mid-channel bars tend to cause convergence of flows into the adjacent near bank regions accelerating the tendency toward bank erosion. Any features which concentrate stress in the near bank region will increase the tendency toward bank erosion. The nature and size of materials in the channel will affect the channel's resistance to this stress and, in the case of stream types with well developed flood plains, vegetation on the banks will have a great deal to do with how they resist this stress.

RESULTS

Stemmer's Run was analyzed using the Rosgen Classification system. The Rosgen Classification System is a system of classifying stream channel reaches on the basis of measurable morphological characteristics (Rosgen, D., 1985). It classifies stream channels using delineative criteria that integrate the eight variables that shape and maintain channels. The delineative criteria are width, depth, confinement ratio, sinuosity, particle size of materials in bed and bank, and slope. The Rosgen Classification System enables one to predict how a river will respond to certain stresses such as a change in flow regime or sediment supply. and to select appropriate restoration approaches for streams in disequilibrium.

Understanding the morphology associated with the different stream types will allow prediction of how the channel will respond to an increase in magnitude, frequency and duration of bankfull flows. Streams channels with differing entrenchment in their valleys distribute the stress of high flows very differently. The distribution of stress in the near bank region will be a key to channel stability. Thus, determination of the stream type will indicate how these increased flows will be handled and will permit the selection of appropriatedesign parameters for reducing stress in the near bank regions so that the restored channels can handle the flows. Thomas Hammer reported channels that enlarged more than an order of magnitude in a process that was still occurring several decades after development had increased the runoff regime in the urbanized areas of eastern Pennsylvania and Maryland (Hammer, T.R. 1973).

Erodibility or potential for erosion is primarily a result of the size, cohesiveness, steepness and vegetative state of banks along with the ratio of total bank height to bankfull height. The other factor that determines rates of erosion is the amount of stress in the near bank region during frequent high flows. Rosgen defines this stress as the percent of flow in the third of the width nearest the bank. He categorizes the stresses as low moderate or high depending whether the percent of bankfull flow is less than 30%, 30-50%, or greater than 50% respectively(Rosgen, 1991). Restoration designs are therefore, aimed at reducing the stress in the near bank region or increasing the banks resistance to erosion or both. Stream types that concentrate stress in the near bank regions are mainly "F" and "G" stream types.

Design approaches for restoration were aimed at converting "G" and "F" stream types to "B" or "C" types. Properly restored channels reduce stress in the near bank regions, maintain sediment transport to handle the sediment delivered by the watershed, but significantly reduce sediment supply from stream bank erosion.

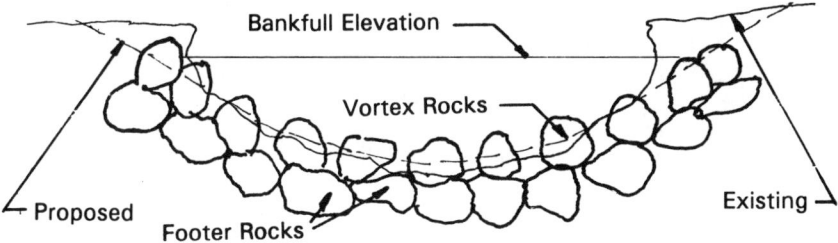

Figure 1. VORTEX ROCK WEIR DESIGN - Typical existing and proposed cross section showing an increase of width depth ratio from 10 to 16 to reduce stress in near bank region. The vortex rock weirs are placed at the upstream and downstream end of each pool.

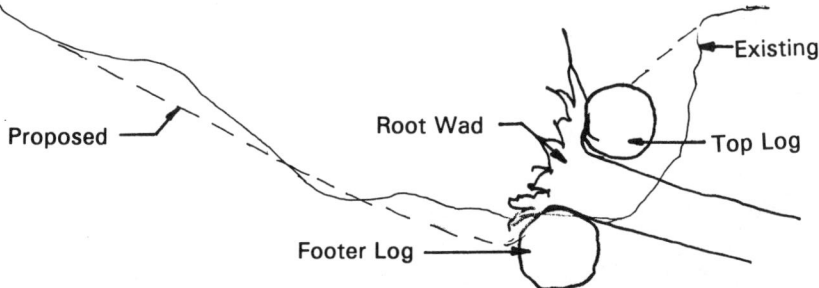

Figure 2. ROOT WAD DESIGN - Typical cross section on an eroding meander bend showing placement of root wad, footer log, and top log with an increase in width depth ratio from 12 to 18. This approach both reduces stress in near bank region and increases erosion resistance of the banks.

APPENDIX - References

Hammer, Thomas R., 1973. Effects of Urbanization on Stream Channels and Stream Flow. Regional Science Research Institute, GPO Box 8776, Philadelphia, PA.

Leopold, L.B., Wolman, H.G., and Miller, J.P., 1964. Fluvial Processes in Geomorphology. W.H. Freeman and Company, San Francisco, CA.

Rosgen, David L., 1985. "A Stream Classification System," Paper presented at the the Symposium, "Riparian Ecosystems and Their Management: Reconciling Conflicting Uses," Tuscon, Arizona.

Rosgen, David L., 1991. Applied Fluvial Geomorphology. Unpublished Textbook used in proprietary short course. Pagosa Springs, Colorado.

Use of Flood Control Features for Environmental Mitigation

J. Craig Fischenich, M. ASCE[1], E. A. Dardeau, Jr[2]., and Kenneth D. Parrish[3]

Abstract

The US Army Engineer District, Vicksburg (CELMK), plans to construct up to 167 water control structures, 52 confined disposal facilities, and 47 borrow pits as part of a major flood control effort known as the Upper Yazoo Projects (UYP). Many of these project features are capable of ponding water and thus can be managed to mitigate for aquatic, terrestrial, waterfowl, and wetland resource losses expected to occur as a result of the UYP. The benefits to be derived will depend upon the land uses and management of the ponded areas. The US Army Engineer Waterways Experiment Station (WES) developed procedures to quantify the cost and habitat benefits of the many management options for these sites. The mitigation strategy was derived by optimizing various combinations of land acquisition, reforestation, land-use change, and site hydrology so that the least-cost mitigation plan could be selected.

Introduction

The Yazoo River Basin extends from Vicksburg, MS, to just south of Memphis, TN, and covers 13,400 square miles. Congress authorized construction of the UYP to provide flood

---

[1] Research Civil Engineer, Environmental Laboratory (EL), US Army Engineer Waterways Experiment Station (USAEWES), Vicksburg, MS.

[2] Geologist, EL, USAEWES.

[3] Civil Engineer, Planning Division, US Army Engineer District, Vicksburg, Vicksburg, MS.

control for 8,900 square miles in the basin. Construction on the UYP began in 1976 but was halted in 1989 due to environmental concerns, with project continuation subject to findings of the UYP Reformulation Study (USAED Vicksburg 1989). From 1976 - 1989, 55.7 miles of channel enlargement and 3.8 miles of levees were completed.

One of the UYP Reformulation Study objectives was to develop strategies to reduce adverse environmental impacts expected to occur as a result of the UYP. This task proved to be arduous both in quantifying impacts and in identifying mitigation strategies. This paper briefly discusses the utilization of project features to reduce and mitigate habitat losses. These topics covered in more detail in the parent study (Dardeau et al. 1992).

Habitat Losses

A Study Team consisting of engineers and scientists from CELMK, WES, the Mississippi Wildlife Fish and Parks, and the US Fish and Wildlife Service was formed to quantify expected habitat losses for the UYP. The team reached a consensus on values for terrestrial and aquatic habitat using the Habitat Evaluation Procedure (HEP). The HEP is based on the concept that habitat value can be assessed in relation to the food, shelter, and reproduction requirements of key species. Losses for each species are determined in habitat units (HUs). Waterfowl habitat was determined based upon available food and energy requirements of ducks. These methods permit site and alternative comparisons for multiple species on the basis of net loss or gain in wildlife resources, rather than strictly on a habitat basis. Wetland losses were based upon affected acreage.

Mitigation and Management Approaches

Because actual habitat loss was computed by species, compensation for lost habitat could be made using various habitat types. For example, the loss to waterfowl of 1 acre of native bottom land hardwoods (BLH) flooded from November to January 15 was calculated to be 91 duck days (DD). This loss could be offset with 1 acre of similar habitat, or with 0.11 acre fallow land (816 DD/acre), or 0.19 acre planted BLH (469 DD/acre), or 0.02 acre soybeans (4,037 DD/acre).

The CELMK/WES/USFWS UYP Study Team first developed mitigation and management strategies that could be applied at large land tracts. The team noted that many of the engineering features that are integral components of the UYP have the potential to provide mitigation for one or more habitat category. Given the potential for reduced mitigation costs, these features were further analyzed. Three types of project features were found to provide the greatest potential for environmental mitigation: ponded areas behind water control structures, borrow pits, and confined disposal facilities (CDFs).

The study team computed habitat potential at 36 ponded areas behind water control structures, 52 CDFs, and 47 borrow pits. For each, the terrestrial, waterfowl, wetland, and aquatic habitat values were computed for 27 Management Plans (MPs) derived from various combinations of land-use conversion and site hydrology. Figure 1 and the associated tabulation show unit habitat values at one ponded area for two MPs. A spreadsheet was developed to calculate these values, and to compare the various site/MP combinations to other mitigation alternatives.

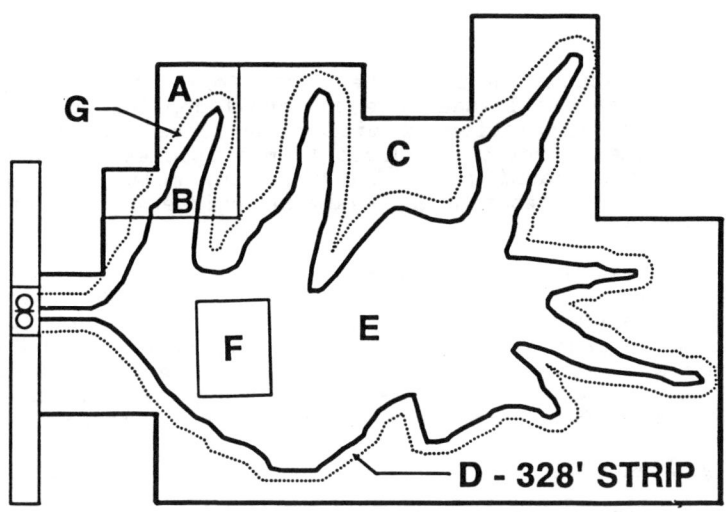

Area	Existing Land Use	Managed Land Use	Habitat Benefit (Plan 1/Plan 2)			
			Waterfowl (DD/Acre)	Aquatic (HU/Acre)	Terrestrial (AAHU/Acre)	Wetland (Acre)
A	BLH	BLH	0/0	0/0	0/0	0/0
B	BLH	BLH	91/121	0/1.04	0/-2.67	0/-1
C	Agr.	BLH	0/0	0/0	1.56/1.56	0/0
D	Agr.	BLH	0/0	0/0	1.56/2.74	0/0
E	Agr.	Fallow	816/1088	0/2.47	0/0	0/0
F	Agr.	BLH	469/121	0/4.47	2.05/0	1/0
G	BLH	BLH	0/0	0/0	0/0.16	0/0

Figure 1. Habitat Values For Two MPs at a Project Feature Flooded Nov. to Jan. 15 (Plan 1), or Nov. to June (Plan 2)

Plan Optimization

To judge their value for mitigation purposes, the project features were compared to acquisition and conversion of large tracts of agricultural land. Economics were considered by normalizing on a benefit-cost basis to a baseline mitigation strategy at a large land tract (ie. $(benefits/costs)_{feature}/(benefits/costs)_{lg.\ site})$. The resulting ratios indicated the value of implementing a given MP at a project feature relative to real estate acquisition. MPs with a ratio greater than one were more cost-effective than the baseline mitigation plan. Nearly all project features provided a management option with a ratio greater than unity, and some offered several options which were compared to select the optimum plan for that feature.

Conclusions

The methods used in the UYP Reformulation Study to mitigate habitat losses have universal application. They permit the formulation of mitigation plans that directly address the affected resources from a species standpoint. With proper planning and coordination, multiple habitat losses can be offset using a common engineering feature or land tract.

By taking full advantage of the project features and the proposed mitigation sites, the UYP Study Team has proposed approaches that reduce impacts and offer several options to satisfy the mitigation requirements of the UYP. Such a systematic ecosystem approach will result in a sound engineering project and may save substantial mitigation costs, depending upon the plan ultimately selected.

Acknowledgement

Information presented herein, unless otherwise noted, was obtained from research sponsored by CELMK in conjunction with the UYP Reformulation Study. Permission was granted by the Chief of Engineers to publish this information.

References

Dardeau, Elba A., Jr., Fischenich, J. Craig., Olin, Trudy J., and Landin, Mary C. (1992). "Environmental Engineering Design and Mitigation Considerations," draft technical appendix to The UYP Reformulation Study, US Army Engineer District, Vicksburg, Miss. Prepared by US Army Engineer Waterways Experiment Station, Vicksburg, Miss.

US Army Engineer District, Vicksburg. (1989). "Flood Control, Mississippi River and Tributaries, Yazoo River Basin, Yazoo Headwater Project, Mississippi," GDM 41, Supplement 1, Vicksburg, Miss.

AN ISOLATED WETLAND USED FOR STORMWATER TREATMENT

Betty Rushton, Ph.D. and David Carr[1]

Abstract
A marsh used for stormwater management showed gradually increasing levels of pH, dissolved oxygen and conductivity in the flow path. The system was especially effective in removing suspended solids though somewhat less so for selected metals. Relationships are discussed to explain significant differences in water quality.

Introduction
Numerous isolated wetlands and rapid population growth in southwest Florida make wetland treatment of stormwater an attractive alternative. Uncertainty exists, however, in their ability to tolerate the changed hydrology and higher levels of pollutants associated with urban runoff. For example, many marsh plants have adapted to low nutrient concentrations, while high nutrient and sediment loads have caused cattail and algae colonization. Also values for pH, dissolved oxygen and redox potential, are often much lower in native wetlands which may affect pollutant removal.

To address these concerns, a native herbaceous marsh is being studied to determine the treatment efficiency of the system and to document the effect of stormwater on the integrity of a natural wetland. This report presents results from the first year of a two-year study. Since hydrology calculations are incomplete, emphasis was placed on processes which may affect constituent concentrations in the water column. Future reports will address the efficiency of the system to remove pollutant loads as well as analyses of the hydrology, rainfall, vegetation and sediments.

Methods
The study site, located in a corporate park in Tampa, Florida (Fig 1), is an isolated wetland covering 1.2 ha receiving runoff from a 6.2 ha drainage basin. The sedimentation basins at both inflows provide some pre-treatment of stormwater before it enters the marsh. The east basin drains a roadway and the west basin collects stormwater from parking lots and rooftops. The south rim of the marsh and the two basins are surrounded by well-maintained landscape plants and lawn; the north rim of the marsh is protected by a buffer of native vegetation.

---
[1]Southwest Florida Water Management District, 2379 Broad Street, Brooksville, FL 34609.

The two inflows and the outflow are instrumented to automatically collect and refrigerate flow weighted water quality samples during storm events at stations 1, 6 and 8 (Fig 1). Stormwater overflows into the marsh during storms greater than 12.7 mm, but the marsh does not discharge except during July, August and September. Therefore, at the outflow, grab samples were taken after rain events until the marsh discharged; samples were then collected every two days. Field parameters were measured using a Hydrolab™ while statistics were performed using the Statistical Analysis System (SAS Inc. 1990). When water quality values were below the laboratory detection limit (mostly for cadmium and manganese) one half the detection limit was substituted. For correlations, two non-parametric tests, Spearman and Kendall, were used as well as the Pearson Correlation coefficients. Most results discussed were significant for all three tests ($P < 0.05$).

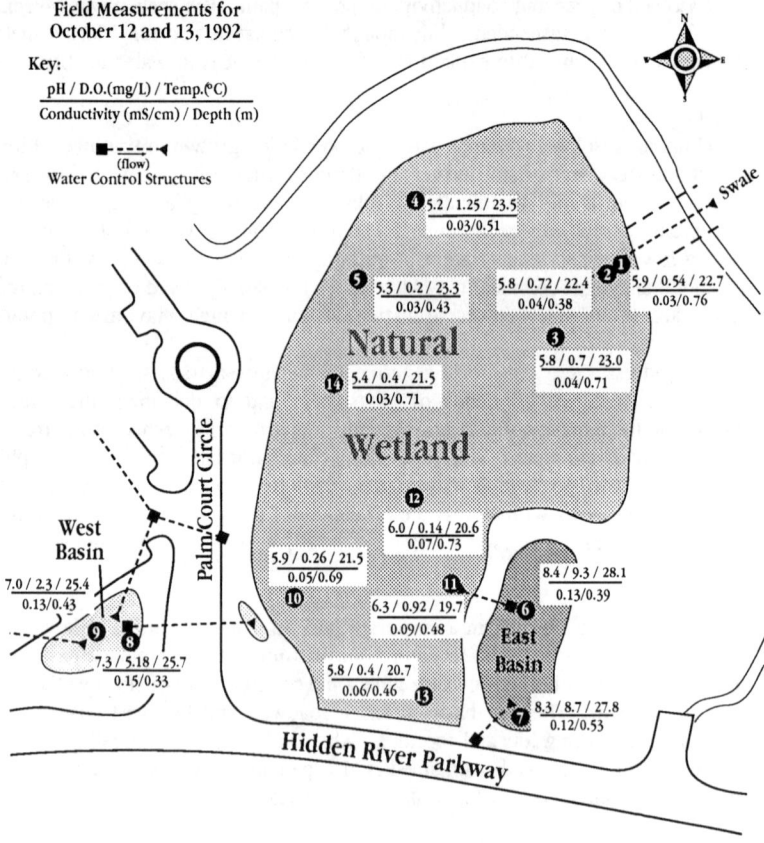

Figure 1. Site plan and field measurements.

Results and Discussion

The field parameters measured during one sampling in October 1992 were typical for averaged values measured with continuous recorders (Fig 1). The spatial distribution showed the dissimilar characteristics of water in the marsh compared to stormwater in the basins. The east basin was a turbid highly productive pond dominated by phytoplankton. It always contained water and had the highest readings for pH, dissolved oxygen, and temperature. In contrast both west basins were choked with emergent vegetation (predominantly cattails), and held water only after storm events. Field measurements were about the same or somewhat lower than the east basin. The marsh, 95% covered with vegetation, had much lower readings for all field parameters. Dominant species measured in the marsh were maidencane, pickerelweed, water lily and arrowhead. Data showed slightly higher values for pH and conductivity nearest the inflow stations in the marsh.

Significant differences were noted in the three systems for some constituent concentrations (Table 1). The marsh performed well at removing suspended solids with average concentrations at the outflow at least 78% less than at the inflows. Concentrations of inorganic nitrogen and zinc at the west basin were significantly higher than the other two locations; and iron concentrations were much lower. The total phosphorus levels in the east basin were over twice that of other stations.

Some processes which might account for these results were investigated with correlation analyses. Correlations were strongest at the west inflow where runoff had little opportunity for treatment before being measured. In this basin ammonia was positively related to ortho-phosphate, suspended solids and nitrate plus nitrite. Also zinc and iron showed a tendency to be positively correlated with suspended solids. These results can be explained by the fact that metals and phosphorus often find attachment sites on particles.

In the east basin total phosphorus increased as suspended solids increased suggesting that phytoplankton had taken up ortho-P which was measured at the discharge point as organic phosphorus compounds and suspended solids. This would help explain the high ratio of total-P to ortho-P in this basin.

At the outflow total phosphorus and zinc were positively correlated with organic nitrogen. Low pH conditions are reported to increase plant uptake of zinc and to make phosphorus more available to biota which may explain these relationships.

An analysis of the data (Table 1) indicated the discharge water from the marsh was acceptable for most parameters. Although the only constituent that was significantly reduced at the outflow was suspended solids, there is reason to believe that the system is efficient at removing most pollutant loads since preliminary hydrology data showed much more water entered the system than left it. Except for phosphorus in the east basin, nutrient concentrations, both entering and leaving the system, were lower than 60 to 80% of Florida lakes though some metals, especially cadmium and zinc, exceeded state water quality standards.

Table 1. Summary statistics for constituents (mg/ℓ) found at the two inflows and the outflow of a natural wetland used for stormwater treatment during the summer of 1991.

Sample	Location	N	Mean	*	Std. dev.	Min.	Max.	Median
Ammonia	East In	17	0.023	B	0.014	0.001	0.067	0.019
as N	West In	16	0.079	A	0.075	0.015	0.291	0.042
	Outflow	22	0.042	B	0.031	0.015	0.120	0.035
Organic	East In	17	0.829	B	0.373	0.000	1.633	0.793
Nitrogen	West In	15	0.369	C	0.334	0.067	1.169	0.231
as N	Outflow	21	1.079	A	0.223	0.844	1.593	1.004
Nitrate+	East In	17	0.062	B	0.042	0.001	0.160	0.048
Nitrite	West In	15	0.186	A	0.107	0.015	0.422	0.173
as N	Outflow	22	0.026	B	0.034	0.008	0.169	0.015
Ortho-	East In	17	0.039	A	0.022	0.001	0.100	0.038
Phosphate	West In	14	0.034	A	0.024	0.005	0.104	0.032
	Outflow	22	0.024	A	0.021	0.002	0.089	0.016
Total	East In	17	0.152	A	0.065	0.001	0.294	0.151
Phosphorus	West In	15	0.044	B	0.032	0.016	0.138	0.036
	Outflow	22	0.050	B	0.027	0.011	0.138	0.047
Total	East In	16	15.630	A	7.149	0.025	30.070	15.735
Suspended	West In	11	11.703	A	10.445	0.840	40.280	11.600
Solids	Outflow	21	2.552	B	1.431	0.580	6.210	2.360
Total	East In	17	0.027	B	0.027	0.009	0.120	0.011
Zinc	West In	17	0.104	A	0.073	0.023	0.319	0.096
	Outflow	21	0.026	B	0.017	0.008	0.067	0.021
Total	East In	17	0.007	A	0.006	0.001	0.019	0.006
Cadmium	West In	16	0.008	A	0.008	0.001	0.027	0.005
	Outflow	22	0.005	A	0.007	0.001	0.022	0.004
Total	East In	17	0.649	A	0.424	0.500	1.900	0.590
Iron	West In	16	0.168	B	0.175	0.020	0.790	0.135
	Outflow	22	0.605	A	0.188	0.040	0.840	0.595
Total	East In	17	0.015	A	0.012	0.005	0.042	0.013
Manganese	West In	16	0.015	A	0.011	0.002	0.038	0.014
	Outflow	22	0.021	A	0.014	0.005	0.050	0.019

* Means of each constituent with the same letter are not significantly different (P<0.05) Duncan Multiple Range Test.

Acknowledgements

Special appreciation goes to Hidden River Corporate Park for their cooperation and use of the site. This research was funded in part by USEPA 205J grant number WM434 administered through the Florida Department of Environmental Regulation.

References

SAS Institute, Inc. 1990. SAS/STAT User's Guide, Version 6, Fourth Edition, Volume 1 and 2, Cary, NC:SAS Institute Inc.

Innovative Wetlands in Urban Settings
Using Dredged Material

M. C. Landin, Ph.D., and T. R. Patin, P.E., M. ASCE

Abstract

As a result of its work in the waterways of the nation, the U. S. Army Corps of Engineers (USACE) has restored or constructed wetlands using dredged material in a number of urban/suburban areas. Habitat restoration, creation, enhancement, and protection that also offer recreational potential and human resource values within projects have been and are currently being used to help meet the demands of complex urban water resource planning and management.

Introduction

The USACE works in virtually all waterways of the United States. As a result, the agency often finds itself facing issues and challenges in urban settings in both its flood control and navigation missions that involve waterway channel maintenance, structures that protect lives and property, port and harbor maintenance and expansion, marina improvements, cultural resources, socio-economic factors, fish and wildlife habitats, water supplies and quality, and legal considerations for wetlands and other protected areas. Most Corps projects now are partnered "compromises" that offer the best way of dealing with all of the above items occurring in a project or group of projects.

Research Biologist and Research Civil Engineer, U.S. Army Engineer Waterways Experiment Station (WES), 3909 Halls Ferry Road, Vicksburg, MS 39180-6199

Where navigable waterways and urban areas coincide, the USACE must find ways to place dredged material that is acceptable to urban citizens and that will provide multiple purpose use. This often includes restoring or constructing a wetland or other habitat with the dredged material, but will also include other planned uses as park facilities, visitor centers, nature trails, biking/hiking, birdwatching, boating, and other activities.

While a number of habitats in urban settings have been or are being constructed, four examples of various types of wetlands are presented in the paper to give the reader an indication of the potential for natural resource beneficial uses of dredged material in urban areas.

Kenilworth Marsh, Washington, DC

A 30+-acre freshwater intertidal marsh is being constructed in Winter 1992-93 in Washington, DC, adjacent to the Anacostia River on National Park Service property. A shrub/marsh wetland was destroyed on this site prior to World War II, resulting in a lake that has partially filled with sediment. The lake is tidally connected to the Anacostia. It is being filled to an intertidal elevation conducive to fresh marsh establishment and survival, and will be planted in Spring 1993 with selected plant species that will provide both cover and food for wildlife and small fish.

Custom-designed geotubes filled with water have been temporarily placed at the tidal entrance to the lake to hold maintenance dredged material from the Anacostia in place until it has consolidated. The geotubes will be deflated and removed prior to planting. The site is being filled in three sections, with berms temporarily isolating each section. Baseline data have been gathered for the site, and a pilot study conducted. Site data will be collected by the USACE and NPS after the project is completed for several years to determine success of the restoration project. It is jointed funded by the USACE and the NPS, and partnered with the DC Council of Governments and the U. S. Fish and Wildlife Service.

Kenilworth Marsh is near NPS's Aquatic Gardens exhibit in Northeast Washington. The existing degraded lake was used for educational talks and tours. The NPS plans to use the restored wetland for more extensive educational classes on wetlands and for public canoeing in a natural setting, and canoe trails and pull-outs, piers, and other recreational features are being constructed as part of the project.

Pointe Mouillee, Rockwood, Michigan

Pointe Mouillee is a 4600-acre wetland restoration project in western Lake Erie, the only wetland left in that portion of the Great Lakes. The 3700-acre degraded and eroded wetland was restored by the construction, completed in 1983, of a 900-acre confined disposal facility (CDF) on the footprint of an eroded barrier island. The island is being filled with material from the nearby shipping channel, and is considered contaminated. This was a consideration in the overall design and management plan, which calls for the CDF to be completed with "clean" material and planted after the project is completed in another two decades.

The management plan, which was completed in 1979, was developed by an interagency, multi-disciplinary group. The plan being implemented includes a visitor center, piers, a marina, hiking/biking trails, nature paths, hunting, fishing, birdwatching, and other recreational features. At the same time, the site is home to numerous species of songbirds, has nesting colonies of black-crowned night herons, ring-billed gulls, and common terns, nesting by Canada geese, mallards, and mute swans, and an abundant muskrat and other small mammal population. White-tailed deer have been found on the site. There is also considerable fish abundance and diversity. Pointe Mouillee site is a major migratory stopover for shorebirds, raptors, waterfowl, and songbirds.

The 15-year-old restored wetland is being compared by the USACE to Pointe Pelee, a 4500-year-old natural Lake Erie wetland protected by a natural berm, in Ontario, Canada. Since sediments are still being placed in Pointe Mouillee, initial comparisons are of larval and other fish, wildlife, and aquatic and emergent herbaceous and woody vegetation, and do not include soils and water quality.

Times Beach, Buffalo, New York

Times Beach is a 200-acre freshwater wetland in Buffalo, NY, on Lake Ontario. The site was originally a CDF constructed by the USACE to hold contaminated dredged material from Lake Ontario. After initial construction and dredged material placement, it received such abundant use by fish and wildlife that the USACE was requested by the National Audubon Society and local groups to leave it as a natural area. The USACE constructed another CDF, and developed Times Beach with nature paths, birdwatching, observation points, and other recreational features.

This wetland includes forested wetlands, fresh

marsh, and floating aquatics. It has been studied by the USACE for a number of years, and contaminants are not readily translocating to plants and animals. The site is an urban birders' hot-spot, and is used for educational classes and tours.

Riverlands, Grafton, Illinois

The Riverlands is mitigation for Lock and Dam 26, the southernmost lock on the Mississippi River. It is a several thousand acre site separated from St. Louis, MO by the river, but is still highly accessible to a large urban population. The site is being restored as wet prairie, small ponds and streams, and nesting islands, and is planned to have some bottomland hardwood restoration take place in the near future.

The Riverlands contains driving trails, nature paths, some boating opportunities, and observation points. It will ultimately have a visitors center and numerous other recreational features. The site is being used as a wetland ecology study site by the USACE, the University of Missouri and the Missouri Botanical Gardens. Islands provide nesting for both waterbirds and waterfowl, and peregrine falcons live on the wet prairie. Riverlands is a prime birders' spot along the Mississippi River.

Summary

Other wetlands built of dredged material designed for recreational use include Muzzi Marsh in Marin County, CA; Vancouver Lake, Lincoln Avenue, and Jetty Island, WA; and a number of other manmade wetlands located in U. S. lakes, rivers, and estuaries.

Multiple purpose habitat restoration projects built in urban areas using dredged material are more practical, are less expensive per habitat unit gained, gain greater public acceptance, and provide public recreational outlets. At the same time, they provide for scarce fish and wildlife habitats that are frequently either absent or degraded in urban areas. A perception expressed in the past that dredged material is a waste product, not a resource, has been proven to be unfounded. Such beneficial uses of dredged material are win-win situations, and should be encouraged wherever possible.

Acknowledgments

The projects described herein were developed and are being studied by the U. S. Army Engineer Waterways Experiment Station and the Baltimore, Detroit, Buffalo, and St. Louis Districts. Permission was granted by the Chief of Engineers to publish this information.

Innovative Wetland Re-creation in Urban Setting

C. Lynn Miller, M. ASCE[1]
George G. Feher[2]

Abstract

The project site is a 106 hectare, mixed-use, commercial development east of Tampa, Florida and at the westerly limit of the town of Brandon, Florida. Delaney Creek traverses the site from east to west. The majority of the site is within the 25 year floodplain. The flow length across the site was increased by rerouting the creek and providing low-flow, meander channels within the creek. Approximately 39% of the site area is committed to surface water management or wetland mitigation design. More than 90,000 wetland-species plants were included in the final mitigation design. The paper describes the difficult design conditions on the site and the final, permitted and constructed wetland system.

Project Setting

This project is the site of Brandon Town Center, a mixed-use, commercial development located east of Tampa, Florida and is within the western limit of the town of Brandon Florida. The total developed area is approximately 106 hectare. Intended land uses include commercial, residential, office areas and a regional shopping mall. The site is located in the southeast quadrant of the intersection of Interstate Highway 75 and State Road 60 (SR 60). SR 60 is the main commercial corridor through Brandon, Florida. Apartments occupy a site adjacent to the southeast corner of the project site (1992). A channelized reach of Delaney Creek crosses the site from east to west. There are several off site inflow points to the creek system and there are additional wetland areas adjacent to the site. Figure 1 illustrates pre-project conditions and Figure 2 illustrates project (developed) conditions.

---

[1] Manager, Water Resources Division, Greiner, Inc., 7650 West Courtney Campbell Causeway, Tampa, Florida 33607-1462
[2] Senior Environmental Scientist, Greiner, Inc., 7650 West Courtney Campbell Causeway, Tampa, Florida 33607-1462

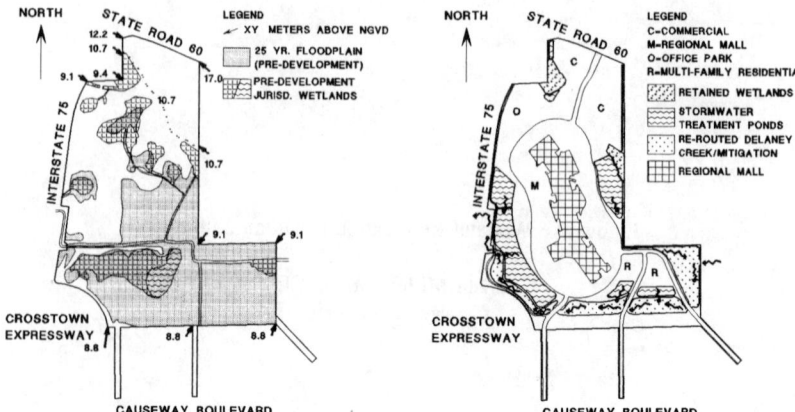

Figure 1. Pre-Project Condition   Figure 2. Project Conditions

Historic Conditions

The historic floodplain of Delaney Creek covered most of the project site, with the exception of the northern 14.6 hectare near S. R. 60 (Figure 1). Approximately 57 hectare (or, 56% of the total site) of the project site is at or below the 25 year flood elevation which had previously been determined for this reach of Delaney Creek.

The project is underlain by the Myakka-Bassinger-Holopaw soil association, which is nearly level, poorly to very poorly drained and has a sandy or loamy subsoil. The water table in this soil association generally ranges from standing water to a depth of 76 centimeters below the surface. The combination of the seepage slope near S. R. 60, the poorly drained soils, underlying clay lenses, and the broad floodplain of Delaney Creek provided the necessary hydrologic conditions for a series of bayheads and shallow grass ponds.

By the early 1900's, this part of Hillsborough County was converted from natural cover - mainly slash pine and longleaf pine - to land uses associated with agriculture and cattle grazing. This included realignment and channelization of creeks and their tributaries and the draining to the greatest extent possible of adjacent wetlands.

By 1982, when this project was initiated, the site had undergone various land uses such as tomato, watermelon and sod farming. Delaney Creek was nothing more than a narrow, approximately 6 meters wide, channelized water conveyance facility in the middle of a 106 hectare cattle ranch. The historic wetlands had become fragmented, drained or otherwise impacted to the point where they provided minimal habitat value. The entire watershed of Delaney Creek has been altered through farming and development practices.

Project Development

Site design was initiated in 1982 with permitting and regulatory agencies having jurisdiction over wetlands, wildlife resources and surface water management on the site. These included the Hillsborough County Environmental Protection Commission (HCEPC) for Waters of Hillsborough County, the Florida Department of Environmental Regulation (FDER) for Waters of the State, the Southwest Florida Water Management District (SWFWMD) for isolated wetlands and surface water management, and the U. S. Army Corps of Engineers (ACOE) for Waters of the United States. Ultimately, the United States Environmental Protection Agency (EPA) would also review and comment on the ACOE permit. This resulted in several different jurisdictional lines and areas, varying from approximately 4 hectare (FDER) to 12.5 hectare (SWFWMD and ACOE). There were a total of 2 hectare that were considered "forested" and 10.5 hectare of "non-forested" wetlands, including a 2.4 hectare man-made wet prairie that was created several years ago when a haul road was built to service two borrow pits in the southern portion of the site.

For this project to be feasible, the design had to maintain the pre-development discharge rate for a 25 year, 24 hour design storm event, have no net loss of the 100 year floodplain storage, limit the import of fill material, and mitigate for all wetland impacts. During the design and permitting of the project, the Delaney Creek watershed was designated as a "volume sensitive" watershed and the 100 year, 24 hour design storm event discharge and volume became design issues. Permits were submitted in 1986. Negotiations continued until 1989 when the final permit was issued. Construction started in 1990.

Characteristics of the Post-Project Condition

The relocation of Delaney Creek and the construction of the stormwater treatment system was completed in 1992, in accordance with the approved permits.

Unique features include five stormwater management ponds which total 9.3 hectare of surface area at normal water level with 41% of the surface area planted as shallow littoral shelves (maximum depth of 1 meter of water). This provides not only the required water quality treatment but also mitigation for non-forested wetland impacts. Two percolation-type, stormwater treatment areas (not ponds) have be constructed to provide pre-treatment for rainfall runoff to the preserved wetland in the northwest corner of the site. A manifold-type, flow distribution system (using three flumes) has been constructed to provide a redirection of off site runoff to the preserved wetland along the east property line.

Approximately 250,000 cubic meters of floodplain storage was detained on the site through the use of a tie-back levee (or, berm), along the west property boundary, which was set at the 100 year design flood elevation for the site. This was achieved by flooding a portion of the south vehicle parking area to an approximate depth of 15 centimeters beginning at about a 50 year, 24 hour design rainfall event. The regional mall and residential areas were graded to be about 61 centimeters above the design 100 year flood elevation and are accessible from the north access roadways.

The rerouted Delaney Creek provides 6.9 hectare of wetlands in a 61 meter wide floodplain that is designed not only to contain the 25 year flood event (which formerly would overrun the site); but also to provide mitigation for forested impacts. Two low pools have been incorporated into the system that will provide feeding areas for wading birds during times of drought. Eight trees were transplanted from the impacted wetlands to serve as cavity trees or perching sites.

The rerouted creek's floodplain has been further diversified by adding low-flow, meandering channels within the created, herbaceous wetlands adjacent to the creek. The total flow length (for flood flows) was increased from 1,180 meters in the pre-project condition to over 1,380 meters in the developed condition. The low-flow, meandering channels in the relocated section further increase the effective flow length for the non-flood events. The total channel flow length under project conditions nearly equals the flow length of all channels that once existed anywhere on the site (2,070- and 2,320 meters, respectively).

The total right-of-way that is dedicated to surface water management systems and wetlands preservation was approximately 41.3 hectare or 39% or the project site. Using 20 different species, a total of 86,000 herbaceous and 4,300 woody plants were installed.

Quality control of the mitigation area grading was performed by cross-section surveys of areas prior to the placement of a 15 centimeter organic muck layer and final grading. Since most of the area was excavated under wet conditions, a tolerance of 15 centimeters was considered acceptable. In cases where finished grades could not support the originally specified plant material, species and zonations were modified to be compatible with the actual finish grades. However, the species diversity was not changed to ensure compliance with the conditions of the approved permits.

The concurrent use of the mitigation area for stormwater attenuation allows for habitat diversity within an urbanized setting, increases wetlands utilization by wildlife and provides attractive, minimal maintenance, entrance landscaping for the development. To date, observed wildlife include hawks, herons, egrets, wood storks, sandhill cranes (including a nest), woodpeckers, otters, alligators and numerous other species. No other project in the vicinity of this site provides the habitat diversity that this site offers.

WATER TEMPERATURE ISSUES IN THE 90'S AND BEYOND

S. F. Railsback[1], M. ASCE

Introduction

Water temperature issues are expected to receive increasing attention in the 1990s. Temperature impacts are among the most common and most expensive environmental issues requiring mitigation at water projects, but few changes in mitigation technologies and little research have occurred in the past decade. Several emerging environmental and regulatory concerns and issues are likely to focus additional attention on temperature. The purposes of this paper are to review current water temperature issues and mitigation methods, to identify new and future temperature issues, and to identify research needs.

Current Temperature Issues

Temperature objectives. Water projects alter water temperatures because the heat balances in reservoirs and in streams with altered flows are significantly different from natural. Water temperature objectives often are to provide sufficiently cold summer temperatures to support populations of cold- or coolwater game fish. Water temperature mitigation is designed to avoid the effects of high temperatures on such fish. These effects can include mortality, but temperatures that are not high enough to kill fish can cause reduced growth rates and degraded physiological condition. Another important effect of altered temperatures is alteration of the timing of migration and spawning. Migration and spawning activities of a number of fish, including anadromous (e.g., salmon and steelhead) and non-anadromous species, are triggered by temperature changes; projects that change water temperatures can therefore change when, and therefore how successfully, these fish reproduce.

A common reservoir temperature concern is cumulative effects of high temperatures and low dissolved oxygen (DO) concentrations on fish. The so-called temperature-DO squeeze effect occurs when the upper levels of a stratified reservoir have summer temperatures too high for fish and the lower levels have cooler temperatures but insufficient DO concentrations. Severely restricted habitat and fish kills can result

---

[1]P.O. Box 8248, Stanford, CA 94309.

from the temperature-DO squeeze.

Water temperature may also be managed for water quality reasons. Because the solubility of oxygen varies significantly with temperature, rivers that suffer low DO concentrations (as a result of reservoir releases with low DO or of waste discharges) could be adversely affected by temperature changes that limit aeration and reduce DO concentrations. Other concerns arise when thermally altered rivers and reservoirs are used for cooling water. Waters with elevated temperatures reduce the efficiency of power plants that use them for cooling and may make it difficult to meet thermal discharge limits. As an example, one of the Tennessee Valley Authority's primary objectives in managing the Tennessee River reservoir system is to keep temperatures low enough to (1) allow efficient operation of their steam-electric power plants that withdraw cooling water from the reservoirs, (2) avoid exceedances of thermal discharge limits at power plants, and (3) avoid impacts to fish due to the combined effects of high temperatures and low DO concentrations.

Study methods. Water temperature issues are usually studied using reservoir or stream temperature models. [Doughty (1993) discusses stream temperature modeling.] These models have been shown in several tests to be generally adequate for prediction of the effects of projects or mitigation on water temperatures, if applied carefully. Temperature predictions by reservoir models, however, sometimes are highly dependent on variables that are difficult to measure or model. For example, temperature predictions are often highly dependent on wind speeds, which drive evaporative cooling and reservoir mixing, but these processes are difficult to model and appropriate wind speed data are often unavailable. However, predictions of water temperatures under various conditions are usually relatively reliable. Water temperatures are easily and inexpensively monitored using commercial instruments.

Temperature criteria for fish are used to predict how a given water temperature affects a particular species and life stage (i.e., incubating eggs, juveniles, and adults within a species; each of which can have different temperature needs) of fish. Temperature criteria are usually a temperature above which either physiological impacts or mortality is expected. Criteria are generally based on laboratory studies, which have been conducted using a wide variety of methods and assumptions, and few of which have been conducted in recent years. Criteria are usually available for recreationally and commercially important species but vary in quality among species and life stages. The temperature criteria currently in use do not consider the variability in temperature tolerance that may occur among different populations of the same species (e.g., rainbow trout native to California are expected to have evolved higher temperature tolerances than have rainbows from northern states). The use of temperature criteria also does not consider the observed ability of fish to respond behaviorally to high temperatures (e.g., to find and use thermal refuges such as pockets of cool water in deep pools). Methods to link temperature criteria (which evaluate the effects of temperatures on individual fish) to effects on fish populations have not been applied, although a few models that eventually may be able to address this issue are being developed. Although the methods used to predict the effects of water projects on water temperatures are fairly reliable, the biological effects of temperature changes are complex and poorly understood. Except for the work of the Environmental Protection Agency's (EPA's) Duluth Research Lab, where researchers are attempting to use field data to define fish temperature tolerances, there is little research on temperature criteria currently underway.

Mitigation methods. There are few ways to control the temperature of releases from water projects. Release temperatures of stratified reservoirs can be controlled by selective withdrawal of releases from various elevations. Selective withdrawal requires an outlet structure that can withdraw controlled amounts of water from selected reservoir elevations. In combination with monitoring of a reservoir's temperature structure, selective withdrawal can be used to release temperatures that meet temperature criteria but conserve sufficient cold water to meet the criteria throughout the warm seasons. [Johnson (1993) discusses the design and operation of new selective withdrawal structures at existing dams.]

Warm (and, occasionally, cold) temperatures that result from reduced instream flows are mitigated by increasing flow rates. Instream flows intended to mitigate temperatures are usually based on conservative and static design conditions (e.g., steady high air temperatures, low wind speeds), and usually are not allowed to vary with actual conditions.

New and Future Water Temperature Issues

Trends in how water temperature impacts of water projects are assessed and regulated, and predicted changes in environmental management objectives, indicate that several new issues will need to be addressed in the 1990s and beyond.

Non-game species and biodiversity management. The traditional fisheries management orientation toward production of sport and commercial fisheries is expected to give way to management for other species for several reasons. First, a growing number of non-game species (as well as native stocks of salmon and steelhead) are proposed or listed as threatened or endangered species; such species are protected by law and assume management priority where they occur. Secondly, there is a clear trend in environmental and fisheries management toward preservation of biological diversity instead of management of only a few species. Biological diversity objectives in fisheries management include preservation of (1) individual stocks of anadromous fish, (2) native strains and varieties of species, and (3) non-game species. A third reason why non-game species will increase in management importance is the increasing application of biological water quality criteria. USEPA policy calls for all states to adopt narrative biological criteria as part of their water quality standards by the end of September, 1993 and quantitative biological criteria beginning as early as 1996. Biological indicators respond to the cumulative effects of such parameters as temperature, toxicity, sediment loads, and flow rate. Application of such indicators, which are often based on the numbers and diversity of fish and benthic invertebrates, would probably increase the regulatory importance of water quality issues like temperature and may require temperature mitigation designed to protect non-game and non-fish species.

A shift of management priority from commercial and sport fish species to non-game species is likely to result in a shift from emphasis on providing cold water in summer to an emphasis on providing more natural temperatures that native species are adapted to. However, where threatened or endangered cold-water species occur (e.g., endangered stocks of salmon), there will be continued emphasis on providing cold water in summer and fall.

Cumulative impacts. In assessing the environmental impacts of, and requiring mitigation for, water projects, resource and regulatory agencies are giving cumulative impacts increasing consideration. There are several reasons for the increasing attention to cumulative impacts. There are steps being taken by the EPA and others toward management of whole watersheds instead of individual river reaches. Concern over nonpoint source pollution and application of biological water quality criteria require more comprehensive and large-scale water management. In addition, tools for evaluating cumulative and large-scale impacts (both ecological theory and mathematical models) are rapidly improving.

Water temperature effects can be cumulative in several ways. Multiple projects in the same watershed can have cumulative impacts on temperature; each project can alter the temperature of water arriving at the next project downstream (the Tennessee River system being a major example). Also, water projects may have effects on temperature that accumulate with the effects of other activities (e.g., timber harvest, grazing) in the watershed. Finally, temperature stress may be one of a number of impacts affecting the aquatic ecosystem; an important example is the cumulative effects of flow changes, entrainment, temperature, and other impacts on Pacific salmon populations.

Assessment of cumulative impacts is often seen by water managers as an increment in the burden of regulatory compliance and mitigation. However, there are numerous cases where the focus of temperature mitigation has been on a water project when cumulative effects from other activities in the watershed have gone unmitigated. The increasing focus on cumulative impacts can be seen as an opportunity to focus mitigation requirements on all the activities contributing to environmental problems.

Climate change. The scientific consensus on the effects of greenhouse warming is that annual average air temperatures in the continental United States will increase by approximately two to four degrees centigrade as carbon dioxide concentrations double in the upcoming decades, and that changes in precipitation and runoff are likely. Climate change would likely accelerate future increases in water demands (including demands for power production and cooling water). The consensus is also that these changes could occur within the life span of many water projects that are currently in operation. Climate change, should it occur as predicted, can be expected to worsen many water temperature problems and complicate the determination of appropriate mitigation for water projects.

Research Needs

Research needs have been determined by identifying (1) important weaknesses in current temperature study and mitigation methods and (2) methods needed to address new and future temperature issues.

Better temperature criteria for fish. Research is needed to develop methods of better predicting the effects of temperature changes on aquatic ecosystems. Methods are needed which (1) account for variability in temperature responses within species, (2) account for behavioral response to temperature stress, (3) include more species, and (4) predict effects on fish populations, biodiversity, and ecosystems. The work of EPA's Duluth Lab is one of the only programs addressing these research needs.

Lower cost mitigation. Methods are needed to reduce the high costs of temperature mitigation. Less expensive ways to selectively withdraw reservoir releases (e.g., Johnson 1993) are needed. Several potential ways to reduce the instream flow releases needed to mitigate stream temperatures need exploration; these include using riparian vegetation to increase shading, taking advantages of fishs' ability to use thermal refuges, and distributing cool releases along the stream instead of releasing them all from a single point. In addition, flexible management schemes to allow temperature mitigation to respond to actual conditions (instead of being fixed over time; real-time instream flow control for temperature management is in use at at least one site, an Idaho Power project on the Snake River) need to be developed.

Analysis of climate change effects. There are numerous questions about the potential effects of climate change on water temperature management that need research. These include (1) how difficult it may become for projects to meet temperature requirements, (2) whether reservoirs could magnify or mediate the effects of climate change, (3) what cumulative interactions with other effects of climate change (e.g., changes in vegetation and water quality) may occur, and (4) what changes in water demands may occur. Many of these questions are being addressed by the electric power industry and by interagency federal research programs.

References

Doughty, K. (1993). "Field comparison of temperature models and study design considerations." Presentation at the 1993 ASCE Water Resources Planning and Management Division Conference.

Johnson, P. L. (1993). "Sacramento and Trinity river systems: A demonstration of river temperature management for the 90's." Presentation at the 1993 ASCE Water Resources Planning and Management Division Conference.

Sacramento and Trinity River Systems: A Demonstration
of River Temperature Management for the 90's

Perry Johnson[1] and Chet Bowling[2]

Abstract

Drought in Northern California resulted in the potential that summer and early fall, 1992, Sacramento and Trinity River temperatures would exceed critical levels for sustaining salmon populations. As a result, extraordinary efforts, both operational and structural, were undertaken to manage cold water releases from the reservoir system.

Introduction

The U.S. Bureau of Reclamation's Central Valley Project (CVP) is a large water project that serves much of California. The project includes 16 storage dams, 3 diversion dams, nearly 1000 km of canals, 39 pumping plants, and 9 powerplants with a combined capacity of 1850 mW. Although developed primarily for irrigation, this multipurpose project also provides flood control, improves Sacramento River navigation, supplies domestic and industrial water, generates electric power, conserves fish and wildlife, creates opportunities for recreation, and enhances water quality.

One part of the CVP, the Shasta and Trinity River Divisions include Trinity and Lewiston Dams on the Trinity River and Shasta and Keswick Dams on the Sacramento River. Water is also diverted from the Trinity River at Lewiston

---

[1] Hydraulic Engineer, U.S. Bureau of Reclamation, Mail Code D-3751, PO Box 25007, Denver CO 80225-0007

[2] Supervisory Hydraulic Engineer, U.S. Bureau of Reclamation, Mail Code MP-2800, 2800 Cottage Way, Sacramento CA 95825

Dam, through Whiskeytown Reservoir, to Keswick Reservoir and the Sacramento River. The reservoir system has a dominate influence on discharge and water temperature in the Trinity River below Lewiston Dam and in the Sacramento River below Keswick Dam.

Salmon populations on both rivers are in decline. The runs on the Trinity River are of major concern to the Hoopa Tribe and are being addressed by a multi-agency task force. The "winter run" of Chinook salmon on the Sacramento River has been listed as threatened under the Endangered Species Act. The "spring run" population has also declined and listing may follow.

California has been in an extended drought. As a consequence reservoir storage has been low and volumes of stored cold water limited. Nineteen-ninety-two was a critical low water year. The potential existed that reservoir release temperatures coupled with in river warming could generate lethal water temperatures for egg incubation and juveniles. As a consequence the Bureau of Reclamation initiated an aggressive program to modify operations and add structural features that would optimize cold water releases.

Reservoir and river system numerical models were used to develop operating guidelines. The models defined release rates that would yield an extended supply of cold water. The models were used to estimate atmospheric warming and tributary influences for predicting the reaches of river over which adequate temperatures could be maintained.

Operational Modifications

Water deliveries to project contracts, both agricultural and urban, were reduced. Water releases to meet project requirements for water quality and the limited deliveries, to the extent practicable, were supplemented from other reservoirs. Project operators coordinated extensively with federal and state resource agencies to develop operations plans to maximize temperature control in both the Trinity and Sacramento Rivers.

As the season progressed it became apparent that colder releases could be supplied from both Trinity Dam and Shasta Dam by curtailing power operations and by making all releases through outlet works. At both dams the power intakes are positioned higher than the low level outlets. Power releases were bypassed at Shasta Dam for 170 days and at Trinity Dam for 110 days in water year 1992. Approximately $10,000,000 in power revenues were lost.

Release Temperature Control Structures

Selective Withdrawal
A retrofit multilevel intake for Shasta Dam has been designed but has not yet been installed (Johnson et al 1991). After consideration of numerous alternatives, the Shasta design consists of a gated steel frame, with sheet pile skin, that is placed over the power intakes and attached to the face of the dam. Studies for the multilevel intake show that to achieve optimum cooling, release must be made from high in the reservoir early in the season saving large volumes of cold water for late summer and early fall. The study shows that in critically dry years there is insufficient cold water in the system to allow both maintenance of historic releases and adequate river temperatures.

Light weight curtain structures for release temperature control were studied as an option for Shasta Dam (Johnson 1990). The curtains show substantial cost benefit over more traditional selective withdrawal intakes. However, because of structure size and structural uncertainties a curtain was not pursued for Shasta. Because of cost advantages and growing need for temperature control, Reclamation initiated a research program to install a field curtain. In the summer of 1992, under a very tight schedule, a curtain was installed in Lewiston Reservoir (O'Haver 1992). The curtain design had been developed with use of numerical and physical models (Brown et al 1992). The 250 m long, 10 m deep curtain was suspended from surface floats and was retained by a cable and anchor system. The curtain was used to exclude epilimnion withdrawal thus cooling both water diverted to the Sacramento River and Trinity River releases. Field performance data have been collected but are not reduced as of this writing.

Pneumatic Diffuser
In early October 1992, a pneumatic diffuser or bubbler was installed to allow access to approximately 1.0×10^8 m^3 of hypolimnion water that was unaccessible with the low level outlet works at Shasta Dam. The diffuser was used to upwell the cold water to the intake. Based on experience of the California Department of Water Resources, garden soaker hose was used as a diffuser line to generate 2.0 to 3.0 mm diameter bubbles. Experiments were run varying air flow rate, diffuser size, diffuser position, diffuser depth, release discharge, and intake velocity to the outlet works. Initial evaluation indicated that the diffuser could be used to reduce released temperatures at least 0.5°C. Again the collected data has not been reduced as of this writing.

Results

The operational measures, particularly the bypass of powerplant releases, have been effective in sustaining reduced water temperatures. Structural alternatives (curtains, pneumatic diffusers, selective withdrawal) show promise for further management of water temperatures.

References

Brown, R.T., Yates, G., and Johnson, P.L. (1992). "Physical and 2-D Computer Models of Skimmer Curtain Effects on Lewiston Reservoir and Outlet Temperatures." Proceedings of the Hydraulic Engineering sessions at Water Forum '92, ASCE, August 2-6, 1992.

Johnson, P.L. (1991). "Hydraulic Features of Flexible Curtains Used for Selective Withdrawal." Proceedings of the Hydraulics Division National Conference, ASCE, July 29 - August 2, 1991.

Johnson, P.L., LaFond, R., and Webber, D.W. (1991). "Temperature Control Device for Shasta Dam." Proceedings of Technical Exchange between the Japan Dam Engineering Center and the U.S. Bureau of Reclamation.

O'Haver, G. (1992). "Temperature Control Curtains - Lewiston Lake and Hatchery." Presented at the Power O&M Workshop, U.S. Bureau of Reclamation, Mid-Pacific Region, October 20, 1992.

Development of Bioretention Practices
For Stormwater Management

Larry Coffman[1], Raymond Green, P.E.[2],
Michael Clar, P.E.[3] and Susan Bitter[4]

Abstract

This paper introduces the concept of bioretention as an innovative stormwater quality management practice. Bioretention is a method to treat the first flush of runoff using a combination of retention, native terrestrial vegetation and soil conditioning. The material presented in this paper is the result of a study to determine the technical feasibility of using bioretention for stormwater management. The study addresses the plant materials suitability, hydrology, water quality, and soil materials aspects of bioretention.

Introduction

Bioretention, attempts to maximize all available physical, chemical and biological pollutant removal or transformation processes found in the soil and plant complex of a terrestrial forested community. Stormwater runoff treatment occurs in many ways: through sedimentation, transpiration, evaporation, infiltration,

---

[1] Assistant Branch Manager, Prince George's County Government, Watershed Protection Branch, 9400 Peppercorn Place, Suite 600, Landover, Md. 20785

[2] Project Manager, Engineering Technologies Associates, Inc., 3458 Ellicott Center Drive, Ellicott City, Md. 21043

[3] Principal, Engineering Technologies Associates, Inc., 3458 Ellicott Center Drive, Ellicott City, Md., 21043

[4] Project Manager, Biohabitats, Inc., 303 Alleghany Ave, Towson, Md. 21204

bio-decay, nutrient cycling and bio-uptake. Similar biological systems have been routinely used in the retention and transformation of nutrients in the land application of sewage effluent. Bioretention areas, when sited properly, also have the potential to improve the site landscaping providing aesthetic enhancement, shade, wind breaks, and noise reduction.

Bioretention Practices For Graded Areas

The first phase of the study centered on the development of details and specifications for bioretention practices in the landscaped or graded green space areas of commercial and industrial sites. Commercial and industrial sites have the most impervious area by percentage, and consequently, the most runoff and pollutant loading potential of the land uses found in urban areas. The bioretention areas were designed to provide retention of the first flush of runoff for pollutant uptake and transformation by the vegetation/soil complex, with the source of water being the sheet flow from grass and impervious areas.

The generation of specifications included the development of plan and section details for bioretention practices in parking islands, parking edge and perimeter areas. A plan and section detail of a bioretention area in a perimeter area is shown in Figure 1. The width of the bioretention area of fifteen feet provides for adequate density of trees and shrubs, creating a microclimate which would be able to tolerate the effects of the pollutants in the runoff, heat stress, insect and disease infestations, and acid rain.

Plant Species

Bioretention area plantings are selected to emulate a typical terrestrial forest community. Overall, the plant species selected for bioretention have been shown to be resistant to stresses from the pollutants in urban runoff, tolerant to wet and dry moisture regimes, and are able to meet local zoning and landscaping requirements. Lists of suitable trees, shrubs, and herbs have been developed as part of the study.

Hydrologic Analysis

A water balance was developed for a bioretention area based on the precipitation, evapotranspiration and the infiltration for a commercial tract for the median rainfall event occurring in the Washington, D.C. area. The median rainfall event is a half inch of rainfall over 6 hours. According to the water balance, the minimum size

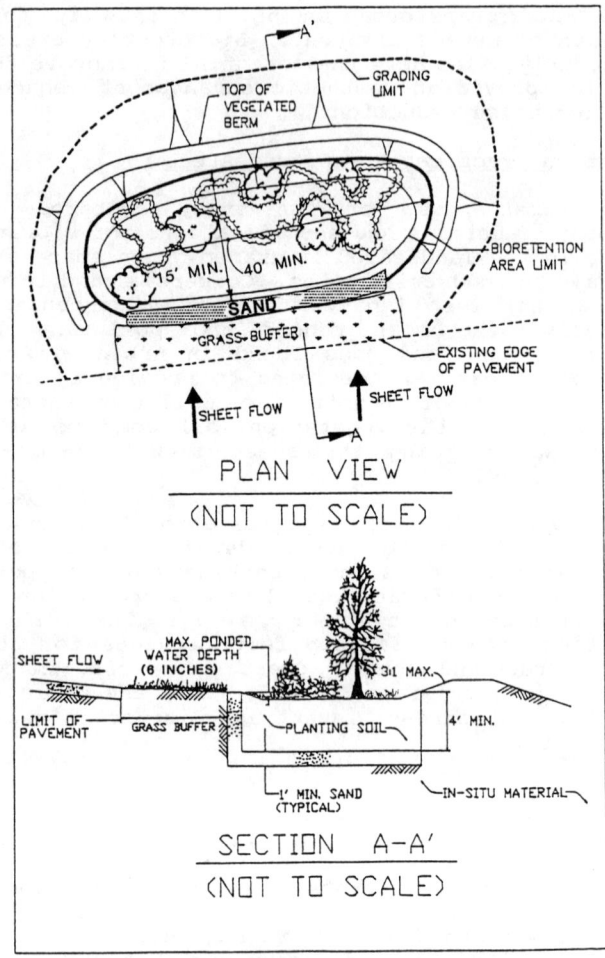

Figure 1. Plan and Section View of a Typical Bioretention Area

bioretention area of 600 square feet (15 by 40 feet), would be able to infiltrate the rainfall event from a site of 0.3 acres having an 80% impervious cover.

Water Quality

The reduction of nitrogen and phosphorus through bioretention can be highly variable. The amount of nitrogen and phosphorus cycled in the leaf fall and leaching to the ground water would vary according to plant species, the age of the plants, and the degree of maintenance. If bioretention practices were applied to six percent of a commercial site, the nutrients cycled by the vegetation would represent from 19 to 38 percent of the nitrogen loading in the storm water runoff, and from 18 to 74 percent of the phosphorus loading.

Soil Materials

It is recommended that the soil used for bioretention have a sandy loam, loamy sand, or loam texture. These soils have a clay content ranging from 10 to 25%. Clays are the most chemically reactive portion of the soil profile. Soils deficient in clay content will not adsorb phosphorus from the runoff and would not be suitable for bioretention. High organic content is needed as a suitable medium for vegetation and microbial growth and reactions.

Conclusions

The results of the study indicate that bioretention can be a viable method for the water quality treatment of site runoff. The treatment of the first half-inch of runoff from impervious areas is the minimum water quality standard required by State of Maryland and Prince George's County for storm water management. The study demonstrated that a bioretention area comprising of five percent a commercial or industrial site would be able to infiltrate the first half-inch of runoff . For a typical commercial and industrial site in Prince George's County, green space is approximately six percent of the site area, and incorporating bioretention practices into the site plan would not modify the area available for parking and buildings.

Where applicable, bioretention promises to be an effective, low cost, and relatively low maintenance management practice. It is an on-site treatment technique which can be easily integrated into new commercial or industrial development landscape schemes and can be used as a retrofit technique for existing development. Because of the use of terrestrial plant materials and the ability of these materials to tolerate greater variation in the hydrologic regime this practice could have wider application potential than typical wetland systems.

Design Considerations Associated
with Bioretention Practices

Larry Coffman[1], Raymond Green, P.E.[2],
Michael Clar, P.E[3] and Susan Bitter[4]

Abstract

This paper presents the design requirements of bioretention practices. Bioretention is a stormwater management practice using a combination of retention, native terrestrial vegetation and soil conditioning. The paper will present the methodology for the sizing, and the preparation of grading and planting plans for bioretention areas. The cost of bioretention practices will also be discussed.

Introduction

The overall goal of bioretention, a stormwater management technique using native plants and soil conditioning, is the treatment of the "first flush" of runoff from impervious areas. The "first flush" is the runoff containing a disproportionally large pollutant load during the early part of a storm event. Conventional storm water management methods such as

---

[1] Assistant Branch Manager, Prince George's County Government, Watershed Protection Branch, 9400 Peppercorn Place, Suite 600, Landover, Md. 20785

[2] Project Manager, Engineering Technologies Associates, Inc., 3458 Ellicott Center Drive, Ellicott City, Md. 21043

[3] Principal, Engineering Technologies Associates, Inc., 3458 Ellicott Center Drive, Ellicott City, Md., 21043

[4] Project Manager, Biohabitats, Inc., 303 Alleghany Ave, Towson, Md. 21204

detention ponds or oil-grit separators, treat the "first flush" of runoff after it is concentrated in the storm drain system, where pollutant removal is costly and limited by space constraints and variable flow rates. The rationale of using bioretention methods is to provide for the management of sheet flow from grass and impervious surfaces before runoff enters into the storm drain system.

Bioretention Area Sizing

Bioretention areas are conceived to capture sheet flow and will typically be limited to small drainage areas ranging from 0.25 to 1 acre. The size of the bioretention area should equal or exceed 5% of the upland impervious area to allow for the infiltration of the first half inch of runoff. The bioretention area should have a minimum width of 15 feet to accommodate multiple trees and shrubs. The length of the bioretention area should be at least twice the width, with a minimum of 40 feet, to allow sheet flow to be adequately dispersed.

Grading Plan Development

Ideally, bioretention should be sited off-line in a location which would allow drainage to be treated before it leaves the site, to prevent untreated runoff from discharging into existing drainage systems. A typical grading plan for a bioretention area is shown in Figure 1. The bioretention area shown in the figure, 975 square feet, was designed to treat the runoff from 0.8 acres of a site which contained a parking lot having an impervious percentage of 70 percent. As indicated in the figure, sheet and gutter flow is diverted into the bioretention area through an opening in the curb. Inlet deflector blocks are located in front of the curb openings to channel the flow into the bioretention area. Water is allowed to pond to a one half foot depth before runoff bypasses the bioretention area and flows into the storm drain system.

The bioretention area in the figure features a grass buffer strip and a sand bed to filter the fine material from the runoff which would tend to clog the planting soil. The sand bed also increases the infiltration capacity, and provides aeration for the plant roots in the bioretention area.

Planting Plan Development

The plant species selected for bioretention should be resistant to stresses from the pollutants in urban

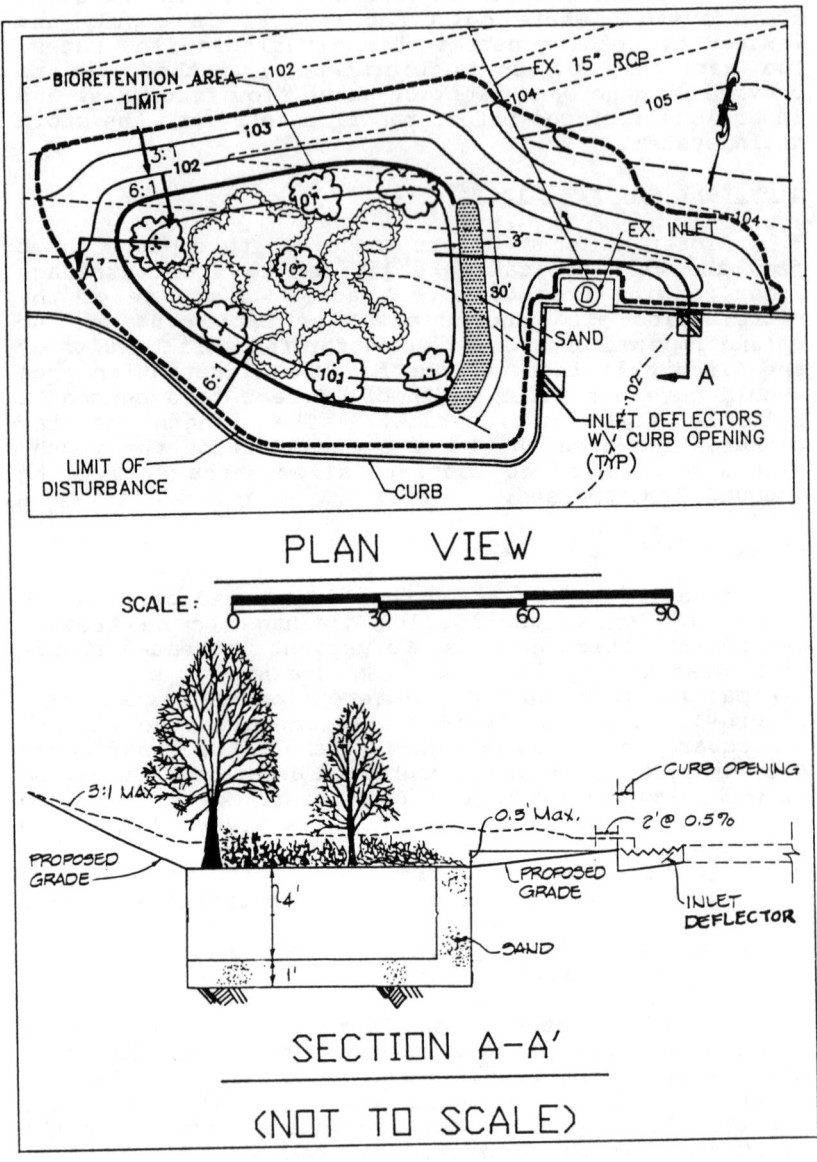

Figure 1. Grading Plan and Section View for a Typical Bioretention Area

runoff, tolerant to wet and dry moisture regimes, and able to meet local zoning and landscaping requirements. The planting plan should use native plants indigenous to the region. Non-native plants tend to be invasive, and to compete with native plants for space. Also, the use of a minimum of three species of trees and shrubs is recommended. Ecosystems that are dominated by a single plant species are susceptible to complete collapse if the plants become weak or diseased.

Estimated Cost

The costs associated with bioretention will vary according to the size of the bioretention area, and the plants used. The estimated construction cost (Prince George's County, Maryland costs) for the bioretention area shown in the Figure 1 is $ 6,500. The construction cost was developed assuming that the planting soil would be imported onto the site. The excavation and disposal of the existing site material, and the import of the planting soil comprised one third of the overall construction cost ($ 2,200). The next major cost, landscaping, was $1,400 or 20% of the total cost. The remaining construction costs included sediment and erosion control, ground cover, and miscellaneous concrete work.

Conclusions

Bioretention is a method used for treating upland sources of runoff. Using the design criteria developed as part of this study[5], bioretention can be a cost effective treatment for the "first flush" of pollutants from impervious areas. The cost for a bioretention area treating 0.8 acres of a parking lot is $ 6,500. An oil-grit separtor treating the same area would cost approximately $20,000, or have approximately three times the bioretention area cost.

Bioretention methods may modify existing site grading practices. In applying bioretention to a site a designer would use sheet flow as a stormwater conveyance rather than inlets and storm drain pipe. The reduction of the number inlets and storm drain will have the benefit of lowering site grading costs.

---

[5] Bioretention Feasibility Analysis, prepared for Watershed Protection Branch, Department of Environmental Resources, Prince George's County, Maryland, prepared by Engineering Technologies Associates, Inc., Ellicott City, Maryland and Biohabitats, Inc., Towson, Maryland.

Pollutant Removal Efficiency of Detention Basins

A. Osman Akan[1]

Abstract

A simple method is presented for quick estimates of pollutant removal efficiency of detention basins under dynamic conditions. The method is based on the generalized solutions to the reservoir routing problem and the settling velocities of pollutants in urban runoff.

Introduction

Detention basins are commonly used for water quality improvement. By detaining stormwater runoff for a prolonged period of time, particulate pollutants can be removed from the runoff before discharging into a receiving water body. Although various processes affect the removal of certain specific pollutants to some extent, detention basins primarily rely upon the settlement processes for pollutant removal. Table 1 summarizes the estimates of particle settling velocities in urban runoff as reported by Environmental Protection Agency (1986).

The approximate method presented in this paper is based on a reservoir routing technique which couples the hydrologic storage equation with an expression describing the hydraulics of the outlet structure. The equations are written in dimensionless form and solved using a finite difference scheme. The numerical results obtained in terms of the governing dimensionless parameters represent pre-determined solutions to the reservoir routing problem. The results are generalized based on the concept of hydrologic similarity and are presented in chart form. These charts are useful to estimate the runoff detention time in a basin given the inflow hydrograph and the reservoir characteristics. Once the detention time is found, the pollutant removal efficiency is estimated on the basis of the settling velocities of different size fractions. In the absence of local data, the EPA estimates of the settling velocities can be employed.

---

[1] Professor, Department of Civil and Environmental Engineering, Old Dominion University, Norfolk, VA 23529.

Detention Pond Charts

A set of pre-determined solutions to the reservoir routing equation are provided in Figures 1 and 2 in chart form for weir-type outlets. Similar charts can be developed for other types of outlets using the same approach. The details of the mathematical model employed to obtain these charts were reported previously by Akan (1991), and they are omitted here for brevity. However, we should note that the inflow hydrographs are assumed to have same shape as the Soil Conservation Service dimensionless hydrograph. Also, we assume that the pond stage-storage relationship can be expressed in the form of

$$s = b \, h^c \qquad (1)$$

where s = storage above the basin outlet, h= depth of water above the outlet, and c,b = constant parameters.

The expressions for the governing dimensionless parameters P, S*, and W are given in Figures 1 and 2. In these expressions, g = gravitational acceleration, i_p = peak discharge of the incoming runoff hydrograph, t_p = time of occurence of the peak runoff rate, k_w = dimensionless weir discharge coefficient, L = weir crest length, w_{90} = time period during which at least 90% of runoff will be present in the detention pond, and s_{max} = maximum volume of water that will be stored in the detention basin.

Example Application

Let the inflow hydrograph for a detention basin have a peak discharge of i_p = 1.5 m³/sec occurring at t_p = 1.0 hour = 3,600 seconds. The stage - storage relationship is s = 18485 $h^{1.22}$, that is b = 18,485 and c = 1.22. The outlet is of weir type with k_w = 0.40, and the crest length is L = 0.38 m. We will determine the solids removal efficiency of the detention basin.

First, from the expression given for P in Figures 1 and 2, P = 0.1. Then from Figure 1, S* = 1.12, and therefore s_{max} = 6048 m³. Next, using Equation 1 h_{max} = 0.40 m. This approximates the depth of water above the outlet for the period during which 90% of the runoff is present in the pond. Also, using Figure 2, W = 6.5, and therefore w_{90} = 6.5 hrs.

To calculate the pollutant removal efficiency of the detention basin the five size fractions listed in Table 1 are considered separately. For the first size fraction, the settling velocity, V_s, is 0.009 m/hr. To settle all the particles belonging to this group, the required detention time is (0.40 m)/(0.009 m/hr) = 44 hrs. The required detention time is greater than the actual detention time of 6.5 hours. Therefore, only part of the particles will settle. Assuming that the particles belonging to this group are uniformly distributed within the runoff, and noting that only 90% of the runoff is detained 6.5 hours, the fraction of pollutants removed is (0.9)(6.5 hrs)/(44 hrs) = 0.13. The calculations for the other size fractions are summarized in Table 2. For size fractions 2,3,4, and 5, the required detention time is less than 6.5 hours. Therefore all the particles contained in the detained portion of the runoff will settle. This will result in a removal fraction of 0.90 for each size group, because only 90% of the runoff is detained 6.5 hours. The overall pollutant removal efficiency is found as the average of the five fractions, that is $(0.13 + 0.90 + 0.90 + 0.90 + 0.90)/5 = 0.75 = 75\%$ The efficiency calculated in this example does not include the particles removed from the 10 % of the runoff that is retained less than 6.5 hours. Therefore when this procedure is used the computed efficiency would be slightly smaller than the actual efficiency.

References

Akan, A.O. (1992), "Storm runoff detention for pollutant removal," *Journal of Environmental Engineering*, 118(3):380-389.

Environmental Protection Agency (1986) *Methodology for Analysis of Detention Basins for Control of Urban Runoff Quality*, EPA 440/5-87-001, EPA Office of Water Regulations and Standards, Washington, DC.

Table 1. Settling Velocities in Urban Runoff (After EPA 1986)

(1) Size Fraction	(2) % of Particle Mass in Urban Runoff	(3) Average Settling Velocity (m/hr)
1	0 - 20	0.009
2	20 - 40	0.091
3	40 - 60	0.457
4	60 - 80	2.134
5	80 - 100	19.812

Table 2. Calculations for Sample Application

(1) Size Fraction	(2) Settling Velocity (ft/hr)	(3) Required Detention Time (hrs)	(4) 90% Detention Time (hrs)	(5) Mass Fraction Removed
1	0.009	44.00	6.50	0.13
2	0.091	4.40	6.50	0.90
3	0.457	0.88	6.50	0.90
4	2.134	0.19	6.50	0.90
5	19.812	0.02	6.50	0.90

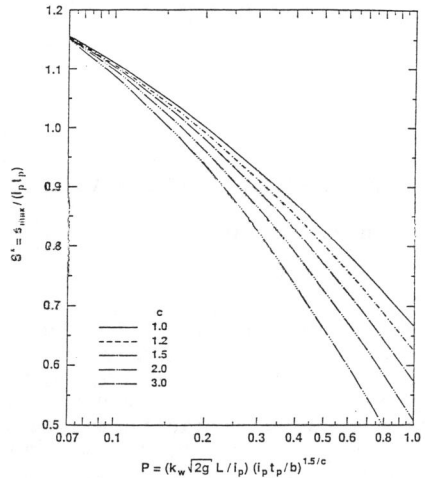

Figure 1. Maximum Storage in Detention Basin

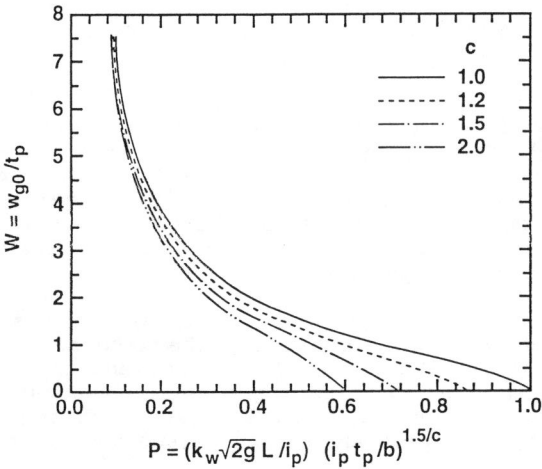

Figure 2. 90% Detention Time

LONG-TERM PERFORMANCE OF WET DETENTION PONDS

by Agustin E. Maristany[1], P.E., M. ASCE

Abstract

Wet detention is a technology which has been successfully applied in recent years to improve stormwater runoff quality in urban areas. Since most systems have a short operational history, their long-term performance has been questioned. There is speculation that once these systems reach a state of dynamic equilibrium, nutrient removal may decline due to the reduced nutrient uptake of a mature ecosystem.

This paper sheds some light on the subject by comparing the performance of relatively new wet detention systems to a case study of Lake Munson, a wet detention system which has received wastewater effluent and runoff discharges for over 30 years. Long-term removal rates for Lake Munson compared favorably with rates reported for relatively new systems.

Introduction

Lake Munson is a shallow man-made lake created in 1950 by impoundment of a cypress swamp. The 3.4-square-kilometer lake, which has an average depth of 1 meter, currently receives runoff discharges from a 311-square-kilometer watershed containing about 51 percent of the City of Tallahassee, Florida. In 1986, its watershed was estimated to be 20 percent impervious.

As a result of over 30 years of treated wastewater and untreated stormwater discharges, the lake has manifested severe water quality and ecologic problems including fish kills, algal blooms, floating aquatic vegetation, high nutrient and bacteria levels, low game fish productivity, and depressed oxygen levels. Although water quality

---

[1]Senior water resources engineer, CH2M HILL, 800 Fairway Drive, Suite 350, Deerfield Beach, FL 33441

improved significantly after the elimination of all effluent discharges in 1984, degradation attributable to stormwater discharges continues to be evidenced.

The Northwest Florida Water Management District initiated an investigation in 1986 (Maristany, et.al, 1988) to assess the current status of the lake, determine the impact of stormwater discharges, and propose restoration alternatives for enhancing lake water quality. The data collection program implemented as part of that study provided the opportunity to assess the long-term performance of an old and heavily impacted wet detention system.

The data collection program was designed to monitor stormwater, lake water, and lake sediments over a 1-year period. Monitoring encompassed lake inflows and outflows, stormwater quality, precipitation, lake level fluctuations, 7 lake water column stations, and 15 lake sediment stations.

History of Nutrient and Sediment Loads

From 1950 to 1984, Lake Munson received effluent discharges from Tallahassee's wastewater treatment plants. From 1977 to 1979 effluent discharges reached their peak and were estimated to contribute 35 percent of the hydraulic load, 66 percent of the biochemical oxygen demand (BOD_5), 88 percent of the total phosphorus, and 91 percent of the total nitrogen loads into the lake. In contrast, stormwater discharges were responsible for 92 percent of the load of suspended solids. Measures of lake water quality for the period 1966-1980 indicated an advanced state of eutrophication. In a 1982 study of Florida lakes, Lake Munson was classified as hypereutrophic and ranked the seventh most degraded lake in the state.

Starting in 1979, effluent was incrementally diverted to spray irrigation fields, and by 1984 all effluent discharges had been eliminated. As a result, by 1987 the water quality of the lake was much improved over conditions observed from 1966 to 1980. The lake's trophic level changed from hypereutrophic to eutrophic over a short period of time, as evidenced by sharp reductions in chlorophyll_a (0.127 to 0.03 milligrams per liter [mg/l]), total nitrogen (4.9 to 0.89 mg/l), and total phosphorus (2.2 to 0.26 mg/l) in the water column. Except for occasional, accidental sewage spills into the lake, stormwater currently accounts for virtually all pollutant loads to the lake.

Lake Pollutant Retention Capacity

Although not designed for that purpose, Lake Munson functions as a wet detention pond with its permanent pool at the elevation of the emergency spillway. The storage capacity of the permanent pool, which equals the volume of the average storm event, is exceeded only by 20 percent of all storms. Therefore, most storm events are captured within the lake, displacing an equal volume of lake water downstream. The captured volume undergoes treatment by settling and biochemical uptake during the relatively long periods of time between storms (averaging 115 hours).

Pollutant retention capacities for Lake Munson were estimated by comparing flow-weighted inflow and outflow concentrations. This approach is valid because the negligible groundwater and atmospheric interactions result in lake outflows that are practically equal to stormwater inflows.

Observed retention capacities for Lake Munson were compared to retention capacities of comparable new wet detention ponds. Performance data from wet detention ponds monitored as part of the Nationwide Urban Runoff Program (NURP) (EPA, 1983) were used for this comparison. NURP data were interpolated to estimate the retention of new ponds having the same storage capacity and stormwater loads as Lake Munson. Table 1 shows the retention capacity of Lake Munson and a comparable NURP pond.

Conclusions

Table 1 reveals that Lake Munson displays removal rates expected from new wet detention ponds of comparable characteristics. The similarity in suspended solids, nutrients, and metals removal is particularly striking. Although lower-than-expected removals were observed for oxygen demand, the overall performance was impressive for a system that has been heavily impacted by wastewater effluent and stormwater discharges for over 30 years, and which has never been maintained.

The results suggest that wet detention ponds provide consistent removal of particulate constituents, but experience declining removal of organic matter (oxygen demand) if not properly maintained. The declining removal of oxygen demand is attributed to the long-term accumulation of organic matter in the sediments which increases sediment oxygen demand and reduces the availability of oxygen for oxidation of organic matter in the water column.

An important conclusion of this study is that wet detention ponds designed with minimum treatment storage capacity and which are not properly maintained are likely to experience significant water quality problems due to eutrophication. Therefore, to prevent the pond from being overwhelmed by common storm events like Lake Munson, it would be advisable to increase storage capacity beyond minimum treatment requirements. In addition, sediment stabilization is recommended as part of a maintenance program to reduce sediment nutrient release and oxygen demand.

Table 1
Lake Munson and Comparable NURP Pond Retention

Constituent	Lake Munson Retention (percent)	Comparable NURP Pond Retention (percent)
Total Suspended Solids	95.4	93
Turbidity	86.6	
Total Chromium	77.5	
Total Copper	72	64
Total Lead	91.3	84
Total Nickel	68	
Total Zinc	84.9	51
Total Organic Carbon	24.3	
Chemical Oxygen Demand (COD)	14	44
Biochemical Oxygen Demand (BOD_5)	20.3	51
Total Nitrogen	31.3	
Ammonia	54.5	
Total Kjeldahl Nitrogen	28.8	38
Nitrate	60	44
Total Phosphorus	64	64
Orthophosphate	-50	

References

Environmental Protection Agency. 1983. **Results of the Nationwide Urban Runoff Program, Volume I - Final Report.** PB84-185552.

Maristany, A. E., Bartel R. B., and Wiley D. 1988. **Water Quality Evaluation of Lake Munson, Leon County, Florida.** Water Resources Assessment 88-3. Northwest Florida Water Management District. Tallahassee, Florida.

Biological Modeling in the Columbia Basin: An Organized Approach to Dealing with Uncertainty

Willis E. McConnaha[1]

Abstract

Development of the Columbia River basin has had a profound impact on its natural resources, particularly species of Pacific Salmon. Passage of the Northwest Power Act of 1980 put in motion an unprecedented regional effort to restore the natural resources of the basin as affected by development of the hydroelectric system. Provisions of the act are compelling an interdisciplinary approach to hydrosystem planning and operations, as well as natural resource management. Symptomatic of this has been the development and use of computer modeling to assist regional decision making. This paper will discuss biological modeling in the Columbia basin and the role of modeling in restoration of large ecosystems.

Background.

Large-scale hydroelectric development of the Columbia River drainage began with the construction of Rock Island Dam in 1933. This was quickly followed by the completion of Bonneville Dam in 1938 and Grand Coulee Dam in 1942. Since then, the river and its resources have been aggressively developed to provide hydropower, flood control, navigation, irrigation, and recreation. The river is now regulated with a series of 75 dams on the Columbia and its major tributary the Snake River.

Although this development has been economically lucrative and contributed to the safety and well being of the Pacific Northwest, it has profoundly affected the biological resources of the basin. This is particularly true for fish species such as chinook, coho, sockeye salmon and steelhead trout that utilize the river and its tributaries for spawning, juvenile rearing and as migratory corridors. The production of salmon and steelhead in the Columbia River prior to development of European civilization has been estimated at 10 to 16 million fish annually (NPPC, 1987). This compares to a present production of around 2.5 million fish.

For most of this century, the region has struggled to balance development with the biological needs of the fishery resource. Until relatively recently, this consisted of replacing naturally produced fish lost to development with large numbers of hatchery fish. While this succeeded in boosting sagging fish numbers and supporting commercial and sport fisheries, it has become apparent that hatchery

---

[1] Systems Ecologist, Northwest Power Planning Council, Portland, Oregon.

production may have diminishing benefits over time (Hilborn, 1992). Society is increasingly concerned with the intrinsic value of resources in their natural state, over and above their economic value to a fishery, as evidenced by legislation such as the Endangered Species Act. The result is that the focus of management has changed to embrace a wider set of values than just those of the harvest community. Restoration efforts involve a broad spectrum of management and interest groups including fisheries agencies, Indian tribes, the hydroelectric industry, agriculture, forestry, recreational and environmental groups.

The focus of these efforts during the past decade has been the fish and wildlife plan prepared by the Northwest Power Planning Council[2]. The Council was created by Congress as part of the Northwest Power Act of 1980. Among other things, the Act attempted to give fish and wildlife equal status to power production and other uses in the management of the Columbia basin hydroelectric system. The Council was directed to prepare a plan to "...protect, mitigate, and enhance" the fisheries resource as affected by development of the hydroelectric system. While the Council can direct funding and action by federal agencies, especially the Bonneville Power Administration, it has limited regulatory or coercive power. Instead its role has been to forge a consensus between the multitude of interests and disciplines.

One of the most significant features of the Act was the requirement that the Council take a system-wide approach. This has meant that all users of the system-- farmers, hydroelectric system operators, fishers and many others-- are recognized as part of the problem and components of the solution. The result has been significant progress in achieving a multi-disciplinary approach to ecosystem restoration.

The Council was directed to rely on the best scientific information available. It was not to await scientific certainty. Although much information exists regarding the biology of Pacific Salmon in the Columbia basin, major uncertainties exist regarding the proper mix of actions. Because most actions have expensive consequences and far-reaching ramifications, these scientific uncertainties have ignited heated regional debate and stymied progress on several fronts.

Development of a rehabilitation program is particularly daunting in light of the very weak condition of many fish stocks. Presently there is one population of sockeye salmon listed as endangered and three chinook salmon populations listed as threatened under the Endangered Species Act. Most other naturally produced populations are at greatly reduced levels (Nehlsen et al, 1991). Rebuilding such populations requires a cautious approach that enhances learning while making needed short term biological change (Volkman and McConnaha, 1992).

Application of Biological Modeling in the Columbia Basin

Computer models are important tools for meeting these challenges. They offer a way to fashion a large body of disparate information into a coherent, if uncertain, picture. They allow the many different theories and opinions to be aired and examined. Models are also fundamental to the multi-disciplinary approach necessary to rehabilitation of a heavily developed ecosystem.

---

[2] Copies of the Council's plan are available from the Council office at 851 SW Sixth Avenue, Suite 1100, Portland, OR 97204.

The use of biological models in the Columbia basin emerged initially as a tool for implementing an adaptive management approach to the Council's program. Adaptive management calls for actions to be implemented as experiments within a formal experimental design (McAllister and Peterman, 1991). It was incorporated into the program as a way to take action while resolving important scientific uncertainties (Lee, 1990). A critical feature is to form management hypotheses that can be tested. Computer models are important tools for bringing information together to make these hypotheses.

The Council initiated development of the first salmonid life cycle model in the Columbia basin as part of its adaptive management strategy in 1985. A series of workshops were held to bring together a wide variety of interests and disciplines to design what became known as the System Planning Model (Anderson, 1992). The model thus early on became a tool to bring together disparate ideas and interests.

The model was initially viewed with considerable skepticism and distrust. Models were seen as unfathomable black boxes that had little connection with reality. Familiarity and comfort with the use of models increased however after the System Planning Model was used in a region-wide process to plan recovery actions in each of the 31 subbasins in the Columbia system (Anon, 1990). The model was used to compare alternative production plans and identify limiting factors and critical uncertainties. Much was learned in this experience and the model evolved to reflect changing needs, new information, and improved computer and software technology.

Recent management needs have resulted in an increase in the use of biological models. The Endangered Species Act, for example, has spurred a region-wide debate over recovery actions. Models are being used to compare alternatives and identify critical uncertainties. They are also used to study innovative modifications to the system that have far-reaching social and economic implications. This includes, for example, drawing down the elevation of mainstem reservoirs to increase water velocities and speed the downstream migration of juvenile fish. Such actions are impossible to field test without major engineering, economic, and social costs. As part of the planning process, models permit us to examine these options and develop strategies for implementation and identify information needs.

Since the initial development of the System Planning Model, several alternative models have emerged. In general, these fall into two categories. Juvenile passage models analyze how manipulation of the hydropower system affects survival of downstream migrating juvenile salmon and steelhead (CQS, 1992, McConnaha, 1992, Weber and Petrosky, 1992). These combine biological information on fish passage with engineering and hydrologic data. The second category is life cycle models, such as the System Planning Model, that synthesize information from all aspects of the life cycle (Lee and Hyman, 1992, Schaller, et al., 1992). These models incorporate the juvenile passage models as sub-components. They are capable of expanding the analysis of downstream survival, for example, to address how actions to improve survival might aid rebuilding of weak salmon and steelhead populations over time.

While different models are being used to address management problems in the Columbia basin, it is not altogether clear that each represents a unique perspective on the questions. Each was developed to address specific needs, but all manipulate the same, limited data. The lack of understanding of the various model systems limits their usefulness to management. Differences in interpretation or legitimate scientific uncertainties lead to perceptions of "model wars" and distrust of the techniques in general. It will be the responsibility of the analysts to foster the notion that limitations in the models reflect limitations in our knowledge that can only be solved through research and adaptive implementation. These limitations cannot be solved by more complex models or sophisticated software.

Applications of biological modeling to ecosystem management

As in the Columbia basin, natural resource management in general, will increasingly become focused on the ecosystem. Natural resource management is expanding beyond its traditional role of harvest management. At the same time, conventional methods of dealing with damage to the resource, such as hatcheries, are proving inadequate over the long term. While they will continue to have a role, they will have to be reassessed in the context of the ecosystem.

Ecosystem management will rely heavily on computer models, data bases, and information systems. Management of ecosystems and the myriad of interests that use and affect their resources will require more information than traditional natural resource management. Ecosystem management focuses on smaller population segments, each requiring information on status, limiting factors, and management needs. Computer models and information systems will be needed to make this information available and synthesize it into usable form.

Models have great potential for bringing together information dealing with all aspects of the ecosystem. Scientific research typically concentrates effort and money on examination of individual components of the system. Each piece by itself does not explain the system or how they contribute to the overall goal of restoring the resource. Models allow us to assemble the various pieces and form ideas and hypotheses about the overall system.

Most importantly, models can help span the gap between science and management. The implications of management alternatives can be explored, while the available information is focused on the problem. By highlighting knowledge limitations, research can focus on critical data needs.

The role of modeling in natural resource decision making continues to evolve. While they can highlight the limitation of existing knowledge, misused, models can add an air of legitimacy to conclusions based on inadequate data or which represent opinions. Decision makers need to appreciate the value and limitations of the techniques. Analysts bear a responsibility to ensure that their work is not misused in the zeal to push political agendas. In the classical scientific method (Platt, 1964) models are simply hypotheses that are designed to be tested through research. Elimination of a model through contrary observations is as useful as its confirmation. Even in management applications, models are best viewed in this light, as hypotheses that will be modified and improved over time.

References Cited

Anderson, D.A., 1992. Documentation for the System Planning Model, version 5.1. Northwest Power Planning Council, Portland, OR.

Anon, 1991. Integrated system plan. Northwest Power Planning Council, Portland, OR, document 91-16.

CQS, Center for Quantitative Sciences, University of Washington, 1992. Columbia River salmon passage (CRiSP) model, documentation for CRiSP 1.4. University of Washington, Seattle, WA.

Hilborn, R., 1992. Hatcheries and the future of salmon in the Northwest. Fisheries, 17(1):5-8.

Lee, D.C. and J.B. Hyman, 1992. The stochastic life-cycle model (SLCM): A tool for simulating the population dynamics of anadromous salmonids. USDA Forest Service, Intermountain Research Station, Boise, ID.

Lee, K.N. 1990. Rebuilding confidence: salmon, science, and law in the Columbia Basin. Environmental Law, 3(1).

McAllister, M.K. and R.M. Peterman, 1992. Experimental design in the management of fisheries: A review. North American Journal of Fisheries Management, 12(1):1-18.

McConnaha, W.E., 1992. The Passage Analysis Model: A spreadsheet model of fish passage in the Columbia basin. Northwest Power Planning Council, Portland, OR.

Nehlsen, W, J.Williams, J.Lichatowich, 1991. Pacific salmon at the crossroads: Stocks at risk from California, Idaho, and Washington. Fisheries 16(2): 4-21.

NPPC, Northwest Power Planning Council, 1986. Numerical estimates of hydropower related losses. Northwest Power Planning Council, Portland, OR.

Platt, J.R. 1964. Strong inference. Science 146: 1879-1889.

Schaller, H., C. Petrosky, E. Weber, and T. Cooney, 1992. Snake River spring/summer chinook life-cycle simulation model for recovery and rebuilding plan evaluation. Oregon Department of Fish and Wildlife, Portland, OR.

Volkman, J. and W.E. McConnaha, 1992. A swiftly tilting basin: The Columbia River, adaptive management, and endangered species. in Watershed Resources: Balancing environmental, social, political, and economic factors in large basins. Oregon State University Press, in press.

Weber, E., C. Petrosky and H. Schaller, 1992. FLUSH documentation. Columbia River Intertribal Fish Commission, Portland, OR.

MANAGING SNAKE RIVER OPERATIONS FOR JUVENILE SALMON MIGRATION

Charles M. Brendecke[1], M. ASCE

ABSTRACT

A simulation model of Snake River hydrology and water management institutions has been developed. The model was used to investigate management changes to enhance flow conditions for migrating juvenile salmonids.

INTRODUCTION

This paper reports on a study commissioned by the National Marine Fisheries Service (NMFS). The principal objectives of the study were to develop a computer model of Snake River Basin hydrology and water management institutions and to use that model to evaluate quantitatively the effectiveness of hypothetical changes in water management practices in augmenting flow conditions at Lower Granite Dam for spring juvenile salmon migration. The model evaluations represent initial efforts to quantify the effects and effectiveness of selected augmentation strategies.

MODEL DEVELOPMENT

The Snake River Operation Model (SROM) represents the various river reaches, reservoirs, diversions, and water management institutions of the basin as a network of 421 arcs and 167 nodes. The allocation of water among arcs of the network is accomplished using network flow programming methods. The model specifically represents the water supply, flood control, and hydropower operations of fifteen major reservoirs in the basin and runs on a monthly timestep over a 1928-1989 hydrologic study period.

The SROM uses inflow and water demand data sets obtained from the Idaho Department of Water Resources and used in the IDWR models of the basin. Additional model data and parameter values were obtained from the IDWR model input data sets, from direct contacts with Idaho Power Company, the Corps of Engineers, and the Bureau of Reclamation, and from secondary sources. The model was calibrated against the results of the IDWR models over the entire study period and against historical reservoir contents and streamflows over the 1980-1989 period.

---

[1]President, Hydrosphere Resource Consultants, 1002 Walnut, Suite 200, Boulder, CO 80302

MODELING APPROACH

The SROM was used in a sensitivity analysis mode to compare several scenarios depicting hypothetical changes in various water management practices in the basin. The first model scenario run was a Baseline Scenario meant to reflect current operating procedures and against which the effects and effectiveness of later scenarios could be compared.

The scenarios were defined to permit investigation of the potential augmentation strategies outlined in the predecessor study. These augmentation strategies fell into three broad categories: 1) modification of flood control and power generation operations at Brownlee and Dworshak reservoirs, 2) use of existing storage supplies to insure refill of Brownlee in return for greater Water Budget releases from Brownlee, and 3) use of water conservation measures to increase the availability of stored water above Brownlee.

The model scenarios were structured to respect the existing physical constraints in the system such as outlet capacities, minimum stream flow targets, and channel capacities. Legal and institutional constraints were relaxed in many scenarios to effect the management changes to be studied. Another important objective in defining the scenarios was to insure that the model had sufficient flexibility to investigate the range of water management measures that would be of interest to the NMFS.

Model results were evaluated in terms of the resulting flow regimes at Lower Granite Dam and the effects on water deliveries, reservoir storage, and instream flows elsewhere in the basin. Lower Granite inflow was characterized both in terms of the monthly flow rates entering the reservoir and in terms of flow velocity through the reservoir at four assumed water surface elevations; these assumed elevations range from normal full pool (738 feet MSL) down to spillway crest elevation (681 feet MSL). This permitted comparison of the relative effectiveness of drawdown versus augmentation from upstream sources.

BASELINE SCENARIO

Scenario 1 represented a "no action" scenario; it assumed current (1989) demand levels in the basin and the continuation of current reservoir operating policies, including the provisional flood control rules for Brownlee and Dworshak reservoirs that have been used for the last few years. No special operations were made for flow augmentation for migrating salmon.

Under these assumptions, modeled inflows to Lower Granite Dam averaged 110,000 cfs (110 kcfs) in May and 107 kcfs in June. There were a few shortages to diversion demands evident in the dry years of 1934 and 1977; these shortages were concentrated on the Henry's Fork and upper mainstem. Flows at Weiser frequently exceeded 25 kcfs and Brownlee Reservoir was frequently required to bypass flows in excess of Hells Canyon powerplant hydraulic capacities during the runoff season.

OPERATIONAL CHANGES AT BROWNLEE AND DWORSHAK

In model Scenarios 2 through 4 the SROM was used to evaluate hypothetical changes in flood control and power generation operations at Brownlee and Dworshak reservoirs. In the case of flood control operations, the hypothesis was that by reducing flood control evacuation from those reservoirs in average-to-dry years, the reservoirs would enter the Water Budget period with more water in storage and would be able to increase their Water Budget releases accordingly. In the case of power generation operations the hypothesis was that reduced winter power generation would similarly lead to greater reservoir storage and Water Budget releases.

The SROM was used to evaluate scenarios depicting changes only in flood control (Scenario 2), changes only in winter power generation (Scenario 3), and changes in both flood control and winter power generation (Scenario 4). In all three scenarios Water Budget period releases were increased to powerplant hydraulic capacity to evacuate the increased storage made available on May 1. These scenarios showed that changes in flood control were the most effective of the two water management modifications; without concurrent changes in flood control operations, most water saved by foregoing winter power generation was subsequently evacuated for flood control purposes.

The combination of both flood control and power generation modifications lead to substantial increases in modeled Lower Granite inflows over the Baseline Scenario. With these assumptions, the average May inflow to Lower Granite increased from 110 kcfs to 135 kcfs and average June inflow increased to 110 kcfs. This translates to an increase in May-June Lower Granite inflow volume of more than 1.6 million acre-feet (MAF). There were only minor increases in modeled spills at the two reservoirs.

UTILIZATION OF UPSTREAM STORAGE SUPPLIES

In Scenarios 5 through 7 the SROM was used to evaluate hypothetical changes in flood control operations and releases of stored water from reservoirs above Hells Canyon. The over-arching hypothesis of these scenarios was that Brownlee releases could be increased beyond those of Scenario 4 if upstream supplies could be used to help assure Brownlee refill after the Water Budget period.

The three scenarios investigated water leasing from the District 1 Water Bank (Scenario 5), water leasing enhanced by water conservation measures (Scenario 6), and changes in flood control and water leasing from the Boise and Payette reservoir systems (Scenario 7). In Scenario 5, leases from the District 1 Water Bank were assumed to utilize only the historically unused consignments to the Water Bank; the amounts leased ranged from 0.5% to 14% of total District 1 storage, with greater amounts leased in wet years. In Scenario 6, District 1 storage and Water Bank consignments were increased by assuming the adoption of water conservation measures at selected diversions; the assumed diversion reductions ranged from 10% to 20%, with greater reductions being simulated for demands which have historically shown lower overall water use efficiencies. In Scenario 7, minor reductions in Payette system flood space requirements were assumed and leasing of up to 100 kaf was assumed from both the Payette and Boise systems.

The scenarios showed that none of these measures by themselves were sufficient to assure Brownlee refill by the end of July. However, Brownlee did refill under most scenarios by September. The increased Brownlee releases lead to average inflows to Lower Granite of about 145 kcfs in May and 110 kcfs in June in all three scenarios The increased draft on Brownlee to achieve these flows resulted in substantial drawdowns in the reservoir during May and June. The increased utilization of upstream storage shifted delivery shortages somewhat from those of the Baseline Scenario but did not increase the overall magnitude of shortages nor did it affect the ability of reservoirs other than Brownlee to maintain minimum storage levels or release targets.

INTEGRATED AUGMENTATION STRATEGY

The final scenario (Scenario 8) evaluated with the SROM assumed the implementation of a combination of all the hypothetical measures examined in the individual model scenarios. This integrated scenario assumed the hypothetical changes in flood control and power operations at Brownlee and Dworshak of Scenario 4, the increased releases and bypasses from Brownlee of Scenario 5, and the releases from upstream storage to assist in Brownlee refill of Scenarios 6 and 7. As a result of these combined measures, the average May inflow to Lower Granite was 146 kcfs, an increase of 35 kcfs from the Baseline case and representing an increase in May inflow volume of over 2.0 MAF. Lesser increases were accomplished in June; average inflow was 110 kcfs representing a volume about 0.2 MAF greater than the Baseline Scenario. Brownlee was refilled by the end of August in most years, and the average end-of-July storage deficit in Brownlee was reduced to 34 kaf (about 3% of active capacity).

CONCLUSIONS

The analyses obtained strongly suggest that the Water Budget goal of 1.19 MAF at Lower Granite Dam is within reach using non-structural approaches to water management. With a broad cooperative effort between Idaho water users, the Bureau, the Corps, and Idaho Power, the available Water Budget volume might reach an average of 2.0 MAF. Furthermore, these inflows appear to be achievable without adversely impacting the water supply functions of upstream storage facilities, although early summer recreation potential at Brownlee would be substantially reduced.

It should be noted, however, that even with this broad cooperation the flow velocities obtained through Lower Granite Reservoir in the integrated operations scenario were still less than those achieved in the Baseline Scenario by assuming only the drawdown of Lower Granite Reservoir to an elevation of 710 feet MSL during the spring migration period.

Columbia River System Operations - Water Quality Assessment

R.G. Willey[1], Bolyvong S. Tanovan[2] and Donald J. Smith[3], Member ASCE

Introduction

In mid-1990, the U.S. Army Corps of Engineers, U.S. Bureau of Reclamation, and Bonneville Power Administration embarked on a Columbia River system operation review (SOR). The goal of the SOR is to establish an updated operation strategy which best recognizes the various river uses as identified through community input. Ninety alternative operations of the Columbia and Snake River systems (see Figure 1) were proposed by various users. These users included the general public, irrigation and utility districts, as well as local, state and various Federal government agencies involved with specific water resource interests in the Columbia River basin.

Ten technical work groups were formed to cover the spectrum of interest and to evaluate the alternative operations. Using simplified tools and risk-based analysis, each work group analyzed and then ranked the alternatives according to the effect on the work group's specific interest. The focus of the water quality technical work group is the impact assessment, on water quality and dissolved gas saturation, of the various operations proposed by special interests (i.e., hydropower, navigation, flood control, irrigation, recreation, cultural resources, wildlife, and anadromous and resident fisheries.

---

[1]Corps of Engineers Hydrologic Engineering Center, Davis, CA
[2]Corps of Engineers North Pacific Division, Portland, OR
[3]Consultant, Resource Management Associates, Lafayette, CA

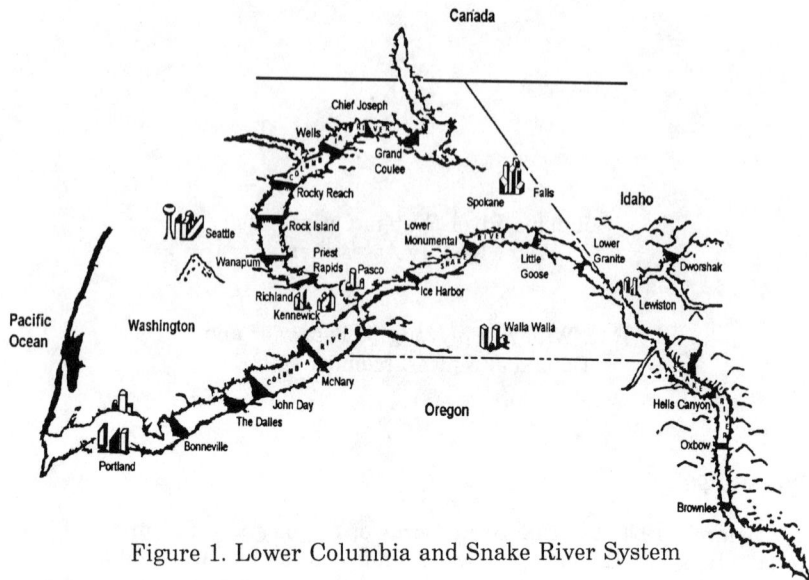

Figure 1. Lower Columbia and Snake River System

System Operation Review

A major finding of the SOR screening of the 90 alternatives was that there is no longer enough water, even in a wet year, to meet all the needs of the Columbia River Basin users. Many of the alternatives heavily favor a single use at the expense of the other uses. The screening results from all the technical work groups provided ten blended strategies for more detailed study in the Environmental Impact Statement (EIS).

Model Calibration and Verification. The water temperature impacts were studied using the HEC-5Q computer model available from the U.S. Army Corps of Engineers Hydrologic Engineering Center (HEC). The calibration for the system shown in Figure 1 was completed using weather data that approximately reflect below average, average and above average weather conditions (i.e., 1985, 1984 and 1990) in the lower Snake River basin. Verification of the model was done with 1991 data, collected in more detail than the other years.

Data availability is a major constraint to analysis and requires estimates of many missing data. All tributary water temperatures were estimated for the calibration years based on either the measured 1991 data for three major tributaries to the Snake River or calculated from a regression equation derived from 1991 Snake River data versus equilibrium temperature.

Other methods first tried for computing these estimates related the inflow temperature to departure from equilibrium temperature. The departure method required fairly extreme inflow water temperature at many locations in order to reproduce the measured scroll case water temperature discharged from the projects. It could also lead to erroneous results when applied to spring runoff that resulted from snowmelt. This method was not considered further for this study.

In general, the simulation results for the four historical years studied provide a water temperature accuracy of about one degree Celsius but in a few cases (in time and space) maximum errors of up to three degrees Celsius exist. An example of typical results of reproduction are shown in Figure 2 for 1985, 1984 and 1990 scroll case discharge water temperatures at Lower Granite Dam on the Snake River.

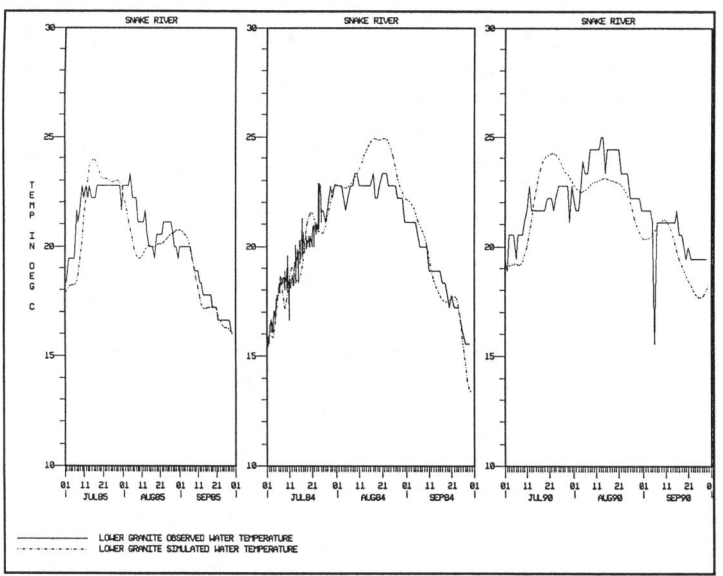

Figure 2. Example of Water Temperature Reproduction Results

<u>Operation Alternatives Analyzed.</u> The calibrated HEC-5Q model was then applied to the 90 operation alternatives. The alternatives were evaluated using monthly required discharges and pool elevations from the projects as determined by a flow accounting model that simulates specific features of the individual alternatives. A hydrologic balance was next achieved by calculating local tributary flows from downstream at Bonneville Dam to upstream for five independent years covering a complete range of conditions from below average to above average flow.

The local tributary water temperatures were estimated as described for the model calibration. The meteorological data for three separate independent years (i.e., those previously used: 1985, 1984 and 1990) were superimposed on the five hydrologic periods to provide 15 scenarios, from wet to dry conditions, for each alternative evaluated.

Each alternative was ranked based on water temperature extremes resulting from the 15 scenarios. The results of the HEC-5Q executions were tabulated and used in a spreadsheet that can be accessed by risk- and-uncertainty software. The software used in this instance was SUPERTREE[4]. For each alternative, expected values of mean and maximum water temperature at selected locations were provided. The probability distribution function of the expected values were supplemented with values at 10 and 90 percent exceedance probability levels.

Miscellaneous Considerations

Other Parameters. A separate empirical spreadsheet model was used to determine dissolved gas concentrations at the same locations where water temperatures were provided. Concentrations were calculated based on regression that takes into account saturation of the inflow, spill amount, air entrainment characteristics of the dams, and gas dissipation within specific reservoirs and river reaches. These results also included the expected mean, and the 10 and 90 percent probability values. They were incorporated with the water temperature results mentioned above to determine the rankings of the alternatives.

Multi-dimensional Concerns. Because HEC-5Q is a one-dimensional model, a two-dimensional model was also applied to the lower Snake River's four reservoir projects shown on Figure 1. This was prompted by the need to simulate the movement of the cool water release from Dworshak Reservoir through the various lower Snake River reservoirs where the water is usually much warmer. Mixing, both laterally and longitudinally, was an issue for consideration. The results of both levels of modeling will be compared and conclusions reached regarding the adequacy of one-dimensional modeling for long and relatively shallow reservoir projects, some of which have two major inflows of sufficiently different density to raise questions regarding mixing of inflows.

Technology Transfer. Training on the HEC-5Q model was provided to local Corps of Engineers offices (i.e., North Pacific Division; Seattle, Walla Walla and Portland Districts), Bonneville Power Administration, Idaho Power Company, National Weather Service, National Marine Fisheries

---

[4]Available from The Scientific Press, So. San Francisco, CA

COLUMBIA RIVER SYSTEM 155

Service and the Columbia River Inter-Tribal Fish Commission. The Inter-Tribal Fish Commission and the Corps of Engineers North Pacific Division are currently applying the model for real-time operation alternatives of the lower Snake River system where thermal blocks are known to have impacted the upstream migration of adult salmon.

Model Expansion. Although water temperature and dissolved gas are the only two water quality parameters analyzed in the Columbia River System Operation Review, the HEC-5Q is being expanded to include pH, suspended solids, dioxins, and heavy metals. These enhancements will provide a more comprehensive water quality assessment of the alternative operations in the future. Auxiliary tools have also been developed to automate flow balancing and for making a series of water temperature model executions with a minimum of user intervention. The enhanced model is expected to have wide application potential for modeling water quality conditions in other regions of the country.

Summary

The modeling effort for the Columbia River SOR of the 90 operation alternatives provided by public input has been tremendously rewarding to understanding of the intricacies of the reservoir control capabilities. The 90 cases were screened for ranking purposes and the ten blended strategies are being analyzed in more detail for input to the EIS.

The expanded HEC-5Q model will no doubt be a significant tool for future analysis as more data is collected to continue this operation review during the next decade. This study will hopefully give other study managers ideas for water quality analysis of their systems of reservoirs to provide for more user benefits from existing projects. As populations increase, water control managers must be prepared to address requests from potential water users for a wide variety of water uses never anticipated when the water projects were authorized for construction.

Hydro Models and Salmon Recovery in the Northwest
Ken Dragoon[1]

Abstract

Hydro regulation models provide extensive support for analyzing the efficacy of salmon recovery plans in the Northwest. Power planners developed these computer programs to help plan and efficiently operate a large multiple use river system. The models represent physical relationships and operational requirements on the system. They also simulate coordinated system operations for efficient power generation.

These models are being pressed into service to provide data for fish recovery plans. They provide important information about hydro system capabilities and responses to recovery programs. However, the models cannot meet all of the analytical needs of fish biologists working toward salmon recovery.

Background

Efforts to analyze coordinated hydro system operations were originally motivated by a number of power needs. Power planners needed to identify the most effective of the new dams being proposed, assess the most beneficial time to install new generators. They needed to know how much power would be generated by new projects, especially in low water conditions. Planners needed the ability to assess how river operating requirements would affect power generation.

---

[1] Chief, Hydro Resource Analysis Section, Bonneville Power Administration, PO Box 3621, Portland, OR 97208.

Engineers began devising computer programs to model the Northwest hydro power system as far back as 1958[2]. The programs were adapted from desk calculator and paper spreadsheet methodologies previously used by hydro planners. In the late 1950s, the coordinated hydro power system included six major reservoirs with 14 billion cubic meters of storage capability and 11 run-of-river projects (dams with relatively small reservoirs). Today's models simulate the operation of about 80 projects with 50 billion cubic meters of storage.

Model Characteristics

The original purpose of the computer models was primarily to assess power generated from the hydro projects. Generally, the models work by starting with naturally occurring inflow (adjusted for irrigation) to the most upstream project. The water flowing out of the most upstream reservoir is the inflow from upstream adjusted by the amount of water stored in, or drafted from, the reservoir. The programs compute how much electricity is generated from the resulting outflow and the average head over a specified time period.

This process is repeated at each next downstream project, using the upstream project's outflow plus any side flows between the projects. The programs continue through the last project before the ocean. A set of rules, or "rule curves", is used to apportion reservoir drafts or storage among the major storage reservoirs. The rule curves are designed to efficiently meet indigenous loads, refill reservoirs, and protect against flooding.

Additional operating requirements can be placed on reservoir operations by limiting minimum and maximum outflows or reservoir elevations. Water can also be diverted from the turbines over the spillway. All of these requirements are represented by hydro models.

---

[2]See "Digital Computer Aids in Power Pool Operation Studies", McIntyre, Blake and Clubb, AIEE Conference paper, May, 1958.

Power planners have been churning out studies for more than three decades with these programs. Improvements in computer hardware over the years prompted improvements in software design. Modern programs feature interactive sessions, and are much easier to use than their counterparts from a generation ago.

Hydro programs consume and produce vast amounts of data. A variety of rule curves for each major project must be input and up to date. Reservoir operating requirements for fish passage, recreation, flood control, and other purposes are continually changing and need to be updated in the model data base. Tremendous volumes of data define physical characteristics of hydro projects. Load and non-hydro resource data are also represented. Modern data management methods and expert system technology are just beginning to be incorporated in our latest generation hydro models.

Enter Fish

Under the *Pacific Northwest Electric Power Planning and Conservation Act* [3] of 1980, the Bonneville Power Administration (BPA) is authorized to fund programs to "protect, mitigate, and enhance fish and wildlife to the extent affected by the development and operation of any hydroelectric project of the Columbia River and its tributaries..." BPA is also motivated by a desire to re-negotiate the Pacific Northwest Coordination Agreement (PNCA). The PNCA expires in the year 2003. The National Environmental Policy Act (NEPA) requires an environmental impact statement (EIS) on any major Federal action with significant environmental impacts. Signing a new PNCA is thought to be such an action.

Other signatories to the PNCA include the US Army Corps of Engineers and the US Bureau of Reclamation. These Federal agencies must also submit an EIS on a new PNCA and have

---

[3] Public Law 96-501, 94 STAT. 2697, section 4(h)(10)(A)

joined with BPA in a joint EIS. This effort is known as the System Operation Review (SOR)[4].

Environmental groups are successfully petitioning the National Marine Fisheries Service (NMFS) to place an ever-growing number of salmon species on the endangered species list. NMFS must propose mitigation actions for the endangered stocks. Interim proposals have already been presented and adopted.

Focus on Hydro Models

The power industry needs to understand how these proposals will affect their ability to meet regional loads. Hydro models, originally designed for just that sort of thing were pressed into service. Some of the proposals stretch the ability of the models to represent the changes.

However, the needs of power planners paled in comparison to the job of assessing environmental impacts. Wildlife interests want to know reservoir elevations to an accuracy of inches. Conceivable modeling techniques fall well short of that level of precision. Recreation interests' needs for daily and hourly information face a similar fate. Most hydro programs are based on monthly or half-monthly periods.

Conclusion

Over time, people have come to assume that analysis of hydro operations can only be done using these complicated computer programs. Relying on our own imaginations to predict what would happen to the system in a world very different from today's has been disconcerting to many. It has been just as troubling to those of us who have relied on these programs for so many years, as it has been to environmentalists, eager to use high technology analysis to evaluate their proposals.

---

[4]The PNCA is one of several motivations for the SOR and its EIS. The SOR also seeks to provide EIS coverage for actions taken pursuant to the Columbia River Treaty between the US and Canada. Under the treaty, the US is obligated to deliver energy to Canada beginning in 1998. The energy will come from BPA, and three public utility districts (Chelan, Douglas, and Grant). An agreement regarding how the energy payments will be allocated to these entities needs to be covered by an EIS.

SNAKE RIVER SALMON RECOVERY PLAN DEVELOPMENT

Peter C. Klingeman[1] and James Litchfield[2] [3]
Members, ASCE

Abstract

The process used by the Snake River Salmon Recovery Team to develop its recovery plan is described. Major conditions affecting fish survival are identified. The Team's approach to evaluate these conditions and prepare recommendations for recovery of fish stocks is summarized. Views expressed are those of the authors about work in process.

Background -- ESA Listing and Recovery Team Formation

Snake River Sockeye Salmon were declared endangered in November 1991 and Snake River Spring-Summer and Fall Chinook were declared threatened in April 1992 under the Endangered Species Act (U.S. Congress, 1988). Species may be listed under the Act (ESA) if any of five factors are found to have affected their decline: 1) habitat or range limitations, 2) overuse, including harvest, 3) disease or predation, 4) inadequacy of existing regulatory mechanisms, and 5) other natural or manmade factors affecting continued existence. All five of these factors affect the Snake River salmonids.

The National Marine Fisheries Service (NMFS) of the U.S. Department of Commerce, National Oceanic and Atmospheric Administration, established a Snake River Salmon Recovery Team in December 1991 to advise and assist NMFS in developing and implementing a recovery plan. The Team is an independent entity with seven members: three fish biologists, one ecologist, one resource economist and two engineers. All have extensive experience with fish problems and expertise in other areas that are complementary to recovery planning. The Team began in January 1992 to develop a recovery plan for reversing the decline of listed species and rebuilding the stocks to numbers that would allow their delisting.

---

[1] Professor, Civil Engineering Department, Oregon State University, Corvallis OR 97331
[2] President, Litchfield Consulting Group, 101 SW Main St., Suite 900, Portland OR 97204
[3] Members, National Marine Fisheries Service Snake River Salmon Recovery Team

SALMON RECOVERY PLAN 161

Process for Recovery Plan Development

The process being followed to develop the recovery plan is one established by NMFS in response to ESA requirements (NMFS, 1990). The initial step is for the Team to submit a draft recovery plan to NMFS. NMFS will review the Team's plan and may either accept it as presented or modify it. A notice of availability of the NMFS draft plan will then appear in the Federal Register, initiating a period for public comments. Comments received will be considered as NMFS prepares the final plan. Once approved, the recovery plan will be implemented as the official position of the agency. Implementation will place a variety of requirements and constraints on activities of many agencies and other groups in the region, until recovery of the listed species either is achieved or is deemed to be hopeless and is abandoned. Many actions toward recovery or to avoid jeopardy have already started through the consultation and conference processes of ESA and through agency obligations under ESA.

General Recovery Plan Requirements

Recovery planning requires problem identification and resolution. The recovery plan must identify, describe and prioritize all tasks and actions needed for recovery. The plan must provide estimates of time and cost to carry out the recommended recovery measures. An implementation schedule must be provided. Objective, measurable criteria must be given that, when met, would indicate recovery of the listed species. The plan may be amended or revised as additional data are developed.

Scope of Recovery Plan Being Developed

The Team formed an organizational structure to address the survival-and-recovery issues and to facilitate recovery planning and plan implementation. The categories used in this organizational structure were: a) Stanley Basin initial measures, b) Sockeye captive broodstock, c) fish habitat, d) fish hatcheries, e) main-stem passage (including relevant basin hydrology, flows and river hydraulics), f) predation, and g) ocean-river harvest.

The recovery plan that is being developed addresses all aspects of the life cycle of Snake River salmonids. The general stages of this cycle are summarized in Table 1. Each stage involves a reduction in the number of surviving fish due to mortalities from a wide variety of causes. This can be described by a survival ratio for the stage. Survival over the full life cycle is the product of all survival ratios for all stages times the initial number of fish available as fertilized eggs. Thus, recovery can be viewed as an effort to increase survival ratios in the life cycle.

Table 1. Salmonid Life-Stage Approach to Recovery Planning

Life Stage of Fish	Nature of Problems and Potential Measures
0. Fertilized Eggs in Gravel	Habitat, Streamflow, Predation
1. Incubation & Emergence	Habitat, Streamflow, Predation
2. Freshwater Rearing	Habitat, Streamflow, Predation
3. Migration to Lewiston	Habitat, Streamflow, Predation, Hatchery Smolts
4. Migration past 8 dams	Main-Stem Passage, Streamflow, Predation, Hatchery Smolts, Transportation System
5. Lower River Migration	Predation, Hatchery Smolts, Water Quality, Transportation System
6. Ocean	Predation, Harvest
7. Upriver Migration	Predation, Harvest, Main-stem Passage
8. Spawning	Habitat

Life-cycle models are available to assess points where recovery measures can be most effective in improving survival ratios. However, survival ratios used in these models are based on estimates. The estimates are highly controversial due to differing interpretations of a limited and poorly documented data base. The numerous factors involved are sensitive to climate variations, streamflows, or human activities and hence vary over time and with location.

Problems and risks occur at each stage of the salmonid life cycle. Table 1 lists these in general terms. Habitat problems may involve degraded or insufficient habitat, over-wintering losses, insufficient streamflow, and/or water quality problems. Hatchery fish may cause crowding and food competition, predation, and/or the spread of disease. Main-stem passage problems involve the eight reservoirs and dams between Lewiston to Portland. Problems occur due to losses in reservoirs, turbine passage, bypassing of turbines into collection facilities, insufficient flows, elevated water temperatures, and/or nitrogen supersaturation. Stresses caused during main-stem passage may amplify problems such as disease or may cause disorientation that make fish more susceptible to predators. Effects may be delayed, such that mortalities go undetected at the place of occurrence but are measured as fewer returning adults.

The recovery plan addresses actions to improve survival for each life-cycle stage. For some stages the opportunities are very limited, but for other stages major changes may be feasible. Table 1 suggests categories for recovery actions

needed to address problems. These are now being developed
and are given in the oral presentation based on this paper.

The Data-Base Obstacle to Plan Development

The limited biological data base available is one of the
biggest problems in analyzing the complex issues involved in
salmon recovery. Most of the data were developed for other
purposes -- not species recovery. Recovery planning raises
questions about species, ecosystems, constructed facilities,
and management/operation activities that may not have been
evaluated previously. This requires very careful scrutiny of
existing information to determine its relevance and the
limits of its applicability. This also leads to delays as
potentially relevant information is sought or new information
is developed to shed light on the questions being analyzed.

The Team has made requests and consulted with NMFS staff
and with many other qualified individuals, organizations, and
agencies to obtain needed information. Examples include data
provision, supplemental data analyses, and new ideas on such
topics as how to better evaluate the survival of transported
fish or conduct a biological test of reservoir drawdown.

The Need for a Flexible Recovery Plan

The recovery plan must set the course toward recovery.
But the proposed plan is not likely to be the ultimate answer
on how recovery may be achieved. The plan must provide for
flexible management because of many uncertainties about fish
behavior and about conditions that can severely impact fish.
Not all outcomes of potential recovery plan measures can be
fully known in advance. Some initial measures may need later
modifications to improve effectiveness. Hence, the recovery
measures may need flexibility to meet varying conditions, to
incorporate new information, and to allow for uncertainties.

Closing Comments

The ESA process is providing an ecological evaluation of what
must be done to keep Snake River salmon from perishing. The
recovery plan will establish the needed actions. Costs and
needed implementation times for potential measures are being
considered in evaluating alternative actions. Once the
recovery plan is adopted by NMFS, its achievement will depend
in large part on use of the sophisticated regional system for
water and fish management and on regional cooperation.

References

U.S. Congress, Endangered Species Act of 1973, Public Law 93-205, as amended through 1988.
National Marine Fisheries Service, Proposed Recovery Planning Guidelines, U.S. Dept. of Commerce,
 NOAA, NMFS, Office of Protected Resources, February 1990.

Columbia River Basin Fish and Wildlife Program Strategy for Salmon

James Ruff and John Fazio[1]

Abstract

Three species of Snake River salmon have been listed as threatened or endangered under the federal Endangered Species Act. In response, the Northwest Power Planning Council worked with the states of Idaho, Montana, Oregon and Washington, Indian tribes, federal agencies and interest groups to address the status of Snake River salmon runs in a forum known as the Salmon Summit. The Summit met in 1990 and 1991 and reached agreement on specific, short-term actions. When the Summit disbanded in April 1991, responsibility for developing a regional recovery plan for salmon shifted to the Council. The Council responded with a four-phased process of amending its Columbia River Basin Fish and Wildlife Program. The first three phases, completed in September 1992, pertain to salmon and steelhead. Phase four, scheduled for completion in October 1993, will take up issues of resident fish and wildlife. This paper deals with the first three phases, collectively known as our Strategy for Salmon.

Introduction

The Strategy for Salmon constitutes a comprehensive recovery program for salmon -- all salmon in the Columbia River Basin, not only those protected under the Endangered Species Act. The Strategy addresses all stages of the salmon life cycle and can be used by the National Marine Fisheries Service as the foundation of its recovery plan

---

[1] Mr. Ruff is a hydrologist and Mr. Fazio is a power system analyst at the Northwest Power Planning Council, 851 S.W. Sixth Avenue, Suite 1100, Portland, Oregon, 97204.

COLUMBIA RIVER SALMON PROGRAM

for listed Snake River salmon runs. The Council worked closely with the Service in preparing the Strategy.

Program goal

The goal of the fish and wildlife program is to double salmon production in the Columbia River Basin from approximately 2.5 million fish returning to the mouth of the Columbia River each year to 5 million fish. We hope to accomplish the doubling goal with no appreciable risk to the biological diversity of fish populations.

Framework

The strategy establishes interim rebuilding targets for naturally spawning Snake River salmon. These numbers are: 1) 50,000 spring chinook; 2) 20,000 summer chinook; 3) 1,000 fall chinook. The Council agreed to review the rebuilding targets in 1993.

To help focus efforts toward the rebuilding goal, the salmon strategy establishes six principles to use in evaluating actions: 1) give priority to weak, upriver runs; 2) cause no appreciable risk to biological diversity among or within fish populations, including resident fish; 3) take a basinwide approach to habitat and production improvements; 4) respect obligations to Indian tribes and other harvesters; 5) focus research on key uncertainties; and 6) use existing hatcheries unless the need for fish cannot be met with existing facilities.

Enhance salmon survival in the rivers

- Increase water velocity in the Snake and Columbia rivers during the spring migration. This is accomplished, in part, by lowering reservoirs behind John Day Dam and the four dams on the lower Snake River to the lowest level at which navigation locks and irrigation pumps can still operate. In addition to this, for the Snake River, release an additional 1.2 million acre-feet of water from the Dworshak, Brownlee and upper Snake reservoirs to speed up the river. This amount of water release is nearly triple the amount called for under the existing water budget operation. On the Columbia River, release up to 3 million acre-feet of water in addition to the 3.45 million acre-feet called for under the existing water budget operation, thus nearly doubling the amount.
During late summer and fall, the Strategy also calls for additional releases to lower water temperatures. Because the immediate measures in this strategy do not appear to be enough, in themselves, to rebuild salmon runs, we call for deeper Snake River drawdowns to begin in April 1995, unless drawdowns are shown to be economically or structurally infeasible, biologically imprudent or inconsistent with the Northwest Power Act.

- Improve and/or install screens to divert juvenile salmon away from turbines at all federal and Columbia and Snake river non-federal dams. Screening should be completed by March 1998.

- Reduce predation of juvenile salmon, including a Squawfish Management Program. The goal is a 20-percent reduction of squawfish populations in the Columbia and Snake rivers within five years. Experts believe this will lead to a 25-percent reduction in predation.

- Accelerate improvements in downstream barge transportation of juvenile salmon past Snake and Columbia dams. We call on the Corps of Engineers, which collects and transports the fish, to evaluate techniques to improve transportation, such as the use of cooler water in the barges, reduced densities of fish in the barges and broader dispersion of the fish when they are released below Bonneville Dam.

- Work with states to conduct water availability studies, establish minimum instream flow levels, deny new water appropriations that would harm anadromous fish and acquire existing water rights on a voluntary basis to improve fish flows.

- Protect endangered Snake River sockeye by allowing no commercial harvest of sockeye below the confluence of the Snake and Columbia rivers. Overall harvest rates on Snake River fall chinook should be reduced to 55 percent of the run from levels of up to 77 percent in recent years. Our strategy recommends substantial reductions in Canadian harvest of U.S. salmon. The strategy also calls for voluntary lease-back and buy-back programs for commercial fishing licenses and development of a compensation plan for fishers. Harvest alternatives, such as live-catch, known-stock and terminal harvest fisheries, should be demonstrated and evaluated. We also call for a review of sport fishing regulations and adoption of catch-and-release rules where appropriate.

Improve hatcheries and production practices

The Strategy encourages improved and consistent basinwide hatchery practices and better coordinated management throughout the Columbia Basin so hatchery fish are better able to survive in the natural environment and do not harm wild fish. Our strategy supports the continued involvement of genetics experts in discussions of how to sustain the diversity of salmon runs and to learn more about naturally spawning fish.

Protect and restore habitat

Our Strategy gives highest priority to habitat protection and improvement in areas of the Columbia Basin where there is low productivity or low survival of adult fish. Priority goes to actions that yield the greatest value for a reasonable cost, and the focus should be on approaches that involve local landowners and governments. Our strategy says that permanent riparian management areas should be identified and protected and that habitat performance standards should be developed.

Costs and economic mitigation

Program expenses are calculated to cost the Bonneville Power Administration about $30 million in 1992 and approximately $36 million in 1993. This is in addition to an estimated $50 million that Bonneville spends to implement previous measures in the Council's fish and wildlife program. We estimate Bonneville's cost to be an average of $40 million to $70 million a year because of the increased river flows. The cost could be as low as $10 million in a very wet year or over $100 million in a dry year. The Strategy will result in about a 4-percent increase in Bonneville's wholesale rates. Retail rate increases will be less. Thanks to support of the Northwest Congressional delegation, U.S. taxpayers will contribute about $100 million in 1993 for salmon recovery work in the Pacific Northwest through the budgets of federal agencies that manage and maintain the Columbia River power system, federal fish hatcheries and federal lands.

Closing comments

The Strategy for Salmon is not a panacea, but a valuable foundation for the salmon rebuilding effort that is yet to be completed. The Strategy calls for aggressive monitoring and evaluation of results. Where changes are indicated, they will be made.

A regionwide cooperative effort is clearly preferable to federal or legal intervention that could lead to extensive and expensive conflict, litigation and economic disruption. Balance is a key word. The Council's overall intent is to have balance so that all uses of the river remain viable.

References

U.S. Congress, Endangered Species Act of 1973, Public Law 93-205, as amended through 1988.

Northwest Power Planning Council, Columbia River Basin Fish and Wildlife Program, Strategy for Salmon, Volumes I and II, September 1992.

BPA Efforts to Protect Salmon and Supply Energy

Jim Geiselman[1]
and
Roger Schiewe[2]

Abstract

The Bonneville Power Administration has sought to increase numbers of anadromous fish in the Columbia River for many years. In spite of these efforts, numbers of some species have continued to decline while others increased. As a result, several species of Salmon from the Snake River portion of the Columbia River basin have been listed as threatened or endangered under the Endangered Species Act. This presentation will identify analytical tools used to assess fish mitigation measures and the changes in power production and marketing expected from implementation of the National Marine Fisheries Service Recovery Plan and the Northwest Power Planning Council's Fish and Wildlife Program.

---

[1] Resource Modeling Coordinator, Division of Fish and Wildlife (PJ), Bonneville Power Administration, PO Box 3621, Portland, Oregon 97208

[2] Hydraulic Engineer, Division of Power Supply (PS), Bonneville Power Administration, PO Box 3621, Portland, Oregon 97208

SNAKE RESERVOIR DRAWDOWN
A BRIEF PROGRESS REPORT

John J. Pizzimenti[1], Kevin Malone[1], Paul Tappel[2] and Brian Sadden[2]

Abstract

The Northwest Power Planning Council has directed the region to employ the use of "Drawdown" as tool to help threatened and endangered salmon stocks in Snake and Columbia rivers unless it is economically unfeasible or biologically imprudent. Harza is providing third-party review of the feasibility studies and biological evidence on Drawdown. It is hypothesized that Drawdown will expedite smolt migration and survival to the estuary. Various options are being studied, from a 33-foot Drawdown of four Snake River reservoirs to complete (100-foot) Drawdown of four Snake River reservoirs plus partial Drawdown of John Day. Cost estimates range from two to five billion dollars, and schedules span up to seventeen years to complete the project.

Harza offers specific suggestions to reduce both costs and schedule of the designs. We also have reviewed the biological literature which indicates that although Drawdown could help salmon, there are numerous causes of mortality that must also be addressed in all stages of the life cycle if we expect to increase the number of adult fish returning.

Introduction

Harza Northwest engineers and scientists were asked to provide a third-party review of the U.S. Army Corps of Engineers plans for "Drawdown". Drawdown is a reduction in the pool elevation of four mainstem Snake River dams plus the John Day pool on the Columbia. Inspired by drought and declining salmon stocks (one endangered and four threatened), Drawdown is proposed to reduce the cross-sectional area of reservoirs and increase hydraulic velocity. The objective is to accelerate outmigrating Snake River juvenile salmon.

By the use of a Drawdown Test in March 1992 and subsequent modeling efforts, the U.S. Corps of Engineers has demonstrated that Drawdown will increase average velocity. However the increased speed of the average water particle is relatively small for most Drawdown options studied, especially at average discharge conditions. For example, a Drawdown of 33 feet would only shorten the average

---

[1]Harza Northwest, Inc., 9600 Southwest Oak, Suite 350, Portland, Oregon 97223
[2]Harza Northwest, Inc., 2353-130th Northeast, Suite 200, Bellevue, Washington 98005

water particle travel time (WPTT) in the Snake reservoirs by about three days at 100 kcfs (Figure 1). If fish traveled at approximately the same speed as water, such efforts would only shorten the journey to the Columbia Estuary by about ten percent. Biological evidence suggests that on average, smolts usually travel about one-third the current velocity, but at times, reach migration rates about equal to the current speed. Thus, many of the proposed Drawdown options may only provide gains on the order of 3 to 5% in fish migration speed. Complete reservoir removal would make more significant changes to WPTT.

It appears that the cost of Drawdown will be significant in both capital costs to redesign the dams as well as operational costs related to lost hydropower, navigation, recreation, irrigation and flood control benefits. Additionally, there are numerous engineering and biological problems that must be overcome to allow it to be done safely and in a biologically prudent manner. One necessity is the redesign of fish passage facilities for downstream migrants and upstream migrating adults. Nitrogen supersaturation is also a problem that occurs under high spill. As a result, the benefits to the salmon fishery must be predictable, not just in more rapid transit to the estuary, but in true reversal of decline of wild stocks if the investment in Drawdown is to pay off. Finally, the Corps estimates that full Drawdown could take between 12 and 17 years to be fully implemented.

Several options exist to help the salmon situation. Drawdown is one of those options. Drawdown is logical and well intended to compensate for drought, increasing consumptive uses, and storage. Drawdown, however, is only one tool in our arsenal to help restore salmon. If it is to be used, other factors that are also significant causes of mortality, must be correctly quantified, assigned priority and abated. For it will do little good to increase migration speed, only to see whatever benefits Drawdown provides, nullified by other causes of mortality.

Turbine Mortality

Turbine mortality is estimated to claim an estimated 15% at each dam of those fish not removed by transportation. The efficiency of removing fish from the river is only 60% at best, thus 40% or more pass through the turbines at most dams. New fish by-pass facilities at Bonneville Dam showed very disappointing results in that mortality appeared to be higher for fish utilizing the by-pass than those that slipped past the fish screens and through the turbines (22% Vs 16%).

Predation

Predation losses are estimated between 9-19% in John Day. Although 14 million fish are released above Lower Granite dam, less than 5 million reach the dam. In recent mark/recapture experiments, it appears that mortality rates in the Snake River upstream of Lower Granite may be as severe as mortality in the reservoirs. Potential causes include sick fish and high predation rates. One optimistic view is that faster moving fish will be more difficult to prey upon. A pessimistic view is that Drawdown will concentrate prey and predators in a smaller reservoir, leading to even higher predation rates. Smallmouth bass switched primary prey items from crayfish to chinook smolts after the Drawdown Test of 1992 possibly because Drawdown killed a large number of crayfish. Predator-prey relationships are complex and currently the data is inadequate to address this quesion.

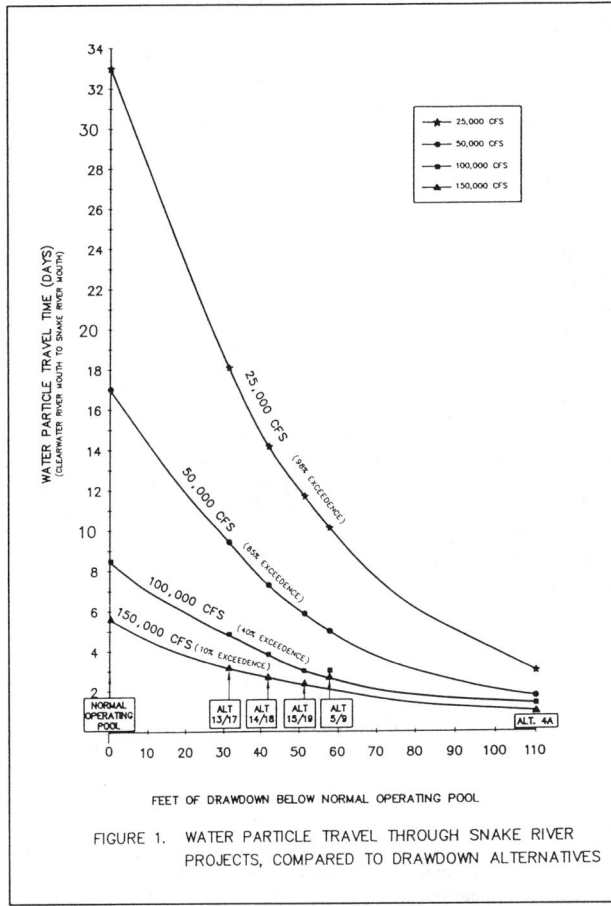

FIGURE 1. WATER PARTICLE TRAVEL THROUGH SNAKE RIVER PROJECTS, COMPARED TO DRAWDOWN ALTERNATIVES

Hatchery Practice

Currently, nearly 30 million hatchery fish are introduced into the Snake River each spring. And nearly 300 million in the Columbia system as a whole. A 3% adult return would yield nearly a million Snake River fish and nearly 10 million adults for the system. Historical record high returns were estimated to be 16 million. Today, most populations are yielding a return of between 0.1% and 0.5%; a few yield 1%. Overall the system is yielding about 0.6% return. Less than 200,000 adults returned to the Snake River recently (0.5% return). Hatchery fish are known to harbor numerous diseases that come from crowding and the hatchery environment. Hatcheries are primarily managed for productivity. Sources of hatchery stock are of highly mixed origins. Their genetic origin may be far from the rivers in which they are reared and planted. They may also suffer from inbreeding which diminishes the building blocks of future generations. At best they are poor substitutes for the native gene pools suffering high mortality. At worst, they further the demise of native fish by dilution of the native gene pool and competition with native fish.

Harvest

On first blush, sufficiently large numbers of fish (about 2 million salmon and steelhead in 1991) return to support recreational and commercial harvest in the Columbia and Snake rivers. However weak stocks are mixed with strong stocks and wild fish with hatchery fish. As a result, if harvest includes 70% of all fish, on average, this will include 70% of stocks that are on the verge of extinction. Traditional fishing practice and economics generally govern harvest and there is a tragedy of the commons occurring. As many as 73% of certain troubled stocks may be harvested in-ocean even though they represent less than 1% of the ocean harvest. Results of the Stochastic Life Cycle Model suggest that reducing adult mortality will lead to faster population recovery than similar improvements to any other life history stage of salmon.

In short, there appears to be numerous sources of mortality of juvenile fish as they come down the river. Delay, if it too is a problem, is only one of many. Thus if Drawdown is an appropriate strategy to speed fish down the river, fish will still suffer a high percentage of mortality from the dams and reservoirs unless we abate mortalities from a variety of sources including turbines, ineffective by-pass systems, predation, disease, transportation, adult migration delay and over-harvest.

What else can be done? There are a variety of engineering proposals to move fish faster and safer through the system. These include canals on the river bank, tubes in the river, levees to narrow the reservoirs, more upstream storage and barge transportation. Barging fish around the dams is the only cost-effective or biologically prudent option among most of these non-Drawdown alternatives. Barging has been practiced for ten years and currently yields about a 1.6 to 1 adult return advantage over fish not barged. But transportation is far from perfect. Return rate of most stocks is far below what fishery biologists would predict for a healthy salmon population. Most of the salmon now in the system come from hatcheries. The number of wild salmon returning to spawn is less than 5% of all returning fish. Some stocks are down to hundreds of adults.

A Proposal

An experimental transportation system located upstream of the existing Snake reservoirs could provide immediate relief to diminished wild stocks, especially if they could be segregated from hatchery fish. Because all hatchery fish will be marked for the first time in 1993, it will be possible to determine the fate of wild salmon. Since the genetic legacy of the wild fish cannot be replaced once gone, wild fish deserve special attention over hatchery fish.

Conclusions

Harza biologists and engineers are currently reviewing the designs, costs and construction schedules of several Drawdown alternatives as well as the question of biological prudence of Drawdown and other alternatives. These alternatives could save hundreds of millions of dollars and years of design and construction. Concepts include (1) use of existing dam as an upstream cofferdam for construction of a new spillway outlet structure; (2) design of a side channel spillway and downstream weir that will allow existing fish by-pass facilities to continue to operate. Finally, we suggest that mortality at all stages of life history— from egg to adult— must be addressed. Use of Drawdown is only one of those potential tools. Depending on progress made in the first half of 1993, we will report on additional new developments concerning the engineering feasibility of Drawdown and the biological advantage it may afford juvenile salmon.

Role of Economics in Endangered Species Act Activities Related to Snake River Salmon

Edwin J. Woodruff[1] and Daniel D. Huppert, Ph.D.[2]

Abstract

The development of recovery actions for the species of Snake River Salmon listed under the Endangered Species Act (ESA) must consider a wide range of actions covering the different life-cycles of the species. This paper examines the possible role of economic analysis in assisting in selection of actions to undertake and draws heavily on similar opinions presented by others in the region.

Background

The ESA does not require the use of economic analysis in the majority of tasks associated with the act. The ESA tasks of listing a species as threatened or endangered, the requirement for recovery plan development, and consultation with Federal agencies, are all determined on an ecological basis. The only component of the act that specifically mentions economic trade-offs is in the consideration of what to include as critical habitat. The act calls for weighing the benefits of inclusion of an area as critical habitat against the benefits of exclusion. This allows for the consideration of economic costs and benefits along with other impacts such as social and cultural costs; provided that exclusion will not result in the extinction of the species. Designation of candidate areas for critical salmon habitat and associated impacts has been done by NMFS and is currently out for public review and comment.

---

[1] Regional Economist, North Pacific Division, Corps of Engineers.

[2] Professor, School of Marine Affairs, University of Washington.

Opportunity For Use of Economics in ESA

Economics, as a discipline, can contribute to the decision making process under ESA in three ways. First, economic analysis can be used to assess the economic costs of designating critical habitat and of implementing recovery actions. Second, once the economic costs have been calculated, the indirect or secondary impacts can be estimated by use of a regional economic model, such as an input-output model. Third, economics can assist in the formulation of priorities within the recovery planning process by application of techniques such as cost-effectiveness analysis. (Huppert, Fluharty and Kenney, June 1992.)

The development of a recovery plan is ongoing along with several consultation activities between the NMFS and Federal agencies to assure that Federal actions are not jeopardizing survival of the listed salmon. The balance of this paper will prescribe methods for which economics can assist the decision process for these remaining ESA activities.

Basis for Economic Analysis

Political forces will ultimately decide what to protect and how to protect it. Nonetheless, diverse values and view-points need to be integrated in the decision-making process through rational application of biological and economic analyses. On the economic side, one of the most common evaluative approaches is to use some form of cost-benefit analysis. This is inappropriate for the task, however, since species survival is given overriding importance and thus is perceived as having infinite benefits. A cost-effectiveness modeling approach avoids the issue of evaluating benefits by setting desired objectives *a priori* and searching for the lowest-cost ways of achieving these. (Hyman and Wernstedt, 1991.)

Of interest to a lot of people, particularly those impacted and politicians, is the impact on jobs and income in specific regions in the Pacific Northwest. However, a comparison of measures based solely on minimizing losses of jobs is considered by many to not be a supportable approach. To illustrate, how does one judge whether an alternative to save salmon by reducing fish harvesting on the lower Columbia/Ocean that puts 500 fishermen out of work is any better than a plan to reduce irrigation in Idaho that will bankrupt 100 farmers and eliminate 500 associated farm industry and processing jobs? Job impacts are important, but they simply do not provide a comparable basis to make recovery plan cost effectiveness decisions over a wide range of alternatives. The economists, however, through the use of regional input-output models can estimate these regional income and job impacts for those that are interested. This task is now underway as part of the ongoing Columbia River

System Operation Review conducted by the Corps of Engineers, Bonneville Power Administration and the Bureau of Reclamation.

Cost Effectiveness Analysis

Cost-effectiveness analysis follows a simple principle: select actions that produce a desired result in a least-cost manner. (Olsen and Peters, Oct 1991.) Care must be undertaken to consistently define the economic costs and to clearly define the desired output. The cost definition approach agreed to by the economic community is to define the costs of measures as "net direct economic costs." These are the costs to people of choosing to use resources in a specific way. They are measured as the net value of goods and services foregone to conserve Snake River salmon (Huppert, et al., 1992.)

An appropriate biological objective must be defined to weigh the alternative costs against. But, to date no specific goal of X amount of smolts or returning adults has been established. That is, the decision makers are unsure of their (specific) objectives or are willing to consider alternative objectives. One way to help visualize the trade-offs is to develop a cost-effectiveness frontier by plotting the costs and level of biological effectiveness of each proposed alternative on a graph (see figure 1.) Points on the frontier represent alternatives that are equally effective but cost the same or less than an alternative not on the frontier (compare points B and A in the figure), or alternatives are more effective but cost the same or less than alternatives not on the frontier (compare points D and C in the figure) (Wernstedt, Hyman and Paulsen, 1992.)

The cost-effectiveness approach discussed above implies a degree of certainty in the economic costs and biological effectiveness that simply does not exist. Points on the figure would be better defined as a box that reflects the uncertainty associated in both the economic costs or the biological effectiveness of the different alternatives.

Closing Comments

Preserving species is a moral imperative that transcends any economic considerations for many people, but that position does not provide any practical guidance for making economic decisions. And there is no way to avoid making these difficult economic decisions (Mannix, 1992.) This paper briefly discusses the need for some economic analysis and outlines a cost-effectiveness approach that could be utilized as valuable information in the decision making process. Cost-effectiveness analysis will not, and should not, be the only factor in selecting recovery actions. It is the hope of the authors that it at least be a consideration.

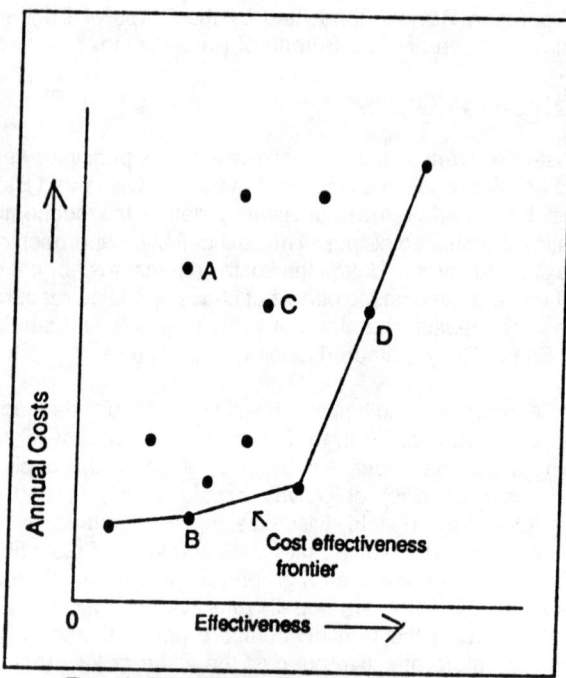

Figure I: Cost-effectiveness of hypothetical Snake River fish recovery measures

References

Huppert, D., Fluharty, D., and Kenney, E. Economic Effects of Management Measures Within the Range of Potential Critical Habitat for Snake River Endangered and Threatened Salmon Species. School of Marine Affairs, Univ. of Washington. 4 June, 1992.

Hyman, J. and Wernstedt, K. The Role of Biological and Economic Analysis in the Listing of Endangered Species. Resources: Resources for the Future. Summer 1991.

Mannix, B. The Origin of Endangered Species and the Descent of Man (With Apologies to Mr. Darwin). The American Enterprise. Nov/Dec 1992.

Olsen, D. and Peters, L. Fundamental Guidelines for an Economic Impact Assessment Under the Endangered Species Act. Prepared for Economics Committee to NMFS. Oct 1991.

Wernstedt, K., Hyman, J., and Paulsen, C. Evaluating Alternatives for Increasing Fish Stocks in the Columbia River Basin. Resources: Resources for the Future. Fall 1992.

Water Quality Impacts from Development of Port Ludlow, Washington

Thomas Smayda[1]; Clain Jones[1];
Martin E. Harper[1]; David Cunningham[2]

Port Ludlow Bay is a tidal estuary located adjacent to Admiralty Inlet in Puget Sound, Washington (Figure 1). A planned development has been under construction along the bay since 1976. The 12,000 acre project includes a marina, 27 holes of golf and currently, 890 residences. Protection of excellent water quality has been a goal of Pope Resources, the developer, since project inception. A 0.64 mgd activated sludge plant and a stormwater control system of wet ponds with swales have been constructed. An NPDES permit for the treatment plant requires bay water monitoring each year, and Jefferson County requires nonpoint source monitoring. Summarized herein are selected data obtained from sea water sampling, from sediment sampling and from runoff sampling in the Port Ludlow area (Figures 2, 3 and 4).

Figure 1. Port Ludlow Bay, Washington

---

[1]Harding Lawson Associates, 1325 Fourth Avenue, Suite 1800, Seattle, WA 98101

[2]Pope Resources, P.O. Box 1780, 19351 8th Avenue NE, Suite A, Poulsbo, WA 98370

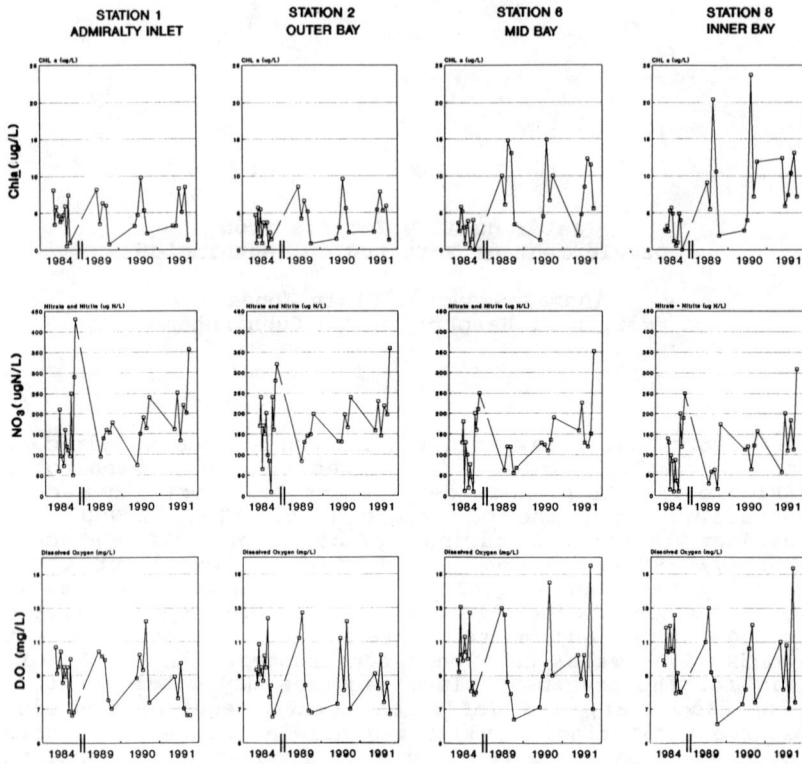

Figure 2. Port Ludlow is a 5.6 km^2 bay with mean and maximum depths of 15 and 30 m. The tidal range is 3.1 m, with strong circulation due to the tidal prism plus currents which enter from Admiralty Inlet. Drogue and drifter card studies were used to select a location for the WWTP outfall location, and indicated an average bay water residence time of 2.3 days. Monitoring has indicated N:P ratios of 7 to 14 by weight, suggesting nitrogen limitation of primary production, but dissolved nitrogen levels remain high even during blooms. Blooms occurred during mid-summer, suggesting light rather than nutrient limitation of growth. Washout of cells from the bay also influenced bloom development and duration as indicated by progressively greater Chl<u>a</u> concentrations as a function of distance in the bay. Dissolved oxygen levels were periodically depressed likely as a result of upwelling from Hood Canal or Admiralty Inlet. Offshore water (Admiralty Inlet) controlled water quality in the bay, but treated wastewater effluent was not found to measurably impact water quality.

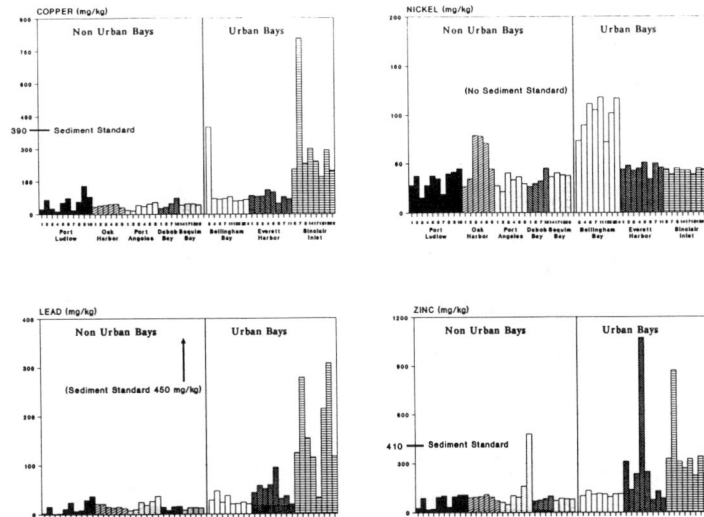

Figure 3. Subtidal marine sediments were collected at the marina and in other quiescent locations of inner Port Ludlow Bay. The sediment was silty with traces of sand, dark gray, anoxic, somewhat gelatinous and contained shell fragments. The samples contained 32 to 76% solids and 1 to 16% volatile solids. Fats, oils and grease, even near the fuel dock, ranged from 51 to 316 mg/kg dry weight, which is within the natural range for Puget Sound.

Copper, nickel, lead and zinc are metals introduced into the water from a range of human activities, including from bottom paint from boats and urban runoff, and then settle to the bottom. Metal concentrations (above) indicate that copper, nickel, lead and zinc were not more concentrated than in other non-urban bays, and were less concentrated than in urban embayments (other data from Battelle[3] and Crecelius et al.[4]). A conclusion is that sediments in Port Ludlow Bay currently have good quality. Shellfish tissue analyses have also indicated good quality but with periodic increases in fecal coliform levels during periods with heavy boater usage.

---

[3]Battelle. 1986. Reconnaissance survey of eight bays in Puget Sound. Volume I. Report to EPA Region 10.

[4]Crecelius, E.A., D.L. Woodruff and M.S. Meyers. 1989. 1988 Reconnaissance Survey of Environmental Conditions in 13 Puget Sound locations. EPA Report 910/9-89-005.

Figure 4. Stormwater runoff enters Port Ludlow Bay via 10 sub-basins. During construction, soil erosion has occurred in some areas which was the principal source of metals in the runoff. Street and parking lot runoff had low metals, but the golf course runoff apparently has slightly elevated copper levels. Pesticides, herbicides and PCBs have not been detected in golf course runoff, and total nitrogen has ranged to 3.2 mg N/L and phosphorous to 0.48 mg P/L which may be slightly elevated over the other tributaries. A constructed two-cell wet pond has had the lowest fecal coliform levels of the 10 discharge points to Port Ludlow Bay indicating good performance, even though solids washout occured during large storms.

Admiralty Inlet water which exchanges with that in Port Ludlow Bay represents the largest source load of nutrients (98%) and fecal coliform (66%). With respect to other sources, (presented here) stormwater is the second largest contributor of fecal coliform and third largest contributor of nutrients to the bay.

TOTAL NITROGEN
(kg/yr)
(31,881 kg/yr)

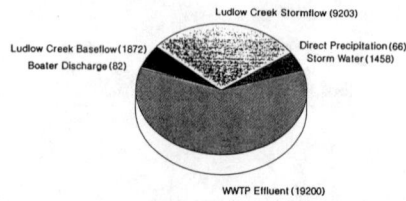

TOTAL PHOSPHORUS
(kg/yr)
(8811 kg/yr)

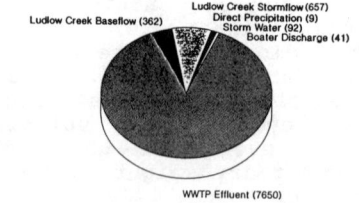

FECAL COLIFORM
(trillion/yr)
(59.8 x 10^{12} /yr)

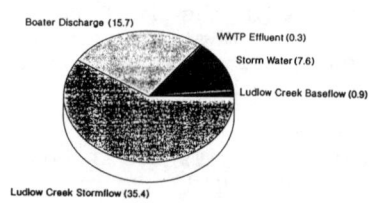

Multiobjective Models for Determining
Fresh Water Inflows to Bays and Estuaries

Ning Mao[1] and Larry W. Mays[2], M. ASCE

Introduction

The estuarine management is defined as a means to manage the fresh water resources for purposes of providing on optimal estuarine environment by controlling the vital components in a estuarine system such as salinity, nutrient, and key organisms. Salinity is an index which has been well-established to indicate ecological conditions in an estuary which is considered as a central element in the models. This paper presents two formulations of multiobjective models for the estuary management problem that are based upon goal programming procedures. The multiobjectives are selected to be minimum fresh water inflow into bays and estuaries while maximizing the commercial fish harvest for the five fish species. Salinity and harvest are expressed as a function of fresh water inflow in nonlinear regression equations and used as the constraints. because of the uncertainty associated with regression equations for salinity and harvest, the second formulation was developed in which the constraints are expressed in a chance-constrained formulation. The nonlinear multiobjective models developed in this study are applied to the Lavaca-Tres Palacios Estuary in Texas. However, these models also can be modified to applied on other bay and estuarine system.

---

[1] Research Assistant, Department of Civil Engineering, Arizona State University, Tempe, AZ 85281.
[2] Chairman and Professor, Department of Civil Engineering, Arizona State University, Tempe, AZ 85287.

Goal Programming Method

The basic concept of goal programming formulation can be stated as follows:

$$\min S_0 = \sum_{i=1}^{p} P_i (w_i^+ d_i^+ + w_i^- d_i^-) \tag{1}$$

subject to

$$x \in X \tag{2}$$

$$F_i(x) - d_i^+ + d_i^- = T_i \tag{3}$$

$$d_i^+, d_i^- \geq 0, \quad i = 1, \ldots, p \tag{4}$$

where d_i^+ is the positive deviation from the target for the ith objective (i.e., overachievement of a goal); and d_i^- is the negative deviation from the target for the ith objective (i.e., underachievement of a goal). w_i^+ is a weight for the positive deviation for the ith objective and w_i^- is the weight for the negative deviation for the ith objective. P_i is a priority factor for the ith objective. T_i denotes the target or goal set by the decision maker for the ith objective function $F_i(x)$, and X represents the feasible region from which the choices of vector x must be effected.

Application of Goal Programming with Deterministic Constraints

The multiobjective problem is formulated to minimize the annual total fresh water inflow and maximize the fish harvest for each of five fish species which can be stated as six objectives subject to the required constraints. In order to obtain preferred solutions directly, the goal programming technique is applied. So, the problem can be stated as:

$$\min S_0 = \sum_{i=1}^{6} P_i (w_i^+ d_i^+ + w_i^- d_i^-) \tag{5}$$

where, d_i^+, d_i^- are positive and negative deviations between target and actual achieved values for fresh water inflow and fish harvest; P_i is the priority factor of inflow and fish harvest deviations; w_i^+, w_i^- are the weights of positive and negative deviations; and $i = 1$, for the fresh water inflow, $i = 2, 3, \ldots, 6$ express the fish harvest for different species. Subject to:
(1) The nonlinear relationship of estuary salinity and fresh water inflow.

$$s_{tj} = \phi_{tj}(Q_{tj}) \tag{6}$$

(2) Upper and lower bounds on the monthly mean salinity at a specified location in the estuary, for each river j.

$$\underline{s}_{tj} \leq s_{tj} \leq \overline{s}_{tj} \tag{7}$$

(3) Upper and lower limits on mean monthly flows in seasons m for each river j.

$$\underline{QS}_{jm} \leq QS_{jm} \leq \overline{QS}_{jm} \tag{8}$$

where $QS_{jm} \equiv \frac{1}{N_m} \sum_{t \in M_m} Q_{tj}$; M_m is the set of months in season m and N_m is the number of months in season m.

(4) The nonlinear relationship between the harvest of organism k and the seasonal inflow in river j.

$$H_k = \psi_k (QS_{jm}) \tag{9}$$

(5) The lower limits on annual fish harvest, $\underline{H}_k$, for species k.

$$H_k \geq \underline{H}_k \tag{10}$$

(6) The upper and lower limits on monthly inflows for each river.

$$\underline{Q}_{tj} \leq Q_{tj} \leq \overline{Q}_{tj} \tag{11}$$

(7) The target value must be equal to the actual achieved value combine the total deviations.

$$F_i (Q_{tj}, H_k) - d_i^+ + d_i^- = T_i \tag{12}$$

(8) Nonnegative of deviations

$$d_i^+, d_i^- \geq 0 \tag{13}$$

The model has been solved by using GAMS/MINOS programming code. For the purpose of doing sensitive analysis of model, the different options with different sets of target value (T_i) have been applied. The results show that different targets will yield very different solutions, therefore, specified reasonable targets is the key step for solving goal programming model. The results of second option are given as example shown in the Table 1.

Table 1 Results of Option II

Objectives	Targets T_i (1000 ac-ft) (1000 pounds)	Positive Deviations d_i^+	Negative Deviations d_i^-	Expected Value (1000 ac-ft) (1000 pounds)
Annual inflow	2035.90			2035.90
All shellfish	6641.21		3606.91	3034.30
Spotted seatrout	5165.78		2138.40	3027.38
Red drum	243.44		176.69	66.75
Penaeid shrimp	4402.12		2125.61	2276.51
Blue crab	1194.86		413.46	781.40

Objective function value = 16922.14

Developed Multiobjective Model with Chance-Constraints

For considering the uncertainty of salinity and fish harvests in the multiobjective programming model, the mathematical formulation can be modified by introducing chance-constraints to replace the deterministic constraints of salinity (Eq. 7) and fish harvests (Eq. 10). The chance-constraints can be presented as following:

$$P_r\left\{ \underline{s}_{tj} \leq s_{tj} \leq \bar{s}_{tj} \right\} \geq p_{tj} \qquad (14)$$

$$P_r\left\{ H_k \geq \underline{H}_k \right\} \geq p_k \qquad (15)$$

where p_{tj} is the required reliability of salinity, for each month t and each river j and p_k is the required reliability of harvest. The deterministic equivalents are not shown for the sake of brevity

The GRG2 programming code was applied to solve the multiobjective model with chance-constraints. The results with different required reliabilities on salinity are shown in the Figure 1. The developed nonlinear chance-constraint estuarine model computes both the minimum deviation of objectives and the achievable reliability for salinity and harvest constraints.

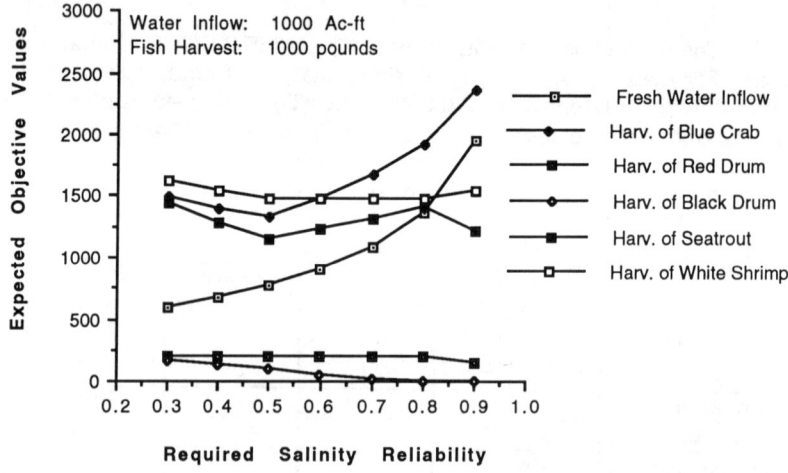

Figure 1 Variation of Objective Values for Different Required Reliability of Salinity

City of Federal Way
Panther Lake Surface Water Study
A Case Study of HSPF Modelling in Urban Detention Basins

Andrew B. Lukas[1] and Cary M. Roe, P.E.[2]

Abstract

This paper presents a case study of using Hydrologic Simulation Program-Fortran (HSPF) as a tool for sizing large urban detention facilities. The Panther Lake drainage basin is a textbook case of the effects of urbanizing a forested area. The study itself combined the elements of urban surface water runoff control, hydrologic computer modelling, detention pond sizing, and environmental impact assessment. In an earlier study, Panther Lake was determined to be a priority in ensuring the stability of downstream natural channels due to its importance with regard to flooding prevention, stream bank erosion control, and habitat protection. Originally the lake was a closed depression, totally in equilibrium with its forest setting. However, with urban development creating impervious surfaces, and the installation of man-made drainage structures that extended the tributary area, the storage of urban runoff from over 526 hectares (1300 acres) was required of the closed depression. Refinements of the tributary area, storage capacity, and outlet hydraulics of the lake were made in order to make an accurate assessment of the amount of additional storage required at the closed depression in the event of a 100 year storm. A first set of alternatives was developed from this data, from which three feasible solutions were chosen and then tested with HSPF to determine their adequacy in a 100 year event.

Key Words: HSPF, urban hydrology, hydrology, detention ponds, computer modelling

Introduction

The modelling portion of the project was performed using the Hydrologic Simulation Program Fortran (HSPF) Version 9.1. HSPF is a continuous simulation hydrologic model written for the United States Environmental Protection Agency (EPA) for the purpose of modelling river basins and developing projects of major economic importance. The use of HSPF in modelling small urban detention facilities has been increasing in recent years.

---

[1]Associate Engineer, Brown and Caldwell Consultants, 100 West Harrison Street, Seattle, WA, 98119

[2]Surface Water Manager, Public Works Department, City of Federal Way, 33530 First Way South, Federal Way, WA, 98003

The model is capable of achieving excellent results with the proper knowledge of soil type, land use, groundwater behavior, topography, and vegetative cover in the study area.

Revisions of the Original Model Input and Calibration

In early work on the Panther Lake drainage basin, King County staff (1991) developed HSPF model input based on regionalized parameters established in a northwest regional study by the United States Geological Survey (USGS) (Dinicola 1990). These parameters have been used in a large number of HSPF modelling projects in King County and are a generally acceptable starting point in developing HSPF input. When adequate field data exists and if determined necessary, model input can be calibrated by adjusting these parameters.

The Hylebos Basin Plan study area encompassed approximately 93 square kilometers (36 square miles), and required a planning-level of detail in developing the HSPF model input. The first steps taken in this study involved taking information used in the plan and refining it to a predesign level of detail. As a result several facets of the modelling were refined. The stage-storage discharge relationship for Panther Lake was improved with new topographic and bathymetric surveys of the lake and a correction to the representation of the outlet hydraulics. The basin's external boundaries, and the delineation of subcatchment areas were improved by reviewing as-built storm drainage maps. The estimated drainage area was reduced by 20 percent, including a smaller reduction in the contributing effective impervious area areas. The basin's subcatchments were also reapportioned. A newly created subcatchment with direct discharge to the lake was also added to the model.

In earlier work (Barker 1990), the USGS regionalized parameters were not adjusted during calibration of the Panther Lake input as the comparison of simulation data to calibration data was considered within desired tolerances. However, with the seemingly drastic changes made in the model input, it was necessary to determine if calibration was still intact. The calibration period for the original model input occurred between October 1986 and September 1988. During that time, rainfall data was taken by a portable King County rain gauge located near Panther Lake, and lake staff gauge readings were taken by USGS field personnel. The entire calibration period was during a very near record-low two-year precipitation period (according to Seattle-Tacoma International Airport rainfall records). In checking calibration, the project team decided to concentrate on data that corresponded to discharge from the lake. Despite the changes in the model input, the resulting fit of simulated lake stage to measured lake stage was good, and the model input was considered calibrated.

Determining the Design Storm

To ensure stability in downstream natural channels the general practice in King County is to provide stormwater detention to restrict flows produced in the fully built-out land use condition to pre-development rates. In this case the pre-developed condition consisted of the present tributary area minus two subcatchments which are diverted to the lake with man-made conveyance structures. The land use conditions existing at the time of the calibration period were used since field reconnaissance (King County 1990) determined that downstream channels were stable at that time.

HSPF runoff was simulated with the Seattle-Tacoma International Airport precipitation record from October 1948 to January 1990. The maximum annual discharges from this simulation were fit statistically with the Log Pearson Type III Distribution (USGS 1981). From this statistical representation of the 42 years of simulation, the 100-year runoff event discharge from the lake under its natural setting was determined to be 0.34 cubic meters per second (m^3/s) (12 cubic feet per second [cfs]).

In order to test solutions, it was necessary to develop a storm that would produce a peak lake discharge equal to that of the statistical 100-year storm. The rainfall period causing the largest peak discharge was chosen as the storm "benchmark." It was "scaled up" with a multiplication factor function in HSPF until the statistical 100-year lake discharge was produced. This "scaled-up" storm was used to test all alternatives under future land use conditions.

Identification of Solutions

Potential solutions were developed in a brainstorming session among the project team members and then screened down to those which could by themselves solve flooding problems of the 100-year event. Due to this evaluation criterion, solutions such as source control, multiple upstream detention facilities, utilization of storage in conveyances, high-flow bypass, lake dredging, and multiple downstream detention facilities were not considered. Some of these eliminated solutions were evaluated on their ability to improve the effectiveness of a larger solution. As a result only three alternatives remained for further consideration.

Alternative 1 provided an addition of 270,000 cubic meters (220 acre-feet) of detention volume at Panther Lake by virtue of raising the existing dam 3.7 meters (12 feet) without changing the discharge structure hydraulics. The peak lake discharge from this alternative was 0.60 m^3/s (21 cfs), a 0.26 m^3/s (9 cfs) increase over pre-developed conditions.

Alternative 2 utilized an adjacent wooded area to detain spilled volumes during certain extreme events. The lake, with its raised dam, and the wooded area, with a its own new dam, in tandem contained the simulated 100-year flood. The wooded area was assumed to have an outlet structure with a relatively low release rate. Due to the small area available in the wooded area, only one foot of lake dam construction was eliminated.

Alternative 3 also consisted of a two-basin scenario. The dam around the lake was raised (less than in Alternative 2) to a level at which the lake contained all simulated runoff produced from the 42-year precipitation record. An adjacent Little League baseball complex would be used to detain the spilled runoff in the simulated 100-year event. A partial-duration series was developed for the entire simulation record to determine the maximum monthly stages for the lake assuming fully built-out conditions. A new dam level, three feet lower than in Alternative 1, was chosen which would prevent spill during baseball season. A new stage-storage-discharge relationship was developed for the lake with its raised dam, and a simulation run was performed using the 100-year storm to determine what volume of detention would be required of the baseball complex. The discharge relationship for the secondary basin was chosen to minimize impacts to facilities due to duration of inundation. Similar facilities operate with a maximum inundation duration of 7 days.

Analysis of annual maximum levels for Alternative 1 revealed that the spill recurrence for Alternatives 2 and 3 were 80 and 50 years, respectively. In all three alternatives, modification of the existing lake outlet was not considered. The model was also used to determine how raising the dam around the lake would increase the frequency and duration of inundation of wetlands species.

Selection of the Recommended Solution

The three alternatives were evaluated using cost and non-cost criteria. The project team developed a ranking system to determine the best alternative. Issues evaluated included wetland impacts, aesthetics, operations and maintenance, dam safety, and possibilities for a multiple-use recreational facility. Each issue was given a weighting factor to reflect its importance relative to others. Alternative 3 was chosen as the recommended solution as it earned the highest ranking and had the lowest estimated construction cost.

References
Barker, B. (1990). Hylebos Creek and Lower Puget Sound Basin Calibration Report. King County Surface Water Management Division.
Dinicola, R.S. (1990). Characterization and Simulation of Rainfall-Runoff Relations for Headwater Basins in Western King and Snohomish Counties, Washington. USGS.
King County Surface Water Management Division (1991). Executive Proposed Basin Plan for Hylebos Creek and Lower Puget Sound.
King County Surface Water Management Division (1990). Hylebos Creek and Lower Puget Sound Basins Current and Future Conditions Report.
United States Geological Survey (1981). Guidelines for Determining Flood Flow Frequency, Bulletin #17B.

Water Quality Model of the
Chicago Waterway

Bill Macaitis[1] (Fellow, ASCE)
Carl Johnson[2] (Member, ASCE)

Introduction

The Metropolitan Water Reclamation District of Greater Chicago (MWRDGC or District) is responsible for wastewater conveyance, water reclamation, sludge disposal and industrial pretreatment programs in Cook County, Illinois. In 1992, the MWRDGC developed a QUAL2E water quality model of the Chicago waterway and Upper Illinois waterway. The modeling effort fulfills the requirements of the District's NPDES permits, and it provides the ability to simulate the water quality impacts of future water pollution control improvements.

The model is capable of predicting $CBOD_{20}$, dissolved oxygen, and ammonia-nitrogen and orthophosphate within 10 to 20 percent of measured values at a 95 percent confidence interval. The model was calibrated and verified under flows ranging from 2,500 to 3,800 cfs in the Chicago waterway at Lockport, and flows ranging from 5,200 to 15,800 cfs in the Upper Illinois waterway at Chillicothe. Simulations were conducted at flows ranging from 1,700 to 5,000 cfs at Lockport and flows ranging from 3,400 to 19,200 cfs at Chillicothe.

Data Acquisition

Data were acquired for the model in the following general areas: channel characterization; flow, time of travel and dispersion; point sources; 1990 water quality; historic water quality; sediment oxygen demand; dam reaeration; and weather data. A major portion of the data obtained for the modeling was provided by the Illinois State Water Survey through field programs conducted in 1989 to 1991.

---

[1]Assistant Chief Engineer, Metropolitan Water Reclamation District of Greater Chicago, Chicago, Illinois

[2]Vice President, Camp Dresser & McKee Inc., Chicago, Illinois

Additional data was obtained from the MWRDGC, the U.S. Army Corps of Engineers, the National Weather Service of the National Oceanic and Atmospheric Administration, the Illinois Environmental Protection Agency, and the U.S. Environmental Protection Agency.

Historic reference material on water quality was based on work by the Illinois State Water Survey, Northeastern Illinois Planning Commission, MWRDGC, and the U.S. Geological Survey. The data acquisition process entailed the review and assimilation of over 50,000 records of time series data on water quality, river flow, point source flows and loads, and weather records, along with about 19,000 records of data on channel geometry. Over 80 previous and ongoing studies on water quality in the Chicago waterway system and Upper Illinois River were reviewed.

Model Development

The modeled waterway system is almost 200 miles long. The upstream Chicago waterway portion is a dendritic, urban waterway, that is routed through relatively deep, narrow, man-made channels. The lower portion of the waterway is the Upper Illinois River, which is mostly natural, wide and shallow outside of the shipping channel, and is controlled by four navigation and flood control dams. The water quality of the Chicago waterway portion must meet secondary contact recreation standards, while the quality in the Upper Illinois River portion must meet general use (water supply) standards.

Because of the differences in its upstream and downstream waterways, two models were created -- one for the upstream Chicago waterway with a 0.5 mile computational element length, and another for the downstream, Upper Illinois waterway with a 1.0 mile element length.

Flow in the Chicago waterway is regulated by federal statute, which allows a total of 3,200 cfs per year to be diverted from Lake Michigan to the Chicago waterway, and ultimately, the Mississippi River. The 3,200 cfs diversion flow includes public water supply use, stormwater runoff, lockage and leakage, and discretionary diversion of Lake Michigan waters to the Chicago waterway for low flow augmentation. The discretionary diversion presently is limited to a 320 cfs annual average flow.

The discretionary diversion flow is used exclusively in the summer; therefore, it has a relatively large impact on the total flow and water quality in the Chicago waterway. Consequently, one water quality model input deck was developed and calibrated for the diversion-flow regime, while a second model input deck was developed for the non-diversion flow regime.

The diversion flow model was calibrated to the Chicago Waterway and the Upper Illinois River for conditions occurring on September 25, 26, and 27 and the non-diversion model was calibrated to October 23, 24, and 25, 1990 conditions.

Model parameters were adjusted to have predicted values of $CBOD_{20}$, orthophosphate, organic nitrogen, ammonia, nitrite, nitrate, chlorophyll-a and dissolved oxygen, agree with actual in-stream measurements. The diversion model was verified for conditions occurring during July 16 - 18 and July 30 - August 1, 1991. The non-diversion model was verified for conditions occurring during May 21 - 23 and June 4 - 6, 1991.

The calibrated and verified model parameters were all within acceptable ranges from literature. The calibrated model can accurately predict, and reflect, the physical, chemical and biological mechanisms occurring in the Chicago Waterway and Upper Illinois River. An overview of the processes occurring in the Chicago Waterway and Upper Illinois River follows, based on the findings of the verified QUAL2EU model.

Findings of Existing Condition Simulations

Existing conditions were represented by average dry weather flows observed in 1991. The model simulations for existing conditions are very close to the conditions observed during the May to August sampling events performed in 1991. Such dry weather conditions are applicable about 85 percent of the time in the Chicago waterway and Upper Illinois River. The results are consistent with the findings of the July and August 1991 sampling runs, which showed that the dissolved oxygen standards are generally met in the Chicago waterway. However, there are times that the standard is not achieved on the Sanitary and Ship Canal below the confluence with the Cal-Sag Channel, in the lower reach of the North Branch of the Chicago River, and in the Cal-Sag Channel, in the lower reach of the North Branch of the Chicago River, and in the Cal-Sag Channel. The dissolved oxygen standards are met in the Upper Illinois River under average dry weather flow conditions.

Nitrification occurs in the Chicago Waterway during the summer, so that total ammonia nitrogen does not exceed the 0.4 mg/l standard at Route I-55 under average dry weather flow conditions. Nitrification is currently (1991) the largest oxygen demanding process in the Cal-Sag Channel portion of the Chicago Waterway. It is followed by sediment oxygen demand, and then carbonaceous decay. The nitrogenous oxygen demand is typically greater than the combined SOD and carbonaceous demand. Algae is not significant in the Channel.

Findings of Future Condition Simulations

Projects underway by the MWRDGC will improve average dry weather water quality markedly in the future. These improvements include completion of most of TARP 1 and completion of sidestream elevated pool aeration station (SEPA) projects.

The completion of TARP will improve the dissolved oxygen
concentrations on the waterways by reducing the volume and
frequency of combined sewer overflows (CSO). This reduction in
CSO discharges will reduce the concentration of $CBOD_{20}$ and ammonia
on portions of the waterway and tributary rivers where TARP has
not yet been completed. Another benefit of the TARP project is
the continued reduction in sediment oxygen demand in the
waterway. Water quality will generally meet or exceed the water
quality standards for dissolved oxygen and ammonia nitrogen. In
the Chicago waterway, unionized ammonia will be less than the
standard of 0.1 mg/l.

TARP 1 improves the dissolved oxygen concentrations on the
Chicago waterway so that the lowest concentration of dissolved
oxygen improves from 3.0 mg/l in 1991 to 4.0 mg/l in 1999. SEPA
adds as much as 6.0 mg/l of dissolved oxygen to the Calumet-Sag
Channel. The spring dissolved oxygen concentration, which is the
worst case for dissolved oxygen, is raised from 2.0 mg/l to 5.0
mg/l at the downstream end of the Calumet-Sag Channel. The 1991
simulations had shown that the dissolved oxygen standard is not
always met on the Chicago Waterway and Calumet-Sag Channel,
particularly in the spring, before discretionary diversion flow
are established. However, in the future, for both the spring and
summer simulations, the dissolved oxygen concentrations are
consistently greater than the standard of 4.0 mg/l for the
average flow conditions in 1999. Hence, the completion of TARP
and SEPA will assure that the minimum dissolved oxygen standard
is achieved throughout the Chicago Waterway, with the possible
exception of a very short reach slightly below the standard on
the lower North Branch of the Chicago River. Ongoing water
quality monitoring programs will determine the need, if any, for
further improvements to protect water quality in this part of the
waterway.

Under average flow conditions, the dissolved oxygen
concentrations always exceed the standard of 5.0 mg/l in 1999 on
the Upper Illinois River. Unionized ammonia is less than the
standard of 0.04 mg/l. The completion of TARP and SEPA improves
the water quality in the Upper Illinois River.

Simulation of Water Quality Processes in the
Chicago Waterway and Upper Illinois River Systems

Bill Macaitis[1], Fellow, ASCE, John Variakojis[2], Member, ASCE and Gary Mercer[3], Member, ASCE

Abstract

The U.S. Environmental Protection Agency QUAL2E-UNCAS (QUAL2EU) computer program was used to simulate the water quality processes in the Chicago waterway and Upper Illinois River Systems. Rates and parameters were developed, calibrated, and verified to six separate sampling events. The processes simulated included nitrification, carbonaceous biochemical oxygen demand (CBOD), sediment oxygen demand (SOD), phosphorus transformation, algal respiration and photosynthesis, and atmospheric, dam, and Sidestream Elevated Pool Aeration (SEPA) reaeration.

Introduction

The Metropolitan Water Reclamation District of Greater Chicago (MWRDGC or District) has developed a water quality model of the Chicago waterway and the Upper Illinois River. The modelling effort was undertaken to fulfill the requirements of the District's NPDES permits. The model provides the ability to simulate the water quality benefits of future water pollution control improvements (CDM 1992). The waterways and rivers simulated include the Chicago waterway and the Upper Illinois River system. The total length of the modeled waterway and river system is 192 miles.

Model

The District selected the QUAL2EU model because of the model's ability to accurately simulate the complex waterway interactions and the model is widely accepted. QUAL2EU simulates the major water quality interactions in the waterways, including the nitrogen cycle, phosphorus cycle, algal production, sediment oxygen demand, carbonaceous biochemical oxygen demand, and atmospheric and dam reaeration. The focus of the study was on the dissolved oxygen concentration.

An extensive data collection program was undertaken by the Illinois State Water Survey (ISWS) under contract to the District. The program included collection of water quality samples and measurements of: flow, SOD, light extinction, time of travel, and dispersion.

---

[1]Assistant Chief Engineer, Metropolitan Water Reclamation District of Greater Chicago, Chicago, Illinois
[2]Supervising Civil Engineer, Planning Section, Metropolitan Water Reclamation District of Greater Chicago, Chicago, Illinois
[3]Senior Water Resources Engineer, Camp Dresser & McKee Inc., Cambridge, Massachusetts

Six intensive flow sampling at fifty stations were taken to calibrate and verify the model.

Processes

CBOD: Biochemical oxygen demand is the utilization of dissolved oxygen by aquatic microbes to metabolize organic matter. Carbonaceous biochemical oxygen demand (CBOD) represents the amount of oxygen required by microbes to stabilize organic matter under aerobic conditions excluding nitrification. The ultimate CBOD values are represented by 20-day values, $CBOD_{20}$

The $CBOD_{20}$ concentration in the effluent of the WRPs were less than 5 mg/l for the Northside and Stickney Reclamation plants and ranged from 5 to 12 mg/l at the Calumet Reclamation plant during the water quality sampling. Calibration of the model to instream $CBOD_{20}$ values was done by adjusting the decay and settling rates (K_1 and K_3). Final CBOD decay rates range from 0.01 to 0.09 /day (Base e) on the Chicago waterway and are 0.01 /day on the Upper Illinois. The low decay rates reflect the highly treated effluent from the District's three large WRPs.

Prior to the start of the calibration and verification process, a mass balance of $CBOD_{20}$ was performed. $CBOD_{20}$ is not a conservative constituent and the total instream mass can be expected to be less than the total mass from the point sources. However, several mass balances indicated a greater amount of instream $CBOD_{20}$ in the lower reaches of the Upper Illinois River than can be accounted for from the known sources. The source of the additional $CBOD_{20}$ instream is hypothesized to be either from incomplete mass flushing from the long residence time (over three weeks) of the system or from decomposition of algal and zooplankton biomass which are high in the lower reaches.

Most of the carbonaceous oxygen demand from the District's plants is met in the Chicago waterway above the Lockport Dam. Though the concentration of $CBOD_{20}$ below the dam were often above 5 mg/l, the decay rates are low, 0.01 /day indicating that little dissolved oxygen demand is present.

Sediment Oxygen Demand: Oxygen demand by benthic sediments and organisms has historically represented a large fraction of the oxygen consumption in the waterways. QUAL2EU represents the demand with a constant rate of oxygen consumption. Measurements of SOD were taken throughout the study area by ISWS and were used as input into the model. SOD rates ranged from 0.4 to 4.4 $g/m^2/day$ in the study area.

Nitrogen: Nitrogen is simulated in four forms, organic, ammonia, nitrite, and nitrate. Nitrification is a two-stage process. The first stage is the oxidation of ammonia to nitrite by Nitrosomononas bacteria. Stoichiometrically, 3.43 grams of oxygen are consumed for each gram of ammonia-nitrogen oxidized to nitrite-nitrogen. During the second stage of nitrification, Nitrobacter bacteria oxidize nitrite to nitrate. Stoichiometrically, 1.14 grams of oxygen are consumed per gram of nitrite-nitrogen oxidized.

The concentration of total nitrogen (organic, ammonia, nitrite, and nitrate) in the effluent of the WRPS ranged from 4 to 7 mg/l at the Northside and Stickney plants and from 9 to 15 mg/l at the Calumet plant. Calibration to instream value of each constituent of nitrogen was done by adjusting the transformation rates, β_3, organic nitrogen to ammonia, β_1, ammonia to nitrite, and β_2, nitrite to nitrate.

As with CBOD, a mass balance of nitrogen was performed for each water quality sampling event. This balance indicated significant ammonia concentrations upstream of the Northside and Calumet Reclamation plants. In natural streams, ammonia can be released to the water column from the hydrolysis of organic nitrogen in the sediment. Past models of the Chicago waterway have included an ammonia sediment source (NIPC, 1981). The source of organic nitrogen, which releases the ammonia, is not necessarily from the WRPs but rather from historic deposition of CSOs discharge solids. A second source of ammonia upstream of the Calumet Reclamation plant, may be CSO discharges to the waterway. Uncontrolled CSOs are present in this reach that have a high frequency of discharge.

Nitrification produces the largest demand of oxygen in the Chicago waterway and is not present in the Upper Illinois River. This is in contrast to studies performed in the 1970s studies which found that nitrification was not present in the Chicago waterway, but was in the Upper Illinois River system (MWRDGC, 1975). Several changes have lead to this. First, the reclamation plants are nitrifying and do not chlorinate and as such discharge the nitrifying bacteria into the waterway. The nitrification process that starts in the plants continue in the waterway. Prior to this it would require a week before nitrification would start and the discharge would be in the Upper Illinois River System.

<u>Phosphorus:</u> Two forms of phosphorus are simulated in QUAL2EU, organic and dissolved (inorganic). The transformation of phosphorus does not consume oxygen, but is simulated as a nutrient for the algal production. As with $CBOD_{20}$, and the nitrogen series, phosphorus is input into the model from headwaters and point sources. The concentration of dissolved phosphorus in the effluent of the reclamation plants discharged typically ranged from 1 to 2 mg/l for the Northside and Stickney plants and 4 to 6 mg/l for the Calumet plant. Calibration of phosphorus was done by adjusting the organic phosphorus decay rate, β_4, the organic settling rate, σ_5, and the dissolved phosphorus source rate, σ_2.

A mass balance of phosphorus was performed for each of the six water quality sampling periods. The balance indicates that a significant amount of phosphorus is 'lost' in the lower reaches of the Calumet-Sag Channel. The removal of dissolved phosphorus is not common in natural river systems. The normal process is to have dissolved phosphorus released to the water column from the bottom sediments by the normal decomposition of organic matter in the sediment. However, the precipitation of phosphorus can be achieved with metals ions, such as iron, manganese, zinc, copper, and others. The precipitation can also be achieved with moderately high Ph and calcium carbonate, which co-precipitates phosphate with carbonates. This was hypothesized to be the mechanism present. The chlorophyll-a values measured in the Cal-Sag Channel indicate that there is not a sufficient biomass of algal to satisfy the imbalance in phosphorus.

<u>Algae:</u> The affects of algae on the dissolved oxygen concentration is the most complex of the processes simulated. Algae can bring significant changes in the dissolved oxygen concentration by several interactions. Algal dynamics and nutrients uptake during algal growth is the main process which removes dissolved nutrients, nitrogen and phosphorus, from the water column. Algal respiration and decay are major components of nutrient recycling. The major source of algae, unlike the other constituents, is not headwater and point sources. Instead, it is instream growth based on the availability of light and nutrients, nitrogen and phosphorus. Analysis of the data indicates that light is the limiting

factor for algal growth in the study area.

Calibration of algae to instream values was done by adjusting the maximum growth rate, μ_{max}, algal respiration rate ρ, and the algal settling rate, σ_1. A small algal population is brought in with the diversion water from the lake at each lock. Light is the limiting factor, and a slow growth of the population starts. The algal population growth remains constant in the waterway at slow rate. A noticeable increase in the growth occurs below Lockport Dam and the algal population becomes large in the Upper Illinois River system.

Dissolved Oxygen: The preceding paragraphs describes the interaction each constituent has dissolved oxygen. Final calibration and verification of the model was done by adjusting the atmospheric reaeration rate and each dams reaeration capacity. Reaeration rates ranged from 0.10 to 0.90 /day in the Chicago waterway and from 0.25 to 0.70 /day in the Upper Illinois River system.

The largest demand of dissolved oxygen in the Chicago waterway is from nitrification, with sediment and carbonaceous demands also being significant demands. The nitrificaeous and carbonaceous demand from the District's Water Reclamation Plants are exerted by the time the flows reach the Upper Illinois River system.

Summary

Improvements to the MWRDGC's water reclamation plants and construction of new facilities have changed the water quality processes that occur in the waterway and River system over the last ten years. Nitrification, which did not occur in the waterway, was found to be the dominate oxygen demanding process occurring. CBOD, formally the dominate oxygen demand, is no longer the major oxygen demand in the waterway. The SOD, long a major oxygen demand in the waterway, has been reduced with the implementation of, TARP, the deep tunnel program to control combined sewer overflows.

Several unusually processes were also found to occur in the waterway and river system. Large amounts of phosphorous co-precipitate with metal ions to the sediments, thus reducing the available phosphorous for algal growth. The concentration of CBOD increased in the Upper Illinois River with no significant source.

The QUAL2EU model was than used to simulate water quality under existing conditions and future conditions to examine the benefits that will be achieved by ongoing construction of TARP and SEPA. Also simulations were made under several extraordinary hydrological conditions to analyze the response of the waterway to point source loads.

References

Camp Dresser & McKee Inc, "Water Quality Modeling for the Chicago Waterway and Upper Illinois River Systems", January 1992.

NIPC, "NIPC Chicago Waterway Model: Verification/Recalibration", 1981.

MWRDGC, Facilities Planning Study, Overview Report, Appendix G, Section IV, January, 1975

OPTIMIZATION OF OPERATIONS - THE MANITOBA HYDRO EXPERIENCE

A.D. Cormie[1] and P.E. Barritt-Flatt[2]

Abstract

As part of a decision support system, Manitoba Hydro implemented a comprehensive optimization model for operations planning in the mid 1980's (Barritt-Flatt and Cormie 1991). Since that time the model has been used on an ongoing basis for scheduling of generation from Manitoba Hydro's two thermal generating stations, eleven hydro generating stations and reservoir system. During this period the system has experienced severe drought, floods in particular basins, and major forced outages of generating equipment. Throughout these events, the optimization model has been used to provide effective strategies for managing the operation of the system.

Why the Optimization Approach?

The optimization approach was chosen for operations planning at Manitoba Hydro because it provides both flexibility in terms of problem formulation and power in integrating the many factors that need to be taken into account in the preparation of an operating plan (Barritt-Flatt and Cormie 1988). For example, within the same problem the complicated hydraulics of the river system, the details of operating thermal generating stations, the complexities of the export/import market and the flow of power in the electrical system can all be considered. All these sub systems can be modeled with adequate levels of detail with the linear programming formulation.

---

[1] Reservoir and Energy Scheduling Engineer
[2] Assistant to the Executive Vice-President
Manitoba Hydro, P.O. Box 815 Winnipeg, Canada, R3C 2P4

How is the Optimization Model used?

As the HERMES (Hydro Electric Reservoir Management Evaluation System) optimization model was being developed, there were limitations on the size of the problem which could be solved. In order to minimize the problem size a good comprehensive understanding of the system was necessary so that only the key elements for each scenario were modeled. With faster computers and more powerful linear programming algorithms, this restriction has been removed. A complete model of the system is formulated independent of the scenario under study.

To develop an operations plan, optimization models representing different possible future scenarios of water supply and load are solved. Scenarios of water supply, lower or higher than forecasted are investigated in order to develop an operating plan which is robust under most conditions. For example, it is important when the system is energy short to know how the power system will respond if there is an extremely cold winter. Or equally, how will it respond if there is a very warm winter. The various scenarios are studied to ensure that contingent actions are in place for conceivable circumstances.

The focus is nevertheless on the expected case and its impacts. Extensive work on the expected case is undertaken to ensure that the data defining the model are as accurate as possible. This work is critical to creating confidence in the results. The results are reviewed with the staff in the control centre, so that they understand what is behind the plan and as a result have confidence in implementing the proposed schedules.

Manitoba Hydro has an obligation to notify others of its plans to regulate the lakes and rivers and provide what can be expected in terms of daily levels and flows. This information is generated by simulation models which utilize the optimization release schedules. These simulation models can calculate in much more detail the hydraulic response of the reservoir and river system to the proposed plan and specific impacts can be identified.

The optimization results for export revenue and generation costs are used in the preparation of the Corporation's annual budget. A "post-process" program uses the optimization results to prepare a report which becomes part of the Corporation's Integrated Financial Forecast. It is a credit to the comprehensiveness of the optimization model that results can be picked up and published in this way.

What is the experience in different circumstances?

HERMES is very powerful in allowing the user to analyze new and unexpected situations. The data can be changed within minutes to reflect new system conditions and new plans are available for assessment shortly thereafter.

For example, at the Corporation's Grand Rapids generating station this last spring, there was a total station shutdown when the headcover on one unit failed flooding the station. Within two days a revised long term operating plan for the power system was available which took into account that there was going to be a long outage.

With a new operating plan in place, we were able to answer questions like: "What is the value of accelerating the repair schedule? and "What is the likelihood of using the Grand Rapids spillway?".

Because the model of the system is comprehensive and the model evaluates the role of any specific site in relation to the whole, system-wide decisions can be made quickly and with confidence. Losing a generating station potentially affects very nook and cranny of the system. Release schedules from reservoirs had to be changed and maintenance schedules for the thermal stations were modified. With a comprehensive model capable of evaluating these changes the financial impacts of the failure were also available.

While much is done automatically, the HERMES system is not in the final analysis automatic. There is an ongoing feedback between the model and the operations planner. Problems are identified by studying the model's results. If necessary additional restrictions may have to be described in the data, for example, on the rate of change of flows from a spillway. In a matter of a few minutes, a new schedule is available recognizing this change.

Managing a drought or a flood, is fundamentally the same experience. The operations plan is based on an analysis of different possible water supply scenarios. As the signs point, for example, to a drought continuing, more attention is paid to the low flow scenario. The operations strategy gradually evolves from maximizing profit to conservation. The level of risk at which the system is being operated is considered very carefully and preparations are made to ensure operations under the worst case scenario are possible.

In the drought of 1987 to 1990, the worst year was 1988. The system had come through 87/88, which wasn't so bad because there was carry-over water in storage at the beginning of the year. However, at the end of that year the carry-over was gone, and in 88/89 it looked like there would be energy shortfalls if the drought continued. Manitoba Hydro made purchase arrangements with adjacent utilities based on the model results knowing when and how much energy could be needed.

Since the model is driven by an economic objective function, it utilizes the most expensive generation sources in an overall least cost manner. When it finally started to rain, the model predicted gradual change from a situation where the load was met by utilizing our generation reserve, to where revenues from export sales were being maximized.

What now?

From the modelling point of view, there is not much more that needs to be done to the HERMES system. There is a risk in going too far in automation which would isolate the user from the decision making process. There will always need to have somebody pouring over the results to say "This is a bad plan!".

APPENDIX I. REFERENCES

Barritt-Flatt, P.E. and Cormie, A.D. (1991). "Implementing a Decision Support System for Operations Planning at Manitoba Hydro", *in Computer Aided Support Systems in Water Resources Research and Management*, D.P.Loucks and J.R. da Costa eds., NATO ASI Springer-Verlag, 357-374.

Barritt-Flatt, P. E. and Cormie, A. D. (1988). "A Comprehensive Optimization Model for Hydro-Electric Reservoir Operations", *Proc. in Computerized Decision Support Systems for Water Managers*, J.W. Labadie, et al., eds., ASCE, 463-477.

IMPACTS OF HYDROPOWER OPERATION ON WATER SUPPLY FROM LOWER COLORADO RIVER IN TEXAS

Quentin W. Martin[1], M. ASCE

Abstract

The Lower Colorado River Authority (LCRA) of Texas is both a water and energy supplier to a large area of Central Texas. LCRA generates approximately 10 percent of its power from hydroelectric power plants on the six dams in the Highland Lakes system of reservoirs. To improve power production, LCRA has investigated alternative operating procedures to increase the winter scheduling of hydroelectric power generation in the upper reservoirs of the Highland Lakes system without adversely impacting available water supplies. A methodology using both optimization and simulation techniques was developed to evaluate the ability of the hydroelectric facilities to meet weather-related winter peaking requirements. A linear programming procedure determined the hourly power generation schedule, over a 24 hour period, that maximized the total amount of power generated over the six hours of peak power demand. The full installed capacity was found to be available during the peak hours without violating system operating constraints including water storage limits at the individual lakes. Based on statistical simulation of daily winter inflows and releases using a LOTUS 1-2-3 spreadsheet, it was found that the full generating capacity could be supplied to meet the weather-related peak winter power demand with no significant impact on water availability.

Introduction

The Lower Colorado River Authority (LCRA) is a water conservation and reclamation district created by the State

---
[1] Manager, Water and Wastewater Engineering Program, Lower Colorado River Authority, P.O. Box 220, Austin, Texas 78767, 1-512-473-4064, Fax 1-512-469-6873.

of Texas in 1934. LCRA also produces wholesale power for a 43 county electric service area in Central Texas. LCRA generates approximately 10 percent of its power from hydroelectric power plants on the six dams in the Highland Lakes system of reservoirs.

LCRA anticipates the need for additional peak power generation during the winter months. To meet this need using existing generating capacity, the author evaluated the maximum hydropower generating capacity and energy from the upper Highland Lakes (Figure 1) available to meet peak power generation needs during a typical winter season, without adversely impacting available water supplies. Only the upper lakes were considered since hydropower generation from Lake Travis can only be made under emergency conditions or for water supply deliveries.

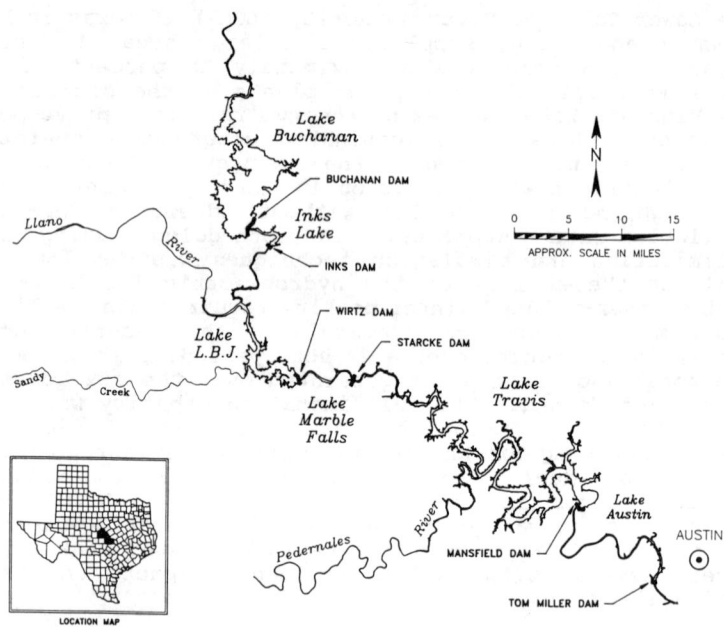

Figure 1. Highland Lakes in Central Texas

Solution Methodology

A methodology using both optimization and simulation techniques was developed to aid in this evaluation. The optimization component used linear programming to determined the hourly power generation schedule, over a 24 hour period, that maximized the total amount of power generated over the six hours of peak power demand. A daily simulation model was then used to evaluate the impacts of hydropower operations on water storage in Lake Buchanan over the three winter months.

Hourly Optimization Model

The hourly operation of Buchanan, Inks, Wirtz and Starke Dams for hydropower generation was formulated as a linear programming problem. Constraints on operation included maximum generation rates, upper and lower limits on the lake levels, and other limitations. The hydropower generation functions were replace by linear functions closely approximating the generation functions at the maximum level of power production for each generating unit. The solution of the linear programming model showed that the full 129 megawatts of installed capacity the four dams was available during the peak hours without violating system operating constraints (Figure 2). A daily release into Lake Travis of approximately 4,000 acre-feet is required to produce the full peak winter power generation.

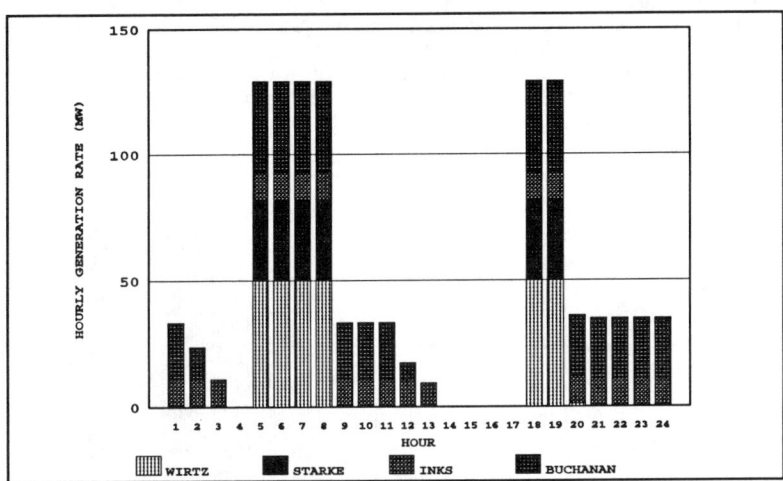

Figure 2. Power Generation at Buchanan, Inks, Wirtz, and Starke Dams Under Simulated Operation to Supply Winter Peak Generation of 129 MW.

Daily Hydrologic Simulation Model

It was necessary to simulation a range of winter conditions in order to determine the impact of hydropower releases on the water storage in Lake Buchanan, which is primarily a water supply reservoir. The PROPS stochastic spreadsheet program (1) (a LOTUS 1-2-3 addin) was used to simulate the daily operation of the four upper Highland Lakes as a generating system. Probabilistic functions were used to represent the timing of peak power demand days and volumes of daily river inflows. The stochastic spreadsheet was solved for a large number of equally likely winter conditions to generate a frequency distribution for the expected net storage change in Lake Buchanan (Figure 3).

Based on these statistical simulations, the full 129 megawatts could be supplied during the average 19 days of peak winter demand, with a net decrease in storage at Lake Buchanan of about 5,500 acre-feet over the winter (Figure 3). Based on a 92% probability level, the hydroelectric generation operation was estimated to cause a net storage decrease in Lake Buchanan of no more than about 24,000 acre-feet over the winter months. This is equivalent to the volume of water in storage in the top one foot of storage in Lake Buchanan. This volume of water may be readily moved into Lake Travis over the winter without having any significant detrimental impact on the water supply of the Highland Lakes.

Further analyses determined that an additional five days of full peak generation could occur without limiting water availability or causing more than an maximum one foot decline in Lake Buchanan's average water elevation. LCRA is currently modifying its power dispatching procedures to take full advantage of this additional power supply.

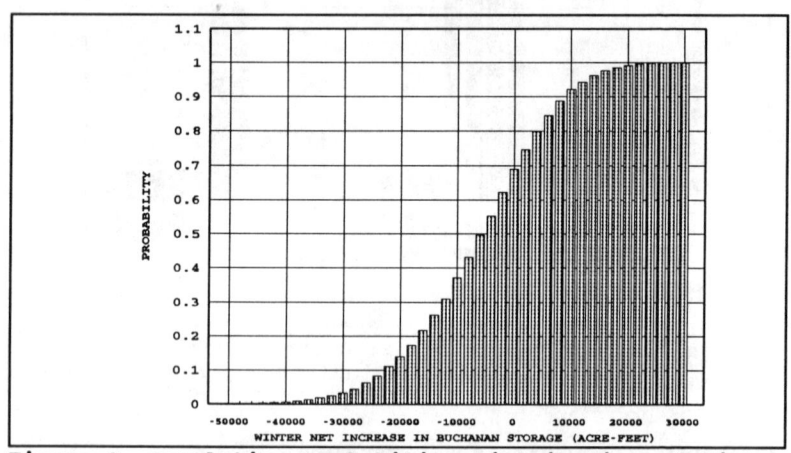

Figure 3. Cumulative Probability Distribution of Winter Net Change in Storage at Lake Buchanan

Appendix: References

1. Peterson, E.R. and A.J. Taylor (1988). "PROPS - Probabilistic Optimization Spreadsheets," Alwington Press, Kingston, Ontario, Canada.

OPTIMAL UTILIZATION OF A MULTI-RESERVOIR WATER DISTRIBUTION SYSTEM

Ali Diba[1],M. ASCE, Peter Louie[2], Manouchehr Mahjoub[3] and William W-G. Yeh[4], M. ASCE

Abstract

A water supply distribution system planning model (DSPM) has been developed using a linear programming approach. Unlike the conventional formulation, the water demand requirements are expressed as a component of the multi-objective function and not part of the constraint set. The model allows the use of preferential weights for the various components. The model has been successfully applied to the water distribution system of Metropolitan Water District of Southern California. It has been used to determine the system's conveyance and storage capacities before and after the proposed facilities; to identify the bottlenecks of the system; to evaluate the timing and amount of additional transfer water to be brought in to meet demands; and to estimate system's reliability by simulating system's disruptions.

Introduction

Given a redundant water distribution system (a network of pipelines and reservoirs in which demands can be met in many different ways), it is often difficult to determine the optimal operations of such a system. This task is further complicated when additional facilities (pipelines, treatment plants and reservoirs) are added to the system whereby the system capacity is altered. In turn, the operations of the system may need to be changed to optimally utilize the added facilities. The analytical tool developed can be used to evaluate the altered system's conveyance and storage capacities and assist in the determination of an optimal utilization of the system for a given

---
1. President, DCSE, 30 Corporate Park, Suite 314, Irvine, CA 92714
2. Senior Engineer, Div. of Planning, MWD of Southern California, Los Angeles, CA 90054
3. Senior Engineer. DCSE, 30 Corporate Park, Suite 306, Irvine, CA 92714
4. Professor, Dept. of Civil Engineering, UCLA, Los Angeles, CA 90024

set of supply and demand scenarios and objectives of the system's operations.

Figure 1 illustrates the distribution network which is supplying more than 60 percent of the total regional water use by the 15 million people in Southern California. The proposed facilities to the system are Inland Feeder and the Eastside Reservoir.

Problem Formulation

Objectives of System Operations

The major goal in operating the water distribution system is to ensure that the demands are met and at the same time the treatment plants and the power plants are not stressed to capacity limits. In addition, target storage levels are to be maintained as closely as possible. In brief the composite objective function contains the following components (Cohon, 1978; Loucks et al., 1981; and Yeh, 1985):

1. Maximization of total system storage.
2. Minimization of total weighted deviation from target storage.
3. Minimization of total shortage.
4. Minimization of uneven distribution of shortage (as a percent of supply to quantity demanded).
5. Minimization of flows exceeding plant capacities.

The decision variables are:

1. Flows from the various sources through the distribution network to the demand points.
2. End-of-period storages in various surface reservoirs and groundwater basins.

The constraint set includes the following:

1. Continuity equations for junctions and reservoir nodes.
2. Inflows from the sources should be less than or equal to the available quantities.
3. Pipe and storage capacities should be within operational limits.

Solution Approach

This problem with the aforementioned objectives and constraints forms a linear programming (LP) problem and can be solved by any

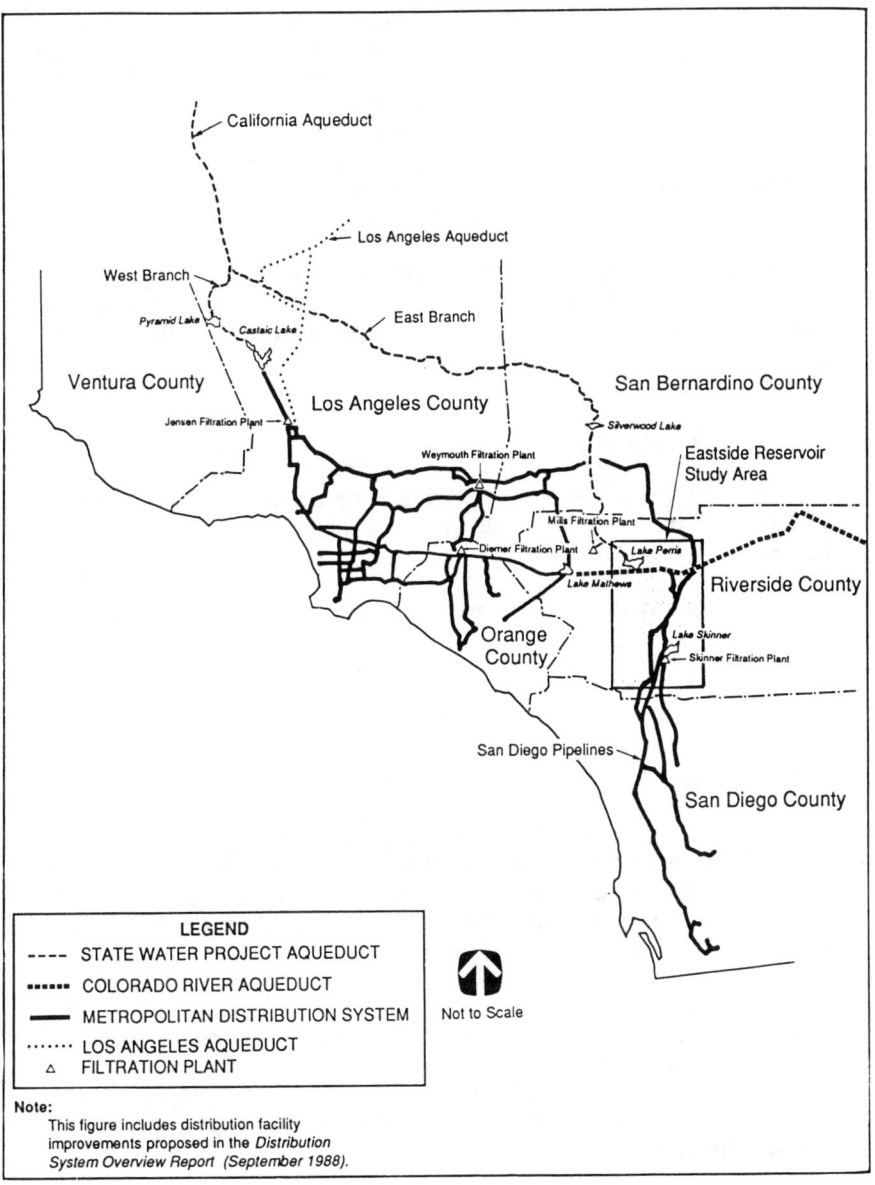

Figure 1 Metropolitan Water District's Distribution System and Eastside Reservoir Study Area

standard LP package (Murtagh and Saunders, 1987). However, unlike the conventional formulation, the objective function contains multiple components which included some of the elements that are traditionally expressed as system constraints (e.g., demands are represented by shortages and flow or storage limitations are given as targets). To express these multiple objective components as a single objective function, a great deal of care needs to be given to compute the proper preferential weights to reach the desired interactions among the components.

In addition, to allow for greater flexibility in reconfiguring the distribution system to reflect addition of new facilities as well as system disruptions, the Direct Graph algorithm is implemented as part of the overall approach.

Results of Case Studies

The DSPM can be applied to solve many different types of planning problems. In this paper however, four types of applications are discussed. These are:

1. Evaluation of system's capacity before and after the proposed facilities;
2. Identification of bottlenecks in the system;
3. Analysis of transfer water timing and quantities; and
4. Evaluation of system's reliability with component disruptions.

System Capacity Evaluation

In determining the beneficial impact of the proposed Eastside Reservoir and the Inland Feeder, the distribution system model was used to determine the additional level of demands that can be satisfied by the expanded system. The system capacity without the proposed facilities was found to be 2.9 MAF ($3.58*10^9$ m^3) per year; whereas the capacity would be increased to 3.3 MAF ($4.07*10^9$ m^3) per year with the additional facilities. The important point in this comparison is that in each case the system was being operated to its utmost for the given system configuration because of the optimization algorithm. It is conceivable that a simulation model may yield an inferior solution that leads to a lower level of system performance even with additional facilities. Therefore it would not be possible to compare the simulation results.

System Bottlenecks

The DSPM, using the Direct Graph algorithms, tracks the flows at various locations of the distribution system whose values are near pipe or treatment plant capacities (e.g., 80 or 90 percent). This tracking capability is extremely useful in determining the need for additional pipes and other facilities.

Analysis of Transfer Water

Transfer water refers to the quantity of water that is purchased from a "water market" at a premium price when there is a lack in water supplies. Therefore the timing and the amount of this purchase need to be considered with greater precision. Working with four different hydrologic sequences, model runs have been conduced to evaluate the timing and the amount that transfer water should be brought into Metropolitan's distribution system for a 15 year planning period. The first two sequences represent the dry hydrologic conditions. The other two represent the normal and wet conditions. The first dry case is characterized by a first-wet-then-dry pattern and the second dry case is reflected by the opposite pattern, i.e., first-dry-then-wet.

The model results indicate that only the dry cases required transfer water whereas the normal and wet scenarios did not call for any transfer water. For the first-wet-then-dry pattern, 3.54 MAF ($4.37*10^9$ m^3) of transfer water would be needed for the 15 year period. For the first-dry-then-wet pattern, only about 1.3 MAF ($1.60*10^9$ m^3) of transfer water was used and all of which came in during the first five years. From the average hydrologic standpoint the two dry sequences may be similar but yet the dry-wet pattern caused a drastically different outcome in terms of transfer water needs (Figure 2).

Evaluation of System Reliability

With the use of the Direct Graph algorithm in the DSPM, system configuration can easily be altered to simulate facility disruptions such as pipe breakages and treatment plant interruptions. For the given altered configuration, the DSPM would search for the optimal resource allocation pattern to meet the system objectives. Therefore by systematically examining various probable disruption scenarios, one can 1) assess the system's reliability in meeting the projected demands at an aggregate level and 2) determine the duration and magnitude of the shortages at various demand points of the distribution system.

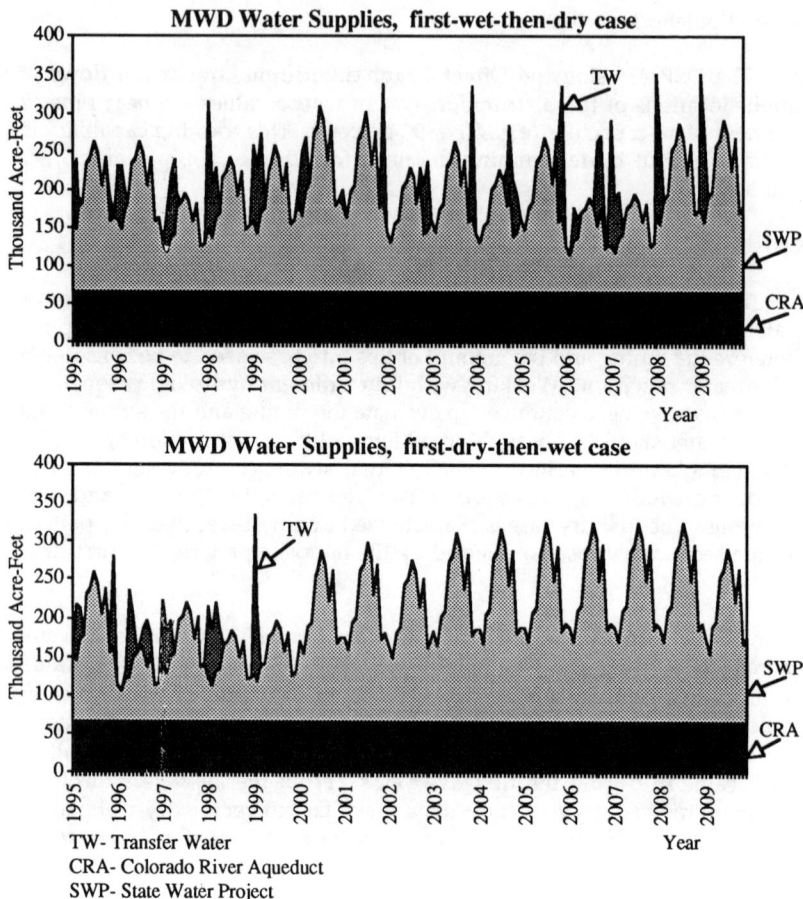

Figure 2 Transfer Water used for two different supply pattern

Final Remarks

The results indicated that the DSPM is a versatile tool for solving various planning problems surrounding a large scale water supply distribution system. These problems range from facility planning to system reliability to supply scheduling evaluations. With the multiobjective function formulation, the various objectives such as maximizing system storage, minimizing total shortage and others can be concurrently achieved. Consequently the resulting solution would reflect the best possible combination of these various operational objectives. Furthermore, the system behavior can be controlled by using the preferential weights associated with each of the objective components. These weights provide the planner with a much greater understanding of the interactions among the various objective components.

References

Cohon,J. L. (1978). *Multiobjective Programming and Planning*. Academic Press.
Loucks, D. P., Stedinger, J. R., and Haith, D. A. (1981). *Water Resources systems Planning and Analysis*. Prentice-Hall, Inc.
Murtagh, B. A and Saunders, M. A. (1987). *MINOS 5.1 User's Guide*. Department of Operations Research, Stanford University.
Yeh, W. W-G. (1985). " Reservoir management and operations models: A state-of-the-are review." *Water Resources Res.*, 21(12), 1797-1818.

Reliability Curves for Reservoir Performance

Loren Jangaard[1], M. ASCE

Abstract

Seattle District, U.S. Army Corps of Engineers, has developed a family of reservoir guide curves that illustrate reliabilities when storage conditions gradually decline from 98% to as low as 50% reliability. The reservoir reliability guide curves are a result of a statistical analysis of calculated reservoir storage values. This paper discusses how the reservoir routing calculation was performed for the storages and time periods of interest.

Location

The Green River is on the west side of the Cascade Mountains in Washington state. After flowing west from the foothills of the mountains, the river meanders along the industrial areas of the cities of Auburn, Kent, Tukwila, and Seattle (figure 1). Portions of the river banks are used for recreation. Many fish spawning sites are along the river's edge for about 30 miles between the dam and Auburn. Instream flows are naturally low during the summer and early fall. Low flows are augmented by releases from just over 24,000 acre-feet of active storage in Howard A. Hanson Dam. The dam is located at about the upper half of the watershed and has a drainage area of 220 square miles.

---

[1]Hydraulic Engineer for the Seattle District, U.S. Army Corps of Engineers, P.O. Box 3755, Seattle, WA 98124-2255. Telephone; (206) 764-3591, fax (206) 764-6678.

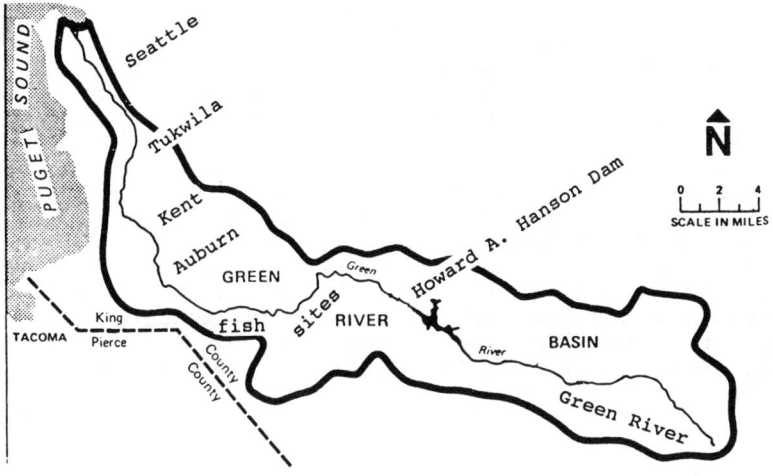

Figure 1. Green River Drainage Basin.

Reservoir Operation

The augmentation storage in the reservoir is intended to be 98% reliable during extreme low flow events. Currently, the reservoir is operated to a single 98% guide curve starting from a full reservoir condition in the spring. When inflow to the dam recedes to less than the low flow demand downstream from the dam, the reservoir level falls below the guide curve. Water resource engineers and biologists then become uncertain about the reliability of the remaining water. It's normal for the reservoir to occasionally drop below the guide curve, but what are the odds that augmentation be maintainable throughout the remainder of the season?

Water Demands

The city of Tacoma diverts 112 cfs for municipal and industrial consumption about ten miles below the dam. Water officials from Tacoma wanted to prepare an action plan to be used by their water division for extreme low flow events. A logical plan was to implement water use restrictions that would increase in severity at a parallel pace with the change in reservoir reliability. To accomplish this, more information was required on reliabilities less than 98%. Similar reliability information would be useful for instream flow. The instream flow demand is 110 cfs which is the design discharge used to size reservoir storage. The demand is usually set greater than 110 cfs when the fish sites are occupied with spawning activity, fish eggs, or emerging smolt. The reservoir has the capacity to supply a portion of the discharge beyond 110 cfs, but at reduced reliability. The normal diversion quantity plus the minimum instream flow together total 222 cfs. An additional 18 cfs was added to the demand because physical gate features and measurement error only allow a precision of approximately 18 cfs per 1% of gate change. A constant demand of 240 cfs was used for this analysis. Semi-monthly discharges were used for streamflow and reservoir routings. Semi-monthly inflows to the reservoir were computed for a 78-year period-of-record, 1914-1991. The outflow depends on the instream flow demand and an offstream diversion demand.

Reservoir Routing

Reservoir routing was iterated at multiple levels of water storage at each time interval. The routing was

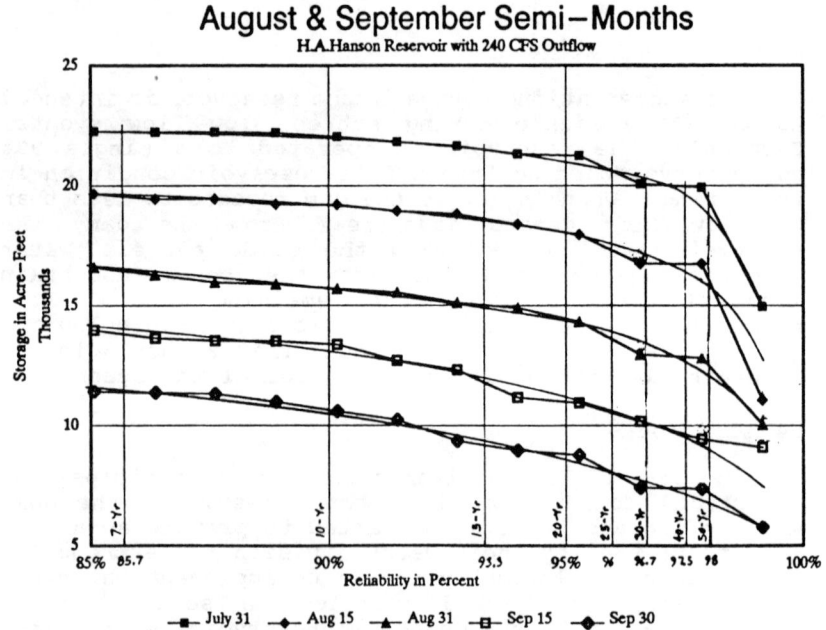

Figure 2. Example Storage Frequency Curves.
(See page 4)

designed in a spreadsheet format to compute reservoir storages from the given time series of inflow and 240 cfs outflow demand. The initial column was moved backward through the year and was changed by increments of about 2,000 acre-feet per iteration. The last row counted the number of failures for each test.

Stage Reliability Curves

Reliability in this study is frequency based instead of using a success ratio or flow-duration. Routed storage values were plotted for each semi-month on a log-probability graph as shown in figure 2 for semi months in August and September. Finally, a family of reliability curves were plotted against time and storage to obtain the reservoir reliability guide curves on figure 3. The additional reliability curves now answer the question, "What is the reliability of 240 cfs outflow given that the reservoir is observed at a particular height, and on a particular *date*?"

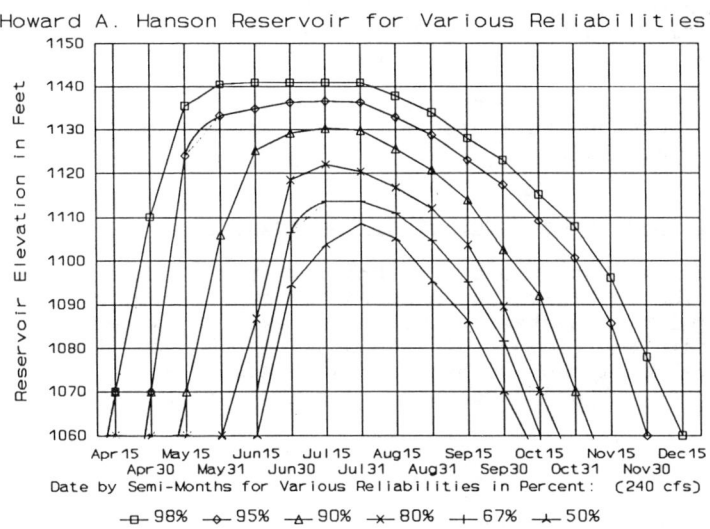

Figure 3. Hanson Dam Reliability Curves.

Geographic Information System Application:
Water Right Places of Use as Polygon Attributes

Tim Reierson, M.ASCE[1]
Hank Riddle[1]

Abstract

A Geographic Information System (GIS) was tested for potential use as a water resource management tool by application to a ground water study area. Because of the limitations of the particular software chosen and the complexity of water rights, emphasis was placed on constructing conceptual models for managing water right data using GIS technology. Implementation will be possible when tabular water right data is managed on a platform which will integrate with GIS software.

A generic method is presented for updating tabular GIS polygon attributes when changes to spatial polygon data are made.

GIS Requirements for Study Application

The Water Resources Program at the Central Regional Office in Yakima administers the state water right permitting system and regulates existing water rights within the region. In response to reported ground water level declines in the Black Rock-Moxee Valley, the department undertook a study of the area, which is

---

[1]Water Right Investigator and GIS Analyst, respectively; Washington Department of Ecology, Water Resources Program, Central Regional Office, 3601 W Washington Ave, Yakima, WA 98903-1164. Phone: 509/575-2800, FAX: 509/454-7830.

GIS WATER RIGHT MANAGEMENT   217

located east of Yakima, Washington.[2] The purpose of the
study was to characterize the aquifers so that decisions
could be made on pending applications for water right
permits. For the purposes of this paper, the method for
addressing wells will not be discussed in detail.

Places of use were more challenging to represent
than wells because polygons are more complex than points.
The major difficulty, however, was representing different
water rights having overlapping places of use.

Results of GIS Pilot

A necessary condition for implementing GIS was that
the spatial and tabular data could be managed easily and
without data conversions. The software chosen, PC
[Personal Computer] ARC/INFO® Version 3.4D,[3] did not
offer a relational tabular data model and therefore was
not an appropriate tool for on-going data management.
Table 1 shows some typical water right data relationships
which would need to be accommodated. An example of a
many to many relationship would be two wells serving two
different water rights.

DATA RELATIONSHIPS		
Relationship Type	Occurs With Water Rights	Handled by PC ARC/INFO®
one to one	✓	✓
one to many	✓	
many to one	✓	
many to many	✓	

Table 1. Water Right Data Relationships.

---

[2]Kirk and Mackie, 1992, Black Rock-Moxee Valley
Ground Water Study, unpublished technical report,
Washington Department of Ecology, Water Resources
Program, Central Regional Office.

[3]Copyright ©1990 Environmental Systems Research
Institute, Inc., 380 New York Street, Redlands, CA
92373.

Figure 1 illustrates how a relational data model would link spatial data (water right places of use, wells, and surface water diversions or pump sites) to tabular data records describing each spatial feature (polygons and points).

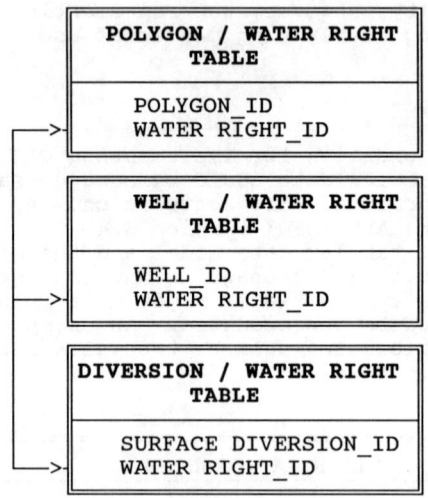

Figure 1.   Data Model Linking Spatial and Tabular Data.

Figure 2 is a generic flow chart which can be used to update tabular polygon attribute data to reflect changes made to polygons. Such a procedure is necessary when dealing with attributes of the same type but different identity which can overlap one another and can change in shape or area. ATTRIBUTE_ID is a polygon attribute identifier. A "displaced polygon" is a polygon whose size or shape was modified because of a spatial polygon update. For our application, the water right number was the attribute of interest. Other examples could be: a zoning designation, a taxing district, or a vegetation type.

GIS WATER RIGHT MANAGEMENT 219

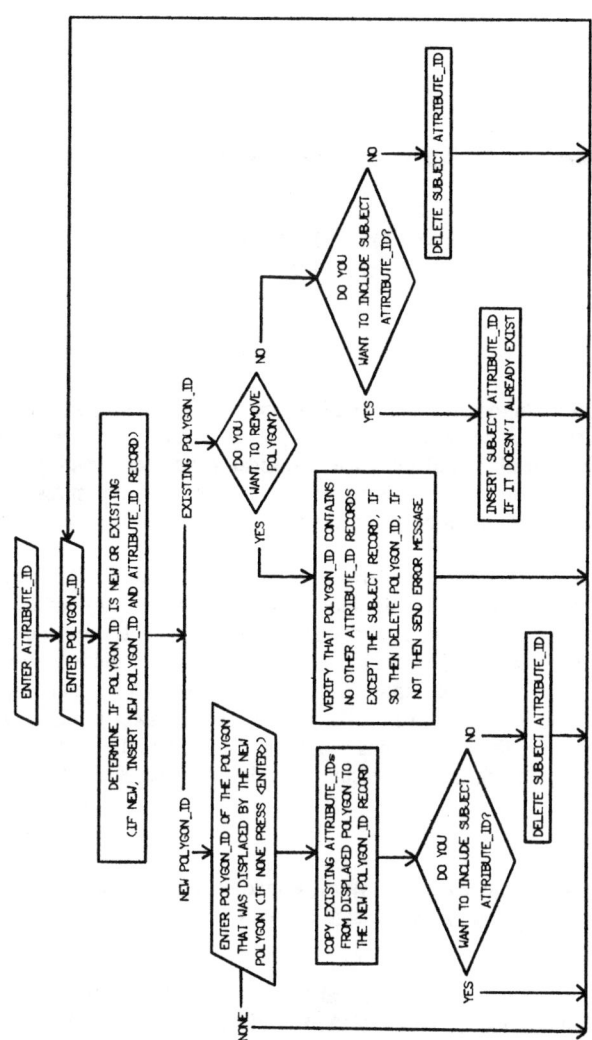

Figure 2. Flow Chart for Making Tabular Record Updates Based on Updates to Spatial Polygon Data.

GIS Reconstruction of Exposure to Water Supply Contaminants

Margrit von Braun, PhD, PE[1]
Ian von Lindern, PhD, PE[2]
Jim C. P. Liou, PhD, PE, M. ASCE[3]

Abstract

Groundwater pollution resulting from improper waste handling practices in the aerospace industry contaminated the water supply in Tucson, Arizona. State and federal health officials concluded that historical exposures could not be quantified at this Superfund site. This paper summarizes how exposures were estimated at the site using a Geographic Information System (GIS).

Using the GIS, cartographic models were developed to simulate i) contaminant release, ii) contaminant plume migration, iii) well pumpage, iv) water distribution system development and v) population growth. The models provided estimates of the degree and extent of contaminated water service over the twenty-five year history of the site. Stratified exposure estimates for the area population and for individual residents were developed. The results are discussed in the context of the site's case history, the implications supporting future health-effect studies, and the benefits of employing GIS in assisting complex environmental studies.

Introduction

Hazardous waste site cleanups typically involve multimedia analysis of contaminants in the atmosphere, groundwater, surface water, and soil. Estimations

---

[1] Assistant Professor, Department of Chemical Engineering, University of Idaho, Moscow, Idaho 83843; Phone: (208) 885-7838, FAX: (208) 885-7462
[2] President, TerraGraphics Environmental Engineering, 121 South Jackson Street, Moscow, Idaho 83843, Phone: (208) 882-7858, FAX: (208) 883-3785
[3] Associate Professor, Department of Civil Engineering, University of Idaho, Moscow, Idaho 83843, Phone: (208) 885-6202, FAX: (208) 885-6608

of the actual or predicted risk to human and environmental receptors associated with exposure to these contaminants are required by the Comprehensive Environmental Response, Compensation, and Liability Act of 1980 (CERCLA or "Superfund") to support site remedial decisions. Geographic Information Systems (GIS) are particularly useful for spatially and temporally integrating the results of environmental models for assessing exposure and risk. These systems can provide an evolving site-specific data base and a common format for the many types of data collected at a waste site to support exposure estimates.

The term "GIS" is used to represent a combination of software, including ARC/INFO©, pMAP©, SAS©, dBASE©, Lotus 1-2-3©, Golden Graphics© and AutoCAD© that access a common spatially-related data base. A variety of external, medium-specific environmental models (such as groundwater models to predict location of a contaminant plume) can be linked to the common format of the GIS.

Site Background

Years of improper solvent, fuel and industrial chemical disposal resulted in a large plume of contaminated groundwater underlying residential areas at the Tucson Airport Area (TAA) site in Arizona. In 1981, as a result of analyses required by the Safe Drinking Water Act, several public drinking water wells were found to be contaminated with trichloroethylene (TCE) and subsequently closed by health authorities. However, initial contamination reports preceded the well closures by nearly three decades. During this period, water containing varying degrees of contaminants was extracted from the aquifer and consumed by the growing community. Despite indications of adverse effects associated with past exposures, government officials cited the lack of adequate historical exposure estimates as precluding quantitative assessment of risks to human health (Caldwell, 1986; Goldberg, 1989; ATSDR, 1988).

Both current data from the Superfund Remedial Investigation and Feasibility Study (RI/FS) and historic information collected from local government, industry, and research institutions were developed and analyzed. Results from a series of external models were integrated using the GIS to infer past exposures. Computerized base maps of 1) Residential Locations, 2) Plume Locations, 3) Well Locations, and 4) Water System Layouts (and others) were created. These maps resulted from specific models of 1) the populations' residential locations over a 30-year period, 2) pollutant transport in the groundwater and 3-4) growth of Tucson's water distribution system, respectively. These base maps were combined and manipulated to create new maps that became part of the site data base. In this way, the data layers could be analyzed simultaneously in the overall model used to estimate exposures to contaminants from the polluted water supply. Intersections of numerous raw data planes were identified in a spatial and temporal "overlay and reduction" technique to arrive at the final combination of individual and population based contaminant exposures. The basic elements of the analyses included:

i) Historical Plume Location (i.e., "Where was the plume when?"): Groundwater flow and contaminant transport models were developed to predict future exposures and determine source contributions as part of the site Superfund effort (CH2M-Hill, 1987). The results of these efforts were used in conjunction with GIS cartographic analyses to project the plume back in time to 1950 in 5-year intervals. Plume locations under two different hydraulic conductivity (K) scenarios were developed to incorporate some of the uncertainty associated with the model.

ii) Contaminant Production by Area Wells (i.e., "How were wells affected?"): Using data from the City of Tucson, the uptake of polluted waters via municipal production wells and development of contaminant levels in wells was accomplished. This analysis used the well production, depth and construction data in conjunction with the historical plume location (from part i). Spatial and temporal variation was significant as different wells pumped different amounts at different depths throughout the 30-year period.

iii) Distribution of Contaminated Drinking Water (i.e., "Where did the contaminated water go?"): Historical water service areas were developed by cartographically combining the contaminant production (from part ii) with community demand and system hydraulics.

iv) Residential Exposure (i.e., "Who drank it?"): The analyses from part iii resulted in annual contaminant concentration estimates for each specific geographic area. These results were cross-indexed with maps of demographic characteristics to develop detailed exposure profiles. Once the conceptual model was applied for a given year, an iterative cartographic modeling procedure was applied to obtain profiles for each year. These results were then combined to provide estimates of the degree and extent of contaminated water service over time. Stratified exposure estimates for the area population and exposure profiles for individual residents were developed and presented for geographic areas, target populations, and individuals.

This analysis is provided in greater detail in (von Braun, 1989).

Discussion and Recommendations

One of the more difficult problems encountered in integrating various pathway models into a multimedia format is assessing and conveying uncertainty. The results can be no stronger than the weakest link in the chain, and uncertainties in one link can be perpetuated or magnified through the system. On the other hand, these integrated systems allow for more sophisticated sensitivity analyses because variable effects on one medium can be assessed in outcome estimates of other media. For example, at the TAA site, assumptions about hydraulic conductivities could be evaluated directly in terms of their effect on household exposures many years later.

The GIS was used to create static representations or "snapshots" in space and time by incorporating the results of specific external models. Although some GIS users favor the development of dynamic and inclusive multimedia models within GIS with 3-D capabilities and enhanced graphics, such models run the risk of implying a false certainty and becoming "black box" applications. Being able to portray the growth of the plume and the transport of contaminants in the Tucson water supply dynamically over the 30-year period might be helpful for visualizing the situation, however it seems that it would imply a false accuracy and precision to the viewers. Exposure and risk analyses are often deliberately presented simply to avoid such misconceptions. The routines employed within GIS applications are often more flexible but less rigorous than those in dedicated external models. The uncertainty of most external models is more easily quantified. The responsibility of quantifying uncertainties in the external models must fall to and remain with the scientist user. As a result, more reliable and credible results can likely be obtained through use of outside dedicated models. GIS techniques can then be used to integrate the results and inputs required. How to convey this uncertainty in GIS systems is an area requiring investigation and research in coming years.

Appendix - Cited References

Caldwell, Glyn. Mortality Rates on Tucson's Southside. Arizona Department of Health Services, Phoenix, Arizona. 1986.

Goldberg, S. J. Human Cardiac Teratogenesis of TCE. Progress Report to the Arizona Disease Control Research Commission. January 8, 1989.

Agency for Toxic Substances and Disease Registry (ATSDR). Health Assessment for the Tucson International Airport Site, Tucson, Arizona. Atlanta, Georgia. 1988.

CH2M Hill. Draft Assessment of the Relative Contribution to Groundwater Contamination from Potential Sources in the Tucson Airport Area, Tucson, Arizona. Prepared for the U.S. Environmental Protection Agency Region IX, San Francisco, California. Contract No. 68-01-7251. 1987.

von Braun, M. C. Use of a Geographic Information System for Assessing Exposure to Contaminants Released from an Uncontrolled Hazardous Waste Site," PhD Dissertation, Washington State University, 1989.

COMPUTERIZED DATA PROCESSING AND GEOGRAPHIC INFORMATION SYSTEMS APPLICATIONS FOR DEVELOPMENT OF A 3-DIMENSIONAL GROUNDWATER FLOW MODEL

Shih-Huang Chieh, Ph.D., P.E.[1], Marc V. Cromer[2], and William R. Swanson, P.E.[3]

Abstract

A 3-dimensional numerical groundwater flow model was developed for a 100 square mile area located in the Santa Clara Valley, California. The model development relied upon data base and Geographic Information System (GIS) processes to evaluate hydrogeologic conditions, define the model structure, and to distribute model parameters throughout the model domain. Approximately three hundred well logs, including location and construction data, lithologic data and water elevation data were available within the study area. In additional to the well logs, numerous groundwater extraction records were available from several decades of groundwater production in the Santa Clara Valley.

GIS applications were developed to display hydrogeologic data along arbitrarily defined cross sections, calculate model parameters at individual wells, distribute the paramters throughout the model domain, and to select wells to be used during transient calibration. Using developed GIS applications, a 3-dimensional flow model structure was defined to reflect the observed hydrogeologic conditions. The model consists of 129 rows and 89 columns, has four layers, and is used to simulate groundwater flow under different aquifer stress conditions and provide boundary conditions for a solute transport model.

1, 2, 3.   Supervising Engineer, Senior Engineer, Supervising Engineer, James M. Montgomery, Consulting Engineers, Inc., 365 Lennon Lane, Walnut Creek, CA 94598, Tel. (510) 975-3400.

Introduction

This paper summarizes the data processing and GIS applications for the development of a 3-dimensional numerical groundwater flow model. The model was developed within a geologically complex alluvial basin in Northern California. Investigations by several parties within the study area have provided abundant, high quality lithologic data from numerous soil borings and well development activities. The groundwater flow model was developed to provide a better understanding of the regional hydrologic framework and to evaluate the effect of local hydrogeologic anomalies on regional flow patterns. A variably spaced model grid was designed to reflect the density and quality of the available subsurface data. Horizontal dimensioning of the finite-difference grid network from 100 to 2000 feet was necessary to capture local structural detail.

GIS technology integrates computerized databases with graphic display systems and enables the evaluation of spatially distributed data. A GIS manages the linkage between the physical location of an object (such as a well or study area) and any information associated with the object. Custom GIS applications may be developed to provides rapid access to specific information that a user requires.

Hydrogeologic Setting

The modeled area is located in the northwestern Santa Clara Valley between the southern edge of the San Francisco Bay and the Santa Cruz Mountains in Northern California. The groundwater basin is defined by the contact between the valley alluvial soils and the bedrock of the surrounding hills and mountains. The bedrock materials, for the purposes of this model, were considered non-water bearing relative to the alluvial soils. The water-bearing alluvium in the model domain consists of a deep confined aquifer and a shallow unconfined aquifer. The separation of these two aquifer systems is based on the areal extent of a regional confining layer.

The structural setting of the Santa Clara Valley provides a foundation for understanding the evolution of the sedimentary basin. Approximately 600 to 800 feet of water bearing alluvium overlies a faulted bedrock surface, which is down-dropped relative to the surrounding uplifted mountains to the west and to the east. Deposition of the alluvial sediments upon this down-dropped block occurred concurrently with subsidence caused by compaction and tectonic activity. A nearly constant rate of uplift and subsidence during a relatively short geologic time period is one of two important factors controlling erosion, sediment transport, and deposition. The second factor is sea-level fluctuation resulting from the effects of global climatic change. Coincident with the period of uplift and alluvial deposition of the basin, sea level fluctuations in the vicinity of the current Santa Clara Valley caused the shoreline to vary significantly.

During glacial periods, fine grain sediments were deposited throughout the area, whereas coarse grain alluvial materials were deposited in stream channels and splay deposits during interglacial periods.

For the purpose of developing a conceptual model of the groundwater flow in the upper portion of the water bearing units, the upper 250 feet of the alluvium was subdivided into four stratigraphic units.

Model Development with GIS

The selected regional model area covers approximately 100 square miles in the northwest and west-central portions of the Santa Clara Valley. The model is discretized horizontally with 129 rows and 89 columns and vertically with 4 layers.

Approximately 300 well logs from various sources were available within the study area. These consisted of drilling logs prepared by drillers or geologists, and a limited number of electric logs. All well location, construction, and lithologic data were entered to a database and linked to a GIS. Specifically, data that were managed through the GIS include:

- geological and stratigraphic data from soil borings, monitoring wells,
- groundwater elevations, stream and river flow, precipitation data,
- well construction data from drillers logs and well construction diagrams,
- groundwater extraction data.

To accomplish the development of the model, several GIS applications were specifically designed to process, display, and update data in the spatial and relational databases. The GIS was used to define the model layer structure and to evaluate the distribution of hydraulic conductivity and transmissivity, specific yield and storage coefficient, precipitation recharge, and groundwater extractions.

The model structure was defined using a series of cross-sections. Lithologic data were projected to the sections and model layers were graphically defined along each cross-section. When the cross-sections were complete, the model structure data were numerically distributed throughout the model domain. Horizontal hydraulic conductivities and transmissivities were initially derived at each boring location using lithologic logs and site-specific aquifer test data. The lithologic data at each well was processed on a layer by layer basis using material specific hydraulic conductivity values derived from pumping test analyses. An inverse-distance weighting interpolation procedure was used to spatially distribute hydraulic conductivity and transmissivity values throughout the model domain for each layer. The specific yield and storage coefficient at each model node was processed using a similar procedure.

Rainfall recharge is a significant driving force in all hydrologic models, and is represented by mass source term directly applied to each model cell. Using historical rainfall frequency data from the San Jose Rainfall Station and an isohyetal map of the San Francisco and Monterey Bay region, a value for direct runoff was calculated at each cell using a rainfall-runoff relationship method developed by the U.S. Department of Agriculture, Soil Conservation Service. The relationship was derived from experimental data for numerous soils, vegetative cover, and land-treatment measures. An algorithm for this relationship was developed and the calculation procedure was designed using a combination of GIS processes and FORTRAN based pre- and post-processing programs.

The Santa Clara Valley Water District provided semi-annual pumping data for approximately 200 wells within the study area. An average frequency distribution was developed chacterize expected pumping patterns during the 4 year simulation period. This statistical processing allowed an for an approximation of monthly activity, which was required by the model. Well construction data, particularly well screen depths, were assessed to properly distribute the pumping contributions expected from each model layer so that mass balance and aquifer stresses could be accurately simulated.

Conclusions

The use of state of the art, low-cost computer technology made it possible to process the hydrologic/hydrogeologic parameters at each of the 45,924 model node locations. Personal computers and UNIX based workstations were used to construct and manipulate data files which often exceeded several megabytes in size. The advantages of these tools were particularly evident in the development of input files for 48 transient time periods.

The interactive GIS and data base applications used during model definition allowed for an efficient assignment of initial parameters for this large groundwater flow model. Both steady-state and transient calibrations were completed with limited adjustment of the model parameters. This is directly attributed to an accurate initial model definition, which was accomplished using the spatial and numerical data processing features of GIS technology.

References

McDonald, M.G., and A.V. Harbaugh, 1988. A Modular Three-Dimensional Finite-Difference Groundwater Flow Model (MODFLOW), Washington, D.C.

Santa Clara Valley Water District, 1977. Mean Annual Precipitation Map for San Francisco and Monterey Bay Region," October 1989.

U.S. Department of Agriculture, Soil Conservation Service, 1968. "National Engineering Handbook, Section 4 Hydrology," March 1985

Stochastic Forecasts in Reservoir Planning and Operations

by

Charles D.D. Howard, M.ASCE
Consulting Engineer
Victoria, BC

ABSTRACT

The factors involved in stochastic forecasts of water supply and demand are discussed in relationship to reservoir planning and operating decisions.

INTRODUCTION

Forecasts of runoff and demands for water are inherently stochastic - they are probabilistic and they take the form of a sequence of partially correlated events. Over the very short-term the correlation is stronger because it is driven by current events. The same is true over the long-term because of the annual cycle of the weather. Between these extremes these correlation is weaker. This complicates analysis.

Sequences may cross over from drought to flood and back again, or from periods of rising demand to intervals of decline. In water resource systems this behavior determines the benefits that can be realized by good water management.

DEMAND AND SUPPLY

The stochastic behavior of supply forecasts has almost always been a major consideration. But in practice, the projections of demands often ignore their sequential variability by selecting alternative high, low, and medium demands for planning, or by using the most recent experience for operations. In fact, demands may be more variable than supplies, and so are just as deserving of statistical respect.

The demand for water is often poorly correlated with supply, tending to be inverse. The degree of correlation and the stochasticity, depends on the scales of time and space that are used in the analysis and the frequency by which decisions are made.

Supply and demand forecasts are preferably generated together to provide a matched set that has a realistic internal correlation. This can be a difficult task because of the factors affecting water demands include societal behavior that is not easily modeled for water resource systems analysis.

For reservoirs the demands and the supplies must both be considered because their intersection determines the plan of operations. Management of demands during drought brings first hand recognition of their jointly intrinsic unpredictability.

PROBABILITY DISTRIBUTIONS

The sequence of a particular series of supply and demand has an undefinable probability, but the totals between specified dates describes an event that can be analyzed. The outcomes from reservoirs operating decisions made NOW include the current deterministic benefits from the recommended decision plus the conditional probability distribution of future volumes of storage, spills, and the future benefits.

STOCHASTIC OPTIMIZATION

Stochastic forecasts are used in stochastic optimization procedures to provide the best bet decision for current reservoir and demand management. In operations purposes the conditional probability distributions are updated at each time increment to account for recent changes in actual conditions of supply and demand. For example, a demand management action may reduce demands significantly, or the weather may change. This will lead to higher storage volumes in previously estimated conditional probability distributions. Current events will now provide new conditional forecasts, and the new initial reservoir volumes. As the time horizon rolls ahead, the conditional probability distributions evolve, and the stochastic optimization procedure recommends the new best bet for the deterministic decision. Time NOW may be reset on a regular schedule, like every Thursday afternoon, or each time that a significant event occurs.

The analysis is more complex for planning. In addition to determining the optimal operating decisions for existing facilities, the planning analysis must anticipate how closely these will be followed. It must consider uncertainty in recommending the timing, scale, and operation of new projects, and their effect on operation of the entire system of supply and demand.

OBJECTIVE FUNCTIONS

Stochastic considerations add the dimensions of reliability and risk, and complicate the discounting that determines the present value of net benefits. At bottom it is the risks that reservoirs are intended to manage. Stochastic forecasts are primary inputs to optimization methods for calculating and minimizing risks that can be quantified or the product of the value lost and the probability of the loss. Both aspects of risk are incorporated in the objective function as departures from prescribed goals of multiple uses for the reservoir.

The optimization method includes an objective function to be maximized within constraints that describe the physical and institutional limitations of the system. The constraints are both stochastic for describing supply and demand, and deterministic for factors that can not vary.

OPTIMAL DECISIONS

Since the key inputs are probability distributions, the risks are reported as a set of probability distributions, one for each goal. These distributions are conditional on the initial conditions at time NOW, and the decision to adopt the recommendation from the systems analysis. If the recommendation is not adopted, the risks will be different and as initial conditions change so do the risks.

Usually the recommendations from the systems analysis will be followed, because their is no better source of advice. But sometimes this is impossible because of equipment failures, or human error. In operations the analysis and the decision are continually updated so this deficiency is easily overcome. But in planning it may not be possible to compensate completely for such misalignments between the analysis and the decisions.

Characterization of Uncertainty in Reservoir Operations

by

Lindell E. Ormsbee
University of Kentucky, Lexington, Ky 40506-0046

Introduction

The efficient operation of complex reservoir systems remains one of the most interesting and yet daunting problems faced by water resource engineers. The development of operating policies for such systems typically requires a consideration of various complex system component interrelationships as well as the prospect of several conflicting operating objectives. However, perhaps the most difficult facet of the problem is the requirement to develop a future operating policy in the presence of uncertain future inputs (or streamflows). This paper will examine the various ways that uncertainty has been incorporated into the development of reservoir operations while at the same time documenting several actual applications.

Real Time Operation With Rule Curves

Attempts to incorporate the uncertainty of future system inputs (inflows) into the resulting operating policy have undergone significant evolution over the past decades. Early attempts to address the problem focused on the development of rule curves for system operation. In most cases such rules curves are expressed in terms of reservoir storage. Rule curves may be derived using mean seasonal inflows or a specific historical streamflow sequence coupled with a deterministic optimization model or associated heuristic. Alternatively, rule curves may be developed by use of a simulation model coupled with multiple streamflow characterizations. In the latter case actual historical streamflows can be used or the potential operating policy can be evaluated using "synthetic streamflow" sequences which are generated using statistical models of the underlying streamflow process. Once the rule curve has been developed, reservoir releases are determined in response to future inflows so as to minimize the deviation of resulting storages from the operational rule curve.

Attempts to develop improved rule curves that consider the impact of additional random hydrologic influences have focused on ways to include probabilistic characterizations of streamflow directly within the associated optimization program. Such approaches have include the use of chance-constrained programming, and stochastic dynamic programming. For a detailed discussion of such methods see the excellent review by Yeh (1985).

Real Time Operation with Forecasts

While the before mentioned methods provide a methodology for generation of a general rule curve, they do not yield themselves to real time operation applications. Methodologies for generation of real-time operation policies will typically rely on the translation of a future hydrologic streamflow forecast into an associated operating policy. Such policies may be characterized as either short term or long term policies depending upon the duration of the policy and the associated streamflow forecast.

Short Term Forecasts

Short term operating policies will normally be associated with durations of between one to three days. Such short term policies can be represented deterministically in terms of a single operating policy or stochastically in terms of a probability distribution of potential releases.

Deterministic policies may be obtained by using an optimal policy algorithm coupled with a representative streamflow forecast. Such forecasts can be obtained by selection of a previous representative historical streamflow sequence or by using a time series model of the underlying streamflow process. Alternatively, the forecast can be obtained using a hydrologic model of the upstream watershed system coupled with forecasts of meteorological data such as temperature and rainfall. Such models may range from the use of correlative models (Stedinger, et al. 1988) or transfer function models (Awwad and Valdes, 1992) to very complex causal models. In applying the causal models, the model is first calibrated using historical streamflow and then adjusted to reflect current conditions using past meteorological data. Once the model parameters have been adjusted to reflect current conditions, a short term deterministic forecast can be made by using forecasted meteorological data. By assigning probabilities to the associated meteorological forecasts, probabilities associated with the resulting streamflow forecasts may be obtained. Such forecasts can then be used in subsequent policy models to obtain a single optimal policy or a range of optimal policies with associated probabilities. Example applications of the use of causal models in real time streamflow forecasting include HEC1F (Charley and Peters, 1988), HSPF (Bradely et al, 1988), SAARR (Rockwood, et al, 1988), and SSM (Widener, 1992).

Long Term Forecasts

Unlike short range operation policies, long range policies may be associated with forecast durations of a week or several months. Since quantitative measures of the causative meteorological parameters associated with such forecasts cannot be accurately predicted beyond a few days, long range streamflow forecasts must rely on the use of historical series as potential samples of future events. Such long range forecasts may be obtained by modification of actual historical streamflow series (Van Do and Howard, 1988) or by conversion of historical meteorological series using a causal hydrologic model (Day, 1985). A deterministic forecast may be obtained by prior selection of a single historical streamflow or meteorologic series (Georgakakos, 1992) while a stochastic forecast may be obtained by processing a complete set of streamflow or meteorologic series and then assigning probabilities to the each resultant forecast (Randall and McCrodden, 1992).

Summary

In recent years, various algorithms and methodologies have been proposed for use in incorporating forecast uncertainty into reservoir operations. While future refinements to such methods can be expected, it should be recognized that the real utility of any methodology will not be necessarily dependent upon its mathematical elegance but its useability by system operators. In the final analysis, the real challenge of systems analysis as it applies to reservoir systems may not lie in the development of more sophisticated computer algorithms, but in the development of more efficient strategies and programs for their implementation and use.

References

1. Bradley, A.A., Anich, L. A., Crawford, N. H. (1988) "Streamflow Forecasting for Hydropower Operation Using Interactive Modeling Systems," *Proceedings of the ASCE Specialty Conference on Computerized Decision Support Systems for Water Managers*, Colorado State University, Fort Collins, Colorado, June 27-30.

2. Charley, W. J., and Peters, J. C., (1988) "Development, Calibration, and Application of Runoff Forecasting Models for the Allegheny River Basin,"

Proceedings of the ASCE Specialty Conference on Computerized Decision Support Systems for Water Managers, Colorado State University, Fort Collins, Colorado.

3. Day, G. N., "Extended Streamflow Forecasting Using NWSRFS", *ASCE Journal of the Water Resources Planning and Management Division*, Vol. 111, No. 2., April, pp. 157-170.

4. Georgakakos, A. P., (1992) "Computer-Aided Management of the Savannah River System," *Proceedings of the ASCE Workshop on Water Resources Operation Management*, Mobile, Alabama, March 16-18.

5. Randall, D., and McCrodden, B. J., (1992) "Modeling the Savannah River System for Improved Operations," *Proceedings of the ASCE Workshop on Water Resources Operation Management*, Mobile, Alabama, March 16-18.

6. Restrepo, P.J. Eom, K., Valdes, J. B., and Jourdan, M. (1992) "Real Time Streamflow Forecasting and Control on the Han River Basin, Korea," *Proceedings of the ASCE Workshop on Water Resources Operation Management*, Mobile, Alabama, March 16-18.

7. Rockwood, D. M., Speers, D, D., and Davis, E. (1988) "Operational Forecasting of a Complex River Basin with SSARR-Micro Versus Mainframe Computer," *Proceedings of the ASCE Specialty Conference on Computerized Decision Support Systems for Water Managers*, Colorado State University, Fort Collins, Colorado, June 27-30.

8. Stedinger, J. R., Grygier, J., and Hongbing, Y., (1988) "Seasonal Streamflow Forecasts Based Upon Regression," *Proceedings of the ASCE Specialty Conference on Computerized Decision Support Systems for Water Managers*, Colorado State University, Fort Collins, Colorado, June 27-30.

9. Van Do, T, and Howard, C.D.D. (1988) "Hydro-Power Stochastic Forecasting and Optimization," *Proceedings of the ASCE Specialty Conference on Computerized Decision Support Systems for Water Managers*, Colorado State University, Fort Collins, Colorado, June 27-30.

10. Widener, D., "Streamflow Forecasting at Alabama Power Company," *Proceedings of the ASCE Workshop on Water Resources Operation Management*, Mobile, Alabama, March 16-18.

11. Yeh, W., (1985) "Reservoir Management and Operations Models: A State-of-the-Art Review," *Water Resources Research*, Vol. 21, No. 12, pp 1797-1818

Hydrologic Forecasting - What are the Issues?

L.E. Brazil[1], D.P. Laurine[2], G.N. Day[3], and J.B. Valdes[4]

Introduction

The purpose of this paper is to briefly describe hydrologic forecasting techniques that are being used today and raise issues that should be addressed as new hydrologic forecast systems are developed and implemented. The issues represent concerns that will become more important as operations begin to take advantage of some of the new technology that is becoming available to make better forecasts. A list of references on hydrologic forecasting is provided at the end of the text.

Hydrologic Forecasting Techniques

Hydrologic forecasting is the estimation of hydrologic conditions for a given time in the future. Forecasting techniques are often categorized into short- or long-term procedures. Short-term forecasts typically range from a few minutes to a few days in the future. They are based heavily on initial conditions and are used for flood warning or real-time hydrologic system operations. Long-term forecasts usually have lead times greater than a week and often are for periods extending several months into the future. Long-term forecasts are used for planning or water management considerations.

Hydrologic forecasting techniques usually involve the use of some type of rainfall-runoff model and may also include models for simulation of snowpack and channel routing. The discussion in this paper will focus primarily on rainfall-runoff modeling. The models may be classified many ways. Rainfall-runoff models are typically physical, physically-based (mathematical or conceptual), or stochastic. The

---

[1] Director, Water Resources Engineering, Riverside Technology, inc., 2821 Remington Street, Fort Collins, CO 80525
[2] Hydrologist, National Weather Service, Colorado Basin RFC, 337 N. 2370 W., Salt Lake City, UT 84116-2986
[3] Director, Water Management and Forecasting, Riverside Technology inc., 2821 Remington Street, Fort Collins, CO 80525
[4] Professor, Texas A&M University, Department of Civil Engineering, College Station, TX 77843-3136

physical models are scaled representations of the true physical processes. Conceptual or physically-based models represent the physical components with mathematical equations describing the hydrologic processes. Stochastic models provide a statistical treatment of the systems on the basis of the stochastic properties of the hydrologic variables. Models also can be categorized according to the types of necessary inputs and processing. Early models were typically driven with lumped inputs. As data processing speeds have increased and data resolution has become more refined, models are being developed to utilize enhanced distributed inputs.

Hydrologic forecasting techniques are also categorized according to the application of the models or systems. The systems may be used for real-time operations or as planning tools. In some cases the same models can be used for both applications. Similarly, rainfall-runoff models may also be categorized as being intended for use in research or operational forecasting. Research models are used to study concepts or provide a better understanding of the hydrologic cycle. Operational forecasting models usually represent mature technology that can be used to provide information about the future occurrence of hydrologic conditions.

The performance of a model is limited by the quality of its input, e.g., precipitation, temperature, evaporation, river stage/discharge, snow water equivalent, and soil moisture. The models often require distributed or areal estimates of these variables, and the estimates are dependent on the ability to observe the variables and to perform spatial interpolation and averaging. Spatial and temporal resolution are improving in many of these measurements; however, great care is necessary to maintain the quality of the data.

Issues

Several issues need to be addressed in the evaluation of a hydrologic modeling or forecasting system.

- What kind of forecast information do users need?
  - Timeliness
  - Frequency
  - Accuracy
  - Uncertainty

- How should procedures, systems, and/or models for hydrologic forecasting be selected?

- Should users (utilities, water supply agencies) invest in better forecast models or better system operations models for optimization and simulation?

- What are the economics of using a particular model?
  - Resources to calibrate?
  - Long-term maintenance?
  - Can simple models provide enough information?
  - How is the hydrologic system defined, and are reasonable inputs available?

HYDROLOGIC FORECASTING

- Who will operate the system?
- How easily can the system be updated to represent current conditions accurately?
  - Continuous simulation versus event-driven?
- What is the value of hydrologic forecast information?
  - How can the value be estimated?
- How should performance in hydrologic forecasting be measured?
- How can the proper resolution be selected to describe the problem?
  - Use of GIS information?
  - Does a better understanding need to be made in varying space-time scales before utilizing models of higher resolution?
- How should forecast information be used effectively in operations?
- How can probabilistic forecast information be integrated into operations?
- How can hydrologic and meteorologic forecast information be integrated?
- How will improved data sources (i.e., NEXRAD) affect hydrologic forecasting?
  - Long-term forecasts - QPF?
  - Estimation of snow data - optimal interpolation?

Future Research

Technology has played a major role in shaping the evolution of hydrologic forecasting techniques. The advent of the digital computer encouraged the development of conceptual models that continuously simulate the hydrologic states of a system. High-powered, UNIX-based workstations have led to the development of forecast systems that provide forecasters with an interactive windowing environment for the graphical display and analysis of real-time hydrometeorological data. Although some data networks are seriously deteriorating, more data are now available for forecasting than in the past. In particular, new automated gage networks, radar, satellite, and airborne data are now available for operational forecasting. These data provide information about the spatial variability of hydrometeorological fields that will improve forecasts as well as understanding of the processes. The availability of these data will undoubtedly influence the future development of hydrologic forecasting models and systems. Several research areas important to the future of hydrologic forecasting are the linkage of hydrologic and atmospheric models, the incorporation of meteorological forecast information, the use of distributed modeling, the estimation of forecast uncertainty, the optimal assimilation of data, the updating of model states, and the use of hydrologic forecasts in water management operations.

References

The following publications will provide the reader with a wide spectrum of information concerning hydrologic forecasting techniques.

Anderson, E.A., "The National Weather Service River Forecast System and Its Application to Cold Regions," Proceedings of the Sixth Northern Research Basins Symposium/Workshop, Michigan Technological University, Houghton, Michigan, January, 1986.

Beven, K., "Changing Ideas in Hydrology - The Case of Physically Based Models," Journal of Hydrology, No. 105, pp. 157-172, 1989.

Bras, R.L. and P. Restrepo-Posada, "Real-Time, Automatic Parameter Calibration in Conceptual Runoff Forecasting Models," Third International Symposium on Stochastic Hydraulics, Tokyo, Japan, August 5-7, 1980.

Brazil, L.E., M.D. Waage, and D.P. Laurine, "Decision Support System for the Denver Water Department: Incorporating Extended Streamflow Forecasting Information into Water Management Operations," Proceedings of the Fourth Operations Management Workshop, American Society of Civil Engineers, Water Resources Planning and Management Division, Mobile, Alabama, March, 1992.

Brazil, L.E., and M.D. Hudlow, "Calibration Procedures Used with the National Weather Service River Forecast System," Water and Related Land Resource Systems, edited by Y.Y. Haimes and J. Kindler, pp. 457-466, Pergamon, New York, 1981.

Burnash, R.J.C., R.L. Ferral, and R.A. McGuire, A General Streamflow Simulation System Conceptual Modeling for Digital Computers, Report by the Joint Federal State River Forecasts Center, Sacramento, California, 1973.

Carroll, T.R., "Cost-Benefit Analysis of Airborne Gamma Radiation Snow Water Equivalent Data Used in Snowmelt Flood Forecasting," Proceedings of the 54th Annual Western Snow Conference, Phoenix, Arizona, April, 1986.

Crawford, N.H. and R.K. Linsley, Digital Simulation in Hydrology: Stanford Watershed Model IV, Technical Report No. 39, Department of Civil Engineering, Stanford University, Stanford, California, 1966.

Day, G.N., "A Methodology for Updating a Conceptual Snow Model with Snow Measurements", NOAA Technical Report, NWS 43, Silver Spring, Maryland, 1990.

Day, G.N., "Extended Streamflow Forecasting Using NWSRFS," Journal of Water Resources Planning and Management, ASCE Vol. 111, No. 2, pp. 157-170, April, 1985.

Duan, Q., A Global Optimization Strategy for Efficient and Effective Calibration of Hydrologic Models, Ph.D. Thesis, Dept. of Hydrology and Water Resources, University of Arizona, Tucson, Arizona, 1991.

Georgakakos, K.P., "Real Time Coupling of Hydrological and Meteorological Models for Flood Forecasting," Recent Advances in the Modeling of Hydrological Systems, edited by D. Bowles and P.E. O'Connell, Reidel Publishing Co., 1989.

Haan, C.T., et al., "Hydrologic Modeling of Small Watersheds," American Society of Agricultural Engineers, No. 5, 1982.

Ibbitt, R.P., Systematic Parameter Fitting for Conceptual Models of Catchment Hydrology, Doctorate Thesis, University of London, London, England, January, 1970.

Kitandis, P.K. and R.L. Bras, Real Time Forecasting of River Flows, Report No. 235, Dept. of Civil Engineering, Massachusetts Institute of Technology, Cambridge, Massachusetts, 1978.

Natale, L. and E. Todini, "A Constrained Parameter Estimation Technique for Linear Models in Hydrology," Mathematical Models for Surface Water Hydrology, John Wiley & Sons, U.K., Chichester, 1977.

O'Connell, P.E. and R.T. Clarke, "Adaptive Hydrological Forecasting--A Review," Hydrological Sciences Bulletin, Vol. 26, No. 2, June, 1981.

Puente, C.E. and R.L. Bras, "Application of Nonlinear Filtering in the Real-Time Forecasting of River Flows," Water Resources Research, Vol. 23, No. 4, pp. 675-682, April, 1987.

Restrepo-Posada, P.J., Automatic Parameter Estimation of a Large, Conceptual Rainfall-Runoff Model: A Maximum Likelihood Approach, Doctoral Dissertation, Department of Civil Engineering, Massachusetts Institute of Technology, Cambridge, Massachusetts, February, 1982.

Sorooshian, S. and V.K. Gupta, "The Analysis of Structural Identifiability: Theory and Application to Conceptual Rainfall-Runoff Models," Water Resources Research, Vol. 21, No. 4, pp. 487-495, April, 1985.

Weeks, W.D. and R.H.B. Hebbert, "A Comparison of Rainfall-Runoff Models," Nordic Hydrology, Vol. 11, pp. 7-24, 1980.

Wood, E.F., "Recent Developments in Real-Time Forecasting/Control of Water Resources Systems," Pergamon, Oxford, 1980.

Wood, E.F. and P.E. O'Connell, "Real-Time Forecasting," Hydrological Forecasting, edited by M.G. Anderson and T.P. Burt, John Wiley and Sons Ltd., pp. 505-558, 1985.

Using Forecasts to Improve Reservoir Operations

Jery R. Stedinger[1], M. ASCE, Mohammed Karamouz, Brian McCrodden, George McMahon, Richard Palmer, and Juan Valdés

Abstract
This review considers methodologies that have or can be employed to incorporate streamflow forecasts into reservoir policies and operating decisions. These include the use of streamflow forecasts in deterministic optimization models, use of multiple stage decision trees, stochastic analytical optimization models with linearized reservoir dynamics, and sophisticated stochastic models that can describe the joint distribution of streamflows and forecasts in the optimization of reservoir operations.

Introduction
Deriving efficient operating decisions for reservoir systems employing realistic forecasts with due consideration of their uncertainty is a challenge. The space rule and other heuristic guidelines are available (Johnson et al., 1991). Alternately, one can use trial-and-error experimentation, or optimization models (Karamouz et al., 1992), to identify reasonable values of the parameters in a given release function. The remainder of this paper describes other ideas.

Deterministic optimization models
Deterministic operations optimization models make use of a nominal set of streamflows for the modelled operating period. These can be historical mean flows or a forecast based upon current meteorological and watershed inputs. Many of these models have been used in practice (Ikura et al., 1986; Kirshen, 1992; Stover, 1992). By employing a critical (wet or dry) appropriately selected inflow quantile, operating decisions can be identified that hedge against adversity at the selected probability level (Randall and McCrodden, 1992).

Multiple scenarios in decision trees
To describe future inflow uncertainty, one can use multiple streamflow forecasts (as in a multi-stage decision tree) to reflect inflow forecast uncertainty. Multi-stage algorithms have been used successfully with the Brazilian system (Pereira and Pinto, 1985) and by the Pacific Gas and Electric Company to describe California snowmelt season runoff-forecast uncertainty. Howard (1992) describes the successful use in

---

[1]Prof., School of Civil and Envir. Engineering, Cornell University, Ithaca, NY 14853-3501.

an LP model for the northwest of a single one-period forecast followed by multiple inflow forecasts for subsequent periods corresponding to specified quantiles of the cumulative inflow distribution; improved operation resulted from use of the optimization model. These models should yield some improvement over use of a single most-likely or critical forecast.

Stochastic optimization models using control theory

Analytic second-moment descriptions of inflow uncertainty can be based upon a current forecast with optimal control optimization algorithms. Because of the efficiency of such algorithms, they can be used to recompute releases within a simulation model using the "latest" inflow forecasts, though discontinuities and constraints on future storage levels cause problems (Soliman and Christensen, 1986; Trezos and Yeh, 1987). Georgakakos used his extended linear quadratic Gaussian control model on the Savannah River (Georgakakos, 1992) and other systems (Georgakakos, 1989; Hooper et al., 1991).

Operations models with dynamic streamflow descriptions

Dynamic programming (DP) algorithms are powerful tools for studying multi-reservoir system operation, because the sequential nature of decisions and the stochastic nature of inflows and forecasts can be modelled explicitly. Several formulations of the reservoir operations problem have been employed.

Model Formulation-- Stochastic dynamic programming can derive release policies that employ different descriptions of the hydrologic state and likely future inflows. One can use either the previous period's inflow or the current period's inflow as a hydrologic state variable. Esmaeil-Beik and Yu (1984) compare the two. Huang et al. (1991) and Estalrich and Buras (1991) indicate that use of the current period's inflow can be viewed as using a forecast of the actual inflow whose accuracy would affect actual operations. Stedinger et al. (1984) note that the issue is more complex because one does not complete a period's release until the end of a period, at which time the current inflow is known.

Stochastic DP models need not be based only upon the current or previous period's inflow. Druce (1990) considers both seasonal variability and forecasts of shorter term flood-control operations. Bras et al. (1983) used last month's inflow as a hydrologic state variable, but in each month solved a new stochastic DP using conditional streamflow distributions based upon other variables.

Models with Dynamic Forecast-Streamflow Descriptions--Stedinger et al. (1984) proposed a stochastic DP model that employed a forecast of the immediate period's inflow as a hydrologic state variable with a model of the stochastic relationship between streamflows and forecasts; this allowed derivation of efficient operating policies which incorporated the best available forecasts of the next month's flow. Employing Bayesian statistical arguments, Krzysztofowicz (1986) developed a similar reservoir decision model which he employed to determine the optimal time to make a water-supply commitment. Karamouz and Vasiliadis (1992) compare the Stedinger et al. and Krzysztofowicz models of the forecast-streamflow processes, as well as the performance of different hydrologic state variables in reservoir operating models. Kelman et al. (1990) used sampling stochastic DP to derive efficient reservoir operating policies that employed snowmelt-season runoff forecasts as a hydrologic state variable. Their sampling DP model describes the complex spatial and temporal correlation structure of streamflows by its empirical distribution.

Value of forecasts
A significant issue is the value of forecasts in reservoir operations. For the High Aswan Dam, Bras et al. (1983) and Stedinger et al. (1984) illustrate use of a SDP model to derive efficient operating policies which incorporate the conditional distribution of future inflows, as did Karamouz and Vasiliadis (1992). Stedinger et al. (1992) consider alternative hydrologic information for use in operating the Shasta and Trinity reservoirs in northern California. Yeh et al. (1985), Johnson et al. (1991), and Kelman et al. (1990) also considered the value of forecasts for reservoir systems in California. The committee is preparing a paper with additional examples.

Conclusions
Incorporating a single forecast into a deterministic optimization model to guide reservoir operation is relatively simple. That methodology has been widely and successfully employed. Models that incorporate streamflow variability and forecast uncertainty are more complicated and have seen less use. We anticipate such models will see greater use in the future. Computing technology has continued to advance, placing greater power at operators' command, while data collection operations have been automated with modern telemetry, bringing current hydrometeorologic information into reservoir control centers on a real-time basis. With access to the necessary data and computing power, the stage is set for reservoir system operators to develop useful stochastic operating models that use available streamflow forecasts and estimates of forecasting precision.

Acknowledgments
The authors appreciate the support ASCE provided for the Task Committee on Streamflow Forecasting: Implications for Reservoir Operations.

Appendix -- References
Bras, R.L., R. Buchanan, and K.C. Curry, Real time adaptive closed loop control of reservoirs with the High Aswan Dam as a case study, *Water Resour. Res., 19*(1), 33-52. 1983.

Druce, D. J., Incorporating daily flood control objective into a monthly stochastic dynamic programming model for a hydroelectric complex, *Water Resour. Res., 26*(1), 5-11, 1990.

Esmaeil-Beik, S., and Y.-S. Yu, Optimal operation of multipurpose pool of Elk City Lake, *J. Water Resour. Plann. Manage. Div., 110*(WR1), 1-14, 1984.

Estalrich, J., and N. Buras, Alternative specifications of state variables in stochastic-dynamic-programming models in reservoir opertion, *Applied Mathematics and Computation 44*, 143-155, 1991.

Georgakakos, A.P., Extended linear quadratic Gaussian (ELQG) control: Further extensions, *Water Resour. Res., 25*(2), 191-201, 1989.

Georgakakos, A.P., Computer-aided management of the Savannah River system, 4th Operations Management Workshop, Am. Soc. of Civil Engineers, Mobile, Alabama, March 1992.

Hooper, E.R., A.P. Georgakakos, and D.P. Lettenmaier, Optimal stochastic operation of Salt River Project, Arizona, *Jour. of Water Resour. Plng. and Mgmt., 117*(5), 566-587, 1991.

Howard, C., Experience with probabilistic forecasts for cumulative stochastic optimization, Am. Water Resour. Assoc. Symposium, Reno, Nevada, Nov. 1992.
Huang, W-C., R. Harbo, and J.J. Bogardi, Testing stochastic dynamic programming models conditioned on observed or forecasted inflows, *Jour. of Water Resour. Plng. and Mgmt., 117*(1), 28-36, 1991.
Ikura, Y., G. Gross, and G. S. Sand, PGandE's state-of-the-art scheduling tool for hydro systems, *Interfaces, 16*(1), 65-82, 1986.
Johnson, S.A., J.R. Stedinger, and K. Staschus, Heuristic operating policies for reservoir system simulation, *Water Resour. Res. 27*(5), 673-685, 1991.
Karamouz, M., M.H. Houck, and J.W. Delleur, Optimization and simulation of multiple reservoir systems, *Jour. of Water Resour. Plng. and Mgmt., 118*(1), 71-81, 1992.
Karamouz, M., and H.V. Vasiliadis, Bayesian stochastic optimization of reservoir operation using uncertain forecasts, *Water Resour. Res. 28*(5), 1221-1232, 1992.
Kelman, J., J. R. Stedinger, L. A. Cooper, E. Hsu, and S. Yuan, Sampling stochastic dynamic programming applied to reservoir operation, *Water Resour. Res. 26*(3), 447-454, 1990.
Kirshen, P.H., Extended experience with a short term hydro scheduling model in New England, *Saving a threatened resource - in search of solutions*, M. Karamouz (ed.), Proceeding of the Water Resource Sessions, ASCE Water Forum '92, Baltimore, MD, August 2-5, 1992.
Krzysztofowicz, R., Optimal water supply based on seasonal runoff forecasts, *Water Resour. Res., 22*(3), 313-321, 1986.
Pereira, M.V.F., and L.M.V.G. Pinto, Stochastic optimization of a multireservoir hydroelectric system-a decomposition approach, *Water Resour. Res., 21*(6), 1985.
Pereira, M.V.F., and L.M.V.G. Pinto, Multi-stage stochastic optimization applied to energy planning, *Math. Programming, 52*(2), 359-375, 1991.
Randall, D., and B.J. McCrodden, Modeling the Savannah River system for improved operations, 4th Operations Management Workshop, Am. Soc. of Civil Engineers, Mobile, Alabama, March 1992.
Soliman, S.A., and G.S. Christensen, Application of functional analysis to optimization of a variable head multireservoir power system for long-term regulation, *Water Resour. Res., 22*(6), 852-858, 1986.
Stedinger, J.R., B.F. Sule, and D.P. Loucks, Stochastic dynamic programming models for reservoir operation optimization, *Water Resour. Res., 20*(11), 1499-1505, 1984.
Stedinger, J.R., J.A. Tejada-Guibert, and S.A. Johnson, Performance of hydropower systems with optimal operating policies employing different hydrologic information, 4th Operations Management Workshop, Am. Soc. of Civil Engineers, Mobile, Alabama, March 1992.
Stover, C., Using real-time forecasts to improve flood control operations, 4th Operations Management Workshop, Am. Soc. of Civil Engineers, Mobile, Alabama, March 1992.
Trezos, T., and W. W-G. Yeh, Use of stochastic dynamic programming for reservoir management, *Water Resour. Res., 23*(6), 983-996, 1987.
Yeh, W. W-G., L. Becker, and R. Zettlemoyer, Worth of inflow forecast for reservoir operation, *J. Water Resour. Plng. and Mgmt. Div.*, Am. Soc. Civ. Eng., *108*(WR3), 257-69, 1982.

The Development and Use of Policy Statements in River Simulation Models

Jon S Behrens[1], Daniel P Loucks[2] and Hussam Fahmy[3]

Abstract

There are a number of ways in which the physical operation of a river system can be expressed in simulation models. Most of the earlier simulation models have incorporated the operating policy within the program itself. This requires changing the code each time the operating policy is to be changed. With the advent of interactive simulation, operating policies are now usually defined in the input data. For example, models that use reservoir release rules or storage target rule curves to determine reservoir releases and storage volumes, and models that use storage target distribution functions of one type or another will have these functions and rules defined by input parameters. Thus the code itself need not be changed to perform multiple simulations of different operating policies.

Work at CADSWES on the development and application of an object-oriented river system simulation model (RSS) has included the use of English--like policy statements as part of the input data. These statements are written in a specific format and allow more flexibility than when specific policies are written directly into the code.

This paper discusses how external policy statements are written and how they interact with the object-oriented simulation engine to provide the river basin modeler with a flexible and powerful tool.

Introduction

Planning for the management and operation of river systems involves a constant re-evaluation of system objectives and the identification of feasible alternatives for satisfying those objectives. It requires prediction of the multiple effects of such alternatives on

---

1. President, Jon Behrens & Associates, Inc., 5575 Bowron Pl., Longmont, CO, 80503.
2. Professor, Civil Engineering, Cornell University, Ithaca, NY, 14853.
3. Research Associate, CADSWES, University of Colorado, Boulder, CO, 80301.

each of the groups having an interest in the river. It involves communicating this information to these interest groups and to the public at large, which inevitably generates debate and the need for public participation prior to making many decisions.

Performing all these planning and management tasks, often under pressure of time constraints, budget limitations, political considerations, and occasionally under conditions of hydrologic stress is not easy. No computer-based decision support system is going to change these facts. Nevertheless the use of appropriate tools can lead to more informed and more efficient planning, managing and decision making.

Operating Policy Statements

There are a number of ways in which the operation of a system of engineering facilities in a river basin can be controlled. A common approach for expressing the release rules of reservoirs is through the use of release zones or storage volume targets. Water allocation policies are often expressed as functions of the flow available at the diversion sites. This paper will discuss a direct method of policy expression that allows the separation of the policies desired from the simulation algorithm needed to implement these policies. Keeping these two functions separate means that either can be modified without affecting the other.

In the RSS model, linguistic policy statements can be used to specify when the desired operation is to differ from the default operation. The use of policy statements provides a number of advantages. One is that operating policy statements are easy to understand given their English-like format. Another advantage is the ease with which operating policies can be changed by the user when performing multiple "what if" simulations. Furthermore policy statements offer more flexibility in specifying how a system is to be operated than has been possible with most generic river simulation programs.

Each policy statement has the following form:

```
POLICY   name   TO_DETERMINE   variable FOR   object name
         { any comments desired }
   BEGIN
         IF    premise   AND   premise   OR   premise
         THEN      variable  =  action
   END
```

The policy statement will assign the variable if all of its premises are true. User defined variables might be something like "flood_control_release" or "safe_drawdown_rate". To find values for these variables, the policy statement in which they are defined may have to invoke other statements to determine components of the variable or to identify circumstances that will affect its value.

Premises are expressions that are either true or false. Examples might be "safe_drawdown_rate >= 500", or "Lake_Alpha_Beginning_storage < 19000". In the latter example the premise would require reading the beginning storage data slot of the Lake Alpha object. Thus this premise may be true sometimes and false at other times depending on the storage state at Lake Alpha. In the case of the first example, the "safe_drawdown_rate" would be interpreted as being a variable defined in some other policy statement. In determining its value, the processor would suspend execution of the current policy statement and look for another statement whose evaluation would return the required value. If the collection of individual premises is true, the policy statement executes.

The action portion of the policy statement assigns a value to the variable identified in the header after which the statement exits returning that value to whatever invoked it, either the simulator or another policy. Example actions might be "safe_drawdown_rate = 1000" or "safe_drawdown_rate = channel_capacity - Lake_Alpha. outflow". The latter case might be used to condition the safe drawdown rate at one reservoir on the release from a parallel reservoir and on the maximum non-damaging flow in the channel below the confluence of their outflows.

The three components of RSS policy statements are shown in the three examples in Figure 1. These three policy statements specify a simple multiple reservoir outflow policy. The release from a downstream reservoir is dependent on the storage volume in an upstream reservoir as well as on the storage in the downstream reservoir itself. The release or outflow from a reservoir is to be 30% of the total storage volume in both that reservoir, called ResDown, and an upstream reservoir, called ResUp. The release is to be made solely from the downstream reservoir, if possible. Water is to be released from the upstream reservoir only if the downstream reservoir is empty.

For this example only a single policy need be written, namely the first policy named d_outflow. If insufficient storage exists in the downstream reservoir, the downstream reservoir object will request additional inflow from the upstream reservoir as part of the simulation engine's default behavior.

The single policy, d_outflow, can also be written using two policy statements, named sumsto and d_release also shown in Figure 1. These two statements illustrate the capability of defining new vari-

```
POLICY d_outflow TO_DETERMINE Outflow FOR ResDown
    {This policy specifies the outflow from a downstream res-
    ervoir as a fraction of the beginning storage in the
    downstream and upstream reservoirs.}
    BEGIN
        Outflow = 0.3 * (ResDown.Beginning_storage +
                         ResUp.Beginning_storage)
    END

POLICY sumsto TO_DETERMINE sum_storage FOR ResDown
    {This policy defines the variable: sum_storage.}
    BEGIN
        sum_storage = ResDown.Beginning_storage +
                      ResUp.Beginning_storage
    END

POLICY d_release TO_DETERMINE Outflow FOR ResDown
    {This policy uses the variable sum_storage in determin-
    ing the release for the downstream reservoir.}
    BEGIN
        Outflow = 0.3 * sum_storage
    END
```

Figure 1. Examples of reservoir release policy statements involving two reservoirs in series.

ables (in this case the variable sum_storage) that permits the writing of a set of simpler policy statements in place of a single more complex one.

In the object-oriented RSS model, any policy written for a parent class will apply to all objects defined by the children classes of that parent. If a policy written for a child conflicts with a parent's policy, the child's policy will apply.

Conclusion

This paper has attempted to convey an idea of the flexibility that is possible when the definition of policy to be simulated is separated from the physics of the simulation itself. The modeler may simply tell the model how it is to handle complex situations without resorting to recoding and without worrying that changes in policy will affect something unrelated. This way of dealing with a model is similar to the actual operation of a river basin. Operators set releases, for example, and the structures take care of the physics.

The Law of the River: Policy Expression

Hussam Fahmy[1], Jon Behrens[2], Kenneth Strzepek[3]

Abstract

The need to accurately represent the law of the Colorado river has heretofore precluded the use of any general purpose river basin simulation model. In order to overcome the inflexibility associated with policies coded directly into a model, a new approach for river basin simulation modeling has been developed. The new approach allows the user to capture the operating policy in a series of "policy statements" separate from the basic simulation model. These statements are accessible without writing code and may easily be modified to explore policy alternatives.

Introduction

The Colorado River is one of the most highly regulated rivers in the world, not only from a hydraulic point of view, but also legislatively and institutionally. Because the Colorado River lies in six states and in two countries, its legislative and institutional requirements range from state law to interstate compacts to international treaties. The technical interpretations of all of these requirements has codified in the "operation criteria" or "law of the river"

Over the last decade the US Bureau of Reclamation developed and implemented the Colorado River Simulation System (CRSS) which is a combination of a large data base and a simulation model. As its name implies, CRSS is specifically dedicated to the Colorado River. Its main purpose is to help water resource managers in preforming planning and operational studies. Since both the Colorado river's configuration and its operating policy are written directly into the model's code, the CRSS is not suited to deal with a wide spectrum of uses, dynamic operating policy, or other basins.

Recently CADSWES, under contract with the USBR, developed a new approach to river basin simulation modeling which integrates a georelational data base, an object oriented simulator, and a policy base. The motive behind this new approach is to provide a flexible tool that can be used by the different parties interested in any River

---

1. Research Associate, CADSWES, University of Colorado, Boulder, CO, 80309.
2. President, Jon Behrens & Associates, Inc., 5575 Bowron Pl. Longmont, CO, 80503.
3. Professor, Department of Civil and Environmental Engineering, University of Colorado, Boulder, CO, 80309.

Basin. The first application of this system is to the Colorado. This paper will discuss the process of capturing the Law of the River in this new paradigm.

Colorado River Operating Policy

The provisions of the Colorado River Compact, The Upper Colorado River Basin Compact, and the Mexican Water Treaty have been given engineering interpretation in what are called the "Operating Criteria". The Operating Criteria essentially determine the monthly releases from Lake Powell and Lake Mead in order to achieve the following objectives:

• Equalize the active contents of both Powell and Mead by the end of each water year,

• Limit the total Upper Basin projected end-of-water-year active contents to the storage required by section 602(a) of public law 90-537,

• Implement pre-described surplus or shortage strategy if surplus or shortage conditions apply,

• Implement Corps of Engineers flood control policy.

The determination of Lake Powell and Lake Mead releases ultimately requires evaluation of such variables as end of water year active storage in each lake, amount of surplus or shortage, storage required by section 602(a), flood control release, etc. Clearly, different interpretations of the law or suggested changes to it will lead to different values of these variables and consequently different operating criteria. Also the procedures and the data used to calculate these variables would change.

New River Simulation System Approach (RSS)

RSS consists of three main components: a georelational data base, an object oriented simulator and a policy base. The georelational data base manages the input and output data of the simulator using a relational database management system coupled with geographic information system. The simulator, which is based on an object oriented representation of the river basin's components (reservoirs, diversions, confluences, etc.), handles the physics of the system (mass balance, evaporation, bank storage, diversions, return flows, etc.). In other words the simulator takes care of the plumbing without respect to any operating policies. The policy base, which is the subject of this paper, is a set of policy expressions that can be referenced by the simulator to control the river system's behavior.

Both the georelational data base and the simulator are stand-alone systems, while the policy base must be attached to the simulator model to be meaningful. The separation of the simulator and the policy base allows RSS to accommodate dynamic operating policies or to model different river basins, each with its unique operating policy.

Policy Base and Language

Although the policy base is separate from the simulator and the data base, links among them must be established. Policy statements have access to any river system variable in the simulator. Further, new variables may be defined in the policy base itself. However, the simulator is only able to access the policy base through four common variables: outflow, diversion, end of period storage and the ratio between the two inflows to a confluence. The simulator does not know about any other internal policy base variables. Thus, the user controls the simulator by writing policies that control the values of these four variables.

Each policy statement has a template which consists of a header, a beginning mark, conditions, an assignment expression, and an end mark. The general format of a policy statement is:

```
POLICY name TO_DETERMINE variable FOR object
   BEGIN
      IF any condition or conditions
      THEN
         variable =
            any expression of defined variables
            and/or functions of variables.
   END
```

The words in uppercase letters are keywords forming the skeleton of the policy statement. Additionally, the key words AND and OR may be used to concatenate as many conditions as needed.

From the Law of the River to Policy Statements

As an example of the process by which we converted the Law of the River to policy statements, we will consider 602(a) storage. Section 602 (a) of public law 90-537 requires a certain volume of storage to be in the upper basin on september 30 of each year as a part of the long term river operation strategy. It also requires the Secretary of the Interior to determine the amount of this storage at the beginning of each calendar year based on all relevant factors (including, but not limited to, historic stream flows, the most critical period of record, and estimates of the probable water supply). Figure 1 displays the text of Section 602(a).

September 30, 1968   - 15 -   Pub. Law 90-537

SEC. 602. (a) In order to comply with and carry out the provisions of the Colorado River Compact, the Upper Colorado River Basin Compact, and the Mexican Water Treaty, the Secretary shall propose criteria for the coordinated long-range operation of the reservoirs constructed and operated under the authority of the Colorado River Storage Project Act, the Boulder Canyon Project Act, and the Boulder Canyon Project Adjustment Act. To effect in part the purposes expressed in this paragraph, the criteria shall make provision for the storage of water in storage units of the Colorado River storage project and releases of water from Lake Powell in the following listed order of priority:

(1) releases to supply one-half the deficiency described in article III(c) of the Colorado River Compact, if any such deficiency exists and is chargeable to the States of the Upper Division, but in any event such releases, if any, shall not be required in any year that the Secretary makes the determination and issues the proclamation specified in section 202 of this Act;

(2) releases to comply with article III(d) of the Colorado River Compact, less such quantities of water delivered into the Colorado River below Lee Ferry to the credit of the States of the Upper Division from other sources; and

(3) storage of water not required for the releases specified in clauses (1) and (2) of this subsection to the extent that the Secretary, after consultation with the Upper Colorado River Commission and representatives of the three Lower Division States and taking into consideration all relevant factors (including, but not limited to, historic stream-flows, the most critical period of record, and probabilities of water supply), shall find this to be reasonably necessary to assure deliveries under clauses (1) and (2) without impairment of annual consumptive uses in the upper basin pursuant to the Colorado River Compact: *Provided*, That water not so required to be stored shall be released from Lake Powell: (i) to the extent it can be reasonably applied in the States of the Lower Division to the uses specified in article III(e) of the Colorado River Compact, but no such releases shall be made when the active storage in Lake Powell is less than the active storage in Lake Mead, (ii) to maintain, as nearly as practicable. active storage in Lake Mead equal to the active storage in Lake Powell, and (iii) to avoid anticipated spills from Lake Powell.

This text has been interpreted in technical engineering terms and translated into several policy statements. The resulting policy base was then applied successfully to the Colorado River simulation. One policy statement, in which 602(a) is determined is shown below.

```
POLICY six02a TO_DETERMINE sixo2_storage FOR powell
  BEGIN
    IF   now () = january AND
         now () = start_month () AND
         now_year () = start_ year ()
    THEN
         sixo2_storage =
              (ub_expected_depletion +
              ub.average_annual_evaporation x
              ub.critical_period_length) x
              (1 - ub.shortage_precent) + 8.23 x
              ub.critical_period_length +
              lees_ferry.critical_natural_flow
  END
```

Where now(), now_year(), start_month(), start_year() are functions that get information from the simulator about the current time step, current year, and starting time of the simulation. Variables such as ub_expected_depletion are determined by other policy statements within the policy base. It is the collection of these other policy statements along with the one shown that constitute the model's understanding of how and when to compute 602a storage.

Conclusion

As the roles and operational objectives of water management agencies such as the Bureau of Reclamation and the Corps of Engineers change to reflect society's changing perception of the benefits to be derived from water projects, it is necessary to develop new analysis and operation tools. Continual modification of tools developed in the 1970's to allow their use in studies of environmental quality and aesthetics can become prohibitive. The approach described here is to develop a tool that can accommodate changes in policy and intent by changing the data stream rather than by recoding the system.

Advanced Decision Support Systems for Environmental Simulation Modeling

Steven C. Chapra,[1] Member, ASCE, Jean M. Boyer[2], and Robert L. Runkel[1]

Abstract

Some new computer approaches for integrating simulation models into water-quality management and planning are described. A system for integrating models, databases and decision-support functions is outlined. The role of object-oriented programming in water-quality modeling is also explored.

Introduction

Transport and fate models provide a linkage between point and non point pollutant discharges and receiving water quality. As such they are an important component of cost-effective management of water pollution problems (Thomann and Mueller 1987). Unfortunately, early computing environments often limited the effectiveness of such models in management contexts.

Most early models were originally written in algorithmic languages like FORTRAN and designed for a batch mode of implementation. Input was via punched cards or flat files and output usually consisted of tables of numbers. As depicted in Fig. 1, the systems typically placed the user at the center of the model application process. This had two impacts on their application: (1) the user had to be an expert, and (2) the user became an intermediary between the decision maker and the decision making tool.

Since the advent of personal computers and engineering workstations, efforts have been made to modify the modeling environment to make it more accessible to managers and planners. The initial efforts focused on making the input/output process more user friendly. Menus were used as

---

[1]Professor and Research Scientist, respectively, CADSWES, Dept. of Civil, Environ. and Architect. Engr., University of Colorado, Boulder, CO 80309-0428.
[2]Environmental Engineer, U.S. Bureau of Reclamation, Denver, CO.

DECISION SUPPORT SYSTEMS

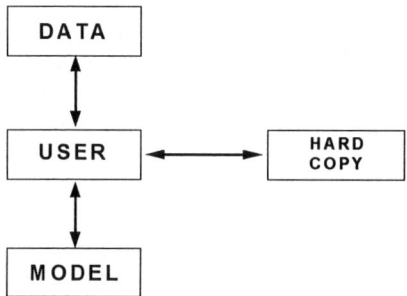

Figure 1. Traditional water-quality modeling placed the user at the center of the model development/application process.

the principle vehicle for user-model interaction. Data input employed spreadsheet-like tables and output was displayed graphically.

Although these modifications have certainly made some water-quality models easier to use, the fundamental structure depicted in Fig. 1 has been maintained. Additionally, most of these frameworks still require considerable expertise. In the present paper, we will describe an alternative environment that is intended to bring the decision maker closer to simulation models and data.

AN ADSS WATER-QUALITY MODELING ENVIRONMENT

Figure 2 shows an alternative framework for water-quality modeling that represents a major departure from the traditional approach outlined in Fig. 1. Rather than placing the expert user at the center of the process, this advanced decision support system, or ADSS, employs electronic linkages to blend the key tools into an integrated package. These tools are:

- **Database**--All problem specific information is centralized in a database management system. This includes spatial (GIS) and temporal (time series) information.

- **Simulation Model**--The model provides a means to simulate the behavior of the natural water.

- **Decision Support System**--This module contains analysis and decision making algorithms needed to support the decision process. Examples might be optimization or statistical tools.

The electronic integration of these tools allows many of the tasks formerly handled by the expert to be taken over by the computer. In addition, the

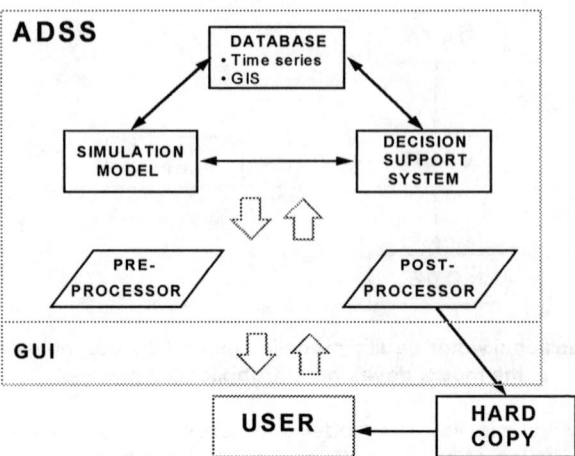

Figure 2. An advanced decision-support system integrates several tools so that users can automate aspects of the management and planning process that were traditionally in the hands of experts.

ADSS also includes a **Graphical User Interface** (GUI) consisting of a pre- and a post processor. This serves to facilitate the analysis by making input and output convenient. Note that the interface provides both visual and hard-copy communication. The latter is designed to support a number of different functions including report generation.

In addition to ease of use, the system can also be designed to incorporate imbedded artificial intelligence. Such capability can provide the user with expert guidance. Such guidance can allow the non-expert to perform valuable "what-if?" computations with the framework. Although such analyses must of necessity be constrained, managers and planners can directly interact with the system to explore options.

OBJECT-ORIENTED PROGRAMMING

Aside from an intelligent interface, another component of an ADSS that can enhance user interaction relates to the choice of the development environment and language. As stated above, most early modeling systems were programmed in algorithmic languages such as FORTRAN. Today, object-oriented programming (OOP) provides an alternative that should facilitate both user interaction and scientific applications.

OOP is a technique to facilitate the development of large computer programs. In contrast to algorithmic languages which organize programs into procediures and control structures, OOP treats the system as a collection of objects. From the perspective of water-quality modeling, it has a variety of promising application areas. Two of these areas involve systems that are directly object-oriented (lumped systems) and others that would not usually be thought of in this way (distributed systems).

1. Lumped-Parameter Systems--These are systems that are made up of distinct entities. For example, large river-basins typically consist of networks of streams and impoundments. Such systems are ideally suited for an object-oriented approach. In particular, OOP facilitates the graphical development of model applications. Users can create simulations by graphically linking icons representing system objects. One such system is presently being developed at CADSWES to model water-quality variables such as temperature, salinity and nutrients in the Colorado River Basin.

2. Distributed Systems--These are systems that are continuous in nature. An example is the vertical water column of a lake or reservoir. Temperature and other water-quality variables change gradually with depth for such systems. Although these variables form a continuum, scientists and engineers typically model them using discrete approximations. Research is presently being conducted to simulate temperature variations in impoundments and tailwaters of the Sacramento River system in California using such an approach.

Advanced decision support systems along with object-oriented programming provide a means to automate functions previously controlled by expert users. Application of these approaches will free expert users from a variety of computational burdens. Thus, they will be allowed to focus more on the validity of the simulation models rather than on mundane programming and data analysis chores. Aside from benefits to experts, ADSS and OOP offer a great opportunity to make models more accessible to decision makers. In particular, advances will facilitate decision analyses as well as allowing managers and planners to rapidly develop and assess simulation based scenarios.

References

Thomann, R.V. and J.A. Mueller, *Principles of Surface Water Quality Modeling and Control*, Harper and Row, New York, 1987.

Solute Transport Modeling under Unsteady Flow Regimes: An Application of the Modular Modeling System

Robert L. Runkel[1]

Pedro J. Restrepo[1]

Abstract

Many existing solute transport codes rely on the assumption of steady (i.e. time-invariant) flow. This assumption severely limits such models, as accurate simulations often require realistic descriptions of the hydraulic flow regime. Small mountain streams, for example, are known to undergo daily fluctuations in volumetric flow rate and groundwater inflow. These changes in flow may act to dilute or concentrate a given solute, thereby affecting a contaminant's ultimate concentration.

In response to this shortcoming, a solute transport model is presented that allows for the specification of time-varying lateral inflows, in-stream flow rates, and cross-sectional areas. This allows the model to be used in conjunction with watershed-scale hydraulic routing models that provide the time-varying parameters. This linkage between the routing component and the solute transport model is greatly facilitated through the use of the Modular Modeling System (MMS). MMS allows the two components to be represented as separate modules that are called at each time step. For a given time step, the routing module computes the flows and cross-sectional areas. These values are then made available to the solute transport module, and a new time step is initiated.

Introduction

A number of mathematical models are used to simulate the behavior of solutes and contaminants in small streams and rivers (e.g. Brown and Barnwell, 1987; Ambrose et al., 1988; Runkel and Broshears, 1991). Although these models have a variety of uses, they all are based on governing equations that describe the physical transport of mass throughout the system. These mass transport equations contain advective and dispersive terms that include hydraulic state variables such as volumetric flow rate and cross-sectional area. In the simplest case, these variables are constant in time and a 'steady' flow regime (Henderson, 1966) is assumed. Unfortunately, many practical problems involve flow fields in which the hydraulic state variables are time variable (i.e. unsteady flow). For this latter case, it is often desirable to link the solute transport model with a hydrodynamic model that provides the time-varying state variables.

---

1. Research Engineer and Senior Research Engineer, respectively, University of Colorado, Center for Advanced Decision Support for Water and Environmental Systems, Department of Civil Engineering, Boulder CO 80309.

In the remaining sections of this paper, we present a solute transport model and a watershed-scale routing model that can be linked to provide a realistic description of the physical mechanisms affecting solute transport. This linkage is facilitated through the use of the Modular Modeling System (Restrepo et al., 1992; Leavesley et al., 1992), a modeling framework that is ideally suited for such applications.

The Solute Transport Model

The solute transport model presented herein is based on the familiar advection-dispersion equation (Thomann and Mueller, 1987), with additional terms to account for the effects of lateral inflow and transient storage. Lateral inflows represent water that enters the stream via groundwater inflow, interflow, overland flow and small tributaries. These flows act to dilute (or concentrate) solutes in the stream channel if they carry solute concentrations that are lower (or higher) than the stream solute concentration. Transient storage occurs in small streams when portions of the transported solute become isolated in zones of water that are immobile relative to water in the main channel (e.g. pools, gravel beds). A model incorporating the effects of lateral inflow and transient storage is presented by Bencala and Walters (1983).

Conservation of mass for the stream and transient storage-zone is given by (1) and (2), respectively (Runkel and Broshears, 1991):

$$\frac{\partial C}{\partial t} = -\frac{Q}{A}\frac{\partial C}{\partial x} + \frac{1}{A}\frac{\partial}{\partial x}(AD\frac{\partial C}{\partial x}) + \frac{q_{LIN}}{A}(C_L - C) + \alpha(C_S - C) - \lambda C \quad (1)$$

$$\frac{dC_S}{dt} = \alpha\frac{A}{A_S}(C - C_S) - \lambda_S C_S \quad (2)$$

where A is the stream channel cross-sectional area [L^2]; A_S is the storage zone cross-sectional area [L^2]; C is the in-stream solute concentration [M L^{-3}]; C_L is the solute concentration in lateral inflow [M L^{-3}]; C_S is the storage zone solute concentration [M L^{-3}]; D is the dispersion coefficient [L^2 T^{-1}]; Q is the volumetric flowrate [L^3 T^{-1}]; q_{LIN} is the lateral inflow rate [L^3 T^{-1} L^{-1}]; t is the time [T]; x is the distance [L]; α is the storage zone exchange coefficient [T^{-1}]; λ is the in-stream first-order decay coefficient [T^{-1}] and λ_S is the storage zone first-order decay coefficient [T^{-1}].

The Routing Model

Inspection of Equation (1) reveals that the solute transport model depends on three hydraulic state variables: Q, A and q_{LIN}. These variables may be provided as output from a watershed-scale hydraulic routing model such as DR$_3$M (Alley and Smith, 1982).

DR$_3$M (Distributed Routing Rainfall-Runoff Model) is a watershed model that routes flow through contributing areas of a drainage basin and through the stream channel. This flow routing is based on a solution of the continuity equation and a kinematic approximation of the momentum equation:

$$\frac{\partial Q}{\partial x} + \frac{\partial A}{\partial t} = q_{LIN} \quad (3)$$

$$Q = \gamma A^m \quad (4)$$

where γ and m are constants that depend on the geometry, slope and roughness.

Model Linkage

Solute transport models and hydrodynamic models are typically linked so that simulations are conducted in two steps. First, the hydrodynamic model is run for the entire simulation period. During the course of the run, computed flow rates and cross-sectional areas are written to a flat output file. In the second step, the solute transport model is executed, and the hydraulic variables are read from the file at the appropriate times. One example of this approach is the union of the WASP water quality model with the DYNHYD hydrodynamic model (Ambrose et al., 1988).

The approach taken here is to link the models within the Modular Modeling System (MMS). MMS is a modeling framework that is specifically designed to link models of this type. Within MMS, each 'model' or process is implemented as a computational module that is executed for each time step.

As depicted in Figure 1, the hydrodynamic model and the solute transport model are linked as modules within MMS. During each time step, the hydrodynamic module is called to compute the flows, areas and lateral inflow rates. The solute transport module then obtains the values of the hydraulic state variables and computes the solute concentrations. A new time step is then initiated and the process is repeated.

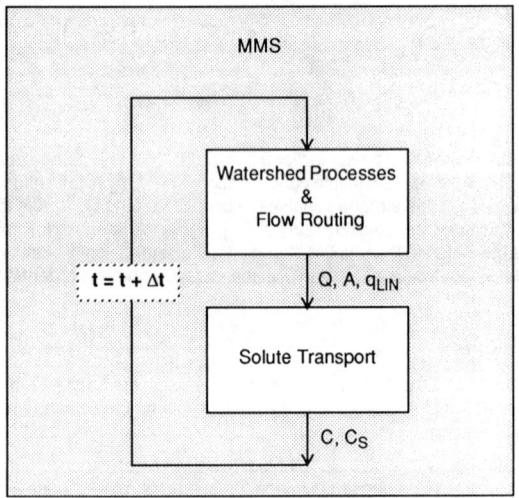

Figure 1. Linkage of Flow Routing and Solute Transport Modules within MMS

Discussion

When modeling solute transport in dynamic systems, it is often necessary to rely on a hydrodynamic model for the specification of hydraulic state variables. This has traditionally been implemented as a two-step process wherein an output file from the hydrodynamic model is used as input for the solute transport model. This paper presents an alternate scheme which makes use of the Modular Modeling System. Linking of solute transport and hydrodynamics under the MMS framework has several noteworthy advantages.

First, each model is formulated as a module that is called at each time step. This eliminates the need for flat files, as the hydraulic variables are made available to the solute transport module during each time step. In addition, the modeling framework provides a graphic environment that allows a model user to display output from each of the modules as they execute. For the problem described herein, the model user can request plots of flow and concentration, thereby identifying a potential relationship between flowrate and solute concentration.

Second, MMS is flexible modeling framework that allows modules to be added and interchanged. For example, if kinematic routing is not appropriate for a given situation, the DR_3M module could be replaced by a more sophisticated hydrodynamic module. Given the modular nature of the system, this change would not affect the structure of the solute transport code. Another example is the addition of an ecological module to assess the effect of a given contaminant on fish populations. Such a module would be placed after the solute module so that solute concentrations are available from the transport module (see Figure 1).

Finally, MMS provides a graphical user interface with which to view, edit and interpret data. The interface includes a spreadsheet that facilitates data entry, run-time graphics, sensitivity analysis features and parameter calibration tools.

Acknowledgments

This work was partially funded by United States Geological Survey (USGS) Toxic Substances Hydrology Program. The Modular Modeling System was developed through a cooperative agreement with the United States Geological Survey and the Upper Colorado Office of the United States Bureau of Reclamation.

References

Alley, W.M. and P.E. Smith, Distributed Routing Rainfall-Runoff Model, Version II, Computer program documentation, User's Manual, *U.S. Geol. Surv. Open File Rep.*, 82-344, 1982.

Ambrose, R.B., T.A. Wool, J.P. Connolly and R.W. Schanz, WASP4, A hydrodynamic and water quality model -- Model theory, User's Guide and Programmer's Guide, *Rep. EPA/600/3-87/039*, U. S. Environ. Prot. Agency, Washington, D. C., 1988

Bencala, K.E. and R.A. Walters, Simulation of solute transport in a mountain pool-and-riffle stream: a transient storage model, *Water Resour. Res.*, 19(3), 718-724, 1983.

Brown, L.C. and T.O. Barnwell, The enhanced stream water quality models QUAL2E and QUAL2E-UNCAS: Documentation and User Manual, *Rep. EPA/600/3-87/007*, U. S. Environ. Prot. Agency, Washington, D. C., 1987.

Henderson, F.M., *Open Channel Flow*, Macmillan, New York, 1966.

Leavesley, G.H., P.J. Restrepo, L.G. Stannard and M.J. Dixon, The modular hydrologic modeling system - MHMS, paper presented 28th Annual Conference & Symposium of the American Water Resources Association, November 1-5, 1992, Reno Nevada, 1992.

Restrepo, P.J., Leavesley, G.H., M.J. Dixon and L.G. Stannard, The modular hydrologic modeling system - MHMS, 1992 Spring Meeting of the American Geophysical Union, Montreal, Canada, 1992.

Runkel, R.L. and R.E. Broshears, One dimensional transport with inflow and storage (OTIS): A solute transport model for small streams, *Tech. Rep. 91-01*, Center for Adv. Decision Support for Water and Environ. Syst., Univ. of Colo., Boulder, 1991.

Thomann, R.V. and J.A. Mueller, *Principles of Surface Water Quality Modeling and Control*, Harper & Row, New York, 1987.

Object-oriented River System Simulation

Jon S Behrens[1] and Daniel P Loucks[2]

Abstract

This paper describes an object-oriented approach to river basin simulation modeling that is event driven rather than time driven as are most river system simulation models. At the heart of this method are autonomous software objects which replicate the physical behavior of the components of an actual river system. These objects react to stimuli (changing values of their inputs, outputs or state) in much the same way as do the analogous components of the actual river. The simulation of river systems is the result of the coordinated actions, in any order, of the various objects representing the entities within the system.

An object-oriented approach to modeling a river has a number of characteristics that make it worthy of consideration. The model's architecture makes it especially suited to graphic user interfaces and database operations. It allows the easy integration of artificial intelligence techniques that can aid in expressing complex policies governing system operation.

Introduction

The river simulation modeling approach to be discussed in this paper was developed under the primary sponsorship of the US Bureau of Reclamation as a way to allow flexibility in the modeling of the Colorado River. The approach taken here to preserve flexibility while allowing expression of the complex operating policies of rivers like the Colorado is to separate policy definition from the program that performs the simulation. The emphasis in this paper is on the so-

---

1. President, Jon Behrens & Associates, Inc., 5575 Bowron Pl., Longmont, CO, 80503.
2. Professor, Civil Engineering, Cornell Univ., Ithaca, NY, 14853.

lution process of the simulator, which can operate independent of policy statements through the usual method of guide curves and demand schedules.

Object-Oriented Structure

River systems in most simulation models are represented as networks of nodes and links. The nodes of the network represent components of the river system such as reservoirs, diversion and confluence sites, hydropower plants, etc. The links of a node-link river system network usually represent river reaches, i.e. the flow of water from one component to another. Using object-oriented jargon, all these components of a river system, both nodes and links, are called objects.

Objects are organized in a hierarchical class structure. The difference between classes and objects is important. A class may be regarded as a template. The class specification determines what data "slots" an object can contain and what operations an object can perform on the data in its slots. Figure 1 shows part of the class structure for a river simulator. Figure 1 reads from top to bottom with each class inheriting the capacity to store data and to perform functions from all classes above. In each class definition, the data slots that are added by that class are in bold type while those that are inherited are in regular type. For example, the water class provides data slots for inflow, outflow and diversion. The reservoir class, which inherits from the water class, simply incorporates those data slots without further elaboration and adds data slots for storage, evaporation, etc.

All of the above is abstract until we actually make an object using the template provided by the class. The object unifies the template with data appropriate to the entity being modeled. For example, to create a particular reservoir object called Lake Mead we would load the template's data slots with information about Lake Mead. We would then have an object that was prepared to represent that reservoir during a simulation. In other words, an object is the combination of a class definition and some specific data. We could continue to make objects until we ran out of memory in the computer.

In addition to the capacity to store data, the class template provides functions that can operate on that data. Thus, the node class provides functions that deal with the graphic user interface and communicate with other objects. The water class adds to this the ability to perform mass balance on an object. The reservoir redefines, or overloads, the mass balance function to handle storage and losses.

As noted in Figure 1 all water class objects are able to perform a mass balance operation and a communication operation, the lat-

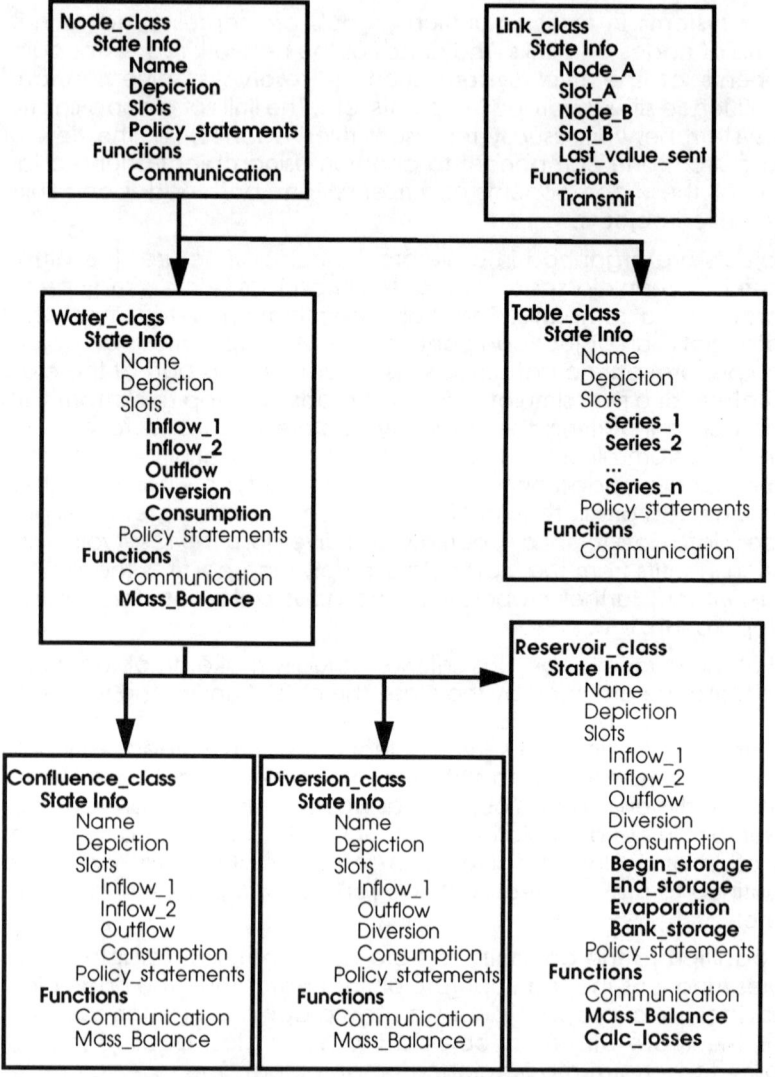

Figure 1. Typical Hierarchical Class Definitions for a River Simulator. Plain-text data slot names indicate inheritance from higher classes.

ter inherited from the parent node class. The mass balance operation involves the beginning (BS) and ending (ES) storage volumes, the inflow (I), the diversion (D), the consumption ratio (C), the outflow (O), and the evaporation and bank storage losses (L), as applicable.

$$ES = BS + I - C*D - O - L \qquad (1)$$

The beginning and ending storage volumes of an object cannot exceed the object's storage capacity. These capacity limitations are, of course, different for each object, and may differ in various time periods. This mass balance equation can only be solved when there are sufficient data, i.e. when all but one term is known. That single unknown term is calculated, possibly with the aid of routing functions in the case of reach objects, and with other functions and policy statements in the case of reservoir, hydropower, confluence and diversion objects.

As new data values are obtained through the performance of this mass balance operation, they are placed in the appropriate data slots of the object. These data are also transferred to other objects' data slots through user-defined communication links (instances of the link class).

As the links move data from those objects that have solved equation 1 to those which have not, the receiving objects gradually build up enough information so that they may also solve that equation. This solution--communication process continues until all objects are in balance, at which point, the next time step can start.

Conclusion

This paper has presented a brief description of an object-oriented simulation scheme which makes use of autonomous software objects to replicate the behavior of a river basin. This scheme promises to yield dividends in improved flexibility and maintainability of the resulting models. The full power of this scheme only becomes apparent when considered in combination with the use of policy statements. These statements give each object the ability to consult an expert rule base in the course of the simulation.

RSS: A Construction Kit for Visual Programming of River Basin Models

Rene Reitsma, Andy Sautins, and Steve Wehrend[1]

Abstract.

With the ever increasing pressure on environmental resources, there exists strong interest in river basin models which can be used to simulate proposed policy alternatives. Traditionally, the development of these models has been a time consuming and expensive endeavour. In addition, these models have typically addressed a limited range of problems and were often not readily extensible. Combining an object-oriented network representation of river basins with principles of visual programming, however, allows for the rapid construction of modifiable river simulation models.

RSS is a construction kit for the visual programming of river basin models. It provides the basic tools necessary to construct a variety of river basin models from a set of primitive water and information objects. These objects can be linked together to form a network which models a desired river basin. In addition, these objects can contain data and/or methods for determining their behavior as they interact with objects to which they are connected.

Introduction.

Decision Support Systems (DSS) are computer based systems characterized by an integration of three elements: state information, process information and analysis. State information is the representation of the state of the problem one tries to solve. For instance, if the problem concerns the monthly operation of a set of storage reservoirs on a river basin, the state information consists of all the "state variables" which uniquely describe the river basin at particular moments in time. Process information, on the other hand, describes the system dynamics over time. For instance, information as to how a reservoir evaporates water, how much electricity is generated by a power plant or how salt is transported through the river system are all system dynamics.

Combining state and process information allows for simulation modeling, since simulation involves the transition of a system from one state into another through a set of dynamics. The addition of the analysis component makes a simulation system into a decision support system. Analysis tools are used to process data generated by the simulation and transform it into a form that supports decision making. Examples range from simple plotting and graphing functions to sophisticated multi-criteria evaluation modeling (Fedra 1990, Reitsma 1990, Reitsma & Sieh 1992).

These three components of DSS can be integrated in a number of ways. In dedicated approaches, state and process information integrate into a single software framework, tailored

---

1. Research Associate, Professional Research Assistant and Professional Research Assistant, respectively, CADSWES/CEAE, University of Colorado, P.O. Box 428, Boulder CO 80309-0428

for the specific decision problem under study (USBR 1987, Fedra 1991). In more generic approaches, state and process information are separated so that different states and processes can be combined arbitrarily (USACE 1982). What these approaches have in common is that the simulation (state and process information) is kept completely separate from the analysis. They also share flaws, which are discussed separately (Reitsma & Sieh 1992).

DSS architectures typically employ separate analysis and simulation utilities which can be applied to data. Although this approach is useful for utilizing existing software applications, it requires the creation of additional software necessary to link these applications. This is a result of a rather awkward operator-operand model. In this model, the operators are the processes and analysis to be conducted and the operands are the state information. Thus, the operators and operands remain separate and the state information must be transformed into a formats that are usable by the different processes.

Object orientation offers a solution to this problem. It allows process, state and analysis information to be stored on a single software object. For example, a reservoir object will contain data that describes its current state and the methods that determine the object's behavior, such as evaporation. An information object will also contain data and may contain methods to display this data for analysis purposes.

Object-oriented modeling.

In an object-oriented world objects themselves know about the organization of their information. Typically, state information pertains uniquely to a specific object. Thus even though Lake Mead and Lake Mohave might have the same organization of state information (they both have variables such as "current storage", and "spill"), the values of the variables are unique for each object. Unlike state information, process information is not uniquely identified for specific objects. Instead, it is defined for entire groups of objects known as "object classes". Thus, the algorithms or methods for evaporating water for both Lake Mead and Lake Mohave are defined on a class called Reservoir. Both Lake Mead and Lake Mohave inherit these methods because they are instances of the Reservoir class.

Object oriented applications typically involve various sets of objects which maintain communications with each other. In the case of RSS, for example, Reservoir and Power Plant objects communicate with each other by sending quantities of water through physical links. The sender of such water might be an upstream reservoir ready to release water to a reservoir downstream, or a downstream reservoir "requesting" water from upstream. Figure 1, for example, shows a network of objects communicating water through links.

Object oriented models have a distinct "discrete event" or "stimulus-response" characteristic. Individual objects behave only in response to stimuli exerted on them by other objects. The system as a whole behaves by virtue of the behavior of the individual objects and the communication links between them.

Apart from technical advantages, object orientation also allows for a more natural way of communicating a decision or analysis problem to a DSS system. For instance, reservoir operators may need to know the best way of running the upstream part of a river basin. The most natural way of posing the problem to a DSS is by describing each object in the system and how they interrelate. These objects, since they are the nexus around which the decision problem is organized, carry around all the necessary state, process and analysis information needed to perform a specific task.

This bundling of information on an object can only work satisfactorily if users have direct access to these objects. In essence, this means that the user functions as an external object, connected to the other software objects through a user interface. This requires construction of a user interface which allows access to object information in a manner similar to the way in which objects communicate among themselves. In other words, if a user wants a reservoir object to set its end-of-period-storage variable to a series of 12 values, the user should construct that series and send it as a message to the reservoir. The reservoir receives the message, inspects

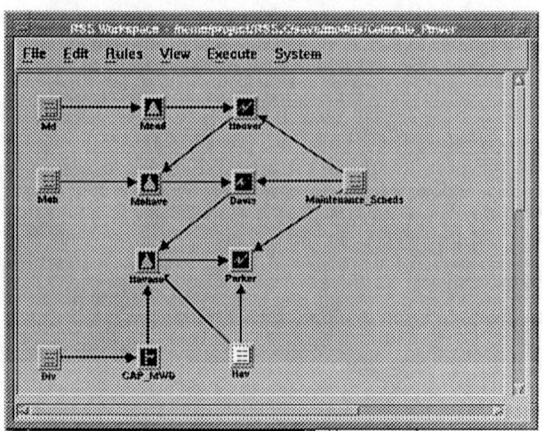

Figure 1: Objects in RSS communicating water through links.

it, and stores the values for later use.

RSS is equipped with a graphical user interface which allows users to do exactly that. Users select one or more objects on the screen (by clicking on them with a mouse pointer), then specify a message to be sent to the object. For instance, objects on the RSS Workspace (Figure 1), may be sent the open message, telling them reveal their state and process information (See Figure 2). In addition, some objects may be sent a more complicated message, such as a request to generate a plot of the object's data. Likewise, objects can be told to consult a rule, or policy, base containing conditional statements about how they should react given information received from other objects.

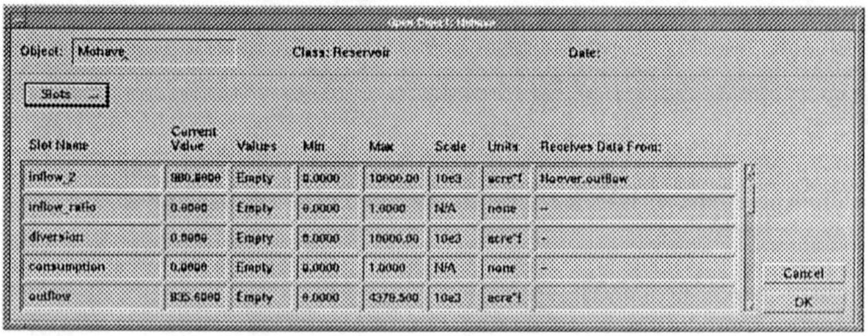

Figure 2: Opening RSS objects.

Visual Programming in RSS.

As argued earlier, object orientation allows for very tight coupling of simulation modeling, analysis, and user interaction. By the same token, RSS can be considered as an example of what is generally known as "Visual programming" (Pong and Ng 1983, Grafton and Ichikawa (eds.) 1985, and Myers 1986). A visual programming environment is one in which the user is able to specify a program in a graphical manner. This often implies using a palette of objects to draw a picture that defines the program.

In RSS this concept of visual programming has been translated into a graphical procedure for specifying river simulation models. In RSS models are programmed by simply generating objects from a palette and then graphically connecting these objects. This connecting, or linking, consists of a simple mapping of one or more variables of one object onto one or more variables of the other object. For instance, a link is created between a reservoir and a power plant, the linking consists of three one-to-one variable mappings: the reservoir's outflow, beginning elevation and ending elevation map to the power plant's inflow, beginning elevation and ending elevation, respectively.

Conclusion.

The object oriented approach applied in RSS lends itself to a visual programming environment. Since each object contains all of the information necessary to determine its state and to communicate with other objects, a program may be written by instantiating and combining these objects from a palette and loading input data. The interrelationship between the objects coupled with their internal data and methods determines how the system will behave when it is executed. While this is not programming in the traditional, or textual, sense, it is nonetheless programming. This approach has been used to build models of existing and hypothetical river systems and is well suited to allowing users to graphically express and solve river system simulation problems without having to learn or know a traditional programming language.

Acknowledgments.

This work was partly sponsored by the U.S. Bureau of Reclamation under Contract No. 8FC-81-12480. References to any products either stated or implied are for example or clarity to the reader and do not represent endorsement by the U.S. Government. Opinions are those of the authors and do not necessarily represent an opinion or position of the U.S. Bureau of Reclamation.

References.

Fedra, K. 1990 "Interactive Environmental Software: Integration, Simulation and Visualization." IIASA RR-90-10; IIASA, Laxenburg, Austria.

Fedra, K. 1991 "A Computer-Based Approach to Environmental Impact Assessment." IIASA RR-91-13; IIASA, Laxenburg, Austria.

Grafton, R.B. & Ichikawa T. (eds.) 1986 "IEEE Computers: Special Issue on Visual Programming; vol. 18(8).

Meyers, B 1986 "Visual Programming, Programming by Example and Program Visualization: a Taxonomy." *CHI'86 Proceedings*; pp. 59 - 66.

Pong, M.C. & Ng, N. "Pigs, a System for Programming with Interactive Graphical Support." Software - Practice and Experience; vol. 13; pp. 847-855.

Reitsma, R.F 1990. "Functional Classification of Space; Aspects of Site Suitability Assessment in a Decision Support Environment." IIASA RR-90-2, Laxenburg, Austria.

Reitsma, R.F. & Sieh, D. 1992 "Bootstrapping Models Using Existing Databases and Object Orientation." *Proceedings of the Eighth Conference on Computing in Civil Engineering*; Dallas, TX.

US Bureau of Reclamation 1987 "Colorado River Simulation System Documentation: System Overview." Denver Colorado.

US Army Corps of Engineers 1982 "HEC-5 Simulation of Flood Control and Conservation Systems. users Manual." Hydrologic Engineering Center, Davis, CA.

STOCHASTIC RESERVOIR OPERATION UNDER DROUGHT WITH FUZZY OBJECTIVES

E. PARENT[1] , L. DUCKSTEIN[2]

Abstract

Biobjective reservoir operation under drought conditions is investigated using stochastic dynamic programming. As both objectives (irrigation water supply, water quality) can only be defined imprecisely, a fuzzy set approach is used to encode the decision maker (DM) 's preferences. The nature driven components are modeled by means of classical stage-state system analysis. The state is three dimensionnal (inflow memory, drought irrigation index, reservoir level); the decision vector elements are release and irrigation allocation. Stochasticity stems from the random nature of inflows and irrigation demands. The transition function includes a lag one inflow Markov model and mass balance equations. The human driven component is designed as a confluence of fuzzy objectives and constraints after Bellman and Zadeh (1970). Fuzzy numbers are assessed to represent the DM's objectives by two different techniques, the direct one (Bardossy and Duckstein, 1992) and indirect pairwise comparison (Saaty, 1980).The real case study of the Neste river system in southwestern France is used to illustrate the approach; the result are compaired to a classical sequential decision theoretical model derived earlier (Parent et al., 1991) from the viewpoints of ease of modeling, computationnal efforts, plausibility and robustness of results.

Introduction

Water resources management constitutes a fertile ground for developping and applying decision theoretic techniques for it may be considered as an engineering attempt to control the dynamic evolution of a complex system under imprecision and randomness.

---

[1]Department of Applied Mathematics and Computer Science, ENGREF, 19 av du Maine, 75015 PARIS, France.

[2]Department of Systems and Industrial Engineering, University of Arizona, Tucson, 85721 USA

In the last decade, much light has been cast on the mathematical techniques to deal with imprecision provided by the fuzzy set theory. In the water resources field, decision processes in which fuzziness enters in one way or another have been studied for example in Bardossy and Duckstein (1992). In the papers of this avenue of thought, a fuzzy set approach has been developed to yield a multiobjective decision making that takes into account qualitative and imprecise environmental objectives. In the present paper we will focus on both imprecision and randomness. Fuzziness is incorporated to reflect the vagueness of human operational objectives. Randomness herein describes natural uncertainty concerning the unknown future states of the stochastic water system under control.

Nature-Driven system description

The Lannemezan Plateau in the Gascogne basin is bordered by the Pyrenees Mountains in the South and the Gascogne and Adour rivers in the Est and West . Consequently, the Gascogne basin rivers are real french "washes" that used to dry out in summer when no precipitation occurs. As early as 1850, a 29 km long canal (the Neste canal) was built in order to convey water from the mountain river Neste to the top of the Gascogne watershed. For more than a century, this abundant new facility has allowed for the development of 30,000 ha of intensively irrigated corn agriculture . A second purpose of the Neste canal is to supply continuous minimum levels in the Gascogne rivers for the purposes of water quality control and domestic supply. In order to fulfill increasing irrigation demand, new dams of a total capacity of 42 hm^3 have been built in the hills of the Gascogne basin in addition to the 48 hm^3 reservoir in the Pyrenees Mountains that already supplies the river Neste. Nowadays, water supply and irrigation demand compete for the same limited resource. This system has been modelled in a lump manner in Parent (1991) and can be represented by a stage-state stochastic discrete time model. Let t be the stage in the decision process, the influence of the decision vector u and the stochastic input ω on the evolution of the state x is expressed in the form of a transition function :

$$x(t+1) = f(t, u(t), x(t), \omega) \dotfill (1)$$

Human-Driven System Description

There exists a minimum constraint that must be kept in the system which can be represented as a contract between the system manager and the water users. The system manager assumes the responsibility to set this level as high as possible , for water quality and aquatic life purposes. In normal periods , the reservoirs release enough water to satisfy irrigation demand and other uses , and there is a general agreement that it is of no use to set this level too high. Considering the climatological and agronomical conditions, irrigation is a necessary means to get maximum corn yields. Nevertheless, in case of water shortage, domestic use and water quality become higher priorities. In this case, the farmers and the system manager agree on a certain amount that will not be furnished to crops. When the system enters this restrictive operation mode, because of agricultural behaviors and administrative reasons, it will practically take some time to switch back in the following periods. An other aim of the system manager is to end the 20 weeks of summer irrigation period with a sufficient amount of water in the reservoirs so as to prevent a very dry automn. Due to the vagueness of these three requirements, after many discussions with the system manager, we modelled objectives and constraints using fuzzy number theory by a fuzzy constraint will express his care for pushing the salubrity level as high as possible for each step t of operation..A terminal goal at the end of the operation period T will indicate that the irrigation period should be as long as possible. A second terminal goal is reflected by a fuzzy goal membership function for setting the reservoir level at T. Using fuzzy cartesian product definition, these last two fuzzy goals can be expressed by a fuzzy vector membership function $\mu_G(x(T))$.Such membership functions are constructed by means of thresholds (Bardossy and Bogardi,1989) or by pairwise comparisons between a discrete number of possible values to assess their relative grade of membership to a desired set after Saaty (1977).

Derivation of operation rules.

Recalling Equation 1 describing the nature-driven system component and associating the human-driven fuzzy objectives and constraints , a rule of operation can be obtained using the classical framework of stochastic dynamic programming. We will state like in Bellman and Zadeh(1970) that the concept of decision is a *confluence* between fuzzy goals and fuzzy constraints. More specifically let $\mu_{c(t)}(u(t))$ be the membership function of the fuzzy constraint for the decision vector and $\mu_G(t+1)(x(t+1))$ be the membership function of the fuzzy goal at time t+1 we can obtain the recurrence equations which are simple instances of dynamic programming applied to membership functions.When dealing with stochastic dynamic systems of the form of Eq 1,the reccurence equations are replaced by

$$\mu_{G(t)}(x(t)) = \max_{u} \{ \min [\mu_e(u, x(t)), \mu_{c(t)}(u)] \} \quad \ldots \ldots \ldots \ldots (2)$$

with $\mu_e(u,x) = \int_{-\infty}^{+\infty} \mu_{G(t+1)} (f(t,u,x,\omega)) \, dP(\omega)$

Conclusions

In the conventionnal approach to stage-state modeling, one should take into account a set of constraints and a performance function which associates with each operation rule a gain resulting from the choice of that policy. Researchers frequently face a lack of economic data to assess such performance functions, or fail to resolve the difficulties that arise when attempting to establish accurate indices due to multidimensionnal human perception, which is often qualitative, always subjective, never precise.The vagueness of the decision maker naturally appears in the fuzzy assertions of the previous sections such as:" the reservoir level should not be *lowered too early*" or " the river streams should be maintained *reasonnably high* so as to ensure a *satisficing* value of *water quality*" and so on. Despite the imprecision of the italicized words, the previous sentences contain much usable information and both explicit and precise decisions are to be made to manage the system. Those fuzzy statements can not be easily reduced by statistical means: in a way they are fundamentally imprecise by nature, with no hope to wait until the exact value is revealed or no means to estimate statistics of the variable of interest. The concepts of fuzzy goals and fuzzy constraints can help to make implementable precise decision rules. The theory of fuzzy numbers is much simpler to explain to decision makers because the notion of probability measure is replaced with the concept of membership function which often obey simpler arithmetical rules. For the system analyst, choosing to treat the two distinct concepts of fuzziness and randomness with distinct mathematical techniques may appear more appealing than embedding the whole process in the conceptual framework of probability theory. However there does not exist an axiomatic background to translate human attitude towards fuzziness into fuzzy numbers. The operational decison rules derived by the fuzzy approach should therefore be compared to classical techniques of decision theory such as the Bayesian approach (Berger, 1985) or utility based techniques (Keeney and Raiffa, 1976).

References

Bardossy,A., Bogardi,I, (1989),Fuzzy fatigue life prediction in *Structural Safety*(6):25-38 Elsevier.
Bardossy,A. and L. Duckstein, 1992. Analysis of a karstic aquifer management problem by fuzzy composite programming. *Water Resource Bulletin*,28(1):1-11.
Bellman, R., E., Zadeh, L., A., 1970. Decision Making in a fuzzy environment. *Management Science*, 17(4):141-161.
Berger, J.O, (1985), *Statistical decision theory and bayesian analysis*, Springer Verlag.
Keeney, R, Raiffa,H (1976), *Decisions with multiple objectives*,Wiley.
Parent, E, Lebdi, F., Hurand P. , 1991. Stochastic modeling of a water resource system : analytical techniques versus synthetic approaches. *Water Resources Engineering Risk Assessment*:415-434. Ed GANOULIS, Springer-Verlag.
Saaty, T. L., 1980. *The Analytic Hierarchy Process*, McGraw-Hill.

Seasonal Water Quality Management Given Sparse Data

Andrews K. Takyi[1], and Barbara J. Lence[2], Associate Member, ASCE

Abstract

   An approach for designing robust seasonal water quality management programs for river basins with sparse or short flow records is presented. This approach uses Markov chain modelling and linear regression relationships between low flow records at adjacent gauging stations to calculate transition probabilities for low flow states of the water quality system. The robustness of the solutions derived from the Markov chain approach is demonstrated for a seasonal uniform waste treatment management model for BOD waste discharges on the Willamette River in Oregon. The results of this application show that management decisions based on the Markov chain approach may be more successful at maintaining the acceptable water quality goal than existing seasonal waste management programs, especially for river basins with sparse historical flow records.

Introduction

   Mathematical programming models for water quality management are generally used to generate information about a river system and to identify alternative management strategies. Under uncertain circumstances, however, the information produced by these models may not accurately represent the water quality system and waste load allocations based on such poor information could result in failure to meet management objectives. Poor information generated by such models can be attributed to uncertainties arising from two main sources, uncertainties in the model structure or formulation, and uncertainties in the input data. Uncertainty in model structure may arise from the lack of complete understanding of the relationships between parameters, variables, and model output, the poor choice of assumptions on which the model formulation is based, and the presence of elements of the water quality system that cannot be represented mathematically. Uncertainty in input data is caused by the stochastic nature of water quality data, the limited amount of input data that are available for most river basins, and the errors made in measuring, processing, and recording data. In this paper, we consider uncertainty in low flow data and develop a seasonal water quality management model that is robust to the sparseness of the historical low flow record.
   The model assumes uncertainty in the low flows at the gauging stations and maintains a specified maximum probability of water quality violation given

---

[1]Graduate Research Assistant, Civil Engineering Department, University of Manitoba, Winnipeg, Manitoba, Canada, R3T 2N2
[2]Assistant Professor, Civil Engineering Department, University of Manitoba, Winnipeg, Manitoba, Canada, R3T 2N2

probabilities of low flow states. A first order Markov process for low flow data at any two adjacent gauging stations is used to model the uncertainty in the flow data. The Markov chain approach is capable of incorporating the relevant statistical relationships in the low flow data even if these data are sparse or limited in length. The robustness of the model output is assessed by the sensitivity of the output to the sparseness and the length of the historical low flow data. The model and the sensitivity analysis is demonstrated for the management of BOD waste on a typical river and results are compared with those obtained from an existing management program, the Risk Equivalent Minimum Average Uniform Treatment (MAUT) program, developed by Lence et al. (1990).

Seasonal Markov Chain Model (MARK)

Under the MARK model, the transition probability matrices for low flows at adjacent gauging stations are developed and used to determine the probability of all possible critical states of the water quality system for each season of the year (i.e., the combination of low flows for gauging stations throughout the stream in that season). For each possible state of the system in a given season, a uniform treatment management solution is obtained. That is, for each system state t, of each season s, a linear programming (LP) model is used to minimize the uniform fraction waste removal, ξ_{st}, required to maintain the desired water quality standard at all locations along the river. The ξ_{st} values are used to obtain a probability distribution for the waste removal levels for season s, ξ_s. Then, a non-linear programming (NLP) model is applied which selects the design seasonal uniform fraction removal levels for each season s, η_s, that minimize total cost of waste treatment and maintain an acceptable probability of water quality violation. The NLP uses the probability distribution for the uniform fraction removal level, ξ_s, obtained from the LP models and limits the joint probability of non-violation of the water quality for all seasons. Let P_s denote $P[\xi_s > \eta_s]$, where $P[\xi_s > \eta_s]$ is the probability that ξ_s is greater than η_s. The formulation for the NLP portion of the MARK model is:

$$\text{Minimize} \sum_{s=1}^{S} f_s \sum_{d=1}^{D} C_d \mu_{ds} \tag{1}$$

subject to

uniform treatment requirement:

$$\mu_{ds} - \eta_s = 0 \quad \forall \ s=1,...,S; \ d=1,...,D \tag{2}$$

limit on the probability of water quality violation:

$$\prod_s (1 - P_s) + R - 1 \geq 0 \tag{3}$$

definition of P_s:

$$P_s = \sum_{h \in G_t} P[Q_h^s] \tag{4}$$

limits on the allowable treatment level for a given season:

$$\mu_{ds} \leq \mu U_{ds} \quad \forall \ s=1,...,S; \ d=1,...,D \tag{5}$$

$$\mu_{ds} \geq \mu L_{ds} \quad \forall \ s = 1,..., S; \ d = 1,..., D \tag{6}$$

where, f_s is the fraction of the number of days of the year in season s; C_d is the annual cost per unit of fraction removal for discharger d ($/fraction removed); μ_{ds} is the design fraction waste removal for discharger d in season s; S is the total number of seasons per year; D is the total number of waste dischargers; μU_{ds} is the upper limit on design fraction waste removal for discharger d in season s; μL_{ds} is the lower limit on design fraction waste removal for discharger d in season s; R is the maximum acceptable probability of water quality violation in a given year; $P[Q_h^s]$ is the probability that the system is in state h in season s; G_t is the set of all system states that result in ξ_s being greater than η_s; and h is the system state which produces a fraction waste removal level, ξ_s, that is greater than η_s.

The objective function minimizes the total annual cost. The first constraint set, *Equation 2*, represents the uniform treatment requirement for all dischargers in any season, s. *Equation 3* ensures that the probability of water quality violation in any given year does not exceed an assigned maximum value and *Equation 4* defines the probability of water quality violation in each season, s. *Equation 3* assumes that the uniform fraction removal levels obtained from the linear programming solutions are independent from one season to the next. This is a reasonable assumption since the spatial dependence of low flows is more important than seasonal persistence for predicting states of the water quality system if the number of seasons is small (e.g., two to four depending on the application).

A MARK Model for the Willamette River

The MARK model is applied to a two season BOD waste management program for the Willamette River in Oregon and compared with an existing seasonal management model, the MAUT model (Lence et al., 1990). A description of the cost data, waste load characteristics, river velocity, reaction rates, background water quality conditions, and flow and temperature data used for analysis are given in Lence et al. (1990). The models are evaluated for a 7.50 mg/l DO standard and a seasonal program with an eight month winter (October through May) and a four month summer (June through September) season. The probability of water quality violation is 10%.

Thirty years of low flow data at Springfield (Station 1), Harrisburg (Station 2), Albany (Station 3), Salem (Station 4), and Portland (Station 5), Oregon, and the highest monthly mean temperature at Harrisburg for a given season are used in this analysis. Each pair of adjacent stations, from Station 2 to 5 inclusive, fit a linear regression model. The low flows at Stations 1 and 2, did not fit a simple linear regression model, however, the difference between the flows at Stations 1 and 2, and the flow at Station 2, fit a simple linear regression model. Therefore, conditional probabilities were determined between the low flows for the inflow between Stations 1 and 2 and for Station 2. A lognormal distribution was assumed for the low flows of the inflow between Stations 1 and 2, and used to estimate unconditional probabilities for these flows (see, e.g., Takyi, 1991). Seventeen flow intervals were assigned to the low flow data for the inflow between Stations 1 and 2 and for Stations 2 to 5.

To analyze the robustness of the MARK and MAUT models for cases where the data are sparse, the models are evaluated for flow scenarios that represent 30 years of record in which 10 years of data are missing. Each scenario of the low flow data is developed by randomly selecting 20 years of low flow data from the actual 30 years of historical low flow data for the five gauging stations. One hundred low flow scenarios are simulated and treatment levels for each scenario are evaluated. Also, solutions are evaluated for these two models based on all 30 years of historical low flow record. These are referred to as Actual MARK and Actual MAUT solutions.

To analyze the results, we refer to Solution A as dominating Solution B if and only if the waste removal levels from A in both seasons are equal to or greater than the corresponding removal levels from B. Thus, if Solution A dominates Solution B, then the former is either as conservative or more conservative than the latter. In such cases, management decisions based on Solution A would have the same or higher probability of meeting the acceptable DO standard than those based on Solution B. Also, Solution C is inferior to Solution D if Solution D dominates Solution C and the two solutions are not equal. Generally, an inferior solution is more vulnerable to failure to meet a specified water quality standard.

The results of the 100 simulations indicate that the MARK model generally designs higher winter and summer waste removal levels than the MAUT model. Of the 100 simulations, 83 sets of waste removal levels produced by the MARK model dominate the corresponding solutions from the MAUT model, while none of the MAUT solutions dominate the corresponding MARK solutions based on the same simulated low flow records. Thus, results from the MARK model are expected to perform better than those from the MAUT model with respect to maintaining acceptable water quality. Despite the general dominance of the MARK model solutions over those of the MAUT model, the differences between these solutions are not very large, so the MARK model results do not indicate large over-design of waste removal levels.

Comparisons of the Actual MARK and MAUT solutions with the 100 simulations indicate that 48% of the simulated MARK solutions dominate the Actual MARK solution and 42% of the simulated MAUT solutions dominate the Actual MAUT solution. Although these two percentages for the MARK and the MAUT models are similar, by contrast all the remaining 58% of the simulation results for the MAUT model are inferior to the Actual MAUT solution while only 36% of the simulation results for the MARK model are inferior to the Actual MARK solution. For the MARK model, 16% of the simulations produce removal levels that do not dominate and are not dominated by the Actual MARK solution. For this 16% of the simulations, although the winter removal levels for the MARK model are lower than that for the Actual MARK solution, the summer removal levels are higher. The under-design in winter may be compensated for by higher removal levels in summer. Therefore, the ability of solutions from a MARK model to meet the acceptable DO standard is less sensitive to sparse low flow data than those from the MAUT model, for the case study presented. Consequently, the MARK model may design more reliable water quality management decisions for sparse low flow data than the MAUT model. This is an attractive attribute of the MARK model since low flow data in some seasons are sparse for many river basins.

References
Lence, B. J., Eheart, J. W. & Brill, E. D. Jr. (1990) Risk equivalent seasonal discharge programs for multidischarger streams, *J. Wat. Resour. Planning and Manage. Div. ASCE. 116* (2), 170-186.
Takyi, A. K. (1991) *Uncertainty analysis for environmental management models: Application of the generalized sensitivity analysis.* M.Sc. thesis submitted to the Faculty of Graduate Studies, University of Manitoba, Winnipeg.

Imprecise Probability and Water Resources Decisions

W. F. Caselton1[1] M. ASCE and W. Luo[2]

Implementing Imprecise Probability

Our present perspective on quantitative decision making under uncertainty has been shaped, to a large degree, by the Bayesian approach. When compared with the classical alternatives, the Bayesian approach offers greater freedom of inputs, allowing both subjective and statistical information sources. Yet there are other perspectives that suggest that these inputs have to be expressed in an unnaturally precise and constrained fashion so that, when the input information is particularly weak, some important uncertainties are inadvertently being suppressed or ignored.

In water resources engineering weak information would correspond to circumstances when the record is extremely short relative to the mean rate of occurrence of important design events, and when the subjective information has been derived from a very limited experience of such events. It is, of course, not uncommon for some of the most important water resources decisions to be made under these conditions.

The term imprecise probability has been coined [Walley 1991] to describe a concept of uncertainty which attempts to recognize these neglected uncertainties due to weak inputs. A number of quite different imprecise probability implementation schemes , which still involve conventional probability to some degree, have been proposed. The nature of the quantitative results produced depends upon the implementation scheme adopted. The two imprecise probability methodologies discussed here have the considerable advantage of retaining a close link with the established Bayesian framework. These are Bayesian Robustness [Berger, 1985], and Dempster-Shafer [Shafer, 1976] methods.

---

[1] Associate Prof. Dept. Civil Eng. University of British Columbia, Vancouver, V6T 1W5, Canada
[2] EIT, Hydroelectric Engineering Division, BC Hydro, 6911 Southpoint Drive, Burnaby, V3N 4X8, Canada

In conventional **Bayesian Analysis** the prior distribution, sample likelihood function, and posterior distribution, are all "precise". That is, they express, quite unequivocally, whether one possible value of the state of nature θ is more likely, equally likely, or less likely, than any other value. This precision cannot be avoided even when the prior information approaches ignorance and the sample data is sparse.

In order to diminish this artificially imposed precision, the **Bayesian Robustness** approach formally introduces vagueness into the specification of a prior by adopting a set of possible prior distributions. The choice of these priors is discretionary. A conventional precise sample likelihood is still adopted. A conventional Bayesian analysis is performed for each of the priors and the final output is expressed in the form of the two posteriors which yield the highest and lowest expected utilities.

The **Dempster-Shafer (D-S)** scheme involves a more significant departure as it permits an explicitly possibilistic expression of the prior and sample information in the form of Basic Probability Assignments (BPA's). This is achieved by allowing probabilities (or proby. densities) to be assigned to intervals and these intervals may overlap. How the probability (or probability density) is distributed within an interval is not specified. Prior BPA's can be elicited and the D-S inference procedures detect imprecision arising from the sample limitations and yield an inference based BPA. Dempster's combination rule is used to combine the prior and sample based BPA's and this yields a result, also possibilistic, in the form of a resultant (i.e. posterior) BPA on θ. Unlike the conventional Bayesian scheme, both prior and sample based inputs are treated symetrically. Furthermore, if there is no prior information then a prior BPA is not required. A unique and universal ignorance prior BPA does exist and has the attractive characteristic of not affecting the resultant BPA even if included in the analysis.

An introduction to some of the relevant D-S theory and its implementation in a water resources decision making context can be found in [Caselton and Luo, 1992]

No decision analysis procedure yet exists which utilizes the resultant BPA directly and it must be interpreted for that purpose. One useful interpretation of the resultant BPA is as a prescription for the set of conventional distributions which fall within the possibilistic limits established by the BPA. These are known as "compatible" distributions. Amongst these there exist two easily extracted conventional distributions which produce the highest and lowest expected utilities when the utility function is a monotone function of θ. These distributions, identified here as $f(\theta)$ and $g(\theta)$, resemble the two posterior distributions produced in the Bayesian Robustness approach in some respects but, technically speaking, are not equivalent.

In spite of its radically different framework, the D-S scheme does retain a connection with the conventional Bayesian scheme. Whenever the prior is a conventional precise Bayesian distribution, then the D-S scheme and the Bayesian scheme produce identical single distribution results for the same sample information. Dempster therefore claimed that the conventional Bayesian analysis is a special case of his more generalized scheme [Dempster, 1968]. Also, as the sample size gets larger, $f(\theta)$ and $g(\theta)$ both approach the Bayesian posterior.

Quantitative assessment of imprecise probability.

The D-S scheme offers advantages in a quantitative investigation of the effects of imprecision. It is able to analytically detect imprecise probabilities arising from sample inadequacies and, in addition, it does not insist on a prior. In contrast, the Bayesian Robustness approach confines all of its imprecise inputs to the prior information, that is, to the subjective inputs. Any results obtained from an investigation of Bayesian Robustness would throw no light on imprecise probabilities arising from sample inadequacies.

The Type I Extreme value distribution is commonly used in Water Resources engineering and its parameters must often be estimated using inadequate record lengths. It was therefore considered worthy of investigation from an imprecise probability point of view.

Some typical numerical results.

A random event was assumed to be modelled by the Type I Extreme (largest) value distribution with known skew parameter but unknown location parameter. Sample values were first simulated using this distribution with a specified value for the location parameter. The location parameter was then inferred using the D-S scheme, with ignorance prior, and the nearest equivalent Bayesian approach utilizing the informationless prior. Plots of the distributions $f(\theta)$ and $g(\theta)$ and Bayesian posterior $\pi(\theta)$ are shown in Figure 1. for sample size 2 and sample size 12.

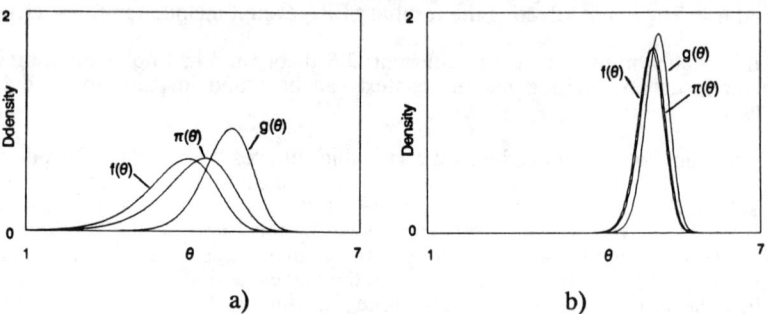

Figure 1. Type I Extreme model. Inference of unknown location parameter. $f(\theta)$, $g(\theta)$ and Bayesian posterior $\pi(\theta)$ for sample sizes a) 2 and b) 12.

Influence of imprecise probability in decision analysis.

In decision analysis, comparisons of decision alternatives are made in terms of their expected utilities. The $f(\theta)$ and $g(\theta)$ distributions are a useful measure of imprecision as they produce the extreme upper and lower expected utilities for monotone utility functions. Without specifying an actual utility function, any assessment of the above results must be confined to cases where the utility function is linearly dependent on θ and where utility is measurable in monetary values. The spread of the upper and lower expected utilities for a single decision is then reflected in the distance between the mean values of $f(\theta)$ and $g(\theta)$. The size of this interval reflects the component of uncertainty arising from imprecise probability considerations. It can be compared with the uncertainty of θ reflected in a posterior distribution produced by the nearest equivalent Bayesian analysis.

The above results are from an individual sample set but are quite typical of those obtained from the numerical experiments carried out so far. Within the assessment limitations outlined above, the results indicate that the influence of imprecise probability is highly significant for very small samples, but declines rapidly as the sample size increases. The interval between the mean values of $f(\theta)$ and $g(\theta)$ declines approximately exponentially with the size of the sample. Other numerical experiments indicate that the size of this interval is also influenced, but to a much less extent, by the population variance. It also varyies randomly from sample to sample.

The ability of a D-S based decision analysis to identify the best decision is impaired when the expected utility intervals from the decision alternatives are large enough to overlap. Thus larger expected utility intervals for individual decision alternatives will reduce the ability of the decision analysis to resolve differences between decisions. Any resulting ambivalence between decision alternatives is not necessarily bad as it may more accurately portray the true decision situation than does a comparable "precise" Bayesian decision result.

In most practical water resources decision situations the utility function will not be linear in θ but may be monotone, in which case the results of a D-S decision analysis will still be governed by $f(\theta)$ and $g(\theta)$. But the size of the expected utility intervals will depend upon the specific utility functions involved and each case would have to be investigated on its own merits.

References

Berger, J.O., "Statistical Decision Theory and Bayesian Decision Analysis" Springer-Verlag, New York, 1985.

Caselton, W.F., and Luo, W. "Decision Making With Imprecise Probabilities: Dempster-Shafer Theory and Application." Water Resources Research. In press, Oct. 1992.

Dempster, A.P., "A Generalization of Bayesian Inference" J. Royal Statistics Soc., Ser. B, 30, 205-247, 1968.

Shafer, G., "A Mathematical Theory of Evidence." Princeton University Press, N.J., 1976.

Walley, P., "Statistical Reasoning With Imprecise Probabilities", Chapman and Hall, London, 1991.

Risk Reduction Approaches for Hydrologic Forecasting System Development

Lynn E. Johnson [1]

Abstract

Development of computerized decision support systems requires that the systems be designed for successful implementation. Risk reduction refers to a suite of activities directed to testing and evaluation of prototype DSS systems to provide feedback leading to refinement of the system design. Risk reduction issues relate to the accuracy and reliability of the hydrological data and forecasting subsystems, as well as use of the forecasting tools by the forecasters in an operational environment.

Risk Reduction for Hydrological Forecasting

The National Weather Service is embarking on a modernization effort in which expanded and new data collection, processing and display capabilities are being developed for implementation at their local forecast offices. A significant component of the modernized NWS forecasting environment is new capabilities for hydrological forecasting. In addition to rainfall data display products produced by WSR-88D radar reflectivity processing, the workstation will allow tracking precipitation and river flow amounts, as well as for forecasting flood magnitude and timing at the headwaters scale (Johnson, et.al., 1990).

Risk reduction activities derive from concepts of software productivity and evolution (e.g. Boehm, 1988). Originally developed for defense system contracting for software development, the so-called "spiral model" for software development (Figure 1).

---
[1] Associate Prof., Civil Engineering, Univ. of Colorado at Denver, P.O. Box 173364, Denver, CO 80217-3364.

The potential impacts of the technological influx on the organization were recognized early on by the NWS. The risk reduction program is being conducted in several phases, each phase designed to validate selected AWIPS-90 system capabilities, data integration and display techniques, and advanced operational concepts. This is being done to (Walts, 1989):
(1) reduce the risk of mis-specifying a system today which will be the backbone for NWS services for the next decade;
(2) gain insight into the broad range of operational transition issues; and
(3) reduce the risk associated with full, nationwide implementation of AWIPS and NEXRAD.

The hydrologic functionality provided in the AWIPS workstation provides new tools for the NWS forecaster which have not been available before. Thus there are a variety of issues relating to the forecaster's acceptance and use of the hydro functions to support expanded forecast responsibilities, amount and mode of training required, and assessment of the hydro message issued. Results of the evaluation activities provide feedback on the success and appropriateness of the enhanced workstation functions and new products.

Aspects of the risk reduction model are being addressed in the NWS modernization program. For the Norman emulation, the scope of evaluation activities includes performance testing of the workstation hardware/software, databases and data feeds; as well as usage patterns of the system by the forecasters. There is interest in the physical performance of the system as well it acceptance by the forecasters in the operational environment. Interface issues are a primary topic.

Various risk reduction data collection and analysis activities are being conducted involving questionnaires, product usage logs (Figure 2), evaluation log (a computerized "diary" for forecaster comments), observations of operations, message coding for "quality" and "completeness", and forecast verifications. In concert the data collections provide a basis for determining the performance of the system and its utility for operational use.

Conclusions

Risk reduction is integral to system design and implementation for large projects of the sort undertaken by the National Weather Service. Risk reduction is being accomplished using existing prototypes of many of the data collection, processing, and display devices; and a phased implementation of key elements of the modernization program into the operations of the Denver and Norman WSFOs.

Lessons learned from these activities contribute directly to the AWIPS-90 system definition, as well as to a better understanding of the future NWS requirements for operations, operations support, training, education, staffing, logistics, facilities, etc. These lessons also contribute to the basic modernization transition planning process at all levels of the agency. Experience with the system for meteorological forecasting has proven to be very positive. Applying this experience to the hydrological package will provide powerful tools to support enhanced flood and flash flood forecaster decision support.

Acknowledgments

Acknowledgment is made to the National Weather Service, an agency of the National Oceanic and Atmospheric Administration, and NOAA's Environmental Research Laboratory Forecast Systems Laboratory where the risk reduction research activities occur. Opinions expressed in the paper are those of the author and do not necessarily reflect policies of the agencies or other individuals involved.

References

Boehm, B.W. 1988. A Spiral Model of Software Development and Enhancement. in Computer, Journal of IEEE. pp 61-72, May.

Johnson, L.E. S. O'Donnell and R. Tibi. 1990. Next-Generation Local Scale Hydromet Forecasting System. ASCE Hydraulics Division Specialty Conference, San Diego, CA. June.

Walts, Denny. 1989. PROFS Role in Modernization of the National Weather Service. in Computerized Decision Support Systems for Water Managers, Proceedings of the 3rd Water Resources Operations and Management Workshop, ASCE, Colorado State University, (Labadie, Brazil, Corbu and Johnson (eds)). June 1988.

RISK REDUCTION APPROACHES

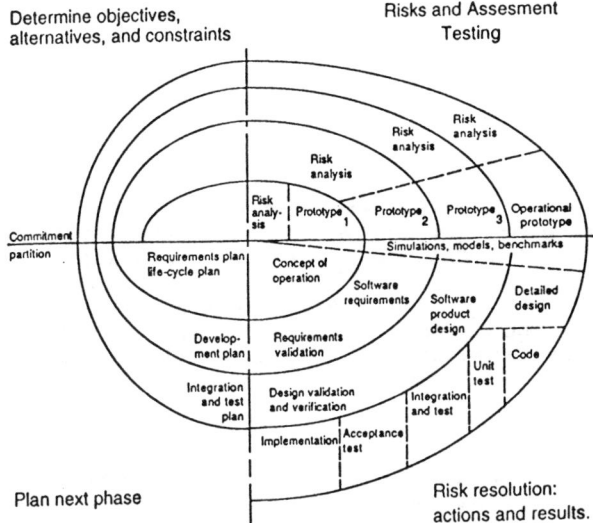

Figure 1. Spiral model of software and system development process (after Boehm, 1988).

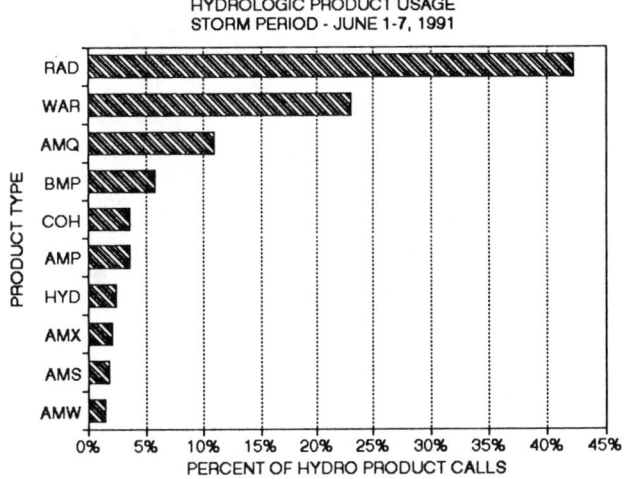

Figure 2. Product usage logs provide profile of workstation information usage.

Development of a Raw Water Master Plan Model to Assist Long Range Drought and Water Conservation Planning for the City of Louisville, Colorado

Noel Hobbs[1], P.E., Member, ASCE
Thomas Charles[2], P.E., Member, ASCE
Jeffrey Carpenter[3], E.I.T, Associate Member, ASCE
Brad Eaton[4], E.I.T

Abstract

The City of Louisville, a municipal corporation charged with meeting the water demands of almost 5,000 water accounts and faced with a diminishing supply of available senior water rights on South Boulder Creek, determined the need for a Raw Water Master Plan to define future water needs for its service area. The Master Plan investigated the City's water rights, conveyance and storage facilities, current and future water demands, and conservation techniques for reduction of demand. To analyze the effects of potential drought conditions, it was necessary to study stream hydrology and develop forecasting techniques and a computer model (SYSTEM) to simulate historical yields from water rights through operation of the City's conveyance and storage facilities.

Approach

The SYSTEM model was developed as a planning tool for examining system yield, storage, and conveyance capacity when compared to historical runoff conditions within the South Boulder Creek Drainage. Due to extensive input requirements for the historical database and water rights portfolio, and the complexities governing operation of individual water rights, storage decrees, and lease

---

[1]Project Manager, Camp Dresser & McKee, Inc., 1331 17th St. Denver, CO 80202
[2]Project Engineer, Camp Dresser & McKee, Denver
[3]Systems Engineer, Camp Dresser & McKee, Denver
[4]City of Louisville Staff Engineer, City of Louisville, 749 Main Street, Louisville, CO 80027

RAW WATER MASTER PLAN   285

agreements concerning Marshall Reservoir, it was determined that a scientific-based programming language would provide the most flexibility in achieving the study goals. Project team familiarity and time constraints favored selection of the Microsoft Version 5.1 FORTRAN compiler to develop the SYSTEM model.

Hydrology

A comprehensive hydrological analysis was performed on the South Boulder Creek watershed and stream flow records to determine the correlation, if any, between snowpack readings and seasonal runoff levels. The United States Soil Conservation Service (SCS) records produced the best correlations. High correlations should assist the City in utilizing SYSTEM to assess potential yields during future irrigation seasons.

Another goal of the hydrological analysis was to identify drought periods, ranging in duration from 1 month to the period of record for return period frequencies ranging from 3 months to 100 years. A computer simulation was performed using Partial Duration Series and Log Pearson III distributions. The historical period of March 1963 through February 1965 was identified as the 24-month duration, 50-year frequency event and was selected as the design target period to utilize in model simulations. The ultimate goal was to develop the capability to meet supply needs during a similar event from the City's portfolio of water rights.

Water Demands and Conservation

Existing water demands for the City of Louisville were analyzed for the period 1979 through 1991 to evaluate base year demand and average monthly distribution. Future demands were based on a 1989 treated water distribution study conducted by another consultant. Present, midpoint and buildout annual demands were analyzed. It was anticipated that at Louisville's historic growth rate, the buildout demand would be reached in approximately 30 years.

Techniques analyzed and approved by City Council for water conservation included: evapo-transpiration reduction programs, installation of low flow plumbing devices, pressure and leakage reduction, increasing block rate pricing, public education, and emergency water rate measures. Indoor water savings of 5 percent and outdoor savings of 6 percent were projected based on the success of past meter installations, leak detection programs, and existing increasing block rate pricing programs combined with the council approved measures.

Model Input

Data requirements for SYSTEM consist of detailed background data input files and data that are input by the user interactively. Three different sets of input files were developed; initial system conditions, limit restraints, and historical flows for each of the 38 decrees in the City's water rights portfolio.

Background information on the City's water rights portfolio, the physical delivery and storage system and specific water rights were obtained from individual agreements, decree stipulations and from the database established for the City's water rights accounting program. In addition, it was necessary to obtain historical data from the database maintained by the State Engineer's Office for the irrigation ditches which influence the City's yield and from Farmers Reservoir and Irrigation Company, for the operation of Marshall Reservoir.

Model Structure

The model was developed as a volumetric simulation of the raw water system based on monthly time increments. This time increment was chosen to minimize the complexity of handling large amounts of data inherent with shorter time steps, while avoiding significant reduction in accuracy associated with longer periods.

Figure 1 depicts the basic structure of SYSTEM and the relationship between the main program and each of the various input/output operations and subroutines. The main program receives and organizes input from the user and data input files and manages operation between the various subroutines which perform specific functions. Physical allocation and operation of the conveyance elements within the raw water system and incrementing of the monthly time steps are also functions performed by the main program. Subroutines address complex issues of water rights yield analysis, Marshall Reservoir operation, storage incrementing, and application of conveyance and evaporation losses.

Future Use

The present system model was used to assess the ability of the City's current portfolio of water rights to meet interim and ultimate water demands under normal and design drought hydrologic runoff conditions. These simulations quantified surpluses and deficits in supply yields for the City for the hydrologic conditions and water demands specified.

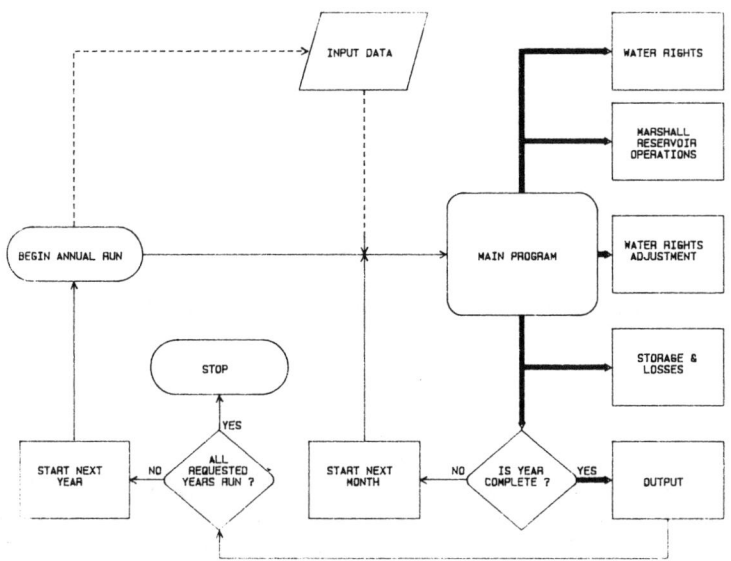

Figure 1 - SYSTEM model

The model was then run for various combinations of direct flow and storage rights to see if deficits could be eliminated and, if not, what practical combination of water rights purchases would best minimize deficits. A water rights purchase program was identified and recommended. Since this program could not eliminate all deficits, an examination was made of other raw water supply options available to the City. This included additional water storage, which was eliminated due to excessive cost; Colorado Big Thompson (CBT)/Windy Gap Project water; local groundwater and Denver Basin Groundwater. Participation in the CBT/Windy Gap water projects was the next most financially attractive alternative for the City, and negotiations are being pursued for this option.

Use of the model to evaluate various operational scenarios will aid the City in developing the most cost-effective method of meeting the future water requirements of the City. In addition to master planning, the SYSTEM model will also be used as a forecasting tool to simulate yield from hydrologically similar water years of record. Model results will then be used in decision making to help optimize the use of the City's water rights.

Application of Computer Support for Water Quality Management

John F. DeGeorge[1], Stephen A. Breithaupt[1], Robert R. Klamt[2], and Theresa Wistrom[2]

Abstract

A Computer-Aided Support System for operation of the QUAL2E-UNCAS river water quality model has been developed for the California Regional Water Quality Control Board, North Coastal Region. It is expected to enhance water quality management of the Russian River, the principal stream of the region. Initial application of the CASS as a management tool and its relationship to institutional issues in the basin are discussed.

Introduction

The Russian River is one of the largest river systems in northern California, providing both urban and agricultural water supplies and a popular recreation resource for a rapidly growing population. Water quality management issues in the basin include the effects of treated urban wastewater discharges, urban and agricultural non-point pollution loads, industrial point loads, gravel mining operations, and excessive algal growth. In recognition of the need for advanced management technologies the California State Water Resources Control Board, parent agency of the California Regional Water Quality Control Board (RWQCB), contracted with the Department of Civil and Environmental Engineering at the University of California Davis (UCD) to construct a water quality model of the river. The UCD Water Resources Modeling Group, after review of available models, selected QUAL2E, the widely used model supported by the U. S. Environmental Protection Agency, calibrated and verified it against historic data, and adapted it for management studies.

---

[1]Doctoral students, Department of Civil and Environmental Engineering, University of California, Davis, CA 95616
[2]Environmental Specialist IV and Sanitary Engineering Associate, respectively, California Regional Water Quality Control Board, Santa Rosa, CA 95403

In order to make the model most useful to the staff of the RWQCB, it was incorporated into a Computer-Aided Support System (CASS) developed under Microsoft Windows™ for IBM compatible microcomputers (DeGeorge and Orlob, 1992). A CASS can be considered a form of Decision Support System (DSS), such as described in the business management and water resources planning literature (Keen and Scott-Morton, 1978; Loucks, et.al., 1985; Arnold and Orlob, 1989). A DSS involves coupling of databases and mathematical models through an interactive user interface which allows users who are not programmers to gain rapid access to information generated and managed by the system. The Russian River CASS supports three principal functions: a modeling environment that facilitates development and execution of model runs; an information management system that contributes to flexible display and analysis of both model output and observed data from the river; and a presentation tool for creation and display of advanced graphic visualizations of the results of model runs.

This paper examines the experience of the RWQCB staff in their initial use of the CASS and some of the institutional issues related to the use of modeling results in the management process.

Application of the CASS

The principal design objective of the Russian River CASS was to make the model QUAL2E accessible to the RWQCB staff in routine assessment of water quality monitoring activities and management alternatives. Normally, without the aid of a CASS, implementation of the model would have been restricted to a single specialized staff member who would have to devote virtually full time in mastering and maintaining the model, and instructing others in its use. Utilizing the CASS, many staff members have the opportunity to exercise the model. To develop the necessary level of familiarity with model, the database, and post processing tools through the CASS a series of training sessions were conducted. The first of these, presented to 15 staff members, consisted of a general orientation on the fundamental structure of the model and how it could be used in management decision making. A second session, conducted for 8 staff members at the UCD computer laboratory, provided "hands-on" experience with the model exercised through the CASS. It covered operation of the CASS and the fundamental tasks of organizing data sets and running the model. Two additional sessions, focusing on more detailed aspects of data set creation, calibration, and uncertainty analysis, were conducted with senior staff who will direct modeling activities for the RWQCB.

Initial uses of the CASS involve coupled evaluation of results derived from the calibrated model and the efficacy of the field monitoring program. Uncertainties evident in both model output and observed data must be critically evaluated to determine sources of discrepancies between the two forms of information. For example in the case of the Russian River, steady state temperature results given by the model for a particular low flow condition were found to be significantly lower than actually observed at a specific location on the river. RWQCB staff noted that the conveniently accessible river location does not always provide a sample representative of the average conditions simulated by the model. They also noted from personal experience that the existing meteorological data set used for modeling may not adequately represent conditions prevailing for that river reach. As a result the sampling point will be moved to a more suitable location where the river is well-mixed and more complete meteorological data will be acquired for the site. This interactive process is expected to reduce uncertainties in the model and its input data, leading to improved calibration.

Another important use of the CASS is in performing comparative analysis of alternative management scenarios. A typical example is an on-going evaluation of the potential effects of increased wastewater loading from the Santa Rosa wastewater treatment facility, the largest discharger in the Russian River basin. Scenario data sets of future loading conditions can be efficiently generated by making incremental changes in base data sets from model control dialog boxes and from map-based representation of the physical system. Results from multiple runs are managed by the CASS to provide for rapid comparative visual analysis. Results may be displayed as graphs, tables, or color contour plots. State variables, as well as internally computed rates and coefficients, can be rapidly displayed (almost instantaneously), allowing a thorough exploration of model results at the user's discretion. The highly interactive structure of the CASS make it an ideal tool for workgroup situations where scenarios are to be created and evaluated during the course of group discussions.

Institutional Issues

Simulation model results are not the only decision factors in the water quality management process. The RWQCB must consider many variables, including feasibility of advanced wastewater treatment technology, conformance with state and federal standards, and public reaction and input. One of the challenges of modeling in the management process is to present results so that appropriate consideration is given to both the capabilities and limitations of the model, while still maintaining its credibility as a useful management tool. For example, it is often more instructive to focus on relative changes in specific state variables between scenarios rather than absolute predictions. This is especially important in communicating results in public forums where participants may be uncertain concerning the appropriate use of models. The RWQCB staff have already met with public interest groups to demonstrate the use of the CASS, thus acquainting the public sector with a new tool to enhance knowledge of water quality processes and improve the management decision process.

Consequences of this exposure of the CASS to the public have been requests to make it available beyond the control of the RWQCB. Reasons for these requests are: 1) that the RWQCB has an advantage over public interest groups in evaluation of water quality management alternatives; 2) that the RWQCB will place more emphasis on model results than on other non-model information; and 3) that the public should be allowed to perform independent calibration of the model. Legal counsel for the Board does not consider the CASS per se to be a part of the public record (although the model QUAL2E is in the public domain). The staff's position is that providing the CASS could lead to misapplications and misinterpretations, counter to the public good. The concern is that public interest groups might present model results to the media or others, giving the perception that they were derived by the RWQCB staff. Changes in input parameters and data could lead to competing results, requiring additional staff effort to analyze and interpret. RWQCB staff credibility and impartiality to present unbiased evaluations could be compromised. The RWQCB seeks to instill public trust in the technical performance and objectivity of its staff. Rather than distributing the CASS to the public, access will be provided through public workshops supervised by RWQCB staff.

Conclusions

A Computer-Aided Support System for the Russian River in California has been developed as a tool for the analysis of water quality management alternatives, providing capabilities for mathematical modeling and enhanced presentation of results to decision makers. The Regional Water Quality Control Board, under its mandate to achieve compliance with federal and state water quality standards, must consider many variables in making its management decisions. Modeling results represent one important input to the decision making process. The CASS allows the RWQCB staff to evaluate control strategies that affect the dynamics of the river's quality. Its value in facilitating adjustments in monitoring design has already been demonstrated by focusing efforts toward a more efficient long-term data collection program. Use of the CASS database and graphics are being used to provide timely summaries of water quality conditions for the RWQCB and the public. Analysis of model runs will help to locate unknown discharges and suspected critical water quality problem areas. The interactive and user-friendly interface of the CASS increases the efficiency of staff effort by creating and evaluating realistic discharge scenarios. The ability to create new scenarios and to obtain tabular and graphic results virtually on demand ensures that the Russian River model will be used and updated on a regular basis. The CASS has the potential to become an integral component of the overall water quality management strategy of the Board.

Acknowledgments

Development of the model and the CASS was supported by the California State Water Resources Control Board. The work was carried out by the Water Resources Modeling Group of the Department of Civil and Environmental Engineering of the University of California Davis for the California Regional Water Quality Control Board North Coast Region.

References

Arnold, U., and G. T. Orlob, (1989), "Decision Support for Estuarine Water Quality Management," J. Water Res. Plan. & Mngmnt, ASCE, Vol. 115, No. 6, pp 775-792

DeGeorge, J. F., and G. T. Orlob, (1992), "Computer-Aided Support for Water Quality Modeling of the Russian River," Water Resources Planning and Management, Proc. Water Forum '92, ASCE, pp 182-187

Keen, P. G. W., and M. S. Scott Morton, (1978), Decision Support Systems: An Organizational Perspective, Addison-Wesley, Reading, MA.

Loucks, D. P., J. Kindler, and K. Fedra, (1985), "Interactive Water Resources Modeling and Model Use: An Overview," Water Res. Research, Vol. 21., No. 2, pp 95-102

CASS for Evaluating Hg Contamination in Clear Lake, CA.

A.E.Bale[1], P.L. Shrestha[2], and G.T. Orlob, F. ASCE[3],

Abstract

Characterization of ecosystem contamination and evaluation of potential mitigation schemes represent two of the most immediate challenges to management of water resources. Heavy metals in the aquatic environment are of particular concern and pose a special problem to water managers. Contamination by heavy metals can range through all levels of aquatic ecosystems as metal species are dissolved in the water column, adsorbed to fine sediments, and transported with biota. Management of these water resources is complicated by the variety of interactions among the various transport mechanisms and the environmentally dependent speciation of heavy metals.

Numerical models are being utilized to describe these aquatic systems, simulate different natural scenarios, and estimate the effects of imposed mitigation schemes. The task of manipulating the data necessary for model implementation and evaluation can be daunting. To assure effective modeling and management of water resources, field data must be coordinated with model requirements and correlated to model results during analysis, calibration, and review. Computer Aided Support Systems (CASS) provide innovative approaches to solving problems of managing, analyzing, and displaying field and simulation data for decision making. This paper describes the development and preliminary application of a CASS for evaluation of mercury contamination and eutrophication in Clear Lake, California.

---

[1] Doctoral Candidate,
[2] Post Doctoral Researcher,
[3] Professor Emeritus
Department of Civil and Environmental Engineering
University of California, Davis, CA.

Introduction

Clear Lake, located in rural northern California, is a recreational resource that supports boating, swimming, and sport fishing. At one time in recent history, the lake supported California's only commercial inland fishery, supplying large numbers of carp and blackfish to the San Francisco Bay Area. Largemouth bass is a very popular sport fish and an important recreational resource of the lake. However, loss of littoral habitat has resulted in dramatic changes in the composition of the fishery, a decline in the largemouth bass population, and may even be impacting avian species.

Mercury contamination is of vital concern to water quality managers and ecologists at Clear Lake. Sampling campaigns over a number of years have revealed elevated levels of mercury contamination of lake sediments apparently emanating from the recently active Sulphur Bank Mercury Mine, a U.S. EPA Superfund site located on the western bank of the lake. Contamination is evident in the sediment, water, and biota of Clear Lake, accumulating in organisms at lower trophic levels and concentrating to potentially toxic levels high in the food chain, thus posing a threat to virtually all levels of the lake ecosystem. An intensive effort on the part of several government agencies seeks to characterize the extent of lake contamination, model the fate of mercury and the extent of contamination, and evaluate the benefits of possible mitigation schemes.

Marked eutrophication of the lake is also of immediate concern. The value of Clear Lake as a recreational water resource is substantially compromised by frequent and periodic blooms of algae that may be associated with addressable, man-made changes in nutrient loading of the lake. Clear Lake is an old, relatively shallow water body that may be naturally eutrophic. However, the character of the lake's watershed has been greatly modified in recent years by forest clearcutting, vegetation clearing, gravel mining and residential and commercial development. Such modifications have adversely affected the quality of inflows, exacerbating eutrophication. The extent of damage done by man-made changes, and possible mitigation schemes are issues of current research that call for new management techniques.

Computer Aided Support Systems

Application of computer aided support systems (CASS), introduced by Loucks and da Costa (1991), holds promise for Clear Lake. The general components and structure of a CASS have been described in previous work by the authors (Bale and Orlob, 1992) where the concept of a data manager-numerical model core accessed from within a graphical interface was demonstrated. A user-friendly interface employed computer graphics to present data and model results in a variety of formats for the purpose of facilitating calibration, verification, and application of numerical water quality models. In the Clear Lake study, application of CASS has been extended to include data analysis and visualization not only for water quality modelers and managers, but for use by researchers in environmental studies, including ecologists, limnologists, biologists, and toxicologists, as well as engineers and planners.

Clear Lake Application

A variety of unconnected studies over the last thirty years, or so, have produced a relatively large, but unwieldy and difficult-to-manage, water quality data set for Clear Lake. This set contains standard water quality parameters, such as dissolved oxygen, temperature, suspended sediment, and salt content as well as laboratory assessments of mercury concentration in the water column and sediment. Coordinated sampling programs now in progress are providing even more data for analysis. Government agencies such as the U.S. EPA, state and regional water quality control boards, Lake County Mosquito Abatement District, the Clear Lake Clean Water Program, and university researchers like those of the U.C. Davis Ecotoxicology Program require an organized framework from which to access, review, and compare all data acquired concerning water quality in Clear Lake.

Numerical models are being developed to analyze proliferation of nuisance algae and the potential distribution of mercury contamination in Clear Lake under future, hopefully mitigating, environmental conditions. Two and three-dimensional models are designed to simulate both the complex hydrodynamics of the lake and a variety of interrelated water quality constituents including dissolved oxygen, temperature, suspended and deposited sediment, particulate and dissolved mercury species, nutrients, and chlorophyll-a. Efficient calibration and verification of these models are enhanced by the organized framework provided by a CASS to compare available data and model results and assess model reliability.

The framework being developed for field data visualization, model building, and managerial assessment - the Clear Lake CASS - is adapted to the Macintosh computer for use by field scientists, modelers, and water quality managers during site characterization, model development, and evaluation of alternative mitigation schemes. Analysis tools like finite element hydrodynamic (RMA2, RMA10) and water quality (RMA4Q) models, together with a linked data base form the core of the Clear Lake CASS. The finite element model, RMA4Q, driven by steady and unsteady flow regimes, has been modified to include bio-uptake and speciation of mercury into four compartments (Hg^0, Hg^{II}, monomethyl mercury, and particle-bound mercury) along the lines of the MCM Lake Mercury Model (Hudson, Gherini, and Munson, 1991). Other analysis tools developed specifically to aid environmental scientists include statistical analysis routines, post-processing graphics, and non-mechanistic models, such as eutrophication-nutrient loading extrapolations (Sas, 1989). Access to analysis tools and data is provided through a user interface that utilizes the standard Macintosh menus, dialog boxes, and onscreen selection routines. Through an interpretive display, data and analysis results are presented in a number of forms including standard graphs, tables, and mapped contour plots. A typical screen display of two contour plots, sampling stations, and a location-sensitive pop-up menu is illustrated in Figure 1. Animation of both graphs and contour plots illuminate changes in constituent concentrations through time and highlight not only anomalies in both collected data and model results, but also the importance to management of temporal changes in water quality .

Acknowledgment

The University of California EcoToxicology Program (Ecotox) provided support for research and development of models for simulating the fate and transport of mercury in the Clear Lake environment.

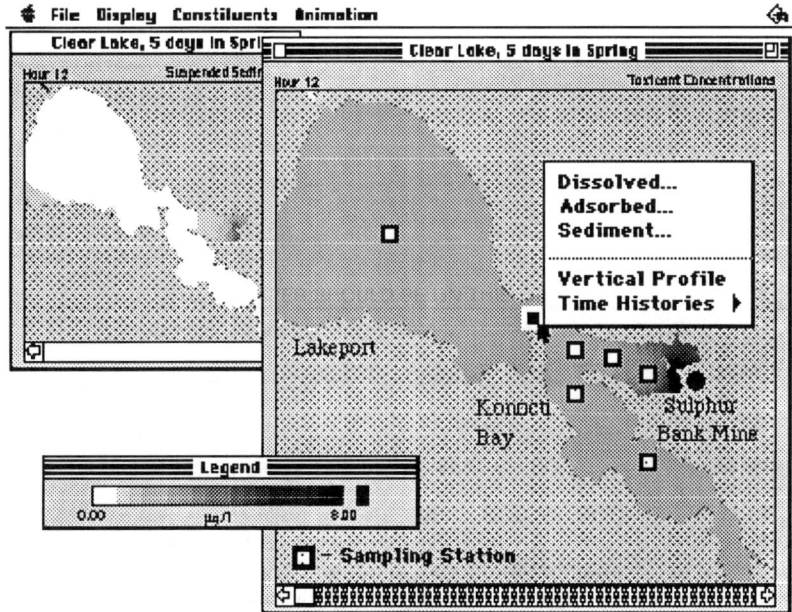

Figure 1. Typical View of Clear Lake CASS

References

Bale, A.E. and G.T. Orlob (1992) Computer Support for Water Quality Management in San Diego Bay. Proceedings of Water Forum '92: Annual ASCE Water Resources Planning and Management Division Conference. Baltimore, MD.

Hudson, R.J.M, Gherini, S.A., and R.K. Munson (1991) The MTL Mercury Model: A description of the model, discussion of scientific issues, and presentation of preliminary results. from Annual report for 1990 to Electric Power Research Institute and Wisconsin Dept. of Nat'l Resources. Ch. 8.

Loucks, D.P. and Joao R. da Costa, editors (1991) Decision Support Systems: Water Resources Planning. Nato ASI series G: Ecological Sciences. v 26. Springer-Verlag.

Sas, H. (1989) Lake Restoration by Reduction of Nutrient Loading: expectations, experiences, extrapolations. Sankt Augustin : Academia Verlag Richarz.

Graphics User Interface for Water Quality Model Calibration

Stephen A. Breithaupt[1], John F. DeGeorge[1], and Gerald T. Orlob[2], F. ASCE

Abstract

The UC Davis Water Resources Modeling Group developed a unique Graphical User Interface (GUI) in the Microsoft Windows environment to facilitate use of the stream model QUAL2E for the Russian River, California. Use of the GUI enhances interaction between model and user, thus streamlining the processes of calibration and verification, and saving time and staff resources in model applications for water quality management. This paper focuses on the use of the GUI for calibration and validation.

Introduction

The Russian River flows through approximately 160 kilometers of California's north coastal region before discharging to the Pacific Ocean at the town of Jenner, north of San Francisco. It experiences climatic conditions varying from that of the arid headwaters to the cool and often foggy Pacific Coast. Precipitation occurs primarily during the winter and spring months; it is almost nil during the summer and early fall period when river flows are lowest. Summertime air temperatures range from 100°F in the upper basin to 70°F along the lower reaches of the river.

During the summer, flows are controlled at Lakes Mendocino and Sonoma, sustaining agricultural, municipal, and industrial consumptive water use, and intensive use of the river for recreation. Four temporary dams maintain recreational impoundments along the river during the summer low flow period, modifying hydraulic conditions for a few miles upstream of each. Flows are generally steady during the recreational season and wastewater discharges

---

[1] Doctoral students Department of Civil and Environmental Engineering University of California Davis CA 95616
[2] Professor Emeritus Department of Civil and Environmental Enigineering University of California Davis CA 95616

are excluded. During the late spring and summer blooms of algae, dominated by Oscillatoria and Cladophora, occur along the river.

To evaluate the response of the river to alternative water quality management measures, including the regulation of wastewater treatment and disposal, the Regional Water Quality Control Board (RWQCB) contracted with the University of California Davis, through its Department of Civil and Environmental Engineering, to model the river and to develop appropriate mathematical tools for decision support. The Water Resources Modeling Group of the department selected the EPA model QUAL2E as most appropriate for the Board's purposes and adapted it to the Russian River system.

Need for a Graphic User Interface

Modeling exercises, including calibration and verification of the model against field observations, requires considerable manipulation of large data sets, a process often very demanding of time and human resources. In the case of QUAL2E, which considers some 13 state variables, including temperature, dissolved oxygen, dissolved solids, primary nutrients, algae, etc., the data requirements can be formidable. The traditional method of operation QUAL2E has been (1) to construct an ASCII input file for the physical system and boundary conditions, (2) to run the model using a text terminal interface, and (3) to extract selected results for post processing from the ASCII text output file. To improve the efficiency of file manipulation and to facilitate the processes of calibration and verification of the model using these data, a unique graphical user interface (GUI) was constructed using the DOS-based Microsoft Windows™. The design objectives and structure of the Computer-Aided Support System (CASS) that includes the GUI has been described previously (DeGeorge and Orlob, 1992).

In structuring the CASS, the GUI together with the QUAL2E computational code were incorporated so that the model runs made through the interface produced results identical with the standard application of the model. The standard text input file was replaced in the CASS by a map display of the Russian River Basin on which graphic objects represented various model components. Also, the input data file was made to be accessed through menu selections. The user is allowed to manipulate the model characteristics by selecting objects on the map display and editing related attributes through a dialog box. Editing of input data can also be accomplished by editing a dialog box, as with the map editor, after selection of an appropriate edit menu option.

Use of the Graphics User Interface

Implementation of the model entails initial discretization of the river system and description of its hydraulic properties. These processes occur outside the interface and are necessary for all versions of QUAL2E, regardless of interface type. Necessary information includes topographic maps, channel cross-sections, and hydraulic parameters defining depth and velocity as functions of discharge. Hydrologic and climatologic data are also required for implementation of the model. Finally, since quality simulation is the primary goal, data from the field on as many quality variables as possible must be assembled and organized into a water quality data file that can be drawn upon for calibration. Spatial coverage should be sufficient to cover the reaches of particular interest in management, and temporal coverage should be such as to represent the range of hydrologic and climatological conditions most likely to test the capabilities of management schemes, e.g., low flows and extreme temperatures.

The initial step in preparation for a model run is to perform a water balance, such that continuity is satisfied at all locations and within all reaches and segments of the model. In the case of the Russian River QUAL2E model the process continues with an application of mass continuity for a conservative constituent, e.g., total dissolved solids. A combination of these steps leads to determination of incremental inflows or outflows along the river and estimation of irrigation efficiency of agricultural zones contiguous to the river. Agricultural withdrawals (balancing flows) are varied until observed TDS concentrations are matched by model output. This assumes that the quantity and quality of agricultural withdrawals and return flows are consistent with assumed irrigation efficiencies. The GUI is utilized iteratively in the calibration process until satisfactory agreement between model and prototype is attained, considering realistic efficiencies. This process is repeated for each hydrologic condition considered in the calibration process.

Next, the other water quality variables of primary interest are calibrated, using appropriate rate coefficients. The sequence usually includes, in order: temperature, BOD, nutrient species, algae, and dissolved oxygen. The order of calibration follows the order of linkages between constituents. Iterative modification of parameter values is carried out until in the judgment of the user the results are considered satisfactory. The user is able to test results and compare with actual field data on the monitor, and to contrast several different model outputs. Observed data can be accessed from an ASCII or a dBase data file. The dBase data can be sorted by two different criteria and can also be plotted independently of model results. In the GUI the results of model test runs can be displayed almost as fast as decisions are made to modify coefficients. (Note: As currently configured, the system runs interactively on an IBM PC with a 486 processor, with run times of the order of two seconds for a system of 189 computational elements.) The GUI greatly accelerates the user's decision process in changing coefficients, boundary conditions, etc., that may lead to an improved calibration.

In addition to plots of state variables, the rates of various processes can be presented graphically to examine trends and sensitivities. Physical properties, such as depth, velocity, and travel times, and biochemical processes in terms of rates of transformation are also accessible to the user through the interface

Concluding Comment

An interactive user interface in Microsoft Windows for an IBM PC 486 has proven to be a very useful tool to enhance the model calibration process. It has been demonstrated in the adaptation of the EPA model QUAL2E to the Russian River System in northern

California. The interface allows the user to interact almost instantaneously with the model, with immediate graphical display of the consequences of his decisions to adjust or modify model performance. It provides insight to the modeler on the idiosyncrasies of the model, leading to improvement in representing the real system. It serves as a medium for the non-modeler user to build confidence in the utility of the model as a management tool.

Acknowledgments

The authors wish to thank Jan S. Hauser and Theresa Fong for their contributions in the implementation of the model, Robert Klamt and Theresa Wystrom of the North Coastal Regional Water Quality Control Board for their thoughtful input in development of the user interface, and the California State Water Resources Control Board for its support of the project.

References

Brown, L. C.,and T. O. Barnwell, Jr., (1987), The Enhanced Stream Water Quality Models, QUAL2E and QUAL2E-UNCAS: Documentation and User"s Manual, U. S. EPA Environmental Research Laboratory, Athens, GA, EPA/600/3-87/007

DeGeorge, J. F., and G. T. Orlob, (1992), "Computer-Aided Support for Water Quality Modeling of the Russian River," Water Resources Planning and Management , Proc. Water Forum '92, ASCE, pp 182-187

Research Needs in Water Resources Planning and Management

Neil S. Grigg[1]

Abstract

When the WRPMD was formed, expectations about research were high, but in this new era we must reexamine the role of research. The greatest policy need for the water industry is institutional: better planning and coordination, an issue that belongs within the Division's mission. To achieve them, the needed ingredient is leadership of problem solving teams and in unraveling complexity through systems analysis. The WRPMD must recruit and build the leadership needed for these tasks, and this can happen if the Division's research activities are made relevant to the nation's water policy and management agendas.

Introduction

In the twenty years of the WRPMD, discussions about "research needs" have been considered both routine and productive, and I hope this paper is the latter. The only reason to look at applied research needs at all is in the context of problems we need to solve. In the spirit of the 20th Anniversary celebration, we are supposed to look backwards and forwards at research needs.

Looking Backwards

Twenty years ago, the Division embarked on a search for methods to apply interdisciplinary knowledge to solve the nation's water problems. The climate for water resources research in the US was exciting. The Water Resources Research Act, the Water Resources Planning Act and the Clean Water Act were getting started. The National Water Commission was debating policy needs, and there was an air of anticipation that planning and research had much to offer the nation.

---

[1]Research Needs in Water Resources Planning and Management

We had many lists of research needs in those days. I reviewed them and the status of research policy in a 1980 paper (Grigg, 1980). The background of US water resources research was traced from the establishment of the first science and engineering schools, through the land grant colleges, through the development of US science policy after World War II, and then through the Water Resources Research Act. This makes quite a history, especially when linked with the nation's political history, and the development of water and environmental policy.

The list of references for the 1980 paper includes a review of the "great books" of water research development. Perhaps the most thorough review was Viessman's review of the Water Resources Research Act (Viessman, 1976).

As far as subject matter is concerned, the early reports on research needs were laundry lists of research needs and topics by scientific area. In Water Resources Planning and Management they were concerned with computer models of water resources systems, both simulation and optimization, with diverse policy topics relating to institutions, economics and policy, with data management systems, and with natural science issues of various kinds.

Perhaps the best way to summarize the kinds of topics is to repeat a list of issues Viessman presented in his paper following the Julian Hinds Award. His list of issues dealing with institutions includes providing information, modernizing institutions, providing national and regional perspectives, paying for water management, uniting technology and society, improving state planning/management capability, modifying agency roles, and defining beneficial use. His list of resource issues includes global climate change, nonpoint discharges, solid waste management, water for natural systems, allocating water resources, dealing with extreme events, water renovation and reuse, protecting and enhancing water quality, restoring water-related attributes, transboundary water management, ground-water protection and management, and land-water management (Viessman, 1990).

Obviously with such a list of issues, there is no end to the research topics that can be listed, and over the years we have pursued them to one degree or another.

Looking Ahead

The important question now is not what is on the laundry list, but what contributions to national advancement can be made by research in this era with a different national mood concerning water development, environmentalism, planning processes, financial capacity, federal-state relations and science policy.

I summarized the status of the nation's water problems in another paper (Grigg, 1992). My hypothesis was that the greatest need for the water industry is better planning and coordination, and I cited the last few national policy studies to verify the hypothesis. My conclusion was that the agenda items needing attention are to reduce tension between regulatory programs and water service providers; to use regional forums to find integrated solutions to regional problems; to create a coordination group to deal with water policy at the federal level; to study the high cost and bleak prospects of water project development; to develop policies for research and education to serve the water industry; to educate the public about water problems and policy; to coordinate decision information and water data; to mitigate conflict over "western water problems"; to stop the uncoordinated rise in utility rates; to find ways for the private sector to contribute more; to find ways to consider the environment as one of the water supply customers; and to review the contributions of the federal water agencies.

Now, neither water resources researchers nor policy analysts are as optimistic about solving policy problems as they were twenty years ago. I would like to summarize points I made about this at the AWRA Symposium on Water Policy in June 1992 (Grigg, 1992).

I observed that the most critical need to solve problems of water resources management is leadership, including leadership in the integration of problem solving teams involving different disciplines and points of view. This is an issue for the WRPMD, because by definition, it is a division that is intended to bridge the disciplines. The next point related to research is unraveling complexity with better research and clearer identification of issues and win-win solution scenarios. This requires systems analysis to show how complex, interacting systems work. Research will be needed at the problem solving level to discover how natural and man-made systems interact. The contributions of each discipline on the problem solving team, engineering, economic, legal, ecological, financial, political, are needed to find the best solutions. Research can help to reduce conflicts to those that require policy debate and political action by providing information, unraveling complexity and pointing out how to reach integrated decisions.

Conclusions

We are here to review the twenty-year history of the Division. With regard to applied research, the only reason for it is to solve problems. Twenty years ago, there were high expectations about the promise of science to solve policy problems. We had long lists of research needs, but the question in this different era is not about the lists, to ask what contributions can be made by research and by the Division.

Resolution of water conflicts will lie in legal, financial and political arenas. These institutional issues will continue to dominate the policy agenda. The clash of values between groups, and issues surrounding property rights in water, will block the visionary dreams of planners, and water conflicts will not be resolved in reasoned, negotiation sessions between interest groups. Water conflicts will be resolved in elections, court battles, agency rulemaking and decisionmaking and in water right purchases. This is the realpolitik that we must deal with in the future.

To deal with this reality, the greatest policy need for the water industry is the institutional issue of better planning and coordination, and this issue is squarely within the Division's mission. To improve planning and coordination, the most critical need is leadership, both leadership of problem solving teams and leadership in unraveling complexity through systems analysis.

Some have criticized the Division for neglecting important policy issues and concentrating on technical issues too much. The research agendas presented in this paper show that the opportunities are great for the WRPMD to make a real contribution to water policy in the US, but it may be possible that "the fields are ripe, but the workers are few". If that is the case, and if the main need is leadership, then we need to recruit and build that leadership. How is that a research need? It is a research need in the sense that if the Division's research activities, our journal, task committees and conference sessions, can be made relevant to the nation's policy agenda, then we will attract and build the leadership needed to make contributions to solving the critical problems of water policy and management.

References

Grigg, Neil S., Management of US Water Research, Journal of the Water Resources Planning and Management Division, ASCE, Vol. 106, No. WR1, March, 1980.

Grigg, Neil S., Engineering Perspectives on Water Management Issues, Symposium on Water Policy, AWRA, Washington DC, June 28, 1992.

Grigg, Neil S., A New Paradigm for Coordination in the US Water Industry, submitted to Journal of the Water Resources Planning and Management Division, ASCE, August 1992.

Viessman, Warren Jr. and Caudill, C.K., The Water Resources Research Act of 1964: An Assessment, Congressional Research Service, United States Government Printing Office, Washington DC, 1976.

Viessman, Warren Jr., "Water Management: Challenge and Opportunity," Journal of the Water Resources Planning and Management Division, ASCE, Vol 116, No. 2, March/April, 1990.

Remarks on Computer-Use History in Water Resources

Leo R. Beard, Hon. M. ASCE[1]

Abstract

Over the past 50 years, use of electronic computers has radically changed water resources engineering practice. It has enhanced capabilities, reduced costs and improved reliability. It has not eliminated the need for a competent professional to make engineering design and management decisions.

Introduction

In considering the history of computer usage in water resources planning and management, there are two aspects of primary interest: the phenomenal growth in computer capability, and the impact of this capability on advancing the technology and on the manner of applying water resources technology.

As one who has been directly involved from the beginning, I can say that this first aspect has been most amazing - almost unbelievable. Starting with the earliest electronic computers used in water resources technology about 1950 (computers that were programmed with jumper wires), computational capability has increased at a rate of about an order of magnitude every five years. This would give us a capability ratio of 100 million to one from 1950 to 1990.

Since I have difficulty visualizing quantities larger than 100, I might elaborate by saying that, by the mid-sixties, the card-input, card-output computer that I had access to performed in 3 minutes one iteration

---

[1] Senior Consultant, Espey, Huston & Assoc. and Professor Emeritus, The University of Texas at Austin.

in an optimization routine. By hand computation, the iteration would take many days. When the same operations were performed on the fastest computers available at the time, the three minutes became two or three seconds. I could not believe it knowing all of the computation sequences that were involved.

But today, on my 25-megahertz desk computer, the operations take a small fraction of a second. Time required on a super-computer would be infinitesimal, and I feel that nobody could fully appreciate the phenomenal performance of these machines.

Early Impact

Despite this fantastic growth of computer capability, its impact on water resources technology was small at first. This was due to difficulties of programming and delays in accessing computers. Engineers were not programmers, and administrators couldn't understand why engineers needed turn-around as fast as once a day or even once a week. These factors retarded performance and increased costs, so that organizations were actually investing in the future when they invested in computers. Computers did not generally pay off at the time.

Once that FORTRAN programming was available to engineers and scientists so that they could communicate directly with the computer (without explaining all to a programmer), application effectiveness greatly increased. At this time (early 1960's) in general, computers were used primarily to solve problems in the same manner as they had been solved by hand computation, and programs were custom-made for each application. Soon, however, the concept of generalized computer programs, whose sequences of computations were controlled by the input file, began to spread, and this led to free and effective interchange of planning and design capability. It became easy for one organization to check designs or other determinations made by another organization.

Standardization

These generalized programs led to a high degree of standardization in computer application. An outstanding example is acceptance by the Federal Emergency Management Agency of particular programs that had been widely tested in flood-plain delineation. When these programs were used, the methodology, but not necessarily the resulting determinations, was readily accepted. Any alternative technology used would be reviewed very carefully. The strong tendency, then, is to use the designated computer programs. While this greatly facilitates flood-plain delineations for more than 15,000 cases, it tends to inhibit development of better technology.

But the computer also produced a great spurt in research as well as application. Large masses of data that were previously available in published form or in private files were gradually converted to electronic files such as the WATSTORE file of the U. S. Geological Survey. Acquisition and data processing that had stringently limited research was no longer a major factor in many cases.

Personalized Computer Use

More recently, a great factor in advancing both research and application is the development of personal computers having capabilities comparable with those of main-frame computers 20 years ago. As the cost of such computers approached the earlier cost of mechanical desk calculators, it became feasible for each researcher or designer to have powerful computational capability at hand and continuously available. With a communications modem, data and software could be obtained or exchanged with relative ease.

And I hesitate to mention the so-called advantages of powerful portable computers that allow one to work while on a plane or in consultation at a remote conference room. It makes me recall the "good old days", when commercial transportation was mainly by train and I would welcome a 3-day trip between Los Angeles and Washington, when one could recoup from every-day stress and catch up on reading and relaxation.

Enhancement of Engineering

But, getting back to the main theme, computer speed and programming simplicity led to two areas of rapid advancement - the rapid increase in complexity of problems solvable by computers, including decision processes relegated to the computer, and increasing simplicity of performing complex operations by a user unfamiliar with the mechanics of the program. As a consequence, problems that had previously no practical solution mechanism were easily solvable. These included solution of complex network problems, unsteady-flow problems, sophisticated analyses of the rainfall-runoff processes and of water quality variations, finite-element analysis of hydrologic factors in groundwater, lakes and estuaries, analyses of various stochastic processes, and automatic model calibration.

I can recall solving 5-variable multiple linear regression problems by hand about 40 years ago. It took half a day, and I was lucky if the solution checked so that the whole process need not be repeated. Now, stepwise multiple linear regression solutions for many variables are made in a matter of seconds, and solutions of 10-by-10 matrices are made thousands of times in stochastic simulation programs.

Simplicity or ease of program execution, usually by menus, has great advantage as long as the program has been thoroughly tested. It helps the analyst to concentrate on high-level decisions, relegating pre-programmable decisions and operations to the computer. On the other hand, it makes it possible for analysts unfamiliar with the technology to obtain solutions that can be in error because of fallacious input or fallacious selection of program options. For this reason, use of such programs by non-experts in the field must be carefully monitored. Regardless of availability of computer technology, engineering and science problems must be solved by competent engineers and scientists.

Closing Comment

Looking back more than 50 years, I recall a room full of noisy desk calculators capable of adding, subtracting, multiplying and dividing, and costing far more than personal computers cost today in terms of constant dollars. These were used in conjunction with tables of logarithms, trigonometric functions and square roots, but not random numbers. It must surely be difficult for younger engineers and scientists to conceive of the spectacular revolution in water resources management that has taken place. It is like two different worlds 50 years ago and today.

An Historical Perspective on Water Resources Education

L. Douglas James[1]
Member

Abstract

A professional education in water resources combines courses that apply sound science to meet popular needs with general knowledge on how society uses and views water. The graduate has a specializations, understands social preferences, and is able to serve the public interest while pursuing a rewarding career.

Introduction

People experience floods and droughts, draw water from streams and wells, and enjoy lakes and rivers. A society that saw some people benefit and others suffer began water resources education by teaching engineers how to obtain water and control floods. It moved on to studying why people need water and how they can meet their needs with minimal natural and social disturbance. In this context, water education functions at three levels:
1. **General Knowledge.** Individuals who understand the roles of water in nature, society, and the economy do a better job of husbanding personal water uses and are better informed when they express opinions on resource management issues.
2. **Technical Knowledge.** Learning how to design, operate, and maintain facilities (reservoirs, canals, wells, and treatment plants) to gain human benefits and to protect environmental and social constraints.
3. **Knowledge Building.** Conducting research to solve puzzles so that society becomes better able to manage water over time.
The first level teaches everyone; the second launches careers in water management; and the third trains people for water resources research.

---

[1] Professor, Civil Engineering, Utah State University, Logan, UT 84322-4500

Forces Driving It
History reveals three principal motives for becoming educated in water resources:
1. **Learning.** People study because understanding a) enriches human spirits, b) produces material benefits, and c) establishes a culture of objectivity and precision.
2. **Politics (Theology).** Water resources management is a public good. Individuals must convince the voting public that the cost of the collective good that they want is justified. Building public support transforms the justifications into "theologies" that believers want taught as water education. The earliest theology was river navigation to get farm goods to market economically. Later theologies were reclamation to give people crowded in urban squalor new lives on formerly non-productive lands; multi-purpose river basin development for efficient electric power, soil conservation, water supply, recreation, etc.; water quality to protect human health; and environmental protection. Each theology became "politically correct" so that objective assessment became difficult and resources were spent carelessly until later eras adopted new theologies and priorities.
3. **Self Interest.** Individuals pursue their personal needs. While supporting programs in popular favor, citizens want to get more value from the water they use and suffer fewer flood and pollution losses. Other people see career opportunities in water resources and want to learn skills to get jobs.

Teaching to meet these three needs respectively raises water consciousness, tells people what to believe about water, and trains specialists for management and research. University water education trains specialists in programs that should also look at the larger picture.

Ying and Yang of It
Educational programs are torn between opposing traditions that have long been argued in a ying-yang dichotomy. One is classical and theoretical. The other is practical and empirical. The boundary separates puzzle solving from problem solving and determines whether a science is demand driven or supply driven. The classical-practical opposites have several dimensions:
1. Populist v. Elite. How should education be divided between training people to make personal decisions and specialists to make societal decisions? When does a problem become so difficult that a specialist is needed?
2. Technology v. Science. How should research be divided between adding basic understanding and applying what we already know to give people better lives?

3. **Majority Favor v. Truth.** Should education teach the viewpoints of the majority in a democracy or present unpopular truth?

Developmental Phases

Society has wanted 1) projects to supply human needs, 2) systems of projects coordinated for cost effectiveness, and 3) programs to sustain systems over project life cycles. Education in water resources engineering first taught hydraulic and structural design to make projects function effectively, added hydrology to quantify the flows and natural constraints, and is now probing the limits to environmental and social sustainability.

At the beginning of each phase, professionals possessed classical knowledge, faced a practical problem, and made a professional judgment. Over time, the methodologies were standardized and written into manuals and computer codes. The designs became evaluated for their compliance with the standards rather than for the reasonableness of the basic judgments.

Water education serving the first phase identified what water engineers were expected to know, established curricula, and tested applicant practitioners. Later political concerns expanded the body of expected knowledge to the point where few people are ever trained on the wide range of related topics.

Nine Themes

Professional water education draws from many disciplines, and training requires courses from many of them. Students are unable to take so much and are left understanding some things at the professional level and other things at the political level.

1. **Engineering.** How to build facilities to function properly. This requires training in hydraulic, structural, and geotechnical engineering.
2. **Hydrology.** How to size facilities. This requires training in science to understand the principles, statistics to analyze frequencies, and a way to assess failures.
3. **Water Quality.** How to protect public health from immediate and long-run threats and protect the natural environment. This requires training in the chemical and biological processes that determine water quality and cause water pollution to have impacts.
4. **Engineering Economy.** How to select from mutually exclusive alternatives. This requires a systematic listing and evaluation of consequences and reduction of the results to a single index of merit.
5. **Economics.** How to use water resources management within a larger program of national economic development. This requires training in the theory of

public goods and evaluations of indirect and secondary effects.
6. **Planning.** How to organize a comprehensive examination of all relevant factors in using multiple means to achieve multiple objectives. A primary component was nonstructural measures for demand management.
7. **Systems.** How to reckon with interactions among components and uncertainties. This introduces systems modeling and stochastic analysis.
8. **Environmental Impact.** How to protect natural features in the environment. This introduces the need for cause-and-effect relationships and ecologic modeling.
9. **Social Impact.** How to protect the values that people hold dear. This introduces needs for social relationships and modeling.

At different times, different themes supported by different combinations of other themes have dominated water resources planning. Agencies search for employees trained in popular themes and may neglect others to their regret.

Planning Expectation.
The paradigm used by water resources engineers also has a ying and yang. The planning ideal has been to learn basic principles, define practical problems, collect available information, and select the best alternative objectively. Where quantitative data were limited, professionals learned to make acceptable judgments from basic principles. Teaching how to reason from principles was the essence of classical water education.
Planning practice, however, has been shifting to prepare, institutionalize, and apply computer models that are programmed to represent legal requirements. Engineering firms are hired to make sure that their clients comply with the law and train their employees to run the models. People with incomplete training over the above nine areas and with minimal practical experience find data and infer conclusions by taking models as reality. Water education shifts from teaching principles to explaining models. Administrators equate good planning as running the models correctly. Planning dogma is formed by interest groups that know what they want done before the facts are gathered or analyses performed. Neither approach brings people to common agreement nor is theoretically correct.
History teaches that the essence of water education is to train specialists to know enough about all nine themes so that important considerations are brought to public attention and to train the general public to recognize that there are gainers and losers and that a balance must be reached. Water education should be developed as both public and specialist levels; and improved harmony between the two in an emerging research need that is fundamental to managing resource demand.

Future Directions In Water Resources Education

Stuart G. Walesh[1]

Abstract

Water resources education at the undergraduate level must change to properly prepare professionals for the 21st Century. While few changes are needed in technical content, graduates must be explicitly prepared to function as a part of a much more heterogenous work force and to thrive within a global economy.

Introduction

The purpose of this paper is to suggest ways in which water resources education should change to prepare professionals for the 21st Century. Undergraduate civil engineering education is the paper's focus. A U.S. perspective is taken to emphasize the dramatic changes that American water resources professionals will experience in the next decade.

Who Will be Available to do Water Resources Work?

About two-thirds of the Americans who will be working in the year 2000 are working now (United Way, 1988). But what about the new workers? Almost two-thirds of the new workers will be female (United Way, 1988). Women as a group have much to offer the engineering education environment and the engineering profession of the future (Marcellino, 1992).

About 57% of the new workers will be non-white, female, or immigrant. Nearly one in three will be what is now called an ethnic minority by the year 2000 (United Way, 1988). One in five Americans are fifty-five or older

---

[1] Dean, College of Engineering, Valparaiso University, Valparaiso IN, 46383

and this group will grow (United Way, 1988). The growing group of older Americans, many of whom have chosen early retirement from their first or principal career, provide a largely untapped resource for many organizations including educational institutions. There are many engineering education implications inherent in the more heterogeneous work force with its greater participation by women, ethnic minorities, and seniors. Engineering colleges need to further improve their recruitment and retention of women (Walesh, 1992) and ethnic minorities. Engineering faculty and administrators must make fundamental changes in their language, metaphors, and style. There should be more appreciation of gender-specific professional traits. New educational paradigms are needed with movement away from the traditional individualistic approach that pits individual students against each other in competition for grades. Interdisciplinary team work should be encouraged through mechanisms such as senior projects and inter-college courses and programs. Finally, colleges need to make more creative use of the knowledge, experience, and energy of senior citizens.

Who Will The Future Water Resource Professionals Serve?

Robert Reich (1990) urges Americans to prepare for a global society with emphasis on the global market. If one accepts Reich's general thesis about a growing global economy and his specification of the symbolic analyst as the individual who will thrive in that economy, then civil engineering seems to have a promising future. The symbolic skills he identifies -- abstraction, system thinking, experimentation, and collaboration -- are the "stuff" that progressive civil engineering is made of.
Allen and Sewards (1992) predict that the globalization of engineering practice will result in major changes in employer and employee attitudes toward international assignments. Arango (1991) argues that the U.S. can stop its backsliding by fundamental changes in engineering education and practice including improvements in the direction of internationalization.
Harris (1992) predicts that "the world of engineering will not be dominated by national security issues as in the past, but by the concerns for social security". Civil Engineering has always been the most people-oriented engineering discipline with the water resources subdiscipline being a quintessential example of people-oriented endeavor.
There are many engineering education implications in serving people and organizations around the globe. Engineering colleges should promote international experiences such as international study, cooperative education, work, and travel for more students. A varied

and significant international student population should be recruited. Colleges should recruit faculty members who were born and raised in a variety of countries. Administrators must encourage and enable faculty to participate in international travel, study, conferences, and work. Colleges should seek opportunities to host international visitors. Partnerships should be developed with internationally-oriented organizations. Finally, engineering students should be encouraged to take courses and pursue minors and majors with international themes or content.

What Kind of Work Will Early 21st Century Water Resources Professionals Do?

From the beginning of recorded history and all over the earth, "water resource professionals" have met the basic needs of communal society (Walesh, 1990). Providing water supply, wastewater disposal, irrigation, transportation, and environmental protection are examples of these needs. Looking into the future, fundamental changes are not likely to occur although there certainly will be many changes in tools and techniques (e.g., see Okun 1992, Pearce 1992, and Roesner and Walesh 1988).

However, all of this will have relatively little impact on the technical content of undergraduate education as it relates to preparing future water resources professionals. Undergraduate engineering education must continue to emphasize the fundamentals. In a field of rapidly changing technology, the leaders want the major emphasis on fundamentals, not on ephemeral technologies; on education, not on training (Walesh, 1988).

Recap of Future Directions in Water Resources Education

Numerous implications for undergraduate engineering education follow from the demographic and economic trends set forth in this paper. These changes have relatively little to do with the technical content of undergraduate education. Rather, they call for major modifications to the non-technical aspects of the overall student's education experience.

References

Allen, L. and J. Sewards, "Issues in Human Resources: Managing Talent in the 21st Century", Journal of Management in Engineering-ASCE, Vol. 8, No. 4, October, 1992, pp. 340-345.

Arango, I., "From U.S. Engineer to World Engineer," Journal of Management in Engineering - ASCE, Vol. 7, No. 4, October 1991, pp. 412-427.

Harris, J.G., "Engineering - The Bridge Between Two Cultures", Newsletter, ASEE Electrical Engineer Division and IEEE Education Society, Summer 1992.

Marcellino, P., "Management: The Study of People -- Do Women and Men Really Have Different Styles?, "Women Engineer, April/May 1992, pp. 50-57.

Okun, D.A., "Urban Water Management in the 21st Century", Proceedings of the Conference Saving Threatened Resource - - In Search of Solutions, Edited by M. Karamouz, ASCE, Baltimore, MD, August, 1992, pp. 150-160.

Pearce, J.B., "Perceptions, Sensitivity, and Solutions; Water Quality 2000", Proceedings of the Conference Saving Threatened Resource -- In Search of Solutions, Edited by M. Karamouz, ASCE, Baltimore, MD, August, 1992, pp. 39-43.

Reich, R.B., The Work of Nations, Knopf, New York, NY, 1990.

Roesner, L.A. and S.G. Walesh, "Urban Water Resources Issues in the 21st Century, Journal of Professional Issues in Engineering-ASCE, July 1988.

United Way, The Future World of Work: Looking Toward the Year 2000, 1988.

Walesh, S.G., "Microcomputer Capability: Practitioners Perspective", Proceedings of the Hydraulic Engineering Conference, ASCE, Colorado Springs, CO, August, 1988, pp. 610-615.

Walesh, S.G., "Recruitment and Retention of Women in Engineering: A Small College Perspective", Proceedings of the 1992 Women in Engineering Conference, --Increasing Enrollment and Retention, Washington, D.C., June, 1992, pp. 39-44.

Walesh, S.G., "Water Resources Science and Technology: Global Origins", Urban Stormwater Quality Enhancement, Proceedings of the Engineering Foundation Conference, Davos, Switzerland, ASCE, 1990, pp. 1-27.

Evolving Institutions for Managing the ACF Basin

Andrew Dzurik[1], Wayne Hall[2], Steve Leitman[3]

Abstract

The ACF basin is a case of interstate conflict in resource management. This study looks at the planning underway with regard to mechanisms for managing the basin.

The Situation

The area drained by the Apalachicola, the Chattahoochee and the Flint Rivers (the ACF Basin) encompasses significant parts of Alabama, Florida and Georgia. This basin's waters and related resources are important to the well-being of this area and the entire three state region.

The three states, and the U.S. Army Corps of Engineers Corps, are today in a serious dispute over the management of these resources. The State of Alabama, in 1989, filed a lawsuit against the Corps, the agency which operates the federally funded hydro-electric, navigation and flood control structures in the ACF basin. This action was in response to a federal proposal to reallocate storage in Lake Lanier from hydropower to drinking water supply and a concomitant reluctance by the State of Georgia and Metropolitan Atlanta to negotiate on the proposed reallocation and coordi-nate it with proposals from other basin water interests.

The technical basis of this suit was the Corps' failure to follow adequately the provisions of the National Environmental Policy Act. Florida filed to intervene on Alabama's side while Georgia filed to intervene on the Corps' side. Alabama officials took this action because they saw the reallocation as preempting their rights to the basin's water resources. Georgia filed to intervene because of its perception that Alabama was trying to restrain the economic growth of Metropolitan Atlanta by restricting access to water. They also held that they were entitled to the water as the major-

---
[1,2] Professors, Dept. of Civil Engineering, FAMU/FSU College of Engineering, Tallahassee, FL 32312-2175
[3] Apalachicola Coordinator, Florida Defenders of Environment

ity of the basin's land area and people are in Georgia. Florida's intervention was based on its concern that unfettered growth upstream would inevitably damage the ecology of Apalachicola Bay and concern about the larger question of the management of water and related resources in the basin. Ultimately, a stay was filed and the parties chose to negotiate their differences rather than go to court. In January 1992, the three states, and the Corps, signed a Memorandum Of Agreement to amicably resolve this suit. As a result, there now exists an opportunity to establish an institution which can meaningfully influence, through study, analysis, coordination and dispute resolution, management of the basin.

Purpose and Scope

This report is the result of a study intended to provide background information and guidance, primarily to the State of Florida and secondarily to Alabama and Georgia and to the U.S. Army Corps of Engineers, in ongoing considerations as to whether a basin-wide institutional entity should be created for the management of the water resources of the ACF Basin and, if so, what form(s) this entity might take. Florida's interests in establishing such an entity relate to the fact that the viablity of the Apalachicola estuary is directly linked to the freshwater flow of the Apalachicola River.

More specifically, the purpose of this study was to develop, from today's perspective and in the context of the ACF Basin, an understanding of the strengths and weaknesses of various institutional models that have been used for water resource management in basins where there are both state and federal interests. The scope of the study included a detailed literature review on basinwide efforts to manage water and related resources through a state-federal partnership, and extensive visits and interviews at three federal-interstate commissions: The Delaware River Basin Commission, The Susquehanna River Basin Compact Commission and the Interstate Commission On The Potomac River Basin.

Background

Water wars are invariably about ways of life; maintaining an old one, or starting a new one. This water war is no different. It is about the differing ways of life of several important sets of folks. The first set is the one who started this particular battle of this war. It consists of all those whose star is hitched to the continued growth of the Atlanta metropolitan area . "As Atlanta goes, so goes the South", is their mantra; and "progress", in the form of continuing economic growth through more development and use of land, water and related resources, is at the center of their culture. More water is required to maintain this way of life, and the

Chattahoochee River seems a likely place to get it.

Then there are those whose way of life has been built around the harvesting of nature's bounty, particularly that which comes from the estuary and bay at the extreme southern end of the basin. There, 400 miles from Atlanta, where the Apalachicola River pours the ACF Basin's waters into the Gulf of Mexico, another ideal of "progress" exists; insuring for the future the continuous plenteous yield of this exceedingly rich wetland. For this, a stable environment is needed, and changes in either the amount or the quality of the ACF's waters reaching the estuary could disturb this stability.

The third group to find its way of life threatened by this dispute includes all those who have adjusted their affairs to take advantage of the relatively constant water level in the system's mainstem rivers and reservoirs: e.g., water based recreationists and those who service them; those who have an interest in the system's irrigated agriculture potential. Large diversions from the system threaten their future use and enjoyment of their investment.

This already difficult conflict is further complicated by the multiplicity of jurisdictions that have a hand, or think they do, in the management of these waters. The federal government, Georgia, Alabama, Florida and numerous local and regional governmental and non-governmental entities all find themselves and their futures caught up in this battle.

The current dispute is not the first skirmish in this war. The lawsuit now before the courts represents the third time in fifteen years that the three states and the Corps have disagreed formally over water management in the basin. It is obvious to most that another way must be found for making such decisions; and making them so fairly, so equitably, that their implementation is unlikely to face serious challenge. The fundamental, critical issue in this matter is whether the resources of this basin will be managed incrementally, from afar, by strangers; or by those most affected, according to agreed upon objectives ?

The Setting

The Chattahoochee - The Chattahoochee River rises in Georgia's mountainous northeast quarter. From there, on the high slopes of Brasstown Bald and Tray Mountain, its waters take a 4700 foot plunge, and a 524 mile run, to the Gulf of Mexico, near Apalachicola, Florida. Initially it is a typical mountain stream, used by many for trout fishing and other fastwater recreational pursuits. It flows through Atlanta and from there to its southern terminus. The Chattahoochee drains an area of about 8800 square miles in three states, but most is in Georgia. About 3,000,000 people live in the basin, two thirds of them in Georgia.

The Flint - The Flint's drainage basin probably

extends to about downtown Atlanta, and by the time it leaves the metropolitan area just South of the Atlanta - Hartsfield International Airport, it is a stream of significant size. The drainage basin of the Flint, about 8500 mi^2, is nearly identical in size to the Chattahoochee's, but the Flint delivers less water, on the average, to Lake Seminole; about 8700 cfs annually compared to the Chattahoochee's 12,000 cfs.

The Apalachicola - The Apalachicola River, Florida's largest in terms of flow, is formed by the confluence of the Flint and the Chattahoochee at Lake Seminole, near the state border. From there it flows about 105 miles into Apalachicola Bay on the Gulf of Mexico near Apalachicola. It delivers an average annual flow of some 24,000 cfs into the estuary. The Apalachicola basin is, by all accounts, a unique and important biological resource.

The System - Perhaps the most difficult concept to grasp about a river basin is that it is a system with connectivity between its various parts. Actions taken in one place within the system, or on one part of the system, have repercussions throughout the system. Using water in one part of a basin may preclude its use in another. In the ACF Basin, an area where until recently there always seemed to be enough water for everything, this is just starting to be understood.

Water Resources Management Issues in the ACF Basin

Surprisingly, officials in all three of the Basin states are aware that "crunch time" is here with regard to competing demands for water and they share a common concern about this. They worry about what comes next; for example, who will make the next such request, who will decide on such requests, and on what bases; who will be the present and future manager of this basin's water and related resources, and to what purpose will the management of these resources be directed?

As might be expected, the three states hold quite different views about the "best" use of the Basin's resources. To Florida view, the Apalachicola's waters are extremely important to tourism and fisheries. Alabama's interests include navigation, recreation and water supply. Aside from Atlanta's request, Georgia has interests similar to Alabama for the basin waters in south Georgia. And the Corps, representing the federal government, is committed to operating the river for navigation, recreation,and power production.

In most management issues, things seem to happen in cycles. It appears that the wheel of resource management is about to complete another turn. Water resources management is about to become, once again, an important regional and national issue. As has been attributed to various authors, "it's beginning to feel like de ja vu all over again."

WATER MANAGEMENT BY ENDANGERED SPECIES

George R. Baumli, * M, ASCE

Abstract

In California, water management decisions are being driven by endangered species considerations. Traditional project formulation and operational criteria seemingly are unimportant. The endangered species considerations exacerbate the impacts of six consecutive years of drought and a stressed water infrastructure where water demands significantly exceed reliable water supplies. The result of endangered species constraints is less water at a higher cost for traditional water uses and delay in implementation of needed water projects. Amendment of the Endangered Species Act is urgently needed. In the meantime, water managers must work within the law and strive to achieve as much balance and reasonableness as possible. This paper describes how endangered species considerations have effected water management in California, presents examples of the resultant economic and social costs and describes problems with the Endangered Species Act and the need to amend it.

Introduction:

Environmental regulations and the Endangered Species Act are paralyzing US businesses and government. When the National Environmental Policy Act (NEPA) passed in 1969, most agreed that more attention to the environment was appropriate. There were countless examples where development ran rough shod and environmental damage resulted. NEPA and companion State acts made businesses and government change their mode of operation and the environment improved. Significant costs were necessary to achieve this, but with an expanding economy, the public didn't object. In only rare cases did the public have any idea or, for that matter, care what environmental costs were associated with a product.

A similar situation surrounded the Endangered Species Act (ESA), which became law in 1973. At that time, the public strongly supported protection of endangered species. However, almost no one realized that economic and social considerations could not be considered in the listing of a species and would be given only lip service in development of plans to protect endangered species. The Act requires protection at any cost.

*General Manager, State Water Contractors
555 Capitol Mall, Suite 725, Sacramento, CA 95814

Again, with an expanding economy and a continual media blitz on the importance of the environment, the public didn't object.

The situation is different today. Businesses are failing, unemployment is increasing and economic depression is a reality. One of the reasons is the onerous environmental regulations and their blind enforcement and advocacy by environmentalists outside and inside government. Rush Limbaugh, talk show host, observed that "Too often the focus of U.S. business today is not on making money. Rather, it is to champion such fashionable left-wing causes as saving the spotted owls and other endangered species, stopping pollution and enacting quota policies.."

Effect of ESA on Water Management

California's population has increased by about 10 million since 1969 when NEPA passed and now stands at over 30 million. Water demands have grown in proportion. Notwithstanding this, California's water infrastructure has been at a standstill, with proposed improvements effectively blocked by environmentalists for the last 23 years. As a result, water shortages are commonplace and have led to significant adverse economic and social effects. Now, the Endangered Species Act is taking its devastating toll. It has the Central Valley Project (CVP) and State Water Project (SWP), which together normally provide water to over 20 million people and for irrigation of over 1.2 million hectares of productive farm land, nearly paralyzed and will succeed unless the Act is changed. The ESA must be modified to provide for consideration of economic and social values, otherwise we will have economic and social disaster.

The California Department of Water Resources has spent about $50 million in planning and development of the Kern Water Bank, a SWP ground water recharge and extraction project in the San Joaquin Valley near Bakersfield. In 1991, none of the 120 million cubic meters (mcm) of water stored in the ground water basin could be extracted because of constraints imposed to protect endangered species. The year 1991 was the 5th consecutive year of drought when urban customers of the SWP received only 30 percent of their requested water supply and agricultural contractors got none of their requested supply. As a result of water shortages in 1991, due primarily to the drought, about 100,000 hectares of farmland were idled in the San Joaquin Valley, resulting in the loss of about 9,000 jobs and economic losses exceeding $500 million.

As a result of drought, water project diversions, over fishing and other factors numerous species of fish, including the chinook salmon runs are in trouble. California has four distinct runs of chinook salmon. The runs are distinguished according to the time they move up the rivers to spawn; spring run, fall run, late fall run and winter run. There are salmon in the Sacramento River system every month of the year, migrating upstream, spawning and migrating downstream. The Sacramento River system produces 70 percent of all chinook salmon caught off the California coast. Water projects get most of the blame for the salmon decline, however, commercial fishermen caught 6,350 metric tons of salmon in 1988. At an average of 4.5 kilograms each, that's about 1.4 million fish; no small impact.

It is not surprising that the conflict between fish and water development has exploded in the Sacramento River Basin. That Basin produces about 25 percent of California's surface water supply and is where the major facilities of the CVP and SWP are located. The endangered winter run chinook salmon currently is of greatest concern and having the greatest impact on operation of the CVP and SWP. In February 1992 the National Marine Fisheries Service (NMFS) issued their biological opinion and the Reasonable and Prudent Alternative (RPA) for operation of the CVP and SWP. The RPA called for a greater volume of carryover storage in Shasta Reservoir to provide cool water for spawning winter-run, meeting temperature requirements in the upper river, opening the gates at Red Bluff Diversion Dam from September through May to facilitate fish migration and closing the gates on the Delta Cross Channel from February through April to avoid transport of young fish to the interior of the Delta where they are threatened by the CVP and SWP pumping plants.

Measures to protect the winter-run salmon in 1992 reduced the CVP and SWP water supply by about 580 mcm. These reductions were in addition to already severely curtailed deliveries due to the sixth year of drought. In 1992, the CVP delivered only 25 percent to 75 percent of their normal deliveries and the SWP delivered only 45 percent of the amount requested by its water users.

The economic costs of protecting the winter-run salmon are significant. One measure of the value of the 580 mcm of lost water is assignment of the cost the State Water Bank sold water for in 1992 or $0.05/mcm ($70/AF). At this cost, which is low, the total value amounts to about $33 million. Water shortages cannot always be completely offset by simply buying water. Sufficient water may not be for sale and facilities to convey the water to the user may not be available.

The CVP power users also have sustained a total loss of about $25 million in power revenues in 1991 and 1992 because of measures to protect the winter-run salmon. These losses resulted from bypassing the Shasta powerplant in order to draw colder water from the river outlets for the spawning salmon. The SWP suffered a similar loss. This resulted from the SWP agreeing to release water from Lake Oroville to maintain salinity conditions in the Sacramento-San Joaquin Delta, which were the responsibility of the CVP. The CVP couldn't fulfill the obligation because their water had to be held in Shasta to protect the winter-run. The result was that about $300,000 in power revenues were lost at Oroville because of the changed operations. Irrigation Districts along the Sacramento River sustained considerable additional economic losses, resulting from constraints imposed on them by NMFS to protect the winter-run.

Consultations on the winter-run salmon under Section 7 of the ESA are continuing with the objective of developing a RPA for 1993 operations of the CVP and SWP. It is expected constraints on water project operations in 1993 will be greater than in 1992. The economic losses and social impacts likely will be substantially greater however, because 1993 may be the 7th consecutive year of drought and the reservoirs are lower than they were going into 1992.

Problems with the ESA

Uncertainty is one of the most difficult aspects of the ESA to deal with. In 1992, for instance, the CVP and SWP water users had no idea of the amount of water they were going to receive until the middle of February when NMFS issued their plan to protect the winter-run. This posed a serious problem for farmers. Without assurance of a water supply, even a partial supply, they could not obtain financing for their operations and couldn't make decisions of the types of crops to plant or when to plant them. The drought itself imposed a high degree of uncertainty. However, the ESA overlay multiplied that uncertainty and caused significant additional stress. This will be the same in 1993. Some farmers, however, won't have to worry about it. They are being forced out of business.

Another frustrating aspect of the ESA is that the water user is frozen out of the Section 7 consultation process and out of deliberations to develop a recovery plan for the winter-run salmon. Yet, the water users' stake is enormously high. This is an unfortunate and undemocratic aspect of the ESA where affected parties have to stand by in silence while middle-level federal bureaucrats decide behind closed doors what will have to be done to protect the species. Little or no consideration is given to the economic or social costs. There are no adopted common economic rules to follow or a consistent procedure for their application. NMFS and the US Fish and Wildlife Service each appear to make up the rules as they go along. Affected parties have virtually no opportunity to provide economic or other input.

If the ESA were doing an effective job of protecting truly endangered species, a warped mind could rationalize that the economic and social costs might be worth it. However, too often after a species is listed it is determined that the species isn't truly endangered. Another failure of the ESA is that measures to protect one species may put other species in jeopardy. This was the case in 1992 with the winter-run salmon. Cold water supplies from Shasta Reservoir were being depleted to assure survival of the winter-run salmon eggs. However, not enough water would remain for release for fall-run salmon if the operations continued. The NMFS, to their credit, stepped in and modified the required plan to avoid jeopardy to the fall-run. Litigation by commercial fishermen provided motivation.

The US Fish and Wildlife Service is expected to list the Delta smelt as a threatened species. On November 6, 1992, environmental groups petitioned to list two additional species, the Longfin smelt and the Sacramento splittail, that also inhabit the Sacramento-San Joaquin Delta. Protection of these three species requires either virtual shut down of the CVP and SWP exports from the Delta or construction of the Peripheral Canal, which would effectively remove the pumps from the habitat of the threatened fish. Unfortunately, it will take a minimum of 7 to 10 years of acrimony and public debate to bring the canal on line, Environmental groups are opposed to any additional water projects. They contend future water needs can be met by conservation, reuse and water transfers. One significant fallacy in their argument is that substantial water transfers across the Delta will not be possible because with curtailment of pumping from the Delta by the ESA,

The real failure of the ESA is its failure to embrace and adjust for mankind's place in the natural environment. The ESA must be amended to provide for consideration of economic and social values otherwise the burden imposed will kill the economic engine of the US and the world. Numerous efforts are underway to amend the Act. However, the concept of protecting endangered species has wide public appeal and significant changes to the Act will be extremely difficult. Until the ESA is modified, water managers and others have to continue to participate in the undemocratic and unjust process to strive for as much balance and reasonableness as possible. In all likelihood, much of the process will end up in the courts, where everyone will lose, except the attorneys.

"Innovation and Diversification -
The Key to our Future Water Supply"

By Albert Muniz[1], P.E.,
R. David G. Pyne[2], P.E.,
and Sharon M. Trost[3], P.G.

Abstract

Water resources management must modernize, innovate, and diversify to meet the challenges of tomorrow: reliable water supply sources of acceptable quality, a cost-effective approach to total water management, and balancing competing needs. Innovative technologies, such as aquifer storage and recovery (ASR), are helping water management agencies to meet these challenges and are changing the direction of water management policies. ASR, for example, can significantly improve water system efficiency while reducing capital costs by at least 50 percent more than other options.

Introduction

South Florida is faced with several major water management challenges, including:
- Increased water demand and limited availability
- Contamination of resources
- Flooding
- Conflicting demands of end users
- Environmental constraints
- Inefficient infrastructure
- Limited surface storage opportunities

Long-term planning to optimize water resources development is critical to maintaining the future viability of these resources. Existing water storage and conveyance systems

---

[1]Division Manager, CH2M HILL, 800 Fairway Drive, Suite 350, Deerfield Beach, FL 33441

[2]Director of ASR Technology. CH2M HILL

[3]Dept. Dir.-South Florida Water Management District

are inadequate for future needs, and the marginal cost of water is increasing, prompting the consideration of new alternatives. One of the most promising water management alternatives is ASR which, although not a total solution to current needs, can play a major role in affordably satisfying competing demands.

This paper discusses ASR, a new technology that has proven effective both technically and economically, and the role of the regulatory community in the development of policies, initiatives and plans to guide utilities in the efficient management of water supply sources.

Innovative Technology

ASR is a water supply strategy by which treated drinking water is stored underground in a suitable aquifer through wells during "wet" months and/or low demand periods. The stored water is then recovered from the same wells during "dry" months or periods of high demand that may exceed the capacity of existing water systems. Hence ASR augments water supply potential without creating new peak demands on the regional system. Typically all of the stored water is recovered and no further treatment, other than disinfection of the recovered water is needed.

ASR technology allows a utility to meet increasing peak demands without an immediate expansion of water supply, treatment, or transmission capacity, typically at less than half the cost of other alternatives. Figure 1 shows the economic advantages of ASR at some ASR facilities in the United States. ASR can also be used to build up a bank account of stored water to meet future demand, or provide water during severe droughts or emergencies. Typical ASR storage is on the order of hundreds of millions, or billions of gallons. By comparison, elevated or ground storage tanks typically hold only a few million gallons, at higher storage unit costs. ASR requirements are minimal because the stored water is not susceptible to evaporative losses and less susceptible to contamination.

ASR technology has evolved during the past 20 years and is now operational at 14 sites across the United States. Its success and increasing acceptance by water users is indicated by the accelerating rate at which new systems are becoming operational and by expansion of existing systems. Currently over 30 ASR systems are under development in 12 states, with several additional systems overseas.
The following is a summary of ASR benefits:
- Cost savings in excess of 50%
- Seasonal storage/long-term storage, or "water banking"
- Emergency storage
- Maintenance of transmission/distribution pressure
- Prevention of salt water intrusion
- Agricultural supply

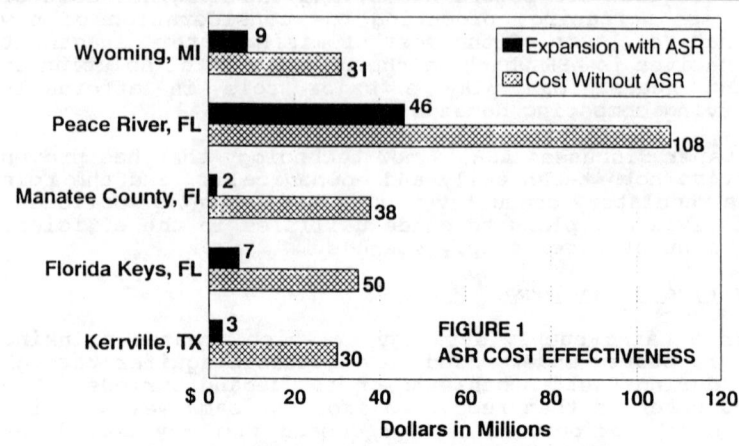

FIGURE 1
ASR COST EFFECTIVENESS

- Nutrient reduction in agricultural run-off
- Defer expansion of water facilities
- Reclaimed water storage for reuse
- Reduce environmental effects of streamflow diversions
- Reduction of evapotranspiration and seepage losses

Implementation

Implementation of any new technology requires the support of the regulatory community. To date, such support has been consistently strong in each new state where ASR has been demonstrated. The South Florida Water Management District (District), which is responsible for the "use of water" in south Florida, has provided leadership in the implementation of ASR. Water Use Permits now recommend and support diverse and innovative water management options such as ASR. The District has included ASR in its Water Use Policy Document (December 12, 1991) and in water supply plans for individual counties. In addition, it has provided financial assistance over the past 6 years to demonstrate the usefulness of and its support for this technology.

ASR technology application has been applied by the Peace River/Manasota Regional Water Supply Authority in southwest Florida. The source of supply, the Peace River, has an average flow of about 1,500 cubic feet per second (cfs) but is a highly unreliable water source in terms of both flow and quality. At times, brackish water from the estuary has approached the intake during droughts. At other times, algal blooms have precluded diversions even though flow was available. As a result, a 2,000 acre-ft offstream reservoir was constructed to store water and supply the system during months when no flow is available from the river.

Faced with the need to expand storage, the utility selected ASR instead explanding the reservoir. ASR wells were constructed and tested under operational conditions, storing offpeak water in a brackish, limestone artesian aquifer beneath the 12 million gallon per day (mgd) water treatment plant. Testing showed that ASR could assist the Authority to meet projected water supply needs at significantly less than half the capital cost of reservoir expansion, and would also defer the need for plant expansion.

Subsequently the system was expanded to include six ASR wells at the plant site. A further expansion to 8 wells with a recovery capacity of about eight mgd is underway. Two zones are used for storing the treated drinking water, one at a depth of 400 to 500 ft and the second at a depth of 600 to 900 ft. A deeper third zone is being investigated. An ASR simulation model was prepared and used to assess the most cost-effective combination of facility components (river intake structure, offstream reservoir, water treatment plant, ASR system) to meet projected demand, confirming the cost-effectiveness of ASR as a water management measure. The ASR system has been operational at this site since 1985.

Summary

ASR technology has been successfully used for the past 20 years. It has been proven technically for the storage of treated drinking water in a variety of geological settings including limestone, sandstone, glacial drift, sand and clayey sand. Experience has also shown that storage zone water quality need not be significantly constrain ASR feasibility. The Florida Keys ASR facility shows that it is possible to store treated drinking water in an aquifer containing seawater. Other sites store water successfully in aquifers containing high nitrates, hydrogen sulfide, iron, manganese, boron, and chlorides.

Economic advantages of innovative water management technologies such as ASR increase the credibility. Cost savings realized through the treated drinking water ASR are only a part of the overall benefits that ASR may offer when expanded beyond the potable water application. New ASR applications with untreated groundwater and/or surface water, and with reclaimed water, may have much greater economic benefits. Support by regulators for innovative but proven water management solutions such as ASR will be essential to ensure we can economically meet future water demands. Efforts by the District are prime examples of how regulators can guide, support, and assist in the resolution of our future water supply needs.

Local Agency Regulation: The Policies and Process of Surface Water Protection

Richard L. Schaefer, P.E.[1] and Jeffrey H. Stern[2]

Abstract

King County, Washington, adopted a Surface Water Design Manual in January 1990. This manual, which has served as a resource guide in developing similar manuals throughout the Pacific Northwest and across the United States, defines the analyses and measures required to control drainage impacts of development projects in the County. Since the Manual's adoption, ongoing research has suggested revisions to facility requirements for improved stormwater quantity and quality control. Subsequent state and federal regulations mandate revisions be made to the manual to further address water quality. Recent growth management policies dictate increased development densities within the urbanizing areas of the County. This paper summarizes the changes to be made to the Manual and describes the technical and political processes utilized to address conflicting concerns of the County, the development community and the general public.

Introduction/History

King County moved to the forefront of stormwater management in the Pacific Northwest region when it adopted its Surface Water Design Manual (Manual) in January 1990. The Manual expanded the scope of development standards addressing impacts to surface water resources, advanced computational methodologies used in analyses and designs, and increased level of protection required of new development.

At the time of adoption, the County understood that such large-scale changes to development regulations would necessitate monitoring and adjustments. Such adjustments are necessary to clarify regulatory language and intent, incorporate new technical knowledge, and respond to state and federal regulations:

---

[1] Project Manager, R. W. Beck and Associates, 2101 Fourth Avenue, Suite 600, Seattle, Washington, 98121.

[2] Environmental Planner, King County Surface Water Management Division, 1111 Third Avenue, Suite 1110, Seattle, Washington 98104.

LOCAL AGENCY REGULATION 329

- As the Manual was implemented in development review, several revisions were indicated to clarify regulatory intent. Improved guidance to development proponents was needed as to the type and quality of information required to evaluate a project. In addition, the thresholds triggering many of the Manual requirements needed review to ensure they properly balanced environmental protection with project feasibility.

- Subsequent watershed planning efforts, research and water quality literature suggested changes to enhance the level of protection provided to water resources and downstream properties. Continuous hydrologic models used in basin planning indicated detention storage volumes needed to be increased substantially over that resulting from the Manual's event model design method.

- It was exceedingly difficult for small commercial projects to comply with the Manual's on-site drainage requirements and with other regulatory and physical site constraints. The cumulative effects of these developments also appeared to be inadequately addressed by the requirements.

- The amount of land area required to accommodate storm drainage facilities interfered with achieving densities promoted under zoning and community plans.

- Compliance with state regulations mandated the Manual be revised in conjunction with implementing the Growth Management Act and that it be consistent with the State Department of Ecology's *Stormwater Management Manual for the Puget Sound Basin*. A provision in the State's manual not addressed by the County's Manual is the application of source control and treatment best management practices (BMPs) to commercial and industrial sites. The State's manual also strongly emphasizes infiltrating stormwater; however, compliance with this provision had to be resolved with the State's groundwater anti-degradation policies and local soil limitations on the use of infiltration.

To address these needs, the County initiated a process to update the Surface Water Design Manual. The balance of this paper describes the process used in developing the Manual revisions and the technical substance of the revisions.

Revision Process

The Puget Sound Basin, with its combination of glacial soils, steep topography and abundant rainfall, is extremely sensitive to development and its impacts. Development regulations, particularly those controlling runoff, must balance the sometimes opposing responsibilities of public welfare, stewardship of resources, and accommodation of growth. The goal of the Manual update process was to ensure involvement of advocates for all of these interests to develop effective yet realistic runoff controls that are flexible in addressing individual site needs and constraints.

Public Process. The process relied on significant public involvement to identify needs, encourage involvement, and to build ownership in the regulations. Involvement was solicited through broad mailings to environmental, real estate and development, community groups, and all current Manual owners. These mailings solicited issues, notified people of County intentions and encouraged further involvement. Public meetings were held to present the rationale for substantive technical and policy amendments along with the proposed changes and the considered alternatives. Review comments and testimony were incorporated into the proposed revisions before presentation to Council.

Advisory Committees. Three groups provided regular advice on policy decisions, review, and help to identify the feasibility and potential implications of proposed alternatives throughout the revision process. One committee was composed of County staff from the development permitting, resource protection and roads divisions. Another committee represented private users of the Manual including developers, engineers and development site designers. The final group was a standing Citizen Advisory Committee, which represents a wide spectrum of private interests and communities across the County. Each group provided a different perspective that proved essential in development approaches that maintained flexibility while achieving protection goals. Their recommendations also carried substantial weight in decision making throughout the update process.

Technical Process. The list of issues was addressed through an analysis of alternatives to develop a proposed package of recommended revisions to the Manual. Issue papers were prepared to document the alternatives considered, their technical and economic feasibility, and the reasons for selecting the preferred approach. These reports provided the technical basis for the three committees and County management to reach agreement on the key decisions behind the proposed revisions. The recommended revisions were subjected to public review and those issues resolved before the final revision packages were prepared for the Council.

A key component of the work on the update was the application of proposed recommendations to various types of development using case studies. Residential and commercial development examples were analyzed to determine if proposed modifications were feasible and addressed the problem as intended, and to determine the cost implications of implementing any recommended regulations. This work was instrumental in ensuring that the Manual could be fairly and properly implemented.

Training. A training program developed for permit writers and private users of the Manual was essential in ensuring its successful interpretation and implementation. The training program used when the Manual was introduced in 1990 was updated to address the new methods and material and to serve as an ongoing, self-supporting program to provide training opportunities at regular intervals. The long-term program is designed for both public and private users of the Manual.

Technical Changes

The revisions developed through the above process addressed a wide range of technical areas including flood protection, application of BMPs, infiltration guidance and enhanced on-site space utilization. Each of these is briefly described below.

Flood Protection. The 1990 Manual utilizes an event-based hydrologic model to determine flows and on-site detention volumes necessary to maintain predeveloped peak runoff rates for events up to 10-year storm intensity. Recent watershed planning utilizing continuous models indicates the level of protection provided by the Manual methods do not provide the 10-year storm level of protection. The sources of the inconsistency are the design storm hyetograph (24-hour duration, single-peak intensity) that does not reflect the characteristics of local rainfall patterns, and the method's overestimation of predeveloped runoff rates that results in artificially high release rates for detention facilities.

Two approaches were considered to resolve this inconsistency: modify the current method to adjust for the shortcomings, or develop an entirely new method based upon continuous simulation results. The recent advent of unit area runoff files based upon continuous modeling provided an opportunity to move toward a more realistic, yet user-friendly, methodology for estimating runoff. Software developed for this method will be distributed with Manual revisions.

Infiltration. Downstream flooding problems historically have resulted from local infiltration facilities having lower percolation rates than designed for. At the same time, the State has emphasized infiltration as the primary means of runoff disposal while simultaneously establishing groundwater anti-degradation policies. The revised Manual calls for improved soil tests to establish more reliable percolation capacity estimates. The Manual also provides guidance on the depths of various-type soils needed to sufficiently treat stormwater prior to contact with the groundwater table.

Space Utilization. Stormwater facilities consume substantial portions of developed tracts and, with larger facilities being required, there is a need to enhance flexibility to accommodate such improvements onsite. Provisions are being developed for joint development of stormwater facilities between neighboring commercial properties and the County. Special designs have also been prepared to address the unique facility siting difficulties for linear developments such as roadways and utility corridors. Guidance has been prepared for "stacking" detention and treatment facilities to reduce their overall footprint. Regulatory relief has been pursued to allow flexible lot sizes and co-location of stormwater facilities in designated open space.

BMP Manual. To achieve compliance with state and federal requirements, the County must develop a program of pollutant source control in stormwater. One component of the County's response was to develop a BMP manual that provides guidance to property owners and County staff in source identification, isolation and treatment. This BMP manual will be used as a resource in stormwater design for new development, redevelopment and retrofitting of existing facilities.

Ag-To-Urban Water Transfer in California: Win-Win Solutions

Lee A. Jacobi, PE[1] and Robert L. Carley, PE[2]

Abstract

Water transfers from farms to the cities are widely viewed as the next major source of supply to urban California. Ag-to-Urban permanent water transfers may have negative consequences to the agricultural sector and to the environment. This paper presents agricultural water use statistics, discusses sources of water for transfer, and suggests sources of water for "win-win" transfers.

Introduction

The current long-term drought in California has generated interest in water transfers. Recent state and federal legislation have loosened restrictions on both temporary and permanent water transfer activity. This paper is concerned with the effects of permanent water transfers. In California, there are three major groups of water users: "Ag," or agriculture; "Urban," including all cities, suburbs, towns and power plants; and the "Environment," where water is purposely or inadvertently applied to support vegetation, fish and wildlife. Ag-to-Urban permanent water transfers may have negative consequences to the agricultural sector and to the environment. The purpose of this paper is to suggest sources of water for transfers that would be "win-win," or at least not detrimental, to all three groups.

Statewide Water Shortage Projection

Water demand in California is now at the limit of existing supply quantities, about 36 MAFY, and is growing due to population growth. Urban water demand is projected to increase by 0.7 to 2.3 MAFY by year 2010 [1], [2]. These increases are for fresh water that will be needed despite projected increases in sewage recycling and conservation measures. At the same time, California's supply from the Colorado River will be permanently decreased by

---

1   Associate Engineer, Boyle Engineering Corp., Newport Beach, CA. Member ASCE.
2   Principal Engineer, Boyle Engineering Corp., Newport Beach, CA. Member AWWA.

about 0.5 MAFY. The current 6-year drought has merely brought the problem to the public's attention sooner.

It is difficult to develop more water in California due to environmental constraints and costs. The agricultural sector at present uses about 75 percent of the developed water supply of the state. It is not surprising that urbanites covet Ag water as a permanent solution to their supply problems.

Agricultural Water Use in California

Net Ag water use (applied water less about 10 percent re-use) in California peaked at about 28 MAFY in the early 1980s and is projected to decrease gradually due to change of land use from Ag to Urban [3]. Nevertheless, Ag will continue to use two to three times as much water as Urban and Environment combined in the near future.

Some statistics of Ag water use in California are presented in Table 1. Of 31.6 MAF of water applied to Ag in 1988, about 10 MAF was used to grow alfalfa and pasture, which are forage for the state's dairy and beef cattle.

Table 1. Applied Irrigation Water in California in 1988, by Crop

Crop	Applied Water (MAF)	Irrigated Land (1) (10^3 Acre)	Average Application Rate (2) (AF/Acre)
Alfalfa	5.4	1,075	5.1
Pasture	4.5	955	4.7
Cotton	4.5	1,380	3.3
Rice	3.0	470	6.3
Peaches & other deciduous fruit	2.2	580	3.8
Lettuce & other truck crops	2.1	915	2.3
Grapes	2.0	760	2.7
All others	7.9	3,155	2.5
Totals	31.6	9,290	

(1) Gross, including double-cropped area. Net area is about 8.9 million acres.
(2) Application rates can vary considerably due to climate and irrigation system.
Source: Calif. Dept. of Water Resources, <u>1988 Annual Water Use - Water Supply Balances</u>, 11/90.

As calculated in Table 1, rice has the greatest unit application rate, averaging over 6 feet of water per year, with alfalfa and irrigated pasture also high at about 5 feet per year.

Sources of Ag Water for Transfer

1. Fallowing

There have already been Ag-to-Urban water transfers in California. In most cases, irrigated crop land has been retired or fallowed. For example, as the City of Fresno expanded in to the surrounding Ag land, the City took rights to the water formerly used to irrigate fields and orchards. This process is occurring haphazardly around almost every town in California. As another example, Castaic Lake Water Agency in Southern California purchased 8,400 acres of Ag land in the San Joaquin Valley and will fallow this land and take its water allotment.

Large-scale retirement of Ag land for the purpose of water transfer can be expected to have many negative consequences. Less irrigated land will result in lower total crop production, which could result in higher food prices statewide and nationally. Higher food prices would result in loss of export earnings for the nation. Less land in production will also result in "third party" losses-- loss of jobs for field workers; loss of sales for suppliers of Ag equipment and fertilizer, etc.; and resulting general loss of sales for rural area shopkeepers. And haphazard fallowing within an irrigation district could encumber the district's operations. For these reasons, permanent fallowing should not be an acceptable source of water for transfer except where the land would have been taken out of Ag use for other reasons.

2. Conservation of "Wasted" Ag Water

Ag water is always applied in excess of actual crop ET because of the difficulty of achieving uniform application rate. However, not all such excess applied water is wasted-- the fate of excess applied water is often re-use. Tailwater is often used by downstream farmers. The Environment is often the unintended beneficiary of excess Ag water, from canal-side weeds and field border hedges to migratory birds' use of rice paddies. Ag drainage water is also re-used by the Environment, with negative consequences where selenium and pesticide concentrations are high.

Many existing Ag water conservation efforts to reduce flows from the farm are detrimental to the Environment, which thrived on the escaped water. Efforts to reduce canal and conveyance ditch percolation will result in less recharge to aquifers. Except in salt sink areas, water "saved" by conventional ag water conservation efforts should not be a source of water for transfer to Urban because it comes at the expense of the Environment.

The only true Ag water wastage is water lost to evaporation and to salt sinks. Ag water wastage could be reduced by:

    a) *Reducing losses to salt sinks.* Lining canals and on-farm ditches, containing and re-using excess deliveries in salt sink areas. The Imperial Irrigation District-to-Metropolitan Water District of Southern California transfer based on reducing flows to the Imperial Valley salt sink area is an example.

b) *Reducing evaporation.* Not sprinkling during mid-day or high wind periods is a simple management change. Replacement of sprinklers with tricklers or sub-irrigation could be funded by a transferee.

3. Reducing Evapotranspiration of Plants

Switching to less consumptive crops may not be feasible due to market constraints. However, development of less-consumptive varieties of the same crop has promise.

Less-than optimum irrigation could be performed in the summertime for certain crops with minimal adverse impacts. For example, it has been suggested that substantial amounts of water could be saved in California if alfalfa was minimally irrigated in summer months. The loss of one cutting of alfalfa could be made up by a substitute item in the dairy cow diet or alfalfa could be imported from other states.

4. Development of Ag Groundwater Basins

The Sacramento Valley contains a large, under-developed groundwater basin. This basin could be used to capture flood water. Local farmers would use the basin as their main supply source, pumping it down to make room for wet period recharge. Water which would have otherwise run off to the ocean would then be available for transfer.

Conclusions

The source of transfer water and where it would have ended up if not transferred should be considerations in permanent water transfers. Water derived from reduced evaporation, reduced losses to salt sinks, reduced plant ET, and development of Ag groundwater basins are suggested for win-win transfers.

Sources Cited

(1) California Department of Water Resources, 1988 Annual Water Use-Water Supply Balances, November, 1990.
(2) Estimated by the author based on state population 44 million in year 2010.
(3) California Department of Water Resources, California Water: Looking to the Future. Bull. 160-87. November, 1987.

Abbreviations and Units Used

Ac	Acre 1 Ac = 0.405 Hectare	MAF	Million Acre-Feet
AF	Acre-Foot 1 AF = 1,233 m^3	MAFY	Million Acre-Feet per Year
ET	Evapotranspiration		

Water Supply Systems Planning with Water Transfers

Jay R. Lund, Associate Professor
Morris Israel, Doctoral Student
Department of Civil and Environmental Engineering
University of California, Davis, CA 95616
(916) 752-5671, -0586

This study examines the recent use of water transfers in California, particularly during the current drought. The emphasis is on how planners and operators of federal, state, and local water systems can integrate water transfers into the planning and operations of their systems. The study identifies many motivations for using water transfers in water supply systems, reviews a variety of transfer types, and discusses the integration of water transfers with traditional supply expansion and water conservation measures. Some limitations, constraints, and difficulties for employing water transfers within existing systems are also discussed. The study focuses primarily on technical, planning, and operational aspects of water transfers, rather than the legal, economic, and social implications which have received extensive attention elsewhere. More complete presentation of this work appears in Lund, et al. (1992).

Transfers in Water Management

There has been little systematic examination of the engineering and operational aspects of water transfers. Instead, the mechanics of economically effecting actual water transfers has fallen on the ingenuity of actual system planners and operators. Their tasks are non-trivial, including the operation of complex storage and conveyance systems to meet multiple urban, agricultural, and environmental demands over a range of uncertain hydrologic conditions. These engineering tasks require rethinking in light of the special opportunities and problems posed by water transfers. The experiences with the many forms of water transfer recently employed in California offer an opportunity for developing these neglected technical aspects of water transfers.

Water Transfers in California

California's water supplies are poorly distributed in both time and space, relative to water demands. The water-related infrastructure, institutions, and legislation which have evolved and the conflicts and antagonisms which have arisen can be traced to these severe imbalances of supply and demand.

Each drought episode has brought new challenges to motivate creative and long-lasting innovations in Californian water management. Response to the 1920s-30s drought focused on construction of storage and conveyance facilities, an approach which continued well into the 1970s. Water conservation became popular in response to its successes in the 1976-77 drought and is now part of almost all California urban water plans and operations.

The current drought also has had significant effects on how water professionals, political leaders, and the public think about water supply. The result has been a refinement of existing water management techniques and experimental implementation of more novel approaches to water management, such as water transfers. The 1991 and 1992 California Drought Emergency Water Banks were the first major State-brokered water transfer programs in the nation. The Drought Water Banks were established to provide water for critical municipal, industrial, and agricultural needs, preservation of fish and wildlife, and carryover storage for potential additional dry years. The 1991 Water Bank was considered successful because it transferred large quantities of water in a short period of time in a state where water transfers have been especially slow to develop, relative to other Western states. Within a few months, the 1991 Water Bank had negotiated over 820,000 ac-ft of water purchases and almost 390,000 ac-ft of sales. The remaining water went to instream flows and overyear storage for the 1992 water year. In the course of arranging these transfers, 348 individuals, firms, and agencies had sold water, 12 agencies had purchased water, and most other major water users and suppliers in the state had become acquainted with the idea and the opportunities of water transfers. Continuation of the Water Bank for a second year demonstrates its perceived success. The 1992 Water Bank illustrates some of the lessons learned from the first year's experience, which included more refined financial arrangements for buyers and sellers, option purchases and sales, an absence of land fallowing as a source of water, and the development of a series of pooled purchases and sales throughout the year.

The Drought Water Banks illustrate the contributions of government to water transfers. State sponsorship firmly demonstrated State support for water transfers as part of overall water management, increased the probability of success for individual transfers, lowered transaction costs, and facilitated coordination of transfers with other water movements in the state. The experience of the 1991 and 1992 Drought Water Banks will likely encourage the independent pursuit of transfers by individual agencies in the future and serve to establish water transfers as a water management technique.

Beyond State-sponsored water transfers, there have been a great number of transfers and exchanges taking place independently of the State. This transfer activity illustrates the diversity of forms and purposes that water transfers can take for both regional and local systems. These types and benefits of water transfers are summarized in the accompanying tables.

TABLE 1: MAJOR TYPES OF WATER TRANSFERS

Permanent Transfers

Contingent Transfers/Dry-year Options
    Long-term
    Intermediate-term
    Short-term

Spot Market Transfers

Water Banks

Transfer of Reclaimed, Conserved, and Surplus Water

Water Wheeling or Water Exchanges
    Operational Wheeling
    Wheeling to Store Water
    Trading Waters of Different Qualities
    Seasonal Wheeling
    Wheeling to Meet Environmental Constraints

TABLE 2: MAJOR BENEFITS AND USES OF TRANSFERRED WATER

Directly Meet Demand
    Use transferred water to meet demand, permanently or just during drought.
Lower Costs
    Use purchased water to avoid higher cost new sources.
    Use purchased water to avoid costly demand management measures.
    Seasonal storage of transferred water to reduce peaking capacity.
    Use drought-contingent transfers to reduce need for storage facilities.
    Wheeling low-quality water for high-quality water to reduce treatment costs.
Improve Reliability
    Direct use of transferred water to avoid depletion of storage.
    Overyear storage of transferred water to maintain storage reserves.
    Drought-contingent contracts to make water available during dry years.
    Wheeling water to make water available during dry years.
Improve Water Quality
    Trading low-quality water for higher quality water
    Purchase water to reduce agricultural runoff.
Satisfy Environmental Constraints
    Purchasing water to meet environmental constraints.
    Wheeling water to meet environmental constraints.
    Using transferred water to avoid environmental impacts of new supplies.

Lessons Learned from California

California's experiences with water transfers suggest several potential lessons:

1. Water transfers can enhance the performance and flexibility of existing water resource systems. These benefits appear in Table 2.

2. Water transfers must be integrated with traditional supply and demand management approaches. Water transfers alone will rarely resolve a region's water supply problems in an economical manner. Typically, a more integrated management approach, employing traditional supply and demand management measures, integrated with water transfers, will provide better results in terms of cost, technical performance, and institutional feasibility.

3. Modification and expansion of infrastructure is often required to take best advantage of water transfers. The operation of existing conveyance, storage, and treatment facilities is likely to require significant changes to facilitate water transfers. In many California cases, the transferred water can only be employed if it is stored for use in dry periods, necessitating new surface water reservoirs or additional use of groundwater storage. Conveyance restrictions, both from physical aqueduct capacities and environmental limitations, are also common.

4. Water transfers can take many forms, each serving a different operational purpose in a water resources system. The California case illustrates the many forms that water transfers can take (Table 1) and the diverse uses for different types of transfer arrangements. Each form of transfer, when utilized for an individual system, can fulfill a different operational purpose and accommodate different legal or third-party considerations.

5. Appropriate use of water transfers will likely vary between systems, reflecting local conditions. Each system is somewhat unique, in terms of its supplies, water demands, costs, and alternatives. Different water supply systems will find somewhat different uses for water transfers. Some water supply systems will not need or be able economically to employ water transfers. This variation in individual water system needs helps explain the diverse ways and degrees that water transfers have been employed in California.

6. Water transfers require a broader scope and scale of thinking about water resources management. The use of water transfers in water management implies a regional and inter-regional integration of different water users and supplies. The differences between the demands of urban water systems and irrigation systems are the reason why transfers can be successful to both parties. The implementation of this broader perspective on water planning will require significant changes in water agencies at the local, state, and federal levels.

7. Environmental, legal, and third-party considerations are important political, planning, and operational considerations in developing and implementing water transfers. Although not the major focus of this study, the environmental, legal, and third-party aspects of water transfers were consistently brought up during our interviews and research. Cases of water transfers, both in California and elsewhere, demonstrate the importance of environmental, legal, and third-party impact issues in the development and implementation of water transfers. While these issues are formidable, they are not insurmountable. There are numerous approaches for accommodating, compensating, or mitigating the real and potential third party impacts of water transfers, many of which involve engineering and operational measures.

8. Government sponsorship is often required for significant water transfer activity to begin. Government has an essential role in accelerating the use of water transfers, reducing the risk and uncertainty involved in water transfers, reducing the costs of completing water transfer transactions, and demonstrating leadership in the legal, technical, and conceptual transitions required for local agencies to implement water transfers.

9. Drought motivates change. Historically, major changes in water management philosophy have been motivated and incorporated as a result of experiences during droughts. Recent water transfers in California are an example of how drought has motivated the exploration of new alternatives in water management.

10. Transfers cannot be avoided only delayed. As increasing demands for water make shortages and droughts more frequent and severe, calls for water transfers are likely to become louder and more forceful. With the current drought, water transfers are now a significant and permanent feature of water resources planning and management in California.

Acknowledgements

This work was funded by the U.S. Army Corps of Engineers Hydrologic Engineering Center. We wish to acknowledge the essential contributions made by those interviewed as part of this study.

References

Lund, J.R., M. Israel, R. Kanazawa (1992), "Recent California Water Transfers: Emerging Options in Water Management," Project completion report, U.S. Army Corps of Engineers Hydrologic Engineering Center, November.

WATER ALLOCATION: MATCHING SUPPLY WITH DEMAND

By John F. Scott[1], Member, ASCE

ABSTRACT: The Colorado-Big Thompson (C-BT) Project was designed in the 1930's as a supplemental water supply for water users in northeastern Colorado. Since that time the water allocation has shifted from primarily irrigation supply to a mix of irrigation supply and municipal and industrial uses. This conversion has necessitated a reassessment of the manner in which the water supply is allocated each year to the project's allottees.

INTRODUCTION

The Colorado-Big Thompson (C-BT) Project was designed and constructed by the Bureau of Reclamation in the 1930's, 1940's and 1950's as a supplemental source of water supply for irrigation and municipal interests in northeastern Colorado. The local organization, acting as the distribution and repayment entity, is the Northern Colorado Water Conservancy District (District), which was formed in 1937, under Colorado state law, to undertake the contractual obligation to the United States. The municipalities of northeastern Colorado have grown beyond the original project design expectations and this growth is impacting the land and water use of the agricultural sector. Consequently, the manner in which the supplies of the C-BT Project are allocated requires review to reflect the change in demand associated with this growth.

COLORADO-BIG THOMPSON PROJECT OPERATIONS

The C-BT Project was designed and has been operated for over 35 years as a supplemental supply of water to all its allottees. There are 310,000 units of C-BT water available for contract by water users within the District boundaries, each unit representing one acre-foot of maximum supply. The District issues unit allotments to essentially three classes of water users: 1) Class B is all municipal water users, 2) Class C is divided into industrial water users and rural domestic water users, and 3) Class D is all irrigation water users. Up to 310,000 acre-feet is available to the water users annually; only a percentage of that supply is usually allotted each April. The District Board of Directors (Board) determines the

---

[1]Supervisory Water Resources Engineer, Northern Colorado Water Conservancy District, P.O. Box 679, Loveland, CO 80539

WATER ALLOCATION

allocation based on the tributary streamflows available to the water users within the District boundaries, the water in storage in both C-BT facilities and local reservoirs, and the needs of the water users. During drought periods the Board sets higher allocations to offset the low tributary streamflow. Conversely, during wet years the allocation is set low so that the C-BT reservoirs can rebuild their storage while the water users rely more on tributary supplies. Over the 35 years of District operations the allocation for delivery has averaged 72 percent or about 223,000 acre-feet. A uniform allocation is being proposed, on an optional basis, for those water users in Classes B and C. The uniform allocations studied are at the 65, 70 and 75 percent levels.

Local water users decide whether to invest in C-BT units based on the supplemental supply nature of the Project, the cost of C-BT units versus the cost of native supplies (including necessary legal costs associated with a change in the water right through the court system) and various other considerations. Since the 1950's when the ownership of C-BT units was approximately 80 percent irrigation and 20 percent municipal and industrial, the municipal and industrial ownership has grown to about 40 percent of the Project and indications are that this trend will only continue. Generally, the irrigation users have only C-BT water as an additional water right to their ownership of water rights in an irrigation company, but the municipal and industrial users have a broad portfolio of water rights. This mix of ownership is necessary to guarantee the municipal/industrial users a firm supply at some predetermined level of hydrologic certainty. It is simply more critical to meet the municipal and industrial needs than it is for the farmer to meet his full irrigation needs in any given year.

The annually varying nature of the C-BT water supply often is contrary to the municipal and industrial user's demand for a firm supply a set level. This conflict has impelled the Board of Directors to investigate alternatives to the allocation process.

UNIFORM ALLOCATION DETERMINATION

To understand the results of this study, some information on the hydrologic basis of the C-BT Project is necessary. The average annual inflows to the C-BT Project are about 280,000 acre-feet, [net system inflows (inflows minus required downstream releases lost to the system and other losses) are about 238,000 acre-feet] the active storage capacity of the three principal reservoirs is about 725,000 acre-feet, and the average annual deliveries are 223,000 acre-feet (72 percent allocation).

A simple calculation from these figures shows that a uniform allocation of at least 70 percent (217,000 acre-feet) is deliverable without injury to the remaining water users. Actually, a 75 percent allocation (232,500 acre-feet) is deliverable, this amount being less than the net system inflows available for delivery. A 75 percent allocation to Class B and C water users, though, would be difficult for Class D irrigation water users to accept as it is greater than the long-term average delivered.

A uniform allocation of 65 percent, on the other hand, has the potential to increase spills from the C-BT Project reservoirs. Spills increased from 28,683 acre-feet under the historical operation of the project to 38,247 acre-feet under the scenario where all water users in Class B and C convert to a uniform allocation of 65 percent.

An interesting result of this study shows a marked benefit to those C-BT water users not opting for the uniform allocation. This is as a result of the operation of the C-BT Project during drought years. During drought conditions, the Board tends to set high allocations to supplement the low water supply of native streams. These allocations are generally over 75 percent for the extent of the drought. During the 1970's drought (years 25 through 29 in Figure 1 below), which during the historical operations of the project was the time when storage in the system was at its lowest point, the allocations were 80 percent, 80 percent, 100 percent and 100 percent. This average allocation of 90 percent for four years (1,116,000 acre-feet) is 224,000 acre-feet more than the average of 223,000 acre-feet if delivered during the four years.

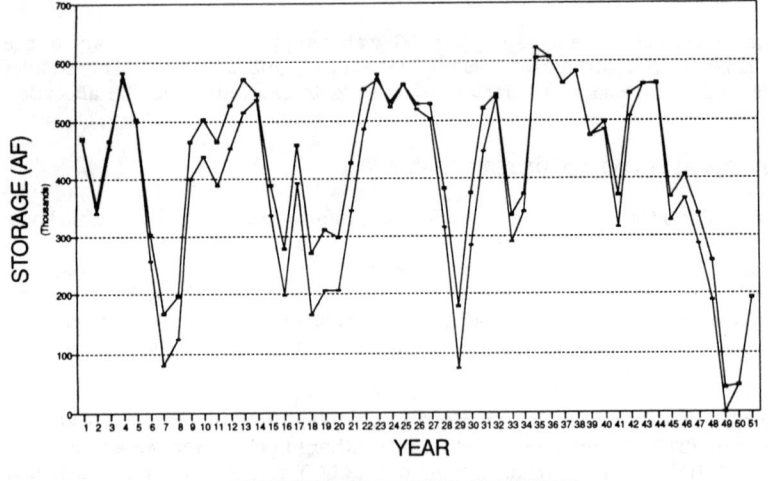

FIGURE 1. Total Active C-BT Storage

The amount of water "saved" for the remaining C-BT water users by allowing certain water users to opt for the uniform allocation is over 100,000 acre-feet as seen in year 29 (November 1, 1977) in Figure 1.

SUMMARY AND CONCLUSIONS

A uniform allocation of 70 percent is suggested to be offered to any water user in Classes B or C. The program will become operational for the 1994 water year (starting November 1993), after presentations to C-BT water user groups. Several contractual matters have to be worked out prior to program initiation. These include: 1) length of time required for participation in the program (at least five years is suggested), 2) termination of participation in the program, if allowed, and 3) how the individual water allocations will be administered during extreme drought conditions.

The management of water supplies in the semi-arid west is not just a cliche trend, but a necessity as populations continue to grow and little or no new water resources are being developed to meet the demand. Pushing agriculture out of existence to satiate the demands of urban centers is not intelligent in the long-term, if the supplies can be managed to adequately meet the needs of both sectors in all but severe drought conditions. This reallocation of the C-BT Project water meets the needs of municipal and industrial users while also benefitting the irrigation water user.

THE ONCE AND FUTURE ASCE MODEL WATER ALLOCATION LAW

Ray Jay Davis, A.M., A.S.C.E.[1]

Abstract

The Water Resources Planning and Management Division during its two decade history has been concerned about water rights allocation and transfer law. It established a Water Laws Committee to foster that interest. With Division approval, the Water Laws Committee in 1990 set up a task committee to prepare a model state water allocation law. The task committee, which now has over a hundred participating engineers and lawyers, has written three drafts of a code. On July 23, 1993, it will hold a one-day code workshop at Park City, Utah, National Irrigation Conference of the Irrigation and Drainage Division, and by September 1994 it will have prepared a fourth and final draft of the model code.

The model code task committee has been disseminating the code to legislators and other stakeholder groups during the drafting process, and after completion of the final draft will continue that effort. A successor ASCE standards committee, working within the Special Standards Division, will periodically update the code and continue the dissemination process. The model code task committee will incorporate in the code provisions from the best state water laws so all states may use it to improve the legal structures for allocation and transfer of water use rights. Through this process, water management in the '90s and beyond will be enhanced. The code process has been, is now, and will continue to be a time for innovation.

Introduction

In a far off land centuries ago, King Arthur gathered his knights in fraternal friendship at Camelot. The bonds of chivalry linking these knights of the roundtable have been celebrated in prose, poetry, and song. This enduring legend

---

[1]Prof., J. Reuben Clark Law School, Brigham Young University, Provo, UT 84602-1029.

of Camelot was recounted in T.H. White's (1958) book "The Once and Future King," a tale of then and now, of the glory of the past, its continuing impact, and the worthy goals of the king. The story of the ASCE Model Water Code Project, like the Arthurian cycle, has a past (the once), a continuing presence (the future), and lofty public policy goals.

Once

The genesis of the ASCE Model Water Code Project may be found in authorization by ASCE of the Water Resources Planning and Management Division two decades ago. According to its purpose statement, the division was established to advance and disseminate knowledge about water resources management by emphasizing interdisciplinary approaches that embrace social, economic, environmental, administrative, and legal aspects basic to water conservation and utilization (ASCE, 1991).

Several technical committees within the division have dealt with the planning, institutional, and legal aspects of water resources development. The purpose statement of the Water Laws Committee states that it is to consider legal problems throughout the water resources development and management systems and to produce reports, recommendations, or position statements related to water laws (ASCE, 1991). In order to further this mandate, the Water Laws Committee launched the ASCE Model Water Code Project.

In the fall of 1989, the executive committee of the Water Resources Planning and Management Division accepted the Water Laws Committee's proposal to establish a Task Committee on a Model Water Code and authorized the task committee to operate for two years from October 1990 through September 1992. Subsequently the executive committee also authorized creation of a follow-up task committee to continue the project from October 1992 through September 1994. The purpose of the task committee is "to develop proposed legislation suitable for adoption at the state level of government for allocating water among competing interests and resolving quantity-related conflicts." (ASCE, 1991).

The Model Code Task Committee, which now has grown to over a hundred engineers, lawyers, and other water resources management professionals, has prepared a draft of the model state water rights allocation and transfer code during each of the three years of its existence (Model Water Code TC, 1991, 1992a, 1993). These drafts were written as "strawmen"--documents which were subjected to internal and external review and criticism. Comments on each draft and other input helped shape the coverage, outline, and substance of the draft that followed. Water resources administrators from federal, state, regional, and local agencies have been given the opportunity to express their views. Water user organizations and environmental groups have been provided with copies of code drafts and their suggestions have been solicited. Legal and engineering academics have been involved with practitioners both in writing and reviewing the drafts.

A workshop on the model water rights code is being sponsored by a Model Code Workshop Committee (1991) and is being held later this summer, July 23, at Park City, Utah in conjunction with the National Conference on Irrigation and Drainage Engineering. Interested persons should register as soon as possible because enrollment is limited to 100 persons. Registrants will be provided with the third draft of the code, and will attend a plenary session and their choices among three concurrent sessions. Topics being covered by the concurrent sessions include establishing water rights in regulated riparian jurisdictions, establishing water rights in prior appropriation jurisdictions, state water rights allocation policy, water subject to allocation, scope of water rights in regulated riparian jurisdictions, scope of water rights in prior appropriation jurisdictions, state water rights administration, special water management districts, status of water rights over time in regulated riparian jurisdictions, status of water rights over time in prior appropriation jurisdictions, rights based upon water supply augmentation and conservation, and state regulation of multijurisdictional diversions. The workshop is a forum for broad input from the engineering and legal professions into the code development process. On the basis of its results and other reviews, a fourth draft of the code will be prepared and readied for publication and general public dissemination prior to September 1994 (Model Water Code TC, 1990).

Future

State legislators, legislative draftsmen, and their advisors and staffs are the principal audience for which the model code is intended. Hence dissemination to them weighs heavily in the future of the project. That can be accomplished only though wide ASCE participation. The model code task committee, accordingly, has sought to acquaint ASCE members with the code project though articles in the *ASCE News*, talks at ASCE specialty conferences, and publications in proceedings of ASCE meetings and other professional outlets. During the first two years of the project, there were nineteen presentations and fourteen publications. Direct legislative dissemination started in April 1992 with three presentations to a Pennsylvania state legislative committee (Model Water Code TC, 1992b). As work on the project progresses, both exposure of professional groups to the code and direct assistance to legislative groups will loom larger in its efforts.

The Special Standards Division of ASCE has approved creation of a State Water Allocations Law Standards Committee. Membership is now being solicited and interested persons are urged to apply either though the model code task committee or directly to ASCE headquarters. The Standards committee will assume primary responsibility for final polishing of the code, dissemination, and updating it as needs change. Unlike the task and workshop committees which have limited periods during which they operate, the standards committee can stay in business as long as the Special Standards Division renews its charter. It is likely that the standards committee will undertake such additional projects as drafting a water quality code for states and preparing a model interstate water compact. The model code project, then, is not intended to have a finite duration, rather it is an open-ended process.

Goals

The purpose of the water allocation and transfer code is to provide a source of legal norms to states seeking to improve their laws. To that end, code drafters have adopted the best of legislation from throughout the country--every state has water rights provisions which are worthy of emulation in other jurisdictions. There is no intent by the task committee to "reform" a state's water code by having it adopt the model in totality (Davis, 1991). The text of the code will be the best effort of water law experts to express what water rights law ought to be--consistent with prior law and political realities.

In addition to a text, the code project is producing commentary to the text. It has been suggested that this is more important than the actual text (Shabman, 1992). The commentary explains what is the issue that a textual provision deals with, what the options are for resolving it, why the code committee chose this particular text, and how the issue has been treated throughout the country. Lawmakers seeking to deal with the important issues of water resources planning and management will be able to learn from the code what their public policy choices are. Through the code process, water management in the '90s and beyond will be enhanced. The code process has been in the past, is now, and will continue to be a time for innovation. The model code is indeed the once and future ASCE model state water allocation law.

References

American Society of Civil Engineers (1991). Official Register 1992, pp. 258, 262. ASCE, New York, NY.
Davis, R. (1992). "Preparing a Model Water Code: Reform or Restatement?" Conf. on Water Resources Planning and Management and Urban Water Resources at New Orleans, p. 390. ASCE, New York, NY.
Model Code Workshop Committee (1991). Workshop Proposal.
Model Water Code Task Committee (1993). Model State Water Rights Allocation Code (Third Draft).
_____ (1992b). Final Report.
_____ (1992a). Model State Water Rights Allocation Code (Second Draft).
_____ (1991). Model State Water Rights Allocation Code (First Draft).
Shabman, L. (1992). "What Should the ASCE Model Water Code Committee Do?" Proc. of Water Resources Sessions at Water Forum '92, p. 237. ASCE, New York, NY.
White, T. H. (1927). The Once and Future King. Putnam, New York, N.Y.

From Riparianism to Water-Use Permitting: Issues Confronting State Government

William E. Cox[1]

Abstract

Many of the original riparian-doctrine states have replaced the common-law allocation approach with administrative water-use permitting, and other states will likely take similar action in the future. This transaction is a major change in a state's water-management institutions. This paper discusses the basic issues that must be resolved during the transition to water-use permitting.

Introduction

Many of the states originally accepting the riparian doctrine have replaced the doctrine with administrative programs that allocate water through a permitting process. This trend will likely continue among the remaining riparian-doctrine states as increasing demands on the resource increase conflict and create the need for more proactive water-allocation systems. Transition from a riparian-doctrine allocation system to administrative permitting is a major institutional change. This change requires decisions on many details regarding the nature and operation of the permit program. Although other issues can be important in certain cases, the following list contains essential considerations important to the typical state confronting the riparian/permitting transition:

- Timing of the transition
- Geographic scope of control measures
- Hydrologic scope of control measures
- Exemptions from permitting
- Treatment of instream water uses
- Duration of permitted water rights
- Relations among permit holders
- Provisions for water-use changes

---

[1]Professor, Department of Civil Engineering, Virginia Tech, Blacksburg, VA 24061.

Timing of the Transition

Determination of the proper time to replace the riparian doctrine with administrative allocation requires consideration of both the costs of operating alternative programs and the consequences flowing from continuance of the existing system. Operation of the riparian doctrine, since it utilizes judicial proceedings, generally will be of lower direct cost than will operation of a permit program. In addition, the doctrine is less intrusive into water-use decision making since prior governmental approval of water use is unnecessary. Thus, premature replacement of the doctrine will impose unnecessary costs.

But the doctrine can impose costs on society under conditions of water scarcity in the form of inefficient water use. For example, the limitation of water use to riparian land and the inherent uncertainty of water rights under the doctrine may impede socially desirable development. Retention of the doctrine beyond some point in the evolution of water use therefore may impose costs greater than any savings associated with the low direct costs of its operation.

Geographic Scope of Control Measures

Most state water-use permitting programs apply statewide, but several states have adopted programs that apply to designated geographic areas smaller than the entire state (see, e.g., VA Code Ann. secs. 62.1-44.83 et seq.). The management-area approach arises because water supplies and demand tend to be unevenly distributed causing water-use conflict to be limited to parts of a state. Limitation of water-use permitting to such areas can be an efficient way to improve allocation where needed without changing the method of allocation where conflict is minor.

Hydrologic Scope of Control Measures

The primary issue regarding the hydrologic scope of an administrative permitting program is whether it applies to both surface and ground water or to just one of these sources. Common-law water-allocation programs traditionally have been limited to one water source and have often been criticized for ignoring the continuity of the hydrologic cycle. Adoption of administrative allocation provides an opportunity for a more comprehensive approach that allows the resource to be managed in a more holistic manner.

Yet, several states have adopted allocation programs limited to a single source, usually ground water (see e.g., VA Code Ann. secs 62.1-44.83 et seq). This approach may be workable where hydrologic interconnections are limited. But such programs contain inherent limitations and require special coordination efforts.

Exemptions from Permitting

Exemptions are a common aspect of administrative permitting programs. Exemptions for small water uses are typical and are justified because of their limited impact on the water resource. But other exemptions that go beyond small quantity uses are commonly incorporated into permit programs (see, e.g., Ky Rev. Stats. sec. 151.140). Such exemptions apply without regard to quantity and usually arise as a result of concessions to reduce opposition to adoption of the permitting program. Categorical exemptions have potential to undermine the effectiveness of a permitting program by leaving major water uses beyond the control of the permitting agency.

Treatment of Instream Water Uses

Creation of an administrative water-allocation program provides the opportunity to enhance the standing of a class of water uses traditionally given limited recognition in common-law allocation systems: instream water uses that depend on maintenance of minimum streamflows. The limited recognition of instream water use in riparian-doctrine jurisdictions arises from restriction of water rights to riparian landowners. These landowners in some cases may not represent the interests of the general public in instream water-use activities. While some water-use permitting programs are silent with respect to instream activities, other permitting programs contain explicit recognition and protect minimum flows from diversion (see, e.g. FL Stats. Ann., sec. 373.223(3)).

Duration of Permitted Water Rights

Water rights established through administrative allocation systems can be limited in duration by two primary means: (1) provisions for termination if water use never begins or ceases at a later time and (2) establishment of time limitations for the life of the permit itself. Essentially all permitting programs employ the first measure while the second, although in common use, is less universal.

The use of time-limited permits can enhance the power of the state to direct water use. Time limitations allow systematic reallocation where program administrators believe such change is in the further interest. Programs with time-limited permits may contain a preference for renewal to the previous party in the absence of evidence that the public interest would be advanced by change. Adoption of a program with provision for such determinations is therefore based on the premise that program administrators can determine the relationships between competing water uses and the public interest. This approach rejects private transactions among water users as the preferred mechanism of reallocation.

Relationships among Water-Rights Holders

A basic distinction among alternative water allocation systems is the relative standing of individual water-rights holders to one another. Permit holders generally have equal status regardless of the time when the permits are acquired. An alternative to the equal-standing principle is a preference system based on type of use to be imposed during water shortages. For example, Florida allocation legislation provides for classification of uses into categories based on their relation to the public interest. During water shortages, restrictions can be imposed on the exercise of rights in one or more classes (Fla. Stats. Ann. sec. 373.175).

Provisions for Water-Use Changes

The above discussion of water-rights duration indicates that water-use changes can either be (1) directed by the state through use of time-limited permits and a formal reallocation process or (2) allowed to occur through private transactions subject to state oversight. The best model for observing private water-use changes is the appropriation doctrine of the western United States. Appropriative rights traditionally have been viewed as existing in perpetuity unless lost through the common-law concept of abandonment or statutory forfeiture. Transfer of appropriative water rights is permissible only when other parties will not be harmed. This restriction is usually interpreted to mean that return flows from the existing use must not be affected by a water-rights transfer. As a result, the amount of water available for transfer is often less than the nominal amount of the water right being transferred.

Conclusion

The decision of a state to replace the riparian doctrine with administrative allocation requires choice with respect to each of these issues. Some of these choices involve functional effectiveness -- for example, decisions with respect to exemptions. But many of the choices also involve value judgments. For example, the choice between time-limited permits and permits of unlimited duration depends on perceptions of the proper role of government in resource allocation.

Because of this dependence on values, water-law changes are usually controversial. Different interest groups support opposing positions with respect to various issues. The final form of a proposed water allocation law results from compromise by opposing interests within the political process. The process sometimes produces flawed results as shown by the existence of administrative programs with ineffective and inefficient provisions. But the political process is the only forum for resolving value conflicts and formulating fundamental public policy expressing preferences for certain values over others.

INSTREAM FLOWS ACCORDING TO
THE ASCE MODEL STATE WATER CODE

by

George William Sherk, A.M. A.S.C.E.[1]
and
Berton L. Lamb[2]

Abstract

The ASCE Model Water Code provides two mechanisms for protecting instream uses of water: reservations and water management areas. Implementation might be difficult unless the relation between the two approaches is unambiguous.

Introduction

The concept of "instream flow" includes the use of water for a variety of purposes. Navigation, water supply and sanitation, as well as fish and wildlife requirements, have been recognized as requiring minimum streamflows. Recreational, aesthetic, ecological and environmental uses are coming to be recognized as equally important uses of water. Historically, there was little recognition or protection in appropriations doctrine states for instream flows or the uses that required such flows. Even less recognition or protection was afforded in the eastern United States under the riparian doctrine.

Recognition of the importance of instream flows has resulted in both eastern and western states enacting legislation to protect those flows. For example, the prior appropriation doctrine's diversion requirement has been eliminated in Colorado, Idaho and Arizona. Legislation authorizing the reservation or withdrawal of water to protect instream flows has been enacted in

---

[1] Attorney at Law, 66 East Ninth Street, Suite 2307, St. Paul, MN 55101-2276 and Virginia Institute of Marine Science, College of William and Mary, Gloucester Point, VA 23062.

[2] Policy Analyst, National Ecology Research Center, U.S. Fish and Wildlife Service, Fort Collins, CO 80525-3400.

Alaska, Oregon, Montana and Utah. Instream flows have been protected by caselaw or statute in Washington, Wyoming, North Dakota and Nevada.

While the western states integrate the protection of instream flows into the prior appropriations doctrine, at least 16 eastern states have enacted legislation to protect such flows in the context of legislation amending (or superseding) the riparian doctrine. The preferred approach in the eastern states has been to authorize a branch of state government to establish minimum streamflows or water levels. This approach has been adopted in at least 10 eastern states.

Virginia, for example, enacted legislation authorizing the establishment of surface water management areas in order to protect both instream and offstream beneficial uses. Offstream beneficial uses may be restricted if the State Water Control Board determines that such uses threaten to reduce streamflows (instream beneficial uses) in a management area. Lawmakers in Massachusetts and South Carolina enacted legislation protecting instream flows by restricting the amount of water available for transbasin diversions. Conversely, Wisconsin allows transbasin diversions for the purpose of maintaining streamflows or lake levels.

At least 28 eastern and western states have enacted legislation or have caselaw protecting instream flows. Protection of instream flows is increasingly a driving force behind state legislative activity. Given the growing recognition of instream values, especially in purely economic terms, it is certain this trend will continue across the Nation.

The ASCE Model State Water Code

One of the goals of the ASCE Model State Water Code project was to provide guidance to state legislatures that are addressing instream flow issues. Such guidance may be needed both in those states that have yet to address the issue and in those states that are revising previously enacted legislation.

The approach taken in the Model Code is unusual in that it integrates two alternative approaches. The first approach is to reserve water from allocation This approach is expressed in Chapter 3, which provides that the minimum quantity of water needed "to preserve the biological integrity of the waters of the State" is reserved from allocation. The purpose of this approach is to mandate that the portion of the State's waters needed to ensure the biological survival of rivers and lakes is not subject to allocation and use. Difficulties in drafting the instream flow section of the Model Code were encountered in focusing on the definition of "biological integrity" and in determining the quantity of water to be reserved.

The second approach, establishing water management areas to protect instream flows (Chapter 4, Part 4) authorizes a state administrative agency to establish a water management area when offstream uses will adversely affect reasonable and beneficial instream uses. To protect instream values, the administrative agency is authorized to require a permit for all existing and proposed uses within the management area. The administrative agency is also authorized to impose conditions on water use permits including conditions to protect instream flow.

Different quantities of water may be reserved or protected under the two approaches. The first approach reserves the quantity of water needed to ensure survival of the water resource. The second may preclude the diversion for a wide variety of other uses. Such a two-pronged approach has the potential to protect the integrity of the water resource and balance public interests with those of private stakeholder groups.

Implementation Issues

A model of implementation (Sabatier and Mazmanian 1980) included tractability of the problem, statutory construction, and nonstatutory factors. First, tractability of the problem refers to the availability of appropriate technology and the extent of behavioral change required of those who would implement the statute. Technologies are available to address instream flow problems. Simple instream flow issues may be dealt with through a number of techniques (e.g., Tennant Method). More complex problems may be analyzed by the Instream Flow Incremental Methodology. Because more than 28 states have instream flow programs, experience is available to inform those who would adopt the ASCE model. On one hand, many of the changes necessary to implement instream flow programs have been made. Implementing a new program or improving an existing one may be easier because there are examples to follow. On the other hand, managers often have difficulty agreeing on appropriate assessment methods; moreover, agreements on flow are elusive, and some States involve so many agencies that coordination is difficult.

Second, the factor of statutory construction includes such elements as clarity of objectives and the unambiguous delineation of procedures and authorities. Because a number of states have implemented instream flow programs, a wealth of practical administrative knowledge exists. A great deal of this information is available in the published literature. Examples include MacDonnell, et al. (1989); the "Opportunities to Protect Instream Flow" series (e.g., Bingham 1992); and journal articles. Additional resources include proceedings published by the American Water Resources Association and the ASCE (e.g., Woessner and Potts 1989). Any new statute could build upon this base of experience.

One of the lessons learned early in the history of instream flow protection is the importance of easily distinguished processes. The integration of two approaches within the ASCE Model may violate this rule. Although combining a water reservation with water management areas--including conditions on water rights--seems logical, ambiguous combinations are problematic. One example is the confusion over Washington's statutes regarding minimum flow (WCA 90.22) and base flow (WCA 90:54). The ASCE Model Code seems likely to at least raise questions about whether the two approaches are additive. If they are additive, the required level of technical analysis is increased.

Third, nonstatutory factors include public support, socioeconomic conditions, media attention, and support from powerful interests in society. A common adage is that change in water policy results from threatening conditions. For example, Iowa's path-breaking program developed out of a drought in the 1950's, Montana's 1973 approach was related to drought and the threat of massive out-of-State diversions, and Wyoming's policy passed after threat of a referendum. Legislative change seems to occur after galvanizing events.

The quality of ASCE's Water Code is itself not sufficient to ensure implementation. Media attention to environmental issues, conflicts over water allocation, low water years, population growth, and Federal land use policy could galvanize State legislatures. However, these same factors must be on-going to prod administrators. When socioeconomic conditions or public support are lacking it is difficult to achieve the needed administrative change for successful implementation.

APPENDIX

Bingham, J. L. 1992. Opportunities to protect instream flows and wetland uses of water in Nevada. U.S. Fish and Wildlife Service, Resource Publication 189.

MacDonnell, L. J., T. A. Rice, and S. J. Shupe (eds). 1989.Instream Flow Protection in the West. Boulder: Natural Resources Law Center, Univ. of Colo. Law School.

Sabatier, P. and D. Mazmanian. 1980. The implementation of public policy: a framework of analysis. Policy Studies Journal. 8:538-560.

Woessner, W. W. and D. F. Potts (eds.). 1989. Proceedings of the Symposium on Headwaters Hydrology. Bethesda, MD: American Water Resources Association.

WATER MARKETING IN WASHINGTON THE NEXT STEP ?

Donald Phelps, P.E., M.ASCE[1]

Abstract

The state of Washington has been blessed with an abundance of water resources for its use, including the Columbia and Snake rivers, but with the listing of the Snake River Sockeye Salmon as an endangered species it must make some difficult decisions. One of these decisions is how much additional water can be appropriated from these river systems. Some streams are already over allocated during low flow years and the salmon recovery plan will make future appropriations even more difficult to obtain. Water marketing has never been prevalent in the state but as various interest groups compete for a shrinking resource it will become the primary source of obtaining water. This paper discusses how the state would benefit from the adoption of that portion of the ASCE Model State Water Code addressing water rights transfers.

Introduction

The state of Washington has long been perceived by most of the nation, particularly those living in the eastern states, as a land of plentiful water with rain being a frequent event. While this perception is generally valid nine months of the year for that portion of the state lying between the Pacific Ocean and the Cascade mountains, more than two thirds of the state, lying east of the Cascades, actually enjoys a much drier, desert like climate. The average annual precipitation for the area between the foothills of the Cascades and the Idaho

---

[1] Senior Engineer, Hammond, Collier & Wade - Livingstone Associates Inc., PO Box 571, Chelan, Washington, 98816

border ranges from 4 to 36 inches. The central Columbia river basin contains the vast majority of the states' agricultural lands and the precipitation it receives varies from 4 to 20 inches annually. The annual precipitation by itself is only sufficient to support dry land wheat farming, but the development of an extensive Columbia Basin irrigation project, operated by the US Bureau of Reclamation, has allowed agriculture to become diversified and contribute billions of dollars per year to the state's economy. The water supply for this extensive irrigation system is removed from the Columbia river at the federally owned Grand Coulee and Chief Joseph dams. Those lands which are not included within the Columbia Basin Project rely on direct withdrawals from the Columbia and a number of its tributaries, including the Okanogan, Methow, Chelan, Entiat, Wenatchee, Yakima, and Snake rivers and, to a very minor degree, ground water.

Competing Uses

Agriculture is only one of the user groups competing for the waters of the Columbia river: hydro power, fisheries, recreation, wildlife, and municipalities also seek to use these waters to satisfy their needs and they are often in conflict with one another. The situation in Washington is further compounded in that those lands served by the Columbia Basin Project do not hold their own rights but purchase water from the federal government, which restricts transfers from one use to another.

Cause, Effect and Options

On December 20, 1991 the Snake River Sockeye Salmon was placed under the protection of the Endangered Species Act. In addition to the Snake River Sockeye several other salmon stocks indigenous to the Columbia river system have been identified as being in jeopardy and are currently being reviewed for possible protection. The listing of the Snake River Sockeye has initiated an examination of the water allocation process for not just the Snake river but the entire Columbia river system within the state of Washington. As an interim step, the Washington Department of Ecology (DOE) issued a moratorium on the issuance of water rights permits which withdrew directly from, or were in direct hydraulic connectivity with, the mainstem Columbia and the Snake rivers.

Although the moratorium on new permits is promoted as an interim measure to allow the DOE time to determine the amount of water available for allocation out of the rivers, it is certain to heighten interest in the concept of water marketing in the region. The lands adjacent to the Columbia and Snake rivers have significant potential as agricultural lands but they are also highly sought after for water based recreational development, driven by the rapid population growth in the western portion of the state. Lands fronting on the Columbia have escalated in price over the last 5 years to the point where developed frontage can cost in excess of 2000 dollars per front foot while land without water rights is virtually worthless. The situation as it now exists is that an individual owning land adjacent to the river for which no water rights have been secured has extremely limited options unless an existing right can be purchased from another landowner. This scenario is new to the state since in the past transfers generally occurred only when land was sold, with all new users seeking and being granted their own permits. With the development of the salmon recovery plan required under the Endangered Species Act it is very likely that future development will depend on the purchase of existing water rights to obtain water.

At issue is whether the transfer process is adequate to protect the interests of all parties. The Revised Code of Washington allows for a water right which was developed and perfected on one piece of property for one use to be transferred to another property and use, providing there is no detriment or injury to other existing rights. The question that must be raised is whether the test of detriment or injury to other existing rights is the only test that should be satisfied before a transfer of right is allowed. It is becoming more and more apparent that there can be significant third party consequences when a water right is transferred from its original point of application and use. To further complicate the issue these third party impacts are often quite difficult if not impossible to quantify since they are frequently related to long term usage patterns or cultural heritage.

The state of Washington has a number of "tools" on the books which may be applied in an effort to assess the impacts of a proposed change in water use. The state Department of Ecology (DOE) is charged with reviewing all requests for transfer of water rights and applying the test of whether the proposed transfer is detrimental or injurious to other existing rights. If the water is to be applied to a new use requiring development the State Environmental Policy Act (SEPA) provisions are implemented and the proponent must complete a checklist identifying the impacts to allow a responsible agency, such as a county planning department, to determine if an environmental impact statement should be required of the project. While the SEPA process allows for review by any and all affected state agencies, this review is often done after the water right transfer has been approved rather than before as the proponent wants assurance that water is available for the project before initiating the design and review process. Transfer requests in the past have most commonly been associated with the sale of property and the intended use is generally listed as being the same as the prior use. A request for change of use is generally not filed until sometime during the planning process, after the transfer has been approved. Thus the full impacts (including third party) which might be associated with a transfer of water right, are not reviewed in the initial processing of the transfer application. In the case of a water right being sold without the land, the DOE has the responsibility to determine the appropriateness of the request and to see that no detrimental or injurious impacts shall be enacted upon other existing rights.

The current statute, while protecting "existing rights", is vague as to what constitutes existing rights deserving of protection and provides the DOE with no guidance as to what public interest issues should be considered when reviewing a request for water rights transfer. It is in this area that the state of Washington could benefit from the incorporation of portions of the Model State Water Code being developed by the Water Laws committee of the American Society of Civil Engineers. The model code contains language requiring that in addition to protecting other water rights any transfers must protect the interests of the public as well. Included within the public interest are issues such as minimum flows, protection of riparian and aquatic habitat, water quality, economic impacts, aesthetics and other non-pecuniary impacts.

Conclusion

The state of Washington, faced with the prospect of increased water marketing to allocate resources, would benefit from adoption of portions of the ASCE Model State Water Code. Specifically that portion addressing water rights transfers and what public interest issues should be addressed before a change of use is approved.

Optimal Planning of Combined Sewer Rehabilitation

Santiago M. Reyna[1], S.M. ASCE, and Jorge A. Vanegas[2], M. ASCE

Abstract

The need to rehabilitate many of the existing sewer systems to keep them operational is evident. The development of a rehabilitation plan that optimizes the use of funds is the first step in this rehabilitation process. This paper presents a methodology that, starting from an assessment of the structural and hydraulic conditions, and data for the different rehabilitation methods, builds an optimization model to select the segments to be rehabilitated and the methods to be used.

Introduction

Many of the sewer systems in the nation cannot hide the pass of time and they are demanding to be rehabilitated. The Nat. Council on Public Works Improvement (Giglio et al., 1988) gave a grade of "C" to waste water on the report's card on the national public works. According to Thomson (1991), the US should renew about 6,000 Mi. of pipes yearly (a $2.5 billion job). Considering the magnitude of the investment that is needed, it is apparent that new methodologies that help optimize the use of the funds allocated for this purpose are going to be required to support the decision maker.

The design of the sewer rehabilitation plan needs to assess the physical and hydraulic conditions as the first step. The physical condition can be determined through a structural inspection program, while the hydraulic condition requires the building and verification of a hydraulics model. The selection of the rehabilitation plan to follow is done with the help of a multiobjective optimization model. We consider the minimization of annual maintenance and claims costs, the minimization of disruptions, and the maximization of overall hydraulic performance and structural

---

[1]Doctoral Candidate, School of Civil Engineering, Purdue Univ., W. Lafayette, IN 47907, reyna@ecn.purdue.edu.
[2]Assistant Professor, School of Civil Engineering, Purdue Univ., W. Lafayette, IN 47907, jorgev@ecn.purdue.edu.

conditions (the latter affected by a failure impact coefficient) as the objectives to be optimized. A database with data of the most used rehabilitation methods is constructed and used. The final results of the model are a sequence of sewer system segments to be rehabilitated and their corresponding rehabilitation methods to be used.

Hydraulic Condition:

The hydraulic condition assessment requires building and verifying a hydraulics model. The Storm Water Management Model (EPA, 1989) has been selected for this task, for it is widely available and of known reliability. The Extran block of this program is able to model backwater and downstream control effects which are of primary importance to evaluate the hydraulic condition of the system.

The Runoff, Transport, and Extran blocks of SWMM are used to generate the flows and route them (Combine is used to overcome size limitations). Hydrologic data, such as rainfall, imperviousness, infiltration parameters, and subcatchment geometry are entered into Runoff. The layout of the pipe system is entered into the Transport and Extran blocks.

Among the results given by the model, three are selected as factors to be considered to compute the Hydraulic Condition Indices (HCI_{ij}). These are: the maximum to design flow rates, the lengths of surcharge, and the lengths of flooding. From these three factors, a single index is given to each reach. These indices are modified accordingly when a reach is selected to be rehabilitated.

Structural Condition:

To obtain the Structural Condition Indices (SCI_{ij}), a Structural Inspection Program is needed. The Program is applied to plan the inspection of the system using the different methods available (i.e. man-entry, CCTV, smoke, thermography). From the inspection program the pipes can be given indices considering their structural condition. The indices rank from 0 (perfect condition), to 10 (structural failure). A modified methodology based on the one suggested by the Department of Housing and Urban Development (1984) is used for the purpose of ranking the pipes in the system. These indices are adjusted in accordance to the expected conditions the pipe would reach once rehabilitated with a particular method.

Failure Consequences Impact:

Also following the Department of Housing and Urban Development, Consequences of Failure Impact Indices are given to each pipe ($CFII_i$). These indices are used to affect the Structural Condition Indices in the optimization process.

Annual Costs:

The non rehabilitated reaches, and the same reaches once rehabilitated with a particular method, have different Annual Maintenance and Claims Unit Costs ($AMCUC_{ij}$). These costs are assumed proportional to the diameter of the pipe.

The lengths have particular investment costs when different methods are applied. To account for the widely varied life spans of the rehabilitation technologies, they are all reduced to annual costs. The Annual Investment Unit Costs ($AIUC_{ij}$) are proportional to the diameters, and linear models are used as a first approach.

Disruption Levels:

Each rehabilitation methodology has a Disruption Level Index associated to it (DLI_j). These are also ranked from 0 to 10 from less to more disruptive. A maximum rank of 10 is given to excavation and replacement and a 0 to no rehabilitation.

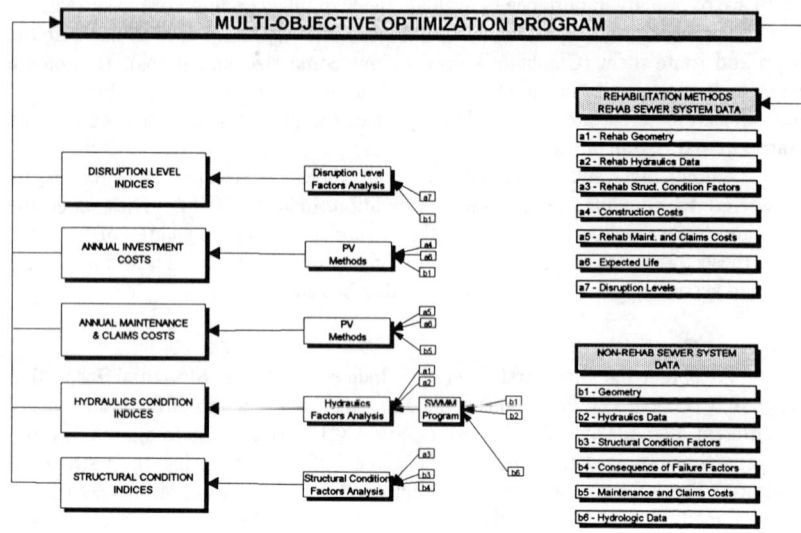

Figure 1

Methodology

Two databases are created, one with the sewer system information, and a second with data regarding the different rehabilitation methods (i.e. costs, applicability, life). Using Workbook and Macros capabilities of Excel, SWMM input files are generated automatically. Also, its results are added into a file with the hydraulic conditions factors. All files are linked in such a way that changing any of the data will bring its consequent modifications automatically (exception made for SWMM that has to be run independently). Figure 1 shows a general overview of the model.

All the different indices mentioned above are link-input into the final optimization file, where What's Best? (a linear optimization package) is used to help solve the following optimization problem.

min Overall Disruption Level = $\sum_i \sum_j DLij * Li * xij$

min Overall Failure Impact = $\sum_i \sum_j SCIij * CFIIij * Li * xij$

max Overall Hydraulic Performance = $-\sum_i \sum_j HCIij * Li * xij$

min Annual Maintenance & Claims Costs = $\sum_i \sum_j AMCUCij * Li * xij$

s.t. Annual Investment Costs = $\sum_i \sum_j AIUCij * LCFi * Li * xij \leq$ Maximum Allowed

$xij \in \{0,1\}$

$i = 1$ to NPIPES $j = 0$ to NMETHODS

In this model, L_i is the length of each pipe and x_{ij} are the 0-1 decision variables, that correspond to each pipe i and rehabilitation method j (j=0, no rehabilitation is done).

Conclusions

A mathematical model is shown that treats the sewer system and its rehabilitation alternatives jointly. It helps the decision maker to choose an optimal set of segments of the sewer system to be rehabilitated first according to the level of investment selected. Tradeoff relations can be developed from this model, too.

The validation part of the present methodology is currently under way (Nov. 92), and it consists of an application to a sewer system in the City of Indianapolis.

Appendix - References

U.S. Environmental Protection Agency (1989). SWWM4 & Extran Addendum. EPA/600/3-88/001a,b (NTIS PB88-236641/AS, PB88-236658/AS).

Giglio, J.M. et al (1988). "Fragile Foundations: A Report on America's Public Works." National Council on Public Works Improvement.

Thomson, J.C. (1991). "Pipeline Rehabilitation: Underground Options." Civil Engineering, ASCE, May 1991, 64-66.

U.S. Dept. of Housing and Urban Development, Off. of Policy Development and Research, Building Technology Division (1984). Utility Infrastructure Rehabilitation. Report prepared by Brown and Caldwell.

COLLECTION SYSTEM
INSPECTION AND REHABILITATION PROGRAM

Bill Macaitis[1], F. ASCE
Amreek Paintal[2], A.M. ASCE

Abstract

The Metropolitan Water Reclamation District of Greater Chicago is a regional wastewater and water resources agency responsible for serving an area of about 875 square miles. Recently the District developed a program for the inspection and preventive maintenance of its 535 mile collection sewer system to increase the reliability of the aging system and to reduce the incidents of cave-ins. The purpose of the inspection program is to determine maintenance and rehabilitation needs, and consists of the physical inspection of manholes and sewer routes, television inspection of the inside of the sewer, infrared or ground piercing radar inspection for locating adjacent cavities, and flow monitoring. Inspection data are analyzed using an computer-aided mapping and data system.

Based on the analysis of inspection data, the preventive maintenance and rehabilitation needs are developed and prioritized. The program is coordinated through an Underground Advisory Committee consisting of expert personnel from the design and maintenance departments of the District.

INTRODUCTION

The Metropolitan Water Reclamation District of Greater Chicago owns and operates seven wastewater reclamation plants and 535 miles of intercepting sewers and force mains. Additionally 85 miles of relatively new deep rock tunnels ranging to 33 feet in diameter have been built to capture combined sewer overflows.

The intercepting sewers and force mains range in size from 8 inch diameter to a 27 by 24 feet horse-shoe section. The first intercepting sewer was constructed in 1906. Fig. 1 shows the age of the intercepting sewers, which have a present worth of $3.8 billion. Sewers 50 or more years old have a total length of 170 miles and a present worth of $1.5 billion.

----

1 Assistant Chief Engineer 2 Principal Civil Engineer
Metropolitan Water Reclamation District of Greater Chicago, Engineering Dept., 111 E. Erie, Chicago, IL 60611.

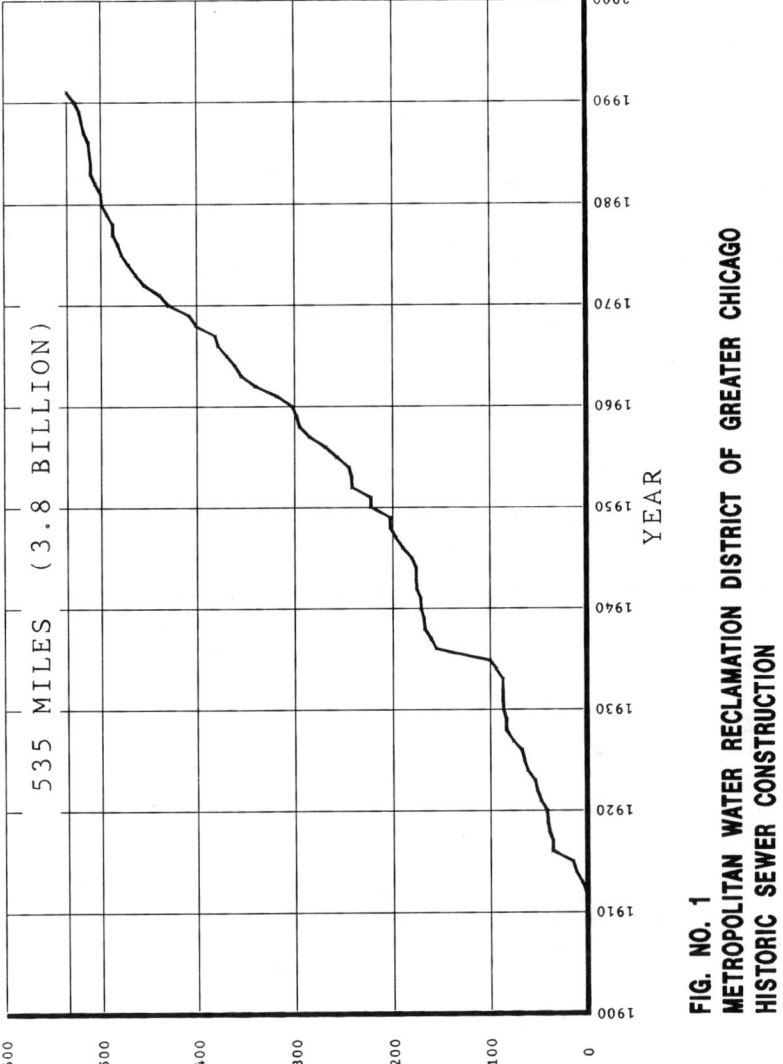

FIG. NO. 1
METROPOLITAN WATER RECLAMATION DISTRICT OF GREATER CHICAGO
HISTORIC SEWER CONSTRUCTION

During the last 10 years, the District has spent approximately $3 million on the emergency repair of sewers and related structures.

INTERCEPTOR INSPECTION AND REHABILITATION PROGRAM

An Interceptor Inspection and Rehabilitation Program (IIRP) has been instituted by the District to comprehensively document the condition of the intercepting sewer system over a period of five years, to evaluate the hydraulic and structural adequacy of the system, and to rehabilitate the system as necessary to preserve its integrity. An objective of the program is to minimize the need for inspectors or workers to enter the sewers or associated structures. The following elements constitute the program and reflect the estimated percentage of the system involved:

1. Physical inspection (100 %)

2. Closed circuit television inspection (50 %),

3. Infrared thermographic inspection (25 %),

4. Flow monitoring (15 %).

5. Inspection mapping and data documentation (100 %)

6. An Underground Advisory Committee.

Physical Inspection: The purpose of the physical inspection is to determine the general condition of all manholes, structures and sewer pipe; and identify segments of the system in need for further study. The inspection consists of surface and manhole inspections. The surface inspection involves walking along the route of the sewer for 'telltale' signs which may indicate potential problem areas.

Closed Circuit Television: The television inspection is to determine the condition of the sewer interior and document cracks, holes, pipe or joint separation, misalignment, root intrusion and illegal connections. The television inspection provides a positive and reliable means of observing and documenting the conditions in the sewer.

Infrared Thermography: The infrared thermographic investigation is designed to locate cavities in the soil adjacent to the sewer. An infrared scanner measures the temperature differences in the ground surface or pavement above the sewer. Voids change the energy flow more than the soil and rock surrounding the sewer pipe and produce a different image on the scanner.

Flow Measurement: Flow measurements are undertaken to find possible obstructions and recirculation of flow. The flow measurement may also identify cracks, open joints, and partial cave-ins, which prevent the intercepting sewer to operate as designed. The flow measurement units are capable of measuring and recording water depth, and magnitude and direction of mean velocity of flow.

Underground Advisory Committee: An "Underground Advisory Committee" has been constituted to support the IIRP program and provide information and advice regarding emergency repairs. The committee consists of six of the most experienced District personnel: three from both the Engineering and Maintenance Departments of the District.

ANALYSIS OF INSPECTION DATA

An interactive computer-aided Interceptor Mapping and Data Management System, IMADS, is being developed specifically to compile and analyze the inspection data to determine the condition of the intercepting sewer system. IMADS is written around Auto-CAD, thereby, making it compatible with PC-compatible computers.

The physical (structural) condition of the sewer is evaluated in the form a matrix which rates the severity and extent of the various conditions found during the inspection. The structural condition matrix will be used to yield a single numerical value representing the structural condition of the sewer, a Structural Condition Index. The Structural Condition Index is being developed. The use of this index will allow comparison of the condition of various segments of the sewer for aid in prioritizing rehabilitation needs.

Failure Consequence Analysis

The structural condition of the sewer as defined by the structural condition index will aid in prioritizing rehabilitation needs. But the rehabilitation priority will be a function of an analysis of the consequences of failure. The service consequences of the failure are different depending upon such considerations as the size of the service area.

The failure consequence analysis is performed in a matrix form. The number of failure impact categories can be increased or decreased depending on the associated problems to be encountered during the failure. The matrix is completed for each problem segment of the intercepting sewer system. The cumulative score of the matrix defines the "Failure Consequence Index". Repairs which cannot deferred through the next budgetary cycle will require emergency action, and all other repairs will be appropriately scheduled.

Estimating Maintenance Cost of Existing
Stormwater Retention Ponds

Robert E. Molzahn[1]
Member ASCE

Abstract

The 280 acre Kensington Business Park includes 11 stormwater retention ponds that were also designed to provide aesthetic enhancement. Maintenance of these ponds became a concern when the 10-year agreement between the Village and the park developer expired. Of particular concern was the estimated short and long term cost to maintain the stormwater detention function as well as the aesthetic enhancement of the ponds.

Introduction

The Kensington Business Park was developed over a ten year period on approximately 280 acres within the Village of Mount Prospect , Illinois,. As part of the general overall development plan, storm water retention ponds were constructed as the park developed. At the end of the development period there were 11 such ponds, some interconnected; but no entity responsible for the maintenance of the aesthetic aspects of the ponds. The Village has responsibility for only the maintenance of the stormwater detention aspects of the ponds.

This study involved the assessment of the physical status of the ponds and provided an estimate of the cost to maintain their effectiveness for storm water retention and for aesthetics.

----

[1]Vice President, Camp Dresser & McKee, Inc., The Sears Tower, Suite 450, 233 South Wacker Drive, Chicago, IL 60606-6306.

Pond Characteristics

The 11 ponds all have permanent pools, ranging in size from 2.8 acres to 0.4 acres. Pond No. 1 is the oldest and deepest pond, about 10' deep, 2.8 acres of permanent pool, and 10 years old. The other ponds were all designed to be from four to 5.5' deep. These permanent pools with clay bottoms provide an aesthetic quality as the retention capacity was designed 2.5' to 4.5' above the normal water surface. The inlets to the ponds are all set with their inverts at the normal water surface elevation. The outlets were all sized to restrict the outflow to that of a 3-year storm prior to development. Side slopes are 4:1 or 5:1 (h:v), except for Pond No. 1 which has much steeper side slopes.

The ponds support aquatic vegetation and aquatic life, such as frogs, and small fish have been observed. Historically, several ponds are eutrophic and require chemical treatment twice each year to kill algae. Pond No. 1 is sufficiently deep and a with a low enough nutrient loading to rarely require chemical treatment.

Field Survey Results

Several field investigations were undertaken in the fall of 1991 to investigate sediment depth and quality in the ponds and to inspect the inlet and outlet structures of the ponds. Measurements were taken of the sediment depth and thickness and sediment samples were collected from several ponds for laboratory analysis.

Each inlet and outlet structure was inspected and photographed. In general the physical condition of these structures was good. Where deterioration was noted, it was limited to small areas of concrete spalling on several of the flared pipe ends. The worst condition was observed at Pond No.1 where the ground around three of the 11 inlets was eroded adjacent to the flared end section or where it joined the first pipe section. Follow up inspection in the fall of 1992 indicated there were two more inlets at Pond No. 1 were the flared end section was pulled away from the first joint, and ground was eroding into the open joint.

The outlet pipes from most of the ponds were partially blocked with weeds and debris. The debris had backed the water into the ponds and raised the water surface .5 to 1.0 ' above the normal water surface, reducing the storage capacity for storm water retention.

Sedimentation in the ponds was checked by comparing the pond bottom elevations against the construction record drawings and by using a rod to locate the top of the sediment and the distance to a firm bottom. Spot measurements were general made over each retention pond on a 50' grid pattern, except near outfalls, where a 25' grid pattern was followed.

Results of the sediment thickness measurements indicated that much of the sediment was located about 25' from the inlet pipes. The measurements of sediment thickness ranged from 0 to 1', with the thickest sediments in the oldest pond. Based on the limited data available, the rate of sediment accumulation was estimated to be 0.05' per year.

Sediment quality was checked in Ponds Nos. 1,4A, and 7. The samples appeared to be organic detritus with little grit. The source of this material was assumed to be leaves, grass clippings, aquatic vegetation, and algae killed by the yearly chemical treatments. The gritty nature was assumed attributable to the streets and parking lot runoff tributary to these ponds.

The three samples were analyzed for polyaromatic hydrocarbons, lead, phosphorus, zinc and cadmium which are commonly found in parking lot runoff. The polyaromatic hydrocarbons were below detectable levels and with the exception of zinc, the metals were typical for concentrations expected in soils and sediments. The zinc concentrations were about 0.200 mg/l as compared to typical zinc concentrations in soil samples of between .005 and 0.180 mg/l. Phosphorus concentrations were between 0.004 and 0.006 mg/l in the sediment.

Storm Water Retention Costs

Costs were estimated to maintain and repair the ponds so that their function for storm water retention would be not be compromised were. These costs reflected the need to repair the inlet pipes and periodic removal of debris around the outlets. Erosion control and prevention of the outlet pipes from moving were considered as the primary objectives of the repair activities.

The Village had already assumed responsibility for the removal of the debris around the inlets and this cost was estimated on the basis of 100 person hours per year for the 11 ponds. This provided for weekly inspections of two hours from May through October and once per month debris removal by a two person crew. The cost of debris disposal was not estimated. At an effective cost of $25 per hour, the annual maintenance cost for these activities was estimated to $2500 for the 11 small retention ponds.

Repairs for five inlets to the oldest pond were designed and bids were taken in the fall of 1992. The repairs for these 18"and 21" inlets(with flared end sections) consisted of removing the end section, preparation of a new pipe bedding and construction of a retaining wall, buried at the joint and fitted to the bell of the pipe for restraint. The buried wall provides restraint on the last pipe section. The low bid for the five inlet pipes was $11,400 or about $2200 per repair. The cost of the repairs was affected by the steep side slopes and the need to restore the pond banks.

Maintenance Cost of Pond Aesthetics

The aesthetic quality of the ponds is of critical concern to the tenants of the business park. This area attracts large numbers of migratory water fowl and the visual appearance has been well maintained. Nuisance aquatic vegetation would spoil the ponds' aesthetics if left unchecked.

The ponds undergo a natural eutrophication process that is accelerated by the migratory water fowl population (contribution of phosphorus) and the shallowness of the ponds. Effective reduction of the phosphorus into the ponds was judged infeasible if the water fowl were allowed to be in the vicinity of the ponds. A resident population of only 40 water fowl around the ponds would contribute enough phosphorus to keep the ponds at an average phosphorus concentration of 0.05 mg/l. Consequently, the use of chemicals to control algae was estimated for use in all but Pond No. 1 at an annual cost of $7500. This cost was obtained as a quote form the licensed chemical application firm that had historically provided this service to the developer of the business park. Chemical treatment has not been necessary in the oldest pond because of its 10' depth and its higher volume relative to the drainage area served.

Removal of sediment was judged to be critical to the long term maintenance of the ponds aesthetic value. If the ponds become too shallow, they will become like a marsh with different vegetation and a different appearance. The shallower ponds were estimated to require sediment removal so that the pond depth did not become less than 3' in 20 years. At a cost of removal and disposal of $20 per cubic yard, the cost in 20 years would be $250,000 for sediment removal.

WYNOOCHEE LAKE AND DAM
Flood Storage Reevaluation Study

Christopher J. Lynch[1]

Abstract

With the desire to increase the revenue generating potential of the Wynoochee Lake and Dam Project the cities of Tacoma and Aberdeen, Washington, have pursued the potential for retro-fitting a hydropower plant at the dam. The feasibility of the hydropower plant is dependent on higher average head for power generation. This paper discusses the Corps of Engineers reevaluation of the winter flood control requirements with the aim of raising the elevation of the winter operating pool.

Introduction

The Wynoochee River Basin is located in Washington State on the southwest corner of the Olympic Peninsula about 25 miles inland from the Pacific Ocean with its headwaters originating in the Olympic Mountains (figure 1). Wynoochee Dam controls about one forth of the basin area and about one third of the basin runoff.

Wynoochee Lake Project, with the sponsorship of Aberdeen, Washington, was built by the U.S. Army Corps of Engineers and in operation by October 1972 as a multi-purpose project primarily for flood control and water supply, although irrigation, fish and wildlife, and recreation were included benefits. Currently 35,000 acre-feet (ac-ft) of Wynoochee Lake's total storage of 69,405 ac-ft is reserved for flood control between 1 October and 1 March. This corresponds to a normal operating flood control pool elevation of 762.6 feet (figure 2).

Motivated by years of fiscal stress and inability to realize the projected benefits of the Wynoochee project the City of Aberdeen became interested in retro-

---

[1]Hydraulic Engineer, Seattle District, U.S. Army Corps of Engineers, P.O. Box 3755, Seattle, WA 98124-2255. Ph (206) 764-3591; fax (206) 764-6678.

fitting the Wynoochee project with a hydropower plant. It was desirable to investigate raising the normal flood control operating pool to improve the feasibility of the hydropower plant.

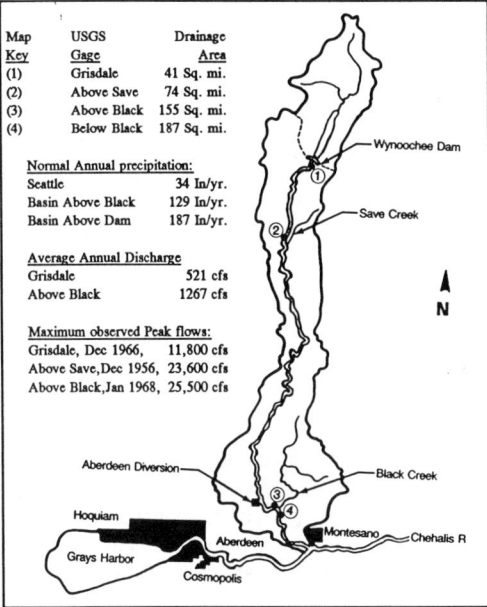

Wynoochee River Basin Map/Information
Figure 1

Purpose

The purpose of this study was to reevaluate the flood control requirement and determine if the amount of space reserved for flood control could be reduced with an acceptably small decrease in flood control benefits. The criteria used in this study was to insure that any change in flood control operation and/or storage would not result in an increased regulated discharge for frequencies up to and including the 200-year event at the Wynoochee River Above Black Creek streamgage (Above Black Creek).

Improving Streamflow Database

A primary justification in reevaluating the flood control storage requirements was the addition of 24 to 28 years of new streamflow data since the project was designed and the original flood control storage requirements were determined (see Table 1).

Wynoochee River Streamgage	Old Record	New record
Below Black Creek	1943-58, 16 Yrs	No new data
Above Black Creek	1957-62, 6 Yrs	1957-90, 34 yrs
Above Save Creek	1926-62, 37 yrs	1926-90, 65 yrs
Near Grisdale	No old data	1966-90, 24 yrs

Table 1. Streamflow Period of Record Summary

This study involved analyzing the largest floods of record. Therefore streamflows from the entire period of record, including both regulated and pre-project years, were examined to obtain hourly discharge data for the largest flood events for each year. All regulated floods were deregulated using an HEC5 model of the basin. The resulting flows at each of the key stations reflected natural pre-project conditions.

Updated Natural Frequency Curves

The natural maximum annual discharge frequency curves for the Wynoochee River near Grisdale (Grisdale), Above Save Creek (Save Creek), and Above Black Creek were updated using the maximum annual peak discharges. The Save Creek frequency curve and statistics were computed first and used to extend the statistics for Above Black Creek and for Grisdale, because Save Creek has the longest uninterrupted period of record, 65 years, 1926 through 1990.

Updated Regulated Frequency Curve.

The Above Black Creek regulated frequency curve was of primary interest because it is the downstream control point currently in use for flood control operations. An updated regulated Above Black Creek Frequency Curve reflecting existing conditions (flood control storage of 35,000 ac-ft) was generated using a three step process: 1) Perform a regression analysis to determine the relationship between peak flows and volume flows at each of the three sites; 2) Synthesize natural normal hypothetical floods for Grisdale, Save Creek, and Above Black Creek for the 100-yr, and 200-yr events; and 3) Regulate the hypothetical floods according to standard procedures using the HEC5 Wynoochee Basin Model to determine the regulated peak flows for the regulated frequency curve.

Alternative Flood Control Storage Test.

The updated Above Black Creek regulated frequency curve was used as the base case against which any change in flood control storage was evaluated. An acceptable reduction would not cause flooding at Above Black Creek for any flood up to and including the 200-year flood. This analysis focused on the 200-yr flood, although smaller floods were also tested. Using an HEC5 basin model, the 200-yr flood was regulated and routed several times through the basin. Each separate routing used a reduced amount of flood control storage.

With a flood control storage volume of 20,000 ac-ft or greater no increase in the 200-year regulated peak discharge at Above Black Creek was observed. Storage volumes less than 20,000 ac-ft caused increased flooding over existing conditions. A volume of 20,000 ac-ft was also verified to control lesser floods without any increase in flooding.

This analysis assumed perfect flood control regulation. Regulation of an actual flood event may involve conditions which cost additional storage and introduce a measure of uncertainty in our analysis. The data, methods, and tools used in this study also introduce some uncertainty. A sensitivity analysis was performed and a safety factor of 4000 ac-ft was adopted to account for these uncertainties. The proposed modified flood control storage is 24,000 ac-ft, an 11,000 ac-ft, or 30%, reduction in flood control storage. The gain in head represented is 13.5 feet during the flood control season of 10 October to 1 March

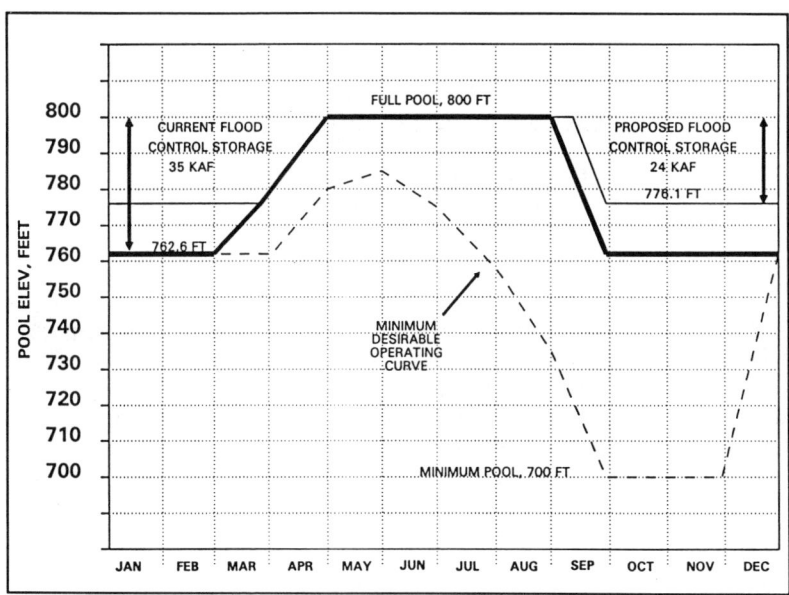

Wynoochee Project Seasonal Operating Guide
Figure 2

Conclusions

This study determined that a reduction in flood control storage from 35,000 ac-ft to 24,000 ac-ft would maintain current flood control protection up to the 200-year event. To implement the proposed reduction three other criteria must be fulfilled: 1) Public acceptance; 2) No increased risk of dam failure; and 3) No net loss of environmental resources. Criteria 1, and 2 have been satisfied and criterion 3 is being worked out among involved agencies. The powerhouse and intake works are under construction and scheduled to be completed by mid 1993. Final implementation of the reduced flood control storage is dependent on an Environmental Assessment Finding of no Significant Impacts.

REAL-TIME WATER-CONTROL SYSTEM FOR THE TRINITY RIVER, TEXAS

By David T. Ford,[1] Member, ASCE, and
J. Russell Killen,[2] Member, ASCE

ABSTRACT

To provide information for improved flood prevention and control in the Trinity river basin in Texas, we developed a real-time water-control system that (1) retrieves rainfall and streamflow observations, (2) processes, files, and manages these data; (3) estimates basin average rainfall and forecasts runoff, and (4) simulates reservoir operation. This system uses custom software for data transmitting and processing, and Hydrologic Engineering Center (HEC) data-management, rainfall-analysis, runoff-forecasting, and reservoir-simulation software. To integrate these disparate programs, we developed a specialized program manager.

FLOODS MOTIVATED SYSTEM DEVELOPMENT

The Trinity river basin includes approximately 17,800 sq mi in northeastern Texas, stretching 300 mi from the Oklahoma border to the Texas Gulf coast. Land use in the basin varies from farmland in the coastal plains to the highly-urbanized central business districts of Dallas and Ft. Worth. Average annual rainfall in the basin varies from 30 to 50 in. Seventeen major reservoirs in the basin are owned and operated for multiple purposes by four government agencies and an electric utility. The total capacity of these reservoirs is approximately 11.08 million ac-ft, including 1.76 million ac-ft of flood-control storage.

Record rainfall in the basin in 1989, 1990, and 1991 caused massive floods that took lives and caused millions of dollars in flood damage. In response, the Texas legislature created a task force that was directed to "... review the floodwater release procedures of lake operators in the Trinity River basin and the coordination of floodwater management efforts within basin boundaries ..." Based on task-force recommendations, the 72nd Texas Legislature enacted Senate Bill 1543, which directed the Texas Water Commission (TWC) and the Trinity River Authority (TRA) to: (1) develop and implement a coordinated basin-wide water release program; and (2) develop a basin-wide flood-warning system.

---

[1] David Ford Cons. Engr., P.O. Box 188529, Sacramento, CA 95818.
[2] Albert H. Halff Assoc., Inc., 4000 Fossil Creek Blvd., Ft. Worth, TX 76137.

SYSTEM INCLUDES FOUR MAJOR SUB-SYSTEMS

To achieve the objectives of the legislation, TWC and TRA commissioned our study team to develop a system "to predict flood flow rates throughout the system for time periods of from a few hours to a few days into the future, using currently-available rainfall and streamflow measurements and short-term rainfall forecasts (Halff, 1992)." Our work focused on development, implementation, and calibration of PC software sub-systems that (1) retrieve data, (2) process and file data, (3) estimate rainfall and forecast runoff, and (4) simulate reservoir system operation.

Data collection, transmission, and retrieval sub-system. The existing data-collection network in the Trinity basin includes 55 automated National Weather Service (NWS) reporting raingages, 192 weather observers, and 44 federal-cooperative automatic-reporting river or lake/reservoir elevation gages. At various time intervals, data are transmitted via telephone or satellite to NWS, U.S. Geological Survey (USGS), and Corps of Engineers (USACE) sites. For forecasting, we developed software to interrogate the databases at these sites, to retrieve data from these databases, to re-transmit them to the forecasting center via telephone modem, and to receive the re-transmitted data.

Data processing and filing sub-system. To store the re-transmitted data, we use HEC-DSS, the HEC's data processing and filing sub-system (USACE, 1990). We developed custom software to convert the data from its raw NWS or USGS format to the HEC-DSS format and to file the data in the HEC-DSS. We use HEC-DSS utility programs for database housekeeping chores.

Rainfall estimation and runoff forecasting sub-system. Software in this sub-system forecasts catchment runoff, given observed and forecasted rainfall, catchment conditions, and model parameters. To estimate catchment rainfall from point observations, we use HEC's computer program PRECIP (USACE, 1989), customized and calibrated for the Trinity basin. To forecast the resulting runoff, we use a PC-implementation of HEC-1F (USACE, 1989), a specialized version of the well-known HEC-1 program. HEC-1F uses a two-parameter unit hydrograph, a two-parameter loss model, a three-parameter baseflow model, and hydrologic routing methods. It updates parameter estimates or forecasts in real-time to match observed streamflow. We calibrated HEC-1F for gaged Trinity basin catchments with data from the 1989 and 1990 floods, and used regional model-parameter relationships to estimate model parameters for ungaged catchments.

Reservoir system simulation sub-system. Forecasting catchment runoff is only the first step to predicting flood flow rates and water levels throughout the Trinity basin. The second step is simulating operation of the system reservoirs to identify how these will affect flooding. To accomplish this, the TRA real-time system uses a customized version of computer program HEC-5 (USACE, 1982). This program simulates system operation following specified release rules, yielding estimates of regulated flows system wide.

SUB-SYSTEMS INTEGRATED WITH DATABASE & PROGRAM MANAGER

The software sub-systems are integrated in two ways. The first is through the data filing sub-system, HEC-DSS. In addition to using HEC-DSS to process and file raw data, we use it as the mechanism for linking the analysis tools. For example, HEC-1F retrieves rainfall data from HEC-DSS, forecasts catchment runoff, and files this unregulated-flow forecast with the HEC-DSS. HEC-5 retrieves this forecast from HEC-DSS, simulates operation, and files the regulated-flow forecast with HEC-DSS. The user then can tabulate or plot forecasted flows and water levels with HEC-DSS utility programs, and can take appropriate action. The flow of data, input, and analysis results is shown in the figure that follows.

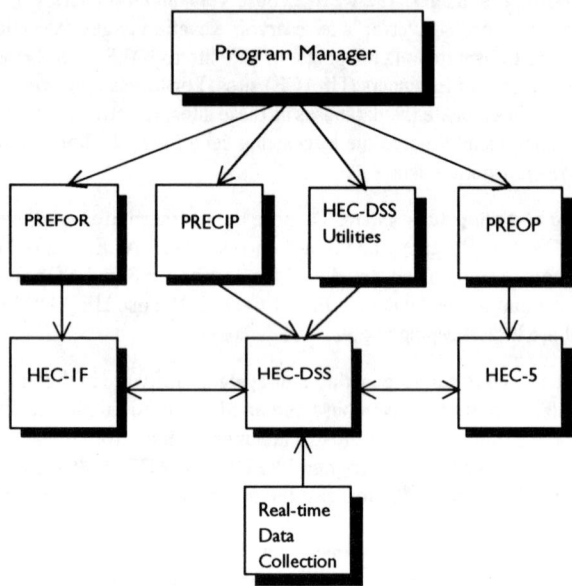

The programs are integrated also through a character-based program manager with pull-down menus, dialog boxes, radio buttons, and all the PC-program features users have come to expect. This program manager, designated TRACE (Trinity River Advanced Computing Environment), serves also as a file manager, input processor, and database interface. The file-manager component names, tracks, and manages the many input files necessary for real-time forecasting and reservoir-operation simulation. The input processor component accepts user input in dialog boxes, such as that shown in the figure that follows, and executes, behind the scenes, programs PREFOR and PREOP to create or modify HEC-1F input and HEC-5 input, respectively. TRACE executes programs PRECIP, HEC-1F, and HEC-5 with the proper input. The database interface component executes HEC-DSS utility program EXTRCT to create working copies of data records, DSPLAY to graph data, and

DWINDO to tabulate and edit data. A comprehensive set of macros insulates the user from HEC-DSS database-management commands.

```
Files  Rainfall  Runoff  Reservoirs  Help
                    Edit Zone Parameters
        Zone No.   BFFCST     RTIOR    STRTL      CNSTL
                   (cfs/mi2)           (inches)   (in/hr)
           1       0          0        0          0
           2       0          0        0          0
           3       0          0        0          0
           4       0          0        0          0
           5       0          0        0          0
           6       0          0        0          0
           7       0          0        0          0
           8       0          0        0          0

              < OK >      < Cancel >     < Help >
```

CONCLUSIONS

System-wide forecasts of regulated flows in the Trinity basin can be made with a PC-based system. Simulated real-time tests with historical flood data yielded reservoir-inflow and local-flow forecasts in 15 minutes and system-wide regulated-flow forecasts in one to two hours, including time for graphing and inspecting rainfall and streamflow data.

The PC-based software is sufficiently flexible to utilize data from additional gages, to change the representation of the catchment runoff or streamflow routing processes, or to simulate alternative reservoir operation rules or system configurations. All this is accomplished without any reprogramming. The value of this flexibility is obvious, considering the sparsity of reliable rainfall data. In one case in the lower basin, we found that a single gage represents rainfall over 2000 sq mi. Additional gages were proposed. When these are available, only minor changes to input files are required to use the new data.

REFERENCES

Albert H. Halff Assoc., Inc. (1992). "Flood prevention and control for the Trinity river basin," report to TWC and TRA, Dallas, TX.
USACE (1982). *HEC-5: Simulation of flood control and conservation systems, program user's manual.* HEC, Davis, CA.
USACE (1989). *Water control software, forecast and operations.* HEC, Davis, CA.
USACE (1990). *HEC-DSS user's guide and utility program manuals.* HEC, Davis, CA.

HYDROLOGIC ANALYSIS OF LEVEED INTERIOR AREAS

Harry W. Dotson, M.ASCE and Michael W. Burnham[1], M.ASCE

Abstract

Flood damage reduction projects that include levees or flood walls usually involve special problems associated with isolated interior areas. Storm runoff patterns may be altered and remedial measures are generally required to prevent increased flooding in the interior area due to natural flow blockage. A new computer program has been developed by the Hydrologic Engineering Center, US Army Corps of Engineers entitled, "Interior Flood Hydrology, HEC-IFH" (HEC, 1992). The program can be used to perform the required hydrologic analyses to characterize the interior area flood hazard and to evaluate the performance of potential flood damage reduction measures and plans. The program is particularly powerful for performing long, period-of-record simulations. The program operates in an interactive IBM compatible Personal Computer environment, allowing full-screen interactive data entry and analysis of results. Annual or partial series interior elevation-frequency relationships are derived directly, using hypothetical event or continuous, historical period-of-record analysis for various measures such as gravity outlets, pumps, and diversions. The paper describes the analysis of interior flood damage reduction measures using the HEC-IFH program.

Introduction

An *interior area* is protected from direct river, lake, or ocean flooding by levees, flood walls, or high ground. The levee or wall protecting the interior area is the *line-of-protection*. The line-of-protection protects areas subject to flooding from the exterior, but may block the natural passage of interior runoff to the main river. Gravity outlets, pumping stations, interior ponding areas (detention storage basins), and diversions are measures commonly implemented within interior areas to reduce flood loss. Gravity outlets (usually culverts) may pass water through the line-of-protection when interior water levels are higher than exterior levels. The flood waters are stored or pumped through the line-of-protection when exterior stages are higher than those in the interior. Flood waters may also be diverted to other areas. Figure 1 is a representation of an interior system.

---

[1]Hydraulic Engineer and Chief, Planning Analysis Division, US Army Corps of Engineers, The Hydrologic Engineering Center, Davis, CA 95616

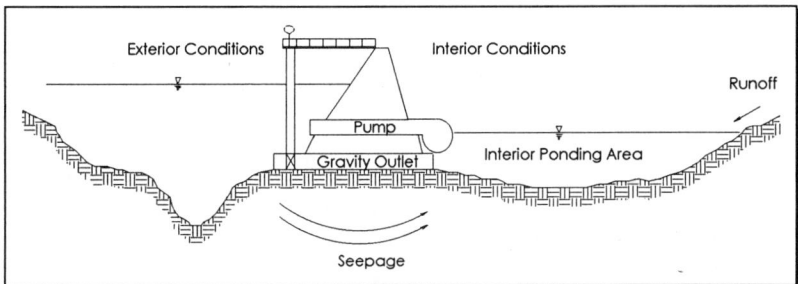

FIGURE 1. Cross-Section of Typical Interior System

Data Requirements

Hydrologic data required for analyses of interior areas include topography, exterior stage data, historic rainfall records and/or hypothetical precipitation data, runoff parameters, and seepage data. Topographic data are required to define watershed and sub-basin boundaries, basin slopes, stream lengths, and elevation-area-storage relationships for natural detention areas. Exterior stage data are required to define tailwater conditions for gravity outlets and pumps, and for determining seepage.

Rainfall data are needed to compute interior and/or exterior basins. If no rainfall gage exists in the basin, records from nearby rain gages can be used in the analysis. Alternatively, depth-duration data for hypothetical rainfall events may be used to determine runoff from a storm of desired duration and frequency. Loss rates may be initially estimated by using values from previous studies or derived through analysis of measured rainfall and runoff volumes at gages.

Runoff transforms provide methods of computing runoff from rainfall excess. Unit hydrographs are common runoff transforms. Initial values for unit hydrograph parameters may be estimated from land use and physical basin characteristics using published values or regression equations, or from analysis of observed events. The importance of volume rather than peak discharge in many studies permits the use of simplified runoff methods with acceptable results. Assumptions should be verified as needed. Interior analyses normally require information on physical and operational characteristics of existing or proposed flood loss reduction measures. For example, information on gravity and/or pump outlet locations, capacity, and operational procedures enable a simulation analysis to reproduce the historic record. Data on ponding areas, storm water collection systems, and any hydraulic controls affecting interior water movement are also often necessary.

Interior Analysis Procedures

There are two general analysis approaches that are used for analyzing interior areas depending on data availability and the characteristics of the area.

1. *Continuous Simulation Analysis* (also called a *period-of-record analysis*), which uses continuous historic precipitation and streamflow records for the

interior and exterior areas. The procedure consists of sequential hydrologic simulation of inflow, outflow, and change in storage. The objective is to derive interior water surface elevations given exterior stages and interior rainfall for the entire period-of-record.

2. *Hypothetical Event Analysis*, is generally applicable when interior and exterior flood events are *dependent*. The analysis can be conducted so that the same series of hypothetical storm events occur over both the interior and exterior areas. This analysis method can also be applied using a constant exterior stage or for any "blocked" or "unblocked" gravity outlet condition.

Continuous Simulation Analysis Using the HEC-IFH Program

The continuous simulation approach involves applying historical rainfall to sub-basin loss rates, runoff transforms, and base flow parameters to yield runoff hydrographs at basin outlets. Hydrographs are combined and routed through the system to yield period-of-record inflows to the ponding area adjacent to the line-of-protection. These data are used with period-of-record exterior stage data to route the flow through the line-of-protection and generate a continuous interior stage hydrograph. The hydrograph is used to derive annual peak stage-frequency relationships. The is repeated for existing conditions and for each set (plan) of interior flood damage reduction measures to be evaluated. Plans are compared based on performance and economic criteria and a recommended plan is selected.

Analysis procedures including data entry, performing an interior area analysis, examining results, and comparing plans are summarized below

1. Data Entry. Data entry is performed using separate data entry *modules*. Each module represents a group of related data. Data entry screens and computational procedures are provided to develop the data for each module. Any entered, imported, or computed data can be tabulated or plotted while within these modules. The data entry modules are described below.

PRECIP - Precipitation gage data is entered or imported from the HEC time series Data Storage System, HEC-DSS (HEC, 1987). Basin average composite precipitation weightings are assigned and composite precipitation is computed. Gage or composite precipitation records are assigned to the upper or lower interior sub-basin, as appropriate.

RUNOFF - Basin runoff data, consisting of drainage area, percent imperviousness, infiltration loss parameters, and unit hydrograph method are entered for computing runoff. The routines employed for computing runoff are the same as in the Flood Hydrograph Package, HEC-1 (HEC, 1990). Parameters may also be entered for routing runoff from the upper to the lower interior sub-basin.

POND - Elevation vs. surface area data are entered to describe the ponding area characteristics. Corresponding pond storage is computed as elevation-area data are entered. If a ditch between the pond and the line-of-protection is to be used, the ditch rating is entered.

GRAVITY - A family of gravity outlet rating curves is computed by entering the culvert shape (box or circular), size, number of identical outlets, length, roughness, and entrance characteristics. Several types and sizes of gravity outlets can be specified for use in subsequent interior plan analyses.

PUMP - Pump unit data consisting of head-capacity-efficiency relationships, and pump start and stop elevations are specified.

EXSTAGE - Exterior stage conditions are specified by entering or importing a stage or a discharge hydrograph and rating curve. The discharge hydrograph may also be determined from rainfall-runoff in the same manner as for the interior sub-basins. Computed or entered stage hydrographs can be transferred to the location of interest by specifying a transfer relationship.

AUXFLOW - Auxiliary inflow for the upper and lower sub-basins, upper sub-basin diversions, lower sub-basin overflow, and pond seepage are specified.

2. Performing Interior Analyses. Plans for interior analysis are defined by specifying the named and saved data set modules. The analysis can be performed in one to five steps consisting of 1) upper sub-basin runoff, 2) upper and lower sub-basin runoff, 3) interior runoff and exterior basin analysis, 4) interior runoff plus exterior basin analysis and pond routing. and finally, 5) all of the above plus determination of pond annual or partial series peak interior elevation-frequency relationships.

3. Hydrologic Analysis Summaries. Twenty-three different hydrologic analysis reports are available for displaying and plotting the simulation results. Variables that can be displayed include precipitation; infiltration; runoff; pond inflow, outflow, elevation-frequency, and duration; gravity outflow; seepage; and pump outflow, operation time, and energy used. The reports are divided into six categories including 1) analysis input summaries, 2) calculation period summaries, 3) monthly summaries, 4) water year annual summaries, 5) entire analysis period summaries, and 6) analysis error and message summaries.

4. Plan Comparisons. Once several plans have been defined and analyzed, plans comparisons can be made. Four types of plan comparison reports can be viewed. The reports include plan summary data and frequency relationships for maximum interior elevation, area flooded, and pond inflow.

References

HEC, *HEC-DSS Users Guide and Utility Program Manual*, U.S. Army Corps of Engineers, Hydrologic Engineering Center, Davis, CA, November 1987.

HEC, *HEC-1 Flood Hydrograph Package Users Manual*, U.S. Army Corps of Engineers, Hydrologic Engineering Center, Davis, CA, September 1990

HEC, *HEC-IFH Interior Flood Hydrology Package User's Manual*, U.S. Army Corps of Engineers, Hydrologic Engineering Center, Davis, CA, April 1992

A Derived Flood Frequency Distribution for Ungaged Catchments

Rafael S. Seoane[1] and Juan B. Valdés[2] M. ASCE

Estimation of extreme flood events is a permanent interest for hydrologists for theoretical and practical reasons. The former is the increase of knowledge about complex processes of the precipitation-runoff transformation and the latter is the relationship with the necessity of civil engineering to use more accurate methods for the estimation of the most important variables of water resources projects. We know that the records of the precipitation variable are longer than runoff series and naturally it appears as a logic necessity the development of technics to use these records to improve the estimation of the extreme floods. This situation is more complicate in developing than in developed countries since the measurements may missing. Therefore it is important to develop new technics with geomorphologic parameters and precipitation register ought to be increased. The area in hydrology that try to resolve these problems has its origins in the theory of instantaneous unit hydrograph (IUH) to calculate the maximum flood and as Dooge(1959) indicated it retained some empirical components and was necessary to relate its variables with geomorphologic characteristics. The experience obtained with practical applications and later investigations could define other goals and therefore, since the late 70's, the investigations could define others goals and therefore, the research concentrated on:

- To represent the influence of the basin structure and the dynamic characteristics on the runoff. This is directly related with the initial concepts of Dooge (1959) and a suggested solution was to define (following Horton and Strahler concepts) the geomorphologic characteristics and the drainage network conduction of a basin that influence in the runoff.

- To represent the influence of the climate introducing the precipitation process. The rainfall-runoff transformation is non-linear. Thus it was required to find a more adequate mathematical representation.The new theory ought to look to the observed variability in the runoff and the maximum time with the intensity of the effective precipitation and the runoff characteristics.

The latter concepts led naturally to the theory GIUH (Rodríguez- Iturbe and Valdés, 1979) and later to the Geomorphoclimatic theory (Rodríguez- Iturbe et al., 1982) and Gupta et al. (1980) studies give the definition of one of the necessary bases for the development of a derived distribution function. Eagleson(1972) developed an analytical expression with an exponential distribution for the intensity and precipitation duration, a model of invariant infiltration in time and as basin response model the kinematic wave theory. This study follows the development in Hebson and Wood (1982), Díaz-Granados et al. (1984), and Raines and Valdés (1992) that incorporated as basin response model the new GIUH and the GcIUH theories respectively. Hebson an Wood(1982) study used a constant infiltration model whereas Díaz-Granados et al.(1984) study used variable infiltration equation. Both models used precipitation represented by the exponential distribution. From the methodological view point it is possible to define the existence of two main research lines from Eagleson theory with two different catchment models. One model used the kinematic wave model: Shen et al. (1990) and

---
[1] Research Associate, National Institute for Water Science and Technology (INCyTH), National Council for Scientific and Technological Research (CONICET) and National University of the Center Buenos Aires. Casilla de Correo. 23 Aeropuerto Internacional de Ezeiza (1820). Argentina

[2] Associate Professor, Environmental and Water Resources Engineering, Department of Civil Engineering and Associate Director, Climate System Research Program, Texas A&M University, College Station, TX 77843-3136; (409)845-1340.

Cadavid et al. (1991). The other research are used the IUH model: Hebson and Wood (1982), Díaz-Granados et al. (1984) and Raines and Valdés (1992) that used the SCS infiltration model. This paper proposes the use of a more realist precipitation model based in a bivariate exponential probability function (Córdova and Rodríguez-Iturbe, 1985) to investigate the effect on extreme floods.

Methodology

This section describes a derived flood frequency procedure based on the stochastic model of point precipitation used with correlated intensities and duration. This model will be used to study the importance of this in the estimation of floods. Our derivation of the probability density function used the results of Raines (1991) and Raines and Valdés (1992). A model for the representation of rainfall events is the Rectangular Pulses Poisson Model (RPPM, Rodríguez-Iturbe, 1984), is used in this work and we consider the relationship between in intensities and duration to be described by a bivariate exponential distribution, e.g. Nagao and Kadoya (1971) and Córdova and Rodríguez-Iturbe (1985), i.e.

$$f_{I_r,T_r}(i_r, t_r) = \frac{\alpha\delta}{(1-\rho)} exp(-\beta_1 i_r - \beta_2 t_r) I_0 \left[\frac{2(\rho\alpha\delta i_r t_r)^{1/2}}{1-\rho} \right] \qquad (1)$$

$$I_0(z) = \sum_{k=0}^{\infty} \frac{(z/2)^{2k}}{(k!)^2} \qquad (2)$$

where $I_0()$ is the zero order Modified Bessel Function and ρ is the correlation coefficient between i_r and t_r.

$$f_{I_r,T_r}(i_r, t_r) = \sum_{k=0}^{\infty} \eta_k (i_r t_r)^k exp(-\beta_1 i_r - \beta_2 t_r) \qquad (3)$$

where

$$\eta_k = \frac{\rho^k (\alpha\delta)^{k+1}}{(1-\rho)^{2k+1}(k!)^2} \qquad (4)$$

$$\beta_1 = \frac{\alpha}{1-\rho} \qquad \beta_2 = \frac{\delta}{1-\rho} \qquad (5)$$

The marginal distributions of i_r and t_r are exponential, where i_r is the mean rainfall intensity and t_r is the storm duration.

The model is built from rectangular pulses associated with a Poisson process. In our approach the SCS Curve Number (CN) method was used to compute the excess precipitation P_e. The same procedure outlined by Díaz-Granados et al. (1983) was used to evaluate the joint PDF of i_{re} and t_{re}. First, the probability of no effective rainfall was obtained by integrating the joint PDF of i_r and t_r over the area where no runoff is produced.

$$i_r = \frac{0.20S}{t_r} = \frac{w}{t_r} \qquad (6)$$

$$\begin{aligned} Prob[i_{re} = 0, t_{re} = 0] &= \int_0^\infty [\int_0^{w/t_r} \sum_{k=0}^\infty \eta_k (i_r t_r)^k exp(-\beta_1 i_r - \beta_2 t_r) di_r] dt_r \\ &= 1 - 2\sum_{k=0}^\infty \eta_k \sum_{n=0}^k \frac{k!}{n!} \frac{w^n}{\beta_1^{k-n+1}} (\frac{w\beta_1}{\beta_2})^{(k-n+1)/2} \cdot K_{k-n+1}(2\sqrt{\beta_1 \beta_2 w}) \end{aligned} \qquad (7)$$

The continuous part of the joint PDF of i_{r_e} and t_{r_e} can be computed as the product of the conditional PDF of i_{r_e} given t_{r_e} and the marginal PDF of t_{r_e}.

The continuous part of the joint PDF was obtained

$$f_{I_{r_e},T_{r_e}}(i_{r_e},t_{r_e}) = 0.776452\alpha S^{0.44161} t_{r_e}^{-0.44161} i_{r_e}^{-0.44161} exp(-1.39047\alpha S^{0.44161} t_{r_e}^{-0.44161} i_{r_e}^{0.55839})$$

$$\cdot 2\sum_{k=0}^{\infty}\eta_k \sum_{n=0}^{k} \frac{k!}{n!}\frac{1}{\beta_2^{k-n+1}} w^m e^{-t_{r_e}\beta_2} \cdot \sum_{m=0}^{n}(C_m^n)(\frac{w\beta_2}{\beta_1})^{(k-m+1)/2} K_{k-m+1}(2\sqrt{\beta_1\beta_2 w}) t_{r_e}^{n-m}[\beta_2 - \frac{n-m}{t_{r_e}}] \quad (8)$$

Integrating the joint PDF of i_{r_e} and t_{r_e} defined by Eqs. 8 and 8, with the approximations proposed in Diaz-Granados et al. (1983) and other assumptions, the expression for the CDF becomes:

$$F_Q(Q_p) = 1 - 2\sum_{k=0}^{\infty}\eta_k \sum_{n=0}^{k} \frac{k!}{n!}\frac{w^n}{\beta_1^{k-n+1}} \cdot (\frac{w\beta_1}{\beta_2})^{(k-n+1)/2} \cdot K_{k-n+1}(2\sqrt{\beta_1\beta_2 w})$$

$$+ \int_0^{Q_r^*}[\int_0^{\infty} f_{I_{r_e}T_{r_e}}(i_{r_e},t_{r_e})dt_{r_e}]di_{r_e} + \int_{Q_r^*}^{\infty}[\int_0^{t_{r_e}^*} f_{I_{r_e}T_{r_e}}(i_{r_e},t_{r_e})dt_{r_e}]di_{r_e} \quad (9)$$

Assuming that baseflow is negligible when compared to the total peak discharge Q_T being considered in flood frequency analysis, the recurrence interval T_E is given by

$$T_E = [\theta[1 - F_{Q_T}(Q_T)]]^{-1} \quad (10)$$

where θ is average annual number of independent rainfall events.

Application of the Proposed Methodology

The four derived distribution procedures and a standard rainfall-runoff simulation model are applied to four catchments in Texas with varying soil and land use characteristics and compared to a flood frequency curve computed from historical data. The watersheds selected vary in size, land use, soil type, slope, and climate to test the derived flood frequency distribution methods. In this study for ungaged catchments where no rainfall-runoff data is assumed to be available for calibration, the SCS Curve Number loss rate and the SCS Dimensionless unit hydrograph were used where the parameters are easily estimated from topographic maps. The design storms were computed using the Texas Department of Highways and Public Transportation Hydraulic Manual (1985) intensity-duration-frequency curves. The log-Pearson Type III distribution was fitted to the annual peak series for the four catchments. Fig. 1 shows the comparison between the derived flood frequency distributions, the new method (with different values of the correlation coefficient), HEC-1 simulation results applied to the Turtle Creek catchment and the comparison with the log-Pearson Type III results. The figure shows a wide variation in the computed flood frequencies. Recall that all four derived flood frequency distribution procedures utilize the same general rainfall model and similar watershed response models based on geomorphologic theory but each model represents infiltration differently. Since the estimated parameters must adequately represent the actual basin characteristics to provide accurate results, the problem may be more the result of limitations with the parameter estimation procedure than model theory. Also, notice that the accuracy is no better with the simulation model as applied to ungaged catchments when no calibration is performed. It is important to realize that for ungaged catchments, there are no rainfall-runoff records to calibrate the model parameters. This work shows the impact that the use of correlated rainfalls has on shape of the derived distribution of annual floods.

FLOOD FREQUENCY DISTRIBUTION

Acknowledgements: R. Seoane express his gratitude to the Department of Civil Engineering of Texas A&M for their cooperation during his stay there. His stay at Texas A&M was funded by INCyTH and CONICET of Argentina. All contributions are gratefully acknowledged.

Selected References

Cadavid, L., Obeysekera, J.T.B., and Shen, H. W. (1991). "Flood-Frequency Derivation from Kinematic Wave." *Journal of Hydraulic Engineering*, vol 117 (4), 489-510. 1991.

Córdova, J. R. and Rodríguez-Iturbe, I. (1985). "On the Probabilistic Structure of Storm Surface Runoff." *Water Resour. Res.*, 21(5), 755-763.

Diaz-Granados, M. A., Valdés, J. B., and Bras, R. L. (1984). "A physically based flood frequency distribution." *Water Resour. Res.*,20(7), 995-1002.

Eagleson, P. S. (1972). "Dynamics of flood frequency." *Water Resour. Res.* ,8(4), 878-898.

Hebson, C., and Wood, E. F. (1982). " A derived flood frequency distribution using Horton order ratios." *Water Resour. Res.*, 18(5), 1509-1518.

Nagao, M and Kadoya, M. (1971). "Two-variate exponential distribution and its numerical table for engineering applications." it Bull. Disaster Prev. Res. Inst.,Kyoto Univ. 20(3), 183-215. 1971.

Raines, T. H. and Valdés, J.B. (1991). "An assessment of derived flood frequency distributions." submitted to the *ASCE Journal of Hydraulic Engineering*, 1992.

Rodríguez-Iturbe, I., and Valdés, J. B. (1979). "The geomorphologic structure of hydrologic response." *Water Resour. Res.*, 15(6), 1409-1420.

Rodríguez-Iturbe, I., Gonzalez, M., and Bras, R. L. (1982). "A geomorphologic climate theory of the instantaneous unit hydrograph." *Water Resour Res.* ,18(4), 877-886.

Rodríguez-Iturbe, I., Gupta, V.K., and Waymire, E. (1984). "Scale considerations in the modeling of temporal rainfall." *Water Resour. Res.* ,20(11), 1611-1619.

Shen, H.W., Koch, G.J., and Obeysekera, J.T.B. (1990). " Physically Based Flood Feature and Frequencies." *Journal of Hydraulic Engineering*, Vol . 116, N 4, 494-514.

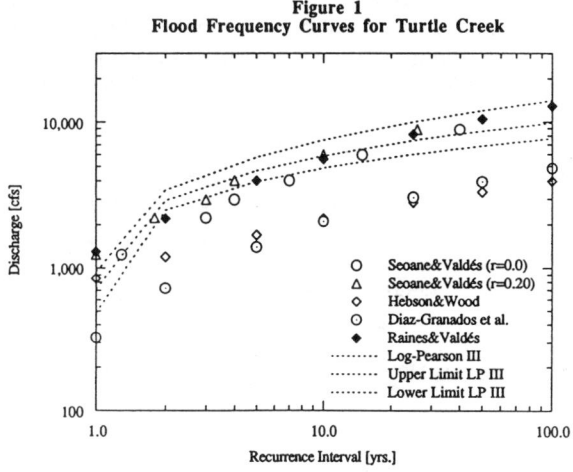

Figure 1
Flood Frequency Curves for Turtle Creek

Global Approaches for the Nonconvex Optimization of Pipe Networks

G. V. Loganathan[1], Member, ASCE and J. J. Greene[2]

Abstract: Because the pipe network problem is nonconvex, two global search schemes, MULTISTART and ANNEALING, are employed to permit a local optimum seeking method to migrate among various local minima. An example problem from the literature is solved using the proposed procedures. The optimal solution has a cost which is significantly smaller than the ones reported by other researchers.

Introduction

Water distribution system design has remained intriguing because of its complexity and utility. Standard hydraulics textbooks treat only the problem of finding flows in a given network layout for specified diameter pipes. The real design aspects of layout selection and choosing suitable diameters for the pipes however, are quite cumbersome. It is not uncommon to find real life problems which have multimodal cost or profit functions defined over a nonconvex feasible region involving multiple local optima and the pipe network optimization problem falls into this group While classical optimization methods find only a local optimum, global optimization schemes adapt the local optimum seeking methods to migrate among local optima to find the best one (Törn and Žilinskas, 1987).

Multistart-local search covers a large number of local minima by saturating the feasible region with randomly generated starting points. From each randomly generated starting point a local optimum seeking method is employed. *Simulated annealing* is an iterative improvement algorithm

---

[1]Associate Professor, Department of Civil Engineering, Virginia Polytechnic Institute and State University, Blacksburg, VA 24061.

[2]Engineer, Engineering Data Systems Corp., Dubuque, IA52002

in which cost of the current design is compared with the cost of the new iterate. The new iterate is updated as current iterate for the subsequent iteration if the cost difference is favorable. To to get out of a local optimum it accepts a worse point (higher objective value for a minimization problem) from the current point with a user specified probability 'p'.

It so happens that for a given layout the nonconvex pipe network optimization problem can be decomposed into a two stage problem comprising of an outer search strategy for selecting link flows and an inner linear program for optimal design of pipes for specified link flows from the outer problem. This decomposition does have merit because by using the aforementioned global search schemes the outer search can be made very efficient and the inner linear programs can be handled with ease.

Inner Linear Program

Let there be P loops, N nodes, and ℓ links. Let M be the number of distinct pipe diameters that are available for the design. By assigning link flows $Q_{(i,j)}$ for each (i,j) the following linear program is formulated:

Problem P(1):

Minimize:

$$f(x) = \sum\sum_{(i,j)m} C_{(i,j)m} x_{(i,j)m} \qquad (1)$$

Subject to:

$$H_s - H_k^{min} - \sum_{(i,j) \in r_k} \pm\sum J_{(i,j)m} x_{(i,j)m} \geq 0 \quad \text{for } k = 1,...,N \qquad (2)$$

$$x_{(i,j)m} \geq 0 \quad \text{for } k = N + 1,...,N + \ell M \qquad (3)$$

$$\sum_{(i,j) \in \mathcal{L}} \pm\sum_m J_{(i,j)m} x_{(i,j)m} = b_{\mathcal{L}} \quad \text{for } \mathcal{L} = 1,2,...,P \qquad (4)$$

$$\sum_m x_{(i,j)m} - L_{(i,j)m} = 0 \quad \text{for } \mathcal{L} = P + 1,...,P + \ell \qquad (5)$$

in which: $x_{(i,j)m}$ = length of the m th diameter segment in link (i,j); $L_{(i,j)}$ = length of link (i,j); $C_{(i,j)m}$ = cost of unit length of pipe segment of the m th diameter pipe in link (i,j); $J_{(i,j)m}$ = hydraulic gradient at the pipe segment; $\mathcal{L}$ = a loop that does not geometrically include any other loops; H_s = head at source node S; H_k^{min} = minimum head at node k; r_k = path through the network connecting the source S and node k. The objective function given by Eq. (1) minimizes the pipe cost. Equation (2) imposes minimum head at node k. In Eq. (2) a positive sign is taken only if the path direction coincides with the flow direction in link (i,j). Constraint (3) requires segment lengths to be nonnegative. Equation (4) sets $b_{\ell} = 0$ for a loop and a positive sign is taken only if the loop orientation coincides with the flow direction in link (i,j). In Eq. (4), b_{ℓ} is the fixed head difference for $\binom{J}{2}$ paths if there are J fixed head nodes. Equation (5) requires that the sum of segment lengths should equal the link length.

Globally Optimal Design

In this section, the global optimization strategies are applied to the solution of a water distribution network. A key observation is that there is a dominating core tree for a network and therefore, it is likely that the flows in the looped layout are perturbed versions of the tree link flows. Commencing with T* as some initial tree, the algorithm first computes v*, the optimal objective cost value via the linear program to obtain a minimum cost design. Then T* is systematically perturbed by adding one link to it from the set of co-tree links to form a loop and deleting that particular link from T* in the loop to form a new tree, T which yields the smallest objective value v. If this latter value is smaller than v*, then the incumbent v* is set equal to this value, and T is recorded as the current best solution T*. Otherwise, the method continues the search with another link in the co-tree which will yield an improving solution. Once the optimal tree network is found, a set of loop-forming links should be chosen to provide sufficient reliability in the case of failure of a tree link. The procedure scans all tree link failures one at a time and finds corresponding sets of reconnecting links from the core tree. The optimal set is found by choosing the cotree links that provide maximum cover with the least total length. The optimal redundant link set augmenting the optimal tree network, yields a looped network that should be optimized for pipe diameters using global search strategies. The example problem is taken from Alperovitz and Shamir (1977). The global procedures yield a cost of $405,301 which may be compared to $412,931 for Loganathan et al. (1990); $415,271 for Fujiwara et al. (1987); $441,522 for Quindry et al. (1979) and $479,525 for Alperovitz and Shamir (1977). The solution is shown in Figure 1.

Summary

The proposed algorithm offers a 'holistic' approach in that it allows for the selection of a core tree layout, selection of redundant links to ensure reliability, selection of initial flows for a search and a means to obtain a global optimum.

References

Alperovitz, E., and Shamir, U., "Design of Optimal Water Distribution Systems," *Water Resources Research*, vol. 13, Dec., 1977, pp. 885-900.

Fijiwara, O., Jenchaimahakoon, B., and Edirisinghe, N. C. P., "A Modified Linear Programming Gradient Method for Optimal Design of Looped Water Distribution Networks," *Water Resources Research*, vol. 23, no. 6, June, 1987, pp. 977-982.

Loganathan, G. V., Sherali, H. D., and Shah, M. P., "A Two-Phase network Design Heuristic for the Minimum Cost Water Distribution Systems Under a Reliability Constraint," *Engineering Optimization*, vol. 15, 1990, pp. 311-336.

Quindry, G., Brill, E. D., and Liebman, J. C., "Comments on 'Design of Optimal Water Distribution Systems' by E. Alperovitz and U. Shamir," *Water Resources Research*, vol. 15, no. 6, 1979, pp. 1651-1654.

Törn, Aimo and Žilinskas, Antanas, *Global Optimization*, Lecture Notes in Computer Science 350, Springer-Velag, New York, N.Y., 1987.

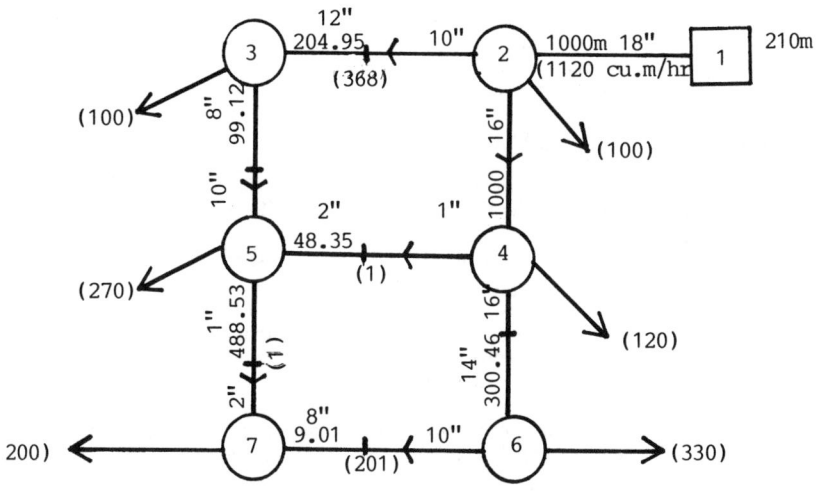

Figure 1. Optimal Pipe Network

Pipe Network Optimisation using Genetic Algorithms

Angus R Simpson[1], Assoc. Mem. ASCE, Laurie J Murphy[2], Graeme C Dandy[3]

Introduction

Optimisation of pipe networks is not used extensively in the design of urban water supply systems by water supply authorities or consultants. Often there are many choices to be made in designs including the size and material of new pipes, whether existing pipes should be duplicated or cleaned, and the sizing of pumps. Once the number of decision variables exceeds about 8 or 10, the number of possible alternative pipe network configurations may easily be in the order of billions. Currently, designers have sophisticated hydraulic simulation tools available for design. A number of trial networks are tested to find if all projected demands under peak loading or fire conditions can be met while maintaining specified minimum pressure constraints at all nodes in the network. The experienced designer uses rules of thumb such as head loss per unit length to eliminate many unrealistic combinations. However, given the many possible combinations, especially for pipe network expansions involving many pipes, it is unlikely that even the most experienced designer will be able to determine the minimum cost network. A new optimisation technique of genetic algorithms has recently been successfully applied to pipe network optimisation. In this paper a parametric analysis is carried out of the genetic algorithm in order to assess the form of the fitness function.

Genetic Algorithms for Pipe Optimisation

The application of the genetic algorithm technique has been developed at the University of Adelaide over the last 3 years (Murphy and Simpson 1992, Simpson et al. 1992, Dandy et al. 1993). A population of pipe network solutions is considered. Each pipe to be sized is represented by a binary sub-string. A linkage is made between each binary sub-string combination and a particular available pipe size with a corresponding cost. Sub-strings are joined together to form a full-length string representing the entire network. There are many variations possible for application of the genetic algorithm technique. In this paper the role of the form of the fitness function is investigated. The genetic algorithm for pipe optimisation involves the following steps:
(i)   **Generation of initial population.** The initial population of solutions (of say, size N=100) is generated using a random number generator.

---
[1]Senior Lecturer, Dept of Civil and Environmental Engineering, University of Adelaide, GPO Box 498, Adelaide, South Australia 5001, [2]Research Officer, [3]Senior Lecturer (both at University of Adelaide).

(ii) **Computation of network cost.** Each sub-string of the 100 strings is decoded into the corresponding pipe size. The total material and construction cost of the network for each of the solutions in the population is then computed.
(iii) **Hydraulic analysis of network.** Each network in the population is analysed for heads and discharges under the specified demand pattern(s). The actual heads are compared with the minimum allowable pressure head and any pressure deficits are noted.
(iv) **Computation of penalty cost.** A penalty cost is assigned to the network by considering the node with the worst pressure deficit. The pressure deficit is multiplied by a penalty factor (e.g. $50,000/metre of head).
(v) **Computation of total cost.** The total cost of each network in the population is the sum of the network cost (ii) plus the penalty cost (iv).
(vi) **Computation of the fitnesses.** For each network in the population, the fitness is taken to be a function of its total cost in part (v). For example,

$$Fitness = \left(\frac{1}{Total\ cost}\right)^n \quad (1)$$

(vii) **Generation of a new population using the reproduction operator.** New members of the next generation are obtained such that the probability of selection of a string to be included in the next generation is directly proportional to its fitness.
(viii) **The crossover operator.** Each pair of strings in the new population are considered in turn. If a random number in the range 0.0 to 1.0 exceeds the crossover probability p_c the crossover operator is not applied. Alternatively, a random crossover point is selected along the m-bit string (e.g. position 5). The digits from 6 to m of the 1st string are moved to the digit positions 6 to m of the 2nd string, while the corresponding digits of the 2nd string are moved to replace the end of the 1st string.
(ix) **The mutation operator.** If a random number exceeds the mutation probability, p_m the mutation operator is not applied to the string. Alternatively, the mutation operator is applied by randomly selecting a digit position along the string and then changing the digit to the opposite value.
(x) **Production of successive generations.** The new generation has now been produced using steps (vii) to (ix). The process is repeated to produce successive generations (e.g. 80 to 100 generations). The least cost strings (e.g. the best 20) are stored and updated as cheaper cost alternatives are generated.

Case Study

The sensitivity of the effectiveness of the genetic algorithm technique is investigated in optimising a network first proposed by Gessler (1985) and shown in Fig. 1. Table 1 gives the available pipe sizes, costs and costs of cleaning existing pipes. Eight choices are provided corresponding to a 3-bit binary string. The network is to be designed for 3 loading cases as shown in Fig. 1. The minimum allowable pressures at each node H_{min} are also given in Fig. 1.

Table 1 Available pipe sizes, costs and cleaning costs

DIAM (mm)	NEW PIPE COST($/m)	CLEANING COST ($/m)	DIAM (mm)	NEW PIPE COST($/m)	CLEANING COST($/m)
152	49.54	47.57	356	170.93	60.70
203	63.32	51.51	407	194.88	63.00
254	94.82	55.12	458	232.94	
305	132.87	58.07	509	264.10	

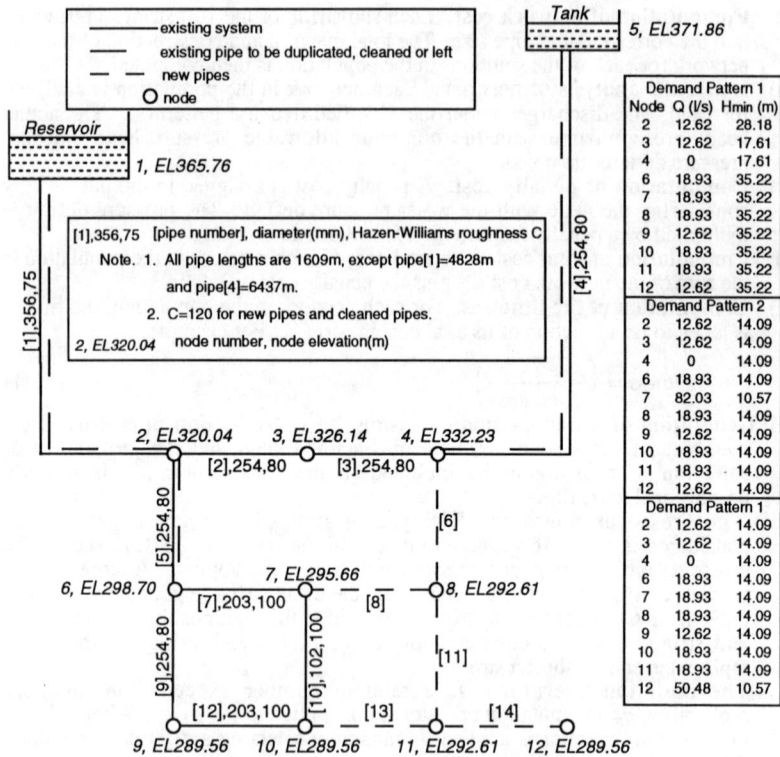

Figure 1 Layout of Gessler problem

Two equivalent optima have been previously identified by a complete enumeration (Simpson et al. 1992). These solutions are shown in Table 2.

Table 2. The two optimal solutions of the problem

No.	Total Cost ($m)	Pipe Selections (mm diameter)							
		pipe [1]	pipe [4]	pipe [5]	pipe [6]	pipe [8]	pipe [11]	pipe [13]	pipe [14]
1	1.7503	leave	356 dup	leave	305	203	203	152	254
2	1.7503	leave	356 dup	leave	305	203	254	152	203

Results

This study considered 5 different fitness functions for the genetic algorithm. The exponent n was allowed to take a constant value for the GA run in the range $n = 1$ to $n = 5$. The 5 fitness function formulations were tested over 8 different combinations of the GA parameters as shown in Table 3. The GA runs were allowed a maximum of 10,000 function evaluations and the runs utilised approximately 6 minutes cpu time on a Sun sparc computer. The results of the GA runs are summarised in Table 4.

Table 3. The GA parameter sets

Parameter	A	B	C	D	E	F	G	H
N	100	100	100	100	100	100	100	100
p_c	0.6	0.7	0.7	0.8	0.8	0.9	0.9	1.0
p_m	0.02	0.01	0.02	0.01	0.02	0.005	0.01	0.01

Table 4. Summary of the GA results

n	Lowest cost solution in $ million and Evaluation number achieved									Avg
	A	B	C	D	E	F	G	H	
1	1.800	1.848	1.848	1.861	1.868	1.839	1.842	1.800	1.838
	3900	7140	9380	2400	10000	5580	6210	9100	6714
2	1.750	1.773	1.750	1.829	1.750	1.750	1.750	1.812	1.771
	4200	8540	3080	4160	4080	6210	2970	5700	4868
3	1.750	1.843	1.750	1.839	1.800	1.839	1.750	1.750	1.790
	4800	3640	2870	8960	4240	4500	1170	2700	4110
4	1.750	1.750	1.750	1.830	1.819	1.812	1.750	1.750	1.777
	660	2730	490	7680	3200	1620	1620	1400	2425
5	1.750	1.839	1.750	1.839	1.750	1.819	1.750	1.750	1.781
	480	3920	910	560	2880	1440	1980	800	1621

The GA using a value of n = 1 did not achieve the optimal solution (= $1.750 million) for any of the runs within the allowed 10,000 evaluations (compared to a search space of >16 million). A value of n = 2 is superior achieving the optimal solution in 5 of the 8 GA runs. Values of n = 4 and n = 5 are equally successful. The averages in the last column of Table 4 suggests the quality of the solution for a value of n = 2 is marginally superior to that for the larger values of n at the expense of a slower convergence rate.

Conclusions

Genetic algorithms are an extremely powerful technique which are capable of finding the minimum cost network in relatively few hydraulic simulations compared to the size of the search space. In addition the genetic algorithm technique provides a number of near-optimal solutions which may be considered by the designer. This new technique is simple to implement and represents an opportunity for water supply authorities to potentially achieve large savings in the capital cost of provision of water supply infrastructure.

References

Dandy, G.C., Simpson, A.R., Murphy, L.J. (1993). "A Review of Pipe Network Optimisation Techniques" *Watercomp '93*, Melbourne, Australia, March.
Gessler, J. (1985). "Pipe Network Optimization by Enumeration". *Proc., Computer Applications for Water Resources*, ASCE, Buffalo, N.Y., 572-581
Murphy, L.J. and Simpson, A.R. (1992). *Pipe Optimisation Using Genetic Algorithms*. Research Report No. R93, Department of Civil Engineering, University of Adelaide, Australia, June, 53 pp.
Simpson, A.R., Dandy, G.C., Murphy, L.J., (1992). "Genetic Algorithms Compared to Other Techniques for Pipe Optimization", submitted to *ASCE, Journal of Water Resources Planning and Management*, July.

ALTERNATIVE DESIGNS FOR PERMANENT PROTECTION OF THE SAN LUIS REY RIVER AQUEDUCT CROSSINGS

Chenchayya T. Bathala, Ph.D., M.ASCE.[1], E. Morris McClung, M.ASCE.[1], Ergun Bakall, M.ASCE[2], W. Jeffery Moncrief, M.ASCE.[2]

INTRODUCTION

The San Diego County Water Authority was organized to provide an imported supply of water to the San Diego region and today imports approximately 90 percent of the total water demand within its service area. The sole source of water to the Authority is the Metropolitan Water District of Southern California (MWD). The Authority operates an aqueduct system composed of five pipelines known as the First and Second San Diego Aqueducts (the First with two pipelines, and the Second with three pipelines). Both aqueducts are buried underground and pass under the San Luis Rey River, between the communities of Pala and Bonsall in north San Diego County. Figure 1 shows the general location of the aqueducts.

A recent study concluded that these aqueducts are severely threatened by erosion caused during floods in the San Luis Rey River. The cause of erosion was primarily linked to the extensive sand mining operations located in the proximity of the aqueduct crossings, although natural river behavior also plays a role. In December 1991, the San Diego County Water Authority retained Parsons Brinckerhoff Gore & Storrie, Inc., to provide engineering services for the evaluation of river behavior, preparation of design alternatives for permanent protection of the aqueduct crossings, and to develop sand and gravel mining guidelines for regulation of mining operations along San Luis Rey River. This paper summarizes the preliminary design alternatives prepared for protection of the aqueduct crossings.

San Luis Rey River Basin

The San Luis Rey River originates in the Palomar and Hot Springs Mountains, both over 6000 feet in elevation and joins the Pacific Ocean near the city of Oceanside. The river extends for about 60 miles across northern San Diego County and drains an area of about 550 square miles. The watershed is mainly unimproved brushland, with forests at elevations over approximately 4000 feet. The upper 206 square miles of the drainage basin is intercepted by the Henshaw Dam built in 1912. Below the Henshaw Dam, the river flows through rocky canyons for several miles. The floodplain consists mainly of alluvium that is primarily sand, gravel, and silt. The channel width varies from approximately 50 feet to more than 500 feet at different locations in a random manner. The low flow channel has dense willow-type vegetation except within active sand mining operations. Historical

---

[1] Respectively, Project Manager and Deputy Project Manager, Parsons Brinckerhoff Gore & Storrie, Inc., 1230 Columbia Street, Suite 640, San Diego, CA 92102, Phone: (619) 338-9376, FAX: (619) 338-8123.

[2] Respectively, Chief Engineer and Senior Engineer, San Diego County Water Authority, 3211 Fifth Avenue, San Diego, CA 92103, Phone: (619) 297-3218, FAX: (619) 692-9356.

AQUEDUCT CROSSING PROTECTION

record of floods in the San Luis Rey River dates back to 1770. The flood of 1916 is considered to have the largest recorded peak flow in the San Luis Rey River, with an estimated peak discharge of 95,600 cfs at Oceanside. Since the construction of Henshaw Dam, downstream floods have been reduced substantially. During the 1977-1978 winter season, several floods occurred on the San Luis Rey River. Erosion and sedimentation occurred in many places, and caused the closing of dip section road crossings. The storm periods of January and February, 1980 were the most severe storms in the County since the great storms of 1916 and 1927. The February 1980 flood had an estimated peak discharge of 20,000 cfs at the Bonsall Bridge. The estimated 100-year peak discharge at this location is 48,000 cfs.

During the past two decades, numerous manmade changes have taken place within the San Luis Rey River Basin. State Highway 76 was built roughly parallel to the river from Oceanside to Lake Henshaw. Major river crossings include bridges for the Santa Fe Railway and Interstate 5 near Oceanside, State Highway 76 at Bonsall, Interstate 15 and old State Highway 395 west of Pala, West Lilac Road, Couser Canyon Road and County Road S6 near Rincon. Considerable residential and commercial buildings have been built along the river near the communities of Bonsall and San Luis Rey east of Oceanside. There are many sand and gravel operations along the river between Pala and Oceanside. In 1969, H. G. Fenton Materials Company was issued the first Major Use Permit for a sand mining operation in the San Luis Rey River. Over time, the number of mining operations peaked at thirteen as development in San Diego County increased. Currently, five operators have County permits to mine in the river.

STATUS OF EXISTING AQUEDUCT CROSSINGS

The SDCWA aqueducts cross the San Luis Rey River at two locations (Figure 1). The First San Diego Aqueduct consists of two 48-inch diameter high pressure pipelines crosses at approximately river mile (RM) 20.8. The Second San Diego Aqueduct which consists of three high pressure pipelines (diameters: 72", 90", and 96"), cross at approximately RM 16.8. At the First Aqueduct crossing, the pipelines are presently about 10 feet below the channel invert. There has been about eight feet of degradation since the pipelines were installed in 1948 and 1954. There are no protective facilities at this location. At the Second Aqueduct crossing, the pipelines are presently approximately five feet below the channel invert. These pipelines were built in 1957 (72"), 1971 (90") and 1983 (96"). This location has experienced approximately eight feet of degradation between 1980 and 1990. A temporary steel sheet piling and rip-rap grade control structure was constructed in 1990 after a flood exposed the westerly pipeline. The alluvium is approximately 100 and 150 feet deep at the aqueduct crossings respectively.

PROTECTIVE STRUCTURES

Several investigations were conducted to arrive at the most appropriate design alternatives for protection of the aqueduct crossings. These include: review of previous study reports, project site inspection, reconnaissance of the entire river including sand mining sites and existing bridges and utility crossings, aerial photogrammetric surveys, interviews with sand mining operators and fluvial modeling studies. A few protective structures built by the U.S. Army Corps of Engineers were also examined. These investigations clearly indicated that the two aqueduct crossings have experienced and will continue to be subjected to major river actions which include channel degradation and local scour. Consequently, it is evident that the pipelines should be protected from further exposure and potential failure during floods.

Proposed Alternatives: A number of alternatives were identified and reviewed. Examples of the various alternatives utilized by other agencies were researched and information gathered. The initial selection was based on a number of factors: 1) Similar structures with a good operating history used by other agencies in similar settings; 2) Reasonable availability of required construction materials; 3) Capability of the structures to handle channel degradation, aggregation, meandering and local scour in an effective manner; and 4) Hydraulic and structural capability to withstand the 100-year design flood and a random sequence of various floods of different magnitude.

From this information, the following five design alternatives were selected for more detailed evaluation:

1) Reinforced concrete drop structure
2) Grouted stone drop structure
3) Elevated pipelines
4) Deeper buried pipelines
5) Precast concrete mat

Alternative 1 is a reinforced concrete drop structure with a weir length of 500 feet. Figure 2 shows an architectural rendering of this alternative. The prototype of this structure was developed by the U.S. Army Corps of Engineers (Los Angeles District) through model tests and has been installed extensively in the Santa Ana River channel (Orange County, California). Alternative 2 involves a grouted stone drop structure with a trapezoidal cross section and a broad crested weir of length 500 feet. The U.S. Army Corps of Engineers, Los Angeles District, built a series of these drop structures along the San Gabriel River near Pasadena. Alternative 3 involves supporting the pipelines on piers above the 100-year flood flow elevation. Alternative 4 consists of burying the pipelines deeper in the river bed, below the potential maximum scour depth. Alternative 5 includes covering the pipelines with cable-tied precast concrete blocks with sufficient anchors on all sides.

Fluvial Modeling: The hydraulic requirements of the above five design alternatives were evaluated by using the FLUVIAL-12 model (Chang, 1993). The stream cross sections data were obtained from an aerial survey and detailed contour maps prepared specifically for this project. Existing sand mining pits data and ultimate mining depths for Fenton and J.B. Unlimited operations were input into the model. A sequence of random flood events occurring in a series such as 10, 30, 20, 40, 15, 100, 20, 15, 70 and 10-year floods, was utilized to evaluate the degradation and aggregation patterns of the river. Local scour around the proposed structures was calculated based on established formulas (Richardson, et al, 1991). The following preliminary results were obtained from the fluvial modeling study:

- Without protective structures, the maximum general scour at the First Aqueduct would be about 15 feet below the top of the pipeline.

- At the Second Aqueduct, the maximum scour would be about 6 feet below the top of the pipeline, if no protective structure is built.

- The maximum general scour downstream of the proposed drop structures (Alternatives 1 and 2) is expected to be 23 feet and 13 feet, respectively, for the First and Second Aqueducts. Additional 10 feet of local scour should be expected due to flow contraction and turbulence downstream of the structures.

- The proposed weir length of 500 feet is satisfactory for both Alternatives 1 and 2. The maximum crest velocity would be 12.3 ft/sec.

For Alternative 3, the total foundation depth for piers should exceed 36 feet at First Aqueduct and 23 feet at Second Aqueduct crossings.

- Based on maximum general scour and potential local scour in the area, the pipelines should be buried (Alternative 4) at depths deeper than 31 feet and 19 feet at First and Second Aqueduct crossings respectively.
- The toe of the concrete mat (Alternative 5) at First Aqueduct crossing should go deeper than 30 feet. At the Second Aqueduct crossing, the toe should be buried deeper than 20 feet.

EVALUATION CRITERIA AND FINAL SELECTION

A systematic evaluation criteria including engineering considerations, overall cost, constructability, environmental impacts, geotechnical considerations, public perception and aesthetics were established. A matrix method consisting of weighing and ranking each alternative based on the evaluation criteria was used. The selection of the preferred design alternative has not yet been finalized.

REFERENCES

Richardson, E. V., Harrison, L. J., and Davis, S. R. (1991). Evaluating Scour at Bridges, Hydraulic Engineering Circular No. 18, Federal Highway Administration, February.

Chang, H. H. (1993) "Preliminary Design Report for Permanent Protection of the San Luis Rey River Aqueduct Crossings," prepared for Parsons Brinckerhoff, Gore & Storrie, Inc. and San Diego County Water Authority.

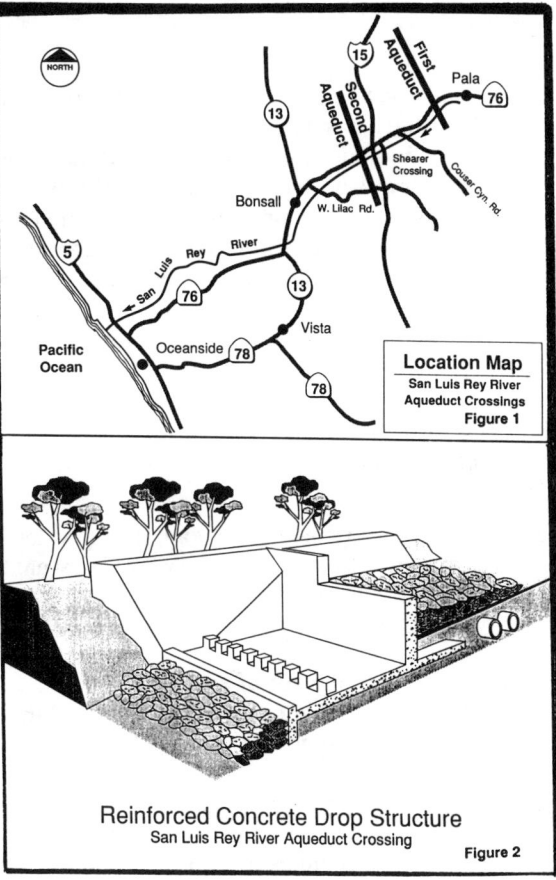

Location Map
San Luis Rey River
Aqueduct Crossings
Figure 1

Reinforced Concrete Drop Structure
San Luis Rey River Aqueduct Crossing
Figure 2

Optimal Rehabilitation Model for Water Distribution Systems

Joong Hoon Kim[1] and Larry W. Mays[2], M. ASCE

Introduction

The objective of this paper is to describe a new methodology which can select the pipes to be rehabilitated and/or replaced in an existing water distribution system and determine how much the pumping capacities should be increased so that the water demand and pressure requirements at all demand nodes are satisfied while the total cost is the minimum. Five cost functions are considered: pipe replacement cost, pipe rehabilitation cost, pipe repair cost, and pumping cost. The model considers the trade-off among decisions: replace pipes, reline pipes, increase pumping power, and leave as is. The decision is made for each pipe.

Model Formulation

$$\text{Minimize (total cost)} \qquad (1)$$

subject to

(1) <u>Demand requirement</u>: The amount of water being supplied to each demand node should be greater than or equal to the required demand, expressed as

$$\sum_{j \in I_i} q_j \geq Q_i \qquad \text{for all demand nodes i} \qquad (2)$$

in which I_i is the set of pipes connected to demand node i, q_j is a continuous variable representing the resulting flow rate in pipe j, and Q_i is the required nodal demand at node i.

---

[1]Assistant Professor, Department of Civil and Environmental Engineering, Korea University, Seoul, South Korea.
[2]Chair and Professor, Department of Civil Engineering, Arizona State University, Tempe, Arizona 85287.

(2) Pressure head requirement: At each demand node of the water distribution system, the pressure head being supplied at the node should be greater than or equal to the minimum pressure head required by the customers. i.e.,

$$\underline{H}_i \leq h_i \leq \overline{H}_i \qquad \text{for all demand nodes i} \qquad (3)$$

where h_i is a continuous variable representing the resulting nodal pressure head at node i and H_i is the required nodal pressure head at node i.

(3) Conservation of energy constraints: For each primary loop, i.e., an independent closed path in a water distribution system, if J_S and K_S represent the sets of pipes and pumps in primary loop S respectively, then the energy conservation equation can be written for pipe sections in the loop as follows:

$$\sum_{j \in J_s} \Delta h_j = \sum_{k \in K_s} PH_k \qquad \text{for all } J_S, K_S \qquad (4)$$

in which Δh_j is a continuous variable representing the resulting head loss in pipe j and PH_k is a continuous variable representing the pumping head of pump k. In case of no pump in the loop, the energy equation (4) states that the sum of the energy losses around the loop equals zero. If there are F fixed grade nodes, F-1 independent energy conservation equations can be written for the pipe sections along the path between any two fixed grade nodes as follows:

$$\Delta E_R = \sum_{j \in J_R} \Delta h_j - \sum_{k \in K_R} PH_k \qquad \text{for all } J_R, K_R \qquad (5)$$

where ΔE_R is the difference in total grade between the two fixed grade nodes, and J_R and K_R represent the sets of pipes and pumps in path R connecting the two fixed grade nodes, respectively. Eq. (4) is a special case of Eq. (5) where the difference in total grade (ΔE) is zero for a path which forms a closed loop.

(4) Decision constraints: These constraints are applied to the model to eliminate the possibility of simultaneously replacing and rehabilitating the same pipe.

$$N_j + R_j \leq 1 \qquad \text{for all pipe j} \qquad (6)$$

in which N_j is a binary variable with with either a value of one representing pipe j is replaced or zero otherwise, and R_j is a binary variable with a value of one representing that pipe j is rehabilitated or zero otherwise.

(5) Hydraulic constraints: The Hazen-Williams equation for each pipe in a water distribution system is given by

$$q_j = \frac{AC_j \; AD_j^{2.63} \; \Delta h_j^{0.54}}{(4.73 \, L_j)^{0.54}} \qquad \text{for all pipe j} \qquad (7)$$

in English unit where AD_j is a continuous variable representing the actual diameter of pipe j after the decision is made and L_j is the length of the pipe j. The actual roughness coefficient for pipe j after decision, AC_j, is given by the following equation:

$$AC_j = CO_j [1 - (N_j + R_j)] + CR_j R_j + CN_j N_j \quad \text{for all pipe j} \quad (8)$$

in which CO_j is the Hazen-Williams roughness coefficient in old pipe j, CR_j is the roughness coefficient in the rehabilitated pipe j, and CN_j is the roughness coefficient in the replaced pipe j. Eq. (8) states that, if N_j or $R_j = 1$, then AC_j equals CN_j or CR_j respectively, otherwise, $AC_j = CO_j$. In other words, if either a replacement or a rehabilitation is performed for pipe j, then the roughness of pipe j takes value from the replaced pipe or the relined value. Otherwise, the value of the roughness is the same as that of the existing old pipe. Similarly, the value of the actual pipe diameter after decision, AD_j, is given by the following equation:

$$AD_j = DO_j (1 - N_j) + DN_j N_j \quad \text{for all pipe j} \quad (9)$$

with a bound constraint

$$DN_j \geq 0 \quad \text{for all pipe j} \quad (10)$$

where DO_j is the diameter of old pipe j and DN_j is a continuous variable representing diameter of the replaced pipe j.

(6) <u>Pump characteristic constraints</u>: For each pump in a water distribution system, the following relation must be satisfied:

$$HP_k = \frac{\gamma \, q_k \, PH_k}{550} \quad \text{for all pump k} \quad (11)$$

where HP_k is a continuous variable representing the useful pumping power of pump k, γ is the specific weight of water and q_k is the flow rate through pump k. Another constraint for pumps is that the constant horsepower, HP_k, can not be smaller than the existing constant horsepower, $\underline{HP}_k$. This can be expressed as

$$HP_k \geq \underline{HP}_k \quad \text{for all pump k} \quad (12)$$

which gives a lower bound on the decision variable HP_k.

(7) <u>Integer (binary) variable characteristics</u>:

$$N_j = [0, 1]; \quad R_j = [0, 1] \quad \text{for all pipe j} \quad (13)$$

SOLUTION PROCEDURE

The proposed model formulation is a mixed-integer nonlinear programming problem. Kim (1992) developed an implicit enumeration algorithm to find the optimal combination for the 0-1 integer variables which represent the optimal rehabilitation plan for the existing water distribution system. Figure 1 shows the incorporation of the program solvers. It shows the connection between the branch and bound (integer) master problem and the nonlinear subproblem. The proposed algorithm solves a nonlinear programming subproblem for each branch node in the enumeration procedure. The integer master problem provides fixed values of the binary variables to the nonlinear subproblem, and the nonlinear subproblem provides the optimal objective values to the master problem. Figure 1 also shows the interfacing of the hydraulic simulation model, KYPIPE by Wood (1980), and the nonlinear solver, GRG2 by Lasdon (1983), in an optimal control framework. Based on a set of initial values for the continuous control variables, DN_j and HP_k, the simulation model is executed and the resulting values of nodal pressure heads, h_i, are transferred back to the nonlinear model.

The mathematical approach which uses the concept of the implicit enumeration algorithm works fine and finds an optimal solution; however, global optimality cannot be guaranteed.

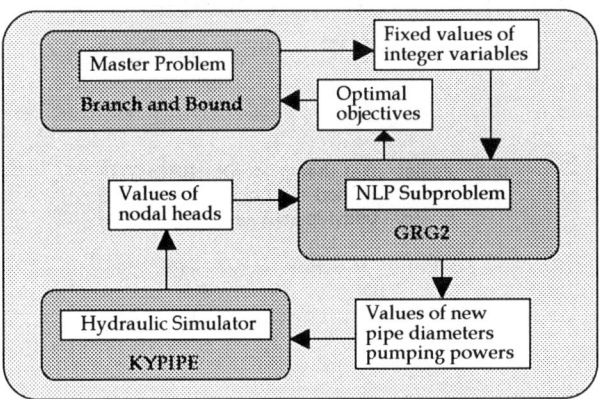

Figure 1   Incorporation of the Program Solvers

APPENDIX. REFERENCES

Kim, J. H., *Optimal Rehabilitation/Replacement Model for Water Distribution Systems*, Ph.D. Dissertation, The University of Texas at Austin, Austin, Texas, December, 1992.

Lasdon, L. S., Warren, A. D., and Ratner, M. S., *GRG2 User's Guide*, Department of General Business, The University of Texas at Austin, Austin, Texas, November, 1983.

Wood, D. J., *Computer Analysis of Flow in Pipe Networks Including Extended Period Simulations*, University of Kentucky, Lexington, Kentucky, September, 1980.

NON CONTACT, REMOTE SENSING
OF
BURIED WATER PIPELINE LEAKS
USING INFRARED THERMOGRAPHY

Gary J. Weil, PE., CPIM.

EnTech Engineering, Inc.
111 Marine Lane
St. Louis, Missouri, U.S.A. 63146
Tel: (314) 434-5255
Fax: (314) 434-3270

ABSTRACT

Subsurface water pipelines, sewer and buried utilities throughout America and the rest of the world range in age from hundreds of years old to brand new. They all have, however, one fact in common: They eventually develop structural failures. Representatives in countries such as Italy and India estimate that they lose 60% to 90% of their fresh water supplies through leaks in their pipeline distribution systems.

This paper describes how a non-contact, non-destructive, remote sensing technique, **Computer Enhanced Infrared Thermography**, may be used to detect both pipeline leaks and erosion voids caused by these leaks, while they are still small enough to be repaired inexpensively, and before they can cause catastrophic system failures. This technology may be used from mobil vehicles, helicopters or man-portable systems and is able to cover several miles or hundreds of miles of pipeline per day. This proven, but relatively unknown technology, will be described in theory, by procedure and by use of case studies based upon successful projects performed throughout the world during the last 10 years.

1. INTRODUCTION

Most pipelines, whether they contain oil, chemicals, water, steam, gas or sewage, have a design life of 20 to 25 years. When they begin to fail, they do so slowly at first through corrosion and small cracks, and gradually progress to a catastrophic ending. This disastrous failure can be expensive in terms of both dollars and lives. But this scenario can be avoided. Planned maintenance can extend the life of all types of pipelines almost indefinitely and regular testing of pipelines form the basis of economically viable restoration. In order for any testing technique to be widespread, it must have the following qualities:

1. It must be accurate.
2. It must be non-contact and nondestructive.
3. It must be able to inspect large areas as well as localized areas.
4. It must be economical.
5. It must not inconvenience the pipeline's users.

One technique for testing in-place, buried pipeline has emerged during the last 10 years that fulfills these requirements. This technique is called **Infrared Thermographic pipeline testing.** During its gestation period, it was used to test pipelines carrying oil, chemicals, natural gas, steam, water and sewage. It has shown itself to be both accurate and efficient in locating subsurface pipeline leaks, voids caused by erosion, deteriorated pipeline insulation and poor backfill, thereby coming into full acceptance as a viable NDT technology.

2. TESTING EQUIPMENT

In order to test areas for pipeline leaks, subsurface voids, and other types of anomalies, surface temperatures must be measured, recorded and analyzed. During the analysis portion, the thermal conductivity and heat capacity of the leaking fluid or air gap in the void will be compared to the same properties in the surrounding soil materials. Since, in even the smallest test area, thousands of readings would have to be made simultaneously in order to outline the anomaly precisely, it is recommended that a high resolution infrared thermographic scanner be used. (See Figures 1 & 2.) This type of equipment allows entire areas to be scanned and the resulting data to be displayed as pictures with areas of differing temperatures designated by differing gray tones on a black & white image or by various colors on a color image. A wide variety of auxiliary equipment can be used to facilitate the data recording, referencing and interpretation.

Figure 1. Infrared scanner mounted on a man-portable system

Figure 2. Infrared scanner mounted on a van for increased mobility

3. CASE HISTORIES

3.1 Buried water pipeline leak detection

In 1983, EnTech Engineering, Inc. performed its second subsurface thermographic leak and void detection on Duncan Street in midtown St. Louis. Prior to the inspection, crews from the Metropolitan St. Louis Sewer District had observed street pavement sinking up to 6 inches along a 600 foot long section of Duncan Street. In-sewer visual inspections using both television cameras and crawl crews had located only 3 dime sized water infiltration points in the 5 foot diameter sewer located approximately 13 feet below the surface. Running alongside the sewer was a pressurized water line.

During the thermographic inspection a cool area was located perpendicular to the buried water pipe. It began at a point beginning at the water line and spreading outward toward the sewer line. It was determined that the cooler surface area was caused by the heat sinking ability of the water plume as it spread out from the water line leak and flowed down the outside of the nearby sewer pipeline. Some of the fresh water was entering the sewer line through the 3 dime sized holes that the crawl crew had located.

In addition to the water leak, the infrared thermographic inspection also located an erosion area above the water line. Evidently the water flowing from the water pipeline to the sewer pipeline was carrying soil which was washing away down the sewer line. This void area had caused some of the pavement sinking and further street collapse was inevitable. The void above the water line was evidenced by a warmer signature in the thermographic image. (See Figures 3A and 3B.)

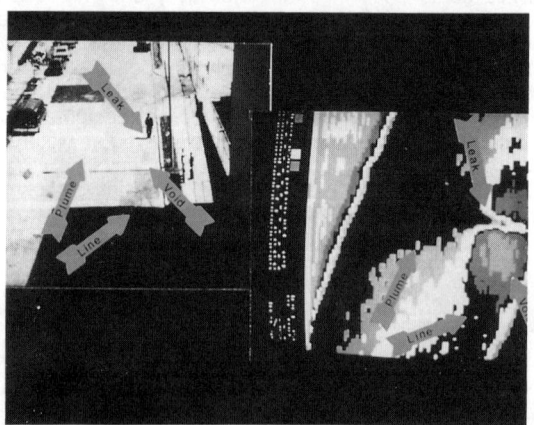

Figures 3A & 3B.   Visual photograph and corresponding thermogram showing subsurface water pipeline, water leak, leak plume and void forming above pipeline

4. CONCLUSIONS

1. Infrared thermographic techniques can be used to detect buried pipeline defects such as leaks, cracks and subsurface voids.

2. Infrared thermographic testing techniques are considered nondestructive.

3. Infrared thermographic testing may be performed during both day and nighttime hours depending upon environmental conditions and what results are desired.

4. Computer analysis of thermal images greatly improves the accuracy and speed of test interpretations.

5. Computer analysis of pipeline thermographic data can improve the ability to set repair priorities for areas in a state of change.

6. Aging water pipeline infrastructure throughout the world is rapidly approaching the end of its design life. This will necessitate more efficient and cost effective methods of testing pipelines under load and in place. Infrared thermography is a nondestructive, remote sensing technique which meets these requirements.

5. REFERENCES

Much of the technical information contained in this article was obtained from the U.S. and worldwide patents and patents pending held by its author, Gary J. Weil and EnTech Engineering, Inc. It is suggested that anyone attempting to use these technologies consult with the patent holders in order to make the most accurate use of this technology.

U.S. and worldwide patent rights will be defended in all cases of unauthorized infringement. Regular U.S. Patent off.

Watershed Management at Three Governmental Levels

Peter E. Black[1]

Abstract

Examples of holistic water resources management at the federal, state, and local levels of government illustrate practical application of basic principles of watershed management and public administration. The examples show evolution *down* to the operational watershed management for a federal agency; *up* for local management, and some fancy footwork where needed in between.

Introduction

Considering the ancient and historical link between medicine on the one hand and forestry and related wildland resources management on the other, it is quite appropriate to apply the word "holistic" to wildland resources management. Watershed management is *the planned manipulation of one or more factors of the environment of a natural drainage so as to effect a desired change in or maintain a desired condition of the water resource.* Described are three situations at the federal, state, and local levels with which I have contact as member of an advisory organization.

Corps of Engineers

The newest mission of the Corps is Environmental Protection and Restoration, and is being applied to wetlands along the Mississippi River where rehabilitation of the aging and undersized locks is taking place. The construction provides the opportunity for re-creation of natural wetland areas approximating those that existed when the nation's settlement reached the river. The St. Louis District of the Corps entrusted to a wildlife

---

1. Professor of Water and Related Land Resources, SUNY College of Environmental Science and Forestry, Syracuse, NY 13210.

biologist the task of creating the ecological framework for the restored area, and the more difficult task of communicating those plans to a large number of groups and individuals. Plans were developed and implemented on the 1200-acre Riverlands Environmental Demonstration Area, where a Prairie-Marsh Complex ecosystem has been restored. It is a part of larger plans and programs for riverine wetlands on the Missouri and Mississippi Rivers. From school kids and citizens' groups to federal agencies and nation-wide transportation companies, people and groups are creatively involved in all aspects of this operation. [Perhaps the most important point concerning wetland management (including restoration) is that the hydrology of wetlands is not well known. This is ramified in the fact that they are, indeed, wetlands: that is, drainage is poor, and watershed boundaries are often ill-defined. In fact, wetland buffer zones — areas that contribute directly and immediately to the wetland — may be the most significant application of watershed concepts.

Could this project have been accomplished at a higher or lower operational/organizational level? The answer is "No!" The ecological scale of operation at the practical watershed limits of the site provided a viable scope. At a higher level, the communication necessary to accomplish the agreed-upon objectives would have been diluted, there would be closer lands perceived to be in need of Corps attention, and energy is more appropriately focused on issues of broader scope. At a lower level, the groups involved cannot meaningfully communicate where there are no appropriate, formal linkages between organizations. Individuals belonging to several different groups may have effective input in each, but the groups cannot establish and maintain interaction. The Corps has met this need by establishing "partnerships"; the process is called "partnering". The Corps used it effectively to involve schools, citizens, interest groups, and local government. Partnering conducted thorough consideration in dynamic, facilitated workshops where all legitimate participants are present and active, can fulfill the same objectives of EIAs and EISs without the confrontational attributes that characterize typically adversarial EIS hearings. With partnering, effective planning and management may be executed at the lowest administrative level, and all legitimately interested groups can participate.

Onondaga Lake Management Conference

The goal of this limited-life entity is to establish a management plan for the lake. The OLMC is organized into a Technical Review Committee made up of representatives from the six official organizations and more than 30 other government organizations, industries, lake-oriented, and numerous local chapter or section interest groups that are associated with larger, often national organizations which were involved in efforts to clean up the central New York lake. Individually these organizations were incapable of achieving the cleanup goal. The organizations were frustrated by lack of coordination, communication, and the ability to get anything accomplished because no one organization had the necessary legal authority, financial resources, or political clout. Here is an opportunity for application of watershed management principles to practical problems such as nonpoint source pollution from urban and rural runoff.

Could the scope of effective lake water quality management have been accomplished on some level other than the lake's watershed? Again, the answer is "No!" No organizational progress of any consequence had been made before creation of the highly-structured, watershed-oriented OLMC. Analysis of the social science concerns reveals that small, ineffective local-government interests had inadequate financial resources or economic power; national organizations had no authority; and regional interests had insufficient political clout.

NYS Soil and Water Conservation Committee

The New York State Soil and Water Conservation Committee (S&WCC) advises two state Departments on soil and water conservation policy. Historically the scope has been mainly agricultural. However, owing to operational connections with (county) soil and water conservation districts, any soil and water concern was addressable, including urban runoff, forestry, recreation, and related land uses.

In 1990, The New York City Department of Environmental Protection (DEP) sought to unilaterally upgrade its long-standing Watershed Rules and Regulations governing land use in the Catskill watersheds from which the City gets most of its water supply. The proposed rules threatened continued profitable farmland use in the Catskills. The action was in response to EPA pressure on the city to filter the water supply and because the watershed rules and regulations had not been changed since 1922; much new information about land and water management practices was available. At the time, a state-wide Joint Legislative Commission on Rural Resources ran a facilitated workshop for all interested groups and individuals at a Catskill retreat. The DEP action stopped the Commission's efforts cold. Reactive opposition by disgruntled landowners suggested the need for coordinated technical assistance from the districts. As the districts' policy coordinator, the S&WCC was called upon for oversight as well as for moral support.

To make a very long story very short, DEP withdrew its proposed rules when the S&WCC and the DEP successfully negotiated a creative watershed management approach with the DEP. The agreement provides a program that will include development of demonstration farms, education and training manuals, and procedural citizen involvement education programs. Protection of riparian zones and development and implementation of BMPs are in the context of the watershed. The approach to water quality control includes pollution prevention, collective responsibility, and watershed planning and management.

Could the city's water quality problems on the Catskill watersheds be accomplished by state authority? Once again, the answer is "No!". Nor could it have been effectively executed by the Rural Resources Commission or peacefully accomplished by the DEP. By focusing on the soil and water conservation districts, attention can and is being directed to the natural Catskills watersheds.

CONCLUSIONS

In the situation of the Corps of Engineers, the operational level came *down* from the federal to the Corps district level where management could be focused on the managed wetland watersheds. Even though created by Congress, effective operation of the local Onondaga Lake Management Conference evolved *up* from numerous ineffective local entities to a regional coordinative organization that could achieve the desired results at a technically sound watershed level. For the NYS Soil and Water Conservation Committee, development of a viable organizational infrastructure was carefully and uniquely *crafted* from county, state, and off-site city interests by complex and delicate political arrangements in order to support watershed-based consideration of the problem. In each case, the capable organization is one that has the legal authority, financial resources, and political clout to efficiently effect and focus policies at the lowest possible organizational level that is accessible to all legitimately-interested publics.

Obviously, watershed definition is a sound way to begin resolution of a water resource problem that is legitimately of interest to numerous groups and individuals. Watershed management is a technically sound and politically viable approach that can be implemented flexibly in a wide variety of situations.

A Watershed Approach to Water Quality Criteria

Scott J. Kenner[1]

Abstract

Water quality has been managed historically using water quality criteria implemented through the NPDES permitting process. Many water bodies have water quality better than that defined by beneficial use criteria while at the same time some parameters naturally exceed defined criteria. Antidegradation policy provides a tool for establishing criteria that reflect ambient conditions in the watershed. An approach to development of water quality criteria that reflect the ambient watershed conditions has been applied to the Suwannee River basin. The criteria developed reflect both time and spatial variability within the basin.

Water Quality Standards and Criteria

The policies established by the state regulations assure that minimum water quality protection for all Florida waters is the water quality necessary to maintain the existing beneficial use classification, Class III, for the Suwannee River. The Suwannee River is also classified as an Outstanding Florida Water (OFW). Additional protection is established for surface waters through antidegradation water quality policy and standards. The antidegradation standard is implemented through the permitting process. More importantly, it establishes March 1978 through February 1979 as the baseline time period for defining the existing ambient water quality for OFW's.

The definition for existing ambient water quality provides the impetus for evaluating available surface water quality and quantity data in a probabalistic framework. The variability of the ambient water quality is based on both

---

[1]Asst. Prof., Dept. Env. Eng. Sci., Univ. of Florida, Gainesville, FL, 32611-2013

systematic (deterministic) and stochastic (random) components. Both of these components are evaluated to define the natural variability of the ambient water quality over time and space.

Database

The database is the essential foundation for development of the ambient water quality criteria. The established surface water quality database for the Suwannee River represents data gathered by seven different agencies at 24 locations along the river. These 24 locations consist of 78 significant monitoring station data sets with periods of record ranging from two years up to 36 years and from 23 to 150 different parameters being measured. An additional component of the database is continuous flow monitoring records. There are eight flow stations on the Suwannee River and correlate to water quality stations. In most cases, the flow data period of record overlaps the water quality period of record and provides for analysis of relationships between concentration and flow.

Comparison with Class III Criteria

As a first step to assessing water quality in the Suwannee River, available water quality data was compared to numeric Class III criteria. This comparison demonstrates how constant water quality criterion do not represent the natural variability of ambient water quality and indicates which parameters merit further evaluation.

The comparisons to Class III criteria can be placed into four groups. The first group consists of parameters that have not historically exceeded the current Class III criteria and include specific conductance, aluminum, arsenic, fluoride, and nickel.

Alkalinity (below White Springs) and dissolved oxygen are group two parameters characterized by mean concentrations (based on the historical data base) that do not exceed Class III criteria, but variability (natural or anthropogenic) causes a fraction of the measurements to exceed the criteria.

The third group of parameters have historical means that exceed Class III criteria. In this situation, improving the ambient water quality condition might be appropriate where the lower water quality is due to anthropogenic impacts. Fecal and total coliform are group three parameters whose concentrations are due, at least in part, to anthropogenic impacts. The parameter that represents this group is alkalinity above White Springs.

All of the metals evaluated (copper, iron, lead, mercury, and zinc) fall into group two or three. The analysis presented here does not provide enough information to identify whether or not the exceedences are due to natural conditions or anthropogenic impacts.

The fourth group are parameters where the current Class III criteria are narrative and are typically stated as some percentage above the background concentration. The parameters that fall into this group are pH, BOD, biological integrity, chlorides, transparency, turbidity, and the nutrients nitrogen and phosphorus.

Data Preparation and Analysis

The objective of the data preparation and analysis is to identify natural cycles, concentration and flow relationships, and water quality trends that must be used to numerically define the ambient water quality. The four major steps in the analysis are; 1) data aggregation, 2) concentration versus flow regression analysis, 3) evaluation of seasonality, and 4) water quality trend analysis. The primary statistical methods used in the analysis are nonparametric. The analysis was applied to specific conductance, total phosphorus, and total nitrogen at eight locations along the river.

Criteria Development

The water quality data, adjusted for trends, are numerically represented by developing the cumulative percentile distributions for ambient conditions and deseasonalized ambient conditions. From these distributions any percentile of interest can be identified and used to represent the ambient water quality. Development of these distributions at several locations along the river provides a description of the watershed characteristics reflected by the water quality. Figures 1, 2, and 3 show the resulting percentile profiles along the river, for specific conductance, total phosphorus, and total nitrogen. The increase in specific conductance below represents the natural contribution of flow from springs. The high total phosphorus percentiles reflect mining impacts in the basin.

The 95th percentile concentration is often used to represent water quality based criteria which focuses primarily on extreme exceedences of water quality. While maintaining extreme exceedences with the 95th percentile, the average or median concentrations can increase or decrease significantly. In water quality management we are concerned with long-term average concentrations of nutrients. Thus, the intermediate percentiles must also be monitored to evaluate changes in average conditions.

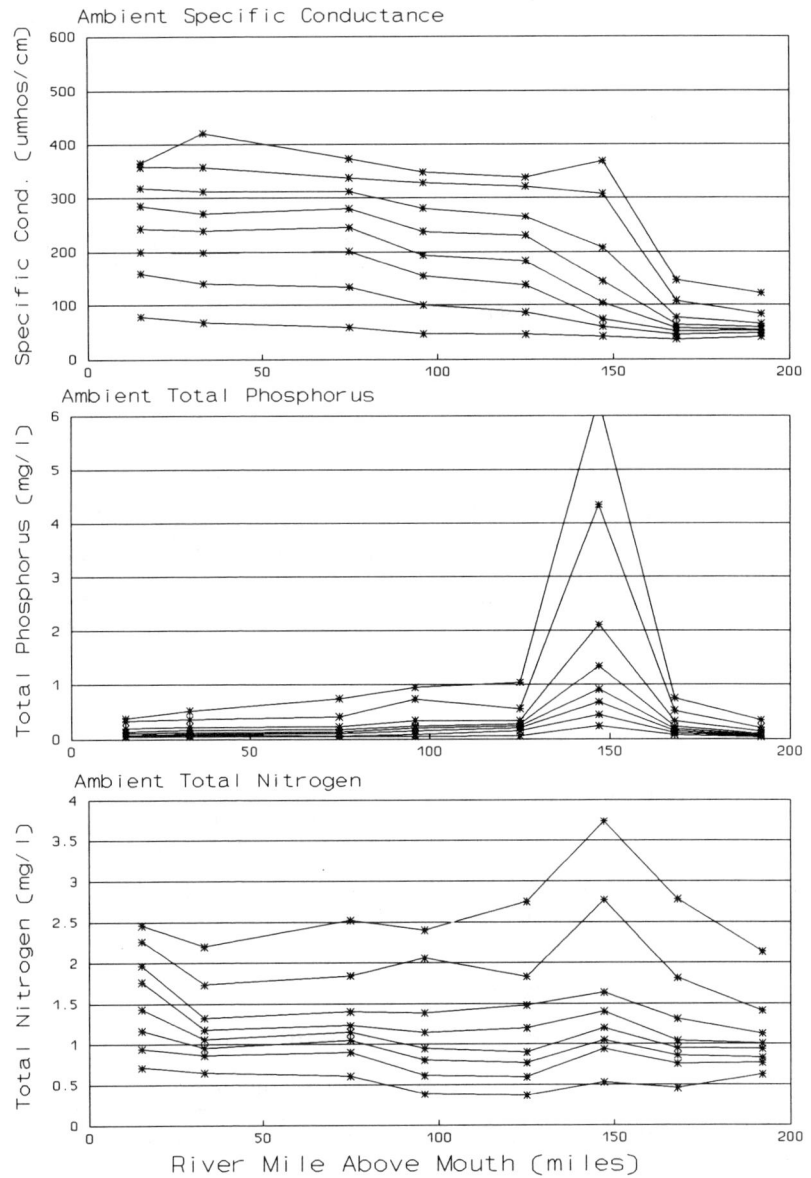

Figure 1. Ambient percentile profiles for SC, TP, and TN representing the 99th, 95th, 80th, 65th, 50th, 35th, 20th, and 5th percentiles from top to bottom.

The Impact of Global Warming on Water Resources:
Overview and Implications for Coastal California

by Jeffrey Garvey[1], Member, ASCE

Abstract

In the past two years, impressive evidence has accumulated that global warming is occurring and is related to the increase in greenhouse gasses. The impact of the warming process on water resources nevertheless remains difficult to identify. One potential approach, which can be applied to California coastal basins, is to analogize from present geographical variation. Impacts projected on this basis range from major along the northern coast to insignificant in the south.

Introduction

Global warming is a popular and controversial topic. This paper reviews indications of present warming trends and their causes, discusses future projections, and examines how resulting effects on water resources can be estimated. One example is presented, relating to projected impacts on coastal drainage basins in California.

Is the world getting warmer?

In the past two years, considerable evidence has accumulated relative to global temperature trends. Much of this evidence consists of improved statistical analysis of actual temperature records (Schonwiese 1992). These analyses show an average global warming of

---

[1]Principal Engineer, Boyle Engineering Corporation, 7807 Convoy Court, San Diego, CA 92111

about 0.5°C in the last 100 years, but the trend has not been seasonally or geographically uniform. North America and Europe, for example, show generally no change or even cooling trends in the winter, while north polar and Pacific Ocean temperatures are markedly increased, especially during the summer. Based on temperature statistics alone, the probability of warming "being real" is greater than 98 percent (Kellogg 1991).

In addition to the direct temperature record, there is also now confirming data from many other fields, including sea level studies, ice season records, and ecological zone boundary changes. Taken as a whole, the evidence of recent worldwide warming is overwhelmingly supportive.

Is it caused by us, or just natural variation?

Of course, the earth's climate is subject to extensive natural variation. In fact, there is even considerable evidence of past major variation on a global scale. An event as extreme as a 5°C worldwide warming occurred naturally during the last deglaciation, for example. What is the evidence, then, that the present warming is not just another example of extensive natural variation? Primary indications are:

- The present rate of average global warming (0.5°C in 100 years) is several times greater than even the extreme average warming rate experienced during the last de-glaciation.

- The present warming trend tracks closely and is statistically related to the gradual increase in CO_2 levels (from 280 ppm to 350 ppm) in the past 100 years. (Schonwiese 1992).

- The present warming trend is broadly consistent with the predictions of climate models, when they are run for "existing conditions." (Kellogg 1991).

The preponderance of the evidence therefore rejects the hypothesis of natural variation, and indicates that the present warming trend is principally induced by increased greenhouse gas emissions.

What are projections for the future?

Indications are that atmospheric carbon dioxide levels will continue to increase, doubling from existing levels by the year 2050 if present trends continue. The implications of this are projected with computerized climate simulation models. What do these models project if a CO_2 doubling occurs?

Primarily, projections are that average global temperature would increase from 1.5°C to 4.5°C, with 3.4°C being a working "best guess" number (Wigley 1992). Precipitation values are also projected by these models, but with much less accuracy: precipitation calibrations have errors ranging from eight percent to 1500 percent (Kuhl 1992). The combined effects on temperature and precipitation of a warmer "$2xCO_2$" scenario are therefore not well quantified.

Then how can water resource impacts be estimated?

There are basically four methods of estimating the hydrological impacts of climate change:

1. Use model results directly. There is some evidence that this may be satisfactory (±20 percent) for certain large watersheds. A study of high-latitude Asian rivers has been done on this basis, for example. (Miller 1992).

2. Model temperature and precipitation outputs can be used as inputs to regional or local hydrological models. Numerous studies have used this methodology. Due to uncertainties in model precipitation results, however, they all need to consider a range of changes from "less than" to "more than" current levels. The end results are thus more sensitivity studies than actual projections.

3. Statistical analysis can be done of actual precipitation trends in warming areas. As one example, an interesting study of Australian trends (Nicholls 1992) shows precipitation increases in some areas, decreases in others, and no change at all in still others.

4. One can analogize from present geographical diversity and documented past changes. This requires (a) a geographically contiguous area with sufficient temperature and precipitation diversity, and (b) historical demonstration over geologic time scales that any relationships hold over wide conditions.

Coastal California is an area with characteristics that allow the use of an analogy model.

Case Example: California's Coastal Watersheds

This example covers 17 coastal basins in California. The basins range from the Smith River at the Oregon border (over 200 cm/yr runoff) to the San Luis Rey River near San Diego (less than 2 cm/yr runoff). There is a relationship between average annual temperature and precipitation in these basins, resulting in a pronounced "colder and wetter" versus "warmer and dryer" curve. (Koltermann 1992). The curve is not linear; it is steeper in the colder area and nearly flat in the warmer area. The implication is that the nature of future hydrological conditions can be estimated by moving a few degrees "warmer" along this curve.

This analogy can be checked by reference to conditions during the last interglacial period (about 120,000 years ago), which was slightly warmer than present temperatures. Lakebed sediment pollen studies indicate that the environment in the San Francisco Bay area during that period resembled present-day Southern California, just as implied by the temperature-precipitation curve. (Koltermann 1992). The usefulness of the curve over long-span climate variation is thus suggested. Consequently, what effects are evident when the curve is used to project future warmer conditions?

Because the curve is steep in the north, a 3°C warming implies reduction of Smith and Eel River basin precipitation by over 50 percent. Because the curve is flat in the south, a 3°C warming has no effect at all on precipitation near San Diego. Intermediate areas have intermediate impacts. Specific projections are therefore quite variable, even within this limited area.

Conclusion

Water resource planners need to begin accounting for potential global warming impacts. Because each region may have a unique climate response, local studies will be required, utilizing one or more of the methodologies discussed here. The specific example presented (Koltermann 1992) may be of significant interest to California coastal communities relying on local basin water resources.

References

William W. Kellogg (1991), Response to Skeptics of Global Warming, Bulletin of the American Meteorological Society, 74,499-511.

Christine E. Koltermann and Steven M. Gorelick (1992) Paleoclimate Signature in Terrestrial Flood Deposits, Science, 256,1775-1782.

Stephan C. Kuhl and James R. Miller (1992) Seasonal River Runoff Calculated from a Global Atmospheric Model, Water Resources Research, 28,2029-2039.

James R. Miller, et al, (1992) The Effects of Climate Change on Monthly River Runoff, Managing Water Resources During Global Change (proc. American Water Resources Association), 175-178.

Neville Nicholls and Beth Lavery (1992) Australian Rainfall Trends During the Twentieth Century, International Journal of Climatology, 12,153-163.

C.D. Schonwiese (1992), Observational Signs of Greenhouse-gas-induced Climate Change, with Special Reference to Northern Latitudes, Journal of Atmospheric and Terrestrial Physics, 54,1101-1106.

T.M.L. Wigley and S.C.B. Raper (1992) Implications for Climate Change and Sea Level of Revised IPCC Emissions Scenarios, Nature, 357,293-300.

Impacts of El Niño in Western Washington

Thomas D. Murphy and V. Bruce Sandoval (member, ASCE)[1]

Abstract

The climatic phenomenon called El Niño is described, and its impact on hydrologic conditions in western Washington is shown for two recent events. Impacts on water management during the 1991-92 event are discussed using management of the Green River in western Washington as an example.

Introduction

In the fall and early winter of 1991, abnormal precipitation and temperatures caused streamflows to drop to very low levels in western Washington. Instead of winter relief from drought conditions, the climatic phenomenon known as El Niño was revealing its impact on the Pacific Northwest once again.

El Niño

The term El Niño was originally associated with a warm ocean current along the Peruvian coast which would appear every few years around Christmastime. It is now primarily associated with ecological and economic disasters in various parts of the world caused by devastating droughts and floods. Consistent with the original connection, an early indication of an El Niño episode is a slight warming of the waters of the central Pacific. At the onset, Pacific Ocean trade winds weaken slightly and the currents that follow these winds also lose some strength. This allows warm water from the western tropical Pacific to slosh back eastward causing a modest warming of the central tropical Pacific. This, in turn, causes a further weakening of the trade winds and further eastward expansion of warm surface

---

[1]Meteorologist and hydraulic engineer, respectively, U.S. Army Corps of Engineers, P. O. Box 3755, Seattle, WA 98124-2255

waters. The warming causes the ocean's storm generating centers to shift and a massive long wave ridge of high pressure to develop and hover somewhere near the west coast of the United States. This ridge builds, decays, and rebuilds throughout an El Niño episode. During the ridge's build phase the polar jet stream is aimed at Alaska, well north of its normal wintertime path across western Washington. Since storms generated in the Gulf of Alaska normally follow the path of the polar jet, these storms are diverted north of the area, carrying away precipitation normally destined for western Washington. During the decay phase the polar jet stream, which storms feed upon and travel along, is near normal, and precipitation returns to near normal.

Water Supply Impacts

Twenty El Niño episodes have occurred this century - ten in the last 50 years. In western Washington typical impacts have been warmer-than-usual winters, dry springs, shrunken mountain snowpacks, and early snowpack melt. Interestingly, the impact of an episode on a global basis is not a good gauge of the severity of its impact on western Washington. In fact, impacts on the region from some recent episodes were opposite of what happened on a global basis. For example, the 1976-1977 episode was a weak one from a global perspective but its impact on western Washington was very severe. Jet stream abnormalities caused significantly warmer and drier than normal weather; rainfall in Washington was well below normal. March and May rains were, however, above normal as the protective ridge of high pressure over the west coast decayed during these months, allowing normal storm tracks to resume temporarily. Washington snowpacks were at or near record low values for the winter of 1976-1977. These conditions were, in part, caused by high freezing levels which frequently meant rain instead of snow in all but the highest basin elevations. In general, the low level snow was gone by 01 April which is 45 or more days early. Soil moisture was very low by late winter because the snow that did fall had melted, leaving the surface bare. Thus, the abundant rains of March and May were mostly absorbed by the soil mantle, leaving streamflows extremely low. Record to near record low runoffs were the result and this led to a significant water shortage during the summer months of 1977.

The recent 1991-1992 episode will probably be categorized as a moderate one on a worldwide basis but a strong one from a western Washington perspective. It was basically mild and dry as the polar jet stream deflected Gulf of Alaska born storms to the north of the region while the subtropical jet stream steered strong storms in the eastern Pacific toward California. With the polar jet well to the north, western Washington was in the mild air normally found well to the south. Consequently, this was the mildest winter in 95 years and was followed by a dry, drought-plagued spring and summer. Rainfall was well below normal, although there were 30- to 45-day periods when the protective high pressure ridge over the west coast broke down and the polar jet stream again focused storms on the region. November 1991, January 1992, and April 1992 had above normal rainfall due to this temporary repositioning of the polar jet stream. Early snowpack buildup from November through mid-December was good and then snow conditions began deteriorating. Warm, dry weather in March was particularly damaging to the pack which peaked about 45 days early at the beginning of March. Streamflows were below to well below average in many river basins throughout western Washington.

Impacts of the 1976-77 and 1991-92 El Niño episodes on precipitation, snowpack, and streamflows are summarized in Table 1 with data from three U.S. Army Corps of Engineers (Corps) projects in western Washington.

Month	1976-77 EPISODE						
	Howard Hanson Project[1]			Mud Mountain Project[2]		Wynoochee Project[3]	
	Precip	Snowpack	Streamflow	Precip	Streamflow	Precip	Streamflow
Oct 76	93	[4]	63	49	70	59	39
Nov 76	44	[4]	52	36	49	30	38
Dec 76	61	17	54	57	38	57	52
Jan 77	37	12	55	27	47	54	49
Feb 77	53	9	41	62	42	85	73
Mar 77	128	15	68	144	67	115	88
Apr 77	67	56	94	75	103	53	83
May 77	150	43	58	176	54	160	103
Jun 77	54	35	63	60	73	56	105
	1991-92 EPISODE						
	Howard Hanson Project			Mud Mountain Project		Wynoochee Project	
Month	Precip	Snowpack	Streamflow	Precip	Streamflow	Precip	Streamflow
Oct 91	44	[4]	33	48	57	26	20
Nov 91	125	[4]	87	155	87	112	92
Dec 91	84	123	123	71	85	82	78
Jan 92	109	75	78	100	73	175	167
Feb 92	59	57	85	84	92	63	118
Mar 92	32	44	62	50	73	15	42
Apr 92	128	26	52	129	65	156	117
May 92	41	9	35	38	54	6	55
Jun 92	82	0	23	66	39	71	35

[1]Green River basin   [2]White River basin   [3]Wynoochee River basin
[4]There is normally no snow during these months

Table 1. Hydrologic Conditions During El Niño as a Percentage of Normal

Water Management Impacts

In 1992 El Niño prompted unique water resources planning and management throughout western Washington. Operations in the Green River basin, where the city of Tacoma draws its municipal and industrial water supply and the Corps owns and operates Howard A. Hanson Dam, are a good example of how the Corps, in coordination with others, dealt with El Niño's most recent appearance.

Coordination and Documentation. Uniting interests in the Green River basin became a primary objective soon after the potential for water shortage was realized. During the course of the year over fifteen persons represented their respective entities, including: the Corps; U. S. Fish and Wildlife Service; City of Tacoma Water Department; Washington State Departments of Ecology, Fisheries, and Wildlife; and the Muckleshoot Indian Tribe. Weekly teleconferences or meetings and frequent one-on-one interaction was initiated in late March 1992 and continued throughout the episode. The coordination identified and helped avoid potential conflicts. All coordination was promptly documented and distributed. This was a key to the success of the coordination. Records served as a confirmation of results and agreements among the parties; if the documentation was not accurate it was amended and/or revisited during subsequent coordination. Records were also important references as issues resurfaced, serving as a forum that avoided conflicts of misunderstanding. Additionally, in some cases documented information was used to make quick decisions that could not be coordinated in advance.

Contingency Planning. Early in the spring much of the effort was concentrated on developing a contingency plan specific to operation of Hanson Dam and Tacoma Water Department's diversion. The plan was composed of drought phases and associated triggering conditions and resultant actions. An example condition is less than 20 days of water supply remaining and an example action is Tacoma Water Department implementing mandatory lawn watering restrictions.

Real-time Water Management. In years of normal water supply micro management of resources on the Green River is not generally required. However, in 1992 summer streamflows were near record lows, so decisions that could normally be made based on limited field data required more extensive, real-time data. As an example, numerous surveys of the Green River fishery were made to measure the depth of water cover over salmon and steelhead redds (nests). Measuring water cover was very important because flows had to be reduced to save water; because real data were gathered, safe streamflows were maintained. Representatives from each agency had the opportunity to participate in a field survey. This enhanced coordination and facilitated a better working relationship among the agencies because better understandings of various positions were gained.

Conclusion

The climatic phenomenon called El Niño impacts western Washington water supply and management by causing lower than normal precipitation and higher than normal temperatures that result in low snowpacks and streamflows. In 1992, impacts on water resource users, especially fisheries, were mitigated due to close coordination between water managers and resource agencies. Special operations, spearheaded by weekly coordination with accurate documentation, resulted in a very successful year for all water resource interests on the Green River.

A Methodology for the Evaluation of Global Warming Impact on Soil Moisture and Runoff

Juan B. Valdés[1], Rafael S. Seoane[2] and Gerald R. North[3]

Global warming is expected to increase the intensity of the global hydrologic cycle (e.g., MacCracken and Luther, 1985; Mitchell, 1989; AAAS report, 1990). Precipitation and temperature patterns, soil moisture requirements, and the physical structure of the vegetation canopy play important roles in the hydrologic system of drainage basins. Changes in these phenomena, because of a buildup of CO_2 and other trace gases, have the potential to affect the quantity, quality, timing, and spatial distribution of water available to satisfy the many demands placed on the resource by society (Callaway and Currie, 1985). In addition, the location and magnitude of many different types of demand for water may be quite sensitive to the effects of CO_2 buildup on temperature and vegetation. A recent comparison study of the five major GCM results by Kellog and Zhao (1989) shows a model consensus that winters will be drier in Texas on all models. Summer is also expected to be drier, except possible along a limited part of the Gulf Coast, where the majority of the models expect an increase in soil moisture. King and Clarkson (1989) and Mays (1989) have reported on the impact of global warming on Texas water resources. The Texas Department of Agriculture (1990) studied the impact on Texas agriculture and natural resources. Valdés et al. (1990) gives results on soil moisture variability. Furthermore, soil moisture variability is tremendously important in the modeling of the earth's weather and climate. This is particularly true in the mid-latitudes and Rind (1982) concluded that "a knowledge of the ground moisture at the beginning of the summer might allow for improved summer temperature forecasts" and Shukla and Mintz (1982) have affirmed that "in the extratropics, with its large seasonal changes, the soil plays a role analogous to that of the ocean by storing of the precipitation it receives in winter and using it to humidify the atmosphere in summer," The recent work of Rodrígez-Iturbe et al. (1991ab) shows that the probability distribution of the soil moisture concentration at large length scales is bimodal with a dry and a wet mode respectively. In their work recycling between precipitation and evapotranspiration was considered.

In our work a methodology for the evaluation of impact on soil moisture concentration and direct surface runoff is presented. The methodology integrates stochastic models of hydroclimatic input variables with a model of water balance in the soil. This allows the derivation of the probability distribution of soil moisture concentration and direct surface runoff for different combinations of climate and soil characteristics, ranging from humid to semi-arid and arid. These PDFs asses, in a comprehensive manner, the impact that climate change have on soil moisture and runoff and allow the water resources planner to make more educated decisions in the planning and design of water resources systems. The methodology was applied to three sites in Texas. To continue in the line of research suggested by Delworth and Manabe we computed the autocorrelation function (ACF) and the spectra of both precipitation inputs and soil moisture concentration outputs for all scenarios of climate change.

Infiltration-Exfiltration Models

To describe the soil moisture concentration, models for the different water balance components were needed. First, the rainfall intensity was assumed to be constant during the duration of the storm. The infiltration and exfiltration models proposed by Eagleson (1978)

---
[1] Associate Professor, Dept. of Civil Engineering and Climate System Research Program, Texas A&M University, College Station, Texas 77843-3136; (409)845-1340
[2] Research Associate, National Institute for Water Science and Technology (INCyTH), National Council for Scientific and Technological Research (CONICET) and National University of the Center Buenos Aires. Casilla de Correo. 23 Aeropuerto Internacional de Ezeiza (1820). Argentina
[3] Distinguished Professor and Director, Climate System Research Program, Texas A&M University, College Station, Texas 77843-3150; (409)845-8083

were adopted here to evaluate the rate of evapotraspiration $e_T(t)$ and the rate of surface runoff $r_s(t)$. Finally, percolation was assumed to be constant throughout the duration of the interval (infiltration or exfiltration), although its magnitude was determined by the initial soil moisture of the corresponding interval. Two types of events were considered in the derivation of the analytical forms of the water balance components in the continuity equation in the paper by Valdés et al. (1990). One case was the conservation of water mass during a storm event and the second case was the conservation of mass during the interstorm periods. In our study since it was desired to model the hourly variations the events were assumed to be equally spaced in time and divided into rainy and dry periods respectively.

Rainfall Event Case

During a rainfall event, it was assumed that no evapotranspiration, $e_T(t)$, occurred. Precipitation, $i_r(t)$, was taken as a rectangular pulse of constant intensity, i_r, and duration, t_r. The difference between rainfall, $i_r(t)$, and runoff, $R_s(t)$, is the infiltration that contributes to soil moisture. The infiltration rate capacity is defined as Philip(1969) and Eagleson (1978):

$$f_i^*(t) = \tfrac{1}{2} S_i t^{-1/2} + a \qquad (1)$$

where the term a is the gravitational infiltration rate (it also takes into account the water table influence), and S_i is the infiltration sorptivity, embodying capillarity. By making the assumption that $f_i^*(t) = i_r(t)$ when surface runoff starts Eagleson (1978) gives an approximate expression for t_o as:

$$t_o \simeq S_i^2 / \left(2(i_r - a)^2\right) \qquad (2)$$

i) No Surface Runoff occurs ($0 \le t_r \le t_o$), thus the soil moisture final state is equal to the value at the beginning of period t plus the precipitation depth minus the percolation losses throughout that period, (Valdés et al, 1990) i.e.:

$$s_1(t+t_r) = s_o(t) + [i_r t_r - K(1) s_o^c(t) t_r]/nZ, \quad 0 \le t_r \le t_o \qquad (3)$$

ii) Surface Runoff occurs ($t_r > t_o$), thus the soil moisture concentration, $s_1(t + \Delta t)$, at the end of time interval t is equal to the value at the beginning of the interval plus the infiltration amount minus the percolation losses thoroughout the storm (Valdés et al, 1990), i.e.:

$$s_1(t+t_r) = s_o(t) + [i_r t_o + S_i(t_r^{1/2} - t_o^{1/2}) + a(t_r - t_o) - K(1)s_o^c(t)t_r]/nZ, \quad t_r > t_o \qquad (4)$$

where s_1 is the final value of the degree of saturation for the infiltration process, t is a relative index of the location of the interval in time, and Z is a measure of the penetration depth of the wet and dry fronts during storm and interstorm periods, respectively.

Exfiltration Event Case During an interstorm period, there is no precipitation and, hence, no runoff or infiltration. The evapotranspiration occurs at a maximum potential rate or at a smaller soil-controlled rate. To simplify the integrations, the potential rate of evapotranspiration $e_p(t)$ was replaced by its long-term climatic average value $\bar{e}_p$ in the previous work by Valdés et al (1990). In this study the actual potential evapotranspiration, $e_p(t)$, rate for the interval was used. Extending the infiltration equation of Philip (1969), Eagleson (1978) has represented the soil-controlled exfiltration rate by:

$$f_e^*(t) = 1/2 S_e t^{-1/2} + a_e \qquad (5)$$

where S_e is the exfiltration desorptivity, a_e is the asymptotic exfiltration capacity. Similar equations were found for the exfiltration case.

Applications of the Methodology to Selected Sites in Texas

The South-east part of the US climate ranges from arid to humid. In the most arid part of the South-east the moisture deficit can not be satisfied and there is little contribution from groundwater. The methodology presented in this work was applied to three sites in Texas, which is part of the South-east region described above: Amarillo, Temple and San Antonio.

The three sites show significant differences in climate and soil characteristics to measure the impact of climate change in hydrologic parameters, in particular soil moisture. Eight climate change scenarios (see Table 1) were designed to reflect the GCM's and experts variabilities in their estimates of precipitation and temperature changes. The most likely scenario is the one representing an increase of two degrees Celsius in the mean annual temperature.

Analysis of Results at the Daily Level The autocorrelation function for the daily precipitation values was computed at all three sites and it resembles white noise. The delaying effect of soil moisture in the ACFs is easily seen. The red noise decay is smoothed when the records with global warming are used since the drying of the soil reduces the ability of the soil moisture to filter the higher frequencies and a more rapid dampening of the ACF is observed. To further study the impact on soil moisture of climate chane a spectral analysis was carred out on the daily values of precipitation and soil moisture at all study sites. This analysis is similar to the one carried out by Delworth and Manabe (1988, 1989) in their work. The precipitation spectra show that the only frequency with strong power is the one associated with the annual cycle. No power is observed for the lower frequencies in the three study sites. The soil moisture spectra shows significant power in the lower frequencies for the base case and all scenarios for the three study sites. Delworth and Manabe indicate that a stochastic model of soil moisture with a white noise input precipitation will produce a red spectrum controlled by a damping term (potential evapotranspiration). The degree of persistence increased in the higher latitudes due to a decrease in the potential evapotranspiration rates. We obtained similar results in our work.

Probability Distribution Functions of Soil Moisture The hourly values of soil moisture concentration were analyzed and a Beta distribution was fitted to the records. These PDFs show definite seasonal variability and three months were selected to be shown in our Figures, January, June and October. In all of the scenarios the result of global warming, i.e. an increase of the mean annual temperature, has been a decrease of the mode of the distribution and a tightening of the PDFs around the mode. Changes in the mean annual precipitation may give a decrease or increase in the standard deviation of the mean annual soil moisture concentration values. In Amarillo the PDF changed its shape for all of the months indicating the vulnerability of this site to climate change.

Analysis of Results at the Annual Level The annual average results are shown in Figure 1. The mean annual value of the soil moisture concentration decreased, as expected in all of the scenarios where the temperature increased and there was a decrease in precipitation (scenarios 1,2,3,5 and 7). It even decreased for scenario 8 where the increase in precipitation was not sufficient to compensate with the increase in mean annual temperature. Scenarios 4 and 6 show an increase in the mean annual values of soil moisture concentration resulting from the 10% increase in mean annual precipitation. It is interesting to point out that soil and climate characteristics make this variability to have significant differences. Amarillo is significant more sensitive to climate changes than the more moist climate of Temple. The results here are more mixed showing increases and decreases within the same scenario reflecting the impact of local soil and climatic characteristics. These results show consistency with those found by the GCM models.

Acknowledgements R. S. Seoane stay at Texas A&M was funded by INCYTH and CONICET of Argentina. R. Bras and P. S. Eagleson of MIT gave invaluable comments. C. W. Richardson kindly provided us with his program and gave valuable advice. C. Fernández of TAES–Texas A&M helped us in the definition of the climate and soil characteristics of the Texas sites. All contributions are gratefully acknowledged.

Selected References

Callaway, J.M., and Currie, J.W., "Water Resource Systems and Changes in Climate and Vegetation", Publication DOE/ER-0236, U.S. Department of Energy, Washington, D.C., 1985.

Scenario Number	Precipitation Change	Temperature Change
0 (base case)	–	–
1	NC	$+2°C$
2	NC	$+4°C$
3	-10%	NC
4	+10%	NC
5	-10%	$+2°C$
6	+10%	$+2°C$
7	-10%	$+4°C$
8	+10%	$+4°C$

NC = No Change

Table 1: Climate Change Scenarios.

Delworth, T.L. and S. Manabe, "The Influence of Potential Evaporation on the Variabilities of Simulated Soil Wetness and Climate", *Journal of Climate*, Vol. 1, pp.523-547, 1988.

Eagleson, P.S., "Climate, Soil and Vegetation 1-7 ," *Water Resources Research*, 14(5), 705-776, 1978.

Eagleson, P.S., "Ecological Optimality in Water-Limited Natural Soil-Vegetation Systems, 1. Theory and Hypothesis," *Water Resources Research*, 18(2), 325-340, 1982.

Mays, L.W., "What Will a Change in Global Climate do to Texas' Water Resources?", Watermarks, publication of the Center for Research in Water Resources, University of Texas at Austin, vol. 24(4), 1988.

Mitchell, J.F.B., "The Greenhouse Effect and Climate Change", Reviews of Geophysics, Vol. 27(1), 115-139, 1989.

Richardson, C.W., "Stochastic Simulation of Daily Precipitation, Temperature and Solar Radiation", *Water Resources Research*, vol.17(1), 182-190, 1981.

Rind, D., "The Influence of Ground Moisture Conditions in North America on Summer Climate as Modeled in the GISS GCM," *Monthly Weather Review*, Vol. 110, 501- 526, 1982.

Shukla, J.M. and Y. Mintz, "The Influence of Land-Surface Evapotranspiration on Earth's Climate," *Science*, Vol. 215, 1498-1501, 1982.

Valdés, J. B.; M. Díaz-Granados and R.L. Bras, "A Derived PDF for the Initial Soil Moisture in a Catchment," *Journal of Hydrology*, 1989.

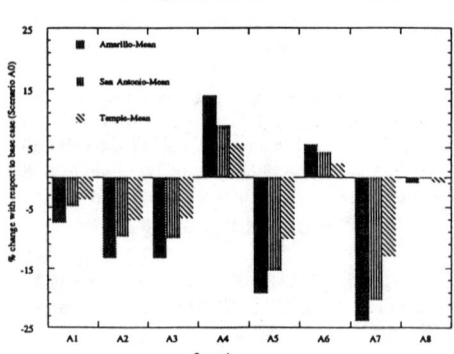

Integrated Water Resource Planning—
"A Northwest Case Study"

Jane Evancho[1] and Kelly Lange[2]

Abstract

The City of Tacoma Water Division recently incorporated integrated resource planning into its comprehensive water system planning process. This method of planning considers demand-side management programs as an alternative to, or in combination with, the development of additional supplies to meet future demands.

Why Tacoma Undertook a Regional Planning Process

As competition for use of our limited water resources has grown in the Puget Sound region, in Washington State it has become necessary to plan the use of these resources with a broader perspective. The Puget Sound area has realized the need to address long-term water resource issues cooperatively and with a broader perspective. Other factors, like droughts, state regulations, the Growth Management Act, existing cooperative efforts, and high project development costs, have also pointed to the need for a regional planning effort.

Tacoma's Past and Present Planning Practices

To assist in long-term resource planning and to comply with state requirements, Tacoma has prepared a series of comprehensive water system plans. The first document was developed in 1970; it was updated in 1980 and 1987.

---

[1] Water Resources Planning Coordinator, Tacoma Public Utilities, P.O. Box 11007, Tacoma, Washington 98411.

[2] Project Manager, CH2M HILL, 777 108th Avenue NE, Bellevue, Washington 98004.

The utility used a "traditional supply planning approach."[3] For the most part, these plans examined source alternatives specifically for Tacoma. Regional supply options, such as purchase of water from Seattle or other neighboring utilities, were evaluated initially, but not considered feasible for plan recommendation. Tacoma's water demand forecast for its designated service area did not include demand reduction resulting from conservation actions. Public comment was taken during the environmental review; two appeals were filed regarding the adequacy of the environmental documentation for the 1987 plan.

This plan recommended construction of an anticipated second transmission pipeline from Tacoma's surface water source to ensure both system reliability and transmission capacity to meet long-term demands. However, the water demands to justify this expensive transmission capacity and associated supply have not been realized within the existing service area. Although Tacoma's service area population projections will gradually bring pressure on the existing water supply sources, the 1992 drought was the catalyst for Tacoma to pursue a regional planning effort with its nearby neighbor and fellow regional water purveyor, the City of Seattle. Needs that became apparent during the summer of 1992 expanded this planning effort to include the Seattle Water Department's demand requirements and all feasible resource options available to the region.

Issues and Concerns Addressed in the Planning Process

Two primary issues and concerns were identified early in the planning process: public involvement and establishing criteria to evaluate supply- and demand-side alternatives fairly.

Public Involvement in Planning Process. Having expended much staff time and expense in responding to the environmental appeals and tribal concerns over the 1987 Comprehensive Plan Update, Tacoma developed an external relations approach that was designed to improve communications and include input from the public and interested agencies throughout the planning process. Citizens, government, recreational, business, fisheries,

---

[3] Janice A. Beecher, and John D. Stanford, "Integrated Water Resource Planning: Discussion Paper," a paper prepared for the Water Industry Technical Action Fund, December, 1992, p. 1-4.

tribal groups and environmental representatives were asked to participate in the planning process.

Evaluation Criteria to Assess Resource Options. Tacoma compiled a list of feasible demand- and supply-side options. These options included wastewater reuse and suggested resource options proposed by the public. Criteria were developed to prioritize options and included water supply yield (either as a constant or peaking source), reliability, operational flexibility, water quality and treatment requirements, environmental concerns, project timing and cost, and water rights. Each criterion was weighted according to its relative importance in ensuring reliable water supply.

The Planning Process

Integrated water resource planning incorporates three quantifiable components: water demands, demand management programs, and supply options.

Water Demand Forecast. Tacoma prepares a long-term (50-year) annual forecast of water consumption and customers to project water resource requirements for the service area. This forecast reflects trends in population and employment and includes estimates of additional Pierce and King County demand for water systems not currently connected to the Tacoma system. The forecast was revised to include projected growth in urban areas as a result of the state's recently enacted Growth Management Act. The forecast was also revised to include on-going demand reductions resulting from a conservation program. Available water supplies were first designated to serve this forecast population. Any surplus waters were then identified as available for transmission to the Seattle area.

Update Water Conservation Program and Strategies. Tacoma revised its water conservation program in 1990, with a commitment to a more active customer program. Concurrently, the utility has developed a maintenance program that conducted a leak survey of approximately 600 miles of main and addressed commercial meter testing and replacement as well as hydrant and valve operations. Records were compiled to quantify reductions in non-revenue water associated with these actions. These savings were incorporated into the water demand forecast as existing conditions.

Tacoma staff then prepared a list of potential demand-side measures for future consideration. The same criteria used to evaluate source options were applied to prioritize the demand-side alternatives. These criteria included water supply yield (either as a constant or peaking source), reliability, operational flexibility, environmental concerns, fisheries, recreation, water quality and treatment, public acceptance, and project timing and cost. Particular attention was given to identifying useful life and saturation potential for each measure based on recent experience from similar utilities and communities.

Demand-side measures recommended for evaluation in the integrated resource planning model included those measures that would best address the type of deficit expected by the region during a summer peak demand period. The types of measures considered included landscaping policies and ordinances, ultra-low flow toilet rebate, conservation rates, and wastewater reuse.

Evaluate Water Supply Options for Regional Needs.
Supply-side options were evaluated as a key component of the integrated resource planning process. The historical list of supply alternatives was expanded to identify options to meet the near-term needs of the region. Supply options including surface water options, additional storage projects, groundwater opportunities including aquifer recharge, reuse, and interties were evaluated for their regional benefits.

A preliminary screening process was used to reduce the number of supply options examined for further detailed evaluation. A life-cycle cost comparison was made to eliminate the most expensive options. The ten highest ranked alternatives were further evaluated.

Summary

Tacoma found this methodology as a useful planning tool. The recommendations that emerged from the integrated resource planning approach tie back to the issues and concerns previously stated. To that end, a number of regional demand scenarios were evaluated to account for the unknowns in the forecasting process. Other uncertainties such as future capital costs and conservation technology were also considered in the integrated resource modeling effort. In the interest of political and consumer preference to maximize conservation opportunities a solution with the least impacts and not the least cost was developed. Three alternative programs evolved from the plan-the least cost, the least impacts, and a compromise combining minimal impacts and costs.

Ground Water Management Standards and Protection Tools

Jon D. Witten[1]

Which Management Approach?

Once local goals and objectives for ground-water protection have been defined, wellhead protection areas delineated, sources of contamination inventoried and assessed, management techniques can be developed. Counties, cities, and towns have many options available to manage existing sources of contamination and to ensure that future land use activities do not pose a threat to ground-water quality.

At this stage of the development of a local ground-water protection program, it is time to think about the types of management controls that may be required. Health regulations, zoning, general ordinances, or voluntary controls are some of the options available to local government. Health regulations can address both proposed and existing development and their impacts on ground-water quality. Zoning controls are limited in that they can only apply to future development and not to existing activities which are exempt or "grandfathered". General police powers are available under a community's home rule powers to protect the public health, safety, and general welfare. Voluntary (or non-regulatory) controls may include educational efforts, monitoring, the adoption of certain best management practices, and land acquisition.

---

[1] President, Horsley & Witten, Inc., 3179 Main Street, Barnstable, MA 02630, Washington, DC and Seattle, WA.

The type of control that a community may be considering will also help to determine who in the local community should be involved. For example, if underground petroleum storage tanks are considered a threat, the local fire official should be involved in the planning of any regulatory measures. It would also be valuable to gather support from local businesses that may be affected by new underground tank controls.

The following table lists many of the techniques or "tools" that are available for ground water protection.

SUMMARY OF WELLHEAD PROTECTION TOOLS

	Applicability to Wellhead Protection	Land Use Practice	Legal Considerations	Administrative Considerations
Regulatory: Zoning				
Overlay GW Protection Districts	Used to map WHPA's. Provides for identification of sensitive areas for protection. Used in conjunction with other tools that follow.	Community identifies WHPA's on practical base/zoning map.	Well accepted method of identifying sensitive areas. May face legal challenges if WHPA boundaries are based solely on arbitrary delineation.	Requires staff to develop overlay map. Inherent nature of zoning provides "grandfather" protection to pre-existing uses and structures.
Prohibition of Various Land Uses	Used within mapped WHPA's to prohibit known ground-water contaminants and uses that generate contaminants.	Community adopts prohibited uses list within their zoning ordinance.	Well recognized function of zoning. Appropriate technique to protect natural resources from contamination.	Requires amendment to zoning ordinance. Requires enforcement by both visual inspection and on-site investigations.
Special Permitting	Used to restrict uses within WHPA's that may cause ground-water contamination if left unregulated.	Community adopts special permit "thresholds" for various uses and structures within WHPA's. Community grants special permits for "threshold" uses only if ground water quality will not be compromised.	Well recognized method of segregating land uses within critical resource areas such as WHPA's. Requires case-by-case analysis to ensure equal treatment of applicants.	Requires detailed understanding of WHPA sensitivity by local permit granting authority. Requires enforcement of special permit requirements and on-site investigations.
Large-Lot Zoning	Used to reduce impacts of residential development by limiting numbers of units within WHPA's.	Community "down zones" to increase minimum acreage needed for residential development.	Well recognized perogative of local government. Requires rational connection between minimum lot size selected and resource protection goals. Arbitrary large lot zones have been struck down without logical connection to Master Plan or WHPA program.	Requires amendment to zoning ordinance.
Transfer of Development Rights	Used to transfer development from WHPA's to locations outside WHPA's.	Community offers transfer option within zoning ordinance. Community identifies areas where development is to be transferred "from" and "to".	Accepted land use planning tool.	Cumbersome administrative requirements. Not well suited for small communities without significant administrative resources.

	Applicability to Wellhead Protection	Land Use Practice	Legal Considerations	Administrative Considerations
Regulatory: Health Regulations				
Underground Fuel Storage Systems	Used to prohibit underground fuel storage systems (UST) within WHPA's. Used to regulate UST's within WHPA's.	Community adopts health/zoning ordinance prohibiting UST's within WHPA's. Community adopts special permit or performance standards for use of UST's within WHPA's.	Well accepted regulatory option for local government.	Prohibition of UST's require little administrative support. Regulating UST's require moderate amounts of administrative support for inspection follow-up and enforcement.
Septic Cleaner Ban	Used to prohibit the application of certain solvent septic cleaners within WHPA's, a known ground water contaminant.	Community adopts health/zoning ordinance prohibiting the use of septic cleaners containing 1,1,1-Trichloroethane or other solvent compounds within WHPA's.	Well accepted method of protecting ground water quality.	Difficult regulation to enforce even with sufficient administrative support.
Septic System Upgrades	Used to require periodic inspection and upgrading of septic systems.	Community adopts health/zoning ordinance requiring inspection and, if necessary, upgrading of septic systems on a time basis (i.e. every 2 years) or upon title/property transfer.	Well accepted purview of government to ensure protection of ground water.	Significant administrative resources required for this option to be successful.
Toxic and Hazardous Materials Handling Regulations	Used to ensure proper handling and disposal of toxic materials/waste.	Community adopts health/zoning ordinance requiring registration and inspection of all businesses within WHPA using toxic/hazardous materials above certain quantities.	Well accepted purview of government to ensure protection of ground water.	Requires administrative support and on-site inspections.

GROUNDWATER MANAGEMENT 437

	Applicability to Wellhead Protection	Land Use Practice	Legal Considerations	Administrative Considerations
Cluster/PUD Design	Used to guide residential development outside of WHPA's. Allows for "point source" discharges that are more easily monitored.	Community offers cluster/PUD as development option within zoning ordinance. Community identifies areas where cluster/PUD is allowed (i.e. within WHPA's).	Well accepted option for residential land development.	Slightly more complicated to administer than traditional "grid" subdivision. Enforcement/inspection requirements are similar to "grid" subdivision.
Growth Controls/Timing	Used to time the occurence of development within WHPA's. Allows communities the opportunity to plan for wellhead delineation and protection.	Community imposes growth controls in the form of building caps, subdivision phasing or other limitation tied to planning concerns.	Well accepted option for communities facing development pressures within sensitive resource areas. Growth controls may be challenged if they are imposed without a rational connection to the resource being protected.	Generally complicated administrative process. Requires administrative staff to issue permits and enforcement growth control ordinances.
Performance Standards	Used to regulate development within WHPA's by enforcing predetermined standards for water quality. Allows for aggressive protection of WHPA's by limiting development within WHPA's to an accepted level.	Community identifies WHPA's and establishes "thresholds" for water quality.	Adoption of specific WHPA performance standards requires sound technical support. Performance standards must be enforced on a case-by-case basis.	Complex administrative requirements to evaluate impacts of land development within WHPA's.
Regulatory: Subdivision Control				
Drainage Requirements	Used to ensure that subdivision road drainage is directed outside of WHPA's. Used to employ advanced engineering designs of subdivision roads within WHPA's.	Community adopts stringent subdivision rules and regulations to regulate road drainage/runoff in subdivisions within WHPA's.	Well accepted purpose of subdivision control.	Requires moderate level of inspection and enforcement by administrative staff.

	Applicability to Wellhead Protection	Land Use Practice	Legal Considerations	Administrative Considerations
Private Well Protection	Used to protect private on-site water supply wells.	Community adopts health/zoning ordinance to require permits for new private wells and to ensure appropriate well to septic system setbacks. Also requires pump and water quality testing.	Well accepted purview of government to ensure protection of ground water.	Requires administrative support and review of applications.
Non-Regulatory: Land Transfer and Voluntary Restrictions				
Sale/Donation	Land acquired by a community within WHPA's, either by purchase or donation. Provides broad protection to the ground water supply.	As non-regulatory technique, communities generally work in partnership with non-profit land conservation organizations.	There are many legal consequences of accepting land for donation or sale from the private sector, mostly involving liability.	There are few administrative requirements involved in accepting donations or sales of land from the private sector. Administrative requirements for maintenance of land accepted or purchased may be substantial, particularly if the community does not have a program for open space maintenance.
Conservation Easements	Can be used to limit development within WHPA's.	Similar to sales/donations, conservation easements are generally obtained with the assistance of non-profit land conservation organization.	Same as above.	Same as above.
Limited Development	As the title implies, this technique limits development to portions of a land parcel outside of WHPA's.	Land developers work with community as part of a cluster/PUD to develop limited portions of a site and restrict other portions, particularly those within WHPA's.	Similar to those noted in cluster/PUD under zoning.	Similar to those noted in cluster/PUD under zoning.
Non-Regulatory: Monitoring	Used to monitor ground water quality within WHPA's.	Communities establish ground water monitoring program within WHPA. Communities require developers within WHPA's to monitor ground water quality downgradient from their development.	Accepted method of ensuring ground water quality.	Requires moderate administrative staffing to ensure routine sampling and response if sampling indicates contamination.

	Applicability to Wellhead Protection	Land Use Practice	Legal Considerations	Administrative Considerations
Contingency Plans	Used to ensure appropriate response in cases of contaminant release or other emergencies within WHPA.	Community prepares a contingency plan involving wide range of municipal/county officials.	None	Requires significant up-front planning to anticipate and be prepared for emergencies.
Non-Regulatory: Public Education	Used to inform community residents of the connection between land use within WHPA's and drinking water quality.	Communities can employ a variety of public education techniques ranging from brochures detailing their WHPA program to seminars to involvement in events such as hazardous waste collection days.	No outstanding legal considerations.	Requires some degree of administrative support for programs such as brochure mailing to more intensive support for seminars and hazardous waste collection days.
Legislative:				
Regional WHPA Districts	Used to protect regional aquifer systems by establishing new legislative districts that often transcend existing corporate boundaries.	Requires state legislative action to create a new legislative authority.	Well accepted method of protecting regional ground water resources.	Administrative requirements will vary depending on the goal of the regional district. Mapping of the regional WHPA's requires moderate administrative support while creating land use controls within the WHPA will require significant administrative personnel and support.
Land Banking	Used to acquire and protect land within WHPA's.	Land banks are usually accomplished with a transfer tax established by state government empowering local government to impose a tax on the transfer of land from one party to another.	Land banks can be subject to legal challenge as an unjust tax, but have been accepted as a legitimate method of raising revenue for resource protection.	Land banks require significant administrative support if they are to function effectively.

A Planning Strategy for Safe
Drinking Water Act Compliance

Scott Trusler, Associate[1]

Abstract

This paper presents a simple planning strategy for drinking water utilities that will allow them to move ahead with system upgrades and specific regulatory compliance programs even though pending regulations remain unknown.

Introduction

Passage of the 1986 Amendments to the Safe Drinking Water Act (SDWA) set in motion a complex and lengthy process of rulemaking. Existing regulations were to be reviewed and updated and new rules were to be researched, drafted and promulgated. Several major rules have now been finalized; the Surface Water Treatment Rule, the Coliform Rule, and the Lead and Copper Rule to name a few. To date, regulations covering 83 different contaminants have been finalized. However, other rules loom in future, such as Disinfectant/Disinfection Byproducts and Radionuclides. Many of these rules create interrelated and even conflicting impacts on a given utility and its water supply. As a result, many utilities have concluded that their only option is to stand by until the conflicting rules are resolved. In some cases, this has postponed needed improvements and upgrades.

Planning Strategy Fundamentals

Use of the approach outlined in this paper will allow utilities to move ahead with essential improvements.

---

[1] Project Manager, Water Supply and Treatment, CH2M HILL, P.O. Box 91500 Bellevue, WA 98009-2050

Although there is always risk when dealing with the unknown, this planning strategy allows utilities to make good decisions based on the current information and then move ahead with confidence. The basic planning strategy is quite simple and involves three elements: plan proactively, adopt a holistic view of SDWA, and build in flexibility. Embracing these elements is important to the success of the overall planning strategy. The extent to which each of these elements is implemented will be specific to the given utility, but some activity is required regardless of the size or level of sophistication of the utility.

Proactive Planning

Utilities have the option of ignoring the regulatory process until the final regulations are published and compliance becomes mandatory. The disadvantage of this approach is that it often leaves the utility with inadequate time and/or resources to avoid being out of compliance. Nobody likes to read about compliance problems in the newspapers. A proactive approach to planning will allow utilities to be ready when regulations are finalized as well as to proceed with needed improvements before regulations are completed. Proactive planning includes following the regulatory process. In general, the process is not speedy, which allows one to follow it with a minimal investment of time. Large utilities often have one or more staff members dedicated to tracking specific rules, but even small utilities can follow the process by maintaining regular contact with their state drinking water program liaison.

A key element of proactive planning is advanced monitoring. As proposed limits for contaminants become available, invest the time or resources to collect water samples and have them tested. Granted, there is a cost associated with this testing; however, the cost of delaying a needed improvement or not anticipating possible future requirements is typically much greater than the monitoring cost. Any data collected can also be used by regulators to help set more appropriate standards.

Finally, a conceptual compliance plan can be developed with very little effort. The goal here is to have a plan that can help guide the utilities' decision making process. The plan should list the various possible impacts from pending rules, options for compliance, and a conceptual plan for compliance. In most cases, a simple "best guess" approach is all that is needed. The plan needs only to identify cost in a relative sense (high-low) until details of the rule begin to solidify.

Periodic updates to the plan are important to absorb new information as it becomes available.

Holistic Viewpoint of SDWA

Many of the current and future rules have conflicting objectives. Perhaps the best illustration of this is the Surface Water Treatment Rule aimed at reducing risks of microbial contamination via filtration and disinfection and the pending Disinfectant/Disinfection Byproducts rule that will set limits on dosages of disinfectant and disinfection byproducts. Compliance with one rule may lead to problems with another. Before the compliance plan for a specific rule is selected, impacts on other rules and system operations must be evaluated to avoid future problems.

One technique to quickly screen compliance options is to use an impact flow chart. The impact flow chart is intended to flag potential problems with other rules or related water quality issues. As a compliance option is developed for a rule, the chart is used to determine possible side effects of a given action. A portion of an impact flow chart is presented in Figure 1. In this case, the water provider is developing options to address a lead/copper corrosion problem. By entering the chart at the top and tracing through a treatment option, the potential impacts become apparent. For example, increasing the pH reduces disinfection effectiveness and increases disinfection byproduct formation. These impacts must be considered, along with initial capital cost, operating cost, and other utility-specific constraints or desires to determine the optimal compliance option.

Flexibility

Building flexibility into physical facilities does not come without added costs. In the long run, providing extra building space, additional piping, or room for a future flow meter or valve may be all that is needed to allow a utility to respond quickly and effectively to future requirements. During facility design, revisit the impact flow chart and assess what types of equipment, monitoring, and so forth, might be needed. Then decide what flexibility features to provide in the planned facility. Accept the cost of these features as insurance in preparation for future requirements. Because flexibility costs usually represent only a fraction of the overall project cost, their impact on project funding is minimal.

Summary

A planning strategy exists that allows water utilities to move ahead with needed improvements even though the requirements of future drinking water rules are unknown. The essential strategy elements, regardless of utility size, include planning proactively, developing a holistic or overall view of the regulations, and building flexibility into physical facilities. Each utility can determine its appropriate level of effort, allowing even small utilities of limited resources full access to this strategy.

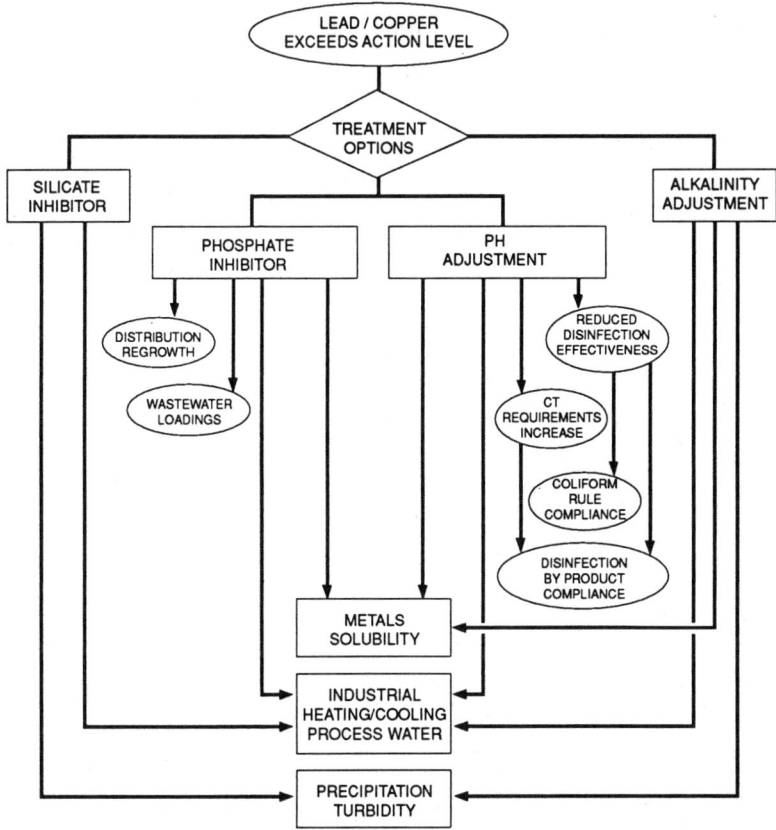

Figure 1. Corrosion Treatment Impact Chart

A Planning Strategy for Safe Drinking Water Act Compliance

1. Safe Drinking Water Act (SDWA)
2. Drinking Water Regulations
3. Drinking Water Utilities
4. Compliance (Drinking Water Act)
5. Corrosion Treatment for Drinking Water
6. Drinking Water Regulations Impact Flowchart

National Study of Water Management During Drought
Results Oriented Water Resources Management

William J. Werick[*]

Abstract. This paper provides an overview of a four year national study of water management during drought and builds a foundation for the other papers in this mini-symposium. The primary objective of the study and the definition of key terms are presented, and the overall study approach is described. The shortcomings of drought planning in the U.S. are enumerated, and the principle products from the study are listed with a brief explanation of how they address these shortcomings. Subsequent papers in this mini-symposium discuss each of the case studies and many of the principal products. The study will conclude this October with the publication of How To Be Prepared for Drought.

Study Background

Study Objective. The National Study of Water Management During Drought (November 1989 - October 1993) was initiated in response to the broad range of conflicts arising from the droughts which plagued so much of the country in 1988. The principal objective of the study is to develop a strategy to improve water management during drought in the United States.

Study Participation. The Drought Study is funded and managed by the U.S. Army Corps of Engineers, but its perspective and conclusions, and ultimately, its effectiveness, come from the breadth and depth of participation by people in other Federal agencies, non-Federal governments, universities, and environmental and public interest groups. Several organizations outside the Corps funded significant amounts of staff time to the study.

---

[*] Manager of the National Drought Study for the U.S. Army Corps of Engineers, Institute for Water Resources, Casey Building, Fort Belvoir, VA 22060-5586.

Definitions in the Study. Different definitions of drought are used in different contexts by researchers. Some authors restrict its use to what others call *meteorological* drought (less precipitation than usual, with "less" quantified or not), others use "drought" to refer to *agricultural* drought (not enough precipitation for crops), or *hydrologic* drought (less water available than usual, whether in the form of precipitation or streamflow)[1].

But surface water reservoirs or well systems can reduce the impacts of lower than normal precipitation, allowing what would have been called a drought to pass without notice. At the same time, adaptation to the new hydrologic patterns - the development of a recreation industry around a lake or reservoir releases, or the creation of new habitat - may introduce new vulnerabilities. In this context, which is the province of the National Drought Study, droughts are times when abnormally low precipitation or streamflow causes significant impacts.

Strategic and tactical responses. *Strategic* water resources planning identifies the appropriate level of investment and the general types of responses to water related problems and opportunities in the long term future, including drought impacts. These responses are based on forecasts of water supply up to a hundred years into the future. The concept of "appropriate" levels of present day investment recognizes the fact that forecasts are uncertain, and that the marginal return on investments to address the negative impacts from successively larger but rarer droughts declines after some point. Thus, strategic planning purposefully leaves residual risks because it not cost efficient to eliminate them. One of these vestigial risks is that there will be times in the future when unconstrained water demands - traditional or new - cannot be met by the available water supply.

Tactical drought contingency planning is used to mitigate the impacts during these periods. During droughts, water managers are expected (within the constraints of laws and contracts) to strike a reasoned balance among the residual impacts which strategic plans do not eliminate. The same broad goals apply in the strategic and tactical planning - economic efficiency, sustainability, and equity. Tactical plans tend to be non-structural adaptations within existing laws and structural capacities.

Study approach. The National Drought Study is managed by the Corps' Institute for Water Resources (IWR), which was established in the late 60's in recognition of the need to develop methods to address environmental and social issues in the study of the feasibility of Federal water projects[2]. IWR serves as a bridge between research and practice in natural resources management. The design of the National Drought Study reflects the nature of IWR in that it is a mix of research and application.

The first year of the study was devoted to analysis of the current state of water management during drought. Each of the 50 Governors wrote to express their main concerns and name a state study coordinator. Three workshops, co-sponsored with the Western States Water Council and the International Drought Information Center, were held to solicit views from water managers, environmentalists, and researchers. Corps managers across the country responded to questionnaires on drought planning, operations, and emergency responses. Existing drought plans and notable case studies with drought management implications were reviewed. A report after the first year of study[3] summarized the problems and proposed the demonstration and testing of a model drought preparedness method, buttressed by research in supporting technical areas, including the development of a National Drought Atlas. The intent of these recommendations was to improve water management during drought by producing drought management guidance which is consistent with the state of the art and which had been applied successfully in practice.

In the second, third, and final year of the study these case studies were supported and supplemented by a variety of studies in component fields, such as law, politics, institutions, and modelling, and additional collaborative efforts such as a demonstration of water allocation gaming for the Colorado River (with the Study of Severe Sustained Drought in the Southwest) and a review of Lessons Learned in the California Drought (based on interviews and focus groups of water agencies, environmental groups, and the media).

Problems and Products

The Problems. Only a little more than half the states have drought preparedness plans[4]. About half of the country's urban water suppliers have drought contingency plans, but fewer than 30% of the urban water utilities have any kind of quantitative data to support decision making during droughts. In many cases, the plans are based on little research and unrealistic expectations of consumer responses[5].

The plans that do exist tend to have functional problems: they may not recognize all the uses of water; they are usually designed for the drought of record, without consideration of the rarity of that event; they often are not understood or endorsed by those who will suffer the impacts of the drought; they may not sufficiently address equity issues or economic differences in the use of water; they are typically triggered by indicators which are not related in a known way to impacts. Finally, almost all such plans are better characterized as documents rather than ways of behaving, and so their effectiveness diminishes as staff changes occur and time passes between plan preparation and drought[6]. In fact, the practice of tactical drought contingency planning has not benefitted much from the research, development and testing that has improved *strategic* water resources planning over the past four decades, although both are meant to assure efficiency and equity in the allocation and use of water and related land resources.

The principal study products were designed to address these problems. They include a model drought planning method which has been tested and refined in four case studies, the National Drought Atlas, demonstrations of the advantages of a new generation of modelling software, a demonstration of the use of IWR-MAIN 6.0 to manage conservation programs in low growth areas, and review and analysis of subject areas important in drought management. These products should help regional managers use scarce resources more efficiently during droughts and minimize the maldistribution of adverse effects.

The planning method developed and tested in the four Drought Preparedness Studies is an evolutionary adaptation of strategic planning methods developed for Federal water resources studies. The development of those methods can be traced through Proposed Practices for Economic Analysis of River Basin Projects (May 1950, revised in May 1958 and referred to as "The Green Book"), Senate Document 97 (1962), Principles and Standards for Planning Water and Related Land Resources (the P&S) (1973) and Economic and Environmental Principles and Guidelines for Water and Related Land Resources Implementation Studies (the P&G) (1983).

The original principles were developed for the purpose of formulating mutually acceptable principles and procedures for determining benefits and costs of water resources projects[7]. Over time these methods have expanded beyond benefit-cost analysis in response to growing public concern about environmental and social issues, public involvement, and the bias towards structural solutions. The Water Resources Council institutionalized the multi-objective planning paradigm which had emerged from the Harvard Water Project in the P&S[8]. The P&G returned the focus to one objective - economic efficiency - with environmental constraints. Although in practice even P&S was used primarily to decide whether or not to invest Federal funds in structural projects, the planning approach that underlies both the P&S and the P&G can easily be modified to allow true multiobjective planning[9]. The model drought preparedness study method does this by the addition of a new planning step - the identification of decision makers, influencers and decision criteria - and the use of a multicriteria, multiobjective, multiparty decision making process.

Public Involvement. The drought preparedness study teams always represent four groups: water users (those who can be hurt in a drought should be involved in preparation efforts), water managers (politicians as well as agencies), issues advocacy groups, and experts outside the first three groups (typically, university researchers). The public involvement strategy in the model drought preparedness method also includes the concepts of *circles of influence* (to increase study efficiency without loss of public involvement), and *alternative dispute resolution.*

In the course of conducting our case studies we were pleased to discover an institutional arrangement which improves public participation over a broad range of water issues, including drought. The Massachusetts Water Resources Authority funds some of the staff and overhead costs of the Massachusetts' Water Supply Citizens Advisory Committee (WSCAC), and allows WSCAC on-line computer access to its data files.

The National Drought Atlas is a compendium of statistics which allows regional planners to determine the probability that a given volume of precipitation or streamflow will occur in an n month period, with n = 1, 2, 3, 6, 12, 24, 36, or 60. The Atlas can be used in may ways, but two uses are considered most important. First, planners can estimate the return interval of the drought of record, and thus can make a more informed judgement about whether this is a suitable design drought for planning purposes. Second, in the event of a drought, operators and managers can better estimate the probability of the drought continuing, or of reservoirs refilling to normal levels.

Modelling. Recent advances in computer software have greatly enhanced the potential for computer simulation models to be used in water resources planning and negotiation. The Drought Study tested one type of an object oriented modelling environment called STELLA in five case studies. As later papers will show, these models allowed an integrated analysis of regional drought impacts under the status quo and alternative plans. These new environments allow simulation models to be built, tested, and modified more quickly and cheaply, and because of their user friendly interface, they can be built collaboratively with water users and advocacy groups. The resultant simulation models are more widely understood and trusted.

IWR-MAIN 6.0. IWR-MAIN is a water use forecasting model which has been widely used in the West to estimate future urban water use in fast growing urban areas. The latest version, IWR-MAIN 6.0, reflects the latest research, adds additional features, and uses a friendlier user interface. The Boston area is not expected to experience rapid population growth, but the Massachusetts Water Resources Authority, as part of the Drought Study, used a beta version of MAIN 6.0 to estimate the effectiveness of each of its many conservation measures. By doing so, MWRA hopes to optimize the investment of conservation dollars.

Conclusions. Most communities could be much better served by drought preparedness efforts. The National Drought Study has collected and tested a comprehensive set of methods and tools to do just that. Regional water managers interested in these measures should write the author or speak to any of the presenters in this Mini-Symposium.

References

1. Dracup, J.A., Lee K.S., and Paulson, E.G. Jr.; 1980; On the Definition of Droughts, Water Resources Research, 16(2): 297-302.

2. Letter from William Cassidy, Chief of Engineers, to the Chairman of the Subcommittee on Public Works, Committee on Appropriations, United States Senate, February 4, 1969: and the response from House Appropriations Chairman Michael J. Kirwan, March 29, 1969.

3. U.S. Army Corps of Engineers, Institute for Water Resources, May 1991: The National Study of Water Management During Drought; Report on the First Year of Study; IWR Report 91-NDS-1; 70 pp.

4. Grigg, Neil and Vlachos, Evan; Drought Water Management, Proceedings of a National Workshop, Washington, D.C.; November 1988; A Report for the Natural and Man-Made Hazards Mitigation Program of the National Science Foundation; 232 pp.

5. Moreau, David and Little, Keith; 1989; Managing Public Water Supplies During Droughts, Experiences in the United States in 1986 and 1988; Water Resources Research Institute Report No. 250; University of North Carolina; 119 pp.

6. U.S. Army Corps of Engineers, Institute for Water Resources, May 1991: The National Study of Water Management During Drought; Report on the First Year of Study; IWR Report 91-NDS-1; 70 pp.

7. Proposed Practices for Economic Analysis of River Basin Projects; 1950, rev. 1958; a Report to the Inter-Agency Committee on Water Resources by the Subcommittee on Evaluation Standards.

8. Lord, William R.; Objectives and Constraints in Federal Water Resources Planning; December 1981, Water Resources Bulletin, American Water Resources Association.

9. Stakhiv, Eugene Z.; Achieving Social and Environmental Objectives in Water Resources Planning: Theory and Practice; Social and Environmental Objectives in Water Resources Planning and Management; 1986; Proceedings of and Engineering Foundation Conference sponsored by ASCE; Edited by Warren Viessman, Jr. and Kyle E. Schilling; p. 107-125.

Empowering Stakeholders Through Simulation in Water Resources Planning

Richard N. Palmer[1], Allison M. Keyes[2], and Selene Fisher[2]

Introduction

During the past two years, researchers at the University of Washington (UW) have had the unique opportunity to facilitate and observe the development of drought planning activities associated with the National Drought Study (NDS) and its Drought Preparedness Studies (DPS) sites as sponsored by the Institute of Water Resources of the US Army Corps of Engineers. Each of the DPS sites is unique, with different study objectives and institutional constraints. However, one uniform requirement of the study is to develop tactical and strategic drought plans that can be *successfully implemented* within the study region.

At the onset of the study, it was recognized that successful implementation is directly related to the active involvement of affected parties and agencies (denoted as "stakeholders") and the degree to which they support the plan's conclusions. Their involvement is also necessary because the problems addressed by the DPS's require the experience and knowledge of a variety of water resource interests in order to arrive at effective alternatives. Their support of the plan conclusions enables regional implementation.

Several techniques were used to encourage stakeholder participation in the planning process. Individuals representing the stakeholders had a wide range of professional backgrounds. This paper concentrates on one specific approach found useful in encouraging comprehensive and meaningful participation by a wide range of stakeholders; the development of object-oriented simulation models for the water resource systems under study.

Simulation models were to develop tactical and strategic drought plans and to ensure the acceptance of the plans by building consensus among the stakeholders. Steps in the modeling process include: 1) identifying affected agencies and groups to involve in the planning process, 2) identifying the important elements in the water resource system, 3) defining measures of system performance, 4) selecting the appropriate level of detail for the study, 5) defining the status quo, 6) identifying alternatives, and 7) ranking alternatives. These activities required training the study teams in the

---

1. Professor of Civil Engineering, and 2. Graduate Research Assistant, Department of Civil Engineering, University of Washington, Seattle, Washington 98195

use of simulation modeling, identifying and incorporating all interested parties in the model building and verification process, and using simulation gaming to identify acceptable solutions to water supply problems. The remainder of this paper describes: 1) how simulation models became a part of the National Drought Study, 2) procedures used to develop the DPS models, and 3) how the model empowered stakeholders.

Background

The successful incorporation of simulation models in the water resources planning process has long been of interest. In their introduction to interactive simulation models, Loucks et al. (1985) discuss several reasons why the use of computer models has been less than totally successful. Factors noted in their discussion include: 1) unless model builders are familiar with both the problem and institutional setting in which the problem is to be addressed, it is unlikely that any model will be effective in obtaining a solution, 2) the model must be complete and able to examine all issues deemed important by the user, 3) the model must be compatible with the mental model users have of the problem, 4) the model must be capable of including subjective information in the modeling and decision process, and 5) the model must be capable of developing and encouraging trade-offs between alternatives. They also suggest that "Perhaps one of the biggest reasons for model solution rejection, even as a basis for discussion in the managing, planning, or policy-making process, has been the lack of adequate communication between the analysts and their clients."

These comments are particularly relevant to the National Drought Study. At the onset, project managers decided that computer models would not be a major focus. This decision was made based on the assumption that insufficient time and funds existed to develop meaningful computer simulation studies and resources would be better spent on developing study objectives and analyses not requiring extensive use of computers. This decision was reversed, however, when participants became aware of recent advances in object-oriented simulation models, creating a significant shift in the modeling paradigm that had been in place since the origin of the project. Specifically, this shift was created by the introduction to a software environment known as STELLA™(High Performance Systems, 1992).

Since the early 1950's, when computers were first used to evaluate water resources planning alternatives, models of water resources have been constructed by a relatively small number of highly trained individuals who straddled the professions of computer programming and hydro-engineering. Many highly respected hydro-engineers in the 1950's and 1960's made little direct use of computers and, when required, worked with other professionals that translated their ideas into computer code. Today most hydro-engineers are trained to some level in computer programming, but many do not make extensive use of these skills after leaving their university.

Given this setting, it is not surprising that the managers of the National Drought Study were extremely uncomfortable with constructing (from the ground up) models of the basins in which the studies were being conducted. Past experience had demonstrated that models sometimes alienated stakeholders rather than encouraging their participation. What changed their minds was the demonstration of computer software that dramatically altered the manner in which water resource simulation models could be developed and significantly empowered stakeholders previously excluded from the modeling process.

Object-oriented programming environments such as STELLA™ (as well as a number of other commercial software packages on the market such as EXTEND™, SimuLink,™ and VisSim™) offer the user the opportunity to construct extremely complex models from a series of basic "icons". Water resources models can be constructed many times faster with these tools than with traditional programming languages and built by planners, engineers, and stakeholders without a high level of programming experience.

Model Construction
Object-oriented environments are not based on traditional procedural coding such as in FORTRAN, but rather are based upon the graphical manipulation of objects or "icons" that have specific characteristics and can perform specific functions. STELLA™ offers a particularly elegant environment as it can present complex systems in terms of four basic icons; stocks, flows, converters, and connectors. These basic icons can be used to represent water resource components such as reservoirs, rivers, economic impacts, or functional relationships.

The figure below illustrates the operation of a simple water resource system composed of inflow to the reservoir, reservoir release, diversion of water to a municipality's pipeline, and flow down stream of the diversion. Stakeholders familiar with the system can quickly recognize key components and understand their relationships.

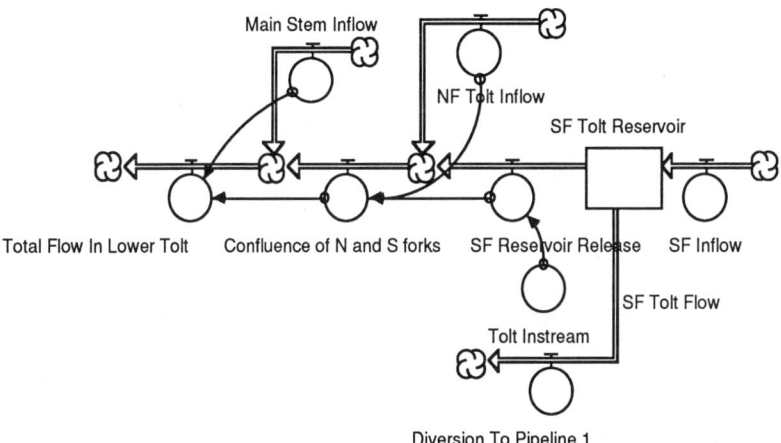

One reason that STELLA™ was chosen over other modeling environments is its relatively short learning curve and the subsequent ease of implementation. Training was provided to the DPS sites in two ways. Initially, researchers from the University of Washington traveled to each of the DPS sites to present a two day overview of modeling philosophy and the use of STELLA™. Some six months later, a workshop held in Seattle allowed all modelers to enhance their models and share their modeling experiences to date.

It is important to note that throughout the modeling process, model developers were encouraged always to keep in mind two questions: who will use the model and how will it be used? Modelers were encourage to prioritize modeling effort by asking: if this feature is incorporated into the model, will it help those responsible for planning and managing droughts to make better decisions to minimize the drought's impact? Because of STELLA's™ interactive characteristics and graphically orientation, modelers were better able to concentrate on these key features rather then on the difficulties typically encountered when coding and debugging computer programs.

Empowerment of Participants
A number of features characteristic of object-oriented simulation environments have been noted that make these tools appropriate for the DPS. However, the most significant benefit of using STELLA™ was its ability to empower those not normally involved in the modeling process. This empowerment was demonstrated in four ways. First, non-programmers recognized their ability to directly impact the modeling process. As a consequence all of the DPS sites actively sought and received stakeholders' input during model development. Typically this came in the form of public meetings followed by in-depth interviews where prototype models were demonstrated and errors or shortcomings of the model were identified and corrected. This approach contrasts sharply with traditional modeling approaches.

Second, stakeholders found that the new model provided previously unavailable information which resulted in a better understanding of the planning and management options available. Stakeholders were better able to recognize the interactions in the system and to appreciate the overall management of the system. Often this increased understanding was accompanied by a willingness to more fully appreciate the concerns and objectives of other participants in the planning process.

Third, they discovered that they could use the models to evaluate alternatives and options without the delays often associated with computer models. Non-modelers also found the models sufficiently clear and self-explanatory to evaluate potential alternatives without the aid of the model developer. Finally, they found that they had confidence in the output of the models because they had been intimately involved in the modeling process from start to finish.

Conclusions
Object-oriented programming has allowed the NDS to empower a wider range of water resource stakeholders in the planning process. This was possible because the models were constructed quickly, allowing the stakeholders to critique the quality of the models and their assumptions and to participate in their development and use. This involvement greatly increases the likelihood that the results of the DPS will be incorporated into meaningful changes in water management and decrease the impacts of future droughts in these regions.

References
High Performance Systems, STELLA II, An Introduction to Systems Thinking, Hanover, NH, 1992, 176 pages.

Loucks, D.P., J. Kindler, and K. Fedra,"Interactive Water Resources Modeling and Model Use: An Overview," Water Resources Research, Vol. 21, No. 2, 1985, pp 95-102.

The National Drought Atlas

James R. Wallis[1]

Abstract

The National Drought Atlas is part of a study of water management during drought, funded by the U.S. Army Corps of Engineers and other state and federal agencies. The Atlas establishes frequency distributions for drought measures (annual and monthly precipitation and streamflow), and includes data on the Palmer Drought Severity Index, and reservoirs for the contiguous 48 states.

The NDA will be published in paper form, and it is hoped that the database will also be made available on CDROMs (with many of the files using database and/or spreadsheet formats with included macros for ease of use). In addition, an interactive PC version of the drought Atlas has been cobbled together using TYDAC SPANS MAP and LOTUS 123/G to serve as a prototype for a standalone PC based interactive National Drought Atlas computer program. The talk is illustrated using screen dumps from this prototype program.

---

[1] Mathematical Sciences Department, IBM Research Division, T.J. Watson Research Center, Yorktown Heights, NY 10598

In 1989 the U.S. Army Corps of Engineers was charged with the responsibility of conducting a national study of water management during periods of drought. During the first year of the study, the nature of the drought problem was investigated through a variety of means, workshops, literature reviews, participation in professional society meetings, interviews and correspondence, and special studies conducted by the Advisory Commission on Intergovernmental Relations. Some of the conclusions that were reached from the study are:

- Much has been done in the United States to reduce drought vulnerability since the 1930's, but many places in the country are chronically ill-prepared for drought;
- No consensus exists within the water management community on a national strategy to improve water management during drought;
- No single conceptual model in law, engineering, economics, social or environmental sciences encompasses the reality of drought;
- Regional differences are substantial in needs, law, climate, and level of investment;
- Communication among regions should be improved;
- The application of water conservation principles is spotty; and
- We have trouble answering some of the most basic questions about drought preparedness such as, "How big a drought should we plan for?"

The first year of the study pointed out that little is known, and less shared, about the probability that droughts of a certain duration or intensity will occur. This ignorance has significant planning impacts. Consequently, a recommendation was made to prepare a National Drought Atlas, NDA, that would provide a common base of data about precipitation, streamflow, soil moisture, and a uniform set of statistical procedures for getting from the data to a drought probability estimate. Floods and high extreme precipitation have been modeled extensively, but there is not a similar source of information for events at the low end of the climatic and hydrologic spectra, a void which the NDA plans to fill.

The NDA team consists of

- G. E. Willeke, Miami University, Ohio, (Chairman);
- N. B. Guttman, NCDC, Asheville, NC;
- W. O. Thomas, USGS, Reston VA;
- J. R. M. Hosking, Math. Sciences Dept., IBM Research, Yorktown Heights, NY;
- J. R. Wallis, Math. Sciences Dept., IBM Research, Yorktown Heights, NY;

with William Werick as point man for the USACE.

In order to accommodate a variety of needs, it was decided to portray drought in terms of monthly precipitation (1119 of the Historical Climatologic Network sites, analyzed by 12 starting months, and for 8 durations, 1, 2, 3, 6, 12, 24, 36, and 60 months). Soil moisture is measured at few stations, and those that exist tend to have short, discontinuous records with sites that are not well distributed across the country. Consequently, proxy measures of soil moisture are commonly used. The Palmer Drought Severity Index (Palmer, 1965), which is based on a calculated water balance using methods devised by Thornthwaite, is the soil moisture proxy for the Atlas. Despite its limitations, it was chosen for the Atlas because it has been widely accepted and used by the water planning community for indicating

NATIONAL DROUGHT ATLAS 457

drought as well as for initiating action for ameliorating the effects of drought. PDIs for 1035 long record sites are included in the NDA.

At the time of writing, (November, 1992), the number of streamflow stations with records suitable for statistical low flow analysis is unknown. DLG boundary vectors, climatic, land use, and geologic information still need to be acquired for almost all of the probable watershed choices. Meaningful clustering of watersheds into homogeneous or quasi-homogeneous low flow groupings is not possible before the necessary data has been acquired. Publication of the Atlas is being postponed. However, interim results continue to be published, and a current bibliography of NDA related papers will be available at the meeting.

The team is currently considering whether or not it would be desirable to make the NDA database available to users on CDROMs (with many of the files using database and/or spreadsheet formats with embedded macros for ease of use). We are interested in hearing from potential users with regard to additional datasets that would serve to make the CDROMs have broader appeal. For instance, it has been suggested that many states have large scale mean annual isohyetal maps that should be digitized and included, (we currently possess only two, WA and CA), while others have suggested that a digitized contour map would be an acceptable surrogate (500' contours?). What about drainage boundary files? or precipitation records for shorter durations than the current 1 month minimum? and if one is to do this would instantaneous flood peak data also be useful? 1009 USGS flood peak streamflow records with little or no diversion or regulation are available, should similar analyses to those planned for the low end of the spectra be extended upwards?

In addition, to the paper Atlas and CDROMs an interactive PC version of the drought Atlas has been cobbled together using TYDAC SPANS MAP and LOTUS 123/G to serve as a prototype for a standalone PC based interactive National Drought Atlas. The remainder of this talk is illustrated by screen dumps taken from this prototype program.

Figure 1, HCN 'Starburst' map showing 1119 HCN precipitation sites grouped into 111 regions. For these annual values three clusters failed the Hosking and Wallis (1992) H_{L-CV} heterogeneity test (marked with triangles on figure 1), while 108 had $H_{L-CV} < 2.0$ i.e. these regions may be considered homogeneous or near homogeneous with regard to their annual data (marked with squares on the figure 1). Interested readers are referred to Guttman, Hosking, and Wallis (1992) for discussion of these regionalization and quantile estimation techniques.

Figure 2, Probability of observing a Palmer Drought Index value < -4 during July for 1035 sites of the NDA database. A PDI < -4 is classified as extreme drought, many sites in the Midwest show high probability of this happening (some as high as 30%).

Figure 3, Collage of windows taken from the NDA prototype PC-based program. The main screen shows a portion of Washington state with the Cedar and Green River reservoir drainage areas marked. The bar graph shows the expected rainfall by months starting in May for HCN site 451237, Cedar Lake. The spreadsheet shows the quantile values for region 60 scaled by the multiplier 15.8 inches. The multiplier was arrived at by assuming the Cedar River watershed to have an average annual precipitation of 121 inches, versus 102.4 for the Cedar Lake HCN site and applying the same ratio to the May-June period as exists for the annual values. The graph in the upper right window shows the median May-July rainfall for the catchment as a whole to be estimated at 15 inches (Note: Mean > Median hence a positively skewed distribution), and with one chance in 50 of experiencing less than seven inches in the May-July period. If the reservoir is at 30% of capacity at the beginning of May how much rain do you need during May-July so as to not have water restrictions by August? You can get the probability of it happening from the NDA.

REFERENCES

Hosking J. R. M. and J.R. Wallis, 'Some Statistics Useful in Regional Frequency Analysis', W.R.R.,[in press 1992].

Guttman N. B., J. R. M. Hosking, and J. R. Wallis, 'Regional precipitation quantile values for the continental U. S. computed from L-moments', IBM Research Report, R.C. 18258, 1992.

NATIONAL DROUGHT ATLAS

459

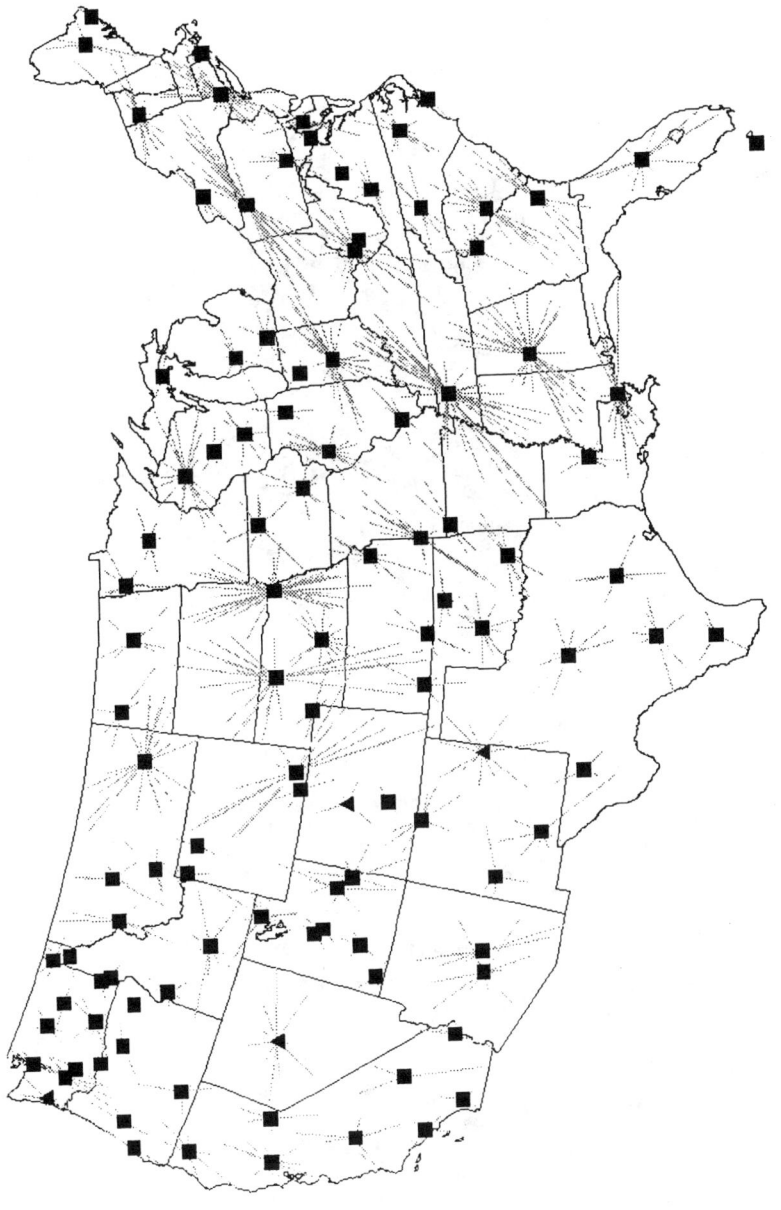

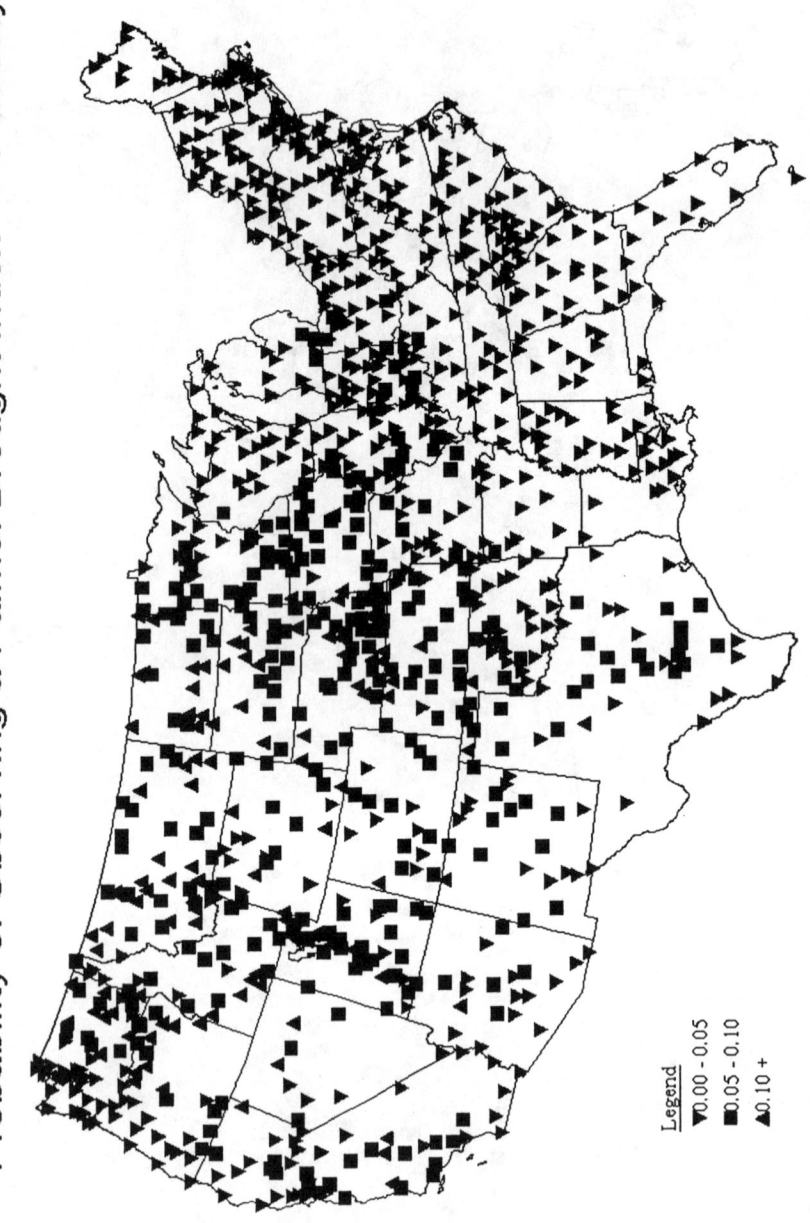

NATIONAL DROUGHT ATLAS

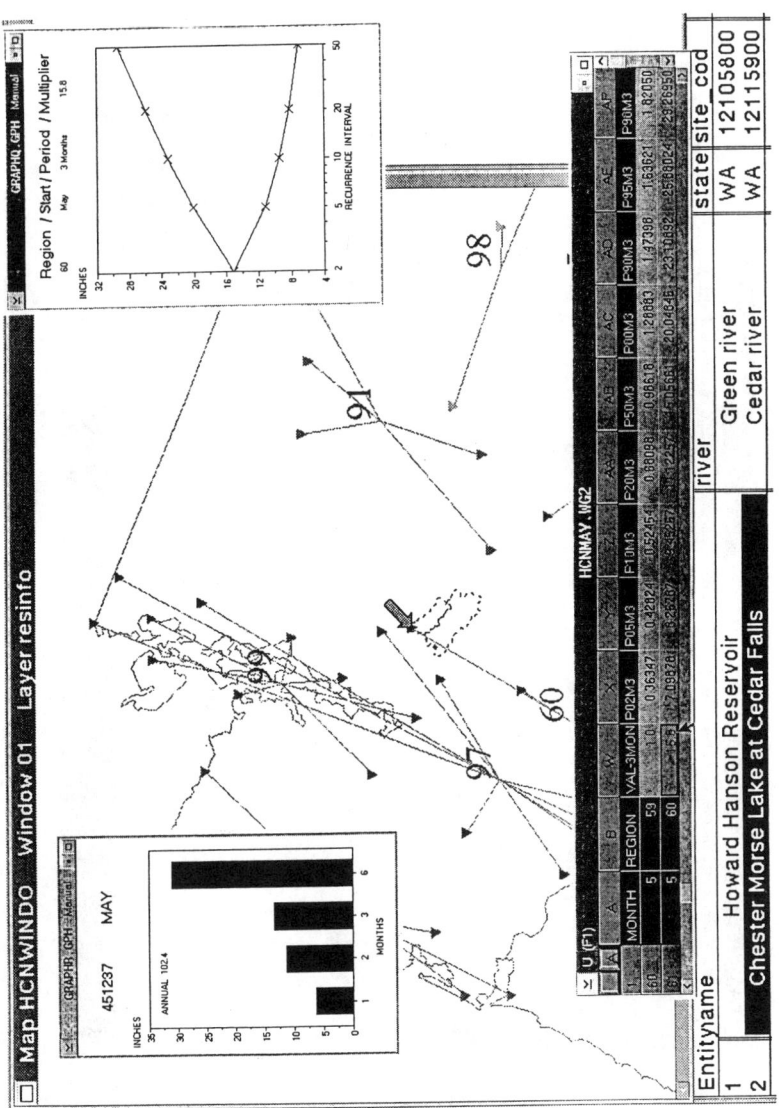

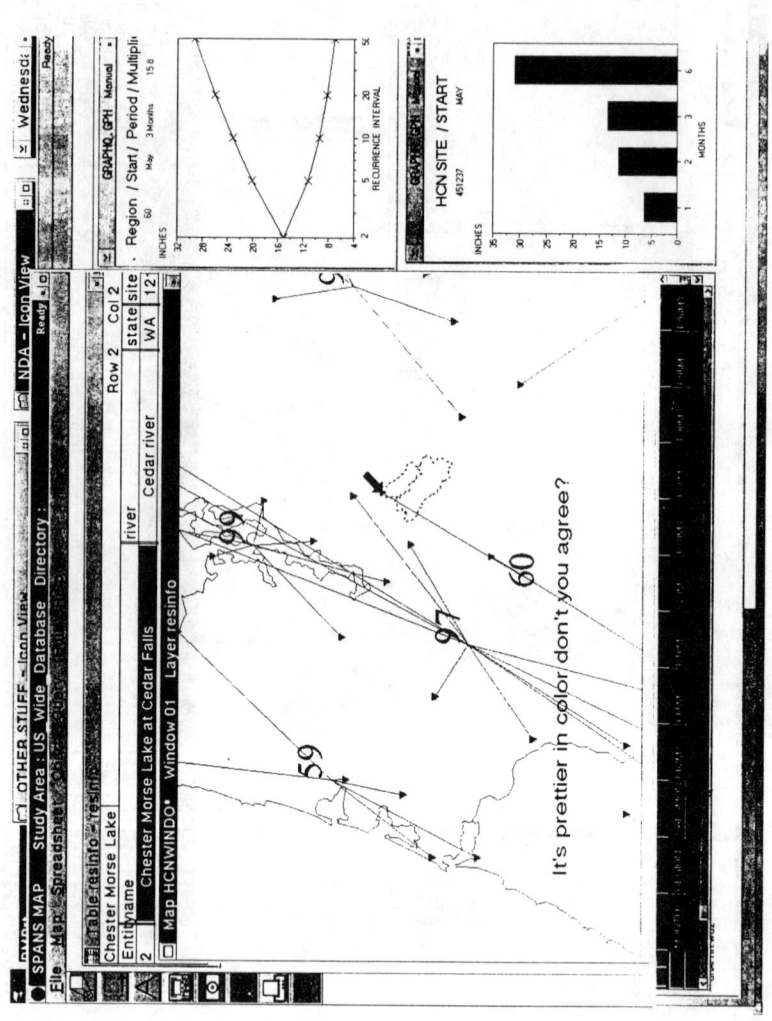

The Potomac Experience - A Forerunner of DPS

Roland C. Steiner[1]

Abstract

Thirty years ago, dramatic events were about to begin in Potomac River basin with the publication of a report of study recommending 16 major reservoirs and hundreds of small ones to solve the foreseeable water supply problems in the basin, especially those of the Washington metropolitan area. The proposed solution was confounded by its shear size, the emergence of an era of more critical thinking, and the growing development of more sophisticated analytical tools. When viewed from the perspective of the current Drought Preparedness Studies, the Potomac Experience does seem in some senses to have been a forerunner. Among the greatest similarities is the simultaneous consideration of demands and resource availability on an operational time step with the aid of computer modelling. Other contributions to the success on the Potomac which are hoped to be included in the DPS include: involvement of local elected officials and citizens groups, technical input from the affected water utilities, creative use of analytical techniques, and the dedicated commitment of the project leadership.

Introduction

Although it might be desirable that specific information be avoided in this presentation because of the potential for distraction, some references to places and dates are included in order to establish context for the Potomac as a case study. The description takes the form of answers to the questions: Where, When, What/Why (the problem), and Who/How (the solution). Each answer describes an aspect or set of aspects of the "Potomac Experience" directly or indirectly alludes to similarities between the Potomac and the DPS. The major elements leading to success in the Potomac River basin are summarized at the end of the paper.

---

[1]Associate Director for Water Resources, Interstate Commission on the Potomac River Basin, 6110 Executive Boulevard, Suite 300, Rockville, Maryland 20852

Where

The subject of this case study is the demand for water in the Washington (DC) metropolitan area and the management of resources to meet that demand. The demand area includes the District of Columbia and the adjacent suburbs in Maryland and Virginia -- three jurisdictions at the "state" level. The resources accessed by the water utilities serving the metropolitan area consist of the Potomac River near Washington, supported by river regulating reservoirs on the North Branch and Little Seneca Creek tributaries of the Potomac, and direct supply reservoirs on Occoquan Creek (a tributary of the Potomac) and on the Patuxent River (adjacent basin).

Throughout their length, the North Branch and mainstem Potomac River form the boundary between adjacent jurisdictions: Maryland, West Virginia, Virginia, and the District of Columbia. Little Seneca Creek and the Patuxent River are wholly in Maryland, and Occoquan Creek is wholly in Virginia. Thus, the demand area and its resources are spread over multiple jurisdictions -- a situation not unlike most of the Drought Preparedness Study sites. In addition, the demand area is served by three major, and several smaller, independent water utilities.

When

In the context of increasing demands, limited resources, and droughts in the mid-1950's and mid-1960's, the U.S. Army Corps of Engineers and others were motivated to examine possibilities for new sources. A notable study (5) concluded in the 1960's that 16 new large multi-purpose reservoirs and hundreds of small ones would meet present and future flood control, recreation, water quality, and water supply needs. The 1970's were characterized by acrimonious objections by the public to new reservoirs, and further studies and negotiations leading to reductions in the number of those proposed new reservoirs. The 1980's saw the culmination of analysis and agreements to implement limited resource development and cooperative management of all available sources for the benefit of the whole demand region. In the 1990's, the agreements are being exercised, and operating rules are being refined.

What/Why (the problem)

The water supply situation in the Potomac River basin was not very different from that in most of the Drought Preparedness Study sites. Demand for water exceeded (or was forecast to exceed) the yield of available resources. The definitive study of the problem at once focused attention on the problem and induced a state of paralysis with respect to its solution. The study emerged at the dawn of the era of effective local opposition to regional projects and the evolution in water quality management from treating the environment to treating the discharge.

For more than 10 years the Corps of Engineers and other agencies examined potential alternative sources, and continued to face stiff public opposition to resource development. The sense of urgency in the need for new sources was heightened during that period by the occurrence of the lowest flow of record in the Potomac River and growing demands in the metropolitan area.

Who/How (the solution)

The beginning of the end of the problem came when, in exasperation, one of the large suburban water utilities (Washington Suburban Sanitation Commission) looked for a solution to its problems within its own service area and harnessed the energies of the public on its behalf (3). A local Bi-County Water Supply Task Force was formed under the leadership of elected officials. A citizens advisory committee was established, and competent imaginative technical support was provided. The result was the identification of a small but adequate reservoir site within the utility's service area. The analysis of proposed operations included the examination of alternate levels of risk of not meeting unrestricted demand, i.e. the imposition of restrictions at various frequencies and for various durations (4).

At approximately the same time, a committee of the National Academy of Sciences recommended (1) to the Corps of Engineers that it consider demand management and conservation scenarios in its forecasts of future demands for the metropolitan area to be met by regional resource development.

The success of the Bi-County Water Supply Task Force was recognized and extended by the establishment of the Washington Metropolitan Water Supply Task Force with comparable political leadership, citizen input, and technical support. A major element of the technical support to the Metropolitan Task Force was a computer model that was evolving to simulate daily operations based on river flows and demands. Through the use of this model, it was determined that with one large new reservoir under construction on the North Branch and the planned smaller reservoir in the WSSC service area, all metropolitan demands could be met with the coordinated operation of all resources through the year 2030. The modelling analysis showed the benefits of the balanced use of near and distant sources, importantly with out regard to ownership, as being far more economically efficient than the sum of new source yields or each utility developing its own sources.

A Corps of Engineers study (2) was published which incorporated the major elements of coordinated operation.

Following the results of the technical analysis, cost-sharing and operating agreements took nearly a year to develop among the previously independent water utilities. Water supply storage in the 2 river regulating reservoirs is owned jointly by the 3 major utilities, and available to each of then through their intakes on the Potomac River near Washington. When storage from the regulating reservoirs is needed, system operation is managed by the Section for Cooperative Water Supply Operations on the Potomac of the Interstate Commission on the Potomac River Basin which is jointly funded by the major metropolitan water utilities.

On reflection, it is the opinion of several of those involved, that the combination of record low flows, increasing demands, resistance to new large reservoirs, computer modeling, pursuit of improved operations, and the dedication of several individuals in leadership roles was critical in successfully meeting the future water demands of the Washington metropolitan area.

Summary

In summary, it can be concluded that success was achieved because:

- Notwithstanding a large study by a national agency, the acceptable solution was not developed until local water utilities, elected officials, and citizens took active participatory roles.
- Departures from conventional analysis included the development of resources and operation rules with acceptable risk of restrictions.
- A specially developed computer model of drought flows and appropriate temporal resolution of demands and operating rules was critical to identifying the efficient solution.

- The coordinated operation of all resources -- irrespective of ownership -- dramatically reduces the need for more storage.
- The selected solution was formalized in new cost-sharing and regional operating agreements.
- Creativity and commitment were demonstrated by several dedicated individuals.

Acknowledgement

The author wishes to acknowledge with thanks William J. Werick of the Institute for Water Resources of the Corps of Engineers for his encouragement to present this material, and to Professors M. Gordon (Reds) Wolman and John J. Boland of the Department of Geography and Environmental Engineering at The Johns Hopkins University for their perspective on the conditions which led to the successful "Potomac Experience."

APPENDIX 1. - REFERENCES

1. Committee to review the Metropolitan Washington Area Water Supply Study, Water for the Future of the Nation's Capital Area 1984, Water Science and Technology Board, National Academy Press, Washington, D.C.

2. Department of the Army, Baltimore District, Corps of Engineers, "Metropolitan Washington D.C. Area Water Supply Study, Maryland ,Virginia and the District of Columbia, Main Report," September 1983.

3. McGarry, R.S., "Potomac River Basin Cooperation: A Success Story," paper presented 1984 at Cooperation in Urban Water Management Seminar sponsored by the National Research Council, Washington, D.C.

4. Technical Advisory Group, "Bi-County Water Supply Study, Montgomery and Prince George's Counties," Bi-County Water Supply Study, April, 1978.

5. U.S. Army Engineer District, "Potomac River Basin Report, "Baltimore, North Atlantic Division, Volume I, Report-Part 1, 1963.

Drought in the Emerald City

Steven D. Babcock[1]

Abstract

This paper discusses a drought preparedness study being conducted for the Cedar River and Green River basins in western Washington state. The study is one of four regional case studies being managed by the U.S. Army Corps of Engineers as part of the National Study of Water Management During Drought. The overriding objective of the drought preparedness study is to leave the region better prepared for drought, through demonstration and test of drought preparedness tools and strategies. The study has served as a vehicle to promote a greater regional focus on drought related water supply problem solving. The 1992 drought in the Seattle/Tacoma metropolitan area provided a unique opportunity for the study team to demonstrate approaches to drought management being researched and tested as part of the study.

Introduction

A drought preparedness study (DPS) is being conducted for the Cedar River and Green River basins in western Washington State. The DPS is one of four regional case studies being managed by the U.S. Army Corps of Engineers as part of the National Study of Water Management During Drought. The goal of the National Study is to improve the way Americans manage water for and during droughts. The overriding objective of the DPS is to leave the region better prepared for drought, through demonstration and test of drought preparedness tools and strategies. The success of the study will be measured by answering the question---how much better is the region prepared for drought as a result of the study?

---

[1]Water Resources Planner, Seattle District, U.S. Army Corps of Engineers, P.O. Box 3755, Seattle, WA 98124-2255

The Cedar River and Green River basins have drainages that extend west from the western slopes of the Cascade Mountains to Puget Sound (Figure 1). The Cedar River provides about two-thirds of the municipal and industrial (M&I) water supply used by Seattle Water Department customers. Masonry Dam is owned and operated by the Seattle Water Department. Howard A. Hanson Dam, located on the Green River, is operated by the Corps of Engineers for flood control and low flow augmentation. The Tacoma Department of Public Utilities diverts M&I water supply downstream of the dam and supplies the principal water needs of Tacoma customers. Both rivers are major producers of salmon and steelhead trout, supporting Indian and non-Indian commercial fisheries and recreational fisheries alike.

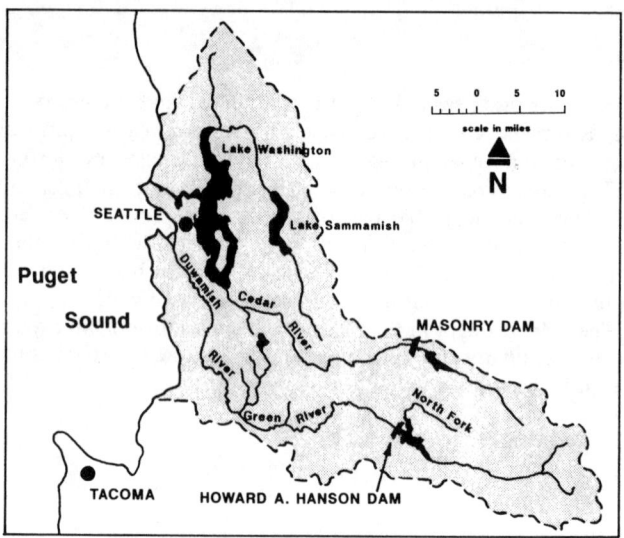

Figure 1. Cedar River and Green River Basins

The Problems

Drought conditions are not unknown to area residents, despite the Pacific Northwest's reputation for green trees and blue water and Seattle's motto as the Emerald City. During the 1992 drought, surface water supply sources were stressed, necessitating mandatory M&I water use restrictions in both basins. The DPS gave the study team the opportunity to develop and test new supply management tools. Also during drought, reductions of instream flows on the Cedar and Green rivers adversely affect migration, spawning, and incubation of salmon and steelhead trout. The DPS gave us the opportunity to develop a reservoir water supply management strategy to ensure that instream flow requirements are met to the fullest extent possible. Drought conditions also make it difficult to maintain the level of Lake Washington, thereby adversely impacting water quality and restricting commercial and recreational navigation through the Hiram M. Chittenden Locks operated by the Corps of Engineers. The DPS provided us the opportunity to improve our ability to manage the system on the Cedar River so as to minimize the chances of receding lake levels during drought.

Planning Objectives

Through workshops, interviews, and discussion, the regions' water managers identified what the DPS could do to reduce impacts and allow the region to endure future droughts with less difficulty. We are pursuing the following planning objectives:

o To ensure the reliability of M&I water service in the Seattle Water Department and Tacoma Department of Public Utilities service areas during drought conditions.
o To maximize Cedar River and Green River instream flow compliance with state criteria during drought conditions.
o To increase the reliability of avoiding drafting Lake Washington below authorized levels during drought conditions.
o To enhance interagency coordination and collaboration and to provide tools for negotiating water management issues during drought conditions.

Decision Makers and Influencers

A variety of decision criteria used by key decision makers and influencers in evaluating the acceptability of drought management alternatives have been identified. Decision makers and influencers in the Cedar River basin include the Seattle Water Department(SWD); Corps of Engineers; Washington Departments of Ecology, Fisheries, and Wildlife; Muckleshoot Indian Tribe; and U.S. Fish and Wildlife Service (FWS). The list for the Green River basin is the same, except that Tacoma Public Utilities is substituted for SWD. Tacoma and SWD are both primarily concerned about reliability of M&I water service, feasibility of alternatives, public acceptance, and impact on instream flows and fish habitat. The Corps of Engineers is primarily concerned with impacts on navigation and lake level and water quality in the Cedar basin. In the Green River basin, the Corps' primary concerns include flood control, reliability and quality of water supply for low flow augmentation, instream flows, fish habitat, and water quality. Ecology's concerns include instream flows, fish habitat, and water quality. The Muckleshoot Indian Tribe and FWS are both concerned with impacts to the salmon and steelhead fishery, instream flows, and water quality. Fisheries and Wildlife share these same concerns, with the distinction that Fisheries is responsible for salmon and Wildlife for steelhead trout. These entities, plus a variety of influencers, have participated in the study process.

Alternatives and Implementation

The region most recently experienced drought conditions in 1987 and again in 1992. Accordingly, awareness of drought impacts and interest in drought preparedness is very high. Tactical alternatives being considered by the DPS study team include adoption of a computer based decision support tool to aid in negotiating water management issues during drought, revision of agency drought contingency plans, adoption of interagency coordination guidelines, and development of indexes based on snow and precipitation observations to predict the onset and progression of drought conditions. All of these measures are intended to leave the region better prepared for drought.

Summary

The Cedar River and Green River basins drought preparedness study differs from previous drought planning efforts in the region. The study team has focused on interagency problem solving and shared responsibility. In the context of the 1987 and 1992 droughts, the study has served as a vehicle to promote a greater regional focus on drought related water supply problem solving. A concerted effort is being made by all agencies to enhance regional cooperation and collaboration. The DPS and the National Study of Water Management During Drought have provided an opportunity to demonstrate and test a variety of tools and experiences from across the country and from our own backyard here in the Emerald City.

The Challenges of Interstate Water Planning and Management

Michael J. Bart[1]
Christopher R. Erickson[2]

Abstract

This study is one of four primary basin studies as part of the U.S. Army Corps of Engineers National Study of Water Management During Drought, a four year study initiated after the Drought of 1988. The Marais des Cygnes - Osage River Basin, located in Kansas and Missouri, is examined in this study. A fundamental goal of this study effort is to identify alternatives to the status quo and thereby better prepare the basin for future droughts. Drought preparation is accomplished with the participation of the Corps of Engineers and the States of Kansas and Missouri throughout the entire study. The obstacles of dissimilar water laws, competing interests for water, and multi-organizational planning participation are examined within this study. By fully involving all parties, the study becomes a collaborative planning effort and traditional barriers to progress can be examined and potentially reduced or eventually eliminated.

Introduction

The Marais des Cygnes - Osage River Basin originates in east-central Kansas and flows easterly to its confluence with the Missouri River, in central Missouri. The entire basin drains an area of about 15,300 square miles. The basin (see Figure 1) contains six Corps of Engineers reservoirs: Melvern, Pomona, and Hillsdale in

---

[1]Supervisory Civil Engineer, Planning Division, Kansas City District, U.S. Army Corps of Engineers, 601 E. 12th Street, Kansas City, MO 64106-2896

[2]Civil Engineer, Special Studies Branch, Planning Division, Kansas City District, U.S. Army Corps of Engineers, 601 E. 12th Street, Kansas City, MO 64106-2896

Kansas; and Stockton, Pomme de Terre, and Harry S. Truman in Missouri. Additionally, Lake of the Ozarks, owned and operated by Union Electric, lies immediately downstream of Harry S. Truman.

The Kansas City District U.S. Army Corps of Engineers, the Kansas Water Office and the Missouri Department of Natural Resources comprise the primary study team. The purpose of this team effort is for these agencies to work collectively in achieving the fundamental goal of the entire study-- "leave the basin better prepared for drought." During drought, four objectives include:

1. Increase the reliability of water supply service to municipal and industrial water services;
2. Increase the recreation opportunities at six federal and one private reservoir;
3. Increase the reliability of power generation at the two federal reservoirs, one private reservoir, and a thermal power plant; and
4. Maintain streamflow at critical checkpoints for environmental and water quality standards.

If these objectives can be fully or partially met during a drought situation the basin will be left better prepared for drought. Within this basin competing water usage during drought includes hydropower, cooling water supply for thermal power generation, recreation, in-stream flow requirements, irrigation, and municipal and industrial water supply.

In this study the development of a model to simulate the entire basin during drought scenarios is utilized. The software used is a graphical, object oriented simulation language. This type of user friendly software allows both programmers and non-programmers alike to participate in the formation of the basin model.

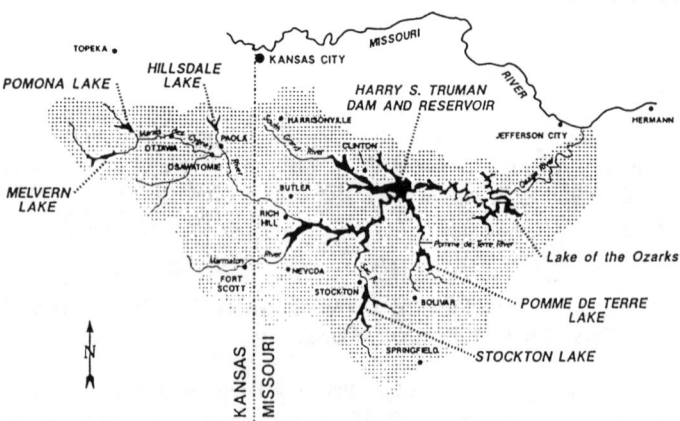

Figure 1

Problems and Opportunities

1. During drought conditions there are potential negative impacts to municipal and industrial users in Kansas and Missouri. As a result of additional environmental statutes, it is becoming increasingly difficult for small water suppliers to provide the level of treatment required by federal and state officials. Therefore, small water districts are likely to call upon the larger suppliers for water from the Marais des Cygnes - Osage River. Industrial users include, but are not limited to, Kansas City Power and Light, a thermal generating power plant, and the hydro-power producers of Union Electric and Southwestern Power Administration.

2. During drought there are impacts to reservoir and stream recreation interests. Economic benefits resulting from recreation opportunities at Lake of the Ozarks and Federal reservoirs are significant to both states.

3. During drought, power losses will likely occur at Bagnell Dam (Lake of the Ozarks), Harry S. Truman, and Stockton Dams in Missouri. Additionally, losses may occur at thermal power producing facilities in which cooling ponds are used for thermal generation.

4. During drought, it is a common occurrence that instream flows within the Marais des Cygnes - Osage River are often depleted or even absent. This lack of flow causes damages to both the agricultural and environmental interests.

The opportunity does exist to reduce negative impacts to the preceding list. This will be accomplished by examining the formation of water assurance districts, water marketing programs, changes in reservoir operation, conjunctive management, institutional change, and demand modification techniques.

Interstate Water Challenges

1. <u>Differing Water Laws</u>. This basin is impacted by differing water laws in each state. Appropriation doctrine is practiced in Kansas whereas riparian doctrine is the focus in Missouri. A main challenge in this basin is the limited availability of information in Missouri. Under appropriation laws, Kansas allocates, distributes, monitors, and grants water rights, therefore this state is relatively well informed of the activities which pertain to its waterways. Conversely, Missouri has a riparian system of water law which is characterized by very little state administration resulting in sparse water use and consumption data.

2. <u>Political Boundaries</u>. In the holistic sense, the preference is to examine a basin as a whole. This ideal philosophy does not negate the fact that the state line divides the basin into two distinct political regions. In the drought period of 1988 the flow at the state line was near zero. The state line area is of close observation

and much concern. Within Kansas there are three federal reservoirs which regulate downstream flows. During drought as water is released from these reservoirs, Kansas will use most if not all streamflow to meet the appropriated rights. This can leave the downstream state, Missouri, with little to no streamflow. The conflicts with this scenario are obvious and lead to much debate.

3. Management Philosophies. The differing water laws dictate the management philosophy each state employs in upholding the users' normal water rights or supplies. Under appropriation doctrine, Kansas is continually seeking methods to protect water rights. Currently, Kansas is trying to implement an Assurance District within the basin. Under this program, storage will be purchased from Federal reservoirs and the stored water will be available to water right holders during periods of drought. Missouri, under the riparian system, has little authority to protect water availability.

4. Basin Geography. The physical arrangement of the basin contributes to potential conflict. Three of the six Federal reservoirs are located in Kansas which can impact the flows that reach the state line with Missouri. Additionally, one large private reservoir which generates as much as six million recreation dollars of revenue per week in the summer months is located on the downstream end of the basin. This large private reservoir, Lake of the Ozarks, is located immediately downstream of Harry S. Truman reservoir, a large Federal multi-purpose structure. The main operation for Lake of the Ozarks, downstream of Harry S. Truman, is for recreation and hydropower purposes, therefore Lake of the Ozarks is seeking a constant pool elevation. An institutional conflict does occur when the public perceives a large Federal reservoir with ample water supply just upstream of their reservoir when their boat docks are well away from the waters edge.

Conclusion

The U.S. Army Corps of Engineers, Missouri Department of Natural Resources, and the Kansas Water Office are employing multiple decision makers in the development of a drought contingency plan for the basin. Throughout this process a series of workshops are held in which the decision makers and influential parties are invited. These workshops help to identify the concerns and problems which the basin is encountering. In this facet of collaborative interstate planning the importance of constructive working relationships and information exchange between all entities is of paramount importance. It is through this multi dimensional planning process that varying ideas, thoughts, and practices are brought together for a unified approach in better preparing the basin for drought.

The James River Case Study

Thomas J. Lochen[1]

Abstract

The James River was selected as a case study for the National Drought Study because it provided an example of an Eastern (riparian) river basin with relatively little storage, increasingly complex uses, an increasing potential for conjunctive use of surface and groundwater and interbasin transfers, and water quality problems during drought.

During a drought no worse than the 1988 national drought, cities in the basin had to implement mandatory curtailment plans with resultant job losses. In a bad drought, losses of cooling and processing water would cause shutdowns in industries which have already sharply reduced water use. Curtailment of outdoor uses of water would cripple many commercial enterprises. Our national security would be threatened by drought due to the presence of the largest naval base in the world, its ancillary facilities, and large installations representing the other branches of the defense sector. Finally, a severe drought would have serious impacts on the groundwater aquifer, leaving it susceptible to saltwater intrusion.

This paper describes the process by which the James River Drought Preparedness Study (DPS) team developed a tactical drought contingency plan which defines roles and responsibilities, makes a judgement about how serious a drought to plan for, refines indicators and decision triggers, estimates the balance between water supply and demand, and explores strategic alternatives such as additional storage and a more comprehensive state water policy.

Introduction

The James River Basin is located almost entirely within Virginia; less than 0.1 percent is in West Virginia. It originates in the Allegheny Mountains on the Virginia-West Virginia border and flows southeasterly through the central portion of Virginia into the Chesapeake Bay at Hampton Roads. The drainage area of the basin

---

[1] Study Manager, Norfolk District, U.S. Army Corps of Engineers, 803 Front Street, Norfolk, VA 23510-1096

consists of approximately 11,000 square miles, which is one-fourth of the state of Virginia. The mainstream is 340 miles in length and descends a total vertical distance of 988 feet to sea level at Richmond, Virginia, which is located at the Fall Line. The actual length of the river basin from northwest to southeast is 240 miles, and it varies from 10 to 90 miles in width. The James River is formed below Clifton Forge near the community of Lick Run by the junction of the Jackson and Cowpasture Rivers. The James crosses more geologic regions and has more tributaries than any other river in Virginia. Its major tributaries include the Maury, Rivanna, Appomattox, and Chickahominy Rivers. This basin was chosen as one of the case study areas because it is very representative of riparian, coastal-piedmont basins with many issues including water quality, salt water intrusion, and groundwater usage, and competing uses such as major metropolitan municipal and industrial (M&I) water supply, hydropower, anadromous fisheries, irrigation, and recreation. It was the highest ranked mid-Atlantic area, and it offered a few problems not seen in other studies.

Problems and Opportunities

A group of key interested agencies and individuals from Federal, state, regional, and local governments, as well as industrial, commercial, environmental, and academic concerns in the region was called together to act as an advisory group for this study. During a series of workshops, this group identified the major problems in the basin and agreed on a course of action to address those problems. The main issue to arise was the lack of M&I water supply in the lower portion of the basin and the need for a state water policy. At one point during the recent drought in the early 1980's, there was only a 30-day supply of water remaining for the Southeastern Virginia area. The many competing uses within the M&I sector absorbed extreme impacts to their operation. There is concern that in future years the middle part of the basin may also develop the same problem. There is no comprehensive state water policy and until now water management has been left to the local municipalities. Increasing restrictions on water use and growing needs due to a swelling population are placing ever greater stress on the existing water supply. Other issues concerning groundwater contamination through salt water intrusion in the lower basin were also raised at these workshops.

Planning Objectives

Our study team agreed that we would pursue the following objectives during this DPS:

1. Insure water service to the domestic sector in the middle and lower James River Basin during drought conditions.

2. Insure water service to the industrial and commercial sectors in the middle and lower James River Basin during drought conditions.

3. Insure water service to medical facilities in the middle and lower James River Basin during drought conditions.

4. Insure water service to military facilities in the middle and lower James River Basin during drought conditions.

5. Enhance surface water quality reliability in the middle James River Basin and the groundwater reliability in the middle and lower James River Basin during drought conditions.

Principals

The principals or decision makers is this study are the Corps, the Commonwealth of Virginia as represented by the Governor and General Assembly, and the local municipalities who are water purveyors. Groups who influence the principals include those who are responsible for regulating water use, local municipalities adjacent to the water purveyors, and various advocacy groups.

Any plans to relieve the situation during drought will be scrutinized according to the respective criteria of the decision makers and those who influence them. The Corps will want to know what impacts may be incurred by existing and future Corps projects. The Commonwealth will want to be sure the plan is within fiscal constraints and in accordance with policy guidelines. The local water purveyors will be concerned with the reliability and level of municipal and industrial water service, fiscal constraints, plan implementability, the ability to fulfill water supply contracts, and impacts on revenue from water sales and on their existing water systems.

The regulating agencies will be concerned with the impact on the environment, public health, sanitation systems, etc. The local municipalities adjacent to the water purveyors will be interested in the impacts on their water supply and their economy. The various advocacy groups will address the impacts to their respective concerns such as the environment, economic growth, agriculture, and water system operation.

Alternatives

Historically, certain areas in the lower part of the basin have suffered serious impacts during drought events. The area in Southeastern Virginia has limited quantities of water available to it and long term water conservation is being practiced. It would appear at this time that a source of water from outside the basin is the best alternative to the area's drought vulnerability.

The James-York Peninsula is also in the lower basin and has much the same problem as does Southeastern Virginia, but not as severe. An additional source of water appears to be needed here also, but coupled with a Drought Contingency Plan which includes long term conservation.

The middle basin area centered around the city of Richmond and Henrico and Chesterfield Counties may be seriously affected by drought in the future. This area lends itself to a more diversified approach to reducing the impacts caused by drought. A Drought Contingency Plan for the area could consider long term water conservation, modifying upstream use, and augmenting flows by additional releases from Federal and local reservoirs. Additional offstream storage is a strategic option.

Innovations and Differences

This DPS is not a reallocation study, reconnaissance study, or Congressionally authorized implementation study. It is based on traditional format, but has been given the latitude to go beyond these boundaries to develop alternate solutions to drought problems, rather than to determine just Federal interest and involvement in a project.

The study is being administered on the national level by the Institute for Water Resources (IWR), Water Resources Support Center, Fort Belvoir, Virginia. IWR has drawn together several professional disciplines such as law, engineering, economics, alternative dispute resolution, risk assessment, planning, and the environmental sciences to assist in this endeavor. Studies will be conducted in each of these fields with the intent of publishing a manual at the end of the national study entitled How To Be Prepared for Drought. Each of the studies will lead to the development of a separate chapter in the manual. The manual will be designed for use by regional planners and decision makers and will combine the analysis from the topical studies with the practical experience from the DPS's. A National Drought Atlas will also be developed and published separately from the manual. The manual will also explain how the National Drought Atlas can be used as part of an overall planning process. In addition, each DPS has access to this national pool of expertise and innovative methodologies and state-of-the-art analytic tools. There is little question that a broader application of the best current analytic techniques in an integrated, customer-oriented study would improve drought preparation at a minimal cost.

The Role of Object Oriented Simulation Models in the Drought Preparedness Studies

Allison M. Keyes[1], Student Member and Richard N. Palmer[2], Member, ASCE

Introduction
One of the major activities in the US Army Corps of Engineers Drought Preparedness Studies (DPS) has been the development of water resource simulation models for each study site. These models were created to provide decision support to DPS participants engaged in strategic and tactical drought planning. STELLA II^{TM}, an object oriented simulation tool, was selected as the environment for model development. This choice of an object oriented environment has had several significant impacts on both the process of model development and the effectiveness of the model as a planning tool. It has enabled the DPS groups to approach model development in a very different way than would have been possible had a conventional programming language been used. It has also enabled the model to be integrated into the planning process more effectively. This paper describes the role of object oriented simulation in the DPS. It sites the many benefits gained by its use and the challenges that have been faced in the development process.

Features of Object Oriented Simulation Environments
Commercially available object oriented simulation environments share several features. Perhaps their most salient feature is that models are constructed in a graphical environment, using objects as building blocks. During model construction the developer defines objects with specific attributes representative of physical or conceptual system components and indicates the functional relationships among objects. This mode of construction is analogous to drawing a flow chart or schematic of the system to be simulated. Thus, these modeling environments are often more intuitive and easily learned by novice users than traditional programming languages. Furthermore, since extensive training is not required, models can be developed by those with a wide variety of backgrounds.

Because they are easily constructed and modified, object oriented simulation environments facilitate rapid prototyping and greatly reduce "programming" effort. Instead, modeling effort can be directed to important tasks such as system conceptualization, data collection, calibration and implementation.

---

[1]Graduate Research Assistant, and [2] Professor of Civil Engineering, Department of Civil Engineering, University of Washington, Seattle, Washington 98195

Another advantage of these graphical environments relates to the ease with which model assumptions can be conveyed to its users. Since relationships between objects are graphically indicated, users can readily discern the interactions among model components. Thus, they are more likely to recognize when invalid assumptions exist, or where greater detail is needed.

Impacts on DPS Model Development

To effectively complement the DPS drought planning process, the simulation models created by the DPS groups must be trusted by stakeholders. They must clearly communicate information to an audience with a diverse background in water resource management and widely differing concerns (Loucks 1992). In addition, they must help stakeholders gain a common appreciation for the scope and source of the problems posed during drought and demonstrate the trade-offs among interests implied by different mitigation alternatives. Finally, they must be sufficiently flexible to assist planners throughout the planning process from problem identification through implementation. Ideally, all these attributes will be packaged in an environment that is interactive, user friendly and easy to use.

STELLA II^{TM}, has provided the DPS groups with a valuable tool for developing models with these desired attributes and has positively influenced many aspects of the DPS modeling effort. In particular, it has facilitated the process of translating planning objectives into an appropriate modeling framework, allowed a wide variety of impacts to be investigated, and established the groundwork for successful model implementation.

One of the most critical phases in DPS model development occurred when establishing an appropriate modeling framework for each region. In this phase, model functions were delineated and translated into specific requirements that define the model's framework including: appropriate system components, level of detail, time step, measures of system performance, and user interface.

The use of an object oriented simulation environment facilitated this process in several ways. First it enabled the modeling framework to be defined iteratively, through rapid prototyping and experimentation. More significantly perhaps, it enabled stakeholders to influence the modeling framework from the outset of the development process. The clarity and ease of use afforded by this environment, enabled users to critique the model effectively and even participate directly in its construction. Misconceptions and inaccuracies could be cited early on, before investing a significant effort in model development. Thus, the models were better customized to meet the needs of their intended users.

One particularly important aspect of defining the modeling framework was the choice of impacts to be represented by the model. To be successful as a decision making tool, impacts relevant to the stakeholders in each DPS region had to be incorporated into the models. In addition, these impacts had to complement DPS planning objectives. Once again, the use of an object oriented simulation tool played an important role in broadening the types of impacts that were expressed in the modeling framework.

A typical approach for soliciting appropriate impacts was through model demonstration. In these sessions, participants were asked to comment on the relevance of impacts shown in the prototype and make suggestions for improvement. Because of the ease of modification and flexibility of the object oriented environment, it was often possible to respond to their suggestions by immediately incorporating new types of impacts into the model. This immediate feedback helped participants gain an appreciation of how their input would impact the model content, and how the model might be useful in their own decision making. As this understanding improved, they were better able to articulate the types of information most relevant to them. When data limitations were encountered, participants were often able to suggest surrogate measures of system performance that could be more easily modeled and come to concurrence on assumptions used to model these measures.

Because the graphical, object oriented modeling environment increased the involvement of stakeholders in model development, a wide diversity of drought impacts were modeled which might have otherwise been overlooked. The range of effects assessed include social and environmental impacts, such as fish losses or political approval. Measures of resiliency and vulnerability were incorporated into many models in addition to the more traditional measure of system reliability.

At the time this paper was written, the degree to which the simulation models will be integrated into the DPS planning efforts has not yet been determined. However, the use of an object oriented simulation environment has clearly helped DPS model developers lay the groundwork for successful implementation. The clarity of the object oriented environment has helped decision makers to understand model assumptions. This clarity, coupled with opportunities to contribute to model development, has enhanced stakeholders' trust in the model and confidence in its results. This involvement in model development has allowed DPS participants to jointly define the model's role as a decision support tool and assure that relevant information is presented in an effective way. Finally, the flexibility that this modeling environment offers will allow the models to easily modified to address changing concerns in the planning process.

Challenges Posed to Model Development

Before learning of STELLA II^{TM}, the DPS groups had decided against building simulation models because of the extensive time and resources required and the limited flexibility offered by non-objected oriented model development tools. But STELLA II^{TM} provided the DPS with a set of tools for creating simulation models that would be potentially effective in the planning process. Thus, it encouraged DPS participants to address several challenges of model development which they were originally reluctant to face.

The task of creating models that are effective planning tools poses a variety of technical, social, and political challenges (Loucks 1992, Delli Priscoli 1989). It has required DPS model developers effectively utilize a wide variety of skills including communication, consensus building, public involvement and creative thinking. Furthermore it has prompted them to actively integrate public involvement into model development.

To be successful in their efforts, the DPS groups have had to recognize how best utilize the tools offered by STELLA II^{TM} and to devise a process for model development that would be as valuable as the final product produced. They have had to devise a forum in which multiple parties could effectively contribute to the model and arrive at a consensus regarding its framework, assumptions, and the context in which it should be used. In addition, they have had to earn and maintain the trust of stakeholders and instill confidence in both the model development and planning processes. In many cases they have had to overcome skepticism, time constraints, and institutional conflicts to encourage stakeholder participation and support. Disagreements over technical issues have also had to be resolved.

DPS participants had varying levels of experience in water resource simulation model development, planning, public involvement and consensus building. They were often unfamiliar with certain aspects of the approach that was recommended for model development. As a result, several workshops were held during the course of the DPS to enhance the skills of project mangers and model developers. The topics covered during these session included developing appropriate planning objectives, using STELLA II^{TM}, developing a modeling framework, interview techniques, improving model clarity, and multiobjective decision making.

Another issue that has arisen in many of the DPS sites is the need for groundrules on model use. This issue stems from the wide range of agencies and interest groups that will have direct access to the models. One fear is that the model will be altered to misrepresent system response to favor a special interest in a region. Another concern is that the model will be inappropriately used to guide decisions made in contexts other than drought planning. The challenge for the DPS groups is to devise groundrules that address these concerns which are supported by all stakeholders

The task of developing groundrules for model use is analogous to structuring protocols for multi-party negotiations. Such rules should define who has access to the model, the context in which its use is appropriate, and how the information supplied by the model will be used. They should also specify conditions under which model modification would be warranted, and how modification would be accomplished. Finally, provisions for future groundrule modification should be included (Cormick, 1989).

Conclusion

DPS model development efforts have greatly benefited from the use of a graphical, object oriented simulation environment. However, the use of object oriented simulation in itself will not guarantee the success of the DPS modeling efforts. The ability of the DPS groups to use this modeling environment effectively and to structure processes for model development and use that complements the planning process will likely determine the value of the models in practice.

References

Cormick, Gerald W. (1989). "Strategic Issues in Structuring Multi-Party Public Policy Negotiations." *Negotiation Journal*, April, 125-132.

Delli Priscoli, Jerome (1989). "Public Involvement, Conflict Management: Means to EQ and Social Objectives." *Journal of Water Resources Planning and Management,* 115(1), 31-41.

High Performance Systems, Inc. (1992). STELLA II Tutorial and Technical Documentation . High Performance Systems, Inc., Hanover, NH.

Hobbs, Benjamin F., Eugene Z. Stakhiv, and Walter M. Grayman (1989). Impact Evaluation Procedures: Theory, Practice and Needs." *Journal of Water Resources Planning and Management,* 115(1), 2-21.

Loucks, Daniel P. (1992) "Water Resource System Models: Their Role in Planning." *Journal of Water Resources Planning and Management,* 118(3), 214-223.

Demonstrating Competition for a Limited Resource
in the Cedar/Green Basins

Christopher J. Lynch[1]

Abstract

An object oriented model building tool is used to develop two river basin simulation models. The models will be used to help manage the water resource during dry years in the Cedar and Green river Basins in western Washington. Alternative management strategies will be tested using the models. Effect on competing uses will be observed and evaluated.

Introduction

The Cedar and Green River Basins are similar but unique neighboring basins on the western side of the Cascade Mountains in Washington state. Each basin experiences very similar precipitation, temperature, and other weather patterns. Historically, management of the water resources in these basins has included M&I water supply diversions, minimum instream fish flows, flood control storage and flow modifications, and Navigation flow. In most years plenty of water exists in the systems to meet the needs of all the uses. During certain dry years, however, shortages to one or all the uses can occur. During these periods extra coordination and careful management practices must be implemented to avert extreme hardship to any of the uses.

An object oriented model of each basin is being developed to enhance both the management of the water resource and the coordination between agencies involved in the management of the resource. These interested and affected agencies are cooperating and contributing to the formulation and development of the models which are intended to be "hands on" tools for all levels of technical staff and management to use and operate.

---

[1]Hydraulic Engineer, Seattle District, U.S. Army Corps of Engineers, P.O. Box 3755, Seattle, WA 98124-2255. Ph (206) 764-3591; fax (206) 764-6678.

General Climatology and Hydrology

The Cedar/Green Basins lie on the west side of the Cascade Mountains about 100 miles east of the Pacific Ocean. During the winter these basins experience relatively frequent, heavy, frontal rains. The summer months are mild, warm, and generally dry. Mean annual temperatures range from 39 degrees F at the mountain passes to 52 F in Seattle. The normal annual precipitation ranges from a low of 35 inches in Seattle to 110 inches at the upper reaches of the basins with around 70 percent of the precipitation arriving between October and March. Snowfall accumulations are appreciable above an elevation of 3,000 feet. Snowmelt runoff is normally counted on between April and June to help supplement receding streamflows as precipitation diminishes.

Streamflow patterns generally follow that of precipitation. Highest streamflows occur from October through March and diminish as Spring and Summer progress. Lowest streamflows generally occur just prior to the return of Autumn rains in September or October. During unusually dry Autumns streamflows have been known to continue in recession into November and early December. This type of low flow pattern is rare and can be particularly hard on the various uses of the water resource.

Cedar River Basin

Chester Morse Lake is located in the upper portion of the Cedar River basin. It is owned and operated by the City of Seattle for conservation and flood control. Lake outflows are controlled by two manmade structures, the main, higher Masonry Dam and the smaller and shorter Crib Dam 1.5 miles upstream. A large amount of seepage is lost to the Cedar River Aquifer between the Crib Dam and the Masonry Dam so this area, known as the Masonry Pool or Cedar Lake, is used on a limited and strategic basis to store water. Downstream 15.6 miles is Seattle's municipal and industrial water supply diversion dam. Between this location and the mouth of the Cedar River at Lake Washington minimum instream flows are maintained for fisheries and other natural habitat. Although the Cedar River ends at Lake Washington its waters contribute significantly to the flood control and navigation operations of the Lake Washington and Ship Canal system before they eventually feed into the Puget Sound. The Corps of Engineers' Hiram Chittenden locks, dam, and fish passage facility are owned and operated by the U.S. Army Corps of engineers and are used to control the level of Lake Washington, and to provide a passage between the Puget sound and Lake Washington for commercial and industrial boat traffic and for fish.

The City of Seattle also owns and operates the South Fork of the Tolt Dam and Reservoir on the South Fork of the Tolt River. Its operation influences the operation and demands on the Cedar River and is therefore a critical element in the model of the Cedar River Basin.

Green River Basin

The Howard A. Hanson project (HAH) is located at the upper half of the Green River Basin. It is owned and operated by the U.S. Army Corps of Engineers for flood control storage and water conservation to supplement instream fish flows during the low flow period. At a diversion dam three and one half miles downstream of HAH the City of Tacoma diverts their M&I water supply. The Green River runs unimpeded for the rest of its length from the diversion dam to its confluence with the Black River. From there the two rivers become the Duwamish River which terminates eleven miles downstream at Elliot Bay.

The Green River provides the majority of Tacoma's M&I water supply. Tacoma also operates a number of wells for water supply and water quality purposes which can influence and effect their diversions from the Green River.

General Management Strategies.

Both Basins follow the same general management strategy. During the autumn and winter flood control season the storage projects in each basin are maintained at or near their flood control operating pool, except for the purpose of reducing floods downstream. Spring refill of the basins' reservoirs begins in early April and is targeted to be complete by 1 June. Minimum instream flows are maintained throughout the year. Municipal and industrial water supply is diverted from both rivers throughout the year but sees a marked increase during the summer and early fall due primarily to increased demands for domestic outdoor uses. As natural flows recede supplemental flows from reservoir storage are essential beginning sometime between late May and early July and continuing until autumn rains return. Infrequent summer rains may reduce the need to supplement natural streamflows.

Competing Uses.

Both River systems serve as the primary municipal and industrial (M&I) water supply source for a major metropolitan area. The Cedar system serves the greater Seattle Metropolitan area, while the Green serves the Tacoma Metropolitan area. Both rivers support a fisheries habitat and a natural environment which is highly dependent on the range and duration of flows as well as the physical qualities of the water. Both basins contain lakes and reservoirs which are used for flood control as well as conservation. The Cedar also provides water for navigational uses. The lower eight miles or so of the Duwamish River are navigable but are influenced primarily by tides and require no special flow demands from the Green River. Table 1 shows a matrix of the competing uses. An X indicates that a tradeoff exists between the use on the left and the use listed at the top. A tradeoff is defined as the potential loss of benefits from one use as a result of a potential gain in benefits by another use.

	RESOURCE USES	CEDAR								GREEN							
		A	B	C	D	E	F	G	H	A	B	C	D	E	F	G	H
A.	WINTER FLOOD CONTROL																
B.	SPRING FLOOD CONTROL		X				X				X				X		
C.	SPRING REFILL	X		X						X		X					
D.	SPRING FISH FLOWS		X			X	X				X			X	X		
E.	SUMMER INSTREAM FLOWS		X			X	X				X			X	X		
F.	M&I DIVERSIONS		X				X	X			X					X	
G.	FALL FISH FLOWS	X		X	X	X				X		X	X	X			
H.	NAVIGATION						X										

Table 1. Tradeoff Matrix of Competing Uses

Basin Simulation Models

A model for each basin is being developed using the STELLA* II programming tool. STELLA* II runs in a Macintosh environment and is easy to learn and operate. The models are object oriented allowing the user to see the basin layout and the interaction between basin elements. "Below" this surface visual level resides detailed basin information such as historical streamflows, reservoir operating guide curves, and minimum instream flow requirements, as well as the mathematical relationships that define how these various elements of the basin effect each other. These data and equations can be easily accessed and modified. This enables interested parties to quickly formulate and test system operating strategies. New elements can easily be added to an existing model and separate models can be joined together.

Objective

The Cedar/Green Basin models will simulate each of the basins and will incorporate the critical elements of the actual systems. Each of the major competing uses listed in table 1 will be represented in the model as well as other elements which are expected to impact or alter a given use. The critical elements were determined through interviews with agencies that are interested in and effected by basin management. The models will be used to demonstrate alternative basin management strategies using the historical period of record. Impacts and benefits to each of the uses listed in Table 1 shall be evaluated as a result of the simulated alternatives. The objective is to minimize and balance negative impacts while improving overall system performance.

INCLUDING EXPERT SYSTEM DECISIONS IN A NUMERICAL MODEL OF A MULTI-LAKE SYSTEM USING STELLA

James M. Stiles[1]
Richard E. Punnett, PhD[2]

Abstract

During the drought of 1988, water users in the Kanawha River, West Virginia, became painfully aware of conflicting uses that have developed since the 1970s. A numerical modeling of the Kanawha River was conducted to evaluate the effects of different operation scenarios on river users -- including water quality flows. The model was created with the object-oriented programming language, STELLA. The model includes user-developed Boolean statements to create an expert system. STELLA, using the expert system, was able to accurately reproduce the water management decisions and river flow for 1988 which included flood releases, minimum flows, lake storage balances, and downstream water quality targets.

Introduction

The Kanawha River in West Virginia has a drainage area of 12,300 mi^2 (31,857 km^2) and includes parts of the states of West Virginia, Virginia, and North Carolina. A total of 5,905 mi^2 (15,294 km^2) of this watershed area is controlled by three flood control projects. Bluestone Lake on the New River, the largest project in the basin, controls a watershed area of 4,565 mi^2 (11,823 km^2).

---

[1]Hydraulic Engineer, Huntington District, Corps of Engineers.
[2]Chief, Reservoir Control Section, Huntington District, Corps of Engineers.

Summersville Lake, the project that makes the whitewater releases on the Gauley River, controls a watershed area of 803 mi^2 (2,080 km^2). Sutton Lake on the Elk River controls a watershed area of 537 mi^2 (1,391 km^2).

This study developed a model of the three lake system with the operational rules being controlled by the two demands: whitewater releases from Summersville Lake on the Gauley River and augmentation flow in the Kanawha River at Charleston, WV, for water quality purposes. During severe droughts, such as in 1988, Summersville Lake could not meet whitewater schedules without releasing storage designated for water quality augmentation. The goal of this model was to facilitate the investigation of the economic impacts of several proposed operational strategies. The design philosophy reflected this goal and required that the model be simple and easy to understand. The objective of this paper was to describe the implementation of expert system decisions in a numerical model of the basin system.

The primary software requirement for the development of a model of the Kanawha River basin was the ability to make a decision making structure that would accurately simulate the current and proposed operational policy. A secondary software requirement was the ability to communicate the decision making logic to those non-technical persons who would be affected by changes in the operational policy.

Modeling Software

STELLA, an object-oriented expert system programming language, provided a relative easy method of implementing the decision making logic. At the beginning of this study, neither of the authors had any experience with either STELLA or the Macintosh operating system. Within a week, a reasonably well behaved STELLA model of Sutton Lake was produced. Within a month, a model of the entire Kanawha River basin was made by fitting together pieces of separately produced models. STELLA's graphical user interface reduced the amount of time required by the authors to become skillful and provided an excellent tool in describing to non-technical persons the program logic. The stock and flow symbols used to make STELLA models can easily be seen as representing lakes and rivers that physically exist.

The four STELLA building blocks that were used to construct the Kanawha River basin model were: converters, stocks, conveyors, and flows. The most numerous building blocks in the Kanawha River model were the converters. In the model, converters served three basic functions: decision making, unit conversion, and graphical function storage. Decision making logic and unit conversion was implemented in individual converters by writing an equation that consisted of Boolean statements and FORTRAN-like formulas. These equations established the functional relationship between the value of a converter and the values of the inputs at each time step. Graphical function converters were used to establish functional relations with time or an input that could not be expressed effectively with analytical relationships. Inflow hydrographs and lake storage rating tables were specified in this manner.

Stocks were used in the model to represent the lakes. STELLA numerically calculated the storage in the lakes as a function of time by integrating the algebraic sum of the inflow and outflow hydrographs. Stocks were also used to combine streams at junctions and local flows with lake releases. The outflow from these stocks was set equal to the sum of the inflows, therefore changes in channel storage were not simulated by this model.

Conveyors are building blocks which lag the inflow by a specified integral number of time steps. Outflows from conveyors are defined completely by the inflow and the specified lag time. Conveyors were used to simulate the routing of water in the model because of their simplicity.

Flow control in the STELLA model is specified with arrows that pass data between converters and flows, and from stocks and conveyors. These arrows define functional dependency.

Model Development

The STELLA model of the Kanawha River basin was constructed from individual models for the lakes and the main stem. All models used a time step of one day and simulated from 1 April to 30 November of the year of interest. The lake models were constructed using a stock to represent the total reservoir storage. Inflow hydrographs to the lakes were taken from historical records. Inflow to the lakes was defined as being the historical hydrograph converted from units of c.f.s. (m^3/s) to ac-ft/day (m^3/day). Outflow from the lake was defined as being equal to either historical records or a converter named "dam tender". The "dam tender" calculated the release from the lake based upon the inflow hydrograph, minimum release, maximum release, downstream water quality targets, pool elevation, guide curve elevation, and lower limiting rule curve elevation. The release logic during droughts was designed to make the lake level follow the guide curve elevation while maintaining minimum downstream flows and under no circumstances allowing the pool elevation to fall below the level of the lower limiting rule curve. During floods, the logic was designed to maintain the guide curve while restricting the outflow to 80 percent of the peak inflow and the channel capacity.

The STELLA model of the main stem of the Kanawha River consisted of releases from the projects combined with local flow hydrographs taken from the historical record. Conveyors were used to route flows through the river channels. Observed releases, gathered from the historical record, by the lakes were input into the main stem model and calculated flows at Kanawha Falls and Charleston, WV were compared with observed flows to calibrate the lag times imposed on the flow by the conveyors between control stations. This calibration was necessarily crude because of the one day time step and the discrete nature of the conveyors.

Model Results

The primary test of the STELLA model was the ability to simulate the hydrological behavior of the Kanawha River during the 1988 drought. The model successfully simulated the Kanawha River during the 1988 drought within the assumptions made in designing the model. These assumptions did not adversely affect the behavior of the model during the actual drought. Some relatively high flows were present in the Kanawha River during the early part of the 1988 simulation and were not accurately modeled by STELLA, but the behavior of interest (during low flow conditions) was not significantly impacted. Additional structures to the Kanawha River basin model were added to facilitate the accounting of whitewater releases from Summersville Lake.

The verified STELLA model of the Kanawha River basin was run with several operational strategies proposed by the major water users. The relative economic value of these strategies were estimated by the amount of storage available for whitewater releases. The model was also run with historical data from other drought periods to determine economic impact.

Multiparty Model Development
Using Object-Oriented Programming

Daniel N. Nvule[1]

Abstract

The Massachusetts Water Resources Authority (MWRA) has taken advantage of recent advances in object oriented programming to build a flexible input and output model of its water supply system. All interested parties were involved in model building. The result has been quick evaluation of the impacts of decisions and timely building of consensus among the parties involved.

Introduction

The MWRA is one of the oldest water utilities in the country, with a history starting in 1652. The MWRA was formed in 1985 to modernize the aging water and sewer system and currently provides wholesale water and sewer services to 2.5 million people in 60 cities and towns in eastern and central Massachusetts. The MWRA is one of the participants in the US Army Corps of Engineers sponsored National Drought Preparedness Study (DPS). The major focus of the MWRA study group is to put in place a planning procedure that emphasizes the long-term strategic aspects of drought planning in order to avert, as much as possible, the necessity of implementing short-term tactical planning when resource levels become critical.

Statement Of The Problem

A crucial component of the DPS study is an input/output model used to evaluate the impact of various decisions on

---

[1]Project Manager, Capital Engineering and Planning, Massachusetts Water Resources Authority, 100 First Avenue, Boston, MA 02129

This article represents the opinions and conclusions of the author and not necessarily those of the MWRA.

the water supply system. Prior to this study, the MWRA had a Fortran model, developed by the late David Hellstrom of the consulting firm of Arthur D. Little, that was inflexible and difficult to use by people not associated with its development. The process of building consensus among interested parties was a drawn out affair due to the time needed to assimilate the Fortran model results.

The challenge therefore is to involve all team members in developing the system model from scratch so that when completed it is transparent to all of them. The model should be flexible enough to explore all aspects of system operations that may vary from strategic measures that have to be taken long before the onset of a drought to tactical measures that get implemented when a drought strikes.

Who Are The Parties And What Are Their Interests?

There are four principal parties involved in model development: the system operator, MWRA, with the objective of reliable provision of water and sewer services; the watershed manager, the Metropolitan District Commission (MDC), with the objective of efficient management of the watersheds and reservoirs; a citizens group, WSCAC, with the objective of minimizing impacts to river basins, and a Federal regulatory agency, the Army Corps of Engineers (ACOE), with the objective of providing planning guidelines and acquisition of new technologies. Indeed, it was the ACOE that introduced object oriented programming software (STELLA[2]) to the group.

Consensus among the parties is key to moving any strategic drought management plan forward. The policy of the MWRA is to exhaust all non-structural measures to balance supply and demand, and with the citizens group, has charted a future course that stays clear of traditional episodic supply planning and puts in place a process that continually monitors supply and demand. Before a crisis develops all parties are familiar with it and with the range of options available to confront it.

MWRA Water Supply System Configuration

The MWRA water supply system harnesses the waters of three watersheds in a synergistic fashion: the total system yield is greater than the sum total of the yields of the individual watersheds. Two principal reservoirs are connected in series. Quabbin, has a maximum capacity of 412 billion gallons and a residence time of about 4 years.

---

[2]STELLA is copyrighted software from High Performance Systems Inc., 45 Lyme Rd, Hanover NH 03755

Wachusett, has a maximum capacity of 65 billion gallons. Water is also diverted during the winter months from the Ware river that lies between the watersheds of the reservoirs. There are minimum release requirements for each source. Water generally flows from west to east by gravity through deep rock tunnels. Whereas the layout of the system is generally considered to be a feat of human engineering, it has the potential to polarize the communities through which it traverses. The easterners are generally considered by the westerners to be beneficiaries of sacrifices of the western communities. Figure 1 shows the STELLA representation of the supply system. The STELLA model runs on a monthly time step.

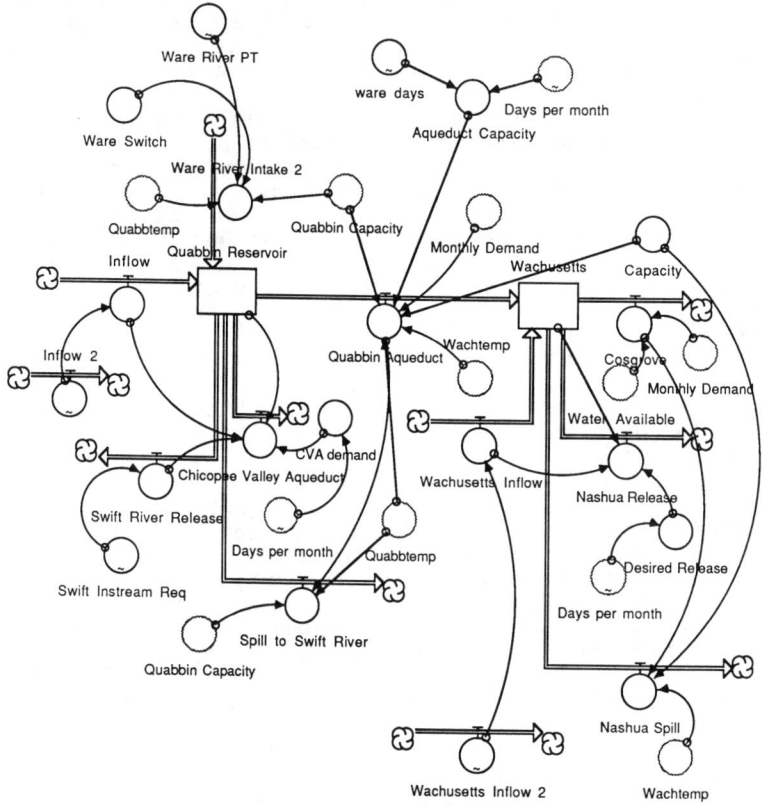

Figure 1: STELLA Representation of MWRA System

Model Validation and Scenario Investigation

Since the inputs to the Fortran model were readily available and its credibility was beyond question, the study team took the approach that if both the Fortran and STELLA models yield the same outputs for the same inputs, then the STELLA model would be reasonably calibrated. The results compared pretty well. The average difference for instance between Quabbin reservoir levels as computed by both models was below 5% and there was no observed bias.

The flexibility of the STELLA Model can be demonstrated by the range of scenarios investigated by the team. Bill Eliott the WSCAC team member was able to demonstrate the rewards reaped by implementing relatively small demand reduction programs even when reservoir levels are in their normal ranges. Along with this work, the team is developing an econometric water demand model. This will considerably increase the team's analytical power. As an example, the group is watching a trend of proliferation of the biotechnology industry. If this trend continues its impact on water demand will be modelled and the results passed on to the STELLA model to evaluate the impact on the supply system. The team also investigated the impact of a plan to re-route Quabbin waters around Wachusett reservoir due to water quality considerations. It was found that the tradeoff of the by-pass is the lowering of system yield. This latter investigation would have been difficult to do with the Fortran code because of changes in the system configuration. In STELLA this is accomplished by simply adding an arrow on the diagram and typing in the transfer rule.

Conclusion

The study team was involved in a wider goal of establishing a new water supply planning framework for the Authority which among other things included preparing the region for a drought. In doing so, the team found that the ability to formulate object oriented I/O models and to demonstrate tradeoffs, facilitated their task. These types of models, due to the ease of their development and use, are inevitably essential catalysts for the process of consensus building. Indeed such is the power of quick model development, that this project has spawned off another one between the MWRA and the MDC that aims at refining the daily operation of the two reservoirs when they are both at high levels. This model will capture the hydraulics of the spillways and the transmission aqueducts and will aim at evaluating various operating rules for water transfers.

BRINGING PEOPLE, POLICIES, AND COMPUTERS
TO THE WATER (BARGAINING) TABLE

RICHARD E. PUNNETT, PhD[1]
JAMES M. STILES[2]

Abstract

Droughts have been defined terms of hydrometeorological terms; however, a Water Control Manager realizes that droughts are also a function of demands. During the drought of 1988, water users in the Kanawha River, West Virginia, became painfully aware of conflicting uses that have developed since the 1970's. Under the National Drought Preparedness Study, the Kanawha River Basin Study was conducted to resolve drought-related issues. A numerical model of the river basin was developed to evaluate the effects of different management scenarios on lake and river uses. The model was developed using STELLA, an object-oriented programming language. The model was used to aid in discussions among private industries, public interest groups, State agencies, and the Corps of Engineers. Since the model was graphically user-friendly, discussions of water management options were greatly enhanced among people of non-engineering backgrounds. This paper focused on the process -- including modeling -- of resolving drought conflicts.

Introduction

A drought could be defined as the extremely dry conditions during which water resources fail to meet normal demands. The 1988 drought in the Kanawha River Basin, West Virginia, had a severe economic consequences -- loss estimates greater than 10 million

---

[1]Chief, Reservoir Control Section, Huntington District, Corps of Engineers.
[2]Hydraulic Engineer, Huntington District, Corps of Engineers.

dollars were given. If the same dry conditions occurred in the mid 1970's, there would have been no significant economic losses. The development of the basin, in terms of population, industrialization, and recreation, naturally lead to greater demands on the water resources. Under normal conditions, all demands (needs) were met. The drought conflicts of 1988 not only revealed the need for new water resource initiatives, but blatantly demonstrated the need for improved coordination, cooperation, and a common understanding of existing needs of all water users.

Traditionally, engineers have not exhibited a propensity for presenting and explaining computer-generated, engineering findings to non-technical groups. The introduction of engineering jargon alone is sufficient to create a communication rift. The presentation of graphs and tables can further preclude useful dialogue. Explaining the hydrologic findings of the modeling results of this study potentially had this kind of communication problem. The study not only required joint understanding of the engineering results, but was strongly dependent upon the input of non-technical groups as well as the final acceptance of the results. However, the model selection and application actually enhanced communication.

The Kanawha River Drought Preparedness Study was conducted to identify some operational schemes that would meet basin needs during a severe drought and be acceptable to all water users. This study was unlike historical drought studies conducted by the Corps in two respects: the model had not been previously used for this purpose, and the drought-resolution process relied heavily upon non-Corps influences for guidance, evaluation, and solutions. The non-Corps influence came from a large group of interdisciplinary people who had a common goal -- to reduce the conflicts associated with drought. The group consisted of representatives from other federal and state agencies, as well as private industries and interest groups.

The Conflicts

The basin resources have not changed since the early 1970's when the newest lakes were completed. The rules of operation, established prior to 1970, have not changed. However, significant changes in demands have occurred in the last 20 years. The recreational use of lake releases for whitewater has grown to a million dollar a day business during a 22-day, fall period. Until the drought of 1988, all water needs were satisfied without conflict. During 1988, the recreational releases were curtailed to store water in order to meet river augmentation requirements. The reality of the conflict, the lack of understanding of each other's needs, and the lack of a formal way to resolve issues lead to the necessity of this study.

The Solution Process

Invitations were distributed to promote participation in an advisory group. Prior to any decisions regarding the direction and extent of the study, the group met twice to determine what were the important concerns during drought, to prioritize those concerns, and finally to shape a plan of study that could provide solutions. Subsequent to those decisions, a modeling program was selected. A user-friendly, object-oriented modeling language, STELLA, was chosen because the object-oriented programming technique provided a graphical link between computer modeling code and the visual representation appropriate for non-technical members. Because the model was constructed using a few graphical building blocks, it is easy to see how the model relates to the real river system. A discussion of the STELLA model technique and the river basin was given in a related paper titled: Including Expert System Decisions in a Numerical Model of a Multi-lake System Using STELLA (Stiles and Punnett, 1993, ASCE Mini Symposium: National Drought Study, Melding Art and Practice).

Once the basic model was complete, the model developers met with representatives from various interest groups separately. The meeting was conducted to ensure the model was understandable and had the support of the representatives, as well as get the input of the representatives for desired modifications and improvements to the model.

After the model developers incorporated the suggested modifications, many different operational scenarios were investigated which included: increasing lake storage, reducing augmentation targets, reallocation of storage, implementing improvements in the monitoring and setting of augmentation releases, combining enhancement for fish habitat with recreation uses of an increased minimum pool, and combinations of various scenarios. Several potential solutions to the conflicts that occurred during the 1988 drought were identified.

The scenarios were then presented to the full group for review, acceptance, and prioritization for implementation. Once the scenarios had been prioritized, the Corps accepted the responsibility for obtaining approval for implementation since the Corps was responsible for the operation of the multi-purpose lakes in the basin. Because the approval process is lengthy, it was beyond the scope of this study. However, the communication, cooperation, and common understanding that was established by this study was expected to continue. The group can be called upon to help resolve new issues in future droughts.

Summary

Because of the cyclic nature of droughts, new conflicts are likely to arise with each successive drought within river basins of limited resources. To effectively resolve new issues, an object-oriented, computer model can be developed to explore new operational scenarios and to aid in discussions between people of both technical and non-technical backgrounds.

Politics and Drought Planning:
Friends or Foes?

Dr. Bruce D. McDowell[1]
Dr. William Blomquist[2]

Abstract

Nothing frustrates the average drought planner more than politics. Yet, droughts cannot be prepared for realistically without reliable political partners, smoothly cooperating government agencies, and strong public support. This paper suggests six rules for linking technical drought planning processes to the political processes and institutions that can implement drought plans.

Introduction

The options for implementing drought plans may include supplementing normal supplies with water from other sources, asking water users to conserve water voluntarily, or imposing mandatory water rationing. Such actions may upset normal patterns of water use, interfere with established water rights, hurt some businesses and industries more than others, and inconvenience many consumers, taxpayers, and voters. The effects may be felt over a large geographic area-- perhaps extending across two or more states. Emergency actions may be required by several different governments, many separate government agencies, private water utilities, and a host of individual water users.

Drought planning is a sub-set of comprehensive water resources planning that provides for developing, allocating, and managing surface and underground supplies, as well as regulating and managing the quality of surface and underground supplies. Periods of flooding as well as normality and drought need to be encompassed so that the whole hydrologic cycle can be drawn upon in the planning. Intra-basin transfers, as well as interbasin transfers, and transfers between river basins and underground aquifers all may offer possibilities for alleviating the effects of drought if the political, legal, and institutional, as well as technical, details can be worked out (ACIR, 1991).

----

[1] Director of Government Policy Research, U.S. Advisory Commission on Intergovernmental Relations, Washington, DC 20575

[2] Assistant Professor of Political Science, Indiana University, Indianapolis, IN 46202

Planning of this scope involves many different "stakeholders," including governments, water planners, water managers, water regulators, and water users, plus those concerned with economic development and environmental protection. From a political and institutional viewpoint, therefore, what needs to be done is to:

1. Know the players;

2. Get all the key players together to plan;

3. Support constructive interactions among them;

4. Make sure they reach decisions that can be implemented;

5. Practice "drought drills;" and

6. Learn from real droughts.

Each of these steps is described briefly below.

Six Rules for Putting Drought Planning into Political Context

1. <u>Know the players</u>. Drought planners usually know most of the important federal, state, and local players within the "water community." If they have a blind spot, it is likely to be for groundwater and water quality regulators. Thorough drought planning needs to encompass more than just a river system and the management of its dams. Where interstate commissions exist, they need to be included.

The affected publics also are probably pretty well recognized. They include the traditional agricultural and urban water users, and major water using industries. The newer publics, needing equal attention, include in-stream users, such as recreationists and the fish and wildlife interests. The general public also has an interest in fairness, equity, and tax rates.

The judiciary's role is likely to become even more important as water demand presses harder on available supplies. Water rights established by state and federal constitutions and laws, and by Indian treaties, are very important, and difficult to change. Legal advice on these issues is needed in the planning process.

Finally, and perhaps most importantly, the federal, state, and local lawmakers and chief executives--and their top aides--are major players. They make water laws, and can change them. They provide for the rules and regulations under which drought emergency steps can be taken. They provide the funds and rules for drought planning, for water resource development and management, for the protection of water quality, and for the protection of environmental quality and endangered species. They also play important roles in determining the economic development and land use policies that influence water demand.

2. <u>Get the players together to plan</u>. Getting all the important players together to plan for drought is a difficult step. Water planners and managers, and the directly affected water interests, may be the easiest to pull in--it's part of their primary job. Others may want to stay on the sidelines to see what develops. Judicial and legislative officials may hold back because of the separation of powers between the three branches of government. Top legislative and executive officials also may hang back because of their busy schedules and reluctance to be pulled into potentially controversial issues before they understand them well and see a personal stake in them. Some environmentalists and Indian tribal governments with special legal standing on the key issues may resist getting involved prematurely for fear of diminishing their rights.

The motivation to do drought planning is stronger when water shortages are occurring, or have recently occurred. This sense of urgency should be used to get key players involved.

The planning group should be convened by an organization having adequate geographic scope and objectivity to gain the confidence of all the parties.

The problem of missing players should be dealt with constructively. They should be sent all the materials prepared for meetings, just as though they were participating, and they should be (a) polled for opinions on the issues, and (b) invited to send close personal representatives capable of reflecting their views accurately. Issues most needing their input should be highlighted for them, and personal briefings should be offered. As the process moves along, special opportunities to involve them should be offered.

3. Support constructive interaction among the players. Intergovernmental and interagency coordination processes yield positive results only with great effort. Too often, protecting turf becomes paramount. Existing laws and procedures may be invoked to close off discussion of potential solutions to problems.

Knowledge, freely shared, should be used to lower the barriers to cooperation. For example, an interactive simulation model of water systems (STELLA) has been found to actively engage a variety of players in a common learning experience that makes them comfortable with the facts of the situation and with each other. Another decisionmaking model (MATS) helps to objectify the group process of choosing among alternative solutions to problems. Professional mediation and negotiation skills also can help to clarify the issues and narrow differences. Formal risk analysis procedures can clarify the likelihood of serious consequences arising from proposed courses of action or inaction. The more open these procedures are, the more confidence they will build among the parties.

Letting the press in on planning exercises can help develop an understanding of the process, the problems, and the proposed solutions. Knowledgeable coverage by the press can help build public support for needed public policies.

4. Make sure that decisions can be implemented. When drought response procedures have been decided upon, there must be a way for them to be carried out. If new organizations or laws are needed for such action, they need to be enacted. If an interstate compact is needed, it should be negotiated with the state legislatures and the Congress. Appropriate trigger mechanisms need to be part of the plan, along with clear responsibility for activating it and clear follow-through responsibilities for each of the players.

5. Practice drought drills. Extraordinary water shortages come along only now and then, plans get old, and the people who put the plans together retire, take other jobs, or forget what the plan calls for. When the drought finally comes, the plan may be obsolete or too unfamiliar to be of much use. This problem has been overcome in the Potomac River Basin by holding drought drills in which all the key players annually go through the motions of managing a drought.

6. Learn from real droughts. During a drought, it is difficult to plan. There is too much else to do, and nerves are too frayed to allow objective reflection. But right after a drought is a perfect time to sit back and reflect on how things could have been handled better.

Conclusion

Water managers and drought planners need the political process and the public support it can bring. They should work as hard at bringing political partners into the planning process as they do at perfecting their technical analysis of hydrologic systems.

REFERENCE

U.S. Advisory Commission on Intergovernmental Relations, (1991) <u>Coordinating Water Resources in the Federal System: The Groundwater-Surface Water Connection</u>, Report A-118 (Washington, DC: ACIR).

CITIZEN PARTICIPATION IN WATER SUPPLY SYSTEM PLANNING AND MANAGEMENT

William G. Elliott, Eileen R. Simonson,
Alexandra D. Dawson, and Robie O. Hubley [1]

ABSTRACT

The Water Supply Citizens Advisory Committee (WSCAC) is a unique citizens group. Although it is funded by the Massachusetts Water Resources Authority (MWRA), it has a paid staff who are independent of the agency, and report only to the Committee. WSCAC was initially organized in 1978 to provide citizen oversight for a proposed large river diversion project. The committee was instrumental in demonstrating that demand management and system repair could meet the projected needs for the planning period. WSCAC presently acts in an advisory capacity to the MWRA Board of Directors and staff on water supply policy and programs. For over a decade WSCAC has provided information and advice to the public, legislators, and state agencies. The Committee has a major impact on state water policy, water management activities, and the passage of major legislation.

During the past year WSCAC has participated with MWRA staff in the U.S. Army Corps of Engineers National Study of Water Management During Drought. As part of this effort, WSCAC has assisted with system simulation modeling, development of system performance criteria, evaluation of alternative management strategies, and "trigger planning." This paper will describe the makeup and functioning of WSCAC and our role in this study.

INTRODUCTION AND HISTORY

For at least 150 years, the metropolitan Boston water system repeatedly expanded westward for additional supply. So long as supplies of pure water were available, treatment of inferior local supplies was not considered. The system was administered by the Metropolitan District Commission (MDC),

---

[1] Water Supply Citizens Advisory Committee to the Massachusetts Water Resources Authority, P.O Box 478, Hadley, MA. 01035

a state agency with a consistent policy of enlarging its water supply and then soliciting system expansion.

Following the mid-1960's drought, the MDC again proposed to augment its system, which has a "safe yield" of about 300 million gallons a day (mgd) of exceptionally high-quality water. The proposal was to use a pumped-storage facility on the Connecticut River at Northfield, Massachusetts to divert up to an annual average of 63 mgd of water from the river through a 10-mile tunnel to the Quabbin Reservoir. The MDC was encouraged in this augmentation plan by the Army Corps of Engineers 1973 study which forecast a prodigious rise in demand from 320 mgd in 1972 to 519 mgd by 1990!

In 1977, in response to public concern, and in recognition of the magnitude of the project and its potential impact on the environment, the Secretary of Environmental Affairs declared the diversion a "major and complicated project" under the Massachusetts Environmental Policy Act (MEPA) regulations. The MDC was told to prepare an Environmental Impact Report (EIR). The Secretary appointed a citizens advisory committee with a "full formal advisory role" to insure informed public input into the review. The "Northfield Citizens Advisory Committee" (NCAC), was established in 1977 jointly by the Secretary, the MDC, and a coalition of Connecticut River cities and towns. It was the first independent CAC to be funded for staff and office expenses by an agency preparing an EIR.

The Secretary appointed 35 committee members from nominees submitted by the parties. The members represented diverse interests, including environmental groups, legislators, water users, agriculture, utilities, and planning agencies. The members hired an executive director in 1978.

EVOLUTION OF WSCAC

One of the Committee's first responses to the proposed Northfield diversion project was to insist, against agency objection, that equal consideration be given to all reasonable alternatives. These included: no action; four river diversions, watershed management, reactivation of water sources in user communities; demand management; and desalination. The Secretary accepted these alternatives as setting the scope of the study, renamed it the Long Range Water Supply Study (LRWSS) and renamed NCAC the Water Supply Citizens Advisory Committee (WSCAC) in 1980. Membership was adjusted to assure representation for each alternative.

The designation of the study as "major and complicated" was amply borne out by its history. During a decade of work, the consultants produced over fifty reports, many including complex analysis and volumes of raw data. To deal with these complexities, WSCAC established a number of task forces, bringing non-members together with Committee members to study such issues as watershed management, supply protection, local water sources in user communities, economic and legislative issues, safe yield and reservoir minimum pool, and population and water needs projections.

WSCAC worked on state-wide policy and legislation in addition to system specific problems. For example, the years 1980 and 1981 were very low in precipitation and streamflow. A state-wide drought task force was created in which WSCAC participated. The work of that group highlighted the need for drought contingency plans and the inclusion of drought planning into state guidelines and regulations. WSCAC spurred the reconvening of the state drought task force in 1985-86 and again in 1988-89. This experience was a factor in the invitation by the Army Corps of Engineers to WSCAC to participate in the National Study of Water Management during Drought.

WSCAC also contributed to changing the legislative and regulatory climate in the Commonwealth. In 1983, after four years of effort, the state passed the Interbasin Transfer Act which declared diversions as measures of last resort, required the receiving area to demonstrate that it had no alternative in-basin sources, and that it had significant conservation programs on line. The 1984 Water Management Act acknowledged the relationship of ground and surface waters and required the registration of withdrawals of 100,000 gpd and the permitting by river basin of all new withdrawals.

In 1985 the Massachusetts Water Resources Authority (MWRA) was formed to replace the MDC water distribution and sewer operations. WSCAC's involvement in drafting the MWRA enabling legislation was solicited, giving the committee an unusual opportunity to help create its future parent agency. WSCAC's efforts resulted in donor river basin representation on the MWRA Board, and conservation and use efficiency being mandated by the legislation.

In 1986 MWRA accepted WSCAC's recommendation to implement demand management which had never before been chosen as the principal means to meet water needs. Now, six years later, water demand has been reduced to 260 mgd, primarily from the success of leak detection and repair, low flow domestic devices, and conservation education.

As part of the earlier EIR, WSCAC developed a comprehensive monitoring, management, and planning program called "Trigger Planning" or alternative futures planning. This is a decision making process which incorporates historic and present operating data, physical characteristics of the system, results of demand management programs, and demographic data. It generates compiled information which guides present management strategies, facilitates the development of projections for timely decision making, and provides for public participation in project planning.

The trigger planning approach is being applied to the Drought Management Study. WSCAC is also assisting with system simulation modeling, development of system performance criteria, and evaluation of alternative management strategies.

WSCAC's most recent legislative initiative resulted in the passage of a comprehensive watershed protection bill which funds land acquisition, regulates land uses along tributaries in the MWRA watersheds, and requires the state environmental agency to upgrade watershed protection throughout the state.

WSCAC presently advises the MWRA Board and staff on water policy and programs. The federal Safe Drinking Water Act has had a major impact on the water supply systems in New England. WSCAC is working with a coalition to amend the Act to emphasize watershed protection and system management as methods of meeting water quality standards.

LESSONS LEARNED ON PUBLIC PARTICIPATION

WSCAC has been in existence longer than most citizens committees. It has grown and changed as its host study grew and changed. A number of points have become clear over the years.

The first is that informed citizen input on a major, long-term study requires independent, full-time staff. Volunteers could never have found time to analyze the numerous reports, farm out tasks, organize responses, set up meetings, technical workshops and conferences, participate in state agency meetings, and make professional presentations. Development of a truly independent critique required a staff answerable only to the citizens committee.

WSCAC believes that citizen participation in public policy is a good investment. The level of support which WSCAC has received is required in other major projects if government is truly committed to meaningful citizen involvement. The total expenditure will always prove--as it has for WSCAC--small compared to the cost of the study and miniscule in proportion to the cost of the proposals being reviewed.

A second point relates to the mandate of a citizens group. WSCAC was given a somewhat unusual mandate: to gather, formulate and represent the public position. Most advisory committees, by contrast, speak only to the agencies they advise. Too often their expressions, especially if they are critical, are buried or lost. Whatever its announced charge, a citizens committee must always be flexible. If blocked in some direction, it should avoid venting its frustration and, instead, seek different audiences for its message. In this way, frustration can actually force the development of independent thinking and a stronger grounding in public support.

A related issue is the credibility of citizen participation. A citizens advisory committee has a dual function: to bring public opinion into a process and to insure that the public is generally satisfied with an outcome. The committee acts as a conduit for public opinion and buffers the government agency from unjust criticism.

A committee can meaningfully endorse a public agency's actions only if it can effectively criticize them. This implies responsibilities for both parties. The committee's criticisms must be pertinent and factual, avoiding broad emotional attacks. This kind of focus requires professional expertise. Obtaining and channeling such expertise requires a lot of energy and time. With staff and facilities, a committee has the resources to analyze facts, criticize when necessary in an effective way, and support good initiatives.

Effective criticism by citizen committees is often curtailed by agency inhospitality and the denial of resources to the committee. Frustration of pertinent criticism has the effect of raising the emotional level and diffusing focus. This forces all parties into entrenched positions and cuts off a vital exchange of ideas. When an agency receives responsible criticism and deals with it professionally, the citizens' work can be truly valuable to improving the project and its public acceptability.

ANALYSIS OF OPERATING CRITERIA: MULTIPLE LAKES AT VOYAGEURS NATIONAL PARK

By Marshall Flug[1], M.ASCE; and Larry W. Kallemeyn[2]

Abstract

An overview of lake and river regulation at Voyageurs National Park, which resides on the Minnesota-Ontario border, is given to demonstrate how water policy agreements can work. In 1905 the United States and Canada authorized private dams with turbines on the Rainy River. The International Joint Commission regulates these dams. The National Park Service is mandated to preserve the natural environment for future generations. State, private, and public sector interests are tourism, flood protection, the pulp and paper industry, native wild rice growth; etc. Rule curves for regulating reservoirs have changed and committee with broad represenattion is cooperating to better manage the waters of Namakan Reservoir and Rainy Lake.

Introduction

This paper describes water policy agreements applied along the Minnesota-Ontario border, where the international boundary between the United States and Canada follows an eighteenth century fur traders' route from Lake Superior to Lake of the Woods. Voyageurs National Park is located along a portion of that network of lakes and rivers. Construction of several private dams and powerhouses began on the Rainy River in 1905. Naturally occurring Namakan and Rainy Lakes, which provide the major inflow to Lake of the Woods, were enlarged. Concerns about changing water elevations were expressed relative to aquatic and riparian ecosystem resources; pollution abatement; municipal water use; navigation; recreation; fish spawning; wild rice growth and harvest by native indians; hydropower generation; and flood control.

These rivers, lakes, and reservoirs are international waters and therefore regulated by the International Joint Commission (IJC). Ownership and operation of the dams remain with private interests for water storage and power generation in nearby pulp and paper mills. In 1971, the US Congress created Voyageurs National

---

[1]Research Hydrologist, Water Resources Division, National Park Service, 1201 Oakridge Dr., Suite 250, Ft. Collins, CO 80525.
[2]Aquatic Research Biologist, Voyageurs National Park, HCR 9, BOX 600, International Falls, MN 56649

Park (United States Code 1971) to preserve the natural environment and native plant and animal life for the benefit of future generations; these management functions are harmonious with the directives in the Organic Act (United States Code 1916) creating the National Park Service. The diversity of water interests, coupled with uncontrolled natural hydrologic variability and regulated control strategies, represents a complex arena for resolving water management conflicts. Annual public meetings are convened by the International Rainy Lake Board Of Control. A steering committee meets to develop a consensus on how to manage the waters of Rainy Lake and Namakan Reservoir for all beneficial uses (Kallemeyn 1992).

Chronology Of Water Policies And Events

1905 - 1909 The US Congress and the Canadian Federal Parliament allowed construction of the International Falls Hydroelectric Project at Koochiching Falls, the natural outlet of Rainy Lake. This dam is privately owned and operated to provide water storage and power for their pulp and paper mills. The rock and masonry dam spans International Falls, Minnesota to Fort Frances, Ontario, with the international boundary crossing the middle of the dam.

1909 The Boundary Waters Treaty between the US and Canada established the International Joint Commission (IJC), to prevent disputes in use of boundary waters, to provide a framework for cooperation, and for regulation of water levels and flows.

1914 Construction of two stop log dams is completed at the outlet of the Namakan Lake, which is actually a series of five natural lakes located upstream from Rainy Lake. This construction creates Namakan Reservoir, with part of one dam on the United States side at Kettle Falls and its Canadian sister dam at Squirrel Falls.

1916 Although the dams at Rainy Lake and Namakan Reservoir are fully operational, high May flood waters and severe street flooding occurred. This inspires an increase in the design outflow capacity from Rainy Lake.

1921 & 1924 Turbines at International Falls and Fort Frances, which generate power and are coupled with grinding stones for turning logs into pulp, are replaced and additions made. The grinders are removed in 1978 during a major turbine upgrading.

1938 - 1940 A convention between Canada and the US, duly ratified in 1940, empowers the IJC to determine emergency conditions in the Rainy Lake Watershed and to adopt measures of control or rules of regulation for dams and control works. A primary objective was securing the most advantageous use of waters for the combined purposes of navigation, sanitation, water supply, power production, recreation, and other beneficial purposes. The International Rainy Lake Board of Control (IRLBC) performed hydrologic studies of the watershed and held public hearings.

1949 The 1940 convention led to the 1949 IJC Order, which set criteria for dam releases, defined emergency conditions, and set monthly water elevations for both Namakan and Rainy Lakes. This Order recognized interests of the pulp and paper mills. Individuals from both sides of the border expressed concerns about impacts from regulated lake levels on riparian lands, shore properties, bank erosion, flooding, creation of unsightly and unsanitary conditions, recreational use, and damaging high flow discharges.

1957   In response to floods in 1950 and 1954, and to complaints expressing dissatisfaction with operations under the 1949 IJC Order, a five year trial period of amendments was implemented. These 1957 Amendments made no changes on Rainy Lake, but established a set of maximum elevations on Namakan Lake for the period October 1 to June 1. Thus a rule band was created between the original 1949 IJC Order and the new maximum values. To accommodate resort owners, full elevation on Namakan Lake was targeted for June 1 instead of July 1 and, to improve fish spawning activity, a maximum lake level was set for the period April 1 to 21. The IRLBC accepted input regarding satisfaction with these amendments and twice extended the original Order and Amendments for additional five year periods.

1970   In 1968 the IJC ordered the IRLBC to further investigate regulation of Rainy and Namakan Lakes. This was in response to heavy rains that caused lake levels above full pool and to low water events that caused lake levels to fall below the prescribed minimum elevations. After public hearings, the 1970 Amendments redefined emergency conditions, allowed a 0.15 m flood reserve, modified rule bands on both lakes, defined minimum elevations on Rainy Lake, and established minimum flow releases for pollution abatement by maintaining downstream dissolved oxygen concentrations that also approximated natural flow conditions.

1970 - 1992   In several instances the IJC amended the rules for reservoir regulation that temporarily modified daily average and instantaneous flow releases. These actions helped overcome operational difficulties at the hydropower plant or responsed to flood or drought conditions that caused both lakes to reach elevations outside the 1970 rule bands.

Environmental Water Management

The above listed rules of reservoir operation were put in place to secure the most advantageous uses of water for the combined purposes of power production, domestic water supply, sanitation, navigation, recreation, and other public beneficial purposes. As evidenced by the needed periodic adjustments to the rules, it is difficult to maintain favorable conditions for a diverse set of water dependent needs, particularly during periods of extreme naturally occurring hydrologic events. This situation is further hampered by a lack of adequate forecasting of runoff and the inability to predict future hydrologic phenomena.

Although concerns regarding biological and recreational impacts were recognized in the early 1900's, the National Park Service (NPS) emphasized preserving the natural environment and determining impacts of regulated lake levels on the aquatic ecosystem. By 1980, research biologists had determined that the IJC Orders created larger than natural fluctuations in Namakan Reservoir and less than natural fluctuations on Rainy Lake (Cole 1982). A hydrologic analysis (Flug 1986a & 1986b) confirmed that Namakan Reservoir's average annual fluctuation is 2.7 m, about 0.9 m greater than estimated natural fluctuations, while Rainy Lake's average is only 1.1 m, about 0.8 m less than natural. This hydrologic analysis also documented that peak water elevations arrived about one month later (i.e., July) than under a natural regime (i.e., late May), and remained stable over the summer with winter drawdowns of 1.8 m on Namakan Reservoir, compared to 0.6 m pre-dam.

The NPS conducted an extensive research program to assess impacts of regulated lake levels on the park's aquatic ecosystem (Kallemeyn and Cole 1990). Flug et al. (1985) developed and used a network simulation model to help analyze alternative rule curves for Rainy and Namakan Lakes and impacts on hydropower. Kallemeyn and Cole (1990) used results from the research program and from the scientific literature to develop ranking factors for resource attributes including: hydropower generation; ice damage to boat docks; lake level changes on beaver, otter, wild rice, aquatic vegetation; fish spawning, and nesting birds. Ranking factor values were assigned for alternative rule curves using outputs from the network simulation model. The resulting impact analysis matrix can facilitate selection of best overall water management alternatives.

Presently a steering committee with international, federal, state, public, and private representation is cooperating to develop a consensus on how to best manage the waters of Namakan Reservoir and Rainy Lake (Kallemeyn 1992). The objective of the committee is to analyze the existing rule curves and determine if there is a more ecologically sound management program that also provides the needs of other water users. In addition the International Falls Hydroelectric Project was relicensed in 1987 by the US Federal Energy Regulatory Commission (FERC), with the stipulation that a water level management plan be developed for Rainy Lake to ensure protection and enhancement of water quality, fish and wildlife, and recreational resources.

References

Cole, G. F. 1982. "Restoring Natural Conditions in a Boreal Forest Park." *Transactions, 47th North American Wildlife and Natural Resources Conference*, 411-420.

Flug, M. 1986a. "Analysis of Lake Levels at Voyageurs National Park." *Report 86-5*, US National Park Service, Water Resources Division, Ft. Collins, CO.

Flug, M. 1986b. "Regulated Lake Levels and Voyageurs National Park." *Park Science*, 7(1), 21-23.

Flug, M., D. Morrow, and D. G. Fontane. 1985. "Modelling of Reservoir Operations for Managing Ecological Interests." In: *Computer Applications in Water Resources*, H.C. Torno (Editor), ASCE, NY, 703-712.

Kallemeyn, L. W. 1992. "An Attempt to Rehabilitate the Aquatic Ecosystem of the Reservoirs of Voyageurs National Park." *The George Wright Forum*, 7(2), 39-44.

Kallemeyn, L. W., and G. F. Cole. 1990. "Alternatives for Reducing Impacts of Regulated Lake Levels on the Aquatic Ecosystems of Voyageurs National Park." Voyageurs National Park, International Falls, MN.

United States Code. Act of August 25, 1916 (39 Stat. 535), 1916.
United States Code. Act of January 8, 1971 (84 Stat. 1971), 1971.

Water Resources Management Issues
Confronting Developing Countries

Yin Au-Yeung[1]

Abstract

Water resources planning, development and management in most of the developing and rapidly developing countries have been driven by consumer interest. The rapid urbanization, industrialization, promotion of export, need for staple food sufficiency and economic explosion in general, have resulted in meeting the demands of consumers at the cost of limited water resources, both surface and ground water. This mismanaged development approach, if not corrected, will lead to serious consequences in regions having water shortages, water quality and environmental degradation, and conflicts in water demands; thus affecting the sustainability of the project. This paper suggests steps to be taken to promote sustainable development of water resources in these countries.

Introduction

Water resources play a key role in civilization. As the result of population growth and associated urbanization and the need for increased economic productivity, demands on the water resources far outstrips water availability, for both surface and ground water. This is particularly true in most of the river basins in developing countries where urbanization and economic growth has increased at a rapid pace in recent years as compared to past history. Currently, major regions in developing

---

[1] Vice President, ATC Engineering Consultants, Inc., 5660 Greenwood Plaza Blvd., Suite 500, Englewood, Colorado 80111.

countries focuse on water resources development strongly motivated and pressured by "consumer interests". In the rapidly developing countries, the trend of emphasis is shifted to improve operation and maintenance of project facilities to protect the investment. However, sound water resources utilization is yet to be implemented in an effective way so as to maintain the sustainability of the water development project.

Sustainable Development

The developing countries, especially the rapidly developing countries, have recognized the need for sustainable development of natural resources, including water resources in recent years. It is now recognized that natural resources should not only be used efficiently but they must sustain the development for the present generation as well as for the future generation. Sustainable development is not a goal but a process, one to be measured in generations rather than years. Therefore, a change towards sustainable forms of progress will require a change in management polices, from one based on consumption towards one based on conservation and improvement. The path to effective resources sustainability is through sound resources management.

Water Availability and Water Consumption

Presently, water resources development focuses strongly on consumer needs without due consideration of the water availability aspects. The governing institutions concentrate on ensuring the water supply to the consumer through greater emphasis on operation and maintenance of the supply systems and through management of the water distribution in the service area with only minimal or no consideration of efficient use, conservation and improvement of the limited water resources. The prevailing misconception is that water is God given and in-exhaustible.

Although in principle, the planning, development and management of water resources projects are based on integrated water resources management according to hydrologic boundaries. In practice, due to budget considerations and social/political pressures, only those components that are needed to control and convey water to the consumers are implemented. Other components such as management of return flows, water reuse, water quality, recharging potential of ground water etc., which are directly related to the sustainability of the project, are not addressed adequately or even completely ignored.

There are many reasons for the mismanagement of the water resources development projects. The following are some of the major reasons:

- The misconception that water is inexhaustible.

- No cost recovery polices and regulations.

- Lack of implementable and effective regulatory instruments.

- Emphasis on single purpose development due to lack of budget

- Imbalance of upstream and downstream developments with priority given to downstream projects mostly because of political pressures and the lack of total river basin development concept.

- Inadequate emphasis on multipurpose and integration grated aspects of natural resources developments.

As long as there is significant imbalance of water availability and water demands and development polices which only address the user aspect without due consideration of sustainability of the resources, the water management problems of river basins will continue.

In order to slow down and eventually stop the trend of over emphasis on the used aspect of the water resources management policy and practices, and to achieve water sustainability in a river basin or a region, polices and projects should be identified, addressed and managed according to water sources and water needs. The management of water needs covers all activities associated with development and implementation for a specific use of water by the user such as agriculture, aquaculture, industries, municipal, hydropower, navigation, etc. Conversely, while the activities relating to maintaining resources sustainability generally involved conservation, reuse and enhancement of water and ecological systems will come under management of water sources.

With the distinction made between management of water sources and water needs, steps and programs can then be formulated to have balanced policies and projects based on both water availability and water needs through effective river basin management. As a result, an integrated approach to a sound water utilization management can be accomplished, and a sustainable water resources development can be maintained.

The following are other measures which will significantly improve the current water utilization situation:

- ▶ Planning of water resources development project should be based on need and not on demand.

- ▶ Improve integrated development of water, land and other natural resources.

- ▶ Install strong and effective regulatory institutions to oversee and monitor the actual use of water.

- ▶ Institutes an effective river basin management to sustain water resources development.

- ▶ Implementation of effective and practical cost recovery.

Conclusion

It is clear that because of population growth, urbanization and economic development in major regions of most of the developing countries, especially the rapidly developing countries, that the demand of water is outstripping the availability of water. Also, the development policy and management direction is, and has been, emphasised heavily on water uses, without proper consideration or action towards sustainable development of natural resources. Proper management of sound utilization of water has not been seriously addressed or implemented. Strategy and management polices are urgently needed to revert the current water resources development trend so that the development of natural resources, including water, not only emphasise the demand aspect, but also the availability of the resources for sustaining development.

In order to effectively manage the water resources for sustainable development, a unique management unit based on hydrologic boundary such as a river basin should be considered. The river basin management will integrate individual projects for conservation and sound utilization of water along with effective coordination and integration of other natural resources within the basin.

Technical and Organizational Constraints on Surface
and Groundwater Irrigation in Small-Holder Areas of S. Asia

Donald E. Campbell, M.ASCE[1]

Abstract

Variable irrigation supply and the independent character of the S. Asian small-holder combine to present a difficult problem in irrigation management. Attempts to introduce advanced technology in water delivery systems can be frustrated by lack of coordination of demand at the farm level. Groundwater development can provide much-needed flexibility in delivery but can have its own problems, particularly between small individually-owned shallow wells and large deep wells.

Introduction

Expansion of agricultural production in the S. Asian area increasingly calls for improvement in irrigation efficiency. This involves both technical and sociologic factors. Sophisticated canal regulation systems are available, but the end user, the small free-hold irrigator, and his motivations and likely reactions remain central problems. Cultivator organization for operation and water distribution can also be a key question in groundwater development.

Based on some twenty years of experience in the S. Asian area, primarily in association with World Bank assisted projects, the author summarizes the problems and options.

The Small-Holder and Traditional Operation of Village Irrigation

A characteristic of the region under discussion is the small size of holding, ranging up to ten hectares in area but commonly one to two hectares or less, and frequently made up of several smaller parcels located in different parts of the village. The holding is usually owned by the

---

[1] Consultant, ANATECH Research Corp., 5435 Oberlin Drive, San Diego, CA 92121

cultivator. Distribution of canal supply would be greatly facilitated by consolidation of parcels. However, where soils are of varying fertility or depth, or for a number of other reasons, the small-holder is often unwilling to consolidate. This, in fact, is the more common situation in much of the S. Asian region, and the irrigation distribution system must be designed to accommodate the existing pattern of holdings. The area to be irrigated may be of near-uniform topography and soil type, or highly variable. The social structure of the village may also be homogeneous, but more often it is divided by caste, by religious and political affiliations, and by level of affluence (or poverty). In spite of these differences small village irrigation schemes have operated reasonably well in the past, due in part to the traditional structure of village authority, and in part to the imperative of communal action to maintain essential irrigation infrastructure. Such communality is unfortunately undermined by any government intervention, however well-intentioned. The advent of democracy at the village level, replacing traditional inherited authority, has also reduced the effectiveness of operation of such schemes, at least for a transition period. The question of whether the virtues of traditional village-level management can be incorporated into operation of major public irrigation systems is much discussed.

The Problems of Variable Supply and Demand in Public Irrigation Schemes. Options in Canal System Design and Implications Regarding Scope for Increasing Irrigation Efficiency

In a monsoonal climate rainfall and river-flow are highly variable both seasonally and from year to year. Provision of storage can regulate supply to some degree, and major reservoirs are contributing much to maintaining crop production in drought years. However, storage sites are not always available, and in any case ecological factors now weigh heavily against construction of new reservoirs. The supply to canal systems is consequently regulated only partially, or not at all in the case of purely run-of-river schemes, and the often-repeated plea for a fully reliable, predictable, delivery of irrigation to the cultivator simply cannot be met. The design of canal systems in such circumstances gives much attention to the question of operation during periods of deficiency.

Catering to variations in the pattern of demand for irrigation is the other half of the supply problem. Where the service area is uniform throughout, the operation of the system may be simplified by designing it to meet the needs of a single preferred class of crop. Where the service area varies in character, however, and more than one class of crop must be grown to meet particular local soil conditions, operation is more complex. The system must then supply water at different times to different crops in accordance with their seasonal needs. Market pressures to diversify from basic crops also calls for more flexibility in irrigation deliveries.

Additionally, there is the problem of distribution at the farm level. The small-holder in an area traditionally subject to variable rainfall and often unpredictable canal supply understandably places first priority on the security of his own crop and his family. He has little concern for the grand design of the supply system and little understanding of or sympathy for the constraints under which it has to operate. Interference with hydraulic controls and out-of-turn diversions are common, amounting in some situations virtually to operational anarchy.

In the circumstances the designer is faced with the choice between simplifying canal operation, at the cost of inflexibility of supply and limitation in the class of crops which can be served, or endeavoring to cater to greater diversity in demand at the cost of more complex operation and greater likelihood of cultivator interference with the system. In either case the unavoidable variations in the supply situation, modified by reservoir storage if any, must be accommodated. The simplest alternative has continuous flow throughout the system, the flow being divided at each canal junction in fixed proportion to the areas served. This requires no gate operation. A disadvantage, however, is that when the available supply or the demand is small the rate of flow in the tertiary canals and the rate of delivery to the fields may become excessively small for efficient operation. Other systems involve on/off full-flow rotational operation of lower-order canals during periods of reduced supply or demand. The primary canal usually runs continuously, flow being regulated, and rotation begins either at the secondary or the tertiary level. There are significant differences in flexibility of supply and in operational simplicity between the latter two alternatives (Campbell, 1992).

Regulation of supply from the tertiary to the individual farm outlet is a key issue in small-holder areas. The tertiary is usually designed to supply one farm only at a time (in the case of non-paddy crops), thereby simplifying regulation within the tertiary command and ensuring a sufficient rate of flow to the field for efficient farm water management. Each cultivator takes the full flow of the tertiary in turn, the duration of the turn being determined in some cases simply by the size of the holding, and in others by the type and area of crop being grown. Regulation of supply within the tertiary command (commonly 40 to 80 ha, serving 50 to 100 or more cultivators) is in fact a highly contentious subject. It is generally (but not everywhere) considered impractical to extend formal Irrigation Dept management down to the level of the farm outlet, and some degree of cultivator management within the tertiary command is necessary. Where such management, by Water User Association, is effective it opens the way for mutual exchange and virtually a controlled demand type of operation, permitting considerable crop diversification. Unfortunately the record of success with such associations to date is generally poor. However, regardless of whatever technical improvements are made in canal delivery systems the efficiency of irrigation in the small-holder situation is largely determined in the final analysis by the effectiveness of operation within the tertiary command. Other than in the simplest situations, there appears to be no alternative but to maintain an adequate level of appropriately-trained staff to supervise the day-to-day operation of the Water User Associations, for however long is necessary to achieve a reasonable level of group self-management.

Groundwater Irrigation in Small-Holder Areas

Over the last three decades a notable development in irrigation has been the proliferation of small shallow tubewells, owned and operated by the individual cultivator, either as his sole source or to supplement canal supply thereby providing desired flexibility. There is, however, a basic constraint on the continued expansion of such installations, due to the fact that pump and motor are located at the surface (or in a shallow pit). They are "suction-mode" wells, limited in lift to about seven meters. Where the dry-season water-table is deeper, or approaches that depth due to continued exploitation, recourse must be had to "force-mode" wells in which the pump, or pump and submersible electric motor, are located "down the hole". However, such installations are too costly and too sophisticated for the individual small cultivator, although not for his larger neighbor. Ideally, regulation of groundwater use could prevent draw-down below the reach of the small "suction mode" well. Deeper "force-mode" wells do not in fact increase the amount of groundwater regionally available (in most circumstances) and they do increase energy use due to greater lift. However, they permit greater use of the groundwater reservoir in dry years. In any case such regulation has rarely been effective in the area under discussion. The options open to small cultivators, faced with falling water-table, are to purchase water from their larger neighbors with force-mode wells, or joint ownership of such wells, or supply from publicly owned deep wells. With such larger wells organization of distribution can be a problem, similar to that discussed for canal supply. Technically efficient distribution systems are available (Campbell 1992), but cultivator organization remains the central factor in the success of such installations.

References
Campbell, Donald E (1992) "Key Factors and Problems in the Design and Operation of Internationally Assisted Small-Holder Irrigation in South Asia", 160 pp., ANATECH Research Corp.

MONTHLY WATER BALANCE FOR BLUE NILE RIVER BASIN IN ETHIOPIA

Peggy A. Johnson[1] and P. Douglas Curtis[2]

Introduction

The Blue Nile River begins at Lake Tana near Bahir Dar and has a drainage area of approximately 324,530 km^2. The Blue Nile supplies nearly 84 percent of the water to the Nile River during high flow season, with the remainder being supplied by the White Nile and other tributaries. Therefore, the Blue Nile is the main source of water not only to Ethiopia, but to Sudan and Egypt as well. From Lake Tana, the river travels southward, then turns to flow west to northwest to the Ethiopia-Sudan border where it begins a more northerly path. The Blue Nile eventually joins the White Nile at Khartoum in Sudan to become the Nile River. Flow volumes along the Blue Nile range from approximately 4,000 million cubic meters annually at the outlet of Lake Tana to 50,000 million cubic meters at the Sudan border. The majority of this volume passes through the river during the rainy months of July through September.

The objective of this study was to present existing stream flow data for the Blue Nile River and tributaries within the Blue Nile River basin in Ethiopia and to use that data to provide some insight on the temporal and spatial variation of the hydrology of the Blue Nile in Ethiopia through a monthly water balance.

Water Balance Model

A monthly water balance model was developed to represent the hydrology of the Ethiopian Blue Nile River basin based on existing water balance models, particularly those by Van der Beken and Byloos (1977) and Schaake and Chunzhen (1989). The model begins by assuming an initial soil moisture, S. An

---

[1]Assistant Professor, Dept. of Civil Engrg., University of Maryland, College Park, MD 20742 (301)405-1965

[2]Graduate Research Assistant, Dept. of Civil Engrg., University of Maryland, College Park, MD 20742 (301) 405-1970

effective precipitation is computed as:

$$P_{eff} = P - E \qquad (1)$$

where E is a function of the monthly potential evapotranspiration and the current soil moisture (Van der Beken and Byloos 1977):

$$E = E_p(1 - \exp(-C_1 S)) \qquad (2)$$

where E_p is potential evapotranspiration. As in the Van der Beken and Byloos model, C_1 reflects the soil type and will decrease with increasingly sandy soil. If the effective precipitation is less than zero, then the effective precipitation is set at zero. The surface runoff component is then computed as a function of the effective precipitation:

$$Q_s = P_{eff}\left(\frac{S}{C_2}\right)^{C_3} \qquad (3)$$

In Eq. 3, C_2 represents the maximum soil moisture so that S/C_2 is the current fraction of the total soil moisture. C_2 is a function of elevation and soil type; at very high elevations, the soil mantle tends to be thinner, so that C_2 should be less than at lower elevations. C_3 is an exponent that allows greater surface runoff when the soil moisture and average watershed slopes are relatively high. If S is greater than the maximum soil moisture C_2, then the excess is considered interflow, i.e., $Q_i = S - C_2$, and S is reduced to C_2. Groundwater contribution to the total stream flow runoff is then obtained by:

$$Q_g = C_4 S \qquad (4)$$

where C_4 is a function of the soil type and will increase with increasingly sandy soil.

The total monthly streamflow runoff is:

$$Q = Q_s + Q_i + Q_g \qquad (5)$$

The soil moisture for the next month is updated by:

$$S_{t+1} = S_t - Q_t - E_t + P_t \qquad (6)$$

The initial soil moisture can be determined in various ways. Since data is limited in the Blue Nile region, the initial soil moisture is calibrated as $C_2/2$. Water balance models are sensitive to the initial value of S only during the first six to 12 months, after which the effect of the initial value is overcome.

Data for the Water Balance Model

Data for calibration of the monthly water balance were taken from a US Bureau of Reclamations report (USBR 1964) and from a climate report by Gamachu (1977). For each of the subwatersheds, it was necessary to determine the monthly precipitation corresponding to the monthly flow. Chemoga and Mugher Rivers have precipitation gages within or adjacent to the watershed boundaries. The records from these precipitation gages were assumed to be the precipitation for those watersheds. For the other watersheds, precipitation data recorded at gages outside the watersheds were weighted by distance, location, and elevation.

Calibration of the Monthly Water Balance Model

The monthly water balance (Eqs. 1 through 6) was calibrated for 15 subwatersheds using a nonlinear, least squares optimization. Calibration results and goodness of fit statistics showed that the model was well suited for the Ethiopian Blue Nile River basin. The relative errors of the predicted discharges tend to be highest in the dry months; this is of little concern since the dry months produce such a small percentage of the total annual runoff.

Conclusions

This study has provided insight into the hydrology of the Blue Nile river basin in Ethiopia through the use of a monthly water balance. The monthly water balance model created for this project provided a good fit for the hydrologic conditions of this region. Since the Blue Nile River is the source of such a large portion of the water for the Nile River, it is important to have an understanding of the rainfall-runoff relationship for forecasting purposes.

The results of this project will be useful in forecasting flows into the Roseries reservoir at the Sudan border and subsequently into the Aswan reservoir in Egypt. The water balance model can also be used to determine the effect of global climatic changes on the volume of water that will enter the Nile river basin from Ethiopia. Since the Ethiopian Blue Nile River provides the majority of the water to the Nile River, such an analysis could be valuable toward water resource planning and management of the reservoirs along the Blue Nile and Nile Rivers.

Acknowledgment

This work was supported in part by USAID and NOAA as part of the Nile Forecasting System project. The writers wish to express their appreciation for the help and support of Dr. J.C. Schaake, Senior Hydrologist at the National Weather Service.

References

Gamachu, D., 1977. Aspects of Climate and Water Budget in Ethiopia. Technical Monograph, Addis Ababa University Press, Addis Ababa, Ethiopia.

Hurst, H.E. and Phillips, P., 1931. *The Nile Basin, Vol. I.* Physical Department Paper No. 26. Government Press, Cairo.

Schaake, J.C., Jr., and Chunzhen, L., 1989. Development and application of simple water balance models to understand the relationship between climate and water resources. *in* New Directions for Surface Water Modeling (Proceedings of the Balitmore Symposium, May 1989), IAHS Publ. no. 181.

U.S. Bureau of Reclamations, 1964. Land and Water Resources of the Blue Nile Basin. United States Department of the Interior, Washington, D.C.

Hydrologic and Structural Considerations for the Initial Filling of Jiguey Dam, Dominican Republic

Guy S. Lund, M.ASCE[1] and Ed A. Toms, M.ASCE[2]

Abstract: This paper describes the hydrologic and structural considerations of the initial filling of Jiguey Dam which is located on the Rio Nizao about 100 kilometers west-northwest of Santo Domingo, Dominican Republic. The Rio Nizao system consist of three tandem multiple purpose reservoirs: Jiguey; Aguacate; and Valdesia. Jiguey and Aguacate dams were added to the system in 1991 to increase the system's hydropower capacity and stored water for domestic and agricultural use. Jiguey Dam is a 115 meter high, 339 meter long, two-centered concrete gravity arch dam with gravity thrust blocks on both the right and left abutment. A 6,900 meter long power tunnel transfers the water from Jiguey Reservoir to a 98 MW underground powerhouse. Aguacate Dam, a power regulating dam downstream of Jiguey Dam, is a 51.5 meter high concrete gravity dam. A 11,000 meter long power tunnel transfers the water from Aguacate Reservoir to a 52 MW underground powerhouse. Valdesia Dam located downstream of Aguacate Dam is a 82.0 meter concrete gravity dam with domestic and agricultural turnouts and a 54 MW powerhouse.

HYDROLOGIC CONSIDERATIONS

Introduction. The reservoir system analysis for conservation, hydropower and flood control was performed with HEC-5 simulations to develop operating procedures for the projected 12 month initial filling period. Monthly reservoir operations were simulated to estimate the filling duration from the river bed elevation to the maximum operating elevation 541.5 meters for Jiguey Reservoir. The monthly reservoir operations were simulated to estimate the required initial reservoir elevation for Valdesia to meet the water supply demands of two alternative domestic water supplies of 3.15 m^3/s and 6.30 m^3/s and irrigation demands assuming the 12 month filling period will occur within the exceedence probability range of 50% to 95%. The filling of Jiguey Reservoir will be affected by the amount of actual rainfall during the initial filling period, the amount of water demands imposed on Jiguey Reservoir from domestic water supply and irrigation requirements and Valdesia Reservoir elevation at the time of closure. The filling will also be affected by the operating criteria to fill the reservoir to elevation 541.5 meters while meeting the water demands from irrigation and water supply for the different series of historical monthly flows.

---

[1]Civil/Structural Engineer, Engineering Consultants, Inc. (ECI), 5660 Greenwood Plaza Blvd., Suite 500, Englewood, Colorado 80111, Phone:(303)773-3788, Fax:(303)740-8671.

[2]Hydrologic/Hydraulic Engineer, Engineering Consultants, Inc (ECI).

System Operation. The water supply and irrigation demands were the first priorities with hydropower generation the second priority for the initial filling period of Jiguey Reservoir. There is no means of regulating the flows out of Jiguey Reservoir below the low level outlet works elevation 478m. The local inflow between Jiguey Reservoir and Valdesia Reservoir and the stored water above elevation 130.75m in Valdesia Reservoir were used to meet the water demands until Jiguey Reservoir reached the low level outlets. After Jiguey Reservoir has reached the low level outlet elevation, if Valdesia Reservoir fell below the minimum operating level of 130.75m then releases were made from Jiguey Reservoir to make up any deficiencies at Valdesia Reservoir. Valdesia Reservoir was used to supply all the water demands until the minimum operating level 130.75m was reached. Aguacate Reservoir was filled prior to the closure of Jiguey Dam and was maintained at maximum operation level elevation 328m during the initial filling period.

Results. Valdesia Reservoir was drawn down to its minimum operation level 130.75m within the first two to four months of the initial filling. Water demands were met by local inflow and releases from Jiguey Reservoir after this period. The deficiencies occur during the months of January through July. A large percentage of these deficiencies occur during the months of February through April with April being the most critical. The deficiencies for April range from 30% to 68% of the amount of water demanded.

In most cases, Jiguey Reservoir did not reached its maximum operating level 541.5m by the end of the simulated 12 month initial filling period. Maximum operating elevation 541.5m was reached and maintained for the 53.6% exceedance period.

The year with 82.1% probability of occurrence (82-83) is less critical than the year 64-65 with 53.6% probability of occurrence. This is because the critical inflow period is much shorter than the 12 month period used for determining the probabilities of exceedence. It rained heavily the first two months of the simulated initial filling period. This caused water being available during the critical periods of the year so there was less deficiencies.

The 1982-1983 flow year changes from 82.1% exceedence probability for the 12 month critical period to 42.9% exceedence probability for the 5 month critical period. The results of the percent exceedence probability revaluation did not change the outcome of the simulated monthly reservoir operations.

Jiguey Reservoir will reach minimum operating elevation 500m usually in the first half of the simulated initial filling period. Valdesia Reservoir was allowed to fluctuate between the maximum and minimum operating levels of 150m and 130.75m, respectively. The system produced power equal to the flow required to meet water supply and irrigation demands during the simulated initial filling period.

STRUCTURAL CONSIDERATIONS
Introduction Stresses and deflections in Jiguey Dam due to usual static loading conditions and unusual static loading conditions due to the Probable Maximum Flood (PMF) were modeled using a general linear elastic finite element analysis computer program. The results from the analysis were used to study the behavior of the dam during the filling of the reservoir.

Material Properties The material properties used for the concrete in Jiguey Dam have been selected based on field data collected during construction and the engineering guidelines specified by the U.S.

Bureau of Reclamation. Concrete test cylinders collected during placement of the mass concrete in the dam have been used to determine the 7- and 28-day unconfined compressive strength of the concrete. The average 7- and 28-day unconfined compressive strength was compared with an average expected strength curve for type-I cements, as reported in the USBR Concrete Manual, and used to estimate an average 1-year unconfined compressive strength for Jiguey dam of 3500 tons/m^2. The instantaneous concrete modulus of elasticity of 3,000,000 tons/m^2 was computed in accordance with guidelines of the American Concrete Institute (ACI), based on the expected 1-year unconfined compressive strength of 3500 tons/m^2. The sustained modulus of elasticity of 2,100,000 tons/m^2 was used for concrete in all static analyses, based on USBR criteria that the sustained modulus should be taken as approximately 70 percent of the instantaneous modulus of elasticity.

The average Rock Quality Designation (RQD) values of the foundation core samples obtained from eight drill holes at Jiguey damsite ranged from 10 percent to 60 percent with the higher values located near the base of the dam. These values suggest that the foundation deformation modulus would vary from (Deere's Engineering Classification of In-Situ Rock) less than 70,300 tons/m^2 to 422,000 tons/m^2.

Previous geoseismic investigations in the early 1980's indicated that the deformation modulus would vary from 1,500,000 tons/m^2 at the base of the dam to 100,000 tons/m^2 at the dam's crest elevation.

After careful review of the data and consultation with geotechnical engineers it was decided that the range of reference foundation deformation modulus values would vary from 1,054,800 tons/m^2 at the base of the dam to 210,000 tons/m^2 at the dam's crest elevation. The precision of foundation material property data was low; therefore, a range of minimum and maximum foundation deformation modulus values was analyzed. The minimum foundation deformation modulus values were taken as 50 percent of the reference deformation modulus values. The maximum foundation deformation modulus values were taken as 200 percent of the reference deformation modulus values.

Load Combinations The loads used in the structural analysis include gravity (the dead weight of dam), temperature (loads due to the volumetric changes in the dam caused by the corresponding ambient air and reservoir temperatures), reservoir (hydrostatic water pressure applied to the upstream face of the dam due to the various Normal and Probable Maximum Flood (PMF) Water Surface levels), and uplift (internal hydrostatic water pressure applied to the dam and foundation along the contact).

The finite element analysis was used to investigate several static load combinations on the dam due to gravity, various temperature loads, various reservoir water surface elevations, and uplift.

Analysis The general finite element microcomputer program SAP386 was used to perform a linear elastic finite element analysis of the dam due to the usual and unusual static load combinations. The dam was modeled using 216, variable node, isoparametric thick shell elements with two elements through the thickness. A significant portion of the foundation was modeled using 630 eight node, isoparametric thick shell elements with eight elements through the thickness from upstream to downstream edge. The foundation extended at least two dam heights upstream of the dam, and one dam height downstream and into the abutments.

The foundation for Jiguey Dam was over-excavated to remove unsatisfactory material, resulting in the use of dental concrete to shape the abutment underneath the left and right gravity thrust block sections. The finite element model used the average dam/foundation contact between the upstream heel and the downstream toe of the dam to define the dam profile. Portions of the left gravity thrust block section have a significant volume of dental concrete below this assumed dam/foundation contact. These analyses assumed that the dental concrete below the dam/foundation contact would behave similar to the foundation, therefore, the material properties were equivalent to those used in the foundation.

Stress Results Based on the results of the analysis the dam does not entirely behave as a three-dimensional structure. In the upper portion of the dam the horizontal arch stresses indicate that the dam's vertical contraction joints will open due to the effects of temperature loads. In the lower portion of the dam the horizontal stresses show that the vertical contraction joints are closed, thus preserving some three-dimensional behavior at lower elevations of the dam.

All of the compressive and most of the tensile stresses predicted by the finite element model are less than the allowable limits. Some tensile stresses predicted by the finite element model on the upstream face dam near the dam/foundation contact are due to the geometrical irregularities of the foundation. The tensile stresses near the dam/foundation contact are unable to develop and will be greatly reduced in the prototype.

The behavior of the dam is influenced by the assumed values for foundation modulus of deformation. Results show an increase in the foundation modulus increases the stiffness at the dam/foundation contact. The increase in stiffness at the dam/foundation contact results in an increase in the cantilever action of the dam and a decrease in the arch action of the dam.

Deflection Results An effective means to determine the behavior of a dam due to various loading conditions is to measure the dam's deflection.

The deflections output by the microcomputer program SAP386 were resolved into a local element coordinate system corresponding to the elements vertical and radial directions. The radial direction corresponds to the deflection perpendicular to the dam's axis.

The deflections results for the various loading conditions from the analyses are plotted comparing the effects of maximum and minimum values of foundation modulus of deformation. The plots define deflection envelopes of expected maximum and minimum deflections values. The deflection envelope for the usual static load combinations corresponding with full reservoir along elevation 554.5, the dam crest, is shown in figure 1.

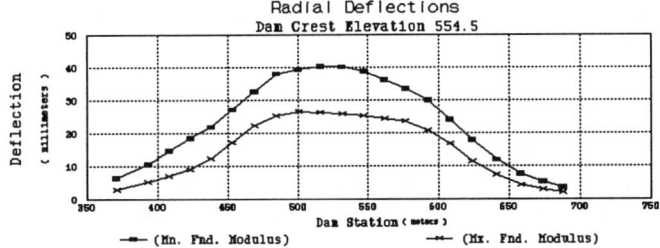

GEOTHERMAL AND HYDROPOWER PRODUCTION IN ICELAND

Duane J. Rosa, PhD, Member ASCE[1]

Introduction

This paper analyzes the impact of current and future development of geothermal and hydropower production on the economy of Iceland. Natural conditions in Iceland favor the increased utilization and development of both of these abundant power sources. The mean surface run-off in Iceland is about 50 l/s/km^2 (liters per second per square kilometer), with a large part of the country consisting of a plateau more than 400 meters above sea level. More than half of the country is above 500 meters above sea level. The technically harnessable hydropower potential is estimated at 64 TWh/year (terawatthours per year), of which 30 TWh/year is considered economically and environmentally harnessable. In addition, Iceland has abundant geothermal energy sources. A quarter of the entire country is a volcanic area. Keeping in mind that geothermal resources are not strictly renewable, it is estimated that the potential power production from this source is 20 TWh/year. Present utilization of these two resources totals only 4.2 TWh/year, or only about 8% of Iceland's aggregate potential.

There are many issues facing Iceland today as it considers development opportunities utilizing both of these abundant power supplies. This paper will first consider the technical aspects of both hydropower and geothermal power production in Iceland. Then, the economic consequences of alternative utilization of these energy sources will be evaluated. The first alternative to be considered will be the direct export of power by HVDC submarine cable to other countries, such as Scotland or the United Kingdom. Iceland could, as a second alternative, concentrate its efforts on bringing in energy intensive industries into the country.

---

[1]Fulbright Professor, University of Iceland, Reykjavik, Iceland and Associate Professor of Economics, West Texas A&M University, Canyon, Texas 79016.

Hydropower Production

The first hydropower plant in Iceland was constructed in 1904, with a power generation capacity of 9 kW, the equivalent of 0.009 MW. Over the years, many additional hydropower plants have been added. The first large-scale hydropower plant was built in 1965, which had a capacity of 210 MW. The total hydropower production in Iceland is now 832 MW. This still, however, represents only 25% of the economically feasible hydropower production for the country. Even though this total amount is small in comparison to world-wide hydropower production, Iceland is second only to Norway in the world in per capita electric power consumption. Even a doubling of Iceland's population would leave the country with usable hydropower potential many times that of its closest neighbor.

Geothermal Power Production

Iceland has enormous supplies of geothermal energy. It is estimated that the total potential for electricity production from the 19 high temperature fields in the country could be as high as 1480 TWh. Fundamentally, there are three types of geothermal energy, hydrothermal, geopressured and hot dry rock, the last being the most plentiful and widely spread. The geothermal areas of Iceland are almost exclusively hydrothermal in nature, and this type of geothermal energy is the most useful for commercial applications. Hydrothermal areas are divided into high and low temperature fields, according to the reservoir temperature. A low temperature field is defined as one where the temperature is below 150 degrees centigrade at one kilometer in depth, and a high temperature area where the temperature is above 200 degrees centigrade. The high temperature fields of Iceland traverse the country from southwest to northeast, with the low temperature fields located on the flanks.

Geothermal energy currently provides 32% of the gross energy consumption in Iceland. The single most important use of this energy source is for space heating (800 MW), with about 85% of all housing in Iceland being heated this way. In addition, geothermal energy is used for greenhouse cultivation of vegetables. The principal industrial uses of high temperature geothermal energy involve drying applications using flashed steam and/or hot water and in sea-minerals production. The largest drying application is the pre-process drying of dredged diatomaceous deposits at the Kisilidjan Plant in the north of Iceland, which annually produces 24,000 tons of diatomaceous earth filter-aids for export. Other drying processes benefitting from geothermal steam or water include fish drying, a seaweed plant, and a wool washing plant.

In contrast to low temperature space heating, power production is only possible in the high temperature fields situated in the volcanic zone which lies diagonally from southwest to northeast through the country. Electricity production by geothermal energy conversion is currently utilized in four power plants in

Iceland. The Krafla Power Plant in the northeast part of the country is powered by a dual pressure condensing turbine generator unit of 30 MWe capacity, with the powerhouse designed for a second 30 MWe unit. The Bjarnarflag Power Plant is a back pressure turbine of 3 MWe capacity. The Svartsengi Power Plant produces 12 MWe of power and 125 MWth used for heating. The Nesjavellir Power Plant, which currently supplies hot water to Reykjavik, can supply 300 MWth of hot water and up to 60 MWe of power. Iceland has found that the most efficient way of using the high temperature fluid is to cascade its use by, for example, using the heat first for the production of electricity, and the remainder for providing water for space heating and swimming pools.

Currently, only 6% of electricity production in Iceland utilizes geothermal power. The reason for this is the competitive advantage of available hydropower projects. This situation may change, however, as Iceland considers plans to develop a submarine cable for export of power to Europe, or to attract energy intensive industries to the country.

Economic Analysis of Energy Utilization

Iceland is a country of many contrasts. While it is rich in energy supplies, Iceland is a barren country. The Icelandic climate is too cold for any significant growing of grain. Therefore, a considerable part of the needed agricultural products such as grain and many vegetables must be imported. It is not economical to compete with more fertile countries in agricultural exports. However, the nation is self-sufficient in meat and dairy products and certain types of vegetables. The principal industry in the country is fishing. This industry, however, has been in a period of significant decline in recent years due to reduced total catch. The principal industrial firms in the country include an aluminum smelter with 90,000 ton production capacity and a ferrosilicon plant. The economy of Iceland is currently in a major recessionary period.

Because of the economic conditions facing the country, Iceland has been considering alternative plans to develop its energy sector. Feasibility studies completed by Landsvirkjun, The National Power Company of Iceland, have concluded that it would be technically feasible to export electricity through submarine cable to Scotland or England. Given the distances involved (a minimum of 950 km), transmitting electricity through cables can only be done effectively by high voltage direct current (HVDC). When electricity is transmitted with HVDC, the alternating current is changed to direct current by rectifying it in a converter station at one end, and then changing the power to AC by inverting it in converter station at the other end.

Though a submarine cable may be technically feasible and, as studies have shown, the price of Icelandic energy competitive, Iceland still faces significant economic and political obstacles. The project will require an enormous expenditure of funds, and this will probably require a sharing of costs by the receiving countries. Also, the cost effectiveness of the project will depend on market prices of energy in the future. From a political perspective, it may also be difficult for Iceland to negotiate an optimal price, given competition from other power producing countries such as France. For the receiving countries, however, the greatest advantage may be that Iceland's electricity does not result in any environmental pollution.

The other development option being considered by Iceland is to attract energy intensive industries to the country. Several international companies have recently been looking at Iceland as a potential location for a 200,000 ton aluminum smelter. The electricity demand for this size of plant is about 3,000 GWH/year. This option has both positive and negative impacts on Iceland. It would bring additional tax revenue into the country, provide additional employment during construction and final operation, and create positive multiplier effects on the other sectors of the Icelandic economy. From an environmental perspective, however, additional industrial development could result in increased pollution levels, and waste disposal problems. Also, increased utilization of geothermal resources can cause surface disturbances, physical effects due to fluid withdrawal, noise, thermal effects, and emission of chemicals into the atmosphere.

In summary, Iceland faces some significant decisions regarding its future hydropower and geothermal power development. Given the decline in its fishing industry, Iceland must promote the use of this valuable energy resource in order to insure a sound economic future.

REFERENCES

The National Power Company, 400 kV HVDC Submarine Cable Link, Iceland-Faroe Islands-Scotland, 1988.

The National Power Company, Export of Electricity, and The Competitive Position of Iceland's Electric Power, November 1991.

Gudmundur Magnusson and Eystein Gjelsvik, Optimal Use of Electrical Power in the Nordic Countries, 1992.

The North East Wisconsin (NEW) Waters For Tomorrow Initiative

Paul E. Thormodsgard and Harold J. Day
Green Bay Metropolitan Sewerage District
Green Bay, Wisconsin

ABSTRACT

A study to improve the understanding of opportunities for better water quality management practices in the Fox/Wolf River, Green Bay watershed in Wisconsin is reported. This investigation was initiated by a local citizen group concerned about inadequate funding for the implementation of the Green Bay Remedial Action Plan, RAP, a part of the International Joint Commission, IJC, sponsored effort thoughout the Great Lakes. They created an organization called "NEW Waters For Tomorrow" and obtained support from a number of public and private organizations, including the University of Wisconsin and the Wisconsin Department of Natural Resources, to fund a one year study.

The goal of the study is to learn more about alternative methods, especially cost effective methods, to continue improvements in water quality in the region. The alternatives, which include various combinations of point and non point source pollution abatement practices throughout the 6,600 square mile drainage basin, are under investigation by a three person interdisciplinary team including a biologist, a resource economist and a civil engineer.

A brief historical review of the region and its water resource management challenge, is the introduction. The framework for the first year effort, based upon the integration of ecology, economics, technology and institutions, follows. Some preliminary results are presented next. Plans for the study completion and presentation of the results follow. Some comments on the possible transferability of the study approach to other areas with similar challenges end the paper.

Developing An Ecosystem Approach for
Restoration of the Bay of Quinte

Murray German

Introduction

The Bay of Quinte was designated a "problem area" by the
International Joint Commission (IJC) in 1975 because of
excessive nutrient enrichment and bacterial contamination.
These problems were apparent well before that time. Beach
closures, nuisance algal production, low dissolved oxygen,
taste and odour problems in drinking water and major changes
in composition of the bay's biota had already prompted
abatement efforts to overcome the bay's bacterial
contamination problem and reduce inputs of phosphorus from
point source loadings. As successful as these efforts were,
restoration of the ecosystem was deemed incomplete in 1985
and the Bay of Quinte remained on the IJC's list of 43 Great
Lakes "Areas of Concern". In 1986, a Work Team was estabished
and the process of developing a comprehensive Remedial Action
Plan (RAP) to complete the ecosystem's restoration was
initiated.

Description of Study Area

The Bay of Quinte is part of the northeastern shore of Lake
Ontario separated from the main lake by Prince Edward
County. The bay is a Z-shaped body of water approximately 64
km long with an area of 254 km2. Water depths increase from
west to east, ranging from as shallow as 1 m in the extreme
western end to a maximum depth of 66 m near its eastern end.
The bay receives drainage from a catchment area of 17,315
km2. Major tributaries include the Trent, Moira, Salmon and
Napanee rivers. The Trent River watershed at the western end
of the bay is the largest and accounts for 67% of the total
drainage area.

Background

Settlement and clearing of the Bay of Quinte watershed
commenced 1784. Two centuries latter, cultural impacts had
shifted the Bay of Quinte from early mesotrophic to early
eutrophic conditions. During the second and third quarters of
this century, the process was greatly accelerated. Commencing
in the 1930s, the bay's aquatic community has undergone a
continuous process of species change and ecosystem
instability characteristic of a hyper-eutrophic ecosystem.
By the 1950s, walleye had replaced the indigenous northern
pike and small mouth bass as the bay's top predator species.
This shift from the indigenous sight feeding species to the
shade seeking walleye marked the onset of highly turbid
conditions resulting from excessive algal production. Algal
densities peaked in the late 1960s. The shading effect of

algae greatly restricted macrophyte production and
contributed to a collapse of the bay's walleye population. By
1970, white perch, an invader had become the bay's dominant
piscivorous fish species. Beginning in 1978 (coincident with
a major reduction of phosphorus inputs from sewage treatment
plants and severe winters in 1977 and 1978) there was a
resurgence of walleye stock and a collapse of white perch.

Causes of Eutrophication

Cultural eutrophication commenced at the start of the 1800s
with the process of converting the watershed from forested to
cultivated land and accelerated during the early 1900s as
intensification of agricultural activities increased.

Ironically however, the onset of bay's hyper-eutrophic
conditions coincided with pollution abatement measures taken
during the period of 1940 to 1960 to overcome severe
bacterial contamination problems. During this period, six
sewage treatment plants with discharges to the bay and a
number of sewage treatment plants with discharges to
inflowing tributaries were constructed. These facilities were
not originally equipped to remove phosphorus. As a result, by
1965, the six sewage treatment plants with outfalls to the
bay were discharging about 215 kg/d of phosphorus.

Point Source Phosphorus Control and Ecosystem Response (1978 to 1990)

Measures to reduce point source loadings of phosphorus
commenced in 1978. The six on-bay sewage treatment plants
were requested to reduce their effluent phosphorus
concentration to 0.5 mg/L. Between 1972 and 1990, this action
reduced point source phosphorus loadings from 215 to 32 kg/d.

Monitoring results for the pre (1972 to 1977) and post (1978
to 1990) phosphorus control periods, while somewhat
encouraging, have revealed a lesser that anticipated
ecosystem response (Table 1). The bay remains eutrophic.

Remedial Action Plan

The disappointing response to point source phosphorus control
measures prompted the RAP Team to undertake a more in-depth
evaluation of the ecosystem and factors preventing the bay's
recovery. Technical workshops, refined substance budgets and
development of substance simulation models have given the RAP
Team a more holistic understanding of the Quinte ecosystem
and factors inhibiting the bay's recovery.

Two internal factors are now known to be exerting a significant trophic influence in the bay. Studies of substance flux from the bay's sediments have produced estimates of 85 to 140 kg/day of phosphorus feedback to the water column. The magnitude of this internal load, when compared with the current phosphorus load of 35 kg/day from sewage treatment plants, explains in part why the bay has not shown the anticipated linear recovery to point source phosphorus control measures. It is believed that disappearance of the bay's macrophytes during the 1960s triggered the onset of phosphorus recycling from the sediments.

Compounding the sediment flux problem, it is now recognized that a food chain influence is operable in the bay. Statistical analyses of survaillance shows no correlation between algal biomass and phosphorus concentration but a strong relationship between algal biomass and the density of alewife and white perch in the bay. The explanation is that heavy predation by alewife and white perch has cropped off the bay's algae grazers (e.g. large zooplankton species and macroinvertebrates). In essence, this "Pac Man" relationship has eliminated segments of the aquatic food chain that eat algae. Invasion of the bay by white perch coincident with a collapse of bay's walleye population is believed to have initiated this 'top down' the food chain influence.

The Quinte RAP has established as an objective, restoration of the bay to its' 1930s trophic status. Clearly, the ability to accomplish this objective depends on our ability to minimize the these internal influences. This will require a return the bay's macrophyte and an recovery of the bay's phytoplankton grazers. Over the past three years, there have been encouraging signs that these missing components of the ecosystem are making a recovery.

The Bay of Quinte RAP will provide a helping hand in the form of further control of phosphorus inputs from the point source discharges and a 25% reduction (approximately 95 kg/day) in phosphorus supply from inflowing tributaries. Other options examined and rejected included the diversion of Lake Ontario water into the upper bay via the Murray Canal, experimental food chain manipulations and alum treatment of the sediments to reduce phosphorus flux.

Plant Modernization and its Role
in the
Efficient Use of Water Resources

Michael J. Haynes, P.E.[1]

ABSTRACT

Engineers and hydrologists alike are challenged with the task of developing hydro projects which satisfy a broad spectrum of criteria. Specific criteria employed by resource agencies and the Federal Energy Regulatory Commission (FERC) dictate many of the features and operating parameters which ultimately become a part of final design.

INTRODUCTION

This paper reviews the aspects of upgrades and rehabilitation of existing projects which are driven, in large part, by resource availability. Plant modernization is one mechanism through which existing projects are evaluated both physically and economically in order to justify improvements. The level of feasibility ultimately defines a particular scope for each individual plant, with all plants being unique. Specific examples will be drawn from the recently completed license documentation for the Snoqualmie Falls Project, located in western Washington.

BACKGROUND

Puget Sound Power and Light Company is an investor owned electric utility which has been operating and maintaining hydro electric projects in Washington State since 1898.

---

[1] Senior Engineer, Engineering Services Department, Puget Sound Power and Light Company, P.O. Box 97034, OBC-11S, Bellevue, Washington 98009-9734

PLANT MODERNIZATION

The concept of modernization is an evolving philosophy which is being developed to provide a proactive approach to some specific hydro issues (and constraints) faced by Puget. Primarily these include: 1) Turbine and generator upgrades, 2) The potential impacts of future (license driven) operational and environmental requirements, and 3) Benefits associated with improved efficiency, unit dispatch and improved control and monitoring through state of the art automated control technology.

The Snoqualmie Falls Project is a diverse hydroelectric system which has been in operation since 1898. The first phase of the project included 4 multi-wheel Pelton style turbines driving stationary field generators. The Project underwent various stages of expansion in 1905, 1910 and 1956. As it stands, Snoqualmie consists of two powerhouses, 7 turbines, penstocks and a tunnel, all with a combined hydraulic capacity of 2,500 cfs. Water rights obtained for the overall Project also equate to 2,500 cfs.

PHILOSOPHY

The path to any study of resource utilization and allocation requires a methodology which is consistent within a given region. Parameters established by State and local governments and resource agencies dictate the basic guidelines which must become the baseline of any study. In some cases certain parameters have been made into law, while others may result from negotiations between project owners/developers and various agencies. Once regional significance has been established, the study must also focus on the more broad national perspective, which employs consistency and balancing. The concept of balancing has developed into the foundation for the hydro licensing process. In this venue, balancing looks at the relative importance of power and non-power project aspects.

Non-power project aspects include recreation, fisheries, aesthetics, wetlands, water quality, water supply, historical/cultural, flood control and air quality, to name a few.

Instream flow requirements, while embedded in the heart of most non-power issues, may ultimately define the baseline for power related project considerations. In this context, most minimum flow requirements start at the bottom of the river or waterway under study. Water resource utilization for power begins, therefore, after minimum flow criteria have been accounted for.

The Snoqualmie system under the existing water right operates at an exceedance of approximately 40 %, while the proposed system will utilize water up to approximately 20 % exceedance (PSPL 1991).

RELATED STUDIES

One aspect of an existing project is the vintage of installed equipment. Generally speaking, hydraulic turbines manufactured prior to 1960 will exhibit reduced capacities and efficiencies when compared to modern designs for the same conditions. Efficiency improvements can be especially attractive since more generation can be obtained with the same amount of water. The difficult part of this type of assessment is establishing economic feasibility. Additional generation due to efficiency improvements alone may not prove economically viable. Utilities, therefore, must consider less tangible aspects such as reliability and operation and maintenance costs.

Puget has conducted studies of generating units throughout its hydro system. The purpose of these studies has been to establish capacities and efficiencies for upgraded turbines and generators. Upgrade capabilities for the units studied include capacity increases up to 35 % and efficiency increases up to 20 %, depending on the particular turbine. A recent turbine upgrade at Snoqualmie attained an increase in turbine horsepower from 28,000 to 34,500.

Optimization:

The concept of optimization, for the purpose of this discussion, will include economic feasibility of a given water resource. In evaluating the Snoqualmie Falls Project from an optimization perspective, Puget developed models for evaluating a range of project flow capacities. The intent of these evaluations was to determine an optimum flow for a given set of economic criteria. Puget's standard financial model provided most of the parameters utilized for the subject study.

Once developed, the optimization model evaluated various project configurations (assuming 300 cfs instream flow) in an effort to attain the most efficient operation in terms of water usage. Key features of the model included turbine efficiency curves and river flow data. By including operating characteristics of the various units within the system, the model could take advantage of the most efficient combination of operating units (PSPL 1991).

The model ultimately correlated river flow data with available energy. In conjunction with this, engineers also considered the costs associated with each alternative configuration. These costs were then levelized over the life of the project, or the term of the new operating license (40 years in this case). In knowing the associated annual costs and the annual generation based on water availability, the installed generation costs were put in terms of mills per kilowatt-hour (PSPL 1991).

Generating resources are typically evaluated in comparison with a utility's avoided cost model. The avoided cost establishes a constraint for the developed cost of a new resource. The various scenarios developed for Snoqualmie Falls were each compared against Puget's avoided cost such that, the last increment of generation would be at the avoided cost. Since the avoided cost represents the best estimate of future generating resource economics, the incremental assessment is a means of comparing levelized benefits and costs. In general, if an evaluated resource option falls at or below a company's avoided cost, that resource meets the long term strategic goals for that company (PSPL 1992).

SUMMARY

The role of modernization in establishing feasible water resource options represents an evolving technology aimed at addressing environmental and economic accountability. It is the responsibility of the hydro system owner or developer to satisfy sound engineering principles, while responding to imposed constraints related to water resource development.

Modernization is essentially a tool which, when properly applied, allows the best technical optimization of a resource in conjunction with allowing designers to factor in environmental parameters. Such issues as instream flow, ramping rates, reservoir operating curves, flood control and aesthetics can be effectively managed through proper modernization planning.

REFERENCES

Puget Sound Power and Light Company,"Snoqualmie Falls Hydroelectric Project - Application for New License", November 1991.

Puget Sound Power and Light Company, "Integrated Resource Plan 1992 - 1993", May 1992.

Developing a Workable Public Input Process
for Aesthetics and Recreational Needs
During Hydropower Licensing

Deborah Howe and Mike Stimac[1]

Abstract

Aesthetics and recreation are becoming increasingly important issues during hydropower licensing. A variety of regulations and legislation mandate the protection of instream flows for aesthetic and recreational resources. These have provided impetus for determining the effects of instream flows on recreation and aesthetic resources. A public survey designed for a proposed small hydropower project, located in a heavily used recreation area, attempts to determine the aesthetic and recreational preferences for instream flows. The major components in designing the survey are discussed. The public input process is still underway, however, preliminary results indicate lower flows in the river are generally preferable by visitors of the area. This is likely because of the types of users and the recreation activities performed.

---

[1]Senior Environmental Planner and Senior Project Manager, HDR Engineering, Inc. Suite 1200, 500 - 108th Ave. NE, Bellevue, WA 98004

Introduction

Agency consultation required under Federal Energy Regulatory Commission (FERC) guidelines, in addition to other federal and state laws, provide protection of instream flows. Several states have passed legislation that specifically provides for the protection and maintenance of instream flows. Oregon state was the first to establish instream flow legislation for the protection of recreation and aesthetics (ORS 536.325 and Rev. ORS 587.322-360). The Washington State Department of Ecology is mandated to provide base flows for "scenic, aesthetic, and other environmental values and navigational values" (RCW 90.54).

The FERC requires that license applications for hydropower projects include considerations for adequate facility bypass flows for instream flow protection. Under the Federal Power Act (Public Law 1082), the FERC is mandated to achieve a balance of potential resource values in its licensing decisions. For example, flows for aesthetic quality, recreation and maintenance of fish populations must be balanced against costs such as revenue lost from reduced power generation.

FERC guidelines for the study of aesthetic resources state that "Standard methods include not only an assessment of physical changes in landscape but also an analysis of viewer response to the changes" (FERC, 1991). To date, however, there are no specific standard methods to assess how recreation and aesthetic resources are affected by flow. Likewise, agencies have not yet taken official positions on how to determine instream flow needs for recreation and aesthetics.

Recently numerous researchers and government agencies have been conducting studies on the relationship of instream flow to recreation and aesthetics. A variety of methods have been used including expert judgement, computer modeling and surveying of recreation users on- and off-site. The following describes a public input process that was developed for a proposed hydroelectric project to determine instream flow needs for recreation and aesthetics.

Project Background

The proposed hydroelectric project is located at one of the most heavily used recreation areas in Snoqualmie Pass. The project would divert stream flows in a reach of the river that includes a scenic waterfall, recreation cabins, a public campground and hiking trails. A public survey was developed to attempt to determine how important instream flow is to visitors aesthetic and recreational experience and to what extent they would feel uncomfortable with low water flows.

Survey Development

During initial consultation with agencies, it was recommended that a user survey be conducted to evaluate the recreation use in the area and the public's attitude toward flow levels. It was determined that the best approach would be a combination of a verbal and visual instrument. Two professors from the University of Washington (a research professor of urban forestry and recreation and another professor of environmental planning) developed the survey instrument. The survey instrument was developed to include two parts.

The first part of the survey uses a questionnaire and photographs to survey visitors at the project site. Components of this survey include:

-Background information/demographics
-Frequency of visitation to the project area
-Recreation activities during visit
-Attributes that make the area attractive
-Photos of landscapes with different stream flows
-Importance of stream flow to recreation experience

A questionnaire was developed for the first part of the survey and distributed to interested agencies. Following agency review, and incorporation of their comments, a pre-test was conducted where the survey was administered on-site to actual recreation visitors.

The greatest problem encountered during development of the first part of the survey was with the photos. Photographs were taken of the river at different flow levels and at different locations on the river (i.e. at the waterfall, upstream of the campground, and at the campground). The intention of the photos was for people to identify stream flows that they preferred. The difficulties were in eliminating the background 'noise' from the photos. For example, rather than the public preferring pictures for reasons of stream flow, they may prefer a picture because the color is better, there are more leaves on the trees, or the sun was brighter, etc. Photographic processes were used to attempt to make the pictures as similar as possible in terms of lighting, color and cropping. During the pre-test it was determined that the public was identifying with the instream flows.

The second part of the survey has not yet been developed. It is anticipated that it will consist of a meeting either on- or off-site with an attempt to include a cross section of all recreation users in the area. This part of the survey will consist of a questionnaire using video tapes. Components of this video meeting will include:

-Background information/demographics
-Primary recreation pursuits
-Video of different stream flows of same sites used in photo survey

Status of Survey

Surveys of visitors in the project area at the campground, on the trails, and at the waterfall were conducted during August and September of 1992. Additional surveying is proposed during the winter of 1992/93 and spring of 1993. The video meeting, (part two of the survey) is planned during the spring of 1993.

Although the analysis of the data obtained during August and September 1992 is not complete, preliminary review of the data indicated that the people that use the area prefer lower flows. The exact flows that the public prefers has not yet been determined. The area visitor's preference to lower flow is most likely due to the types of people that use the area and the types of recreation activities. People that generally use the project area consists mainly of families with children. The types of water activities they engage in include wading, fishing and swimming. It is theorized that the public at the recreation area prefer the lower flows for the safety of their children who enjoy playing in the river. If, for instance, the majority of users at the site were experienced river rafters, higher flows may be preferred. Although results from the initial surveying showed that public responses were fairly predictable, the results are inconclusive since more on-site surveys are planned. Additionally, the video meeting may reveal different public responses related to aesthetic and recreational preferences for instream flows then the on-site survey.

APPENDIX

(References)

Federal Energy Regulatory Commission, 1991. <u>Hydroelectric Project Licensing Handbook.</u> Office of Hydropower Licensing, Washington D.C. p. C-29.

Identifying the True Issues of Hydropower
Resource Development in an Era of
Public and Regulatory Policy Transition

Neil Macdonald and Mike Stimac[1]

Abstract

The last several years have produced changes in public and regulatory policy that have provided for increased roles and participation of resource agencies, Native Americans, and the public in the licensing of hydroelectric facilities. The resulting involvements can significantly affect the schedule, design, and operation of a hydroelectric generation project. However, successful completion of the licensing, relicensing, or license amendment process is usually dependent upon sound engineering design and thorough scientific research underlying the basic viability of a proposal. The true issues remain economics, environmental impacts, and the appropriate priority in the order of electrical energy resources.

---

[1] Vice President/National Director Hydropower and Senior Project Manager, HDR Engineering, Inc., 500 - 108th Avenue NE, Suite 1200, Bellevue, WA 98004

Introduction

The last decade has produced changes in public and regulatory policy that have provided for increased roles and participation of resource agencies, Native Americans, and the public in the licensing of hydroelectric facilities. These changes may be attributed to the perceived need to provide broader review of project proposals and increased protection of natural resources brought about by the rapid pace of filings with the Federal Energy Regulatory Commission (FERC) for new hydroelectric facilities during the 1980s and the anticipated numerous relicensing submittals expected in the early 1990s. While the changes have achieved the intended purpose of greater participation and responsibility, the successful completion of the licensing, relicensing, or license amendment process by an applicant or licensee remains a direct function of the siting, engineering design, and scientific research supporting the basic proposal. Good projects get approved - bad ones don't. The true issues remain economics, environmental impacts, and the appropriate priority in the order of electrical energy resources to be developed.

The Gold Rush of the 1980s

Economic incentives brought about by the Public Utility Regulatory Policies Act of 1978, coupled with the perceived increasing need for electrical power, produced a "gold rush" effect in terms of the investigations for and filings on new hydroelectric sites. Electric utilities, municipalities, and private developers all entered the race. It seemed that if left unchecked, virtually every reach of stream in the Northwest would some day have a dam and/or powerhouse located on it. Free-flowing and white-water would become descriptions of the past. In some cases, preliminary permit applications were filed with the FERC on sites unseen by the proponent of a proposed project. The developers came from everywhere - coast to coast - it didn't matter. All that was needed was a USGS quad sheet and a copy machine. Even advisory organizations, such as the Northwest Regional Power Planning Council (Council) got involved. The Northwest Conservation and Electric Power Plan developed by the Council in April 1983 established a priority listing of electrical resources with hydroelectric second only to conservation. The Council estimated that 1500 average Megawatts (MW) of new hydropower potential existed with 1080 MW at new sites in the four Northwest states.

Perceived Need for Regulatory and Policy Reform for Increased Environmental Protection and Agency/Public Involvement

The 1980s were anxious times for those concerned about the proper utilization of the nation's water resources, particularly as they could be affected by the expanded development of hydroelectric generation. The perception for increased protection through regulatory and policy reform produced protective measures through the passage of national legislation such as the Electric Consumers Protection Act (ECPA) of 1986, modification of the FERC's regulations to require an expanded three stage consultation process, and regional directives like the Council's Protected Areas Program.

Although ECPA was primarily focused on the relicensing of hydroelectric projects, it also included provisions for the filing and processing of applications for new licenses. The ECPA amendments to the Federal Power Act (FPA) were incorporated into the FERC's regulations through subsequent rulemakings. ECPA and the revised regulations provide for a three stage resource agency consultation process, attendance by the public at the beginning of agency consultation, and procedures for resolution of disputes between applicants and the agencies.

Concern regarding potential environmental impacts to streams as a result of proposed and anticipated future hydroelectric developments led to the Protected Areas Program by the Council. The program was an outgrowth of the Council's responsibilities under the Northwest Power Act which directs the Council to develop a "program to protect, mitigate, and enhance fish and wildlife, including spawning grounds and habitat on the Columbia River and its tributaries". In 1988, the Council amended its Fish and Wildlife Program to designate approximately 44,000 miles of streams both within and outside of the Columbia River basin in the four Northwest states as "protected areas". Protected areas are reaches which contain important habitat for certain valued species of fish and wildlife.

Factors Determining Good Projects, Bad Projects

Putting aside political and emotional influences, distinguishing good projects from bad must be accomplished by the appropriate consideration of factors in the following areas: environmental, engineering, aesthetics/recreation, and economics.

Proper siting of a new hydroelectric project involves more than just the obvious adequate combination of water flow and head. General site topography, access, proximity to existing load and transmission facilities, absence of threatened/endangered species, and where anadromous fish are a consideration, the existence of a natural barrier to their passage are important considerations. Satisfactorily meeting the associated requirements will help to ensure approval and possibly minimize the overall project schedule.

Engineering factors include siting considerations as well as constructability, operability, and maintainability. Sound preliminary engineering assessments will help to clearly determine a project's feasibility from both an engineering and economic perspective.

Aesthetics, aesthetic flows, and recreation have emerged as significant concerns in the licensing of hydroelectric facilities. Projects which avoid or minimize impacts to recreational uses of the stream have a greater chance of success.

Similarly, aesthetic impacts must be a factor in the engineering design process. Facility operation can influence aesthetics, particularly where flows in the bypass reach and over water course features are important. Recreational uses and observer perception must be included in any evaluation.

The proposed project, whether for immediate use by the proponent or sale of the output to others, must be cost-competitive and meet the economic requirements of the proponent. It should also satisfy any applicable resource priority listing in terms of cost measured in mills per kilowatt-hour or dollars per kilowatt.

The Right Project at the Right Time

The decade of the 80s also saw the evolution of the utility least cost resource plan. Electric utilities, whether public, municipal, or investor owned, were and are developing, individually and/or collectively, resource plans which setforth their respective energy and capacity needs for the future and which resources in what order would be utilized to meet those needs. First has been conservation, followed by renewables (including hydro). Provided the utility's conservation opportunities were being pursued, hydroelectric generation was generally considered to be a right resource of choice. Thus, if the project met the economic and design tests, it seemed to be an appropriate selection.

Conclusion

Electrical energy load growth has resulted in a continuing need for new electrical generating resources. Policy changes implemented in the late 1970s and early 1980s resulted in a renewed interest in hydroelectric generation with particular emphasis on the development of smaller capacity, run-of-river projects at new sites. Both in the Northwest and nationally, this added pressure on water resources for which there were significant competing uses produced legislation and regulations increasing the involvement of resource agencies, Native Americans, and the public in the project licensing and approval process. These changes have probably produced a more balanced utilization of the nation's water resources. However, they may have also provided the means for special interest groups to delay "good" projects and, because of the increased complexity of the process and associated cost, thwarted the underlying intent of PURPA. The viability of a project was, and remains to be, based on thorough underlying research, sound siting principles and engineering design, good economics, and being the generating resource option of choice in a relevant resource plan.

APPENDIX

(References)

Electric Consumers Protection Act of 1986, Public Law 99-495, October 16, 1986.

Federal Power Act, June 10, 1920.

Northwest Power Planning Council, April 27, 1983.
<u>Northwest Conservation and Electric Power Plan.</u>

Northwest Power Planning Council, September 14, 1988.
<u>Protected Areas Amendment to 1987 Fish and Wildlife Program.</u>

Pacific Northwest Electric Power Planning and Conservation Act, Public Law 96-501, December 05, 1980.

Public Utility Regulatory Policies Act of 1978, Public Law 95-617, November 09, 1978.

"SNO WATER" LED COORDINATED WATER SYSTEM PLANNING

PAT BURNAROOS[1]

ABSTRACT

The designation of a Critical Water Supply Service Area (CWSSA) which leads to Coordinated Water Planning evolved from problems recognized much earlier. Regional water suppliers in Snohomish county foresaw upcoming water needs and started meeting as the North Snohomish County Regional Water Association ("Sno Water").

INTRODUCTION

The Tulalip Tribes deserve a special commendation for getting regional water planning going. In looking back to the early 1980's, it was tribal planners who called my office to see if we could meet with them regarding water supply on the plateau which we shared. At first, we were a small group but as time passed our group grew. This was really the first time that water purveyors of the area had ever sat down together and addressed our joint concerns.

ASSESSMENT OF SITUATION

As the months passed, we were united in our assessment of our problems. The four main problem areas were:
1. Quality and quantity of water in the region,
2. The proliferation of numerous small water systems,
3. The duplication of facilities,
4. The lack of system standards and enforcement.

The tribes engaged an engineering firm to study the plateau's water situation and to advise our group how to best accomplish our goals. One item was the need for a

---

[1] Manager, Seven Lakes Water Association, P.O. Box 100, Lakewood, WA 98259

large pipe line to supply water to the town of Marysville and the Tulalip Reservation. At this time the Tulalips were actively seeking industrial development.
We named our group "North Snohomish County Regional Water Association". This was later shortened to "Sno Water". During the ensuing months, we had representatives from the County and state agencies attend our meetings. The state health officials informed us about the Public Water System Coordination Act. Some of our group had grave concerns about this Act so a great deal of time was spent checking with other counties that had the Act in place.
After much deliberation, in 1987, the Sno Water Committee formally requested that Snohomish County Council declare the northern part of the county a Critical Water Supply Service Area. This was something new for the County and a great deal of apprehension surrounded it. One concern that was voiced was "Will this group of purveyors try to limit growth?" Our group continued to push and to educate the County as to our concerns. The County assessed all the water systems and their capabilities and recognized the needs we had enunciated.

COORDINATED PLANNING UNDER THE ACT

Finally in mid 1988, the County took the big step and passed the motion that put in place the Water Coordinating Committee and the Coordinated Water Planning Process. The Committee would be members from the State and local agencies as well as a representative from each water purveyor with 50 services or more, The study area was expanded to almost double our original size. Our committee began the task of writing a WAter System Plan that each of us would be comfortable with. The first task was to hire a consulting firm familiar with drinking water planning. The committee chose Economic and Engineering Services (EES) with Bob Wubbena and Glenn Fiedler. Both men had worked with the Coordination Act and drinking water officials.

Now we had a goal, a committee and an engineer. The next two years of meetings were not always easy. As our small group had grown, we now had cities, districts, associations and agencies. Our biggest problem came when we defined our future service area boundaries. Slowly but surely disputes over claimed areas were resolved. Another problem came when one of the cities needed water and could not wait for the plan to be completed. The conflicting parties met and outlined what was necessary to each and the new pipeline resulted in a three-way sharing agreement.

The plan was completed and submitted to State and local

authorities in late 1991. In February of 1992 the North Snohomish County Coordinated Water System Plan was certified and put into action. After 12 years we had succeeded, we had our plan, we had made a lot of good friends and solved a lot of problems.

IMPLEMENTATION

A few months later we ran into brand new problem. Just how do we implement this plan???

The committee started meeting again. Purveyors were supposed to meet requests for service in areas which they had claimed but did not yet have waterlines in. The committee wrote sample agreements for remote service in an attempt to provide consistency. Our consultant is going to provide a supplement for implementation. The Appeals Sub-Committee held their first appeals hearing to review an issue where landowners felt they ought to have access to water sooner than the purveyor could provide. The Appeals process seemed to function smoothly.

PRIVATE ASSOCIATION'S EXPERIENCE

As a representative of a private water association the past 12 years have been rewarding. I am general manager of a system that has a few less than 2000 services. We have faced problems that are shared by cities and water districts. Our financing must meet different rules than municipally owned systems, but our residents value reliable service and pay rates which enable us to plan for future growth. Our fee structure includes initial hookup charges so growth tends to pay its way. We feel secure that our hopes of uniform standards are now a reality. We no longer have the problem of a small substandard system going in within our defined boundaries.

SUMMARY

We feel very strongly that because of the planning of the past 12 years, the citizens of our area are the real winners. We, the purveyors, are more aware of our natural resource and just how fragile our water supply is.

3 COUNTIES COORDINATED WATER SYSTEM PLANS

JACKIE HIGHTOWER[1]

ABSTRACT

The Coordinated Water System Plan (CWSP) process is a state mandated management tool created to get agencies and utilities to jointly address problems of current and domestic/municipal water supply, based on local needs and resources. The experiences in three counties, Island and Snohomish and King, illustrate how the process works in situations where the water availability and population differ. The Washington State Department of Health (DOH) claims that large water systems dependably distribute higher quality drinking water with lower cost to customers than do small systems.

INTRODUCTION

People from areas which were experiencing or where problems were expected with delivery of safe, efficient and reliable water service assessed their situation and requested designation as Critical Water Supply Service Area (CWSSA). This is the first step of a CWSP.

Island County designated the entire county because it was almost entirely dependent on groundwater. Snohomish County designated the northern two-thirds of the county, also an area heavily dependent on groundwater. King County designated the whole county under four separate areas, including those which relied on groundwater and surface water for source supply.

Each CWSSA set up a committee, the Water Utility Coordinating Committee (WUCC), where consultants and representatives of agencies and citizen groups refine state CWSP requirements. They are:

---

[1] Manager, Wellhead Protection, City of Renton, 200 Mill Avenue South, Renton, WA 98055

1. mapping of existing and future service areas for utilities,
2. developing the process that allows new purveyors to provide service,
3. setting up a Satellite System Management Agency to handle areas where no service area claims existed,
4. planning future regional water supply,
5. reviewing existing water rights,
6. assisting each water purveyor to prepare a water system plan for their claimed future water service area, and
7. setting minimum design standards for construction of new systems.

With the assistance of consultants these procedures were put in draft form, reviewed and approved by the WUCC, the County government and the state Department of Health. At that point a CWSSA is said to have a certified CWSP. These plans are working documents and reviewed at least every five years. Some planning areas have found that the committee continues to meet the first year as implementation of the plan begins.

RURAL ISLAND COUNTY

Island County has one city and a U.S. Naval Station that pipe surface water from the mainland. The other two towns and the rest of the population depend on groundwater sources. There are approximately 660 public water systems. Almost all people are served by privately owned water systems either as homeowners associations or developer owned. Most of the systems have had little refurbishing, have low capitalization and expect problems meeting the Safe Drinking Water regulations. Few systems claimed areas any larger than their existing service areas.

Several areas of the County have a serious problem with saltwater intrusion in water supplies. There is evidence that the incidence of saltwater intrusion is correlated to withdrawals exceeding available groundwater, but the geohydrologic system is complex and other factors may dominate in some areas. Many water rights are not being fully used in areas of present saltwater intrusion.

More than half of Island County's population resides in the North Whidbey census division. The Oak Harbor water system which is supplied from off-island has the potential of delivering additional supplies. The costs of major line extensions would be significant. Strong public sentiment surrounds the issue of whether to import water from off island. Certified in 1989, the Island County CWSP has been challenged legally about its regional water supply plan.

The Island County CWSP opened the discussion about water rights and the present exemption for withdrawals of less than 5000 gallons per day. Regulation of water quality and the right to drill and pump groundwater is not tightly defined nor stringently regulated. Those who don't want more growth identify water as the limiting factor.

The water systems can deal with intrusion by remaining small and carefully metering water, allowing only the use of less than the annual recharge. Or the majority of water used could be piped onto the islands. The existing systems managed by private interests do not have the funding options available for capital improvements that the latter would cost. Development of interties between utilities has not taken place either, perhaps because of reluctance to share today's surplus when tomorrow there may be a shortage.

No satellite management system has stepped forward to take over ownership, coordination or management of the small utilities. The sparse but growing population and distances between houses work against coordination of water services.

SUBURBAN SNOHOMISH COUNTY

Western Snohomish Coordinated Water Supply Plan was certified in 1992. It includes planning for future supplies for the northwestern two-thirds of the county which includes 640 public water systems of which 50 are over 50 connections each. Snohomish County has a number of small towns and areas that are turning from rural to suburban densities. Most purveyors within the CWWSA use groundwater but some get their water from surface supplies through the Everett Water Department.

The larger water purveyors of the CWSSA were willing to designate future service areas bigger than their existing service areas. This indicates a willingness to serve proposed new developments within their area. All purveyors claiming larger areas must update their system's comprehensive water plan by summer of 1993 to describe how they will serve the larger area.

A major portion of the county which was not claimed as a particular service area falls under the jurisdiction of the satellite management system provider which is Snohomish County Public Utility District #1 (PUD). However the PUD is legally limited to only being responsible for those systems it owns. This creates a dilemma for systems which want management but not ownership. The role of Satellite System Management is developing. The PUD is exploring many avenues in its role of assuring water quality and dependable service. At Lake Roesiger the PUD had installed a water line to densely settled lakeshore homes on septic tanks. They started a groundwater/septic system management program to protect water quality in the lake.

The CWSP includes groundwater protection planning as one method of assuring potable supplies. The County has started a Groundwater Management Plan (GWMP) process for the county. The United States Geologic Service (USGS) is gathering data. Their three year project records water levels and water quality records from existing wells. This will give a comprehensive picture of groundwater at this date in time across the county. The information that the USGS gathers will be the foundation for the ground water protection policies proposed in the GWMP.

Snohomish County is continuing to coordinate water supply planning with proposed code changes asking that new development hook up to existing systems and by promoting groundwater protection education.

URBAN KING COUNTY

King County has four Coordinated Water System Plans that cover virtually all of the county. The first is Vashon Island which is a sole source aquifer, similar to Island county. It's CWSP was certified in 1990 and is being implemented through an ongoing groundwater management plan. The second, Skyway, an urban area, needed coordination of the different suppliers as more infill occurred and fireflow needed to be coordinated. It was certified in 1990 as was the third plan - East King County's. The East King County Regional Water Association has taken the lead in implementation and is doing a partial update to bring the state mandated Growth Management Areas (GMAs) in to compliance with the rural and urban categories in the CWSP. Some areas, previously urban now declared rural, have ULIDs in place to build urban density infrastructure. The update of the CWSP will address those concerns brought by GMAs. The South King County CWSP was certified a year later in 1991 and includes some joint planning for future supplies by the Seattle Water Department and the Tacoma Water Department.

SUMMARY

Of the three counties Island and Snohomish have found that new small water systems (less than 8 hookups) are continuing to form even though the official position of the Department of Health supports only the extension of larger systems. As counties become more densely populated as King County is, larger systems find it more profitable to extend lines, thus King has been most successful at limiting the proliferation of small systems. Full development of Satellite System Management techniques will also increase the dependability and quality of public drinking water service by limiting the number of system managers.

A SATELLITE WATER SYSTEM PROGRAM IN SNOHOMISH COUNTY, WASHINGTON

Mark D. Spahr[1]

Abstract

The Snohomish County Public Utility District No. 1 (PUD) has developed a Program for management of small public water systems. The Satellite Water System Management Program (SWSMP) is attracting considerable interest. The challenge is to upgrade and operate small systems to comply with new regulations while assuring the PUD's financial viability.

Introduction

Over the past few years, Snohomish County has ranked among the top three counties in the nation in percentage of population growth. While most of the growth has been served by a regional water system, much of the geographical area is not served by a large water utility. This, when coupled with the availability of ground water and past land use policies that allowed easy creation of new "public" water systems, resulted in the current number of public water systems which exceeds 600.

---

[1] Water Resource & Planning Manager, Public Utility District No. 1 of Snohomish County, 2320 California, Everett, WA 98201

With the emergence of stringent water quality, monitoring, operational and financial standards for water systems, it is estimated that less that one percent of the existing systems can comply. Further, Washington State's recently adopted Growth Management Act requires that a system must be "safe and adequate" before any expansion or permits are allowed. Where parcels are served by substandard systems, this has resulted in the denial of building permits and mortgage loans for new construction or the sale/refinancing of existing homes.

Satellite Water System Management Program

The specter of increased regulations, enforcement and possible impact on property values is generating interest in the SWSMP. The PUD is an electric and water utility, serving over 220,000 electric customers and about 6,600 water customers in four water systems. While the PUD has had its SWSMP in place since 1979, it has generated little interest until recently. Currently, seven systems have initiated formal requests for the PUD to evaluate their system for possible takeover. Further, approximately 20 others are considering the possibility of transferring ownership to the PUD.

The SWSMP works as follows:

1. The owners or customers of a water system inquire about the SWSMP. A community meeting is typically held to discuss the program and answer questions. If there is continued interest, the PUD's engineering staff conduct a free, "no-obligation" evaluation of the system to determine what, if any deficiencies and/or liabilities exist.

2. Based on that evaluation, a cost estimate for preparing an economic, financial and engineering feasibility study is provided to the system owners. Charging for this service covers the PUD's cost of engineering, water quality testing and other services as needed. The cost also separates "casually" interested parties from those who are serious

about a possible transfer.

3. If payment is received, a formal evaluation of the system is conducted, including a cost estimate and financial plan. If needed improvements are modest (less that $600 per customer), they are financed through monthly water rates. If improvements exceed $600 per customer, alternate financing is developed. The common approach is to create a Local Utility District (LUD), to assess the improvement costs to the benefiting parties. An LUD allows the property owners up to 20 years to pay the cost.

4. Once a financial arrangement is made, the transfer takes place, and the PUD assumes responsibility for the water system.

Financial Issues

The SWSMP does not have the resources to "bail-out" small water systems. The PUD's water utility is operated as an enterprise, meaning that operating revenue derives from rates and charges from water customers. Further, with the relatively small customer base of 6,600, there is limited ability and little desire to subsidize the operation of one system at the expense of another.

This represents the greatest challenge to the success of the SWSMP. While economies can be derived through standardization and efficient scheduling of maintenance, there is currently no regulatory method to allow a meaningful reduction in operational costs. Specifically, full regulatory compliance is now based on the number of satellite systems, rather than the total population served. Such a modification would offer savings in monitoring costs.

As a part of its updated water system plan, the PUD will evaluate options for increasing efficiencies and lowering costs. Some options to be evaluated include; contract operations, interties, and the possibly of reduced regulatory requirements.

Partnership with Land Use and Regulatory Agencies

The key to success of the program will be the PUD's ability to approximate the efficiency and per customer cost of operating a large, consolidated water system. This will require the development of a partnership between land use and regulatory agencies.

While the agencies support and encourage the PUD's SWSMP, currently each of the satellite systems is subject to the same regulatory burden as if it remained separate and independent. For the Program to succeed, some credit for a unified operational approach must be developed.

Summary

There is a significant need for a coordinated and unified approach to the issue of small public water systems in Snohomish County (and many other counties in Washington State). One can imagine the problems if 600 electric or telephone utilities existed within a single county. The PUD's SWSMP is a good approach to the gradual unification of these systems, but other agencies including land use and regulatory agencies, which contributed to the existing problem by approving the creation of the water systems, must participate in the solution.

The most significant need is to allow the satellite water system operator to achieve economies through operating multiple water systems, as compared to each community or neighborhood operating independently.

The Department of Health's Role in
Implementing the Coordination Act in
North Snohomish County

Nancy Feagin[1]

Abstract

A coordinated water system plan in north Snohomish County resulted in development of uniform minimum design standards, a decrease in the number of new small water systems and regional solutions to water quantity and quality problems.

Introduction

The Public Water System Coordination Act (Chapter 70.116 RCW), enacted in 1977, establishes a procedure for water utilities to coordinate their planning with adjacent water utilities and local governments. The law states that the Washington State Department of Health (DOH) or the county legislative authority may declare an area within a county as a critical water supply service area (CWSSA). This declaration is based upon the findings of a preliminary assessment, identifying problems related to inadequate water quality, unreliable service or lack of coordinated planning.

Implementation of this law requires that a coordinated water system plan (CWSP) be prepared for the study area. The CWSP consists of a compilation of the water system plans prepared by each expanding water utility and a regional supplement prepared by a water utility coordinating committee consisting of representatives of water systems serving fifty or more customers, county government and DOH.

---

[1]Regional Engineer, Washington State Department of Health, 1511 3rd Avenue, Seattle, Washington 98101

The Coordination Act has been implemented in eighteen Washington counties. In Snohomish County, the county council declared a CWSSA in north Snohomish County on October 19, 1988, and adopted the formal boundaries of the area on July 5, 1989.

Water Supply Problems in North Snohomish County

North Snohomish County has experienced water supply problems related to rapid growth, limited local sources of water, existing water quality contamination and lack of coordination among water purveyors.

The preliminary assessment for North Snohomish County found that the area had experienced above average growth and concluded that all of the larger water systems would need to increase their source capacity within five years.[2] In addition, a number of subdivisions were created in rural areas of the county during the 1960's and 1970's without adequate water supply.

The preliminary assessment found that additional ground water supplies were available in the region, but the areas with the greatest promise in terms of quantity were most likely to have water quality problems. The feasibility of developing surface water supplies was restricted by potential adverse impacts on local fisheries and the costs of compliance with new and impending regulations.

Water quality problems have limited the availability of local water supplies. In early 1987, a series of arsenic poisonings near Granite Falls led to the discovery of high arsenic levels in the ground water.[3] The problem was most severe in an area served primarily by small water systems and single family individual wells. Several small food establishments were closed and individual families forced to buy bottled water or haul water from town. In addition to the arsenic problem, many small water systems currently use surface water or shallow ground water for

---

[2]*Northwest Snohomish County Regional Water Study and Preliminary Assessment*, Rasmussen and Huse Engineering, March, 1987.

[3]*Seasonal Study of Arsenic in Ground Water, Snohomish County, Washington: A report from the Snohomish Health District and the Washington State Department of Health*, Environmental Health Programs, Office of Toxic Substances, June, 1991.

their source of supply. Increasingly stringent regulations stemming from the federal Safe Drinking Water Act, will require additional, and probably expensive, treatment of these sources. Finally, high iron and manganese throughout the area has made development of ground water sources expensive, and sea water intrusion has been a problem in coastal areas.

In addition to water quantity and quality problems, the lack of a coordinated approach to water service has led to the creation of a large number of small water systems, with nearly 400 systems serving fewer than 10 homes each. DOH has identified a number of problems commonly associated with small water systems: a diffuse ownership, often resulting in a lack of responsibility and understanding of water system development and operation; private ownership, resulting in limited access to public financing such as grants and low-interest loans; and lack of an adequate rate base, limiting the system's ability to adequately finance capital improvements, and operation and maintenance.[4] DOH statistics on the compliance status of small water systems during a recent five year period found that more than 40 percent of the small systems in Snohomish County repeatedly failed to meet state water quality testing requirements for bacteriological quality.

Goals of the Coordinated Water System Plan

The goals of the CWSP are to: develop uniform minimum design standards; establish formal service area boundaries and service area agreements between adjacent water purveyors; develop a referral process for those needing water service to ensure that new systems are not created unless existing ones cannot provide service; and develop and implement a regional supply plan. These goals were intended to address the water supply problems identified in the preliminary assessment. Minimum design standards were set to ensure a higher quality, more reliable water supply for the consumer and facilitate future interconnection of systems. Formal service area boundaries and the referral process were established to eliminate costly duplication of service and competition for service between adjacent water systems, and streamline the process for obtaining water service. The regional water supply plan was intended to make economies of scale available to all purveyors in need of additional water

---

[4]*Small Water System Problems: An Assessment*, Drinking Water Program, Office of Environmental Health Programs, Department of Social and Health Services, May, 1989.

supplies and bring in water to areas with inadequate local supplies.

Plan Implementation

The role of DOH in implementing the CWSP is through the review of individual utility water system plans, project reports and construction documents. DOH reviews these planning and construction documents to ensure that they are at least as stringent as the minimum design standards, that they are consistent with the established service area boundaries and that new systems are not created unless the utility service review procedures laid out in the CWSP have been followed. In addition, when a developer appeals the conditions of service, DOH is the final level of the appeal process.

Most problems encountered in implementing the plan have been related to the minimum design standards. The concept of designing a system around the designated land use of the area rather than for specific buildings or developments has been difficult to convey to engineers, developers, water purveyors and county fire officials. Furthermore, developers accustomed to constructing minimal water systems in unincorporated areas have been resistant to meeting the new standards, especially when fire flows are required to support the proposed land use.

On the positive side, the water utility coordinating committee has operated on a consensus basis, resolving most issues without resorting to a majority vote. All service area conflicts were resolved between the parties involved, and action by DOH as final arbiter was not required. Since the plan was approved by DOH in February, 1992, the Snohomish County Public Utility District (P.U.D.) has been acting as a satellite system management agency. As a result, very few new water systems with independent management structures have been created. A 30 MGD regional supply pipeline sponsored by the City of Marysville has been constructed; and a regional project to bring safe drinking water to the Granite Falls area is being sponsored by the Snohomish County P.U.D. and is undergoing environmental review. Although only one appeal has been requested, the appeals process has received favorable reviews from the parties involved.

In summary, the coordination act is an effective tool to solve local water supply problems, but active participation of utilities and local government agencies is essential to the success of the process.

ECOLOGICAL CONSIDERATIONS IN GROUNDWATER MANAGEMENT

Eugene B. (Gus) Yates[1]

A wide variety of potential effects of groundwater management on terrestrial vegetation and aquatic biota are described and illustrated with examples from California.

EFFECTS ON PHREATOPHYTIC VEGETATION

Effects of Rapid Water Table Declines

Water table declines that exceed the rate of root growth can cause decreased plant growth, withered leaves, canopy die-back, early leaf drop, and death. In high desert scrub (Sorenson et al. 1991) and Central Valley riparian vegetation (Dains 1992), effects vary among different species and, if preexisting root development is different, among individuals of the same species.

Effects of Slow Water Table Declines

Roots of individual phreatophytic plants can grow deeper as the water table declines (Sorenson et al. 1991). However, a deep or declining water table may prevent establishment of seedlings. Valley oaks, for example, successfully reproduce only in areas where the depth to the water table is usually less than 10 meters (m), although mature trees survive temporary drawdowns to depths of 35 m (Cooper 1926). Lowered water tables may limit successful reproduction to sequences of wet years, resulting in marked age clustering in the population.

Species succession following a long-term water table decline is often similar to vegetational changes with distance away from an area with a shallow water table. For example, native riparian species gave way to salt cedar and Russian olive along the Owens River (Brothers 1984). If water table declines are large and prolonged, widespread phreatophyte mortality will result, as occurred near a municipal wellfield in the Carmel River valley (Kondolf and Curry 1984).

Hydrologist, Jones & Stokes Associates, 2600 V Street, Sacramento, CA 95818

Effects of Water Table Rises

If groundwater pumping is not continued in areas where imported water becomes available for irrigation, rising water tables often result. Shallow water table conditions damaged orchard crops in the Solano Irrigation District and prompted the installation of drainage wells (Summers Engineering 1988).

Effects of Subsidence

When groundwater withdrawals result in subsidence near streams or rivers, the frequency and duration of overbank flooding can increase and the depth to the water table can decrease. This has occurred along the Sacramento River near Knights Landing. These changes can favor riparian over upland vegetation.

AQUATIC ECOSYSTEM EFFECTS

Effects of Surface Water Depletions

Groundwater pumping can cause substantial decreases in low flows, which can make critical riffles impassable for anadromous fish. This effect played a central role during design of conjunctive use projects on San Simeon Creek (Engineering Science 1992) and the Carmel River (Furst 1991), two waterways with steelhead trout populations on the central coast. As riffles and pools become shallower, predation of fish by birds increases, and crowding increases competition and predation among fish. If the creek goes completely dry, almost all aquatic organisms die and must repopulate from upstream or downstream areas when flow resumes. This occurred along Putah Creek in the Sacramento Valley when increased groundwater pumpage caused long reaches of the creek to go dry (U.S. Fish and Wildlife Service 1992).

Groundwater withdrawals can lower water levels in water-table lakes and coastal lagoons. Statistical analysis indicates that water level declines in Lake Merced in San Francisco during 1977-1988 resulted not only from recent drought conditions but also from chronic groundwater overdraft (Yates et al. 1990).

Lowered water levels in lakes and lagoons often result in encroachment by upland vegetation types. Also, if water depth decreases to less than about 1 m, rooted aquatic vegetation such as tules will often invade. Coastal lagoons are small and subject to variable flow and salinity. Organisms adapted to these conditions are often rare. The 200 m x 10 m lagoon and adjacent salt marsh on San Simeon Creek hosts four endangered or threatened species.

Effects of Surface Water Accretions

Seeps or streamflow accretions can result from groundwater management in areas where surface water is used for artificial recharge, where drainage tiles are used to prevent waterlogging of irrigated soils, or where return flows from irrigation or septic systems seep into nearby surface waters.

Effects of Water Quality Changes

Groundwater inflow can buffer temperature extremes in surface water bodies. Under predevelopment conditions, native fish in Putah Creek survived the summer months in pools in the creek channel sustained by groundwater, which has a relatively constant temperature of 16-19°C (U.S. Fish and Wildlife Service 1992). Groundwater levels are now too low to sustain pools. Surface inflow to the creek in summer warms up to 30°C before it seeps entirely into the creekbed. Extended periods of low flows and warm temperatures in 1990 eliminated all but two of the hardiest fish species. Similarly, groundwater inflow can prevent freezing of open water surfaces in cold winter areas, thereby maintaining oxygen exchange and biological activity.

Groundwater can introduce nutrients or contaminants into aquatic ecosystems. Selenium-laden agricultural drainage water poisoned waterfowl in Kesterson wildlife refuge. Groundwater flowing into Lake Merced contains about 10 milligrams per liter (mg/l) of nitrate (as N) and corresponds to a mass loading rate of about 91 kilograms per year (kg/yr) per hectare of lake surface, which is comparable to agricultural fertilization rates for many crops. All the nitrogen is removed by biota as groundwater flows through the lake, resulting in large amounts of algae and phytoplankton (Hamlin and Yates 1990).

Water quality of groundwater can differ substantially from that in adjacent surface water bodies (Siegel 1988). Thus, the direction of flow between a wetland and the underlying groundwater system could change the suitablility of growing conditions for vegetation and aquatic organisms. Detailed modeling studies have shown that groundwater flow near lakes and wetlands is often divided into local-, intermediate- and regional-scale flow systems (Winter 1976, 1983). The direction of flow is highly sensitive to slight variations in topography, recharge, and pumpage. In some cases, the direction of flow alternates with time, and in others it remains constant (Carter and Novitzki 1988). Groundwater inflow is necessary to maintain brackish conditions in some coastal marshes, such as in Morro Bay (The Morro Group 1991).

REFERENCES

Brothers, T. S. 1984. Historical vegetation change in the Owens River riparian woodland. Pages 75-84 in R. E. Warner and K. M. Hendrix (eds.), California riparian systems. University of California Press. Berkeley, CA.

Carter, V., and R. P. Novitzki. 1988. Some comments on the relation between ground water and wetlands. Pages 68-86 in D. D. Hook (ed.), The Ecology and Management of Wetlands. Timber Press. Portland, OR.

Cooper, W. S. 1926. Vegetational development along alluvial fans in the vicinity of Palo Alto, California. Ecology 7(1):1-30.

Dains, V. 1992. Survey and assessment of riparian vegetation along Putah Creek from 1990 and 1991. Prepared for Putah Creek Council, Davis, CA.

Engineering Science, Inc. 1992. Preliminary design and evaluation of long term water supply projects. Alameda, CA. Prepared for Cambria Community Services District, Cambria, CA.

Furst, D. 1991. Overview of the Carmel Valley simulation model. Appendix 2-A, Supplemental Draft EIR/EIS for the New San Clemente Project. Monterey Peninsula Water Management District. Monterey, CA.

Hamlin, S. N., and E. B. Yates. 1990. Relation of nitrate concentration and groundwater movement through a water-table lake in San Francisco, California. Poster presented at annual meeting of the Association of Ground Water Scientists and Engineers, September 25-26, 1990. Anaheim, CA.

Kondolf, G. M., and R. R. Curry. 1984. Pages 124-133 in R. E. Warner and K. M. Hendrix (eds.), California riparian systems. University of California Press. Berkeley, CA.

Morro Group, The. 1991. Freshwater influences on Morro Bay, San Luis Obispo County, California. Los Osos, CA. Prepared for the Bay Foundation of Morro Bay, Morro Bay, CA.

Siegel, D. I. 1988. A review of the recharge-discharge function of wetlands. Pages 59-67 in D. D. Hook (ed.), The Ecology and Management of Wetlands. Timber Press. Portland, OR.

Sorenson, S. K., Dileanis, P. D., and F. A. Branson. 1991. Soil water and vegetation responses to precipitation and changes in depth to ground water in Owens Valley, California. (Water-Supply Paper 2370, Chapter G.) U.S. Geological Survey. Washington, D. C.

Summers Engineering, Inc. 1988. Solano Irrigation District groundwater resources. Hanford, CA. Prepared for Solano Irrigation District, Vacaville, CA.

U.S. Fish and Wildlife Service. 1992. Lower Putah Creek resource management plan. Draft report to Congress. Sacramento, CA.

Winter, T. C. 1976. Numerical simulation analysis of the interaction of lakes and ground water. (Professional Paper 1001.) U.S. Geological Survey. Washington, D.C.

_____. 1983. The interaction of lakes with variably saturated porous media. Water Resources Research 5:1203-1218.

Yates, E. B., S. N. Hamlin, and L. H. McCann. 1990. Hydrogeology, water quality, and water budgets for Golden Gate Park and the Lake Merced area in the western part of San Francisco, California. (Water-Resources Investigations Report 90-4080.) U.S. Geological Survey. Reston, VA.

Coupled Simulation-Optimization Approach to Wellhead Protection Area Delineation to Minimize Contamination of Public Ground-Water Supplies

John M. Shafer[1] and Mark D. Varljen[2]

Abstract

A loosely coupled simulation-optimization procedure is used to determine the steady-state pumping rates for individual wells in a multiple well municipal wellfield that result in the least total number of potential contaminant sources in the specified time period capture zones of the wells. A conjugate direction search algorithm is employed to solve the formulated unconstrained, nonlinear optimization problem using a penalty function approach to force the total wellfield pumping rate to meet the wellfield demand.

Introduction

Ground-water time of travel has become the preferred criterion by which many states are delineating wellhead protection areas under recommendations of the U.S. EPA pursuant to the 1986 amendments to the Safe Drinking Water Act. Presumably, land use restrictions and special hazardous substances management practices are to be imposed within these areas surrounding municipal water supply wells in an attempt to minimize the threat of contamination of these wells.

In practice, the ground-water time of travel criterion is translated into time-related capture zones of municipal wells. A τ-year capture zone is the bounded volume defined by the set of all ground-water flow paths that intercept the well in τ years. In most cases, the vertical flow component is ignored and the capture zone becomes the planar area described by the set of all horizontal pathlines with τ-year travel times to the well. The resulting time-related capture zone is designated the wellhead protection area.

---

[1]Associate Director, Earth Sciences and Resources Institute, University of South Carolina, 901 Sumter St., Columbia, SC 29208

[2]Senior Staff Hydrogeologist, Hart-Crowser, Inc., 1910 Fairview Ave. East, Seattle, WA 98102

The size, shape, and orientation of a τ-year capture zone are a function of the hydrogeologic properties of the ground-water system (i.e., hydraulic conductivity, porosity, and regional gradient), flow system boundaries, the pumping rate, and well interference. The area of the capture zone is proportional to the pumping rate; i.e., the lower the pumping rate, the smaller the capture zone area. Depending on the hydrogeologic properties, spacing, and pumping rates, well interferences also can significantly affect the configuration of capture zones. This phenomenon can be used to advantage in determining pumping rates that result in capture zones that have a minimal number of contaminant sources within them. Accordingly, the underlying hypothesis of this research was that for multiple well municipal ground-water supplies, pumping rates could be determined that meet the total wellfield demand yet result in capture zones that minimize the total number of included contaminant sources.

Approach

The formulation of this unconstrained, nonlinear optimization problem is

$$\min \; \frac{\sum_{j=1}^{ns} u_j}{ns} + \left[\beta \left| 1 - \frac{\sum_{i=1}^{nw} q_i}{Q} \right| \right]^{\alpha}$$

where

$u_j = 1$ if potential contaminant source s_j is in a capture zone
$u_j = 0$ if potential contaminant source s_j is not in a capture zone
$j = 1,...,ns$ the total number of potential contaminant sources
q_i = withdrawal from well i
Q = total wellfield target withdrawal
$i = 1,...,nw$ the total number of wells (and therefore the total number of capture zones).

The left-hand component of the objective function is the true objective of minimizing the total number of potential contaminant sources in all of the time-related capture zones. The right-hand part of the objective function is the constraint on meeting the total wellfield output requirement in the form of a penalty function. α and β are penalty parameters that are adjusted to force convergence to an optimum solution. Because the number of contaminant sources is a relative small number compared to the total ground-water withdrawal rate, Q, both parts of the objective function are constrained to values between 0 and 1. In this manner, the pumping rates do not swamp the minimization of the number of contaminant sources in the resulting capture zones.

A finite-difference ground-water flow model and a particle tracking code are used to calculate the time-related capture zones for each well, i, included in the analysis. A conjugate direction search algorithm is used to determine the optimum q_i that result in the least number of contaminant sources in the capture zones defined by the q_i.

Demonstration

A hypothetical ground-water system with a regional gradient of 0.002 is used to demonstrate the simulation-optimization approach to wellhead protection area delineation. The fictitious aquifer is nonhomogeneous and isotropic in hydraulic conductivity with a uniform effective porosity of 25%. The wellfield consists of three wells from which a total withdrawal of 2.0 MGD is required. There are 20 potential contaminant sources located throughout the flow domain. Fig. 1a shows the pre-optimization pumping strategy and the resultant 5-year capture zones for each well. A

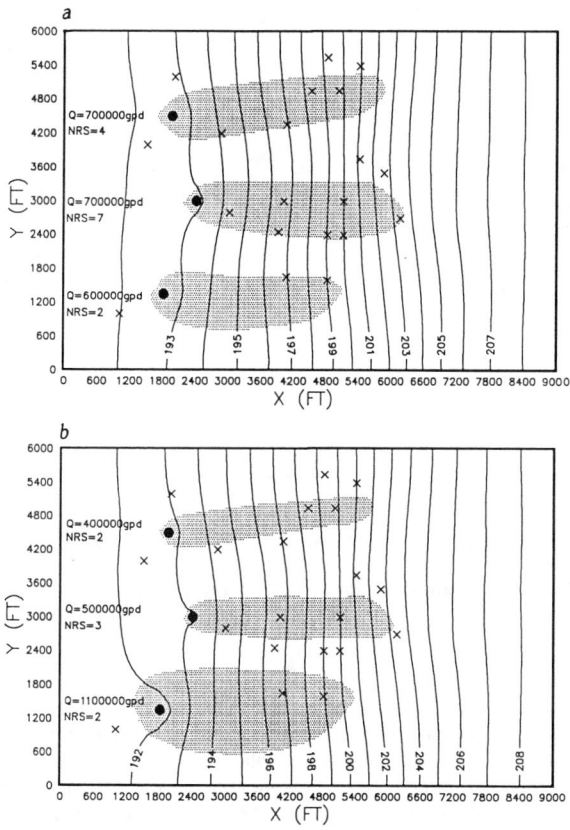

Figure 1. (a) Head Distribution and Capture Zones Resulting from Pre-Optimization Pumping Rates; (b) Head Distribution and Capture Zones Resulting from Optimum Pumping Rates

total of 13 potential contaminant sources are located within the capture zones of the three wells. Fig. 1b shows an optimum alternative pumping strategy resulting from the coupled simulation-optimization procedure. The 2.0 MGD required withdrawal is maintained, but the number of potential contaminant sources included in the capture zones is reduced from 13 to 7.

To confirm the optimum solution, the simulation model was used exhaustively to incrementally (stepsize = 0.2 MGD) evaluate all pumping combinations between 0.2 MGD and 1.2 MGD for each well. Fig. 2 shows the results of this exercise. From Fig. 2 it can be seen that the lowest value of the objective function is indeed achieved at 2.0 MGD and 7 contaminant sources included in the capture zones.

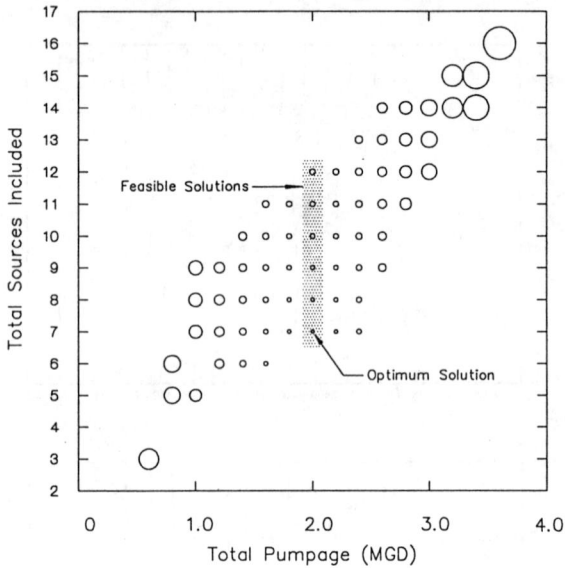

Figure 2. Objective Function Evaluation via Exhaustive Simulation

Conclusions

The coupled simulation-optimization technique for delineating wellhead protection areas allows for changing individual well pumping rates to proactively achieve a more favorable geometry for time-related capture zones. A more favorable geometry is in terms of re-orienting capture zones so that while the total demand from the wellfield is met, the least possible number of potential contaminant sources is included in the capture zones. Future enhancements of this approach may include a more efficient optimization component. Further, potential contaminant sources could be weighted according to perceived threat and buffer zones could be placed around contaminant sources so that they would not be single points that are either in or out of the capture zone regardless of how close they may be to the boundary.

Assessment of TCE Concentration in South Tucson Water Network

Ian H. von Lindern[1], Jim C. P. Liou[2], M. ASCE, and Margrit von Braun[3]

Abstract

This paper summarizes a study that estimated trichloroethylene (TCE) contamination patterns in a portion of the Tucson water distribution network due to several polluted wells. Aspects discussed include a description of the evolution of the network, data requirements for various contaminant estimates, geographic flow balance, and network analysis.

Introduction

Studies have shown that the aquifer providing drinking water to the residents in the southside/airport area of Tucson, Arizona, was contaminated with TCE due to improper waste disposal in the 1950's. Retrospective ground water studies suggest that these contaminated wells, collectively, pumped more than 2,000 gallons of TCE and other solvents into the Tucson public water system from 1952 to 1981. The study area, shown in Fig. 1, is bound by the 22nd Street on the north and by the Country Club Road to the east. The eastern and southern boundaries are natural physical boundaries. Historical distributions of TCE in the study area were needed for assessing residents' drinking water exposure from 1952 to 1981.

Contaminant Distribution During Early Years

The network within the study area evolved from 16 small independent water companies in the 1950's to an integral part of the citywide distribution system by 1982. Examination of the pattern of development in the study area shows that the principal contaminated wells were originally designed to serve local subdivisions. As those subdivisions grew and adjacent areas were developed, these wells were upgraded. As a result, contaminated areas expanded with system growth. Defining the affected area over time for each of the contaminated wells is the main challenge in assessing exposures.

---

[1]President, TerraGraphics Environmental Eng., 121 South Jackson, Moscow, Idaho 83843
[2]Associate Professor, Dept. of Civil Eng., University of Idaho, Moscow, Idaho 83843
[3]Assistant Professor, Dept. of Chemical Eng., University of Idaho, Moscow, Idaho 83843

The most convenient method for describing trends is by the sub-units of the old private water companies. The majority of the study area grew as small service areas with individual wells developed to serve quarter-section units (0.25 mile squares) or contiguous quarter-section groups. A set of sub-units was developed for a geographic flow balance modeling (described later) for the early years of the study period.

The extent that contaminants from an individual well penetrate the distribution system depends on factors such as water demands, hours of operations of wells, the configuration of wells in operation, reservoir levels, and booster status, etc. All of these factors varied with time. In defining long term average characteristic of chronic or sub-chronic exposures, we were mainly concerned with average seasonal and yearly hydraulic conditions.

The first step in establishing the TCE distribution pattern was to identify the area over which a particular source was uniformly distributed. If the annual pattern of production, distribution and demand was dominated by a local production/local consumption scenario, then the area influenced by the well would be small and localized. This was the case when individual private water companies served isolated areas. During that period, local production equaled local demand and service TCE concentration equaled well-head concentrations. In later years, dilution of TCE occurred, both by a inter-connected distribution system and by additional sources of clean water. Through these years, the distribution scenario moved away from local production/local consumption toward more dilution and larger areas of impact.

A geographic flow balance approach was used for several years in the study period. In each sub-zone, water production was allocated over demand. Surplus waters were transferred down gradient (established by well-head pressure and elevation data) to hydraulically connected sub-zones. Deficits were met by importing water from external well fields. Two opposite contamination scenarios emerged as a result of the geographic flow balance. In the southern portion of the study area, contaminated local wells pumped directly into the distribution network. Services hydraulically nearest these wells were the most severely impacted with the effect spreading into adjacent service units. The effect was dissipated as clean sources from either local production or imports from external well fields made up water deficits to the area. To the north, the reverse situation was observed. Local production wells were clean and the imported water was contaminated. Homes hydraulically nearest the local production wells received the greatest percentage of clean water. Those most dependent on the imported Santa Cruz Well Field source experienced the greatest exposure.

Network Modeling

As the sub-areas evolved into an integral part of the citywide distribution network, defining the influence of individual wells in later years (after 1972) by the geographic flow balance becomes progressively more difficult. Network modeling using commerical software packages became necessary.

Network schematics for the lower B and C pressure zones (see Fig. 1) were constructed for 1975 and 1980. The main difference between these two years was that greater local well productions existed in 1975. There were also differences in network connectivity. The blue prints of Tucson Water's distribution systems were the primary source of information. The schematics so constructed did not contain every pipe in the study area.

Some of the pipes with diameters less than two inches were not included. Groups of interconnected small diameter pipes were lumped together and ended at a node which represented a number of houses or buildings fed by a common pipe. The number of nodes within each quarter-section varied from 1 to 49, depending on the flow demand pattern in the quarter-sections. The schematics were detailed enough to show pressure and flow variations within each quarter-section.

Annual average demands were assigned to the nodes established in the network schematics. The study area is primarily residential. Houses associated with each node were counted from the blue prints and the aerial photographs. The flow demands were based on Tucson Water's Flow Demands and Number of Accounts data. Unusual demands caused by large commercial or industrial accounts were noted in Tucson Water's demand data. These users were identified in the blueprints, verified by the aerial photographs, and added to the schematics. The spatial pattern of water usage was captured by this process.

Annual average flow rates of production wells were used. Waters from outside sources were not directly imposed due to questionable flow data. Instead, known HGL elevations at the discharge of the booster stations at the Martin Reservoir and at the Kolb Reservoir (outside the study area) were imposed on the network. Computed HGL elevations at well-heads matched the HGL elevations estimated from periodic well efficiency test data. A similar situation existed for the lower B zone. The division of flow between two 36-inch collector mains feeding B and C zones was unknown. As a result, an average HGL elevation based on efficiency tests on well SC-7 was used. Again, the computed well-head HGL elevation matched those from well efficiency tests reasonably well. Well-head TCE concentrations were obtained from a separate ground water study.

For 1975 and 1980, the B and C pressure zones covered an extremely large area. It was prohibitive, both in cost and in time, to model these zones in their entirety at the level of details described above. Only the lower portions of these pressure zones were modeled. This was not a problem for lower C zone as the Bension Highway, which crosses the C zone from southeast to northwest, was almost a natural boundary. Hydraulically, the lower C zone was connected to the remaining C zone by a 48 inch transmission main. This main crosses the Bension highway to the Kolb Reservoir that was modeled as a node with known HGL elevation. For the lower B zone, we use Ajo Way as an artificial boundary. A mass inventory for the B zone north of Ajo Way indicated that this area imported water from both the south and east from the old Santa Cruz Transmission Main. It was assumed that no flow crossed Ajo Way. This assumption resulted in HGL elevations in the lower B zone higher than that in the Santa Cruz transmission main. As a result, the impact of contaminated flows from the Santa Cruz line was eliminated. This yielded conservative (lower) exposure estimates in the lower B zone.

Figs. 1 and 2 show the HGL elevations and the TCE concentration estimates for 1975. The influence of local wells can be seen in the latter. The computed hydraulic gradients support the local production/local consumption scenario for that year. However, the 1980 results showed some significant differences in TCE concentration estimates between the geographic flow balance and network modeling approaches. This difference is largely related to increased reliance on clean waters from the Avra Well Field outside the study area.

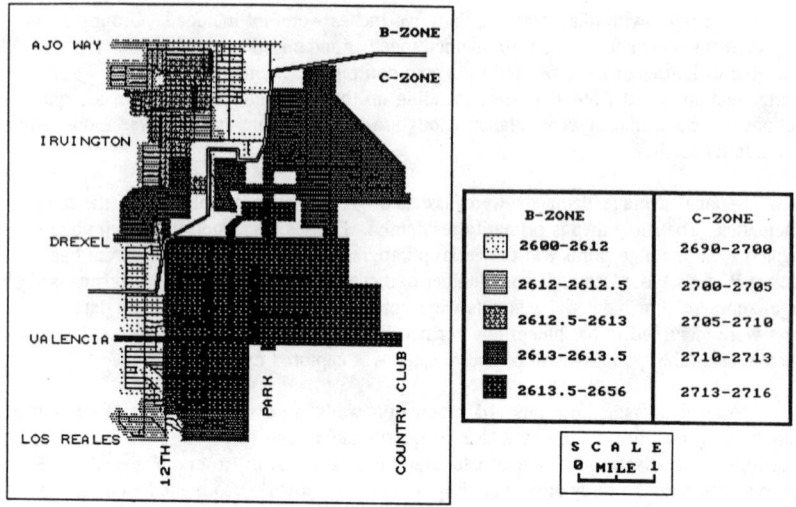

Fig. 1 Study area, pressure zones, and simulated HGL elevations for 1975

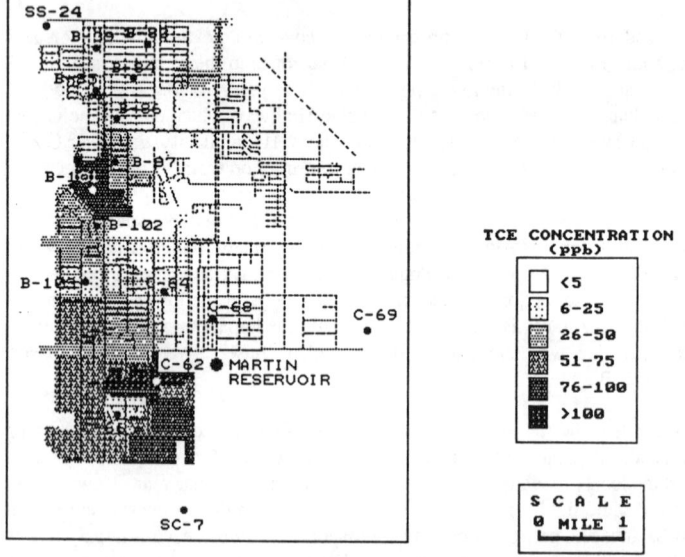

Fig. 2 Well locations and simulated TCE concentrations for 1975

Groundwater Contaminant Transport at a Hazardous Waste Disposal Site, Pullman, Washington

Bill Saur	M. Yavuz Corapcioglu	Kent Keller
Cascade Earth Sciences, Inc.	Dept. of Civil Engineering	Dept. of Geology
3425 Spicer Drive	Texas A&M University	Washington State University
Albany, OR 97321	College Station, TX 77843	Pullman, WA 99164

Abstract: Groundwater contamination at a hazardous waste site near Pullman, Washington is simulated by a numerical model. The aquifer at the site is classified as an unconfined valley aquifer. Solute transport was found to be towards the aquifer outlet, Airport Creek. The model predicts initial trace contaminant, by a nonreactive solute, at the node representing Airport Creek six to 11.4 years after the start of the simulation. This simulation of a nonreactive solute transport represents a worst case scenario. Results are presented and discussed in a context to prevent future groundwater contamination.

Introduction: Hydrogeologic investigation in the Pullman, Washington area was spurred by prospects of abundant supplies of high quality water from the basalt aquifers (Russell, 1897; Smoot and Ralston, 1987). Groundwater flow simulations for the Pullman-Moscow basin were performed by Jones and Ross (1972) and Smoot and Ralston (1987). In the latter study, a three-dimensional computer simulation of groundwater flow and water levels under various pumping scenarios was constructed as a management tool for the major groundwater users in the region. Kuhlman (1986) simulated groundwater flow in the loess at the study site. Kuhlman's and Smoot and Ralston (1987) are the only investigations in which flow in the surficial loess is explicitly considered. The present study will consider solute transport in the loess by employing the USGS solute transport model (Konikow and Bredehoeft, 1978).

Hydrogeologic Setting: The hazardous waste site modeled in this study is located in the Palouse region of southeastern Washington (see Fig. 1). The geologic units present at the site are the Quaternary Palouse formation and the underlying Wanapum and Grande Ronde (Miocene) basalts of the

Columbia River Group. The Palouse formation is an eolian loess, deposited as large dunes which form the present day topography. Grain size analyses show that it is comprised of predominantly silt size particles (Kuhlman, 1986). The aquifer at the study site is the saturated portion of the surficial loess of the Palouse formation. From a classification standpoint, it can be termed a temperate climate, unconfined, valley aquifer. In this type of aquifer, water flows towards the outlets, which are the low points in the topography. The slope of the water table indicates the flow direction in the aquifer. Recharge is assumed to occur, via infiltration, over the entire surface of the aquifer. Figure 1 illustrates the conceptual hydrogeology of the study area. The north end can be considered a groundwater divide and Airport Creek, at the south end, is the low point in the topography and therefore can be considered the aquifer outlet.

Model Data Input: The grid used in this simulation is shown in Figure 2. It is based on Kuhlman's (1986) grid, with the only modification being the addition of enough cells at the southern end to create a rectangle. It is 13 x 15, with square cells measuring 100 feet on a side. The grid is justified by the following: (1) the north end is a groundwater divide with flow towards the south away from this divide; (2) at the southern end these boundary nodes are beyond the aquifer outlet; and (3) the no-flow nodes at the sides of the grid are essentially parallel to flow, so they provide no impediment to flow. The constant head node represents a boundary condition such as that found at Airport Creek. The head at Airport Creek, which represents the contact area between the aquifer and its surface outlet, may be taken to be independent of the flow conditions in the aquifer. Instead, it is determined by the water level of the creek. The other type of special node is the contaminant source node placed to correspond to the hazardous waste and manure dumping areas. The initial water table elevation values were taken from Kuhlman (1986), and represent the estimated average annual water table levels.

The groundwater flow simulation was run under steady-state conditions. The unsteady cyclic fluctuation of the water table in response to seasonal variations in precipitation was compensated for by using the average annual water table values. There are no pumping wells in the loess, and all other possible causes of unsteady conditions, such as land use changes and climatic change, are ignored.

Steady-State Flow Simulation Parameters: The thickness of the loess in the study area varies from about 100 feet at the north central end to eight feet in the Airport Creek area. For modeling purposes, the thickness of the loess was assumed to decrease uniformly from north to south and to remain constant in an east-west direction. In addition, the loess was assumed to be saturated to within five feet of the ground surface except near Airport Creek where it was assumed the aquifer was saturated over its entire thickness. Hydraulic conductivity was determined as a calibration parameter. In this procedure, the value for recharge was taken as known and the value for hydraulic conductivity was manipulated until the model calculated a distribution of hydraulic heads close to the observed distribution. In all calibration cases, the loess was assumed to be uniform and isotropic with respect to hydraulic conductivity. The known recharge value was taken as 1.4 in/yr. The effective porosity value of .33 is taken and represents the value for a silt. The value for longitudinal dispersivity was taken as one foot. A value of 0.2 for the ratio of transverse to longitudinal dispersivity was used in this simulation.

The hazardous waste and manure landfill in this study was an active disposal site from 1970 to 1980 (Budd, 1986). The simulations in this report begin with the cessation of active disposal and run for a 30-year time period. The contaminant concentration at the source nodes was maintained at 100 throughout the simulation, i.e., this was a constant potential source.

Results/Discussion: While groundwater flow in this simulation was steady-state, solute transport was unsteady-state. The hydraulic conductivity calibration procedure yielded a value of 3.28×10^{-6} ft/sec. This value is the mean of the range of values for loess given by Freeze and Cherry (1979). The hydraulic conductivity value of Kuhlman (1986), 2.5×10^{-7} ft/sec was found to be too low to obtain reasonable calibration. Comparison of observed with output head data shows maximum differences of 16-29 feet for 20 nodes in the north central and northwestern part of the grid. All others were within 15 feet or less, and 82% of these were within 10 feet or less. Subsequent to the calibration step, the simulation was run using 50 time steps for solute transport which led to concentration value maps every six years. The concentration values for the concentration plume after 6, 18 and 30 years with 10% and 50% (of initial 100% concentration) contours drawn in are shown in Figs. 3-5. These figures reveal that the model predicts a dominant direction of migration for the leading edge of the plume to be south to southeast. Based on the previous discussion of an unconfined valley aquifer, this is to be expected. The plume is migrating in the direction of the aquifer surface outlet, Airport Creek. The furthest extent of migration for the contamination plume after 30 years (at the C = 10% level) is between 650 and 700 feet in a south-southeasterly direction (Fig. 5). Fig. 6 is a graph which shows the changes in concentration with time at node (7,11), which is marked in Fig. 2. This maximum distance is at node (7,12). It is noteworthy that this node and (7,11), the one immediately adjacent to it on the north, represent a portion of Airport Creek (the creek itself flows across the southeastern corner of cell (7,11), while the node is in the center of the cell.) Fig. 6 shows the changes in concentration with time at node (7,11). A marked change in slope at the 12-year mark indicates the onset of gross contamination. The contaminant concentration in the creek itself would not be expected to reach the levels of the graph owing to dilution by stream flow. Fig. 8 lends insight into the nature of the plume by tracing the movement of the 10% and 50% contours through time. The steeper slope of the 10% line shows that it advances faster than the 50% line. This is due to dispersive effects.

Contamination of Airport Creek represents the beginning of the process whereby contaminant could be disseminated in the environment. Because of its significance as the surface outlet for the aquifer, an imaginary observation well was placed at node (7,11) to closely monitor contaminant concentrations through time. Examination of output data for the observation well reveal that after 30 years simulation, a concentration of 23 (i.e., 23% of original concentration at contaminant source nodes) occurs here. The contamination, with a concentration of 0.1%, first reaches the node between six and 11.4 years of simulation. Depending on the chemical species involved and the original 100% concentration, this seemingly minute amount could be a problem as some chemicals are toxic in trace amounts. This model predicts that the source nodes for the first contamination to arrive at Airport Creek are at the southern tip of the manure burial site and southern tip of the westernmost waste disposal cells (Fig. 5).

The recharge value used was taken from Kuhlman's (1986) study. This value was obtained by calibration and was believed to be reasonable; however, it was not based on any field measurements (Kuhlman, 1986). Smoot and Ralston (1987) propose that the average recharge rate in the Pullman-Moscow area is about 3.6 in/yr, based on results of a groundwater recharge simulation model. An increase in recharge for the simulations would require an increase in calibration hydraulic conductivity and would thus increase the rate of flow in the aquifer. Two other parameters which lacked detailed field data are saturated thickness and hydraulic conductivity. These parameters affect the simulation in that they affect the transmissivity of the aquifer. Owing to the uncertainty in the recharge value for the area, a sensitivity analysis for this parameter was performed. Values of 5, 1.4 (used in this study), .9 and .2 in/yr were used to test the effect of recharge on contaminant arrival time and concentration at the Airport Creek node (7,11). It can be seen that the simulation is sensitive to recharge with the higher values causing earlier arrival times and higher concentrations at the Airport Creek node (Fig. 7).

The results show a dominant south-to-southeasterly migration of the solute plume towards Airport Creek with an initial trace contamination at the creek by an unreactive solute between six and 11.4 years after the simulation begins.

Literature Cited

Budd, W.W., 1986, "A Comparison of Three Risk Assessment Techniques for Evaluating a Hazardous Waste Landfill," *Hazardous Waste and Hazardous Materials*, Vol. 3, No. 3, pp. 309-320.

Freeze, R.A. and Cherry, J.A., 1979, *Groundwater*, Prentice-Hall, 604 pp.

Jones, R.W. and Ross, S.H., 1972, *Moscow Basin Groundwater Studies*, Idaho Bureau of Mines and Geology Pamphlet 153, 95 pp.

Konikow, L.F. and Bredehoeft, J.D., 1978, "Computer Model of Two Dimensional Solute Transport and Dispersion in Groundwater," *Techniques of Water Resources Investigations*, Book 7, Chapter C2, 90 pp.

Kuhlman, J.G., 1986, "Three Dimensional and Quasi-Three Dimensional Modeling at a Hazardous Waste Site," Washington State University, M.S. thesis, 149 pp.

Russell, I.C., 1897, "A Reconnaissance in Southeastern Washington," *USGS Water Supply Paper 4*, 96 pp.

Smoot, J.L. and Ralston, D.R., 1987, "Hydrogeology and a Mathematical Model of Groundwater Flow in the Pullman-Moscow Region, Washington and Idaho," *Idaho Water Resources Research Institute Paper*, 118 pp.

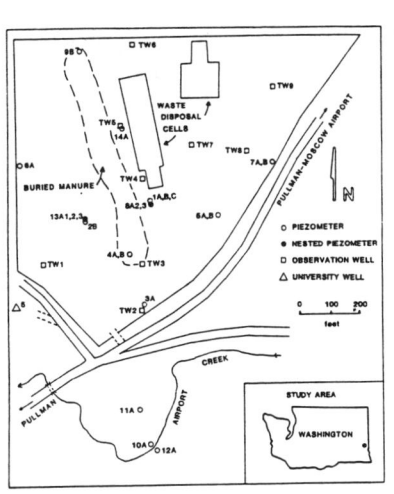

Figure 1. Location of the study area (Kuhlman, 1986).

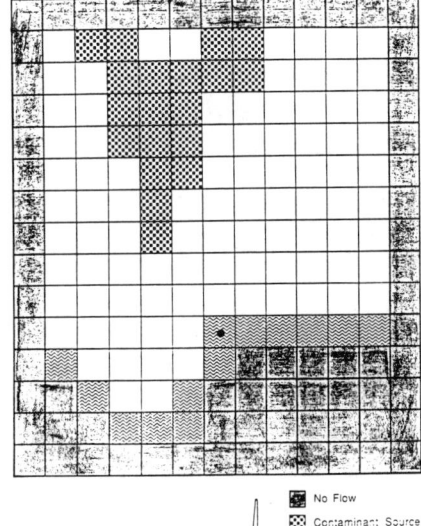

Figure 2. The model grid covering the study area.

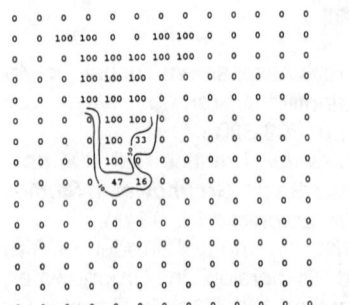

Figure 3. Contaminant distribution after 6 years.

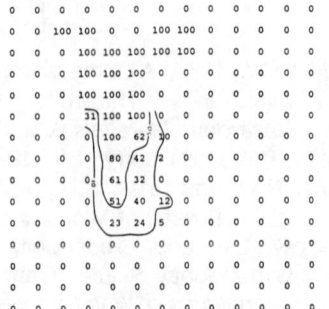

Figure 4. Contaminant distribution after 18 years.

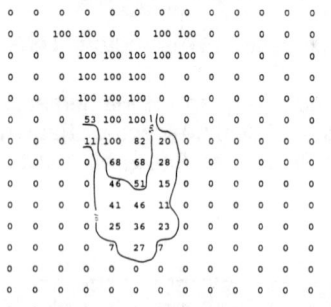

Figure 5. Contaminant distribution after 30 years.

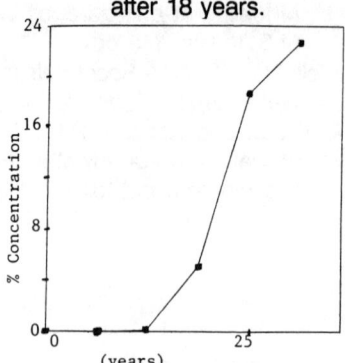

Figure 6. Contaminant variation at the Airport Creek [node (7,11)].

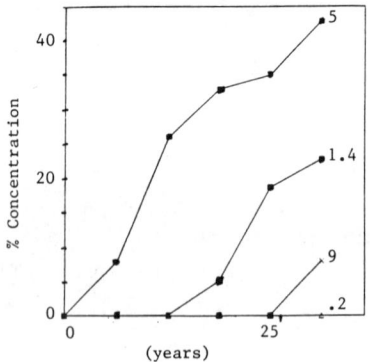

Figure 7. Contaminant variation with recharge (in/yr) at node (7,11).

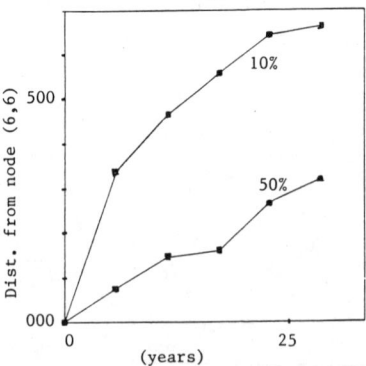

Figure 8. Migration of 10% and 50% concentration contours along the sixth column.

INVESTIGATION AND REHABILITATION OF THE MOSES LAKE LARSON WELLFIELD

David Banton[1] and Robert Anderson[1]

Abstract

In 1988, low levels of trichloroethylene (TCE) were detected in several of the Larson wells operated by the City of Moses Lake. The wells are up to 800 feet deep and are completed in fractured basalt. TCE concentrations in the City wells ranged from 5 to 15 ppb. A hydrogeologic investigation concluded that TCE migration within the deeper basalt sequences was controlled by poor well completion practices which allowed contaminated groundwater to migrate between different aquifers. A program of packer testing of one of the contaminated wells was carried out to evaluate whether well modification could reduce TCE concentration and return the well to service. The well is now back in service at much lower cost than had traditional water treatment techniques been employed to remove TCE from the groundwater.

Introduction

The Moses Lake Larson wellfield supplies water to about 5,000 people. The wells were deeded to the City in the 1960's by the US Air Force following the closure of Larson Air Force Base. There are six wells with a combined capacity of 7,500 gpm. The discovery of TCE in three wells resulted in a loss of capacity of 4,500 gpm.

Physical Setting

Moses Lake is located on the Columbia Plateau in Central Washington. The City wells are generally between 700 to 800 feet deep, cased through 100 to 200 feet of alluvium, and completed open hole across multiple basalt flows of the Wanapum Formation of the Columbia River Basalt Group. The integrity of seals separating groundwater from the alluvium and basalt is uncertain because of the age of the wells and poor records.

---

[1]Associate and Senior Hydrogeologist, Golder Associates Inc., 4104, 148th Avenue N.E., Redmond, WA 98052

Hydrogeology

Evaluation of the hydrogeology of the Larson area focused on the hydrostratigraphic controls to groundwater movement in the basalt sequence. Geophysical borehole logging, water-level monitoring, pump testing and hydrochemical analysis were used to distinguish a number of hydrostratigraphic zones and to estimate the likely extent of TCE in the subsurface.

Basalt hydrogeology is complex. In general, the rubbly lower part of the flows are permeable, while the intervening collonade and entablature structure (containing characteristic vertical hexagonal columns) is of lower permeability. Horizontal groundwater flow occurs preferentially though the rubbly zones rather than the dense flow interior. The flow interiors generally preclude vertical groundwater movement between individual flows, with the exception of limited groundwater movement via vertical fracturing. In some instances, the fractures may contain precipitated minerals from the groundwater which further reduces the vertical hydraulic conductivity.

The Larson wells are cased through between 100 to 200 feet of alluvium, and then penetrate numerous basalt flows all part of the Wanapum Basalt. Based on hydraulic properties, water quality and groundwater level measurements, five hydrostratigraphic zones were identified at Moses Lake. Each basalt zone may contain one or more individual basalt flows.

Alluvium - an aquifer where saturated (generally the lower 50 to 100 feet);

A-Zone - a permeable rubbly basalt horizon (aquifer) over the uppermost 50 to 100 feet of basalt.

B-Zone - a dense basalt aquitard between 100 and 200 feet thick underlying the A Zone.

C-Zone - a number of permeable basalt flows 50 to 100 feet thick underlying the B-Zone aquitard; and

D-Zone - a lower aquitard.

The alluvium and Wanapum A zone appear to be in hydraulic communication. There is a downward component of hydraulic gradient between the alluvium/Wanapum A zone and Wanapum C and D zones in the summer months because of depressed water levels in the deeper zones caused by the pumping of high-capacity irrigation wells. The difference in water levels between the alluvium/Wanapum A and deeper zones is over 100 feet in the summer months. This head difference develops across the Wanapum B zone and demonstrates that the basalt is of low permeability. In the winter and spring, there is little difference in hydraulic head between the individual basalt zones.

Well yields vary depending on the depth of the well and the particular basalt intervals open to the well. The upper portions of the basalt have very high transmissivites, producing yields of several thousand gallons per minute with only a few feet of drawdown. Wells sealed into the deeper basalts generally encounter less transmissive basalt flows resulting in correspondingly lower specific capacities.

Nature and Extent of TCE in Groundwater

The extent of TCE in groundwater at Larson is poorly known. Monitoring wells have been installed by the Corps of Engineers in localized areas but only in the alluvium and Wanapum A zones. Present information suggests that TCE is restricted to the unconsolidated deposits and Wanapum A zone, except where uncased wells have penetrated the Wanapum B aquitard allowing contaminated groundwater to migrate down to the Wanapum C and D zones via the open wellbores. There are therefore localized zones of TCE contaminated groundwater in the Wanapum C and D zones.

Larson Wells

The Larson wells that are contaminated with TCE (ML-21, ML-22, and ML-28) draw water from the four basalt hydrostratigraphic zones because the wells are open hole from the top of basalt. Two of the uncontaminated wells (ML-23 and ML-24), although of similar depth, are completed with a sealed casing at least one hundred feet into the Wanapum B aquitard. These deeper sealed wells have a different inorganic geochemistry from the Wanapum A zone, and show different seasonal water level fluctuations indicative of a separate aquifer system.

The conclusions drawn from the hydrogeologic investigation were that: 1) the TCE was likely restricted to the alluvium and Wanapum A zone; 2) the absence of a deep well seal in wells ML-21, ML-22, and ML-28 resulted in TCE entering the wells and potentially migrating to deeper zones in the basalt via the borehole particularly in the summer when there is a downward hydraulic gradient; 3) that the low-permeability Wanapum B Zone (were not breached by a poorly sealed well) would act as a natural barrier to vertical migration of TCE to deeper zones in the basalt (as demonstrated by the good water quality in wells ML-23 and ML-24); and 4) wells sealed with a casing into the Wanapum B Zone and drawing groundwater from deeper zones should be reasonably well protected from the TCE contaminated groundwater provided there were no wells breaching the aquitard within the capture zone of the sealed well.

Remedial Alternatives

Options considered for well rehabilitation included: blending groundwater with uncontaminated wells; air-stripping, carbon adsorption, or chemical oxidation of the contaminated water; and well modification to seal out TCE contaminated water from the alluvium/Wanapum A Zone. To discriminate between the various remediation alternatives based on the level-of-knowledge at the time of the investigation and uncertainty in remedial costs for each alternative, a probabalistic decision analysis model was developed and implemented in a LOTUS 123 worksheet with a commercially available risk assessment program, @RISK. The program clearly indicated that well modification had the lowest life-cycle cost and highest probable well yield/quality requirements to satisfy the City's objectives.

Modification of Well ML-28

To evaluate well modification on well ML-28, a packer pump assembly was set in the well to draw groundwater only from the Wanapum C and D zones. The well was then pumped for about 30 days at a rate of up to 1,000 gpm. TCE concentrations declined consistently during the test and stabilized at about 1 ppb after 30 days. The pump was then shut off and TCE concentrations monitored for a further 30 days. TCE concentrations remained at 1 ppb. In effect, the well was operated to pump out contaminated groundwater from the Wanapum C and D zones within the capture zone of the well. Post-pumping monitoring indicated no new source of TCE to the C and D zones.

Subsequently, the packer was removed, and a 12-inch diameter steel well casing was cemented in place to a depth of about 430 feet (Wanapum B aquitard). A permanent pump was installed and the well was reconnected to the distribution system. The well is now in use; TCE concentrations are about 1 ppb.

The total cost of the investigation and rehabilitation of well ML-28 amounted to about $200,000 (which included a new submersible pump and fittings ($50,000) versus about $500,000 for a single air stripper (excluding engineering design, and administrative costs).

The probabalistic decision analysis model significantly aided the City in its decision-making process making it possible to complete cost-effective decisions among rehabilitation alternatives in the face of considerable uncertainty. The cost savings associated with well rehabilitation rather than using traditional groundwater treatment technologies demonstrated the successful application of the technique to a real-world problem.

Acknowledgement

Golder Associates wish to thank the City of Moses Lake staff and in particular Brian Hinthorne, for their assistance on the project.

OPTIMAL STRATEGY FOR AQUIFER REMEDIATION

Fethi Ben Jemaa[1] and Miguel A. Mariño[2]

Abstract

A methodology for the restoration and cleanup of existing subsurface contaminated sites and for the containment of pollutants is developed. The remediation problem is posed as an optimization model where the rates and locations of pumping and injections are to be determined given the characteristics and extent of the contamination plume. The solution of the remediation problem is based on the econometric method of feedback control coupled with groundwater flow and transport simulations. The remediation plan is optimized so as to lower the contamination level to a prespecified level by the end of the remediation period while minimizing the cost of pumping and treatment. The objective function of the optimal feedback control model consists of a successive minimization of a weighted sum of squared deviations of the achieved cleanup level at each stage from the desired target level of groundwater quality.

Introduction

The problem of aquifer remediation and water quality management has been the focus of numerous studies. As the general aim in any remediation effort is to improve the groundwater quality and decrease the contamination level, a number of proposed remediation models consist of minimizing the pollutant concentrations and/or the cost of the removal and treatment of the contaminated groundwater.

Gorelick and Remson (1982) used a linear programming superposition technique to manage several groundwater pollutant sources so as to maximize waste disposal rates and not violate the groundwater quality standards. Gorelick et al. (1984) combined groundwater flow and contaminant transport simulations with nonlinear optimization to determine optimal pumping and injection rates for

---

[1]Research Assistant, Department of Land, Air and Water Resources, University of California, Davis, CA 95616.

[2]Professor, Department of Land, Air and Water Resources and Department of Civil Engineering, University of California, Davis, CA 95616.

groundwater quality control. Wagner and Gorelick (1989) examined the effect of uncertainty of the hydraulic conductivity on aquifer remediation strategies. Andricevic and Kitanidis (1990) used a discrete time optimal control methodology for finding optimal pumping rates for aquifer remediation; the methodology used accounts for and reduces parameter uncertainty. Ahlfeld et al. (1988a and 1988b) presented two optimization formulations for aquifer cleanup, a first approach aiming to extract a maximum amount of pollutant over the remediation period, and a second approach aiming to lower the contaminant concentration to a specified level. Jones et al. (1987) presented an optimal control model for groundwater management using differential dynamic programming. Murray and Yakowitz (1979) applied a modified differential dynamic programming to a multireservoir control problem. Makinde-Odusola and Mariño (1989) used the feedback method of control to determine optimal pumping rates necessary to maintain the piezometric surface below a desired level. Andricevic (1990) used optimal control methods to determine withdrawal rates so as to satisfy a given water demand while keeping the piezometric heads close to some prespecified levels.

This paper presents a groundwater remediation approach based on the econometric method of feedback control and coupled with flow and contaminant transport simulations. The approach aims to determine the optimal control rules on the locations and rates of pumping and injection so as to lower the contamination to some target level. The feedback control model used consists of a successive minimization of a loss function describing the deviation of the state variables (pollutant concentrations) from their targets.

Governing equations

The equations used to describe the groundwater system and to predict its evolution under the proposed remediation strategy are the basic groundwater flow and transport equations:

$$\nabla (T \ \nabla h) = S \frac{\partial h}{\partial t} - q \qquad (1)$$

$$v = - K \ \nabla h \qquad (2)$$

$$\nabla (D \ \nabla c) - v \ \nabla c = \frac{\partial c}{\partial t} \qquad (3)$$

where T is the transmissivity tensor, h is the piezometric head, S is the storativity, q is a source/sink term, v is the Darcy flow velocity, K is the hydraulic conductivity, D is the hydrodynamic dispersion, and c is the contaminant concentration.

The feedback control model

In order to formulate the remediation problem as a feedback control problem, some transformations to the equations governing the groundwater flow and transport regimes are needed. Similarly to the transformation of equation (1),

used by Makinde-Odusola and Mariño (1989), and using the finite element formulation, equation (3) can be written in the following form:
$$c_t = A_t c_{t-1} + C_t x_t + b_t + \mu_t \tag{4}$$

in which c_t is the state variable giving the pollutant concentration at stage t, x_t is a control vector, A_t, C_t, and b_t are coefficient matrices of known elements, and μ_t is a vector of random disturbances.

Using the above transformation, the groundwater remediation problem can be posed as a feedback control problem (Chow, 1981). The objective function consists of a successive minimization of a loss function containing the squared deviations of the state variable from its prescribed target value to be reached by the end of the control process, such function can be written as

$$W = \sum_{t=1}^{N} (c_t - a_t)^T K_t (c_t - a_t) \tag{5}$$

in which N is the number of time periods, c_t is a vector containing the state variables (i.e., the pollutant concentrations at time t), a_t is a vector containing the target values of the state variables, and K_t is a penalty matrix for deviating from the target a_t. The minimization of the objective function in a feedback control problem is subject to a constraint of the form (4), where the state variable is expressed as a function of its previous value and of a control vector containing the undertaken decisions. In the case of groundwater remediation problems, the state variable can be taken as the pollutant concentration, whereas the control vector will contain information on the pumping and recharge rates and their locations. It is obvious that the contamination level at each stage will be directly dependent on the remedial decisions embedded in the control vector and on the state of the system on the previous stage. The optimization model can be written as

$$\text{Minimize } W = \sum_{t=1}^{N} (c_t - a_t)^T K_t (c_t - a_t) \tag{6}$$
x_t

subject to:
$$c_t = A_t c_{t-1} + C_t x_t + b_t + \mu_t \tag{7}$$

The remediation problem expressed as an optimization model and posed in its above form can be solved using the econometric method of feedback control. The solution process consists of a stage by stage optimization. The minimization of the loss function (equation 6) will result into a set of feedback control rules given by the control vector:

$$x_t = G_t c_{t-1} + g_t \qquad t = 1, ..., N \tag{8}$$

in which G_t and g_t are feedback coefficients defined below. The control vector containing the optimal decisions for the time period t is basically dependent on the state of the system at time t-1. The feedback coefficients, G_t and g_t, needed to compute the control vector are derived recursively from the following equations:

$$G_t = \left(C_t' H_t C_t\right)^{-1} C_t' H_t A_t \qquad (9)$$

$$H_t = K_t + \left(A_t + C_t G_{t+1}\right)' H_{t+1} \left(A_t + C_t G_{t+1}\right) \qquad (10)$$

$$g_t = -\left(C_t' H_t C_t\right)^{-1} C_t' \left(H_t b_t - h_t\right) \qquad (11)$$

$$h_t = K_t a_t + \left(A_t + C_t G_{t+1}\right)' \left(h_{t+1} - H_{t+1} b_{t+1}\right) \qquad (12)$$

The solution process starts at the end period (t = N) by using the conditions

$$H_N = K_N \qquad (13)$$

$$h_N = K_N a_N \qquad (14)$$

and proceeds backwards in time in a stage by stage optimization using a dynamic programming approach.

References

Ahlfeld, D. P., J. M. Mulvey, G. F. Pinder, and E. F. Wood, Contaminated groundwater remediation design using simulation, optimization, and sensitivity theory: 1. Model development, *Water Resources Research*, 24(3), 431-441, 1988a.

Ahlfeld, D. P., J. M. Mulvey, G. F. Pinder, and E. F. Wood, Contaminated groundwater remediation design using simulation, optimization, and sensitivity theory: 2. Analysis of a field site, *Water Resources Research*, 24(3), 443-452, 1988b.

Andricevic, R., A real-time approach to management and monitoring of groundwater hydraulics, *Water Resources Research*, 26(11), 2747-2755, 1990.

Andricevic, R. and P. K. Kitanidis, Optimization of pumping schedule in aquifer remediation under uncertainty, *Water Resources Research*, 26(5), 875-885, 1990.

Chow, G. C., *Econometric Analysis by Control Methods*, John Wiley, New York, 1981.

Gorelick, S. M. and I. Remson, Optimal dynamic management of groundwater pollutant sources. *Water Resources Research*, 18(1), 71-76, 1982.

Gorelick, S. M., C. I. Voss, P. E. Gill, W. Murray, M. A. Saunders, and M. M. Wright, Aquifer reclamation design: the use of contaminant transport simulation combined with nonlinear programming, *Water Resources Research*, 20(4), 415-427, 1984.

Jones, L., R. Willis, and W. W-G. Yeh, Optimal control of nonlinear groundwater hydraulics using differential dynamic programming, *Water Resources Research*, 23(11), 2097-2106, 1987.

Makinde-Odusola, B. A. and M. A. Mariño, Optimal control of groundwater by the feedback method of control, *Water Resources Research*, 25(6), 1341-1352, 1989.

Murray, D. M. and S. J. Yakowitz, Constrained differential dynamic programming and its application to multireservoir control, *Water Resources Research*, 15(5), 1017-1027, 1979.

Wagner, B. J. and S. M. Gorelick, Reliable aquifer remediation in the presence of spatially variable hydraulic conductivity: From data to design, *Water Resources Research*, 25(10), 2211-2225, 1989.

Wellhead Protection Area Delineation for the
Weyerhaeuser Wellfield, Springfield, Oregon

Mark Cunnane[1], David Banton[1], Jonathan Snell[2]

Abstract

The Weyerhaeuser Wellfield draws groundwater from a shallow alluvial aquifer to serve 18,000 citizens in Springfield, Oregon. The aquifer consists of unconsolidated and partially cemented sand and gravel with average transmissivity of 40,000 ft^2/d. The 10-year time-related wellhead protection area was determined to be approximately 2 miles wide and 4.3 miles long in the up-gradient direction. This protection area was based on calculations using both analytical and numerical methods and considering uncertainty in groundwater flow direction.

Introduction

This paper summarizes the salient components of a wellhead protection area delineation project for the Weyerhaeuser Wellfield in Springfield, Oregon. The Weyerhaeuser Wellfield draws 4 MGD from three wells to supply a population of 18,000 citizens. The wells, penetrating a shallow alluvial aquifer, range in depth from 60 to 90 feet and are located within 300 feet of the McKenzie River. The wells were drilled between 1956 and 1962 and were completed with perforated casing and surface seals. The shallow well depths, high water table conditions, and transmissive aquifer materials leave the wellfield vulnerable to releases of hazardous materials at ground surface.

---

[1] Project Engineer and Associate, Golder Associates Inc., 4104-148th Avenue N.E., Redmond, WA 98052
[2] Senior Hydrogeologist, Golder Associates Inc., 317 S.W. Alder, Portland, OR 97204.

Geologic Framework

The McKenzie River drains the western Cascades in central Oregon. The river valley was formed during late Tertiary to early Pleistocene time. In the Springfield area, the ground surface elevation is 500 feet above mean sea level and the valley width is 2 miles. Upstream the valley extends for greater than 20 miles. Downstream the valley widens and joins the Willamatte River, which flows north to the Columbia River at Portland, Oregon.

The valley fill sediments include two formations, older and younger alluvium, respectively (Frank, 1973). Older alluvium constitutes the majority of the valley fill. Younger alluvium unconformably overlies older alluvium and is restricted to deposits along the McKenzie River channel. Younger alluvium may have been partially derived by reworking of older alluvium.

Older alluvium textures range from silt and clay to boulders, although most of the formation is coarse sand and gravel. The deposits are discontinuous, reflecting the fluvial depositional environment. Many strata are partially cemented, although open gravels also are present. Younger alluvium textures are similar to older alluvium, although no cementation has occurred and there are fewer fine-grained deposits.

In the Weyerhaeuser Wellfield, younger alluvium forms the upper 35 feet of sediment. Below 35 feet, older alluvium continues to bedrock at an unknown depth. The McKenzie river bed cuts through younger alluvium.

Aquifer Hydraulic Properties

Transmissivity and storativity values for the alluvial aquifer have been estimated from well tests at various locations (Frank, 1973). Transmissivity values ranged from 2,700 to 67,000 ft^2/d for older alluvium and 76,000 to 270,000 ft^2/d for younger alluvium. Test interval thicknesses were unknown, but likely range from 60 to 100 feet. Storativity estimates for these tests ranged from 0.002 to 0.06, suggesting semi-confined conditions.

Pumping tests conducted in the Weyerhaeuser Wellfield confirmed the high transmissivivity of the aquifer materials. Transmissivity estimates from 24 hour pumping tests yielded values ranging from 20,000 to 69,000 ft^2/d. The effective test intervals were on the order of 60 feet and crossed from younger alluvium to older alluvium. Storativity values obtained for these tests ranged from 0.01 to 0.05, also suggesting semi-confined conditions.

The McKenzie River was anticipated to behave as a recharge boundary in the pumping tests. Water levels in the wellfield were observed to fluctuate with changes in river stage, implying a degree of hydraulic communication. But a conclusive recharge boundary condition was not observed in the pumping test observations. The absence of a boundary was attributed to the short duration of the tests, the partially penetrating conditions of the river, and the presence of a clogging layer at the river bed. Temporal variation in river stage during the pumping tests was observed to interfere with interpretations.

Groundwater Flow

The groundwater flow direction in the McKenzie River valley is primarily from east to west down the valley. The rate of groundwater flow is fairly rapid for porous media. The average linear pore water velocity is about 6 ft/d. Volumetric groundwater flow rates through the upper 60 feet of the aquifer range from 4 MGD upstream from the wellfield to 10 MGD where the valley widens to the west. Recharge to the aquifer occurs as infiltration where tributary streams enter the valley and from direct precipitation onto the valley floor. Precipitation in the Springfield area is approximately 40 in/yr. The bedrock on either side of the valley is not anticipated to provide significant recharge to the aquifer.

In the Weyerhaeuser Wellfield, groundwater recharge appears to be primarily from the McKenzie River. Major-ion water chemistry analyses distinguish surface water from regional groundwater. Based on these analyses for the McKenzie River, the wellfield, and two wells distant from the river, over 75% of the wellfield supply originates from the river. This recharge may be induced or occur naturally from loosing reaches of the river.

Wellhead Protection Area Delineation

Wellhead protection areas (WHPAs) for the Weyerhaeuser Wellfield were delineated based on time of travel. The WHPAs were evaluated using the two-dimensional numerical model, FLOWPATH (Franz and Guiguer, 19922), and the analytical WHPA delineation software by EPA (1991). Both models track particles in a steady-state groundwater flow field to obtain the time-related WHPAs.

Figure 1 shows the recommended 10-year time of travel wellhead protection area. Results from FLOWPATH are shown by the stipled pattern and conform to a down-valley flow direction forced by the model configuration. The extension of the WHPA to the south side of the valley resulted from analytical calculations using the flow direction (N 50° W) shown on a 1969 water table map (Frank, 1973). The WHPA computed using the analytical model was truncated at the contact between bedrock and alluvium. The recommended WHPA combines the numerical and analytical results. A buffer zone was also recommended to include the bedrock highlands south of the valley were runoff and infiltration to the aquifer is possible.

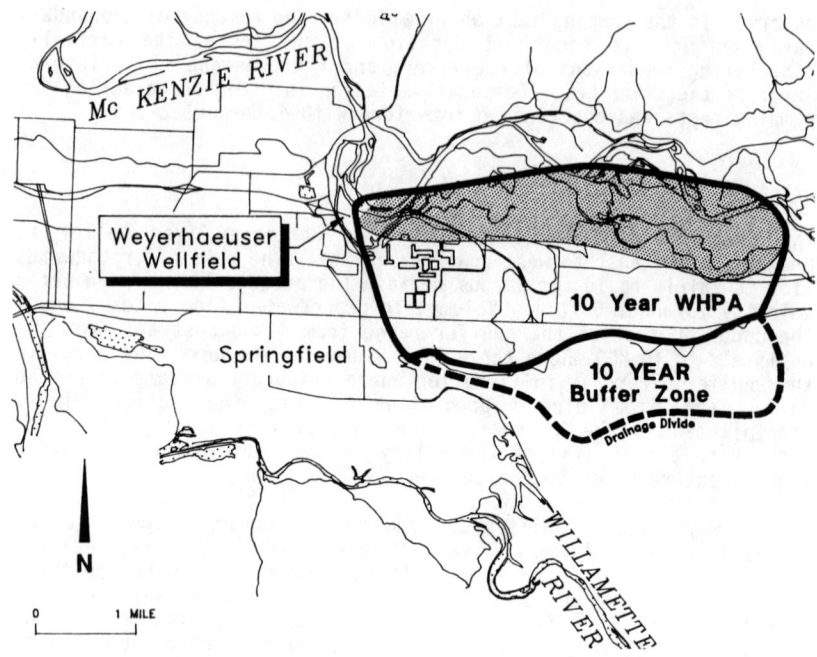

Figure 1. 10-year time of travel wellhead protection area.

Appendix I. References

EPA, 1991, WHPA, A Modular Semi-analytical Model for the Delineation of Wellhead Protection Areas, Version 2.0, U.S. Environmental Protection Agency Office of Ground-Water Protection, March 1991.

Frank, F. J., 1973, Ground water in the Eugene-Springfield Area, Southern Willamette Valley, Oregon, U.S. Geological Survey Water Supply Paper 2018.

Franz, T. and N. Guiguer, 1992, FLOWPATH User's Manual, Waterloo Hydrogeologic Software, Whitby, Ontario.

Appendix II. Unit Conversions

1 m = 3.28 feet (ft); 10 km = 6.2 miles; 86,400 s = 1 day (d).

Optimum Operation of Recharge Basins
Hasan Mushtaq [1], Larry W. Mays [2], M. ASCE, and Kevin E. Lansey [3], A. M. ASCE

Abstract

Mathematical models based on nonlinear programming are developed for operating recharge basin systems. The objective of these optimization models is to determine the operation policy that maximizes the infiltration subject to constraints for continuity, infiltration, groundwater flow, and the physical constraints describing the recharge basins. Two basic physical processes that are involved in the recharge basin systems are the infiltration process and the soil moisture redistribution process. The infiltration process is modeled using both the Philip's (1957) and the Green-Ampt (1911) infiltration models and the soil moisture redistribution process is modeled using the Darcy's equation as well as a kinematic wave model (Charbeneau, 1984). An application is presented.

General Model Formulation

The overall optimization model can be stated as :

Maximize Infiltration (during application time, X, and infiltration time, Y)

Subject to
 (1) Mass balance equation during application time, X.
 (2) Mass balance equation during application time, X, and

---

[1] Research Assistant, Department of Civil Engineering, Arizona State University, Tempe, Arizona 85287

[2] Professor and Chair, Department of Civil Engineering, Arizona State University, Tempe, Arizona 85287

[3] Assistant Professor, Department of Civil Engineering, University of Arizona, Tucson, Arizona 85721

infiltration time, Y.
(3) Groundwater/subsurface flow equations.
(4) Application time constraints.
(5) Cycle time constraints.
(6) Surface water depth constraints.

The objective function is the maximization of the infiltration volume during the application time, X, and the infiltration time, Y (Figure 1). This equation is the summation of cumulative infiltration, F(t), over all the basins. As noted, the objective of the process is to maximize the infiltration volume during the operation period.

Mathematically, this can be stated as following :

Maximize $F = F(X_t, Y_t)$ (1)

subject to

Continuity equation, $\quad g_t(X_t, Y_t) = 0 \quad t = 1, \ldots, T$ (2)

Redistribution equation, $\quad h_t(X_t, Y_t, Z_t) = 0 \quad t = 1, \ldots, T$ (3)

System operation equation, $\quad w_t(X_t, Y_t, Z_t) \geq 0 \quad t = 1, \ldots, T$ (4)

where F is the infiltration which is a function of the application time (X)

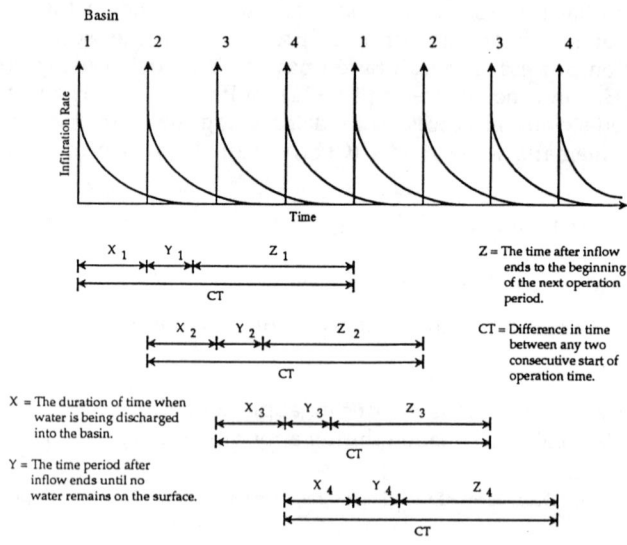

Figure 1 Definition Sketch of Application Time, Infiltration Time, and DryingTime (for four basins only)

and the infiltration time (Y). $g_t(X_t, Y_t)$ are the continuity equations which describe the infiltration process. This is essentially a set of mass balance equations. $h(X_t, Y_t, Z_t)$ are the soil moisture redistribution process equations, which describe the change in moisture content with time and depth and $w_t(X_t, Y_t, Z_t)$ are the physical system constraints including bounds on the decision variables.

Two physically based infiltration models (Philip's and Green-Ampt) and two soil moisture redistribution models (Darcy and kinematic wave) are used in this research. Using these mathematical models for the physical processes, a series of four optimization problems of the form (Equation 1-4) were formulated and solved using a nonlinear optimization code. Following is a hypothetical example.

Application

A hypothetical recharge system consisting of four basins was chosen for the purposes of illustration. The basin areas varied from 0.25 acres to 1.0 acre. This specific model was solved for only one cycle time of 192 hrs. Other preset model parameters are listed in Table 1. The multibasin single cycle time optimization model was solved using the nonlinear optimization code GAMS MINOS5 (Murtagh and Saunders, 1977). The results are compared in Tables 2 and 3.

Results

The results from the hypothetical examples show (1) it is possible to mathematically model these infiltration and redistribution process equations, (2) the model results also show that the Green-Ampt infiltration model is more conservative than the Philip's infiltration model (approximately 12%-14% less total infiltration) (3) kinematic wave model results in about 25%-27% more total infiltration than the Darcy's plug flow assumption model.

Table 1 : Parameter Values for the Hypothetical Examples.

I	=	1.85 cfs.		E	=	0.000356345 ft/hr	=	0.26 cm/day
S	=	0.0984251 ft/hr$^{1/2}$	= 3.0 cm/hr$^{1/2}$	K	=	0.0328083 ft/hr	=	1.0 cm/hr
T	=	48 hrs		N	=	4		
Ψ	=	11.01 cm	= 0.3612204 ft.	L	=	5.0 ft.		
P	=	8.29		θ_r	=	0.048.		
ϕ	=	0.423		θ	=	0.30		
LMIN	=	5.0 ft.		$\Delta\theta$	=	0.375		
DMAX	=	3.0 ft						

Table 2 : Results from Darcy's Plug Flow Assumption.

	A acres	ET acre-ft.	X hrs	FO acre-ft.	Y hrs	FI acre-ft.	Z hrs	D acre-ft.	IN acre-ft.	F acre-ft.
(1)	(2)	(3)	(4)	(5)	(6)	(7)	(8)	(9)	(10)	(11)
Basin 1	1.00	0.017	14.87	0.867	33.33	1.397	143.80	1.409	2.282	2.265
		0.018	13.09	0.671	36.22	1.320	142.69	1.333	2.008	1.991
Basin 2	0.75	0.013	11.15	0.521	37.05	1.178	143.80	1.186	1.711	1.699
		0.013	9.81	0.405	139.49	1.088	142.69	1.099	1.506	1.493
Basin 3	0.50	0.009	7.43	0.256	140.77	0.876	143.80	0.883	1.141	1.132
		0.009	6.54	0.201	42.77	0.795	142.69	0.802	1.004	0.995
Basin 4	0.25	0.004	3.72	0.078	144.48	0.488	143.80	0.492	0.570	0.566
		0.004	3.27	0.062	46.04	0.436	142.69	0.440	0.502	0.498
Sum	2.5	0.043	37.17	1.722	155.63	3.939	575.20	3.972	5.705	5.662
		0.044	32.71	1.338	164.52	3.638	570.76	3.674	5.021	4.977

Table 3 : Results from Kinematic Wave Model.

	A acres	ET acre-ft.	X hrs	FO acre-ft.	Y hrs	FI acre-ft.	Z hrs	D acre-ft.	IN acre-ft.	F acre-ft.
(1)	(2)	(3)	(4)	(5)	(6)	(7)	(8)	(9)	(10)	(11)
Basin 1	1.00	0.024	19.73	1.084	47.27	1.919	125.01	1.936	3.027	3.003
		0.024	17.13	0.828	49.86	1.777	125.01	1.795	2.629	2.605
Basin 2	0.75	0.018	14.80	0.648	52.20	1.605	125.01	1.619	2.270	2.253
		0.018	12.85	0.496	54.14	1.458	125.01	1.472	1.972	1.954
Basin 3	0.50	0.012	9.86	0.316	57.13	1.185	125.01	1.196	1.514	1.502
		0.012	8.57	0.244	58.43	1.059	125.01	1.069	1.315	1.303
Basin 4	0.25	0.006	4.93	0.095	62.06	0.656	125.01	0.661	0.757	0.751
		0.006	4.28	0.075	62.71	0.577	125.01	0.582	0.657	0.651
Sum	2.5	0.060	49.32	2.144	218.66	5.365	500.04	5.412	7.568	7.509
		0.060	42.83	1.642	225.14	4.871	500.04	4.918	6.573	6.513

(Highlighted numbers are from Green-Ampt Infiltration Model)

A = Area, ET = Evaporation, D = Depth of water remained in the basin, IN = Volume of water supplied to the basin, X = Operation time, Y = Infiltration time, Z = Drying time, F = Volume of infiltration, FO = Infiltration during operation time, Infiltration during infiltration time.

References

Charbeneau, R. J., (1984). "Kinematic models for soil moisture and solute transport."*Water Resources Research*, 20(6), 699-706.

Green, W. H., and G. A. Ampt. (1911). "Studies on soil physics, part I, The flow of air and water through soils", *Journal of Agricultural Sciences*, 4(1), pp. 1-24.

Murtagh, B. A., M. A. Saunders, and M. H. Wright. (1977). MINOS, a Large Scale Nonlinear programing System, SOL-77-9, Report, Systems Optimization Laboratory, Stanford University, Stanford, California.

Philip, J. R. (1957). "The theory of infiltration : 1. The infiltration equation and its solution." Soil Sci., vol. 83, no. 5, pp. 345-357.

Artificial Recharge To Manage Groundwater Quality In A Connected Surface Water Groundwater System

Seshadri Suryanarayana[1]
A. Osman Akan[2], M. ASCE

Abstract

Finite difference method is used to formulate the management problem of movement of water and solute in a connected surface water groundwater system. Pumped groundwater quality of conservative substance is controlled by manipulating pumping and artificial recharge of water.

Introduction

Conjunctive-use of surface water and groundwater is often the only solution to meet the growing water demands. On the other hand, disposal of liquid wastes into surface water bodies or through deep injection wells, contaminates the aquifer and the surface water. So, the water quality at the pumping wells is controlled either by (a) optimizing well discharges and surface water diversions and the concentration of waste input into the system such that the water quality at the pumping wells are within the standards specified as discussed in detail by Suryanarayana and Akan (1992), or (b) optimize well discharges and surface water diversions and the concentration of waste input into the system and contain the movement of contaminants into water supply wells by artificial recharge of water with quality of recharge water being either of a superior or of an inferior quality to that of aquifer.

In the development of this model, it is assumed that locations of all withdrawal points and injection points both from and into surface water and groundwater are known. Additionally, artificial recharge rate and its concentration, and disposal flow rate of wastes into groundwater and into surface water are assumed to be known. Further, it is assumed that the constituent dissolves fully in water but the density of the fluid remains constant.

---

[1] Graduate Student
[2] Professor, Department of Civil and Environmental Engineering, Old Dominion University, Norfolk, Virginia 23529-0241

Methodology

The flow and transport model is based on steady, two-dimensional, horizontal flow equation in a heterogeneous and anisotropic medium. Streams and rivers are represented by the continuity equations and the mass transport in streams is treated as a plug flow model. The conjunctive-use of surface water and groundwater optimization scheme is the same as previously reported by Suryanarayana and Akan (1991). Further, the solute transport model coupled with the conjunctive-use management model is the same as previously reported by Suryanarayana and Akan (1992). Here, our objective is to contain the movement of contaminants into water supply wells by means of artificial recharge of water. Thereby, concentration input through injection wells could be further maximized and still maintain the water quality at the pumping wells within the standards.

Management Model

The management model consists of an objective function subjected to a set of constraints. Here, our objective is concerned with maximizing the surface water withdrawal together with pumpage from groundwater. At the same time we are interested in maximizing the waste input concentration into surface water and groundwater and control the movement of the contaminants by a set of artificial recharge wells. To enable a linear programming scheme, we approach the solution of this problem using dual programming algorithms. In the first sub problem, optimal pumping and surface water diversions are found from the solution of the model,

$$Max \ Z \ = \ \sum (Q_{GW})_{iw} + \sum (Q_{SWD})_{js} \qquad (1)$$

where, summation of all groundwater pumping Q_{GW} and summation of all surface water diversions Q_{SWD}, constitute the management objective function, subjected to both hydraulic and management constraints.

After seeking the solution of the first sub problem, we proceed to the next sub problem, in this we optimize the concentration input into surface water and groundwater

$$Max \ Z \ = \ \sum (C'_W)_{nw} + \sum (C'_R)_{nr} \qquad (2)$$

where, summation of all concentrations injected C'_W and summation of all concentration disposed in river C'_R, constitute the management objective function, subjected to mass transport and environmental management constraints.

Illustrative Example

A hypothetical basin shown in Fig.1 is considered. In the figure, GW1, GW2, and GW3 represent the location of pumping wells, IW1 is the injection well for wastewater disposal of 4 m^3/min, SW1 and SW2 represent the location of surface water diversions, SD1 is the site of 50 m^3/min waste disposal into surface stream, injection of 1 m^3/min with a concentration of 10 mg/l is considered in each of the artificial recharge wells. The various management and environmental constraints imposed and its optimal solution obtained is shown in Table 1. The concentrations in the aquifer for optimum solution is shown in Fig. 2.

Appendix - References

Suryanarayana, S. and Akan, O.A. (1991), "Conjunctive Use of Surface Water and Groundwater Management", Proceedings of the International Symposium on Groundwater, ASCE, July 29 - August 2, 1991, Nashville, Tennessee.

Suryanarayana, S. and Akan, O.A. (1992), "Water Quality and Quantity Management in Connected Surface Water Groundwater Systems", Proceedings of the Water Forum '92, ASCE, August 2 - 6, 1992, Baltimore, Maryland.

Table 1: Optimal Solution For the Management Problem

Types of Constraints	Constraints	With Artificial Recharge	Without Artificial Recharge
Demand (m^3/min)	≥ 200.	328.602	328.475
Pumping capacity (m^3/min) - GW1	≤ 40.	38.525	38.524
- GW2	≤ 40.	36.322	36.306
- GW3	≤ 40.	29.305	29.237
Diversion capacity (m^3/min) - SW1	≤ 150.	74.45	74.41
- SW2	≤ 150.	150.0	150.0
Minimum flow requirement (m^3/min)	≥ 400.	400.0	400.0
Input concentration (mg/l) - SD1		1060.0	1060.0
- IW1		423.394	396.264
Pumped water quality (mg/l)- GW1	≤ 50.	41.176	41.143
- GW2	≤ 50.	31.928	31.736
- GW3	≤ 50.	50.00	50.00
Downstream river water quality	≤ 100.	100.00	100.00

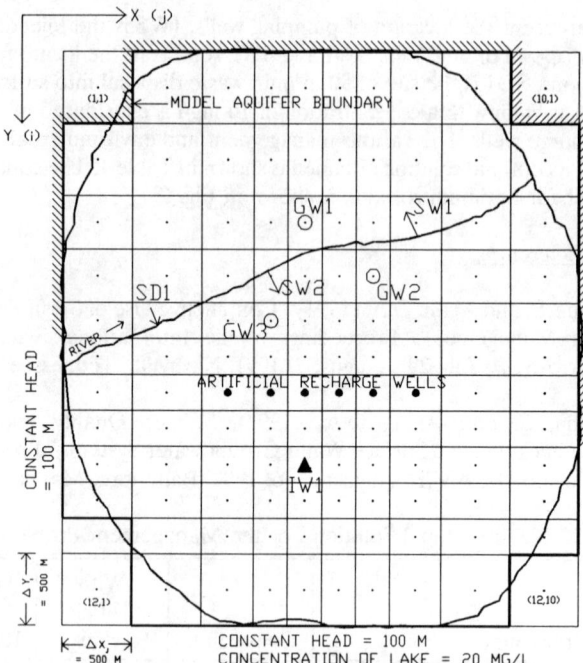

Figure 1. Description of Example Problem

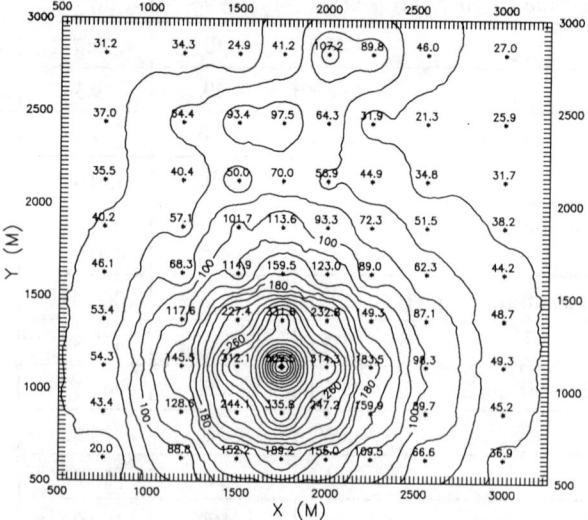

Figure 2. Aquifer Concentrations for the Optimum Solution

The Strategic Development of a Sole Source Aquifer to Improve Water Quality While Minimizing Environmental Impact

Marc V. Cromer[1], Mark J. Abbott[2], and Shih-Huang Chieh[3]

Abstract

A water supply project is currently underway which prevents future degradation of a sole source aquifer while providing a groundwater resource with limited usage restrictions. Salt water intrusion at the coastline and along estuary canals is the principle component currently contributing to this degradation. Projected water demands were expected to exacerbate the current water quality situation at existing coastal wellfields, if an alternative supply was not developed.

The selected water supply alternative provides for the consolidated "buildout" raw water demand of several coastal treated water purveyors through an inland groundwater supply. The specific strategic objectives were to meet year 2030 projected demands without significant cutbacks or degradation during a design drought condition of 210 days.

To realize these objectives, site selection alternatives for the inland systems were determined using specific design and regulatory criteria, and evaluated through aquifer performance testing and numerical groundwater flow modeling. The interaction between ground and surface waters in this environmentally sensitive South Florida region also added to site selection complexity. Numerical modeling was required to assess potential drawdown impacts on the coastal saltwater intrusion front, existing local groundwater users, suface water features, and individual well performance within the designed wellfield.

1,3. Senior Engineer, and Supervising Engineer, James M. Montgomery (JMM), Inc., 365 Lennon Lane, Walnut Creek, CA 94588, (510)975-3400
2. Hydrogeologist, JMM, 2328 10th Ave. North, Lake Worth, FL 33461, (407)586-8830

Introduction

A regional water supply project is currently underway which implements the findings and recommendations from a study on long-term water supply development and water quality protection of a sole source aquifer in South Florida. The regional study provides a comprehensive evaluation of a county's water supply by accounting for average and maximum day raw water demands, as defined by the existing water purveyors, through a projected buildout date in the year 2030. Two centralized wellfields, located well inland from the coastline, are recommended to avoid potential saltwater intrusion and contamination problems and to assure supply to existing large treated water purveyors.

The study proposes a phased approach to the relocation of withdrawals from 10 major coastal wellfields to two inland systems having a total designed capacity of 110 million gallons per day (mgd). To meet water supply objectives, an intensive screening effort was required to evaluate the location selection for these centralized wellfields. Siting of these wellfields considers weighing the inland displacement from the saltwater intrusion front with the economical distance from coastal treatment facilities.

A Threatened Resource

The Biscayne aquifer is the highly productive portion of the surficial aquifer system in southcentral and southeastern Florida. With the unprecedented population growth which has occurred along the coast over the last twenty years, a steady increase in water demand has placed the existing water supply systems in jeopardy. Extensive tidal waterway systems, which allow access to the Atlantic Ocean from many coastal communities, coupled with increased withdrawal at the coastal wellfields, have caused significant saltwater intrusion into the former fresh water aquifer.

Characteristic seasonal fluctuations in rainfall recharge create a cyclic saltwater intrusion problem. During the rainy months of June through October, large amounts of rainfall occur along the coast and supplies ample water to recharge the coastal aquifer. Recharge occurs both by infiltration from the inland canal drainage system and by direct surficial recharge during this period. However, during the dry months between November and May, water is supplied to the coastal wellfields through groundwater storage. Aquifer recharge during this period is limited to sporadic rainfall events. This problem of limited rainfall recharge during the winter months is intensified by increased water demand from a large influx of seasonal population.

Frequent occurrences of chloride levels above the maximum contamination levels (MCLs) during heavy demand periods have historically caused the closure of wells at several existing coastal wellfields. This urgency at specific locations demanded that individual attention to these purveyors be considered with the implementation of the project. Those factors deemed to influence project priorities include the distance from the existing supply to the saltwater interface, local groundwater withdrawal practices, local groundwater levels, and the available remedial alternatives. The critical nature of the situation at some of the purveyors requires that a portion of the water supply project be implemented prior to the development of the centralized regional water supply. This measure provides a local short-term solution which can later be integrated into the overall regional plan.

Design Strategy

All applications for new water supply must undergo extensive review by the South Florida Water Management District (SFWMD). The permittee must evaluate well siting against the potential for adverse impacts to surrounding wetlands, natural water bodies, existing permitted users, and the saltwater front. Well siting criteria were established for the purpose of locating preliminary study sites and selecting individual well sites. Many of these criteria were developed following an extensive environmental audit.

The purpose of the environmental audit was to identify current and past potential environmental liabilities associated with the wellfield site selection. This identification helps to prevent locating proposed production wells in areas which have or had activity which could adversely impact the groundwater. Preliminary well sites were to be compatible with the existing land use, accessible for construction, near transmission pipeline easements, at least 300 feet from surface water bodies, and at least 100 feet from a sanitary sewer. Application of these regionally identified land use, water quantity, and water quality criteria, yielded a combined total of 140 potential well sites. The preliminary sites were further screened using local regulatory guidelines to establish a total of 50 well sites between the two centralized systems. The transmission routes and wellfield configurations were established following criteria applications and site specific testing and analyses.

The SFWMD additionally requires that the projected demand for the wellfield be justified based upon the projected water demand of the urban service areas. To determine this demand, approved population projections are used to estimate the permanent populations for the participating water purveyors. A per capita water use, based upon recorded population and water use over the previous five years, is then applied to the projected population to determine the demand over the allowed permit duration. These demands are translated into a design withdrawal needed to evaluate potential adverse impacts.

The feasibility of achieving the designed, high-volume withdrawal of limited impact, was examined at each of the two centralized systems using long-duration aquifer performance testing and numerical groundwater flow modeling. At each system, a representative well-site was selected for a performance study, as defined by a 72-hour pumping test. These tests yielded valuable information on aquifer hydraulic\hydrogeologic properties, water quality, and observed environmental impact. More specifically, the impacts on local surface water features, neighboring wetlands, and existing groundwater users due to sustained groundwater withdrawal were monitored. The water quality in this region generally varies with depth. Deeper connate waters exists here due to their physical displacement from "flushing" mechanisms active nearer to the surface. An examination of long-term surficial aquifer stress was needed to evaluate the potential for inducing upward flow from this poor quality, deeper system.

Two numerical groundwater flow models were calibrated and verified against the aquifer performance testing. These models were used to predict aquifer hydraulic response and anticipate the resulting local and regional impact associated with various wellfield configurations and withdrawal scenarios. The models also supplied the groundwater flow field for evaluation of potential water quality degradation. Connate water upconing was locally evaluated with the assistance of a particle tracking algorithm, while the impact on the groundwater flow gradients (which restrain coastal saltwater intrusion) was examined at the regional scale.

Conclusions

The wellfields are to be built and operated by a county agency, with water to be sold to the communities based upon operating and maintenance costs. Construction costs will be paid by county-wide, ad-valorem taxes. However, the long term use of water and expense of the system requires that a legal document be established which defines the quantity of water to be purchased from implementation through buildout of the participating water purveyor. Large User Agreements between the county and the individual purveyors were negotiated based upon the amount of water demand projected on an annual basis and peak day conditions. Based upon these agreements, the distribution system was designed to deliver the raw water to the participating Large Users.

References

Fish, J.E., 1988, Hydrogeology, Aquifer Characteristics, and Ground-Water Flow of the Surficial Aquifer System, Broward County, Flordia. USGS Water Resources Investigation Report No. 87-4034.

MacVicar, T. and S.S.T. Lin, 1984, Historical rainfall activity in central and southern Florida: average, return period estimates and selected extremes. Environments of South Florida Present and Past, Miami Geological Society Memoir No. 2, Second Printing, pp. 477-509.

Sonenshein, Roy S., 1984, Altitude of Water Tables in the Biscayne Aquifer, Broward County, Flordia, May 22-23, 1984. USGS Open File Report.

Developing, Managing, and Protecting Urban Aquifers
in the Pacific Northwest

Michael R. Warfel[1]

Abstract

Water purveyors in the Puget Sound region have applied a variety of innovative techniques to effectively manage urban groundwater supplies from a quantity and quality perspective. Two examples are discussed: enhancement of well field yield by the Seattle Water Department through aquifer recharge; and implementation of a comprehensive aquifer protection program by Renton, Washington, to protect a shallow alluvial aquifer from contamination.

Introduction

The recent population growth in the Pacific Northwest has led to an increased awareness among water purveyors and citizens of the importance of groundwater as a water supply resource. As the demand for water has increased, the importance of managing the quantity and quality of groundwater supplies has become a critical component of the overall water supply infrastructure. Examples of effective management of groundwater supplies (aquifers) in urban environments of Western Washington are provided by the Seattle Water Department groundwater recharge program, and by the City of Renton aquifer protection program. The key elements of these two programs are described in this paper.

---

[1]Manager, Hydrogeologic Services, CH2M HILL, 777 108th Avenue NE, Bellevue, Washington 98009

Groundwater Recharge Program, Seattle Water Department

Overview and Hydrogeologic Setting

The Seattle Water Department (SWD) has utilized the Highline well field since the late 1980s to provide summer peaking supply as a supplement to surface water from the City's mountain watersheds. Of the three aquifer zones present in the well field area, the intermediate aquifer exhibits the highest transmissivity and is the primary source of groundwater for the well field. This coarse-grained aquifer (composed of sand, gravel, and cobbles) is 100 to 150 feet (30.5 to 45.7 meters) thick and occurs at depths of 120 to 300 feet (36.6 to 91.5 meters) below land surface in the well field area.

The goal of SWD was to develop the Highline well field as a reliable summer peak-demand supply of 12 million gallons per day (45.4 million liters per day) by maximizing yield and mitigating groundwater level declines during the 4-month summer pumping period. SWD embarked on a program to assess artificial groundwater recharge as a means of augmenting the well field yield and mitigating drawdown impacts. A successful proposal was prepared by SWD and accepted by the U.S. Bureau of Reclamation (USBR) for the Highline project to be included in the USBR High Plains States Groundwater Recharge Demonstration Program.

The project has been implemented in three phases: an initial evaluation, pilot testing, and full-scale testing. The following sections describe the approach and results of each phase of the project.

Initial Evaluation

This phase of the project was completed by September 1990 and was designed to assess technical aspects of artificially recharging the intermediate aquifer with treated drinking water from the SWD Cedar River Watershed through a recharge well, and recovering this stored water at a later time by pumping from the same well. Factors that were assessed included hydrogeology, aquifer mineralogy, water chemistry, and potential recharge impacts to the aquifer (such as chemical precipitation, clay swelling, and bacterial growth).

The initial evaluation did not identify any technical factors that would preclude continuation with the pilot testing phase of the project. Additional monitoring to further assess the potential for subsurface bacterial growth and formation of disinfection byproducts (DBPs) was included in plans for the pilot testing phase.

Pilot Testing

The pilot testing was conducted between January and July 1991 using a test well that had been drilled during a prior tunnel exploration project. Three recharge-recovery cycles were completed, with recharge and recovery volumes ranging from 5.1 to 50.8 million gallons (19.3 to 192 million liters) and 9.3 to 46.8 million gallons (35.1 to 176.9 million liters), respectively. All water recovered from the test well was initially routed to an onsite holding pond prior to discharge to the local stormwater collection system.

The pilot testing results indicated that observed clogging in the vicinity of the well screen was caused by fine-grained solids, rather than other mechanisms, and that surging effectively reversed this type of clogging. Bench-scale testing results indicated that re-chlorination of recovered water creates DBPs in concentrations similar to those measured in recharge water. A full-scale testing program utilizing SWD production wells was proposed, with an emphasis shift to fewer recharge and recovery cycles with greater recharge volumes and longer storage periods.

Full-Scale Testing

The initial cycle of the full-scale test was conducted at the Boulevard Park Production well between December 1991 and February 1992, with recovered recharge water routed to the stormwater collection system. The well clogging mechanism and reversibility observed during the pilot test was confirmed during the initial full-scale test. A multi-well test is scheduled to start by the end of 1992, during which recovered water is planned to be pumped into the SWD distribution system.

Aquifer Protection Program, Renton Water Utility

Overview and Hydrogeologic Setting

The City of Renton is situated at the mouth of the Cedar River Valley on the southeast shore of Lake Union, approximately 11 miles (17.7 kilometers) south of Seattle. The majority of the City's production wells are completed at depths of less than 110 feet (33.5 meters) in the Cedar Valley Aquifer, an unconfined aquifer consisting of coarse sands and gravels. With a prevailing static water level less than 20 feet (5.3 meters) below ground surface, the potential for leaks and spills of hazardous substances to contaminate the Renton water supply is extremely great. Over 90 percent of the City's water supply is provided by groundwater.

An awareness of the susceptibility of the aquifer to contamination led to the initiation by Renton of an aquifer protection program by Renton in 1983. This program is comprised of three major phases: a well field protection study, a well field monitoring study, and ordinance development.

Well Field Protection Study

Completed in 1984, this study included characterization of aquifer hydrogeologic conditions and potential contaminant migration pathways, an inventory of potential groundwater contamination sources, and development of preventive measures for different contaminant source types. The study resulted in a prioritization of potential contaminant sources and preparation of a petition for Federal sole-source aquifer status. The USEPA designated the Cedar Valley Aquifer as a Sole Source Aquifer in 1988.

Well Field Monitoring Study

The objectives of this phase of the program were to assess rates and directions of groundwater movement in the vicinity of the well field, to determine zones of potential capture for various pumping conditions, and to delineate appropriate aquifer protection area boundaries for a City aquifer protection ordinance. The study was completed in 1988 and included installation of an initial network of 11 monitoring wells, analysis of multiple-well pumping tests (at rates up to 15,000 gallons per minute [56,700 liters per minute]). Water-level contour maps were evaluated to estimate groundwater travel times and to delineate aquifer protection zones.

Ordinance Development

The cornerstone of the City of Renton's aquifer protection program is a comprehensive aquifer protection ordinance that was adopted in September 1992. This ordinance establishes a framework for regulating land uses and hazardous substances within wellhead protection areas, and provides incentives for nonconforming uses to relocate outside the aquifer protection area. The aquifer protection ordinance joined existing City ordinances regulating underground storage tanks and hazardous waste sites as groundwater protection measures.

Large-Scale Conjunctive Use
of the San Gabriel Basin: An Environmentally
Beneficial Water Supply Project

Jonathan Harris[1]
and
Timotheus Hampton[2]

Abstract

The 170-square-mile San Gabriel Basin in Los Angeles County, California, has commanded the attention of both water supply and environmental agencies because: (1) it is an idyllic natural water storage facility with almost 10 million acre-feet of storage that is currently used to meet the needs of about one million residents, and (2) it contains one of the largest areas of groundwater contaminated with volatile organic compounds (VOCs) in the world. The Metropolitan Water District of Southern California (Metropolitan) is evaluating a project designed to more effectively use this resource in the management of Southern California's water supply, in a manner consistent with the U.S. Environmental Protection Agency's (EPA's) remedial objectives in the basin.

Introduction

In 1989, VOC contamination was first detected in groundwater from the San Gabriel Basin. Following this discovery, State and Federal agencies embarked on extensive groundwater sampling programs that resulted in the listing of four areas within the Basin on the National Priority List (NPL). EPA's Superfund program has since been used to fund remedial investigations throughout the basin, initiate wellhead treatment programs, and develop a long-term plan for basinwide remediation.

---

[1] Groundwater Resources Department Manager, CH2M HILL, California, 2510 Red Hill, Santa Ana, CA 92705

[2] Senior Engineer, Metropolitan Water District of Southern California, 1111 Sunset Blvd., Los Angeles, CA 90012

EPA's remedial strategy is divided into distinct Operable Units (OUs) that focus remedial efforts on specific problems in limited areas. Currently, one of EPA's highest priority OUs involves controlling continuing migration of contamination out of an area centered on the City of Baldwin Park.

The term "conjunctive use" as used herein, refers to the practice of storing surface water in groundwater basins, for subsequent extraction and use when surface water supplies are in short demand. Concurrent with EPA's efforts over recent years, Metropolitan has been evaluating use of the San Gabriel Basin as a large-scale storage facility for imported water. Current ongoing programs emphasize relatively short-term, small-scale conjunctive use projects to help take better advantage of an abundant winter supply of imported water to satisfy local demand in the summer. The objectives of an expanded, large-scale project are much broader in scope, involving benefits to Metropolitan's entire service area, which exceeds 15 million people throughout Southern California.

Despite the distinct objectives of the projects under evaluation by Metropolitan and EPA, both projects involve extraction of large amounts of water. In Metropolitan's case, this water would be replaced, either before or after its extraction, with recharge of an equal amount of imported water. EPA has considered a variety of water disposal options, including local distribution (displacing groundwater production elsewhere in the basin), and recharge. If the location and design of Metropolitan's extraction system were made consistent with EPA's by centering it on the Baldwin Park area of contamination, Metropolitan's project could be designed to meet the objectives of both agencies.

Water Supply

The San Gabriel Basin contains one of the most productive aquifers in Southern California. Its immense size --almost 10 million acre-feet of storage-- coupled with high permeability and natural recharge make it an ideal natural reservoir providing benefits far exceeding the needs of San Gabriel residents. Through expert management of recharge and extraction operations, the basin has successfully met these local needs for decades, even through some of the region's most severe drought periods. Nonetheless, to date these water management practices have used only a small fraction of the basin's available capacity. If substantial additional imported water could be made available to replace extractions, the working storage of the basin could be increased to help meet the needs of a much larger region.

However, increased use of the basin's storage would result in greater fluctuation in water levels, which would decrease water production costs when levels were high, and increase costs and potentially curtail production at current facilities when water levels were low. Mitigation of these effects would be required to avoid impacting the current uses and benefits of the basin.

From a water supply perspective, the project provides a variety of benefits in addition to the availability of substantial additional storage. For example, the containment and removal of contamination will allow for the continued use of groundwater by local purveyors to meet overlying needs. If contamination were allowed to continue to spread, eventually these needs would have to be met with imported supplies, or through very expensive wellhead treatment. Increasing the use of storage available in the San Gabriel Basin is a step toward maximizing the use of local resources and reducing the region's dependence on direct supplies of imported water.

Groundwater Contamination

VOC contamination in groundwater in the Baldwin Park area extends vertically throughout the horizons currently used for groundwater extraction (up to about 600 ft depth). Groundwater contamination has also been detected to near the bottom of the aquifer, which ranges in depth from 500 to over 2000 ft. The original number of individual sources of VOC contamination is unknown, but could range into the dozens and potentially hundreds of individual facilities at which surface spillage occurred. It is likely that much of the contamination originally introduced remains in an undissolved state that will continue to dissolve and contaminate groundwater for many decades to come. Thus, EPA's current objectives in this area are limited to interim measures that will minimize continued migration of dissolved contamination away from this area into less contaminated or uncontaminated areas. The amount of groundwater extraction recommended by EPA is the minimum amount considered necessary to control migration.

Metropolitan's proposed project involves an eventual extraction rate of 100 million gallons per day (mgd), which is more than twice that required by EPA. Groundwater modeling evaluations indicate that this amount of extraction would exceed EPA's objectives of containment providing a further benefit of removing large amounts of contamination. EPA is evaluating the relative merits of this project along with other projects designed for remedial purposes only, with no water supply benefits.

Proposed Project

The San Gabriel conjunctive use project will involve expanding use of the basin by extracting up to 100 mgd of groundwater from the middle of the most extensive zone of VOC contamination in the basin. This water would be replaced through recharge of imported surface water during times of excess supply (e.g., most winter seasons). This would provide a source of water when imported supplies are limited during periods of peak demand, drought, or natural disasters.

The project consists of ten extraction locations, each made up of several wells. These locations have been situated throughout the contaminated zone, which is approximately six miles long and almost two miles wide. Their locations have been chosen to optimize containment of this contamination, while extracting from the most highly contaminated areas to maximize removal efficiency. Extracted groundwater will be piped to a central treatment facility that will remove contaminants to concentrations below standards for distribution within Metropolitan's service area.

Treatment will probably consist of both air stripping towers and liquid-phase granular activated carbon (GAC) for VOCs. Ion-exchange units will be added as needed for nitrate treatment. Air emissions from the stripping towers will be treated with vapor-phase GAC. The treated water will be delivered through a six-foot, pressurized pipeline to a regional feeder about a mile away.

Although the ultimate project is expected to achieve a capacity of 100 mgd, several smaller projects will likely be implemented as interim phases. Early project phases will probably include significant redundancy in the treatment process to compensate for uncertainty in both influent contaminant concentrations and process train performance. Data gathered during early phases will be used to help optimize the design of subsequent phases. The response of the aquifer to increased pumping will also be monitored to refine conceptual and numerical models of the groundwater system and the design of the 100-mgd extraction system.

Strategies for Developing Major Capital Facilities

Walt Anton(F), Rosemary Menard, Dave Hilmoe(M),
Jay Laughlin[1], Bob Ellis[2], Kristie Langlow[3],
David Every[4], Chips Barry[5], Tim Block[6], Jim Goetz[7]

Abstract

Siting, permitting, and mitigating impacts present many challenges during the development of major capital facilities. As a result of the need for public involvement, environmental review, wetland restrictions, regulatory agency permitting, as much effort is generally required for these activities as for project engineering and financing. By adopting a strategic approach to organizing the project development process, project proponents can expect to encounter fewer surprises and to receive timely project reviews and approvals. Recommendations for collaborative public involvement approaches and for dealing with wetland issues are described. Experiences during the development of four major water and wastewater projects are described.

Introduction

Strategies for effective collaborative public involvement and for dealing with environmental issues, especially wetlands and associated regulatory agency requirements, are as important to the successful development of major capital facilities as are project engineering and financing. This session will include presentations on public involvement strategies and wetlands issues and on the experiences of four public agencies who are in

---

[1]Directors/Managers, Seattle Water Dept., Seattle, WA
[2]Project Manager, James M. Montgomery Con Eng Bellevue,WA
[3]Principal, Langlow/Hall Associates, Seattle, WA
[4]Senior Ecologist, Dames & Moore, Seattle, WA
[5]Manager, Board of Water Commissioners, Denver, CO
[6]Project Engineer, Municipal. of Metropolitan Seattle,WA
[7]Project Manager, CH2M-Hill, Bellevue, WA

varying stages of developing water/wastewater facilities. These projects are

Tolt Filtration Plant (Seattle Water Department)
Two Forks Dam Project (Denver Water Board)
Ware Creek Reservoir Project (James City County)
West Point Secondary Treatment Plant Upgrade (Metro/Seattle)

Public Involvement Strategies

The public expects to be involved in making decisions about projects which affect them. They will not tolerate being notified late in the process. By taking the initiative to seek out and use citizen ideas and interest in the initial problem definition stages, a sponsor can build a coalition which will support the project. In addition to early involvement, stakeholders need the kind of in depth information anyone requires before making a decision of major importance. The more collaborative the approach, the greater the opportunity to create a strong base of support for a major project.

Early in the life of a project, sponsors must begin thinking strategically about how to tap different viewpoints in order to reach a solution that will be acceptable to a broad base of support. To achieve that result requires identifying who the stakeholders are, investigating and analyzing their viewpoints, considering how they might participate in the decision to be made about the project and determining the most useful ways to engage their cooperation. The object is to prevent the polarization that results when proponents and opponents reach an impasse.

The benefits of engaging stakeholders as partners in determining what the problem is and how best to solve is illustrated in the development of the Seattle Water Department's Water Supply Plan. A citizen advisory committee was established to help design a long range water supply planning project. Public opinion surveys and situation analyses of critical issues associated with water supply to neighboring markets, conservation requirements and impacts were conducted. Public forum exercises identified and prioritized water resource alternatives, which helped citizens understand and weigh how water supply options might interact with one another and impact growth management for the entire region.

Wetlands Issues

Wetland protection and preservation has become a national issue. It should be no surprise that this issue affects most major civil projects, especially water projects. The effects range from minor costs and schedule delays to fatal flaws in otherwise apparently viable projects, and sometimes wetlands appear to be the scapegoat issue for broad opposition to a project.

The opportunity for innovation in addressing wetland issues is greatly enhanced by early detection of the problem and by proper diagnosis. The symptoms of impending wetland problems are variable and often not obvious to any but the seasoned specialist. There are many who can identify a wetland and yet have no understanding of the impacts, evaluating their importance, or proposing viable solutions to the conflicts. Some solutions that would be easy if applied during the planning or even the design stages are costly or infeasible late in the project development.

One major hurdle that should be considered early is the Alternatives Analysis that the Corps of Engineers must conduct under a provision of Section 404 of the Clean Water Act for projects requiring a wetland fill permit. They must be provided documentation to determine that alternative locations or designs were not available that would have substantially less impact on the aquatic environment, including wetlands. The importance or even the existence of this requirement is not common knowledge.

Mitigation, in the broadest sense of the word, is the area with the most potential for innovation. This must start with attempts to avoid the impacts, and the real opportunities come during the planning and design phases of a project. To be most effective, a wetland impact and mitigation specialist should work with the planners and design engineers. This is also true for minimizing impacts, the second step in the mitigation process. Compensatory mitigation for wetland impacts that cannot be feasibly avoided is the final step. Here, innovation and creativity are often important in negotiating equitable agreements on mitigation requirements under permits.

Experiences with Tolt Filtration Plant -- Seattle

The Seattle Water Department's proposed Tolt Filtration Plant would ensure that the water supply from the South Fork Tolt River meets all current and reasonably anticipated drinking water quality regulations and would improve the reliability of SWD's overall water supply system serving 1.2 million people in Seattle and surrounding communities.

The State of Washington has a long history of extensive environmental protection. To enforce the numerous environmental regulations, federal, state, and local agencies have complex project review and permitting requirements. In addition to these numerous review and permitting requirements, a comprehensive environmental impact statement is required. Since the Tolt Filtration Plant would provide potable drinking water, regulatory approval is required by the State Department of Health. The project is also subject to public review by the general public, wholesale customers, Indian tribes, and other interest groups.

As a result of the need for environmental review, permitting, regulatory agency approvals, and public participation, as much effort is required for these activities as for the project engineering and financing. In addition, to properly address these issues, special strategies must be applied early in the project development phase.

Special techniques and strategies have been applied to the planning of the Tolt Filtration Plant. Special emphasis was devoted to development of a clear set of project objectives and key issues at the beginning of the project. Potential permits were identified and a flexible public participation strategy was developed. All of these items were documented in a Project Strategy and Schedule Report prepared during the first two months of the planning effort. During project predesign, the engineering effort incorporated sensitive environmental constraints such as wetlands. A range of site layout options and mitigation plans were developed to provide flexibility in responding to agency comments. Techniques for project planning included development of a specific permit and approval schedule separate from the overall engineering and construction schedule. A permit handbook will also be utilized subsequently.

Experiences with Two Forks Dam Project -- Denver

The Two Forks Dam Project had been proposed by the Denver Water Board to meet future water supply needs. The area that would be inundated to create the reservoir includes wetlands. The Environmental Protection Agency vetoed this project during the permit process. The presentation will describe how a large municipal utility and its suburban customers are dealing with a federal veto of their water project from a legal, political, hydrologic,

MAJOR CAPITAL FACILITIES 617

and planning perspective. The situation remains extremely fluid politically and is now in litigation.

Alternate methods of supplying the region's water are under consideration, including ground water withdrawal from a large aquifer. The potential for recharging the groundwater aquifer is under study.

Experiences with Ware Creek Reservoir Project -- Virginia

The Ware Creek Project involves the construction of a water supply reservoir, pipeline, and treatment plant to provide an alternative water supply and to meet the growing water demands within James City County, Virginia. The proposed reservoir would inundate wetlands. This resulted in a Environmental Protection Agency veto of this project during the permit process. James City County took this matter to court where they received a favorable finding based on a determination that the James City County had no other reasonable alternative. EPA appealed the district court's decision to the 4th Circuit Court of Appeals. The 4th Circuit Court affirmed that a no less damaging practicable alternative exists. The Circuit Court did grant EPA's request for a 60 day remand during which EPA had an opportunity to veto the project a second time. EPA issued the second veto in March, 1992. The Federal District Court overturned EPA's second veto; EPA has again appealed the District Court's decision.

Experiences with Metro's West Point Secondary Treatment Plant Expansion Project -- Seattle

The Municipality of Metropolitan Seattle (Metro) is under a court order from the Washington Department of Ecology to upgrade Metro's comprehensive secondary treatment quality standards by December 1995. Compliance with this order from Ecology was complicated by a requirement in the Seattle's Shoreline Management Plan that prohibits expansion of treatment plants in the shoreline zone "unless there is no feasible alternative". Over a 2-year period, an evaluation was conducted to assess the alternatives for phasing out the existing West Point Plant and constructing a treatment facility in a new location. In 1986, Metro decided that no alternative was feasible and expansion would occur at West Point.

The environmentally sensitive West Point site is located on the shores of Puget Sound at the edge of Discovery Park, the largest natural park within the Seattle city limits. Panoramic views of the natural environment and the distant Olympic mountain range characterize the plant surroundings. Only one overland access route exists to the plant and that route is through a major residential area and Discovery Park. Thus siting the plant expansion at West Point has been complex and controversial. Following an integrated design and environmental review process, permits have been obtained and construction of this major project is now under way.

The area of the site is half that normally needed for a plant this size. The construction permit conditions require materials and equipment to be transported by barge or over the single, two-lane access road through Discovery Park. No more than 100 truck trips per day are allowed and workers are bussed to the site. Public access must be maintained to the park and beach surrounding the plant.

Since the project must be compatible with the natural surroundings, Metro is creating a striking new public open-space area for visitors, which extends from the adjacent beaches to Discovery Park. When completed, the plant itself will be largely hidden from view. Secondary facilities will be screened by earthen berms and landscaping so that visitors using the new trails along the beach will be unaware of the treatment plant located only a few dozen feet away. Odor and noise control will be provided to minimize impacts on park and beach users.

The West Point Story
Chapter 1: Overview and Planning

John Lesniak, P.E.[1]

Introduction

The West Point wastewater treatment plant is the Seattle metropolitan area's largest wastewater treatment plant, serving a population of over 750,000. The plant, an existing 125-mgd primary treatment facility, is being upgraded to provide secondary treatment with an ultimate capacity of 159 mgd average wet-weather flow and a peak hydraulic capacity of 440 mgd. The project budget is $610 million and represents the single largest investment ever in the protection of Puget Sound water quality.

The planning, siting, and permitting process leading to the West Point project was complex, difficult, and controversial. The West Point site is located in an environmentally sensitive area on the shoreline of Puget Sound adjacent to Discovery Park (Figure 1). At more than 500 acres, Discovery Park is the largest natural park within the Seattle city limits. Panoramic views of Puget Sound, the Olympic Mountains, Mt. Baker, and Mt. Rainier make Discovery Park and West Point an incredible recreational asset to residents in the region. The site has limited access through Discovery Park and the Magnolia residential community. The 32-acre site is very tight, only half the area normally needed for a plant of this capacity.

Other project constraints include a steep, unstable bluff immediately adjacent to the site, nesting bald eagles on the bluff above the plant, eelgrass beds offshore of the site that are protected by state and

---

[1] Project Superintendent, Metro West Point Project Office, P.O. Box 70716, Seattle, WA 98107-4699

federal regulation, Native American fishing grounds on the site's North Beach, and an archaeological site protected by federal and state law. More than 200 permit conditions established by the City of Seattle and state and federal regulatory agencies affect the design, construction, and eventual operation of the new secondary treatment plant.

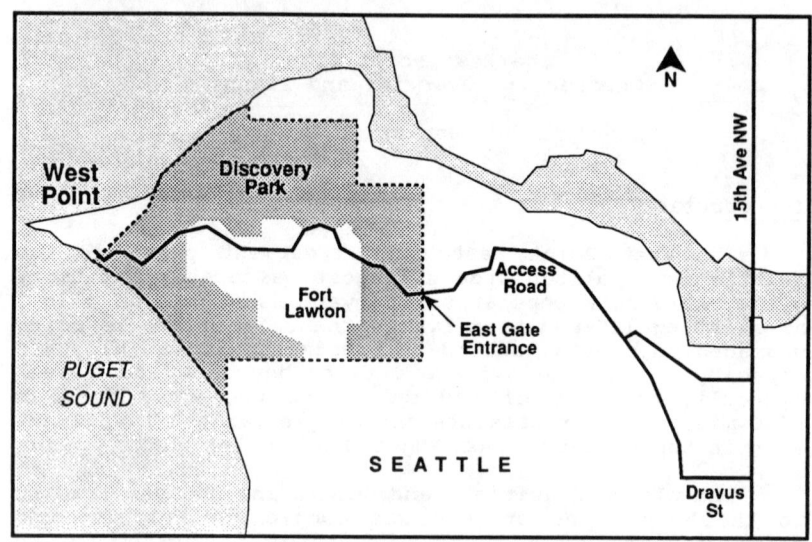

Figure 1. Site Location

Historical Background

The decision to build a treatment plant at West Point was part of a regional grassroots effort to improve water quality in the late 1950s. At the time, the West Point sand spit was part of the Fort Lawton military reservation, and off limits to the public. The Metro Council approved the site as the best location for the plant, not only because of its isolated military location, but also because of ideal currents off the point, a stable soil base, and its proximity to the existing Seattle sewer outfall, the major sewer discharge pipe for the city, built in 1909.

Then, in 1972, when Fort Lawton was surplused by the federal government, most of the land became Seattle's Discovery Park. In that same year, the Clean Water Act

(PL 92-500) became law. Those two unrelated actions set the stage for what was to be Seattle's great land use debate of the 1980s. Figure 2 summarizes the chronology of that debate.

1984 Dept. of Ecology order to secondary treatment	1987-1988 West Point Predesign
1985-1986 Planning Process	1988 Project EIS on West Point
1986 Metro Council Decision to upgrade West Point	1988-1989 Project Shoreline Permit Process (City)
1986-1988 Plan Shoreline Permit Process (City)	1989-1991 Design
1988 City Council Decision on Permit	1990 Construction Begins
	1996 Construction Complete

Figure 2. Chronology

For 10 years, the local scientific community, citizens, and elected officials debated the need and cost of secondary treatment, versus funding needs for what some saw as higher priority water quality projects. Finally, in 1984, all municipal treatment plant discharges to Puget Sound were directed to undergo more advanced treatment. A court ordered consent decree directed Metro to make secondary treatment operational by December 31, 1995.

The Metro Council spent 2 years and more than $3 million studying alternatives. In the planning phase of the project, several alternative sites for new treatment facilities, but, after weighing costs, potential environmental and construction impacts, and other factors such as currents and the proximity of the existing sewer network, the Council voted in 1986 to upgrade the West Point primary treatment plant. That decision was strongly opposed by the Mayor of Seattle, the majority of the Seattle City Council, and citizens and environmental groups who wanted the West Point plant relocated.

Years of controversy followed the decision. Opponents appealed every step of the permitting process. While the design process continued, Metro worked to resolve the permit appeals. The appeals delayed the start of construction by more than a year. Finally, after favorable rulings from the Seattle City Council, the State Shorelines Hearings Board, Superior Court, and the Washington Court of Appeals, Metro and the project opponents began negotiating a final settlement agreement.

While Metro could eventually start construction, the opponents could delay the project for at least 2 more years through continued permit appeals. It was in everyone's interest to settle the differences and proceed with secondary treatment, a goal all parties shared. The settlement agreement and local, state, and federal permits included 200 conditions that affect the design, construction, and operation of the new secondary facilities at West Point.

The conditions imposed on design require that the plant be hidden from view and produce no objectionable odors or noise for park users.

Construction of the West Point project began in May 1991. While construction will continue through the end of 1996, the construction schedule is driven by a court ordered consent decree requiring that secondary treatment be on-line by December 1995.

Conclusion

The layout and compaction of treatment facilities on a small site, the extensive visual screening, park development, and landscaping, and the extensive measures being implemented to control odors and noise represent a new standard for wastewater treatment facilities located adjacent to sensitive neighbors such as a residential community and a heavily used natural park. Metro's experience with innovative approaches to equipment procurement and construction planning and management provides food for thought for others responsible for managing future major public works projects.

Permits have been obtained and construction of the expanded plant is currently underway. Establishment of the goals jointly with the community prior to the start of design and providing a design to meet these goals will allow this major public works project to be completed within 10 years from the start of initial planning. Metro is proud of the program that has facilitated the development of this design process and looks forward to the successful completion of the project in 1996.

The West Point Story
Chapter 2: Project Management Approach
Timothy J. Block, P.E[1]

Facility Planning

In 1985, serving as the program lead, Metro contracted with Lewis Zimmerman Associates to conduct a formal facility planning effort for the West Point wastewater treatment plant. With Metro directing the work, major planning studies followed, including evaluation of more than 200 configurations of treatment plant locations, technologies, and collection system improvements and environmental review under Washington's State Environmental Policy Act. Metro held more than 100 community meetings to discuss the alternatives and the planning process. Seattle's Shoreline Management Plan specifies that expansion of treatment plants in the shoreline zone is prohibited "unless there is no feasible alternative."

To make this determination it was necessary to evaluate alternatives with new treatment plants in detail. The alternatives analysis showed that expansion of the West Point plant to include secondary treatment was far more economical than building a new secondary plant elsewhere, and following an extensive environmental review process, the Metro Council decided in 1986 to upgrade the West Point plant. Although key permitting issues had not yet been resolved, Metro elected to begin preliminary design in order to meet the 1984 EPA consent decree deadline of December 1995. Metro's engineering staff took the lead during preliminary design, directing the work of its engineering consultants, CH2M HILL and associated firms. At the same time, Metro's

---

[1] Engineering Manager, Metro West Point Project Office, P.O. Box 70716, Seattle, WA 98107-4699

environmental planners and scientists began preparing an environmental impact statement to address the numerous environmental issues surrounding the project. At the conclusion of the preliminary design in 1988, the final site layout was selected.

As part of the product deliverables schedule for the preliminary design effort, Metro required a plan-level cost estimate, contract package scenario, and proposed construction schedule from CH2M HILL. To get a second opinion on costs, schedule, and contract packaging, Metro hired a value engineering team that included consultants and contractors from around the United States. The value engineering team issued a report that proposed similar construction sequencing and contract packaging strategies to those proposed by CH2M HILL. Seven basic contracts were suggested by both teams: (1) Site Preparation, (2) Liquids Stream, (3) Solids Handling, (4) Renovation, (5) Administration Building, (6) Marine Outfall, and (7) Landscaping. The contract packages, approximate construction schedule, and contract amounts are shown in Figure 1.

Cost estimates prepared by both groups were $250 million over the budget allocation of $534 million established by the Metro Council in 1986. To bring cost in line, Metro, with its consultants, reviewed everything that was involved in the plant. Major cost reductions were achieved by eliminating two of the eight proposed aeration basins, raising the plant 15 feet out of the ground (which required a new intermediate pump station to pump the wastewater to the secondary process) relocating the chlorine contact channel from along the North Beach to beneath the new secondary sedimentation tanks, and privatization of the oxygen production facility. The completion of this value engineering/cost reduction process yielded a revised version of the site layout, implementation approach, and basis for project-level permits. Subsequently, the Seattle City Council concluded that there was no feasible alternative to upgrading the West Point plant, resulting in the issuance of the plan-level shoreline permit by the City of Seattle and affirmed by the State Shoreline Hearing Board in 1989.

In 1989, Metro began final design of the project, contracting with CH2M HILL for these services. At the peak of design activities, the project reached approximately 200 full-time equivalents (FTEs). Metro staff engineers worked side-by-side with the consultant team throughout the 2-year design period. The Metro engineers were discipline oriented, and provided technical and contractual direction accordingly. During the final design process, there were still extensive permitting response needs and public presentation requirements in which Metro staff took the lead role. Metro then submitted the project-level permit to the City of Seattle. This application demonstrated Metro's Secondary Expansion compliance with the permit conditions established in the plan-level permit. This permit was issued by the City of Seattle in January 1991 and affirmed by the Shoreline Hearing Board on March 1, 1991, with 110 permit conditions.

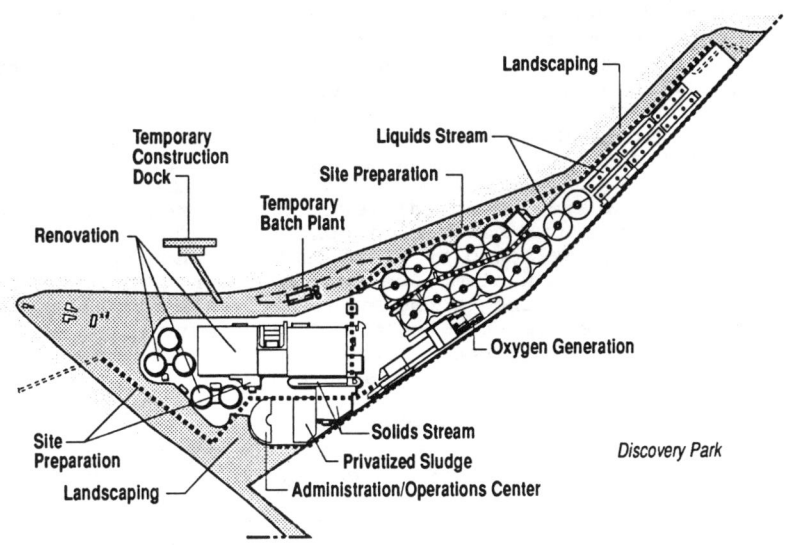

Construction Contracts

Oxygen Generation
$25 million (1994-1995)

Administration/Operations Center
$8 million (1994-1995)

Landscaping
$12 million (1996)

Temporary Construction Dock and Batch Plant
$5 million (1991)

Site Preparation
$75 million (1991-1993)

Renovation
$30 million (1991-1995)

Liquids Stream
$107 million (1992-1995)

Solids Stream
$25 million (1992-1995)

Privatized Sludge
Unit Price (1994-2004)

Not shown

Procurement
$40 Million

Miscellaneous Contracts
$54 Million

Figure 1. Construction Program

Construction

The protracted process for acquiring the various construction permits for the project delayed the planned start of construction, intensifying the intercontractual coordination requirements. In response, Metro pursued a pro-active construction management program that included the implementation of varying contractual methods. Examples of innovative responses to the secondary treatment program's project construction needs are: Complete privatization (design, build, own, and operate) for a portion of the solids handling system; a general contractor-construction manager with a guaranteed maximum price contract for the oxygen generation facility (similar to a turnkey design-build contract); an aggressive owner-furnished procurement program, which consists of 54 separate contracts worth an estimated $60 million; and the traditional lump-sum bid process, of which there are a total of nearly 30 construction contracts that Metro will manage and administer. Metro directs the work of the $30 million renovation contract. The renovation contractor was selected by qualifications and experience, rather than by the lump-sum bid approach because this contractor is working in the existing treatment plant and that it is essential for the existing primary plant to be operational at all times. There are nearly 90 individual contracts worth an estimated $450 million.

Following the acquisition of the project-level permit in 1991, major construction activities began for all work associated with the secondary treatment program. Construction cost average more than $7 million per month. The project's construction manager, construction contract leads, resident engineers, engineering manager, procurement manager, permit compliance officer, and community relations representatives are all Metro staff. The current construction management staff (over 140 FTEs) includes consultant engineering staff for traditional services during construction, and other contract staff for technical and administrative support.

Conclusion

Metro is proud of the program that has facilitated the West Point project to date and looks forward to continued success in its construction program. The project remains on schedule to have secondary treatment on-line in 1995 and construction of the perimeter berms and landscaping elements completed by 1996.

The West Point Story Chapter 3: An Integrated Park and Treatment Plant

Linda Sullivan[1]

Introduction

The West Point Treatment Plant expansion is a microcosm of some of the more compelling elements of today's megaprojects: complex permit requirements, community involvement, extensive mitigation measures to address both short- and long-term impacts, and innovative contracting and project management.

The West Point wastewater treatment plant in Seattle sits on a spit of land that juts into Puget Sound. High on a bluff next to it is Discovery Park, which at 534 acres is Seattle's largest city park. The two were thrown together by geography and circumstance; the bluff and the land above it once housed an Army fort, but were transferred to the city in 1972, with the proviso that they be transformed into a park. During the 1970s, the West Point plant had a delicate relationship with the developing city park. Then in 1984, controversy erupted when Seattle Metro (the agency that provides wastewater treatment as well as public transportation to the Seattle Metropolitan area) recognized that West Point would need to be upgraded to provide secondary treatment to reduce pollutants in the Sound. A court order in 1986 established a tough implementation schedule, but Metro, a design team of more than 200, and an activist local community have managed to come up with a design-construction plan that enhances the park while expanding the plant. Metro carefully considered all treatment options before proceeding with the West Point plant upgrade.

---

[1] Mitigation Manager, Metro West Point Project Office, P.O. Box 70716, Seattle, WA 98107-4699

Sensitive Design

Design goals were daunting. The project not only required fitting additional wastewater treatment processes (from 125 mgd up to peak flow rates of 440 mgd) into an operating plant in half the normal size, but also doing it in a park-like setting.

Construction traffic through the surrounding neighborhood was a critical issue. Many permit conditions contained tight constraints on traffic. Because of public concerns about treatment-plant odors and noise, the project include a complex odor-containment and -control system. Noise in public areas was limited to 55 decibels A-rated (dBA), and the sound level of the plant's mechanical equipment was limited to 52 dBA on the beach.

A major element of Metro's commitment is to create a new 19-acre public open-space park along the beach, doubling the area available to the public today. The new plant will be largely hidden from view by earthen berms, covered by extensive native landscaping. Trails will provide dramatic views of Puget Sound and the Olympic Mountains. Metro has set aside $30 million in a Shoreline and Park Improvement Fund for future shoreline recreation areas in the city and metropolitan area.

Visual Screening and Recreational Improvements

Two major design elements, berms and landscaped covers (lids), were critical to the effort to create a seamless blending of the treatment plant with the shoreline and park. The large earthen berm that will surround the treatment plant will hide it from view of shoreline visitors to the park. By analyzing computer-generated views, the designers were able to "look" along proposed trails to ensure that the berm height, with vegetation, will be sufficient to screen treatment facilities from view. On the treatment plant side, the berm is supported by a reinforced earth retaining wall that averages 18½ feet in height. In areas where horizontal space is limited, the berm will be tall and relatively steep. In wider areas of the site, more level space allows incorporation of wildlife enhancement features such as a one acre fresh water wetland. The berm will incorporate an upland bench wall that will serve as a security barrier and retaining wall. Above the upland bench wall, which will average 6 feet in height, a wide planting area will provide added visual screening from the trails and beaches. Approximately 2 miles of primary and secondary trails will traverse the berms and public area.

Landscaping on the berm is an integral element of the visual screening and effort to blend the facilities with the shoreline and hillside environment at West Point. Eighty-five species of native plants will be used. Low grasses and shrubs will be planted in the zone nearest the shoreline. Shrubs and low trees will be located inland from the shore area. In the "upland bench" area, a dense forest of trees and shrubs will screen the treatment facilities. The native species to be planted were selected primarily for their ability to provide visual screening and provide habitat for wildlife. Metro already has entered into growing contracts with three northwest nurseries to guarantee that the plants are available for landscaping in 1996.

Of the landscaped "lids" and covers that will screen treatment plant facilities, the most critical are at the high-purity oxygen aeration basins, where the site is at its narrowest and is surrounded by a public trail (Figure 1). To maximize both screening and public access to the shoreline near the high-purity oxygen basins, it was essential that the basins have the smallest possible footprint. This was accomplished by designing tanks much deeper than the average, which is typically about 17 feet. By using a 25-foot depth, the designers were able to reduce the footprint and free an extra acre of land for shoreline access. Four feet of topsoil and landscaping will cover the aeration basin tanks, building, pipes, and valves to blend the area into the nearby hillside.

Figure 1. Visual Screening and Landscaped Lids

Traffic Mitigation

Some of the most complex permit conditions attached to the West Point project are those that place constraints on traffic during construction. From the beginning, delivery of supplies to the site and removal of excavated materials has been a challenge due to requirements that minimize heavy truck traffic through Discovery Park and Magnolia. Access to the West Point construction site is limited by a winding two-lane road through Discovery Park. That meant that up to 300 trucks a day might have passed through the surrounding community and park during peak construction periods. Neighbors felt this volume of traffic was excessive. Metro's solution was to build a temporary construction dock at West Point to barge materials to and from the plant, and an onsite concrete batch plant. Presently, those two facilities are reducing construction truck trips by more than two-thirds. An offsite parking area for construction workers, who are then shuttled by bus to the site, further reduces construction traffic through Magnolia and Discovery Park.

Public Participation

To ensure that the mitigation program was responsive to public concerns, Metro and its consultant team met monthly during design with a special Citizens Advisory Committee. Since construction began in early 1991, Metro staff, citizens, and City of Seattle staff have met monthly as part of an official Traffic Management Team to monitor and address traffic issues relating to the West Point project. In addition, Metro participates in a monthly forum with neighborhood activists and staff from the City of Seattle Construction and Land Use, Engineering, and Parks and Recreation departments.

Conclusion

Not until 1996 will the public be able to fully measure Metro's success in delivering on its promise to build a treatment plant the public can't see, hear, or smell. However, the mitigation goals and permit conditions that shaped planning and design of the West Point project influence all phases of construction now underway, and Metro's success at controlling traffic impacts already is noticeable.

The West Point Story
Chapter 4: Design

James G. Goetz, P.E.[1]

Metro's approach to design was one of integrated environmental and plant design. The project team, with guidance from the community and the city's parks department, established goals before design began.

Selection of the Secondary Treatment Process

To ensure that the new plant would meet technical and economic requirements, as well as comply with the various permit conditions, Metro initiated an exhaustive evaluation of technical alternatives. Factors such as reliability, operability, cost-effectiveness, land area requirements, visual impacts, and potential odors were considered. Following initial process screening and detailed layouts, the high purity oxygen activated sludge (HPO) process was selected. The HPO process provides better odor containment and control; allows the basins to be shallower than would be required with the air-activated sludge process; has a low profile that was less obtrusive visually than that of the TFSG process; and features an encapsulating concrete slab over the HPO basins that provides an opportunity for additional landscaping and enhanced visual screening.

Footprints and Rubber Bands

Permit conditions set a limit of 32 acres on the area that the aboveground facilities could occupy. Furthermore, no more than 6.1 acres of that could be in the shoreline zone (a 200-foot strip inland of the high tide

---

[1]Project Manager, CH2M HILL, P.O. Box 91500, Bellevue, WA 98009-2050

line). Specific techniques used in the West Point layouts to minimize the plant footprint were to:

- Maximize the distance of the expanded facility from the shoreline area by constructing the new facilities along the upland hillside property line adjacent to the park.

- Stack process units to the maximum extent practical.

- Minimize space for solids processing facilities by consolidating dewatering and thickening into a single, multi-story building.

Description of Facilities

Key elements of the secondary treatment plant expansion at West Point are shown in Figure 1 and include the following.

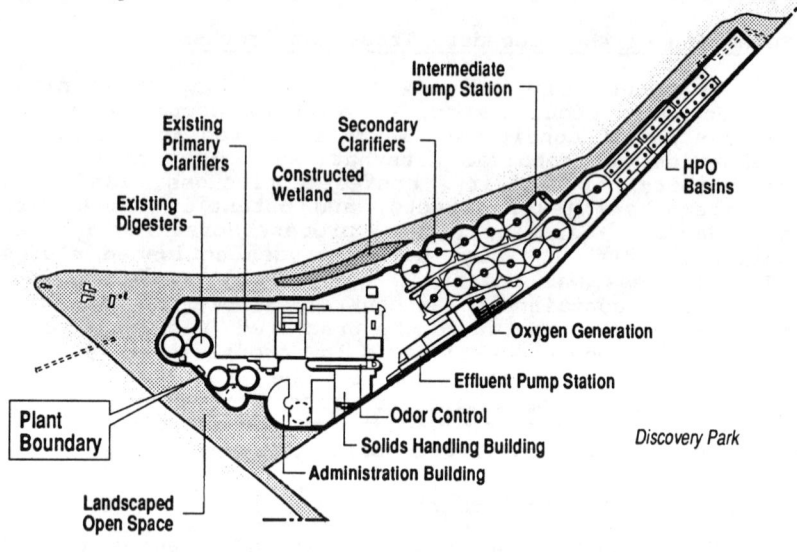

Figure 1. Site Plan and Plant Boundary

HPO Aeration Basins will be located on the panhandle, the narrowest portion of the West Point site. The HPO basins will have 25-foot side water depth, resulting in a footprint saving of 34,850 square feet (or approximately 0.8 acre) over a typical HPO basin of this capacity.

Secondary Clarifiers. Metro considered three secondary clarifier configurations: Stacked rectangular tanks, rectangular tanks, and circular tanks. Circular tanks were selected because of their superior operational characteristics compared to stacked rectangular tanks and their lower capital and operation and maintenance costs compared to rectangular tanks. However, circular tanks typically occupy more space--about 9.6 acres as opposed to 8.1 acres (rectangular tanks) and 6.2 acres (stacked rectangular tanks)--for a plant such as that at West Point. Therefore, considerable effort was given to reducing the area required by the clarifiers to 8.9 acres. This was achieved in three ways: (1) The space allocated for appurtenant structures such as channels, roads, and walkways was reduced as much as practicable, (2) the structural clearance between the tank and channel walls and the gallery was reduced, (3) space in the cleavages between adjacent structures was used for flow splitting structures, pump stations, and blower facilities.

As a result, 62 percent of the clarifier footprint at West Point will be actual tankage, whereas the footprint at a comparable plant (Metro's Renton treatment plant) is 45 percent tankage.

Solids Processing. The selected alternative calls for Metro operation of the existing digesters and onsite processing by a private vendor using a sludge drying technology. If this technology proves successful, Metro will explore the feasibility of processing the entire West Point plant solids in that manner and ultimately abandoning the digestion and dewatering system now in use, thereby freeing up an additional 3 acres of the site for public access.

Other Facilities. Many of the new plant structures and conduits will have common walls, and some will be stacked to reduce the amount of land that they occupy. A raw sewage line, a primary effluent line, and a pipe gallery will be located below the grade of the perimeter berm; this excluded them from the calculations of plant acreage. Some control centers and galleries along the east side of the plant will be buried by extending the existing grade of the hillside. The taller structures such as the solids handling facility, the effluent pump station, maintenance building, chlorination facility, waste gas incinerator, and odor control facilities will be clustered together away from the shoreline. Designers were then able to screen these facilities as a unit.

Hillside Retaining Wall

In order to create public shoreline access and reduce the visual impact of the plant, plant facilities will be clustered against the hillside property line, with the tallest buildings closest to the hillside. This required Metro to demonstrate that stability of the hillside will be improved, but with minimal disturbance of park property. As a result, a 3,000-foot-long tieback retaining wall is being built. The wall will range in height from 35 feet to 65 feet and cost an estimated $16 million.

Odor

Permit conditions require a maximum of three odor units at publically accessible areas around the plant. The approach to odor control incorporated both source control to minimize odor generation and scrubbing of odorous air. Exposed wastewater surfaces in the headworks and primary treatment facilities will be covered to contain odorous air. A centrally located two-stage scrubbing system will provide treatment of the air prior to discharge. Solids thickening, dewatering, and sludge truck loading operations will be covered and the odorous air will be scrubbed using a single-stage chemical scrubber. In addition, air from the anaerobic digester pressure relief vents will be collected and scrubbed in carbon canisters prior to discharge.

Noise

Because public park areas surround the plant, permit conditions specified maximum noise levels at numerous points near the plant. The goal was that no plant noise would be discernible to the park and beach user. To address this requirement, noise levels in similar plants were measured; noise levels of specific plant equipment were estimated; and a computer model was developed to predict sound levels from all potential noise sources. This information was used to modify specific features of the design to reduce noise levels at the source.

Conclusion

Integration of the environmental and technical requirements early in the design process allowed the establishment of sensitive and realistic design criteria. The results are a treatment process, plant layout, odor and noise control that will provide cost-effective secondary treatment and meet or exceed the project goals.

The West Point Story
Chapter 5: Construction

Jim Benedict[1]

Introduction

As design of the West Point secondary treatment plant neared completion, Municipality of Metropolitan Seattle (Metro) staff turned their attention to constructing the facilities. The challenge was enormous: build a complex wastewater treatment plant on a peninsula in a park with limited access and keep the existing primary treatment facilities operational throughout construction, comply with over 200 permit conditions, and complete the work by December 1995. To make this happen, the Metro construction team approached their task with a flexible attitude and a willingness to innovate when necessary. This paper describes some of the strategies used during construction of the West Point plant by focusing on preconstruction activities and on two major contracts: the Site Preparation and the Renovation contracts.

Preconstruction Activities

In late 1989 Metro was experiencing design delays because design cost estimates were well above the project budget. Negotiations for permit requirements and conditions were also taking longer than planned. Meanwhile, the amount of time Metro had to complete the upgrade by December 1995 was getting shorter every day. To lessen the impact of these delays, Metro decided to consider prepurchase of long-lead-time equipment and materials.

First, Metro considered long-lead items or items that would be needed in the front-end of projects, like a

---

[1] Construction Manager, West Point Project Office, P.O. Box 70716, Seattle, WA 98107-4699

temporary administration building, gravity belt thickeners, and nursery stock. Metro also decided to prepurchase equipment like large flow control gates and oxygen dissolution equipment to maintain consistency across contract lines. Finally, Metro decided to preselect equipment and establish prices for contractor use. Examples of equipment in this category included programmable logic controllers and a concrete batch plant.

To inform the bidding contractors of these actions, numerous references were included in the contract documents describing the prepurchased and preselected equipment. The equipment involved in the contract was listed in a document that described the actual equipment prepurchased or identified the contract and unit prices on the preselected items so that contractors could prepare their bids.

Contracting Strategies

With the completion of the permit cycle, Metro had four critical construction contracts to complete to meet the court order completion date of December 1995. Strategies used to keep two of these contracts, site preparation and liquids stream, on schedule are described below.

SITE PREPARATION The original schedule called for starting work in June of 1990. However, the first contract package, the site preparation contract, was not ready to advertise for bid until December 1990. The City of Seattle had not approved the project-level permit in December 1990, so Metro began planning for the possibility that the project-level permit might be appealed. At this point Metro, had three choices: wait until the full appeal process was over to bid the job, use the interim period for planning and start the job as soon as the appeal process was over, or delay the bid.

The project team opted to use the interim period for planning. The effect of delaying the site preparation contract was analyzed, and two strategies resulted: (1) minimize the impact of delayed site preparation activities on another critical contract, Liquids Stream; and (2) have the site preparation contractor ready to begin work when the permit was issued.

First, the two critical contracts-Site Preparation and Liquids Stream-were overlapped. This was done by requiring the site preparation contractor to complete the construction access road within 12 months of notice to proceed. This allowed access to the east end of the site, which was the area farthest away from the access road. The second milestone was to complete the east end of the site excavation approximately 16 months from notice to proceed. This would allow award and construction of the high-purity oxygen building and three secondary sedimentation tanks to begin on time.

Second, to ready the site contractor, Metro developed a planning period to prepare schedules and submittals after selection of the lowest responsive bidder, but before award and notice to proceed. This allowed Metro to share information about the project with the contractor, explore value engineering ideas on how to build the project better, develop the construction schedule, and review critical submittals on the retaining walls, shoring, and dewatering. During this time, the City of Seattle issued the project-level permit. This issuance was followed by a 30-day appeal process during which no appeals were received. Following this, Metro signed a settlement agreement with the opponents, thereby allowing construction to begin.

M. A. Mortenson was the lowest responsive bidder for the site preparation contract. Metro staff went to Mortenson offices in Minneapolis and met with senior staff to discuss this pre-award activity. After discussions about the project and Metro's ideas, Mortenson agreed to participate in the pre-award activity. Mortenson and some of its key subcontractors met with Metro project and design staff to discuss critical areas and develop requirements for the planned deliverables.

This pre-award activity was a success for both Mortenson and Metro: good communication between the owner and contractor was established and the project started on time. The Metro Council subsequently authorized the contract award to Mortenson on May 2, 1991.

RENOVATION was the second critical contract. This contract was to upgrade the retained primary facilities to the requirements of increased plant flow and secondary treatment needs. This project is of long duration-more than 4 years. It is a remodel and requires working around an active process. Some of the items of work were only one-time but happened throughout the contract life. Finally, after design was completed, the plant operators requested additional changes.

To make this contract a reality, Metro modified its method of contracting and issued a combination consultant-construction contract. The team has experience in project management, scheduling, mechanical, electrical, instrumentation, and process engineering of active process facilities. The team provides leaders in each of the six fields who will both work with Metro and direct the mechanical, electrical, and instrumentation work required for the renovation. The team will propose overhead fee which will be added to mechanical, electrical, and instrumentation work. The mechanical, electrical, and instrumentation work will be negotiated by work packages issued by Metro over the life of the contract. The team will also propose overhead fees for all subcontracted work.

Partnering

Metro does its own construction management and is playing the role of general contractor for this project. Formal partnering with construction contractors has set the stage for cooperative rather than adversarial approaches to managing construction contracts on the job. Metro is procuring the owner-furnished equipment and providing engineering support, construction management and start-up services from a team consisting of Metro employees and consultant staff led by CH2M HILL. This team is responsible for managing the 78 procurement and construction contracts that make up the project and is responsible for delivering the project within design quality specifications, on time, and within the project budget.

Conclusion

The Metro construction team spent over 2 years planning, changing, and adapting ideas and sequences into a workable plan that will result in what the politicians call the "World Class Treatment Plant" by December 1995.

During this process Metro will be using the path of teamwork with the contractors so to meet our ultimate goal of secondary treatment by December 1995 and create an extension to Discovery Park that will enthrall visitors who come to enjoy a panoramic view of the United States' newest estuary, Puget Sound.

Upper Guadalupe River Authority, Texas
Innovative Solution to Long-Term Water Needs

Paul D. Thornhill, P.E.[1], Robert Adams, P.E., D.E.[1],
John McLeod, P.E.[1], and B. W. Bruns, P.E.[2]

Abstract

Through conjunctive use of groundwater and surface water, the Upper Guadalupe River Authority (UGRA) has designed a water supply system for the City of Kerrville, Texas that saves 80 percent of the original project cost and provides for a 73 percent reduction to impacts on flows in the Guadalupe River while avoiding Federal permitting. In spite of providing an innovative solution to an age-old problem, with minimal environmental impact, opposition to the project threatens it's implementation.

Introduction

Since 1975, UGRA and the City of Kerrville have been seeking a long-term solution to the water supply problems for the City of Kerrville. UGRA is currently before the Texas Water Commission (TWC) seeking approval of an innovative system combining surface water, conjunctive use of native groundwater, artificial groundwater storage and recovery (ASR), and aggressive conservation and reuse commitments. Further, previously anticipated expansions of the water treatment plant have been indefinitely postponed. Studies indicate the proposed system will have a 73 percent less impact on flows in the Guadalupe River, and no facilities requiring Federal permitting must be constructed. The solution to this age-old problem of providing a firm municipal supply for the City of Kerrville is an innovative example which used a mix of implementable/permittable resources; made use

[1]CH2M HILL, 5339 Alpha Rd., Dallas, Texas 75240; [2]Upper Guadalupe River Authority, P. O. Box 1278, Kerrville, Texas 78028.

of existing infrastructure and avoided expansion of same; provided enormous cost savings - 80 percent; provided a superior alternative over the next-best feasible "conventional" alternative; and, has pointed out the need for water resource planners and engineers to be more creative, especially given the lack of Federal grant funding and while facing extreme regulatory constraints.

The Proposed Project

The proposed project is based on the concept of aquifer storage and recovery (ASR) being used to supplement native, naturally recharging groundwaters in conjunctive use with diversions from the Guadalupe River. The use of underground storage of treated surface waters in lieu of a surface reservoir eliminates a substantial portion of the cost, environmental impact, and all of the evaporation loss associated with the surface reservoir. Conjunctive use of the native, naturally recharging groundwaters further minimizes the net required diversions from the surface source of supply (the Guadalupe River), and also eliminates mining of the native groundwater levels, which had occurred historically.

The proposed project calls for (1) diversion, treatment and underground storage of up to 3,702 acre feet of surface water per year from the Guadalupe River for municipal purposes; (2) diversion, treatment and storage underground of up to 1,058 acre feet of water per year for refilling of the depleted storage after drought periods; (3) continued conjunctive use of 412 acre feet per year of native, naturally recharging groundwater for the City of Kerrville and 1,848 acre feet per year of native, naturally recharging groundwater for Economic Development Corridor (EDC) municipal uses; (4) construction of new injection/recovery wells, or rehabilitation of existing City of Kerrville municipal wells, for use as both injection and recovery points for treated waters; (5) delay in construction of water treatment plant expansions for at least 10 years.

Demands

A planning horizon of Year 2040 was used. Gross demands for Kerrville and the EDC are 12,218 acre feet per year.

Table 1 provides a summary of existing and Year 2040 municipal demands and supplies for the region. Demands for water were allocated such that 100% of the available yield of the regional groundwater aquifers (the Upper, Middle and Lower Trinity formations) was assumed to be used. Remaining unmet demands were assumed to be provided by existing and future surface water supplies to be developed and firmed up using the conjunctive use system. Net new water supplies required to meet Year 2040 demands for Kerrville are 1,489 acre feet per year and EDC demands are 2,213 acre feet per year, for a total of 3,702 acre feet per year.

LONG-TERM WATER SOLUTION

Sources of Supply

Starting with a gross demand of 12,218 acre feet per year for municipal uses in the Year 2040, 1,678 acre feet per year is available from a 15% reduction in per capita usage to be implemented between 1992 and 2030. An additional 975 acre feet per year reduction occurs from the elimination of the off-channel reservoir and the subsequent elimination of evaporative losses. Net demand after elimination of these two items is 9,565 acre feet per year.

Existing surface water supplies available to the City of Kerrville amount to 3,603 acre feet per year firm yield from an existing on-channel dam and reservoir. Groundwater availability includes 412 acre feet per year for the City of Kerrville and 1,848 acre feet per year for EDC users. Note that the EDC use assumes the conversion of 1,027 acre feet per year of existing groundwater use for municipal purposes to surface water use. This conversion allows the shallower aquifers to meet the regional manufacturing and irrigation uses. All municipal groundwater use is expected to come from the deepest aquifer, the Lower Trinity, also known as the Hosston-Sligo Formation.

New firm supplies required to be met from the proposed project are 1,489 acre feet per year for the City of Kerrville and 2,213 acre feet per year for the EDC, a total of 3,702 acre feet per year.

Permitting/Environmental Requirements

The Texas Water Commission (TWC) is the state agency responsible for evaluation and permitting of requests for surface water diversions within the State of Texas. A permit application has been filed with the TWC, and the administrative hearing process is underway at this writing.

No facilities requiring Federal permitting under Section 404 or Section 10 are required.

Groundwater pumpage is essentially unregulated in Texas. The right to pump groundwater lies with the overlying surface landowner. Some minor regulatory powers fall to various districts regarding water quality and pumpage. Such a district has been created in the Kerrville area (Kerr County), but as yet no rules or regulations have been promulgated. The City of Kerrville has ordinance authority requiring permitting of wells drilled within the city limits. This ordinance is silent with respect to potential reasons for denial of such a permit. Thus, groundwater remains and is expected to remain subject to the right of capture in the Kerrville area.

Thus, regulatory requirements covering the diversion of water from the Guadalupe River are solely concerned with surface water impacts, and with the ability to recover any water injected into the ground. Surface water impacts include prior downstream water rights that must be recognized; instream needs for fish, plants and wildlife; recreational uses; bay and estuary inflows; and, water quality impacts. Opposition to the project has developed at the TWC hearing by the actions of intervenors seeking denial of the requested permit for diversion from the Guadalupe River, or the imposition of extremely high pass-through flow requirements. The outcome of this contested proceeding is not known.

Costs

Table 2 provides a cost comparison between the proposed project and the next least costly feasible alternative, an off-channel reservoir. The costs of the proposed project are minimal, especially when compared to the next least costly feasible alternative, i.e., an off-channel reservoir. Total costs for the full Year 2040 demand project include two 2.5-mgd expansions of the water treatment plant, expected to cost $1.45 million each (1992 dollars), and $85,000 per well for rehabilitated wells or $400,000 per well for new wells. Each well has a useful capacity of 0.5 mgd (based on field injection/recovery tests), requiring therefore approximately 17 wells to meet Year 2040 demands 100% from groundwater (as a worst case), at a cost of about $4 million, for a total project cost of about $6.9 million. Permitting and mitigation costs would add proportionately about the same amount to each type of project, such that the proposed project would save 80% of the cost of the next least-costly alternative.

Conclusions

The proposed conjunctive use project involving surface water, naturally recharging groundwater and artificially recharged groundwater provides an environmentally sensitive, innovative solution to the long term water supply needs of the City of Kerrville and the EDC. Evaporation loss that would result from a surface storage reservoir is eliminated. Environmental impacts are minimized or eliminated. Water treatment plant expansion requirements are delayed and reduced. Costs are reduced by 80%. Conservation and reuse play an integral part of meeting future demand. Regulatory requirements are minimized, with the need for Federal permitting eliminated. Texas water law still does not support private ownership of stored waters, which requires creation of special districts or the existence and use of a geographically limited and/or slow moving aquifer for storage. Natural groundwater levels can be restored and maintained without the need for mining of groundwater to meet future needs.

Table 1
Demand/Supply — Kerrville and EDC
(Acre feet per year)

	City of Kerrville	EDC	Total
Gross Demands (2040)	7,445	4,773	12,218
Sources of Supply			
Groundwater			
Existing (1990)	412	2,875	3,287
New (2040)	0	(1,027)	(1,027)
Surface Water			
Existing (1990)	3,603	0	3,603
New (2040)	1,489	2,213	3,702
Total (GW and SW)	5,504	4,061	9,565
Implement Conservation and Reuse (By 2030)	966	712	1,678
Eliminate Evaporation	975	0	975
Total Supplies	7,445	4,773	12,218

Table 2
Cost Comparison

	Off-Channel Reservoir	Proposed Conjunctive-Use Project
Reservoir/Pipeline	$ 30.0 million	$ 0
Treatment Plant	$ 2.9 million	$ 2.9 million
ASR Wells	$ 0	$ 4.0 million
Subtotal	$ 32.9 million	$ 6.9 million
Permitting		
TWC	$ 0.5 million(1)	$ 0.4 million
USCE	$ 1.0 million(1)	$ 0
Mitigation	$ 1.0 million(1)	$ 0
TOTAL	$ 35.4 million	$ 7.3 million

(1) Unknown

Small Area Water Demand Forecasting at the Salt River Project

Robert S. Nichols[1]

Abstract

The Salt River Project Small Area Water Demand Forecast (forecast) is designed to provide water resource planners with long-range water demand forecasts disaggregated by user type and geographic area. Projections are based on small area water consumption factors for types of usage applied to specific land-use acreage. The small area nature of the forecast allows demand forecasts to be disaggregated by lateral and canal, as well as by geographic area.

Introduction

In Salt River Project's (SRP) service area, there are essentially four groups of end-users of water; agriculture, municipal and industrial, urban irrigation and contractual obligations. The demands of these groups are dependent on variables such as climatological conditions, the economic situation, population growth, land-use and agricultural cropping conditions.

The water demand forecast enables SRP to plan more effectively for a number of endeavors, ranging from facility operation and maintenance, to water resource planning which will help ensure an adequate supply of water is available for our customers and shareholders. Demand projections also facilitate enhanced analysis of water issues facing SRP, such as potential demand side programs to encourage efficient use of water.

Methodology

Figure 1 summarizes the process by which the water demand

---

[1]Corporate Economist, Strategic Planning, Salt River Project, P.O. Box 52025, Phoenix, Arizona 85072-2025

1990 Water Demand Forecast Process

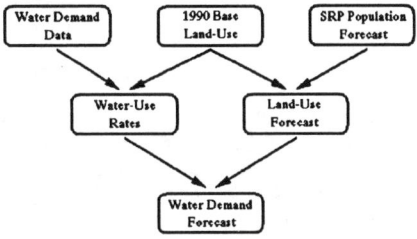

Figure 1

forecast is developed. Small area water demand data is collected and matched with corresponding land-use data to determine base year water-use rates. These water use rates are then applied to projected land-uses generated by SRP's Land-Use Forecast (Borrego, 1992). The details are as follows:

Land-Use Forecast[2]

Forecasted land-uses are derived from SRP's population forecast[3] and base year land-uses that are digitized from aerial photographs. Forecasts are based on land-use relationships between factors influencing development such as access to transportation, proximity to recent growth, land-use restrictions, and compatibility of adjoining uses.

Water Demand Data

Detailed demand data are collected for each class of customer. For the municipal and industrial class, individual customer usage data are collected directly from water providers. These billed consumption data are converted to calendar month usage and assigned to a forty-acre parcel via each customer address.

---

[2] See (Borrego, 1992) for a more complete discussion of SRP's Land-Use Forecast

[3] The population forecast is based on employment and demographic trends such as rising labor force participation rates for women and the aging population.

Data pertaining to agricultural, irrigation and contractual deliveries are taken from SRP's water accounting system. The data provided includes date and amount of delivery and delivery point (canal, lateral, and gate) allowing each agricultural or irrigation account to be assigned to a forty-acre parcel.

Finally, canal, lateral and other distribution loss data are collected and assigned to a forty-acre parcel. The end result is a database of total water demands (including losses) for each forty-acre parcel within SRP's water service area.

Water-Use Rates

Once the land-use and water demand data are collected and geocoded to a forty-acre parcel, average water-use rates are calculated for each land-use using a Geographic Information System (GIS)[4]. Average water-uses rates for each square mile are then assigned to undeveloped land to allow computation of water usage if the land develops.

Water Demand Forecast

Using the GIS, water demands are projected by applying the water-use rates to future land-uses generated from the Land-Use Forecast. The result is a long-term small area forecast (by forty-acre parcel) of water demands for SRP's water service area. The projections are used in a variety of plans and studies. The SRP Water Resource Plan has utilized the forecast to assist in developing recommendations and solutions to ensure that an adequate and reliable water supply is available to meet future SRP shareholder and customer expectations. A 20-Year Well Utilization Plan and a 5-Year Groundwater Action Plan have used the projections to optimize groundwater resources to meet water demands in the long and short term. The SRP Canal Available Capacity Study uses the demand forecast to provide information regarding potential future canal capacity problems.

Future Development

The locational or geographic aspect of SRP's water demand forecast has proved to be very useful for water resource and facility planning at SRP. As currently formulated,

---

[4] The current forecast utilized SRP's in-house GIS, MAPCALC. SRP recently purchased Arc/Info which will be used in future forecasts.

the forecast is based on applying specific usage rates to future growth resulting in a point estimate. To facilitate contingency planning and to incorporate uncertainty into the forecast, it is anticipated that future forecasts will utilize end-use or econometric models for each class. This will allow the development of forecast ranges for each forty-acre parcel.

References

Borrego J. A., "Projecting Future Water Customers Using a G.I.S. Land Use Forecasting Model", Proceedings of a Conference on Water Management in the 90's: A Time for Innovation, American Society of Civil Engineers, May 3-5, 1993.

A Comparison of Short Term Forecast Methods
for
Municipal Water Use

by

Ashu Jain and Lindell Ormsbee
Department of Civil Engineering
University of Kentucky
Lexington, Kentucky 40506-0046

Abstract

Water utilities routinely develop long range demand forecasts for use in the normal planning activities. In recent years, many utilities have sought to develop more short-term demand forecasts for use in daily operations. In the current study three different short-term (daily) demand models are developed and tested for the city of Lexington, Kentucky. Developed models include a lumped response model, a distributed response model, and a neural network based model.

Introduction

Historically, most work in demand forecasting has focused on the development of long term forecasts for use in planning purposes. With the development of optimal control methodologies for water distribution systems has come the need for demand forecasts for much shorter time periods (i.e. daily or even hourly). Most attempts to model short term demand have focused on the use of standard statistical methods (i.e. regression analysis or time series analysis). While such methods will generally provide adequate performance in modeling the cyclic variations observed in a typical water demand series, they do require the prior selection of the underlying model structure and are generally not readily adaptive to changes in future demand realizations. In recent years neural network technology has been proposed as an alternative methodology for use in demand forecasting which avoids these deficiencies.

In the current study, separate demand forecast models are developed using both standard statistical methods and neural network technology. In each case the resulting models are developed using a five year daily data set (1982-1986) for Lexington, Kentucky and then tested using a separate annual daily demand series. The formulation of the three model structures is discussed in the following sections.

Lumped Response Models

The simplest way to forecast daily demand is to develop a functional relationship between demand and those factors which are expected to influence demand such as temperature, precipitation, the demand in the previous day, etc. Such functional relationships may be constructed by using either regression analysis, time series analysis or transfer function models. Perhaps the simplest type of regression model would be to regress daily demand with forecasted temperature. Mathematically this may be expressed as:

$$D_t = \phi_o + \phi_1 T_t \tag{1}$$

where D_t = the demand in day t, T_t = the forecasted temperature for day t, and the ø coefficients represents model parameters.

The simplest type of time series model would be one that correlates the demand in day t with the demand in the previous days. Mathematically this may be expressed as:

$$D_t = \sum_{i=1}^{I} \phi_i D_{t-i} + \epsilon_t \tag{2}$$

where D_t = the demand in day t, D_{t-i} = the demand in day t-i, ø = the model coefficients, and I = the total number of previous days considered in the model.

A third type of lumped response model is a transfer function model. In a transfer function model, the response of the system (i.e. demand) is related to some other independent variable such as temperature. For a simple case of a discrete linear model a potential transfer function for demand may be expressed as:

$$D_t = \sum_{i=1}^{I} \phi_i T_{t-i} \tag{3}$$

where D_t = the demand in day t, T_{t-i} = the temperature in day t-i where I is the number of terms in the model, and the ø coefficients make up the impulse response function of the system. Perhaps a more familiar example of a transfer function is the unit hydrograph. In this case the excess precipitation represent the input, the runoff hydrograph the output, and the unit hydrograph the transfer function.

Distributed Response Models

Because of the inherent periodic (both seasonal and weekly) nature of most demand data, lumped response models tend to provide limited model accuracy. One way to improve model accuracy is to disaggregate the daily series into individual subsets (i.e. based on individual days of the week) and then develop lumped response models for each subset. Alternatively, the series may be detrended and deseasonalized first and the resulting residual series modeled using a lumped parameter model (Steiner, 1984; Maidment, et al., 1985; and Smith 1988). The easiest way to detrend a series is by subtracting the expected demand from the original series where the expected demand can be obtained by fitting a regression model through the historical mean series. The two major approaches for use in deseasonalizing the original series are the use of Fourier Analysis (Maidment and Parsen, 1984) and the use of deterministic sinc functions (Valdes and Sastri, 1989). In the current study a deterministic sinc function approach is used.

An examination of the detrended mean daily water use series for most municipalities will reveal a symmetrical pattern having one peak and two ripples with smaller amplitudes at the two tails. This series can be approximated by a simple trigonometric function called "sinc function". Mathematically, the sinc function may be expressed as:

$$S(t) = a \sin \frac{\left[\frac{2\zeta(t-c)}{b}\right]}{\frac{2\zeta(t-1)}{b}} + d \quad \text{for} \quad t \neq c \quad (4)$$

$$S(t) = a + d \quad \text{for} \quad t = c \quad (5)$$

where: a = amplitude during the summer months, b = period available to half the seasonal cycle, c = horizontal shift of the peak water use, and d = vertical shift of the basic water use. The sinc function is useful for describing the seasonal nature of the daily water use time series because of the nature of its model parameters. The parameters b and c (which have units of time) can be selected explicitly depending upon the observed period of seasonality and day of the annual peak. The remaining parameters (a and d) can then be fit using nonlinear regression.

Neural Network Technology

In recent years several researchers have considered the use of neural network technology in forecasting applications (Crommelynck et al., 1992, Park et al, 1991). Implementation of such technology on a computer involves the development and calibration (training) of the network using historical data and then applying the network in a forecasting application. Conceptually, a neural network model may be thought as a network of nodes and links where the nodes are arranged in distinct sets or layers such that inputs are processed from an input layer through the network to a final output layer. Each node may be characterized as a function which can receive information or output from nodes in a previous layer, operate on that data, and then transmit the response to the nodes of the next layer. In addition, outputs from individual nodes may be adjusted in transmission to other nodes by the use of weighting factors associated with the connecting links. By supplying several sets of independent (input) and dependent (output) data values, the parameter values of the network may be iteratively adjusted until the network can yield the correct response for a specified historical input. Once the network has been calibrated in this fashion, the network can then be used to predict or forecast the response of the underlying physical system for a future set of independent variables. With regard to applications to demand forecasting, the daily demand will be the dependent variable (or output node) and the independent variables (or input nodes) would consist of such factors as temperature, rainfall, previous demand, etc.

References

Crommelynck, V.; Duquesne, C., Mercier, M., and Miniussi, C., (1992) "Daily and Hourly Water Consumption Forecasting Tools Using Neural Networks", *Proceedings of the AWWA 1992 Computer Conference*, Nashville, Tennessee, April 12-15, 1992, pp. 665-676.

Maidment, D. R. and Parzen, E., (1984) "Time Patterns of Water Use in Six Texas Cities", Journal of Water Resources Planning and Management, ASCE, Vol. 110, No. 1.

Maidment, D. R., Miaou, S., and Crawford, M., (1985) "Transfer Function Models of Daily Urban Water Use," *Water Resources Research*, 21(4), April, pp. 425-432.

Park D.C., El-Sharkawi, M.A., Marks II, R. J., Atlas, L.E., and Damborg, M.J. (1991) "Electric Load Forecasting Using An Artificial Neural Network," *IEEE Transactions on Power Systems*, Vol 6., No. 2., May, pp. 442-449.

Smith, J. A., (1988) "A Model of Daily Municipal Water Use for Short Term Forecasting," *Water Resources Research*, 24(2), February, pp. 201-206.

Steiner, R. C., (1984) *Short Term Forecasting of Municipal Water Use*, Ph.D dissertation, The Johns Hopkins University, Baltimore, Md.

Valdes, J.B., and Sastri, T., (1989) "Rainfall Intervention for On Line Applications," Journal of Water Resources Planning and Management, ASCE, Vol. 115, No. 4.

WATER USE PATTERN OF RESIDENTIAL AND COMMERCIAL CUSTOMERS

Pen C. Tao[1]

Abstract

The availability of the Automatic Meter Reading (AMR) devices make it possible to read a large number of meters in a short period of time. Through the use of AMR, we monitored weekly water consumption of several thousand customers. This project, identified who contributed to the base load, and who generated the summer peak demand of a water system.

Introduction

Hackensack Water Company (HWC) and Spring Valley Water Company (SVWC) are two water utilities in the New York-New Jersey Metropolitan area. HWC serves Bergen and Hudson Counties in northern New Jersey, and SVWC serves Rockland County in New York. The served population of HWC is about 0.8 million and that of SVWC is about 0.2 million.

Both HWC and SVWC are equipped with Automatic Meter Reading (AMR) devices to obtain their customers' water consumption data. The AMR system makes it possible to read a large number of customers' water meters within a short period of time and with relatively low costs.

In August 1991, HWC conducted a three-month pilot test program to read 2,500 randomly selected accounts on a weekly interval. The test provided us with information indicating who caused the summer peak demand. Such information is essential for the design of a geographically based water conservation program. It is also needed for the development of a demand oriented rate structure.

Encouraged by the results from the pilot test, we expanded the pilot test program and started a full scale customer weekly water use monitoring program in February 1992. The samples selected for this program consists 5,200 accounts from HWC in

---

[1] Director, System Planning, Hackensack Water Company, 200 Old Hook Road, Harrington Park, New Jersey 07640

New Jersey and 2,700 accounts from SVWC in New York. For both systems, the samples were selected from single family residential, multi-family (2 to 5 units) residential, apartment (more than 5 units) and commercial (business) customers. It is planned to monitor the 7,900 customers' weekly water uses for one full year.

Sample Selection and Weekly Meter Readings

HWC serves about 160,000 quarterly billed accounts. All the quarterly accounts are divided into 56 clusters. A cluster (about 3,000 accounts) is called a Cycle. The meters of all accounts in a Cycle are read on the same day for billing purpose. We randomly selected about 40 samples from each Cycle during the 1991 pilot test. The number of samples were expanded to 80 for each cycle in the 1992 test.

The population of multi-family residential and commercial accounts is much smaller than that of the single family residential accounts. To ensure enough samples for analysis, a total of 1,500 samples of those accounts in HWC were included in the 1992 test.

SVWC serves about 55,000 quarterly billed accounts. Similarly, all the quarterly accounts were divided into 56 clusters or cycles. Each cycle consists of 1,000 accounts. About 40 samples were selected randomly from each cycle for the 1992 test. An additional 500 samples from multi-family residential and commercial accounts were also included.

We used several personal computers to obtain meter readings of the selected samples. Typically, readings began at 10 p.m. and ended about 8 a.m. the next day. For example, during the night of August 3, 1991 we programmed the first set of readings for the pilot test. The readings were completed in the morning of August 4th. The reading date in this case is reported as August 4, 1991.

Normally, an AMR reading contains the first 4 significant digits of the 6 digits on the meter registration for customer billing. The resolution of a normal reading is at 100 CF (cubic feet) or 750 gallons. That amount is too large to be practical for reporting an individual account's weekly water consumption.

A special program was prepared for this project to capture all six digits from the meter registration. When a meter transmits the 6-digit signal to the computer, the last digit (0 to 9 CF) of the signal is transmitted as either an "E" (the last digit of the actual meter registration is greater than 0 but less than 5) or "5" (the last digit of the actual meter registration is equal to or greater than 5). In this study, an "E" is interpreted as 2.5 CF and a "5" as 7.5 CF. Therefore, the maximum error of a reading is at 2.5 CF or 18.7 gallons. That amount is about 1% of the average use of a quarterly billed account in a week.

Weekly Water Consumption

Figure 1 shows the water use distribution of about 2,100 accounts for the weeks of August 27 and October 21, 1991. The first period was a warm week. The average air temperature was 78.7°F and there was no rainfall. The average system draft during that week was at 20% over the annual average. The second period was a relatively low draft week. The air temperature was 56.4°F with no rainfall. The average system draft was 6% below the annual average.

Figure 2 shows the water use distribution of about 400 commercial accounts in the week of February 5 (average temperature at 31.4°F), and July 21 (average temperature at 74.6°F), 1992.

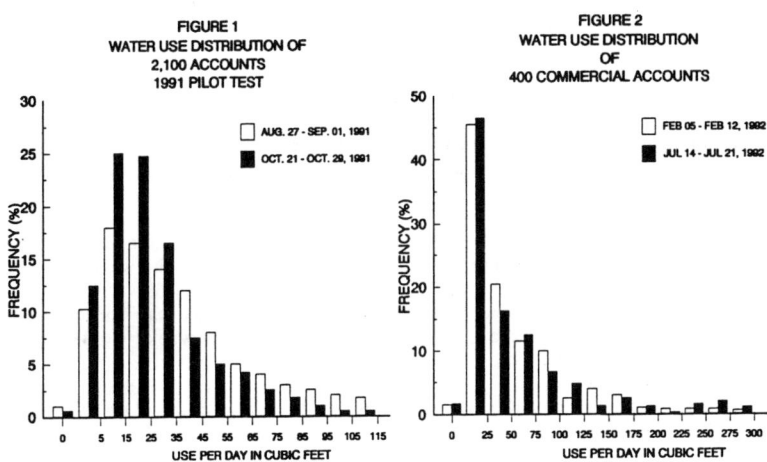

Statistical Analysis

A regression equation was applied to the metered weekly consumptions of the sampled accounts and the total weekly demands of the water system:

$$Q_i = A + B \times D_i \qquad \text{Eq. (A)}$$

where Q_i = metered consumption (CF/day) of the selected samples during the i-th week;
D_i = average system draft (MGD) during the i-th week;
A = constant of the regression equation; and
B = constant multiplier of D_i.

Applying the regression equation to the 1991 pilot test data, the correlation coefficient "R" of Eq. (A) was 0.8974. This indicates that the weekly water use of the sample accounts contributes directly to the weekly variations of the system draft.

Discussion of the Regression Models

In 1991 pilot test the regression model, Eq. (A), was applied to each of the 56 Cycles. For more than 50% of the 56 Cycles, their "R" values are greater than 0.8. Bergen County's less densely populated, affluent communities are the ones with high "R". In contrust, those with low "R" are located in high population density areas.

Among the 2,500 samples selected in the 1991 test, about 2,000 of them are residential, single-family accounts. The "R" value of those 2,000 accounts was at 0.918. On the other hand, the "R" value of the residential, multi-family accounts (a total of 270 samples) was at 0.106.

In conclusion, the residential, single-family customers in the affluent Bergen County communities are primarily responsible for the seasonal load (summer peaking) of the system draft.

Concluding Remarks from the 1991 Pilot Test

The following conclusions were derived from the 3-month, 2,500-point pilot test in 1991. Data collection on a large scale, full year test program is being conducted at this time.

1. The residential, single-family customers in the affluent communities are primarily responsible for the seasonal load (peaking) of the water system.

2. The communities in the high population density areas may contribute a high volume of base load. They cause very little to the weekly variation in the system draft.

3. The residential multi-family accounts may have high demand on base load. They have insignificant contribution to the weekly system draft variation.

4. The telephone cost for this pilot test was $200 per set of readings.

5. The weaknesses of the pilot test were: a) the pilot test period was too short, it lasted only 3 months; b) the test period did not cover the entire warm season; and c) the pilot test sampled about 40 accounts per Cycle. A Larger sample is recommended to make the result more reliable.

6. The framework of the full scale water consumption test, designed in part to correct prior weaknesses, is to be implemented in 1992.

LOW FLUSH PLUMBING FIXTURES AND WASTEWATER SYSTEMS

Thomas P. Konen[1]

Srinivasan Pongavanam[1]

R. Bruce Martin[2]

ABSTRACT

The significance of low flush plumbimg fixtures on the wastewater collection and treatment system serving a small community has been determined. Among the concerns studied were: functional performance of the fixtures, transport of wastes in building drains and laterals, and the operation of the treatment plant. The results of the laboratory and field measurements show 39 percent reduction in water use with no detrimental impact on the collection system or wastewater treatment.

INTRODUCTION

San Simeon, California, is a small town located on the Pacific coast about half way between Los Angeles and San Francisco. Its economy is based almost entirely on tourism - sheltering, feeding and serving visitors to the nearby Hearst Castle.

---

(1) Center for Environmental Engineering, Stevens Institute of Technology Hoboken, New Jersey
(2) W/C Technology Corporation (A Sloan Valve Co.)
    Troy, Michigan

By the end of 1987, the combination of drought and record tourist usage had reduced fresh water supplies to the point that salt water was intruding into the wells. Additionally, peak summer tourism produced flows which exceeded wastewater treatment capacity. Voluntary rationing begun in 1986, had failed. Water consumption had to be dramatically reduced since new (additional) supplies of fresh water and waste treatment would take years to obtain. Thus, faced with having to close (not rent) motel rooms to curb water usage - which would cut directly into their economic base, the Village leaders decided instead on a radical alternate plan - replacement all toilets with low consumption units operating with 1.6 gallons per flush or less.

While other communities had mandated that all new installations be low consumption, flush volumes less than 6 litres (1.6 G), San Simeon was the first community in the nation to require the replacement of existing WC's, as well.

San Simeon, which is unincorporated, has operated since 1961 as a California State Special District under the name "San Simeon Community Services District" ("SSCS"). It has its own water supply and waste treatment systems.

- ** Water is obtained from wells located near the village and about 230 meters from the shortline. These wells are fed by underground aquifers that depend heavily on mountain runoff.

- ** The waste treatment plant is located on the shore and is supplied by gravity flow through 2,500 lineal meters of sewer pipe.

All structures located within the village are connected to both the water and sewer systems.

The conversion to low flush toilets raised concern in three areas:

- Functional Performance of the products

- Transport of wastes in the building drains, laterals and main sewer.

- Operation of the treatment plant with increased concentration of wastes.

FIXTURE PERFORMANCE

All water closets installed in the United States must meet the performance requirements established by the product standards ANSI/ASME A112.19.2 Vitreous China Plumbing Fixtures and ANSI/ASME A112.19.6, Hydraulic Requirements for Water Closets and Urinals. The latter standard provides specific tests and criteria for waste removal efficiency, surface washing and water change.

Since the SSCS was determined to minimize the risk, it was decided to purchase proven products. Accordingly, commitments were placed with area plumbing wholesales for a total of 1,198 Flushmate activated water closets split between American-Standard's Cadet Aquameter^{TM}, Briggs Industries Turboflush^{TM} and Mansfield Plumbing Products Quantum^{TM}. The conversion of San Simeon's toilets was completed in December, 1989.

Subsequent evaluations of this class of product, flushometer tank, have confirmed the performance. This work was conducted under the auspices of the American Society of Plumbing Engineers (1).

EFFECT OF REDUCED FLOW ON THE SEWER LINES

A major concern was what effect would the reduced flow resulting from the new toilets have on waste transfer to the treatment plant? Would blockages occur in the sewer system? To answer these questions, in March, 1992, San Simeon hired Video Inspection Specialists ("VIS") of fresco, California to both clean and inspect (via use of a video camera) the lines. All that VIS removed amounted to only less than 1/2 truckload -- more than two years after installation of the low consumption WC's.

The impact of reduced flush volumes may be discussed best with reference to the discharge curves for the various products, Figure 1. These data shown that although the peak flow rates are similar discharge times are considerable different, fifteen seconds for the nineteen litres (5.0 GPF) units and five seconds for six litres (1.6 GPF) products. The area under the curve is the flush volume. These time dependent flows combine with those from other fixtures and become the driving force in transporting wastes. Calculations (2) have shown the use of low flush fixtures have an incremental effect on the cumulative flow. In addition the variation of flow rate with velocity and depth in circular drains is such that considerable reduction in the flow rate results in lesser changes in both depth and velocity. Analysis are underway to determine the depth of flow and velocity at key points within the system.

RESULTS

While San Simeon anticipated a 20% overall reduction in water usage and the stabilization of its well levels, the actual results of the water closet conservation program exceeded projections:

**   Total community water consumption was reduced by 39% -- almost twice what they had hoped for.

**   Salt water incursion was stopped. Well head levels actually began rising;

**   The economic vitality of the community was preserved. By not having to shut 10% percent of the available rental rooms, the community retained more than $500,000 in annual revenue.

Water usage for the 12 months preceding the conservation (June, 1989) compared with the 12 months following its completion (December, 1991) is given in Table 1.

TABLE 1, WATER USAGE (TWELVE MONTH PERIOD)
(CUBIC METERS)

Segment	Before Retrofit	After Retrofit	Savings Percent
Residential	30.7	19.8	36
Commercial	3.5	0.6	83
Restaurant	14.5	12.5	14
Lodging	100.7	60.2	40
Irrigation	3.1	0.0	100
Totals	152.5	93.1	39

These savings have continued as evidenced by the data in Figure 2 illustrating water and wastewater pumped for the near five year period December 1987 through August 1992.

Besides the immediate results, important longer-term benefits were also obtained:

1. The need to expand the waste treatment plant was able to be pushed back for (at least) seven (7) years -- saving more than $750,000 (capital deferment of $500,000 and an annual savings of approximately $37,500 in debt service or $262,500 over the seven years).

2. A gain in the operating efficiency of t he water treatment plant was obtained because of the higher solid-to-liquid ratio of the incoming effluent. "This resulted in the discharge of a higher quality effluent ...";

The effluent BOD levels for the near five year period are given in Figure 3.

The amount of sludge hauled is shown in Figure 4.

3. Deferral of the immediate need for additional water supplies which meant a significant economic savings to the tax payers -- instead of San Simeon having to expend somewhere between $600,000 and $3,500,000 (cost range of the six alternative plans evaluated), they now have the option to tie on to adjoining facilities currently under construction -- at a fraction of that cost.

REFERENCES

1. American Society of Plumbing Engineers, Research Report 91-01 Water Closet Volume. ASPE Research Foundation, Westlake, CA, 1991.

2. Kannan, Ramesh, Additional Papers, CIB/W62 1992 International Symposium on Water Supply and Drainage for Buildings, Washington, D.C. Available from the American Society of Plumbing Engineers, Westlake Village, CA.

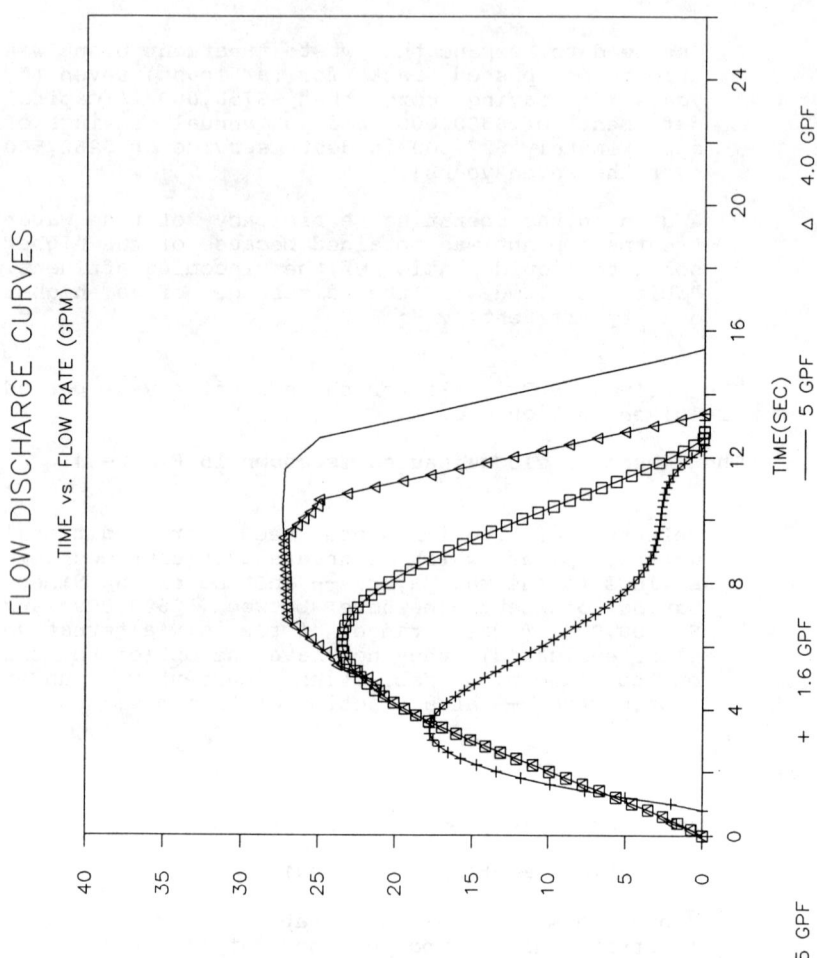

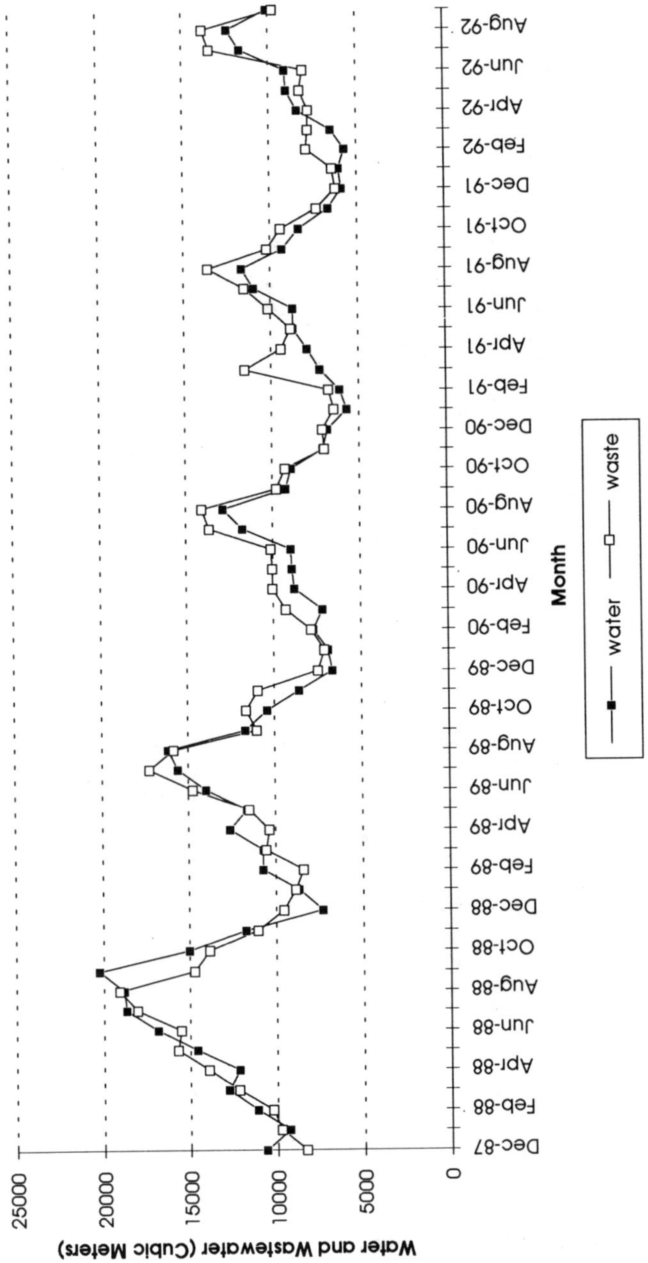

LOW FLUSH PLUMBING 663

Figure 2
Water and Wastewater Pumped

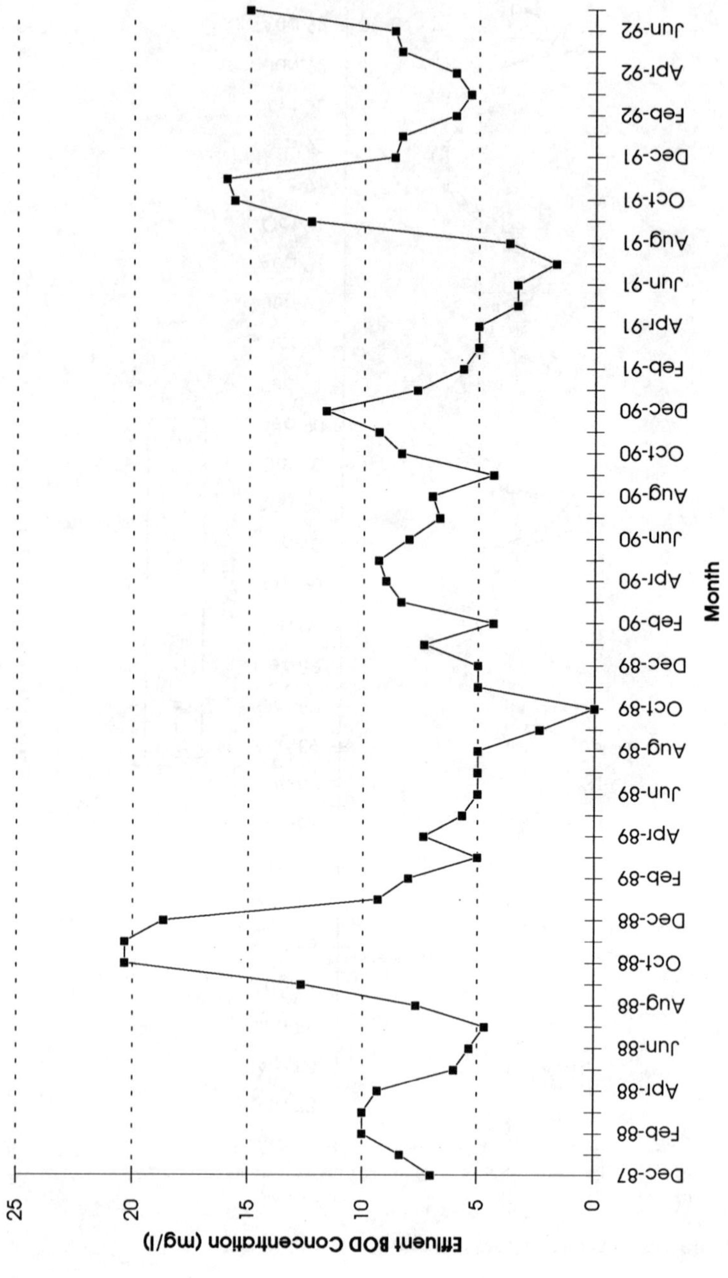

Figure 3 Effluent BOD Concentration

LOW FLUSH PLUMBING 665

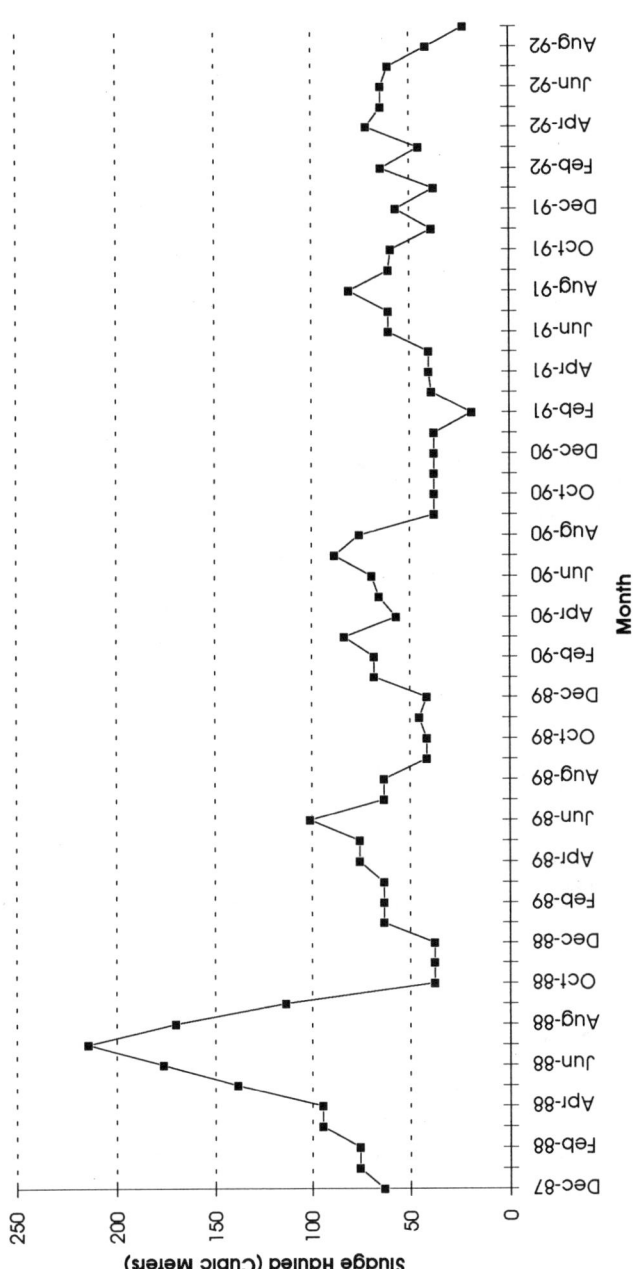

Figure 4

Sludge Hauled

Use of a Statistical Model to
Forecast Future Wastewater Flows

Erick Heath[1]
John Calmer[2]
William Maddaus[2]
Jack Weber[2]

Abstract

Future wastewater flows are generally predicted based upon an assumed rate of population and industry growth within a treatment facility's area of influence. In a community with limited growth, factors outside of population increases may have a greater impact on the quantity of future flows. South Lake Tahoe is a recreational area and experiences flow variations of 2:1 for summer versus winter flows. Therefore a statistical model was developed to evaluate relative impacts of contributions from many factors to changes in flows received at the treatment facility. Conclusions were reached by performing a sensitivity analysis on the factors which contributed significantly to the flows predicted by the model.

A statistical model was created for the South Tahoe Public Utility District (District) using historical data. It was intended that a variable, or a combination of variables, would be found that would have a high degree of correlation with the flows. Anticipated changes in the variables could then be used to predict expected changes in the magnitude of future flows.

Methodology

Prior to setting up and running a statistical model it was necessary to determine input variables and parameters to be used. Variables for consideration were divided into three categories dependent upon the type of flow they were expected to contribute to the WWTP. These categories were:

1. Base flows - Generated from the permanent population.

2. I/I - Generated from sources such as lakes, groundwater, rain, and snowmelt.

3. Non-base flows - Generated from seasonal residents and from visitors.

---

[1]Project Engineer, James M. Montgomery, Consulting Engineers, Inc., 740 University Avenue, Suite 160, Sacramento, CA 95825.

[2]Principal Engineer, James M. Montgomery, Consulting Engineers, Inc. Same address.

None of the flows in the three categories were directly measurable and therefore it was necessary to develop surrogate variables to approximate the magnitude of the flows in each of the categories. Base flows would be related to the number of sewer connections and to the number of permanent residents within the District. I/I flows were not measurable directly as their methods of entry into the system are typically through illegal connections or through defects in manholes or pipe walls. Non-base flows in the Tahoe area were generally related to the number of tourists in the area.

The creation of the statistical model was an evolutionary process. The general approach was to assume that a linear relationship between the WWTP flows and the independent variables existed, and was of the form:

$$\text{Flow} = a + b1 * x1 + b2 * x2 + \ldots + bn * xn + e$$

where "a" is a constant, the "x" terms are the values for the independent variables, the "b" terms are the coefficients that define the change that occurs in the flow related to a unit of change in the independent variables, and "e" is an error term. The error term is the portion of the flow which is not explained by the previous terms.

Transformations were performed on the variables (for example, log and exponential transformations), but in all cases no improvement in the correlation was found. Other variable manipulations were attempted, and the following ones improved the correlation in one or more of the scenarios:

1. Lagged Variables - A variable was lagged by one or more time periods to determine if the variable's impacts were delayed by time. A two time period moving average was also analyzed to evaluate any improved correlation.

2. Seasonal Indices - Seasonal Indices were used to identify any recurring patterns, and were evaluated for both the independent variables and for the dependent variable. It was not necessary to identify the cause of any recurring pattern, only to identify if such a pattern existed.

3. Difference Variables - These were used to evaluate the correlation of the deviations of the independent variables from their normal values, with the deviation of the flow from its normal value.

4. Outliers - An outlier variable was assigned to explain the data points in each of the variables which fell outside the 99 percent probability boundary. It was assumed that these outliers were random and unpredictable.

5. Time Variables - A time variable was used to evaluate the change of a variable with time, and was implemented by assigning a value of 1 to the first period and increasing the value by 1 for each subsequent period. A second type of time variable was used to evaluate the correlation of periods of interest to the flows, for example the periods corresponding to a drought. A value of 1 was used in the time period represented by the variable, and a value of 0 was used for all other time periods.

Case Study

Initially many variables were identified which had potential for a high correlation with one of the categories, and therefore with the amount of wastewater flows. Variables identified were evaluated for availability of data, and the variables for which data was available were tested for independence from both the wastewater flows and from each other. Variables chosen for evaluation with respect to the District's flows were:

1. Precipitation
2. Gross Sales in the City of South Lake Tahoe
3. Inbound Traffic to the South Lake Tahoe Area
4. South Lake Tahoe City Population
5. Number of Rooms Rented in the South Lake Tahoe Area
7. Number of Sewer Accounts in the District

Additional variables were also defined, as described in the previous section, and were evaluated in an effort to achieve a better correlation with the flow data. Dozens of combinations of variables were evaluated and the model which reliably explained the greatest amount of the variation in flow was chosen.

The optimum model developed was based on the average number of gallons of wastewater per day per account (GPDA) on a monthly basis. Other models were developed which explained the variation to a similar degree as the selected model but which did not include a factor for account growth. An account growth factor was deemed necessary for forecasting purposes. The model equation chosen included variables to explain the historical outliers, and an autoregressive variable to explain the serial correlation of errors. A graphical representation of the equation, showing residual, actual, and fitted data, is shown in Figure 1. The equation is expressed as follows:

$$GPDA = 35.70 + 243.08(GPDASI) + 91.90(DP2) + 139.32(LDP2) - 28.39(TM5) + 41.23(OUT10) + 49.89(OUT11) - 29.73(OUT12) + 0.36(AR1)$$

GPDASI is the seasonal index derived for GPDA.
DP2 is the change in precipitation from normal for the current month.
LDP2 is the change in precipitation from normal for the prior month.
TM5 is the dummy variable in flows for the period January 1991 to date. This variable explains the decrease in GPDA which is attributed to conservation.
OUTXX are variables which explain the outliers found in the historical data.
AR1 is an autoregressive term which explains the serial correlation of the errors

A shortened version of the equation was used for forecasting because of the assumption that no outliers and no serial correlation of errors would be present. The forecasting equation is as follows:

$$GPDA = 35.70 + 243.08(GPDASI) + 91.90(DP2) + 139.32(LDP2) - 28.39(TM5)$$

Flow forecasting was done by changing the four variables to bracket the expected range of account growth, anticipated changes in precipitation from normal, and anticipated levels of conservation. For example, if precipitation is expected to be normal, and if conservation levels are expected to remain the same, then GPDA is only related to expected account growth.

Conclusions

It was determined that approximately 80 percent of all variation in WW flow could be explained by a defined normal seasonal pattern, precipitation deviations from normal, a variable to remove serial correlation, and a variable to remove major outlier variations. Including a time variable improved the explained variation to as much as 85 percent.

The seasonal index defines the long term pattern of consumption by month. The coefficients for rooms-rented, east-bound traffic, time, and population suggest in some of the scenarios that a change in flows has occurred during the drought period. The seasonal index has implicitly picked up these changes without defining their cause. The seasonal index should be reviewed periodically, especially if the drought conditions change.

Changes in precipitation cause a relatively minor change in the flows. For example, a large change in precipitation, say a 30 percent increase over average for four months in a row, only increases the flows by three percent.

The conservation variable is a dummy variable which has a value of one in months where it applies. The value should be left at one since some measure of conservation is expected for all the forecast years. If more or less conservation is expected, the coefficient of -28.39 should be changed. The -28.39 coefficient represents a 9.4 percent reduction from flows without conservation. It is evident that conservation, possibly coupled with economic conditions, is the prevailing cause of change from normal flow.

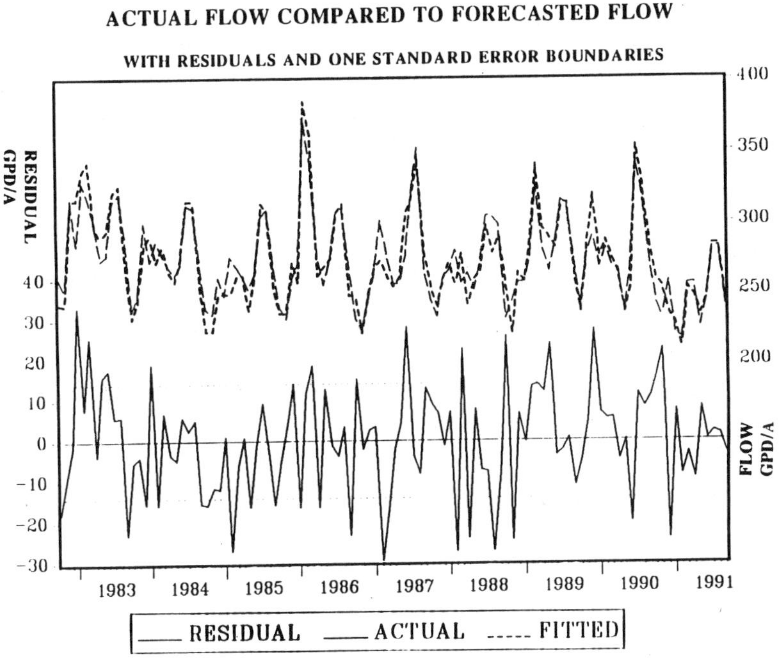
ACTUAL FLOW COMPARED TO FORECASTED FLOW
WITH RESIDUALS AND ONE STANDARD ERROR BOUNDARIES
RESIDUAL GPD/A
40
30
20
10
0
-10
-20
-30
400
350
300
250
200
FLOW GPD/A
1983
1984
1985
1986
1987
1988
1989
1990
1991
RESIDUAL
ACTUAL
FITTED

Figure 1

The Water and Waste water Savings Achieved by
Ultra Low Flush Toilets in Santa Monica, California

Craig Perkins and Susan Munves[1]

Abstract

Since its implementation in December 1989, the Bay Saver
Residential Fixture Rebate Program has replaced over 30,000
water-wasting toilets and shower heads in Santa Monica with
1.6 gallon per flush ultra low flow toilets and 2.5 gallon
per minute shower heads. Upon anticipated completion of the
Bay Saver Program in April 1994, the City will have achieved
a permanent reduction in water usage and waste water flows
of 1.9 million gallons per day (MGD). These 1.9 MGD of
reduced water usage and decreased waste water flows
represent a 15 percent reduction in Santa Monica's average
total daily water demand and a 19 percent reduction in
average total daily waste water flows. It is estimated that
the projected 1.9 MGD permanent sewage flow reduction will
save the City of Santa Monica approximately $9.5 million in
avoided sewage treatment capacity purchases alone by the
year 2000.

Program Background

Implemented in December 1989, the Bay Saver Program
initially targeted a 25 percent retrofit of residential
households within three years, or 19,200 ulf toilets and
shower heads. The 25 percent target was achieved in less
than 2 years. This accelerated pace of retrofits was due,
in part, to the City's implementation of a mandatory 20
percent water cutback with penalty surcharges for exceeding

---

[1] Craig Perkins, Environmental Programs Manager, and
Susan Munves, Conservation Coordinator, work for the
City of Santa Monica, California.

water targets in response to the fifth year of drought. During 1991, the City's water and waste water flows were at their lowest levels in over 20 years. In April 1992, after mandatory cutbacks had been lifted, the City implemented Phase Two of the Bay Saver program which targets an additional 25 percent of residential households (19,200 additional toilets) and an initial 25 percent of commercial/business customers (4,088 toilets) over a 2 year period.

Analysis of Savings

Figures One and Two represent the City's actual water usage and waste water flows from January 1986 through November 1992, and projected usage and flows through April 1994. (Projected completion date for Bay Saver Program is April, 1994.)

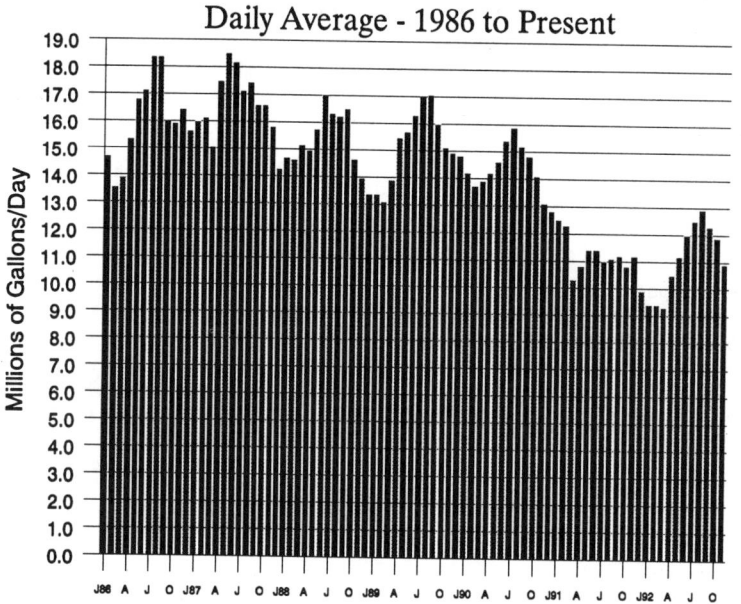

Figure One.

SANTA MONICA WASTEWATER FLOWS

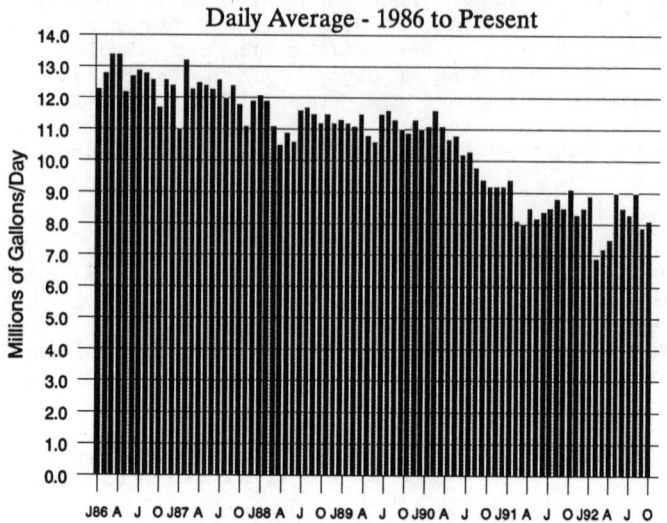

Figure Two.

When a comparison is made solely between the cost of imported MWD water and the cost of water saved through the Bay Saver Program, it becomes clear that when viewed over a 10 year frame of analysis, the retrofit option will result in a significant cost savings to the City. For example, by FY 2000 the projected cost for an acre foot of imported MWD water is $684. By comparison, the cost for each acre foot of water saved under the Bay Saver residential program will be $339, a savings of $345 per acre foot.

Conclusion

The water saved by ultra low flush toilets represents an important alternative for meeting future water needs. Water conservation programs have traditionally been used to reduce short-term demand during drought emergencies, but have only

recently been perceived as permanent water supply
alternatives. As the financial and environmental costs of
developing traditional water supplies become prohibitive,
conservation is now the most cost-effective strategy to
increase water supplies.

In addition to the direct financial benefits accrued through
the Bay Saver Program, significant environmental benefits
have been obtained through a reduction in the use of natural
resources, a decision in the amount of waste water
discharged into the ocean, and a lessening in energy demand
for pumping and treatment. These environmental benefits are
increasingly used in evaluation of demand side management
programs in the energy field. They must likewise begin to
be included in evaluation of water and waste water
conservation programs.

Challenges in Implementing the Use of Reclaimed Water

David B. Parkinson, P.E.[1]
James V. Wodrich, P.E.[2]

Abstract

The State of Washington is currently in the process of adopting standards for use of reclaimed water on land. CH2M HILL recently has finished two feasibility level studies for reuse projects in western Washington. These studies investigated the feasibility of using reclaimed water for irrigation of golf courses, median strips on roads, parks, and landscaping in and around business parks. This paper presents the results of the two studies in terms of economic and technical feasibility and addresses the following questions:

- Why is reclaimed water needed in the wet, cool Pacific Northwest climate?

- What is the future of reuse in Washington?

- How does it fit into the water supply picture for the state?

Introduction

Recent water shortages and concurrent population increases in the Pacific Northwest have raised interest in the use of reclaimed water. In the Pacific Northwest,

---

[1] Regional Water Supply Planning Manager, CH2M HILL Northwest Inc., P.O. Box 91500, Bellevue, Washington 98009-2050.

[2] Project Engineer, CH2M HILL Northwest Inc., P.O. Box 91500, Bellevue, Washington 98009-2050.

many competing uses--hydroelectric power companies, fisheries, wildlife, aesthetics, recreation, and irrigation-vie for the area's water resources along with potable water suppliers. Projections indicate that King, Kitsap, Pierce, and Snohomish Counties in the Puget Sound area will approach their existing water production capacity by the year 2000 if no other measures are enacted. Such meassures could include conservation, the use of interties (pipeline connections between utilities), and the use of reclaimed water (commonly referred to as reuse).

There are economic reasons why communities and industries consider reclaimed water a viable option. The use of reclaimed water for nonpotable purposes can be an economically viable alternative to developing a new water supply source or, in some cases, a means for reducing transmission costs. An example of the latter is a treatment plant located where there are few or no wastewater disposal receiving streams available.

Recent Reuse Feasibility Studies in the Puget Sound Area

In 1992 CH2M HILL completed two feasibility studies investigating the use of reclaimed water for landscape irrigation in two Puget Sound communities: Semiahmoo at Blaine, Washington, and at two master planned developments (MPD's) in Redmond, Washington. These studies evaluated use of reclaimed water to reduce the demand on drinking water supplies and wastewater discharges.

At Semiahmoo we considered using reclaimed water as a source of irrigation water for the expansion of resort facilities, including one existing and three future golf courses. The local communities could not provide the amount of water required to irrigate the facilities and pumping water from the nearby Nooksack River was also deemed infeasible. A new source of water was required, and the idea to use reclaimed wastewater from one or both of the two local wastewater treatment plants serving the Cities of Blaine and Birch Bay was proposed.

The second feasibility study, a combined effort with the Municipality of Metropolitan Seattle (Metro), considered reuse as a solution to the wastewater and irrigation water supply needs at two MPDs in the Redmond area. Metro was interested in reducing wastewater discharge quantities and reducing peak waste flows. The City of Redmond and the Seattle Water Department, wholesalers of water in the area, were interested in studying reuse as a means for reducing the demand for developing new water supplies.

Methodology and Results

Each of the projects was approached in a similar manner, addressing the following aspects:

- Regulatory Treatment Requirements
- Water Budget "Availability Versus Need"
- Storage and Distribution Facility Alternatives
- Economic Feasibility

Regulatory Treatment Requirements. Regulations for the use of reclaimed water in the state of Washington are being developed at this time by a combined effort of the Department of Health, the Department of Ecology and the public. A final set of standards, procedures, and guidelines is due by August 1, 1993. For the purposes of recent feasibility studies, the reuse standards adopted in California and Oregon were used as a basis for the development of treatment requirements. The two standards are similar with a few exceptions regarding buffer distances and public contact issues.

The regulations in California and Oregon are written in terms of both water quality and treatment level requirements, depending upon the type of use for the reclaimed water. The highest level of treatment would consist of oxidation, coagulation, clarification, filtration, and disinfection. California and Oregon allow lower water quality and treatment if the reclaimed water is to be used for restricted access irrigation or irrigation of orchards and some crops. In some instances, primary treatment only or oxidation and disinfection are the only treatment requirements. Additional site requirements generally are required for the lower treatment level facilities such as signs, buffer zones, fences, and night irrigation.

Water Budget. The water budget enables the prospective reclaimed water user to evaluate the amount of acreage that can be irrigated, the storage requirements, and the amount of wastewater that can be treated and reclaimed. A water budget was used as a tool to determine the local irrigation demands, compare these to the available quantities of wastewater in the communities, and determine the amount of storage and treatment required.

Storage and Distribution Alternatives. Storage requirements play a significant role in the development of a reuse project if the wastewater supply is too small to meet the need for reclaimed water on an as-needed basis. An abundant supply of wastewater allows for a decrease in storage requirements. Storage facilities can be taken into account during the design of many facilities. For example, golf courses can be designed to incorporate additional storage needs in their hazards and lagoons.

Distribution system piping can also play a significant role in the feasibility of any reuse project. Locating the treatment system close to the areas that need irrigation water decreases distribution costs. Feasibility level distribution system cost estimates ranged from $6 to $8 per inch-diameter-foot for both studies.

Economic Feasibility

The economic feasibility of reuse projects in the Pacific Northwest depends upon many variables: the extent of upgrade required, treatment level requirements, land availability, proximity to users, land costs, density of user sites, distance from source of wastewater to users, and cost of other options. Typically, new projects are compared to the cost of water to an end-user. However, for reuse the comparison should be between the marginal cost (the cost of the next proposed source of supply) of water minus the savings accrued to the wastewater agency. By completing the comparison in this manner a true cost to the region will be derived.

Conclusions

In conclusion, it appears that under some conditions in the Pacific Northwest it may be feasible to use reclaimed water as a cost-effective alternative to meet some of the future needs for water. In the case of the Semiahmoo study an economic comparison could not be completed because, at this time, there is not other alternative for providing irrigation water to the proposed facilities. For the Redmond study the cost of reclaimed water varied between $1.50 to $5.50 per 1,000 gallons as compared to a preliminary marginal cost of water equal to $2.74. Because of low quantities of available wastewater, in order to irrigate three golf courses in the Semiahmoo area, a total of 180 acre-feet of storage would be required. One-fifth of this storage could be provided in the golf course hazards with the remaining storage allocated to storage lagoons. Water supplies in the Northwest can consider reclaimed water as a viable option in our struggle to provide enough water to everyone. People in the wastewater business can reduce the amount of wastewater discharged to our water bodies in the Pacific Northwest by reusing our wastewater for such purposes as irrigation—a win-win solution.

The City of Los Angeles Gray Water Pilot Project
Shows Safe Use of Gray Water Is Possible

Bahman Sheikh[1]

ABSTRACT

A year-long study of eight residential gray water systems was conducted in the City of Los Angeles. Soils irrigated with gray water were compared monthly with soils irrigated with municipal potable water. All of the soils and the gray water samples tested negative for *Salmonella*, *Shigella* and *Entamoeba hystolitica*. Occasional *Ascaris* positives were reported for samples of soil from control and gray-water-irrigated areas. Fecal coliform and Enterococci (indicators of possible human fecal contamination) were not significantly different in the soils compared. It is concluded that gray water used below the surface for irrigation of landscaping does not pose a significant health risk to the public.

INTRODUCTION

The Gray Water Pilot Project consisted of eight gray water test systems installed at various types of residences, sampled monthly and monitored over a year-long period for safety and water savings. Samples of soils and water were tested for indicator bacteria and pathogens, to compare areas receiving gray water with those irrigated with potable water.

MATERIALS AND METHODS

The test sites were selected from over thirty residences volunteered by owners for participation in the Gray Water Pilot Project. Eight sites were required for the pilot

---

[1] Director, Office of Water Reclamation, City of Los Angeles, 200 N. Main Street, Room 570, Los Angeles, California 90012.

project. Therefore, a selection protocol and criteria were developed to choose the most appropriate locations. Criteria used included diversity of neighborhoods, plant materials, system complexity, topographic conditions, and size and type of homes.

At each of the eight selected sites, a gray water system, usually donated by a manufacturer, was installed. A variety of types of systems were installed to maximize the opportunity to learn about the available variety of systems on the market. One site was a home that had had a gray water system in use for about eleven years.

Drip irrigation was the primary method of application of water in all but two of the sites. The type of drip irrigation system used was the tortuous-path emitter systems, which allow for a fairly wide flow path to minimize clogging. The drippers in each system were monitored to assure proper functioning and to prevent clogging.

For the results of the study to be reliable, it was critical to incorporate certain precautions to avoid bias on the part of project participants, particularly the laboratory analysts. Therefore, a sample numbering system was adopted that was at once logical and cryptic. The sample numbering protocol was not revealed during the course of the project. It encoded site identity, type of sample, and date. A quality assurance/quality control procedure was in place at the analytic laboratory, complete with chain-of-custody procedures for sample handling.

RESULTS

A statistical analysis of the data was performed. For each site, values obtained for each parameter are compared in each of the 12 sampling rounds between control- and gray-water-irrigated areas. In addition, at each round of sampling, values of the same parameters are compared across sites for control- and gray-water-irrigated areas. A summary of the results of statistical comparisons between soils irrigated with gray water and those with tap water appears on Table 1.

Three of the Disease-causing organisms monitored in the sampled soils--*Salmonella, Shigella,* and *Entamoeba histolytica*--were negative at all sites in all sampling rounds, in gray water and in soil--both control and gray-water-irrigated. Apparently, neither the gray water nor the soil carried any of these particular organisms. The fact that throughout the year, none of the samples yielded a positive for any pathogens tested is encouraging for the possibility of safe use of gray water--even where total adherence to hygienic handling of gray water in not assured.

Table 1. Summary of Gray Water Data Statistical Analysis

Parameter	Mean Diff'nce (Xd)	Standard Deviation (s)	Nr of Data (n)	Stand'd Error (Xe)	Test Statistic (Xd/Xe)	Reject* Null Hypothesis? (90 % Conf.)	(95 % Conf.)
Total Coli.	157,321	610,569	98	61,677	2.55	Yes	Yes
Fecal Coli.	16,192	376,566	90	39,694	0.41	No	No
Enterococci	19,080	172,509	96	17,606	1.08	No	No
pH	0.14	0.76	96	0.08	1.75	Yes	No
Sodium	23.52	98.38	97	9.99	2.35	Yes	Yes
Chloride	38.09	266.77	94	27.52	1.38	No	No
Calcium	590.73	4,636.96	96	473.26	1.25	No	No
Magnesium	-11.47	231.66	96	23.64	-0.49	No	No
Spec. Cond.	448.99	3,001.21	90	316.36	1.42	No	No
S. A. R.	0.33	1.33	94	0.14	2.36	Yes	Yes

* If the test statistic is larger than 1.66 or smaller than -1.66, then one can say--with 90 % confidence--that the two sets of data are indeed different. If the test statistic is >1.98 or <-1.98, then the same can be said with 95 % confidence. If the test statistic falls outside this range, then observed differences are probably due to chance. The numbers 1.66 and 1.98 are derived from statistical tables based on the statistical design of the project.

Total coliform was significantly higher in gray-water-irrigated soils than in control soils, at the 95 percent level of confidence. On the other hand, fecal coliform--a measure of possible human fecal contamination--did not appear to be significantly different in the two soils. Sodium concentration and sodium adsorption ratio (SAR) are both significantly higher in gray-water-irrigated soils than in control soils at the 95 percent level of confidence.

pH, sodium, chloride, calcium, magnesium and total salts were measured in gray water and in the soil extract to determine if any of the agronomic characteristics of the soil might be affected by gray water irrigation. For the same purpose, sodium adsorption ratio was computed for each sample from the basic data. As expected, sodium and sodium adsorption ratio were both significantly higher in gray-water-irrigated soils than in the control soils. This points to the importance of using low-salt detergents in homes where gray water is put to irrigation use.

Water savings from the use of gray water systems was estimated based on the potential demand for gray water use at each site. To estimate the water savings, the volume of gray water actually used at each site was calculated as a percentage of the total household water use and extrapolated to the entire landscaped area. This method assumes that enough gray water is available to irrigate the entire landscape area--a valid assumption on the average, but not necessarily for any given household. Calculation of water savings indicates that potential demand for gray water ranges from 13 percent to 65 percent.

CONCLUSIONS

From the results presented above, including baseline data, it is clear that backyard soils are contaminated, whether they are from the control areas or from gray-water-irrigated areas. Therefore, the general sanitary practice of washing soiled hands with soap and avoiding direct contact with the dirt in the yard are as valid for sites irrigated with tap water as those irrigated with gray water. It appears that use of gray water at sites similar to the pilot project sites does not pose a significant risk to the users or the community. Since pilot project sites were controlled, inspected, and repaired as needed, broad generalization of this conclusion may be premature.

However, certain more specific generalizations appear inescapable, e. g.: (1) Indicator bacteria (total coliform) in the soil seem to increase with gray water application. However, the soil is already so heavily contaminated with animal fecal matter that the additional contribution of gray water may be irrelevant. (2) Disease organisms, normally capable of surviving in the soil for a few days, were not present in gray-water-irrigated areas. Neither were these organisms detected in gray water in storage. (3) The water savings potential of a gray water system to an individual home can be significant--about 50 percent of all the water used. However, the savings accrued to a community may be considerably smaller. This is because it appears unlikely that a large enough number of people will install such systems, due to the maintenance requirements, complications with permitting, and cost.

Bellingham Frozen Foods
Spray Irrigation System Operations

Martin Harper, Ph.D, P.E.[1]; Clain Jones[1]; David B. Green[2]

Introduction

Bellingham Frozen Foods (BFF) is a major vegetable processor located in northwestern Washington state. Vegetable products include peas, beans, corn and carrots. Wastewater generated during the processing of vegetables historically averaged about 1.5 million gallons per day (mgd) and was discharged to the City of Bellingham wastewater collection system for treatment. The City of Bellingham treatment system provided primary treatment prior to discharge to Bellingham Bay, a marine water body located in north Puget Sound.

During the mid-1980's, the City of Bellingham began the process to upgrade the treatment plant to provide secondary treatment of wastewater. Because of the potential high costs for treatment at the upgraded City facilities, BFF decided to develop a land application system for treatment and disposal of its wastewater. An application site was located approximately six miles from the BFF processing plant adjacent to the Nooksack River, which discharges to Bellingham Bay. Several environmental issues and regulatory challenges existed to implement the conveyance and treatment/disposal system. These issues and challenges are presented along with the results from one year's operation of the facilities.

---

[1]Harding Lawson Associates, 1325 Fourth Avenue, Suite 1800, Seattle, WA 98101

[2]Bellingham Frozen Foods, Inc., P.O. Box 1016, Bellingham, WA 98227

Wastewater Quality. Wastewater samples were collected weekly during 1989 and analyzed for BODs, COD, TSS, pH, specific conductance, and nutrients. Metals and pesticides applied to crops processed were analyzed once per vegetable type. The wastewater was found to have high BOD and TSS (1,400 and 700 mg/L, respectively) and moderate levels of TN (30 mg/L) and TP (8.5 mg/L) as compared to domestic wastewater. Levels of metals and pesticides were low and below applicable primary drinking water standards, although some metals were above fish toxicity levels.

System Design. A design report provided a comparative analyses of three alternatives deemed to be the most appropriate for the type and quantity of BFF's wastewater: complete mix activated sludge; activated biofiltration; and land treatment.

Technical and economic analysis demonstrated that BFF should withdraw from the municipal treatment system and apply the wastewater to 360 acres of agricultural land that they own along the Nooksack River, north of the food processing plant. This method of treatment would provide the greatest reliability with the least cost and was predicted to achieve 98 percent removal of suspended solids and five-day biochemical oxygen demands. Working closely with the Department of Ecology and local regulatory agencies, intensive surface and ground water quality investigations were undertaken to determine the effects of land application of wastewater on the underlying groundwater and an adjacent salmon spawning stream.

After waste characterization and environmental water quality studies were completed, the wastewater management system was designed consisting of a 2-million-gallons-per-day pump station located at the processing plant in Bellingham; six miles of 14-inch pipeline to convey wastewater to the treatment and disposal site; a 215-acre agricultural irrigation system for treatment and utilization of wastewater; and a 20-million-gallon lagoon for treatment and storage of wastewater generated during rainy periods of the processing season and during the non-irrigation season.

Permits Required. The project required a number of permits including a State Waste Discharge Permit. The Waste Discharge Permit was the first granted under new state groundwater regulations and therefore received considerable scrutiny.

Loading Rates and Water Quality Impacts. Loading rates of water, nitrogen, and BOD during 1992 were approximately 1.2 in./week, 155 lb. N/acre, and 10 lb. BOD/acre-day. These application rates are all well below EPA guidelines based on soil permeabilities and crop uptake rates yet there were still problems with surface ponding and mounding groundwater. Certain portions of the field could not be sprayed on for short periods during September and October because the State Waste Discharge Permit does not allow spraying if there is less than 3 feet to groundwater. Nitrate levels at the downgradient groundwater station were consistently at least five times lower than the 3 mg/L enforcement limit due to the low nitrogen loading rate.

Surface water metal levels were also well below the enforcement limits which were set at chronic toxicity levels. The monitoring effort also found no significant impacts to worms, larvae, and rodents due to metal bioaccumulation although the number of worms did increase substantially, likely due to an increase in soil moisture and organic matter. Only 1 of 16 pesticides applied to BFF vegetable crops was detected in the aerated wastewater. This pesticide, linuron, had a concentration of 2 ppb in the October wastewater sample. Linuron will be tested in surface water bimonthly throughout 1993 and in groundwater during 1993 and 1994 due to this detection. In summary, low loading rates of organics and nitrogen have been readily assimilated by the crop/soil matrix, and no water quality impacts were detected during the first year of operation.

Operations

To ensure successful operation and permit compliance, daily monitoring is conducted using an all-terrain vehicle for easy access. The spray site is monitored for ponding, runoff and overall conditions. Groundwater levels are checked on a daily basis to anticipate changes in level and allow the application area to be adjusted accordingly.

The use of computer-controlled, center-pivot spray irrigation equipment provides optimum flexibility and control. This system allows the application area and frequency to be adjusted quickly and accurately to compensate for changing conditions.

Activated carbon gas scrubbers placed at the last two air release valves in the pipeline mitigate odor caused by released gas. The addition of three 75 hp aerators to the 20 million gallon lagoon eliminates odors during times when the water level is high and heavy loads are entering the lagoon. Because wastewater is not allowed to pond in the field, no odor problems have been detected.

Water Conservation. The operations and maintenance costs for the conveyance, treatment and disposal facilities were estimated to be $107,000 per year. At an average annual flow rate of about 0.6 MGD, the cost for conveyance, treatment and disposal amounted to approximately $0.006 per gallon of wastewater. BFF recognized that this cost could be reduced by the implementation of water conservation measures and that these measures could be cost-effective.

Conservation measures that have been or are planned to be implemented are listed in the following table. Three of the four measures involve the reuse of water rather than direct discharging the water to the treatment system after one use. The carrot line capacity increase resulted in shortening the processing season by 10 days.

Water Conservation Measure	Water Reduction (mil. gal./year)	Capital Cost	Annual Savings
Pea Line Transport Water Recirculation (1989)	9.0	$160,000	$14,850
Carrot Line Capacity Increase (1992)	7.5	78,000	12,375
Tunnel Defrost Water Recirculation (proposed)	22.0	83,000	36,300
Bean Line Water Recirculation (proposed)	8.0	68,000	13,200
	46.5	$389,000	$76,725

The benefits of the water conservation measures include a significant reduction in water use and therefore, wastewater generation and a significant savings in annual operations costs. The cost of the individual projects are recovered over periods ranging from nearly 2 to 11 years.

Summary

An aggressive project schedule allowed construction to be completed one year ahead of schedule. The initial plans for the municipal wastewater treatment plant previously accepting BFF's wastewater were to accept the plant's effluent through the summer of 1993. However, the city could not accept the facility's wastewater effective July 1992 and had the wastewater treatment system not been able to accelerate its construction schedule, the company would have not been able to discharge or treat its wastewater.

The annual cost savings will exceed $1 million relative to other alternatives available to BFF. The total facility capital cost is $5 million less than the next least expensive option of complete mix activated sludge treatment.

EBMUD's Approach to Demand Reduction

by John B. Lampe[1] and Jacqueline A. Millet[2]

Abstract

The East Bay Municipal Utility District (EBMUD) supplies water to more than 1.2 million people in the East Bay area of California. The District has developed a comprehensive plan to deal with its long-term water supply needs. Increased conservation, reclamation and rationing during drought will be relied upon more heavily in the future to assure adequate water supply.

Introduction

EBMUD is a publicly owned utility formed under the Municipal Utility District Act passed in 1921. EBMUD's primary function is to provide a reliable, high-quality water supply to residential, commercial, and industrial water customers, while responsibly managing the environmental resources the District relies upon to provide the potable water supply. The District's source of water is the Mokelumne River watershed located on the west slope of California's Sierra Nevada mountain range. The District's future supply of water will be reduced due to three factors: increased use by Senior Water Rights holders on the Mokelumne River, increased water allocation for protection of in-stream resources, and increased growth in the District's service area.

In 1989, the District began preparation of a Water Supply Management Program EIR/EIS (WSMP) to identify and evaluate the actions and projects needed to continue providing its customers with high quality water in a reliable and environmentally acceptable manner. The year 2020 was selected as the planning horizon. The WSMP effort included the development of a comprehensive demand-management strategy as well as the identification and evaluation of supplemental water supply alternatives. This paper outlines the WSMP planning process, the

---

[1]Manger of Water Planning, East Bay Municipal Utility District, 375 11th Street, Oakland, CA 94607

[2]Associate Civil Engineer, Ibid.

DEMAND REDUCTION APPROACH

demand-management programs already in place, and the District's future commitments to conservation, reclamation, and rationing.

Existing Drought Management, Conservation and Reclamation Programs

October 1992, marks the sixth consecutive water year in which precipitation and runoff have been substantially below normal in the Mokelumne River watershed. It also marks the sixth consecutive year in which a Drought Management Program (DMP), including rationing, has been imposed on District customers. The District's rationing goals have ranged between a 12% voluntary program and a 25% mandatory program. Currently, a 15% mandatory program is in place. District customers have exceeded their rationing goals in all but two years and residential customers have reduced summer use (June through September) by as much as 38%. In addition to rationing, other elements of the DMPs have included an increasing block rate structure, water use limits, and water use prohibitions.

Conservation has been an integral component of the District's water supply operation since the early seventies. By the year 2020, the District expects to save approximately 20 million gallons of water per day (MGD) (based upon 1990 usage) from existing, adopted, and state-mandated conservation programs. Replacement of conventional toilets with ultra-low-flush toilets and modification of landscaping practices will yield the largest savings. Other elements of the program include interior and exterior water audits, distribution of water saving devices, construction of demonstration gardens, development of model landscape standards for new construction, leak detection and pipe replacement, and public education.

In addition to its conservation efforts, the District has long recognized the benefits of reclamation as a means of conserving its fresh water resources. The District has been investigating the feasibility and implementation of reclamation projects since 1967. By the year 2020, the District expects to save approximately 8 MGD (based upon 1990 usage) from existing and adopted reclamation programs. The most ambitious single program is the Chevron Oil Refinery project which will save over 5 MGD of potable water, or enough water to serve 20,000 residential households.

The WSMP Planning Process and Future Demand Management Programs

The initial phase of the WSMP planning process was devoted to identifying the District's need for water. The future "need for water" was defined as the quantity of additional water required at projected 2020 levels of customer demand to limit drought rationing to District customers to a maximum 25% of normal water demand levels. There are a number of factors which affect this future need for water. Some of these factors can be influenced by the District (customer demand and the drought management program), while others are largely outside District control (future Mokelumne runoff and precipitation, releases for protection of instream resources, operation and diversion by Senior Water Rights holders.

Using the 70 known years of Mokelumne River hydrology, the District's reservoir operations model, EBMUDSIM was used to calculate the future water need given input assumptions such as customer demand, fishery releases, demands of other agencies and a worst-case drought scenario. In order to provide the required input, the District conducted extensive analyses of the flow and habitat requirements for key fish species in the Mokelumne River and customer demand within its ultimate service boundary. The District also had to identify the worst-case drought it would use in the planning effort. A three year drought planning scenario was selected which represents the lowest two consecutive Mokelumne River runoff years on record followed by a third year which was an average of those two years. Finally, the District had to decide how much rationing it should plan to impose on its customers during the drought scenario. The District chose the 25% limit as identified in the definition of the "need for water".

Once this "need for water" was established, the District developed a range of potential alternatives which would provide the required water. Since 25% rationing, and existing and adopted conservation and reclamation had already been taken into account, elements of the potential solutions included: additional conservation and reclamation, reservoir storage, groundwater storage/conjunctive use, and supplemental supplies. Each potential alternative was compared against a list of criteria developed by the District to evaluate its engineering, operations, legal, public health, public safety, sociocultural, economic, wildlife, and biological implications. Given the environmental impacts and the regulatory and institutional uncertainties associated with many of the structural solutions, the District recognized that it must continue to maximize the efficient use of its existing supplies before turning to additional sources of water. Consequently, demand management must to play an increasingly prominent role in the District's future operations.

One of the most difficult decisions the District had to make in the WSMP preparation was how much rationing it should plan to impose on its customers during the worst-case drought planning sequence. In order to establish this limit, the impacts from District-wide rationing levels between 15% and 50%, were evaluated. Individual goals were established for each customer category based on the District-wide goal, projected 2020 demands, water use characteristics, and past rationing performance. A typical customer response scenario was developed for each goal and customer category. For example, the 40% District-wide rationing goal required a 60% reduction from residential customers. In order to meet this goal, residential customers would have to reduce toilet flushing by 50%, bathing by 40%, appliance use by 20%, and outside use by 100%. (In general, it was believed that inside water uses would take precedence over outside uses.) Obviously, individual customers would respond to the goals differently, but this analysis allowed the District to evaluate the level of effort that would be required from each customer category. The analysis concluded with the selection of a 25% rationing limit. The following goals would be established for each customer category: 32% for residential customers, 50% for major irrigators, 30% for commercial and institutional accounts, and 15% for industrial accounts.

In addition to future rationing levels, the District also evaluated what future conservation and reclamation programs it should implement. The District developed five increasingly stringent "levels" of conservation. The levels ranged from a moderate program with little impact to District customers to an intensive program which required mandatory compliance from District customers. The conservation levels were rated against each other to determine which would best meet the District's future needs and a proposed conservation program was selected. The proposed conservation program expands on the District's current efforts by significantly increasing the coverage of existing programs and including incentives to improve landscaping practices. It also includes investigation of a toilet replacement program. The future conservation program would triple the number of staff currently working in the District's Water Conservation Office and would conserve an additional 13 MGD above the 20 MGD savings expected from the District's current program. Thus the 2020 savings from conservation would equal 33 MGD over the 1990 base demand. The fixed annual cost for the program is $3.9 million (1992 dollars).

Thirteen potential reclamation alternatives including local, non-potable, reuse alternatives, regional export alternatives, and potable reuse alternatives were identified in the WSMP. Again, the criteria were used to rate these potential alternatives against each other to identify which would best meet the District's future needs. The alternative which rated the best became the proposed reclamation program. It would double the District's current reclamation savings from 8 to 16 MGD by the year 2020. The total capital cost for the proposed reclamation program is $74 million (1992 dollars). It would provide reclaimed water to large outdoor users, typically parks and golf courses, and selected industries within five miles of sources of reclaimed water.

The economic and environmental consequences of the current six year drought have made Californians painfully aware that water is a precious, scarce resource. It has also reinforced the push for state review of long-held water rights to ensure protection of the public trust resources. The combination of these two events has, in turn, forced water districts to carefully evaluate how they are using water today and how they will provide water service in the future. In order to achieve some balance between the three competing water interest groups (agriculture, urban, and environmental), the agricultural and urban water suppliers must first demonstrate that they are making reasonably efficient use of existing supplies. These actions have reconfirmed the importance of and the District's commitment to demand management. The Draft WSMP EIR/EIS, which includes a full discussion of the demand management programs, was released to the public in late December 1992.

Global Distribution of Water Through the Oceans

E. Robert Winter, P.E.[1]

Abstract

The paper is a conceptual investigation of a pipeline and bladder system for conveying and storing fresh water in the oceans. Modeling is recommended to provide background on what could become an industry, capitalizing on the global supply and demand of water over distances and time.

Introduction

Rudyard Kipling wrote of a water supply problem resolved, to a degree, by Gunda Gin. Perhaps we can take a lesson from Gunda Gin's humble sack to address some of our current water supply dilemmas. So invoked, here goes a conceptual investigation of fresh water distribution through the oceans. The purpose is to provide alternate storage facilities and better distribute a portion of the fresh water that currently flows to salt. The oceans typically receive excess water, so they provide a convenient corridor. The problem of water supply that can, to a degree, be so addressed is neither unique to a particular region nor is it temporary.

The elements of a classic trading regime are represented by the abundance of water frequently available where sunlight is limiting and the abundance of sunlight where water is frequently limiting. Serving those in need of a product

----

[1] APWA member, District One Hydraulic Engineer, Washington State Department of Transportation. Home address: 19526 Ashworth Avenue North, Seattle, Washington 98133.

and those in need of a market has some interesting ramifications, for instance:

1. Fresh water is lighter than salt water; so lines may, with certain provisions, float.

2. The sea level pressure differential will reflect the depth at which fresh water is moored.

3. Deeply moored pipes can be fully flexible, with the fresh water confined only the heavier salt water. These flexible tubes or bladders are potentially suitable for both storage and conveyance. The conduit becomes the reservoir.

Harvesting

Routing water through the oceans, if considered, must first provide for harvesting that water on land. Provisions for harvesting may complement provisions resulting from the emerging National Pollution Discharge Elimination System permit requirements. The legislative process tells us that, although the costs are becoming uncomfortable, those requirements are not going away. Issues include maintaining clean ground, surface, and municipal water flows as the land becomes increasingly impervious.

Collection is a challenge. The resource is only available intermittently and it falls with enough energy to dislodge any soil that lacks artificial or vegetative stabilization. An adequate exploration of erosion types and recommended remedial prescriptions far exceeds the goals of this discussion. Suffice to say, collection of a quality product must include good erosion control practices and effective management of collection and storage facilities.

Storage

The reservoiring of storm flows is important for stream bank protection as well as for pipe line conveyance economies. Off shore storage will provide a productive destination for secondary conveyance systems, systems that supplement existing pipe and stream channels. They will reduce storm water flows and reservoir requirements.

Availability and timing of markets for water vary, but not as radically as supply. Thus a need not only exists to attenuate flows but to either store water or economically produce it. Production from salt water is a function of energy supply, and is usually expensive. The current residential rate for water in Seattle is $5000 per 32,500 cubic meters. That volume will fill a one meter increment of a bladder extended to a diameter of 203 meters.

Such a flexible bladder can be moored 3.5 kilometers deep off the continental shelf. The ambient pressure will support a column of fresh water 60 to 80 meters high at sea level. That pressure is modest, and it will be derated for working conditions to perhaps 40 meters. Still it is adequate to serve lower areas and to supplement other supplies.

Joints

Effective joints are essential for implementation. Pipe diameters will be large and walls thinner than conventional practice. Two concepts are displayed in figure 1. The first is the use of negative pressure to lock opposing gaskets on inclines against a pressure pipe interior. A receiver engages pipe lugs for restraint. The other concept involves negative pressure to secure non-pressure sheets to anvils as longitudinal clamps make mechanical connections to thickened sheet edges.

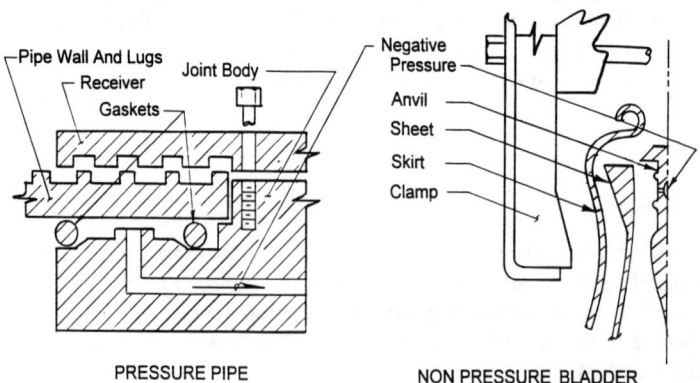

Figure 1

Damage Control

Provisions will be required, in moored pressure pipe systems, to prevent catastrophic failures following line breaks or excessive draw down. Either condition will collapse and damage pressure pipes. Check valves are recommended to allow salt water intrusion until repairs are made. Salt can then be purged by passing balloons, followed by fresh water, down the tube.

Conclusion

The schematic proposal of moving fresh water through salt water, a fluid through a fluid, is physically possible. Modeling the technology will test details and help prove the economics. The concept will allow dry season use of wet season flows as flexible conduits double as reservoirs.

RESIDENTIAL WATER CONSERVATION AND REUSE DEMONSTRATION: CASA DEL AGUA AND DESERT HOUSE

Martin M. Karpiscak[1], Richard G. Brittain[2], and Mark A. Emelity[3]

Abstract

Occupied single-family homes can provide factual data as well as create an active real-world setting for education of the public. The installation of water-efficient fixtures, rainwater harvesting, and systems for graywater storage and reuse can potentially reduce potable water use by over 50 percent. Casa del Agua and Desert House demonstrate that efficient resource utilization can become an integral part of a comfortable and quality living environment.

Introduction

Arizona receives erratic and highly variable annual precipitation throughout the state. For example, the Sonoran Desert City of Yuma in the southwestern part of the state receives about 75 mm of rainfall per year while the City of Tucson, also situated in the Sonoran Desert in the southcentral part of the state, has an average annual rainfall of approximately 275 mm. Rainfall in the Phoenix metropolitan area averages approximately 175 mm per year while surrounding mountainous areas receive 250-500 mm. The highest mountain elevations receive greater than 625 mm of precipitation annually. Rainfall is not a dependable water supply source.

---

[1]Office of Arid Lands Studies, University of Arizona, Tucson, AZ 85719
[2]College of Architecture, University of Arizona, Tucson, AZ 85721
[3]Senior Planning Analyst, Water Planning Department, Salt River Project, Phoenix, AZ 85072-2025

Primary water sources in Arizona include surface water from in-state rivers and streams, an allocation of Colorado River water delivered via the Central Arizona Project (CAP), groundwater, and reclaimed water. Surface water including the CAP, currently accounts for 54 percent of total water use within the state. Groundwater use is approximately 43 percent. Reclaimed water use makes up the remaining 3 percent (Eden and Wallace, 1992). Currently, groundwater withdrawals in southern and central Arizona are exceeding recharge by approximately 74 X 10^7 m^3 per year (Olsen, 1992).

The City of Tucson, Arizona's second largest city, has a population of approximately 600,000. Projections indicate that by the year 2020, Tucson will grow to more than a million people. Prior to the 1992 delivery of CAP water, the Tucson metropolitan area was the largest urban area in Arizona almost entirely dependent upon groundwater. An estimated 53 X 10^7 m^3 of water is being pumped from deep underground aquifers annually, while only 19.2 X 10^7 m^3 is being recharged, thus creating an overdraft situation (Karpiscak et al., 1991).

The City of Phoenix, located centrally in the state, is Arizona's largest city and has an estimated population of about one million people. Phoenix currently receives most of its water supply from the Salt River Project (SRP), an organization formed in 1903 primarily to provide water for agricultural irrigation in the Salt River Valley. The City of Phoenix continues to grow and has expanded its water service area beyond the original SRP water service territory boundaries (Karpiscak et al., 1991).

Conservation

In 1980, the Arizona legislature passed the Groundwater Management Act (GMA) and created the Arizona Department of Water Resources (ADWR) to administer the GMA. The GMA requires all major cities to balance supply and demand by 2025. The GMA identified areas within the state where the overdraft situation was most severe and either identified them as Active Management Areas (AMAs) or Irrigation Non-Expansion Areas (INAs). Phased management plans are being implemented for the various AMAs. These plans include many augmentation and conservation strategies including building codes which now require installation of water-saving fixtures in all new construction.

Residential water-efficiency programs can potentially save substantial amounts of water along with energy and money. For the homeowner, efficiency programs can reduce water, sewer, and energy bills. For water providers, reduced water demands can potentially defer the need to expand water and wastewater treatment plants for several years. Reduced water demand can also mean that less water needs to be pumped creating an energy savings. Operating costs can also be decreased because of a reduced need for chemicals and reduced equipment operating time.

Casa del Agua

Casa Del Agua (CDA) is a single-family residence in Tucson, Arizona, which was retrofitted in 1985 with water-conserving fixtures, rainwater harvesting, and graywater reuse systems. A family of three occupies the house which allows for the monitoring of real-time data. A six-year study shows that municipal water use was reduced by 50 percent over that of a typical Tucson residence. This reduction in demand was accomplished without a reduction in the residents' quality of life. The success of CDA in the area of education and research led to the development of an expanded demonstration project (Karpiscak et al., 1990).

Desert House

Desert House was designed to build upon the knowledge acquired and still being gathered from CDA. The overall concept is to create an educational exhibit at the Center for Desert Living at the Desert Botanical Garden. The exhibit will refine the residential water use concepts employed at CDA and will incorporate various aspects of residential energy efficiency. This project will also have residents who will generate the data for analysis and dissemination to the visitors (Karpiscak et al., 1991).

The **prime goal** of Desert House is to raise the awareness of the public, building industry, agencies and financial institutions as well as government officials to the commercially-available and cost-effective technologies that will save residential water and energy.

The **water efficiency** features of Desert house include low water-use plants, irrigated with an efficient drip irrigation system using harvested and stored rainwater and graywater. The front, rear, and side yards are contoured and all patios and walkways are used to direct rain to plants. Discharge of water to the sewer system is limited to wastewater from the toilet, kitchen sink, and dishwasher. The toilets are 4 liters-per-flush, faucets equipped with aerators use 6 liters-per-minute, and showerheads use 9 liters-per-minute.

The **energy efficiency** of the residence is achieved by applying passive solar concepts which include but are not limited to proper orientation, window sizing, shading, optimal insulation, and thermal mass walls and flooring.

Additional efficiencies are achieved by using a high-efficiency 3-function heat pump (space heating/cooling and water heating). Appropriately sized and located dual pane windows and clerestories provide passive solar gain during winter, reduced heat gain during summer, and natural daylighting. The landscaping pallet provides for multi-season color, enhanced habitat for wildlife, and is designed to provide lush outdoor living spaces adjacent to the home and maximum summer shade for the structure.

The **education program** for the Desert House Project includes an adjoining structure featuring an interactive computer tour of the home which is available at all times to the exhibit visitors. Computers will enable visitors to access real-time data on household water and energy use and to explore alternative strategies. Informational brochures describe simple, practical approaches in a how-to format.

Conclusions

We look forward to the acquisition of new water use data from Desert House and comparing it to the CDA six-year database as well as the new data on energy use. New systems will be investigated and many visitors will benefit from these demonstration projects.

Acknowledgements

Casa del Agua is supported by Tucson Water. Desert House is a cooperative program of the Arizona Department of Commerce Energy Office, Salt River Project, City of Phoenix Water Services Department, Desert Botanical Garden, and The University of Arizona.

Appendix: References

Eden, S. and Wallace, M.G. (1992). Arizona Water Resources Research Center (WRRC), Arizona Water: Information and Issues. Issue Paper Number Eleven.

Karpiscak, M.M., Brittain, R.G., Gerba, C.P., and Foster, K.E. (1991). "Demonstrating residential water conservation and reuse in the Sonoran Desert: Casa del Agua and Desert House." Wat. Sci. Tech., 24(9), 323-330.

Karpiscak, M.M., Foster, K.E., and Schmidt, N. (1990). American Water Resources Bulletin, American Water Resources Association. Residential Water Conservation: Casa Del Agua.

Olsen, S. (1992). Arizona Department of Water Resources (ADWR). Personal communication.

Agricultural Water Conservation Programs to Improve Water Use Efficiency

Baryohay Davidoff[1]

Abstract

California's growing population poses an ever increasing demand for water from the existing limited and scarce water resources of the State. In addition the continuation of six years of drought has made water shortages a daily reminder of the scarcity of this commodity. The State provides technical and financial assistance to the irrigation districts and growers to use irrigation water as efficiently as possible. Programs such as California Irrigation Management Information System, (CIMIS); Mobile Laboratory; Agricultural Drainage Reduction; and Training and Education activities and how these programs are being implemented will be discussed.

Introduction

The limited water supplies of California coupled with a population increase of over 600,000 per year and continuous and recurring droughts have made irrigation water management a necessary and integrated part of the farming operation.

Solving water problems is a complex issue involving an even more complex physical, legal, and institutional set-up. Social, economical, and environmental factors have interrelated in a competing fashion. The economic viability of the State depends on the availability and

---

[1] Chief, Agricultural Water Conservation Section, Water Conservation Office, California Department of Water Resources, 1416 9th Street, Sacramento, CA 94236.

reliability of water resources without which economic losses, damage to rural communities and the environment can be enormous and irreversible.

Agriculture has taken up to 100 percent cuts in water deliveries leaving thousand of acres fallow. To minimize the adverse effect of water shortages and water cuts, the State provides assistance to the agricultural community through many on-going water conservation and newly developed water bank programs.

The State water bank has helped meet emergency needs of cities, industry, and permanent crops and wines. The State water bank purchases water from agriculture and sells it to where the emergency need is.

In addition, managing continued water shortages and the drought has made the cooperation of agricultural and environmental water users an essential and imperative element in overall management of agricultural water. Representatives of many interested parties involved in water issues are working toward development of efficient agricultural water management practices and an implementation process to further advance agricultural water management in California.

Water Conservation Programs

The technical and financial assistance to the farming community to help further advance on-farm irrigation efficiency is being achieved through the following programs:

1. California Irrigation Management Information System (CIMIS);
2. Mobile Laboratory or Irrigation System Evaluation Program;
3. Training and education, drought workshops, flexible water delivery system; and,
4. Agricultural Drainage Reduction.

1. California Irrigation Management Information System, (CIMIS)

CIMIS is an integrated network of computerized weather stations located at key agricultural and municipal sites throughout California. The weather data is used to estimate reference evapotranspiration or "ETo". Growers, landscapers, consultants, managers etc. can call a centralized computer and obtain ETo data along with available crop coefficients. ETo and crop coefficient information helps determine when to irrigate and how much water to apply, the two fundamental parameters of good irrigation scheduling. Currently, over 1,300 direct users access the system regularly.

2. Mobile Laboratory, Irrigation System Evaluation Program

Irrigation efficiency is directly related to how good an irrigation system distributes the applied water. The distribution uniformity often is determined by catch-as-catch-can uniformity. The Mobile Laboratory or Irrigation System Evaluation program provides on-farm irrigation system evaluation and practical recommendations on how to improve distribution uniformity. There are seven mobile labs on a cost-sharing basis working in cooperation with resource conservation districts and performing irrigation system evaluation on thousands of acres annually. A recent analysis of data collected from field evaluations indicates significant improvement in distribution uniformity and irrigation efficiency when recommendations from the program were implemented. The results from over 1,500 field evaluations indicate an average irrigation efficiency of 72 percent.

3. Training and Education Program

Thousands of growers, irrigation district board members and engineers, irrigation specialists, consultants, etc. have benefitted from the training and education program. The activities are in the form of short courses, seminars, conferences, and symposiums. Topics of this program are diverse and tailored to the needs of a specific area. The topics include short courses on irrigation evaluation; drought workshops; deficit irrigation; AgWater software training and demonstration; flexible water delivery; water delivery control systems; water measurement technology; and irrigation scheduling. Recognizing that the majority of the irrigators and foremen speak Spanish, a bilingual program is under development to provide training in Spanish as well.

4. Agricultural Drainage Reduction Program

The objective of this program is to help growers and local agencies reduce their drainage water and conserve water at the same time. In the west side of the San Joaquin Valley, shallow ground water is rising, resulting in salinization of the soil. In addition to this problem, the discovery of naturally occurring selenium and other toxic elements in the shallow ground water and drainage water is threatening food and fiber production in the Valley. The deep percolation, particularly during prorogation is the main source of excessive drainage. According to a recent ruling by the State's regulatory agency, the State Water Resource Control Board, in areas where saline shallow ground water is not usable, the annual deep percolation should not exceed 0.4 acre-feet per acre. This is almost about 50 percent of current deep percolation. The program activities include demonstration of state-of-the-art irrigation and management techniques. New and emerging irrigation technologies such as low-energy precision application (LEPA), subsurface drip irrigation, automated gated pipes, etc. are being demonstrated and studied.

Conclusion

Thousands of California growers, irrigation managers, irrigation district board members and engineers are benefiting from existing assistance programs. The results of the implementation of these programs are manifested in increased awareness toward more efficient use of water, improvement in irrigation systems and irrigation management throughout the State. Irrigation management now is becoming an integral part of the irrigation districts operations.

Impact of Groundwater Management Act and
CAP Water Supply on
Agricultural Water Conservation Programs

Thomas Carr[1]

Abstract

Irrigation districts in central Arizona have increased the tax assessments for their farms to pay for distribution systems to deliver Central Arizona Project (CAP) water. Many farms must also invest in water conservation improvements to comply with mandatory reductions in annual water allotments. Changing economic conditions in the last few years has severely limited the ability of the farms to pay for these new water supply and regulatory costs.

Introduction

The objective of water management programs in Arizona is to reduce and eliminate regional groundwater overdraft in central Arizona. Over 85% of the historical overdraft has been caused by agricultural pumpage. The long-term strategy to reduce overdraft has been to increase renewable water supplies by importing CAP water from the Colorado River, and to reduce demand through mandatory conservation programs. Both strategies require large investments by individual farms. Many areas which expected to take CAP water had to invest in distribution systems to deliver the water. The loans for these investments came from private bonding companies and the federal government. These same farming areas are also

---

[1]Assistant Deputy Director, Arizona Department of Water Resources, 15 S. 15th Ave, Phoenix, AZ 85007

expected to reduce their maximum water use by 25% over the next decade by investing in state - of - the - art irrigation systems. These investments were determined to be feasible if crop prices and yields increased moderately. However, crop prices and yields have declined during the last five years. Crop prices have declined as a result of world market conditions, while yields in Arizona have been reduced by unfavorable weather, pests, and a reduction in crop rotation. The increased costs associated with the CAP water supplies has decreased water demand significantly.

On-farm Conservation Investments

The Department of Water Resources projected in 1987 that investments in level basin irrigation systems would result in water savings of about $50 per acre for a typical cotton farm in central Arizona (ADWR 1987). Cotton yields were also found to improve substantially after installation of these systems. Total increased profits to a farm from these two factors amounted to about $175 per acre total. Although the water savings are still accurate, the estimate of yield increase has been offset by adverse environmental conditions. Although many farms adopted level basin irrigation technology in the mid-1980s, at a cost of hundreds of thousands of dollars per farm, their ability to recover investment costs may have been reduced significantly (Wilson, 1992).

CAP Water Supply Costs

The Bureau of Reclamation (BOR) prepared estimates of the feasibility of constructing CAP distribution systems for the irrigation districts in the late 1970s and early 1980s. The BOR estimated that the farms in central Arizona could pay $65 to $85 per acre-foot for all CAP water costs. New estimates indicate that the ability to pay is less than $40 per acre-foot (Wilson, 1992). The current cost of CAP water is slightly over $50 per acre foot. The result is that water sales to agricultural CAP contractors has been declining during the past few years. Some districts and the Central Arizona Conservancy District have used reserve funds to lower the price of water during the last two years. This action has increased CAP water sales somewhat, but the reserves are limited and cannot continue indefinitely. Also, the CAP will be declared complete in 1993 at which time the federal government will expect project repayment to begin. Total repayment costs will increase to the farms.

Conclusions

Recent trends in the farm economy have severely reduced the ability of farms in central Arizona to pay for investments in improved irrigation technology and to pay for renewable water supplies from the CAP. Some researchers project that farms which have made these investments, and their irrigation districts may become financially insolvent. In the short run, a reduction in total water use is expected, since the production may be reduced as farms are unable to obtain financingas a result of high irrigation district assessments and the higher cost of water(Dedrick, et al. 1992). In the long term, if the irrigation districts are able to relinquish their CAP investments, groundwater use may increase since there will be no economical water supply to substitute for it use.

REFERENCES

1. Arizona Department of Water Resources. 1987. "Economic Analysis of the Areas of Similar Farming Conditions with the Pinal AMA." Unpublished report.

2. Dedrick, A.R., et al. 1992. The Demonstration Interagency Management Improvement Program (MIP) for Irrigated Agriculture in the Maricopa-Stanfield Irrigation and Drainage District (MSIDD). U.S. Department of Agriculture, Agriculture Research Service.

3. Wilson, Paul N. 1992. "An Economic Assessment of Central Arizona Project Agriculture." The University of Arizona.

AGRICULTURAL WATER CONSERVATION PROGRAMS IN THE LOWER COLORADO RIVER AUTHORITY

Jobaid Kabir, M ASCE [1]

INTRODUCTION

Rice irrigation is the largest user of water within the area served by the Lower Colorado River Authority (LCRA), accounting for approximately 75 percent of total annual surface and ground water demands. In an average year, about 30 percent of surface water supplied to rice irrigation is satisfied with water released from the storage in the Highland Lakes located at the upstream reaches of the Lower Colorado River and its tributaries. During a severe drought, the demand for stored water could be as much as 70 percent of annual rice irrigation demand.

LCRA owns and operates two irrigation canal systems which together supply water to irrigate 60,000 acres of rice each year. These irrigation systems are the Lakeside and Gulf Coast Irrigation Divisions. The Lakeside system is located in Colorado and Wharton Counties and the Gulf Coast system is located in Wharton and Matagorda Counties.

In the 1987 and 1989, the Lower Colorado River Authority Board of Directors authorized implementation and funding for Canal Rehabilitation Project and Irrigation Water Measurement Project respectively. These two projects are key initiatives to agricultural water conservation goals established in the LCRA Water Management Plan and Water Conservation Policy. In addition LCRA participated actively in agricultural water conservation research projects and technology transfer activities.

Senior Engineer, LCRA, P.O. Box 220, Austin, TX 78767

PROGRAM ELEMENTS

The LCRA Agricultural Water Conservation Program consists of four interrelated elements:

Irrigation Water Measurement

During 1990, LCRA began field testing a water measurement system that is intended to provide the capability for more efficient operation of LCRA water distribution systems and provide information for more precise on-farm water management. Ultimately, a full-scale water measurement system will also enable LCRA to implement water pricing incentives for on-farm water conservation.

The field test involved measuring the flow of water within certain canal segments as well as measuring the flow and volume of water delivered to each field within the test area. In-canal water measurement involved flow rating and modifying water control structures and installation of electronic equipment to monitor fluctuations in canal water level. Farm-level water measurement was accomplished using either modified concrete water delivery structures, which have been calibrated by the U.S. Bureau of Reclamation for flow measurement, or by using current meters.

The initial field test of the measurement system was designed to resolve questions regarding the technical and economic feasibility of full-scale project implementation. The results of the field test were favorable and the project is proceeding toward full implementation by 1992. Water pricing incentives for on-farm water conservation (e.g., volumetric rates) are expected to go into effect during 1993.

Canal Rehabilitation

LCRA is currently in the fifth year of a major rehabilitation project on the Gulf Coast canal system. This system consists of approximately 370 miles of earthen canals and laterals serving an average of 37,000 irrigated acres. Due mostly to a lack of preventive maintenance, the operating efficiency of the canal system was allowed to deteriorate to the point that total water diversions reached nearly 10 acre-feet per acre in 1986.

Much of the inefficiency of the Gulf Coast system is attributed to an overgrowth of undesirable vegetation on the system. Literally, forests had become established on the canals such that routine maintenance had become impossible on large segments of the system. Studies indicated that plant evapotranspiration accounts for the largest share of canal water losses. Correcting this problem has required mechanical clearing and selective use of herbicides. To date, over 100 miles of canal have been mechanically cleared and roughly 160 miles of canal have been treated with herbicide. Once cleared, the canals are routinely mowed or treated to prevent re-establishment of nuisance vegetation.

In addition to vegetation removal, the LCRA canal rehabilitation project also includes measures to improve the hydraulic performance of the system. For example, many canal segments have been narrowed by as much as 50 percent and re-graded to improve flow velocity and reduce the volume and surface area of the water in the canal. Approximately 100 water control and delivery structures have been replaced as well.

On-Farm Water Conservation Research

In 1982, the USDA Soil Conservation Service (SCS), local Soil and Water Conservation Districts, Texas A&M University, the Texas Agricultural Experiment Station (TAES) and the Texas Rice Research Foundation initiated an in-depth study of water use and management practices throughout the Texas rice belt. The objective of the study was to develop and evaluate irrigation water management practices to improve on-farm water use efficiency and reduce rice production costs. In 1987, LCRA contributed $90,000 to the study in order to prevent its premature demise due to federal and state budget cuts. LCRA has also provided a significant contribution of in-kind assistance.

Fourteen water management practices were identified and evaluated for their potential to reduce on-farm water use. These practices can be grouped into two major areas - shallow water management and improved water delivery. Shallow water management is defined as maintaining the flood depth on a rice field at three inches or less. Recommended practices include precision land leveling, water leveling, closer levees, flushing, and maintaining a shallow continuous flood. Improved water delivery system and maintenance to reduce losses from customer-owned laterals.

Results of the research program have shown readily attainable reductions in on-farm irrigation water use of 25 to 30 percent and significant increases in crop yield (an average increase of 17 percent). Significant additive effects of various management practices have also been observed. For example, one field monitored in 1986 exhibited a 61 percent reduction in inflow relative to the average inflow for surface water irrigated fields. This reduction is attributed to the combined effects of precision land leveling, closer levee spacing, multiple inlet water delivery, and "intensified" water management by the producer.

Technology Transfer

LCRA has used a variety of methods to educate rice farmers about on-farm water conservation practices. Methods have included dissemination of fact sheets and videos, assistance with extension service seminars, field demonstrations, and one-on-one consultation. LCRA has also organized an Agricultural Water Conservation Task Force to obtain farmer input on issues relating to LCRA programs and policies.

The success to date of the "technology transfer" process is seen in the results of a survey of LCRA irrigation customers. That survey showed that the majority of LCRA customers were familiar with the "less water, more rice" practices and that a majority have used or plan to use one or more of the practices.

PROGRAM COSTS AND BENEFITS

Funding for the Agricultural Water Conservation Program has been provided primarily by LCRA. From inception to program completion in 1994, LCRA will have committed approximately $2.5 million to the program. In addition, the U.S. Bureau of Reclamation has provided a significant level of technical assistance through a four-year federal appropriation Two grants for water measurement equipment purchases has also been provided by the Texas Water Development Board (TWDB). TWDB also funded a policy research project to the University of Texas Lyndon B. Johnson School of Public Affairs for the evaluation of LCRA's agricultural water conservation program.

A benefit-cost analysis has been performed using the estimated total LCRA costs for canal rehabilitation and water measurement. Direct economic benefits included in the analysis were reduced energy costs for water diversion and reduced demand for stored water. These benefits were calculated on a range of water-savings estimates (reflecting varying acreage levels at the LCRA irrigation districts. The low estimate is water-savings of 64,900 acre-feet per year, the base-case estimate is 100,800 acre-feet per year, and the high estimate is 120,500 acre-feet per year. Under the base-case scenario, which reflects average acreage levels in recent years, the projects were shown to return $1.87 for each $1.00 invested. Under the base-case scenario, the projects show a net economic benefit of $3.5 million over 20 years.

The impact of the overall program on gross water diversions at LCRA's Gulf Coast Irrigation District has been impressive. Total gross diversions per irrigated acre have decreased from 9.86 acre-feet in 1986 to 5.23 acre-feet in 1990. Some of this decrease is attributed to above average rainfall during 1990 and refinements in pump curves used to calculate diversions. Nonetheless, the trend toward reduced overall water diversions and reduced water diversion per irrigated acre is clearly established.

Agricultural Water Conservation Technology Transfer

Gerald W. Buchleiter[1]

Abstract

Efforts to implement computer programs which encourage water and energy conservation in the Columbia Basin, are described. Irrigators use the irrigation scheduling program to receive recommendations on the timing and amount to irrigate. The selected fields to be irrigated are entered into a pump selection program to obtain recommendations for operating pump combinations which minimize the total daily energy cost.

Introduction

In the early 1980's, many irrigated farms along the Columbia River faced escalating electric energy prices which greatly increased the cost of pumping and distributing water through center pivot irrigation systems. In these cases, energy use is directly related to water use so water conservation efforts are interrelated with energy conservation efforts. Electric bills for irrigation are based on rate of use or demand and amount used or consumption. Demand charges are based on the peak 30 minute demand at each billing point which typically is each pump station. Typically the demand component is about one third and the consumption component is about two thirds of the total energy bill. Many of these large privately owned irrigation systems were constructed in the 1970's when energy prices were much lower. Since the installed irrigation system is a fixed capital cost, the only alternative for reducing energy costs is to redesign the pump stations and pivots for lower operating pressures and to improve

---

[1]Agricultural Engineer, USDA-ARS, AERC, CSU Foothills Campus, Fort Collins, CO 80523

system management to reduce electrical consumption and demand. Some agencies such as the Bonneville Power Administration and local electric utility companies implemented cost sharing improvements to upgrade the system hardware which can reduce electric demand and consumption.

Approach to the problem

In 1984, ARS researchers began to look at system operational procedures, seasonal water application and energy usage for a 103 pivot, 5000 ha. farm in northeastern Oregon. This branched pipe network with multiple pump stations is typical of many irrigation systems pumping water from the Columbia River. The complexities of operating these dynamic irrigation systems and the undesirable consequences of a sudden pressure drop encourage the managers of these systems to maintain pressures in excess of the minimum required pressures and discourages water conservation.

Significant savings in the amount of water applied are possible if crop managers would use more accurate information on crop water use and were confident that the system would deliver the required water when necessary. Past experience with irrigation scheduling indicates crop yields can be maintained or possibly increased while significantly reducing water applications. The concept of computerized irrigation scheduling for center pivots is to maintain soil water budgets at two points in a field to determine a range of dates that an irrigation of a specified depth can be effectively used. Climatic data are used to calculate crop water use and rainfall and irrigations are measured to provide input to the soil water budget. Knowing the pivot's flow rate, the program recommends when to irrigate to minimize overirrigation as well as prevent yield reducing water stress. Based on these recommendations, the irrigator decides when to irrigate (Heermann et al., 1976). Reducing overirrigation conserves water and reduces the highly visible pumping costs. Other less visible benefits include reduced leaching of fertilizer and chemicals. These benefits are difficult to quantify but are becoming much more important as society and irrigators in particular become more aware of the costs of environmental degradation.

The approach for improving management of daily operations was to develop software that analyzes the current status of the entire hydraulic network and provides recommendations for operating the most economical pump combinations. The program accepts user input of which pivots to irrigate, calculates required discharges and pressures throughout the system and selects pump combinations at each pump station which gives the lowest total energy cost on a daily basis. These recommendations which satisfy the user specified irrigation schedule, affect both the demand and the consumption components of energy cost.

Two main aspects must be considered when selecting the least expensive pump combinations. For a single pump or pump combination at an individual pump station, the most economical choice is the pump(s) that operates near peak efficiency with a minimum excess pressure. The marginal cost for this demand depends on the peak 30 minute demand incurred to date within the current billing period. If the current demand exceeds the previous peak demand, a demand charge is incurred for the incremental increase in peak demand. If. the current demand does not exceed the previous peak, no additional demand charge is incurred by selecting this pump combination. Most large distribution systems have sufficient flexibility to provide the necessary discharges and pressures throughout the system with a number of different pump combinations. There may be significant savings in choosing pump combinations at each billing point which minimize incremental demand charges even though the pump combinations may not be the most efficient at an individual billing point (Buchleiter and Heermann, 1989).

Experiences in implementation

Identifying the problems faced by the various beneficiaries of an improved management program is important in defining the various parameters and requirements to be considered. Upper level managers are usually interested in the monetary consequences of a decision so including the economic ramifications of various alternatives is important. Lower level managers are more interested in the impact on their workload. Understanding the capabilities of the personnel who actually operate the program is important in designing the computer interfaces for communicating information in an understandable and usable form. Users are likely to be skeptical of the results initially so it is important to involve them in identifying the problems or verifying the recommendations.

The technical feasibility of using these computer programs was demonstrated on two large irrigation systems in the Columbia Basin. A radio telemetry system was installed on the 5000 ha. farm in Oregon to improve the daily management and recording of data. This system continuously monitored and allowed control of all pivots and pumps from a single location. Pressure and flow information used in calibrating the hydraulic model was recorded instantaneously when requested. Annual labor savings were 33%. Savings in water and energy were more difficult to quantify but the estimated total annual savings were $88 /ha. On a 20 pivot farm in south central Washington, the pump selection program ran on a battery powered computer in the irrigator's pickup. Pumps at the river pump station were monitored and controlled using an infrared communication linkage with the mobile base station. This arrangement significantly reduced the time required to change pump combinations but data for calculating monetary savings were unavailable.

The biggest hindrance to implementation of computerized irrigation scheduling was obtaining the necessary data to maintain accurate soil water budgets. Irrigators who believe in water conservation techniques must make the crucial data collection process a priority. For successful implementation, as much data as is economically feasible should be collected and entered in the program either automatically or with minimum effort by the irrigator. The biggest hindrance to implementation of the pump selection program was the necessity of obtaining adequate and consistent data on pressures and flow rates in the pipe network and at the pump stations, to properly calibrate the hydraulic model. Since irrigation systems are dynamic and complex, manual collection of the necessary calibration data was difficult and error prone.

Recommendations for the future

This approach focused on implementing improved water and energy management at the farm level, assuming adequate water resources. In the future there will be more pressure to reallocate the finite water resources which will require a higher level of management and cooperation among competing interests such as power generation, fisheries, transportation, municipal, industrial, irrigation and environmental concerns. Systematic approaches which can quantify different concerns and constraints on a common basis are needed to make objective decisions for allocating water resources.

The data requirements for complex computer models can become burdensome for large field applications. It is imperative that the necessary data be collected with a minimal time commitment. Future directions might include replacing the data intensive numerical models with 'self-learning' approaches such as neural networks to reduce the time and expense of obtaining the necessary data for calibrating and updating the models.

References

Buchleiter, G. W. and D. F. Heermann. (1990). "Management of Multiple Pump Stations". *Applied Engineering in Agriculture* 6(1):39-44.

Heermann, D. F., H. R. Haise, R. H. Mickelson. (1976). "Scheduling Center Pivot Sprinkler Irrigation Systems for Corn Production in Eastern Colorado". *Transactions of ASAE* 19(2):284-287,293.

Price Elasticity and Conservation Potential

Dr. David S. Hasson[1]

Abstract

Water conservation programs usually contain pricing incentives for users to limit water consumption. This analysis uses data from the Portland, Oregon water system as an example of both the reasons for conservation pricing and the limitations of doing so. The results suggest that system planners and policy makers need to have realistic expectations about the effectiveness of conservation rate structures in achieving water savings objectives in order to adequately plan for future supply needs.

Introduction

As part of the preparation of a long term water demand forecast for the current and potential future service areas of the Portland, Oregon water system (CH2M HILL 1991), an analysis was made of the primary determinants of water demand in the service area. Multiple regression statistical techniques were used to estimate these determinants using time series data. Separate demand equations were estimated for the City's retail users and for its wholesale users as a group. The results of this statistical analysis suggest certain limitations on the effectiveness of pricing mechanisms as primary means of achieving significant water conservation targets.

Multiple Regression Results

The regression results indicated that in Portland the variables that best forecast annual water use are population, real (i.e., inflation adjusted) per capita income, and the real price of water. In the surrounding area water utilities, these variables included the number of water accounts and the real price of water. Peak season

---

[1]Director, Water Management Economics, CH2M HILL, 825 N.E. Multnomah, Suite 1300, Portland, Oregon 97232

variables with the best forecasting relationship included the number of residential accounts, nonresidential accounts, and a season indicator variable.

The price elasticities were estimated to be -.3420 and -.224 for Portland and the suburban water utilities, respectively. These results mean, for example, that a 1 percent increase in average water rates in Portland would result in a .3420 percent reduction in water use, if all other things were held constant. These findings are generally consistent with the literature on price elasticity for water use.[2] The overall price elasticity for water is inelastic in the sense that the quantity demanded is not very sensitive to the price. The results suggest that the wholesale customers' water demands are slightly less sensitive to price than are the City's retail customers' demands. Reasons for this might include higher than average household incomes, lesser emphasis on water conservation, differing rate structures, or other such factors.

Conservation Potential

What do these results suggest for achieving conservation objectives? Because the demand for water is inelastic, very large price increases would be needed to achieve large percentages of water savings. In Portland, water rates would need to be increased by more than 50 percent to achieve a maximum of 17 percent water conservation (i.e., .5 x .342 = .171). This would be the maximum potential for this rate increase because the elasticity estimates apply only to a limited range of price changes. After the initial water conservation activities by users are implemented, the remaining water demands are even less sensitive to price changes. A measured or estimated elasticity does not apply to the full range of the demand curve; it is only applicable to a relatively narrow range around the quantity at which the estimates are made. Therefore, the elasticity results of the multiple regression analysis tend to overstate the conservation potential for large price changes.

If one assumes that a 20 percent increase in overall water rates is the maximum acceptable increase, then the potential water savings in Portland would be less than 6 percent in the City and 4 percent in the suburban areas. These savings levels would not defer or eliminate most water supply facility additions that system engineers or planners might consider because traditional supply additions are typically for larger increments.

It can be concluded that very large average rate changes would be necessary to achieve significant water demand reductions unless other nonprice measures are undertaken simultaneously. In most communities the necessary rate increases

---

[2] See, for example, Nieswiadomy and Molina (1989) and Cassuto and Ryan (1979).

would not be accepted by the public.

A similar analysis was undertaken for the peak season water demands in Portland and for its wholesale customers. The results indicated similar elasticity coefficients, but because the monthly demands are larger in the peak season, the absolute amount of monthly savings that can be achieved by water pricing measures is somewhat higher than on an average annual basis.

Implications for Water Supply Planning

As integrated water supply planning is increasingly used in place of more traditional supply planning, consideration is frequently given to conservation as an alternate "supply." The results here suggest that it would be very difficult to achieve water demand reductions of more than perhaps 5 to 10 percent through pricing methods alone because of the very large rate increases that would be required.

Another important consideration is the duration of the elasticity effect. It is possible that the elasticity impacts may be temporary, and over time, users may gradually resume normal usage patterns when they become accustomed to the higher rates and as personal incomes rise. Therefore, continuing rate adjustments might be required to maintain a particular conservation response to water rates.

Specific changes to the water rate structure might be more effective at achieving particular reduction targets than general rate adjustments. For example, seasonal rates may be more effective and applicable for meeting peak season conservation goals than merely increasing all rates on an annual basis, and they be more acceptable to the ratepayers as well. Similarly, excess use rates can be effective at achieving seasonal objectives because they focus the pricing incentive on the particular demand target.

Relationship to a Conservation Program

Although the potential for significantly altering water demands solely through pricing incentives is limited, as indicated by the data for Portland, conservation pricing methods have an important place in a comprehensive conservation program. Rate structures and rate levels that are perceived as being "anti-conservation" by the public, such as declining block rates, may undermine the effectiveness of other conservation measures and programs. Conversely, conservation rates will reinforce the credibility of a comprehensive conservation program and will be an important part of the overall set of activities to achieve water demand reductions. There is no easy method to measure this support role that conservation rates and price elasticity may play as part of a conservation program, but most utility managers recognize the need to send a consistent set of signals to their customers regarding the importance of conservation. For example, the City of Winnipeg is currently considering shifting from declining block rates to a more conservation oriented rate structure to enhance the effectiveness of its conservation program.

Conclusions

Water supply planners and system managers should not have unrealistic expectations regarding the savings that can be realized through conservation pricing. The data for Portland suggest that, at best, general changes in water rates can result in marginal changes in consumption patterns. Large modifications in users demands require additional conservation actions, such as education programs, revised plumbing codes, distribution of water saving devices, landscaping modifications, and other elements of a comprehensive conservation program.

This conclusion does not mean, however, that conservation rate structures are unimportant in a conservation program. Any such program requires public acceptance and commitment in order to be successful. Consistent signals to water users regarding the importance of conservation and the utility's commitment to conservation goals is critical to the public's acceptance and commitment. Therefore, conservation pricing is an important aspect of the program beyond its direct elasticity impacts. Because water rates are sometimes the most visible element in a conservation program, the rate structure should be carefully tailored to the water system and its needs. These needs include revenue adequacy, rate equity, defensibility, public acceptance, and conservation incentives.

References

Cassuto, Alexander E. and Ryan, Stuart. "Effect of Price on the Residential Demand for Water Within an Agency." *Water Resources Bulletin*. April 1979.

CH2M HILL. *Water System Demand Study*. 1991.

Nieswiadomy, Michael L. and Molina, David J. "Residential Water Demand." *Land Economics*. August 1989.

Seasonal Rates - The Pros and Cons:
A Case Study

Frank Gradilone III &
Mark D. Rothenberg (Member ASCE)[1]

Abstract

The Spring Valley Water Company, an investor owned water utility serving over 58,000 customers in Rockland County, New York, implemented a seasonal differential rate structure in 1980. Analysis of summer peaking since the implementation of the rates clearly shows that they have had a significant impact on peak day loads. The ratio of the annual peak day to the average day has decreased by about 15%. On the positive side, the decrease in peaking has allowed the Company to delay the construction of a new, expensive capital project, and based on surveys of customers, has changed the summer water use habits of a significant portion of the company's customers. On the negative side, the rates caused a huge increase in customer complaints (especially during the first year of implementation), has led to customer confusion about rate levels, and highlighted deficiencies in standard meter reading and billing practices.

---

[1] Resources Planning Analyst, United Water Resources, 200 Old Hook Road, Harrington Park NJ 07640; Vice-President & General Manager, Spring Valley Water Company, 360 West Nyack Road, West Nyack NY 10094

Background

The Spring Valley Water Company service territory is primarily a suburban bedroom community for job centers in New York City and northern New Jersey. Residential water consumption accounts for more than 60% of total demand and the system exhibits a strong summer peak use pattern. The decade of the 1970s was the wettest on record and for the most part summer peaking in the system was lower than expected. During the 1980s rainfall was far below average, resulting in two drought periods that affected customers. The first occurred in 1980-81, and was the most severe to be felt in the area since the 1960s. The second drought, which was not as severe, occurred in 1985. In both instances the customers in the service area were exposed to the intense media coverage about the droughts and in the 1980-81 drought, customers were subject to a ban on non-essential outdoor water use and an alternative day sprinkling ban was in effect. The summer winter rate structure was implemented in 1980. Because of the weather patterns that prevailed during the 1970s and 1980s, and the confounding factor of the drought emergencies, assessing the impact of the summer/winter rate structure proved to be very difficult in the early 1980s. A long enough data record has now been established to allow a more accurate assessment of the impact of the rate structure.

The impetus behind the imposition of summer winter rates in 1980 was the pending need for a new capital projection to meet projected peak water demands in the system. The New York State Public Service Commission, mindful of the rate impact the construction of this project would have and interested in seeing if demand could be curbed through rates, ordered the company to perform a marginal cost pricing study in early 1980. The study concluded that a summer winter differential rate structure should be implemented in the system. The summer season was defined as May 1st to September 1st. The study and the review process took longer than expected. The PSC nevertheless decided to implement the new rate structure on June 15th; well into the summer season. For the first year the differential between the summer and winter rate was set at 3 to 1 (the actual rates were $36.90 per ccf during the summer period and $11.61 per ccf during the winter). The late implementation of the rates did not afford the company the opportunity to educate customers about the need for and purpose of the rate structure, and also forced the company to prorate bills based on the number of days before and after June 15th. Customer confusion and anger were expressed in the form of thousands of phone calls and letters; over half of the company's customers contacted the company about the rates before the summer was over. (As an aside, overtime in the company's customer service department increased by 1500% over the previous year.) The differential was dropped to 1.5 to 1 beginning in 1981.

Analysis

On the positive side, the summer winter rate structure clearly has decreased the peak day-to-average day ratio for the Spring Valley system. In the 15 years before the rate structure went into effect the peak-to-average ratio averaged 1.62:1. Since the rates were implemented the ratio has averaged 1.41:1; a reduction of about 13%. This translates into a decrease in expected peak demand of nearly 3 mgd. This, coupled with a slower than expected growth rate and State mandated and company sponsored conservation efforts, has pushed the need for new peaking capacity facilities to beyond the year 2000.

In a 1991 survey of customers, fully 88% said the were aware of the rate structure, and 62% said they didn't like them. On the other hand 48% said they have changed their lawn watering, outdoor and indoor water use habits in response to the rates. The survey also showed that a substantial number of customers now fill their pools before the imposition of the rates; a clear indication that the rates have had the desired effect of shifting load out of the summer season. In short, customers may not like the summer winter rate structure but they clearly have changed customer behavior.

On the negative side of the ledger, many customers interpret the annual implementation of summer/winter rates as a general rate increase. In the three surveys the company has performed during the last decade, rates have been the number one negative issue among consumers. While rates in the Spring Valley system are higher than average, the company believes that the summer winter rate structure and the customer's perception of it has contributed to the pushing rates to the top of the customers negative list.

As noted earlier, in first year the rates were implemented the company simply prorated consumer bills based on the number of days before and after the imposition of summer winter rates. Consumers were quick to point out that this wasn't fair (eg., "I filled my pool after my meter was last read and before June 15th"). The company instituted a customer read program to alleviate the problem. A postage-paid self-read card was mailed to all customers prior to the start and end date for the rates. A high proportion of customers have participated in this program; another indication of the level of awareness and effectiveness of the program. The company has since implemented AMR and has the capability to read customer meters on demand and so the read card program is being phased out.

Conclusions

Summer winter rates appear to have been effective in reducing peak loads in the Spring Valley system and in delaying the need for new capital projects. Customers may not fully understand the rates, but they have changed consumption habits in response to their imposition. On the down side many consumers have misinterpreted the annual cycle of summer winter rates as a rate increase, and this has contributed to rates becoming the number one negative issue about the company among customers. Implementing summer/winter rates, or other innovative rate structures, may also highlight deficiencies in a water system's meter reading and billing system.

THE EVOLUTION OF CONSERVATION RATES IN PHOENIX, ARIZONA

Edward G. Blundon & Jeffrey S. DeWitt[1]

Abstract

Conservation or increasing block rates first implemented in Phoenix, Arizona in 1982 were reexamined by a three year study involving Water Department staff, a Citizen Rates Committee and consultants. The study recommended to replace the increasing block rate structure which consisted of two season, three customer classes and three blocks with a simpler uniform rate structure with a three seasons, and no customer classes in 1990. The new rate structure, which actually enhanced conservation savings, uses a new cost of service concept that allocates extra capacity cost to the months when the capacity is used, This increases the seasonal differential in the rates to enhance conservation savings.

Background

Prior to 1982, the City of Phoenix maintained the traditional "cost of service" or decreasing block rate structure. This traditional rate structure assesses rates at a decreasing cost per unit based on data derived from historical or embedded cost. In 1982, the State of Arizona mandated conservation targets or goals in the form of a gallon per capita per day (gpcd) requirement. In response to this new requirement, the Water Services Department developed a "conservation" or increasing block rate structure. An increasing block rate structure charges a higher per unit charge for the upper blocks as an economic incentive for customers not to use more water. The new conservation rate consisted of three customer classes, residential, commercial/institutional and industrial. The residential rates were substantially higher than the nonresidential rates. In addition, three consumption
----
1. Assistant Director for Administration and Economic Analyst respectively, Phoenix Water Services Dept., 455 N. 5th St., Phoenix, AZ 85004.

blocks set zero at ten hundred cubic feet (Ccf), eleven to twenty-five and over twenty-five Ccf were in the structure design. Finally, the rate design contained a monthly service charge that varied by meter size.

Evaluation Study of the 1982 Conservation Rate

The 1982 conservation rate was reexamined in 1987 by the Water Services Department in conjunction with a newly established Citizen Water Rates Committee and consultants. The goal of the study was to examine the current rate structure to determine if another alternative rate structure might improve water conservation. The goals or objectives set for an alternative rate structure were as follows:

1. Raise sufficient revenue--REVENUE SUFFICIENCY.
2. Apportion charges according to responsibility--EQUITY.
3. Encourage the optimum use of the resource--EFFICIENCY.
4. Be Understandable and accepted by the public--SOCIAL ACCEPTABILITY
5. Be easily implemented and not require constant revision--PRACTICAL FEASIBILITY.
6. Encourage WATER CONSERVATION

Over nearly a three year period, several different alternatives were examined and measured against the rate objectives. The alternatives examined ranged from small modifications to the current rate structure by changing from three to two blocks, to structures where the blocks are determined by the individual customers peaking pattern. From examining the numerous alternatives, different factors or issues were identified that were desired in a new water rate structure. These issues are listed below.

1. Increase the conservation incentive.

2. Simplify the structure for customer understanding and for forecasting annual revenues.

3. Maintain stable bills for low water use customers.

4. Move away from embedded or historical based rates and towards marginal cost pricing (charges based on the incremental cost to purchase the next unit of water).

5. Eliminate subsidies between customer classes unless data supports cost of service differences.

6. Reflect any seasonal cost of service difference in the rates.

The New Conservation Rate Structure

After nearly three years of examining different rate alternatives, a new water rate structure was developed. The rate structure developed used the traditional base-extra capacity method with two major changes. These changes related to the allocation of extra capacity costs and the elimination of customer classes. The customer class issue was address by an analysis of peaking characteristics by customer class. This analysis indicated that there were larger differences in peaking characteristics within classes than between the classes. Based upon this information, the classes were eliminated.

The extra capacity cost allocation issue relates to the supply situation in Phoenix and a philosophy about allocating embedded costs. In Phoenix, more expensive supplies are required in the hotter months and the treatment plants and distribution systems are largely sized to meet summer peak demands. Because of the much higher cost to deliver water in the summer, particularly when one considers the cost to size the plants to meet the peak demands, the rates were developed to reflect this cost differential. Cost were separated into those cost necessary to meet minimum month water demand, the base costs, and those cost to meet demand above the minimum month, extra capacity costs. In a break from the traditional base-extra capacity methodology, extra capacity costs were assigned only to those months when the capacity was used. This cost allocation distinguished three seasons termed the high, medium and low seasons. The high season consisted of the June to September period, the low season the December through March period and the Medium season the months of October, November, April and May. By allocating the bulk of the extra capacity costs into the high season, a large differential was established between the high and low seasons. This differential encourages conservation by assessing the higher rates when the demand is more price elastic or price responsive.

The service charge aspect of the previous rate structure was maintained, although at a lower overall level. The service charge was designed to recover billing and meter reading and maintenance costs that were disproportionate for some meter sizes in the previous rate structure.

A major policy in selecting a new rate structure was to maintain stable monthly water bills for low water users. An additional goal was a rate structure that more closely reflected the marginal or incremental cost to provide an additional unit of water. Whenever rates are based on marginal costs, and the marginal cost are higher than average costs, a revenue surplus occurs. By providing a small block of consumption in the service charge at a zero cost per unit, the cost per unit above this block can be raised closer to the marginal cost without over recovering revenue. This concept also provides stable monthly bills for customers within the established minimum block and an additional conservation incentive for customers consuming outside the minimum block. The minimum block was termed a lifeline block because it provided the necessary monthly consumption for the average sized single-family household with the latest in water conservation technologies. The level of six Ccf or 7,480 gallons per month was raised to ten Ccf in the High Season to provide for minimal landscaping requirements. The resultant rate structure is illustrated in the table below.

Table 1

PHOENIX WATER RATE STRUCTURE (JUNE 1992)

SERVICE CHARGE VARIES BY METER SIZE
FROM $5.43 FOR A 5/8" METER TO $35.25 FOR A 6" METER

VOLUME CHARGE-ALL CUSTOMERS

	0-6 Ccf	Over 6 Ccf
Low Months (Dec,Jan,Feb,Mar)	$0.0	$0.79
Medium Months (Apr,May,Oct,Nov)	$0.0	$0.95

	0-10 Ccf	Over 10 Ccf
High Months (Jun,Jul,Sep,Oct)	$0.0	$1.24

Conclusions

The new conservation rate structure developed by Phoenix has been estimated not only to have maintained the conservation effectiveness of the previous increasing block rate structure, but to have actually enhanced it. Consultant estimates indicate that the rate should decrease annual water demands by an estimated 1.7 percent and lower peak month demands by over 3.6 percent. In addition, the new rate is simpler to understand by the consumer than the more complicated increasing block rates.

The Phoenix Water Services Department believes it has made significant strides over the previous water rate structures by developing a rate structure that more closely reflects the marginal cost of providing water service, maintains conservation effectiveness and is simpler to understand. Finally, stable monthly water bills are provided to low water using customers, and annual revenue estimates and rate updates are simpler.

Optimal State Feedback Estimation in Groundwater: Application to leaky aquifers

Mohamed M. Hantush[1] and Miguel A. Mariño[2], M. ASCE

Abstract

The problem of groundwater flow in leaky aquifer is looked at from a stochastic point of view. Optimal estimation of hydraulic heads are achieved using Kalman filtering. The Kalman filtering recursions are based on quasi-analytical solutions of the governing mean and stochastic flow equations using the Galerkin finite element method and matrix exponential.

Introduction

Groundwater flow instigated by naturally variable recharge is better described within a stochastic framework. The complex nature of natural recharge in space and time and uncertainty due to lack of sufficient information are among the reasons that justify implementing a stochastic framework. Among the endeavors in solving groundwater flow problems subject to stochastic recharge are *Gelhar* [1974], *Sagar* [1978], *Unny* [1989], *Takagi and Harada* [1990], *Cheng and Laffe* [1991], and *Van Geer et al.* [1991].

In the current effort, the continuous time solutions of the finite-element version of the governing mean and stochastic flow equations are expressed in terms of state transition matrices. The nicety of the continuous time solutions is their capability to account for continuous time stochastic recharge process. Furthermore,

---

[1]Research Assistant, Department of Land, Air and Water Resources, University of California, Davis, CA 95616.

[2]Professor, Department of Land, Air and Water Resources and Department of Civil Engineering, University of California, Davis, CA 95616.

their generality to account for continuous time boundary conditions (e.g., fluctuating river stage). Aside from furnishing exact solutions on the time domain, the continuous time solution also reduces considerably the computational effort that would otherwise be required using finite differences. For a white-noise recharge process, the resulting stochastic dynamic system is that of a multivariate Ornstein-Uhlenbeck process.

Application

The mean and the stochastic continuous time solutions have the form (Hantush, 1993)

$$H(t) = \Phi(t,t_0) H(t_0) + B^{-1}A(I - \Phi(t,t_0))A^{-1}U \qquad (1)$$

$$h(t;\omega) = \Phi(t,t_0) h(t_0;\omega) + \int_{t_0}^{t} \Phi(\tau,t_0))A^{-1}K\xi(\tau;\omega)d\tau \qquad (2)$$

in which $H(t)$ and $h(t;\omega)$ are the mean and stochastic head vectors of the finite element nodes, respectively; $\Phi(t,t_0)$ is the state transition matrix from time t_0 to time t; A, B, and K are finite-element matrices; and $\xi(t;\omega)$ is the stochastic recharge vector. if the recharge matrix, K, is defined appropriately, then $\xi(t;\omega)$ can be a stochastic vector of random boundary conditions. The Kalman filtering recursions are based on the continuous time solutions given by (1) and (2).

The robustness of the continuous time solutions in optimal feedback estimation is illustrated by applying it to leaky aquifers such as the one depicted in Fig. 1. The stochastic leakage is attributed to random fluctuations of the water table within the upper leaky unconfined aquifer. The location of piezometric head measurements are assumed to be confined to a few observation wells (Fig. 1.). Multivariate Gaussian and independent water-table fluctuations are synthesized within each simulation time step assuming an exponential spatial covariance structure. Subsequently, they are used to generate piezometric head data.

Figure 2 displays the optimal estimators corresponding to the piezometric head, φ, at the spatial point b. It is conceded at this moment that the optimal estimator consists of the optimal estimate φ_{est} and the reliability of the estimate is given by the 95% confidence interval $\varphi_{est} \pm 1.96 (p_k)^{1/2}$, in which p_k is the variance of error of the estimate at the discrete time t_k. Overall, the optimal estimates of piezometric heads at point b compare favorably well to the actual (generated) values when contrasted to the smoother mean values. Statistical conditioning on available field measurements is quite evident on recouping the actual variability of head not only in space but also in time as shown in Fig. 2.

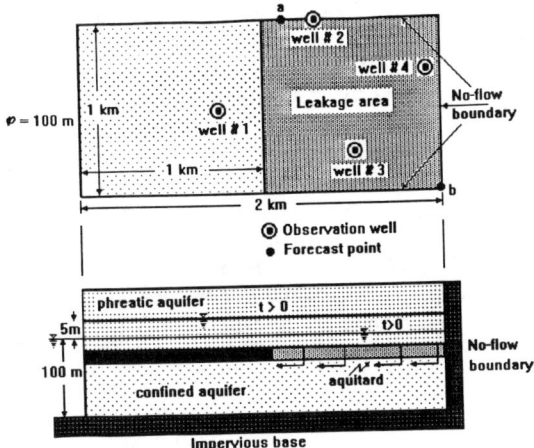

Fig. 1. Plan view and cross-section of the leaky aquifer used in the numerical experiments.

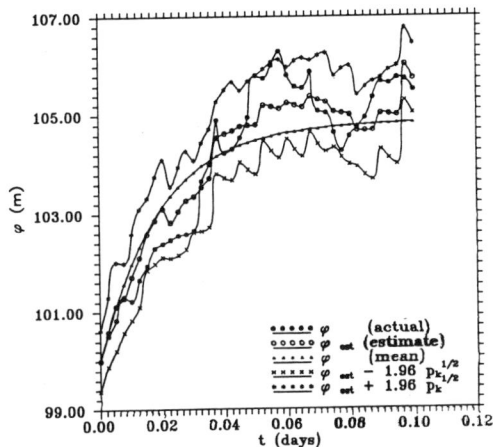

Fig. 2. Comparison of estimated and actual (generated) piezometric heads and the 95% confidence interval for point b.

REFERENCES

Gelhar, L. W., Stochastic analysis of phreatic aquifers, Water Resour. Res., 10(3), 539-545, 1974.

Hantush, M. M., Stochastic Analysis and Optimal Estimation in Groundwater,

Hantush, M. M., Stochastic Analysis and Optimal Estimation in Groundwater, Ph.D. dissertation, University of California, Davis, Calif., 1993.

Sagar, B., Analysis of dynamic aquifers with stochastic forcing function, Water Resour. Res., 14(2), 207-216, 1978.

Takagi, F. and M. Harada, Study on the spatial and temporal behaviors of unconfined groundwater head-field in heterogeneous region, Memoirs of the Faculty of Engineering, Nagoya University, 42(1), 1990.

Unny, T. E., Stochastic partial differential equations in groundwater hydrology, Stochastic Hydrol. and Hydraul., 3, 135-153, 1989.

Van Geer, F., C. B. M. Te Stroet, and Y. Zhou, Using Kalman filtering to improve and quantify the uncertainty of numerical groundwater simulations: 1. The role of system noise and its calibration, Water Resour. Res, 27(8), 1991.

FLOW INVESTIGATION FOR LANDFILL LEACHATE

by

Reza M. Khanbilvardi[1] and Shabbir Ahmed[2]

ABSTRACT

A two-dimensional unsteady-state moisture flow model has been developed in order to describe the leachate flow process in a landfill. The unsteady variation of leachate mound head has also been considered in the saturated zone of the landfill to compute the time-varying leachate flow rates in both the lateral and vertical directions. The contribution of precipitation to the landfill leachate has been investigated by computing evapotranspiration and surface runoff due to side slope. The model was used to simulate the leachate flow rates in Section 6/7 of Fresh Kills landfill, situated in Staten Island, New York. A comparison of the results was made with the Hydrologic Evaluation of Landfill Performance (HELP) model which is based on a quasi two-dimensional approach. Comparisons were also made with the results obtained from previous studies using the Environmental Protection Agency (EPA) water balance model and investigating the real field condition. An underestimate of the surface runoff was observed in the case of the results obtained by the HELP model. The simulated leachate flow rates by the new model were found to be less than those obtained by other methods. The effects of the variation of the boundary condition, which depends on surface runoff and evapotranspiration, were examined to arrive at the better representation of the two-dimensional unsteady mechanism of leachate flow process in a landfill.

---

1. Professor of Civil Engineering and Director of Hydraulics Lab., The City University of New York, The City College, N. Y. 10031.

2. Ph.D. Candidate, Department of Civil Engineering, The City University of New York, The City College, N. Y. 10031.

INTRODUCTION

The estimate of leachate rate in a landfill site is of considerable importance in the design of an appropriate collection system or the treatment alternatives to reduce the offsite migration that might pollute both surface water and ground-water resources. The hydraulics of leachate accretion in the unsaturated zone and the variation of leachate mound in the saturated zone are relatively complicated due to the heterogeneity of the landfill matrix. The nonlinear nature of the governing equations in both the unsaturated and saturated zones of the heterogeneous media of the landfill system, has encouraged researchers to employ numerical solutions for representation of the real field problems. However, some simplification in the computational scheme have been employed by considering the water balance upon the landfill surface. In the water balance method, the amount of water percolating through the solid waste is obtained by subtracting surface runoff, change in soil moisture and evapotranspiration from the total precipitation. The portion of precipitation that remains after surface runoff, change in soil moisture and evapotranspiration is considered to be instanteneously flowing as leacheate through the landfill. The actual process of leachate accretion, retention and accumulation inside the landfill matrix is not considered. The Environmental Protection Agency (EPA) model was developed by Fenn et. al. (1975) is one example of such models using the water balance method.

In our study, a two-dimensional, unsteady state Flow Investigation for Landfill Leachate (FILL) model has been developed. In the FILL model, moisture content is computed at every grid point in a vertical plane. The governing equation was obtained from the mass conservation principle which has been used in the present FILL model to predict the movement of the leachate mound head and to compute the variation of leachate flow rate from the landfill.

BACKGROUND

The mathemathical models representing the flow phenomena in both the unsaturated and saturated zones are used to deal with the specific problems of the leachate flow in a landfill. Some phenomenological solution techniques have been developed in order to avoid the complexities involved in the solution of the nonlinear equations governing the leachate flow process. The complexities exist because of the nonlinearity in the hydraulic conductivity and diffusivity. Also due to the lack of the data to define the initial condition and the characteristics of the landfill matrix, the phenomenological solutions based on water balance procedures were developed. The EPA model is based on this method which was developed by Fenn et. al. (1975).

The EPA model does not take into consideration the actual

process of moisture transport through the refuse. The model also does not consider any percolation that occurs prior to the onset of the computation. However, a significant amount of percolation occurs during the operating life of the landfill that can be defined by the initial values of moisture content and leachate mound head.

Recent mathematical models consider moisture flow through the landfill in both the steady- and unsteady-state flow conditions. The HELP model was developed by Schroeder et. al. (1984) to obtain the leachate flow rates both laterally towards the perimeter of the collection system and vertically downward through the bottom clay layer. In this model, a steady-state solution of the Boussinesq equation is performed in order to compute the lateral flow. The vertical moisture routing is considered unsteady. The three submodels in the main model compute infiltration, vertical flow and lateral flow.

MODEL FORMULATION

Unsaturated leachate flow exists above the leachate mound in a landfill. The two-dimensional unsteady moisture flow in the unsaturated zone can be expressed as (Willis and Yeh, 1987; and Demetracopoulos et. al., 1986):

$$\frac{\partial \theta}{\partial t} = \frac{\partial}{\partial x}\left[D(\theta)\frac{\partial \theta}{\partial x}\right] + \frac{\partial}{\partial z}\left[D(\theta)\frac{\partial \theta}{\partial z}\right] - \frac{\partial K(\theta)}{\partial z} \qquad (1)$$

where,
θ = volumetric moisture content (L^3/L^3),
$D(\theta)$ = soil moisture diffusivity (L^2/T),
$K(\theta)$ = unsaturated hydraulic conductivity (L/T),
t = time (T), and
x and z are lateral and vertical coardinates (L).

An implicit finite-difference expression of equation (1), similar to that employed by Korfiatis (1984) for one-dimensional moisture flow equation, is used for the computation of moisture content at the finite-difference grid points in a vertical section. The upper boundary condition in the finite-difference network of the landfill section is defined by considering the input as the precipitation less surface runoff and evapotranspiration. The moisture gradient along the lower boundary (at the bottom of the unsaturated zone in the landfill) is assumed to be zero which in fact implies free drainage due to gravity. The moisture contents obtained by solving equation (1) in the finite-difference network are used to compute the leachate accretion.

The leachate flow, laterally towards the collection system on the perimeter, and vertically through the bottom clay depends on the leachate mound head variation in the saturated zone at the bottom of the landfill.

The runoff at the upper boundary considers the effect of slope and surface roughness. The runoff submodel is based on the solution of the kinematic wave equation using an explicit scheme at the upper boundary.

The kinematic wave equation is transformed into a system of algebraic expressions using a finite-difference scheme for a two-dimensional grid in x-t plane. Different finite-difference expressions yields different numerical schemes associated with their own stability and convergence criteria.

The runoff submodel runs for 24 hours a day when the temperature is above freezing (32 F) and when there is rainfall and/or snowmelt on the landfill surface(cover).

The evapotranspiration is computed by the modified Penman method which was also used in HELP. The modified Penman method is described in detail by Schroeder et. al. (1984).

FIELD APPLICATION

The FILL and the HELP models were applied to simulate the leachate flow in section 6/7 of Fresh Kills landfill maintained by the New York City Department of Sanitation. The 3000 acre (1,215 hectare) facility has been receiving household waste since 1947. The site area consists mainly of wet land bounded on the south by Arthur Kill Road and on the west by Arthur Kill. The landfill has been developed into four distinct sections which correspond to the areas designated as section 1/9, 2/8, 3/4, and 6/7. The present simulation by the FILL model was carried out only for Section 6/7 to estimate the variation of leachate flow rates.

SUMMARY AND CONCLUSION

The application of the quasi two-dimensional HELP and strictly two-dimensional FILL models to section 6/7 of Fresh Kills landfill helps quantification of leachate flow rates for the conditions in a real field. Refinements in the leachate flow estimates were achieved by comparing the model results with the existing data. The model simulations demonstrate that the computation of surface runoff and evapotranspiration are extremely important to get the real picture of moisture flow onto the landfill leachate mound from which the leachate flow rates occur in both the lateral and vertical directions. The comparison of the results of the FILL model and HELP model with the existing results by water balance method and Darcy's law, indicates that the consideration of leachate accretion, retention, and accumulation are of extreme importance in order to get more refined estimates and variations of leachate flow rates with time.

An interesting conclusion is obtained by the FILL model results for surface runoff. It is observed that a significant amount of surface runoff occurs from the landfill surface because of the side slope and less permeable daily soil cover on the operating landfill in Section 6/7. Comparison of the solution by the kinematic wave equation in the FILL model with the SCS curve number technique in the HELP model, shows the importance of considering the effect of side slope, hydraulic conductivity, and surface roughness for the computation of surface runoff.

The two-dimensional unsteady-state leachate flow model (FILL) indicates a lower value of leachate outflow from the leachate mound in Section 6/7 compared to the values obtained by quasi two-dimensional (HELP), water balance (EPA), and steady-state (Darcy's law) models. The strictly unsteady-state flow conditions describing the accumulation of leachate in the leachate mound for the time-varying boundary conditions in the two-dimensional model (FILL) gives better estimate of leachate flow by representing the field conditions more realistically.

REFERENCES

Demetracopoulos, Alexander C., George P Korfiatis, Efst. L. Bourodimos, and Edward G. Nawy, 1986. "Unsaturated Flow Through Solid Waste Landfills : Model and Sensitivity Analysis". Water Resources Bulletin. American Water Resources Association. Vol. 22, No. 4. pp 601-609.

Fenn Dennis G., Keith J. Henley, and Truett V. Degere, 1975. "Use of the Water Balance for Predicting Leachate Generation from Solid Waste Disposal Sites". Report SW-168. U. S. Environmental Protection Agency. pp 8-11.

Khanbilvardi, R. M., Fillos, J., Ahmed, S., 19902. "Modelling, Monitoring, and Evaluation of Leachate at the Fresh Kills Landfill". Report submitted to New York City Department of Sanitation, and New York State Energy Research and Development Authority, October, 400 p.

Korfiatis, George P. and Demetracopoulos, Alexander C., 1986. "Flow Characteristics of Landfill Leachate Collection Systems and Liners". Journal of Environmental Engineering, Vol. 112, No. 3. pp 538-550.

Schroeder, P. R., Gibson A. C., and Smolen M. D., 1984. The Hydrologic Evaluation of Landfill Performance (HELP) Model. Volume II, Documentation for Version 1. U. S. Army Engineer Water Ways Experiment Station, Vicksburg, M.S. Produced by National Technical Information Service, U.S. Department of Commerce, Springfield< VA 22161-.

Willis, Robert and Yeh Williams W-G, 1987. Groundwater Systems Planning and Management. Prentice-Hall, Inc. Eaglewood Cliffs, NJ 07632.

THE PULSEQUAL MODEL:
USING COMBINATIONS OF SIMPLE MATHEMATICAL EQUATIONS TO EVALUATE COMPLEX STORM EFFECTS ON WATER QUALITY

John K. Marr, P.E. [1]
Robert Eimstad, P.E. [2]

ABSTRACT

Storm discharges (stormwater and combined sewer overflows) have transient and highly variable impacts on receiving waters. Simple water quality equations have been developed to accurately model storm impacts without resorting to complex mathematical formulations. The model (called PULSEQUAL) features a user-friendly environment to model the water quality effects of dozens of outfalls over extremely long periods of record. This provides information, including the expected magnitude, duration and frequency of water quality problems, which can be used to control alternatives. Examples are provided for recent studies on the Lower Willamette River in Oregon and the St. Joseph River in Indiana.

The complexities of storm-related water quality effects often drive engineers and analysts to choose very complex methods and models to evaluate river conditions and the benefits of controls. Unfortunately, complex models often only increase the level of *precision* achieved but not the *accuracy* of model forecasts. Also, these complex approaches are costly, data intensive, and often have limited flexibility to address the many diverse combinations of river flows and rainfall conditions.

---

[1] Associate Vice President, Limno-Tech, Inc. (LTI), 2395 Huron Parkway, Ann Arbor, MI 48104, Member ASCE
[2] Program Manager, Bureau of Environmental Services, City of Portland, 1120 SW 5th Ave., Room 400, Portland, OR 97204.

This presentation introduces an elegant alternative to complex models which offers simplicity, lower cost, higher flexibility and computational reliability for simulating the water quality effects of stormwater and CSO. An engineering solution known as mathematical superposition is used in a model named "PULSEQUAL" to segment transient storm effects into components that can each be described by simple Lagrangian (moving frame of reference) algorithms and combined to simulate complex storm effects on water quality. Below is a flowchart showing the PULSEQUAL model framework.

PULSEQUAL is a user-friendly water quality model developed to implement these concepts as part of the CSO and urban stormwater management programs in Portland, Oregon and South Bend, Indiana. A full color, animated computer graphics presentation is used to present the theory and application of the PULSEQUAL program including the Portland and South Bend examples. The presentation demonstrates (on a large screen projection) how PULSEQUAL was used to evaluate the effectiveness of management options to improve water quality for 10 to 15 year combinations of rainfall and river flows. Pull down menus and interactive simulation capabilities are also featured in PULSEQUAL.

PULSEQUAL MODEL FRAMEWORK

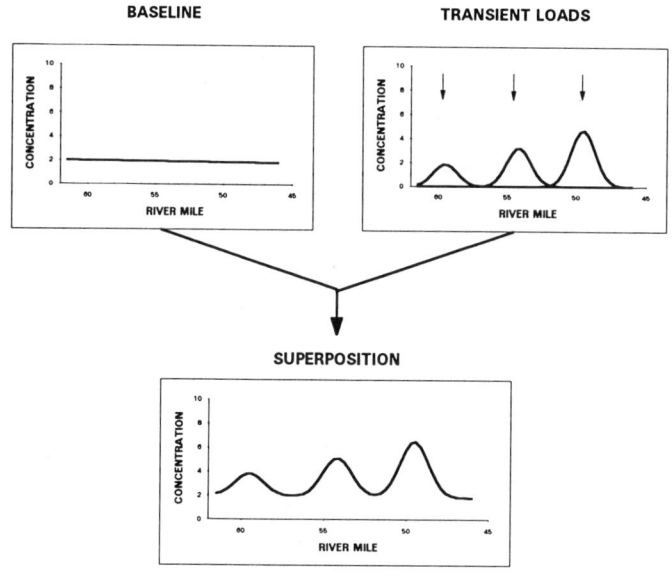

Portland, Oregon Example Results

The most easily understood demonstration of the utility of the PULSEQUAL model is for calculating bacterial effects on water quality. The model dynamically calculates the bacterial concentration in the river as affected by CSO loads, dispersion and bacterial death for each individual storm superimposed on baseline conditions, as displayed in the figure below. Note that for this storm there is a gradual building of bacterial levels during the storm (days 0 through 2) followed by gradual dispersion and die off after the storm (days 3 and 4). By day 5, the effects are flushed out of the system into the Columbia River (river mile 0.0).

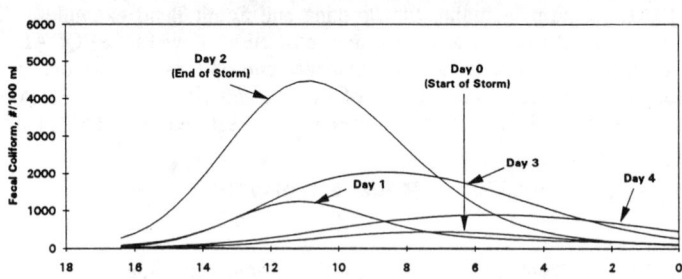

Forecasted Fecal Coliform Concentrations in the Willamette River Due to CSOs During a Large Summer Storm Event

Repetitive application of PULSEQUAL to the storms that occurred over 15 years of record provides very useful information for "bracketing" the potential benefits of CSO controls. Below is a summary figure of the forecasted average time per year that fecal coliform levels in the entire lower Willamette River will exceed various bacterial levels for the current conditions, with full sewer separation and with no CSOs at all. Potential CSO control benefits correspond to the difference between the two purely hypothetical control scenarios and current conditions.

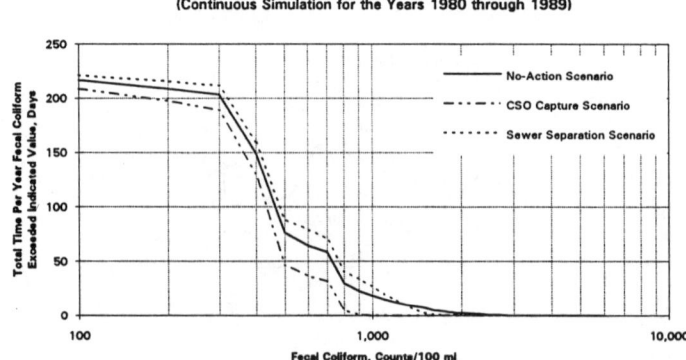

PULSEQUAL Forecasted Fecal Coliform Bacteria in the Willamette River
(Continuous Simulation for the Years 1980 through 1989)

South Bend, Indiana Example Results

A similar PULSEQUAL evaluation was conducted for South Bend CSO discharges to the St. Joseph River. However, the focus was on a slightly different concern than Portland, i.e. whether regional CSO controls were also needed to consistently meet bacterial standards. The figure shown below generated by PULSEQUAL for 10 years of storms confirmed that control of South Bend CSOs alone are insufficient and control of other municipal controls are needed to produce major water quality benefits - especially for bacterial concentrations exceeding 1,000 counts per 100 ml.

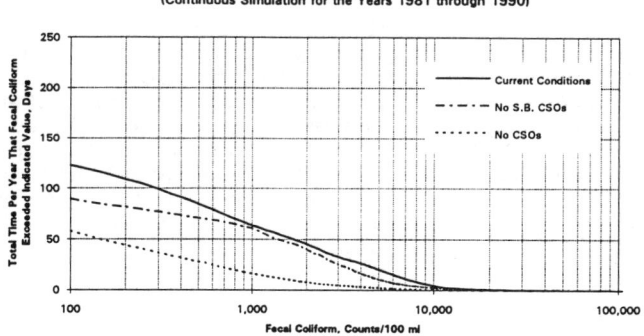

Conclusions

Simple models such as PULSEQUAL can be used to effectively assess the water quality benefits of CSO controls in both tidal and non-tidal one-dimensional river systems. They provide the ability to examine water quality responses to long periods of storm records, they are easy to understand and apply, and they provide comparisons of the magnitudes, durations and frequencies of storm effects due to transient pollutant loads. They are especially effective when insufficient data are available to use more complex model frameworks such as the U.S. EPA WASP model.

Implementing a Watershed Plan for Lake Stevens

Gene N. Williams[1]

Abstract

The water quality of Lake Stevens is declining because of watershed nonpoint pollution and cycling of nutrients from lake sediments. Hypolimnetic aeration and an alum treatment are proposed to address internal nutrient cycling. A watershed plan identified 87 measures to control nonpoint pollution. Successful implementation of the watershed measures has been hampered by the jurisdictional complexity of some measures, the difficulty of changing everyday behaviors, inadequate funding, and vague responsibilities. It is unknown if citizens and agencies can clear these hurdles.

Introduction

Lake Stevens is a 421 ha lake situated 10 km east of Everett, Washington. The lake has a maximum depth of 45 m and a mean depth of 20.5 m. The watershed is 1350 ha in size. Permanent single family residences ring the shoreline and the entire watershed is in rapid transition from rural to suburban uses. Because of its large size and accessible location, Lake Stevens is the most heavily used recreational lake in Snohomish County.

Since the 1970's the water quality of Lake Stevens has declined noticeably. Blue-green algae blooms are common in late summer and winter. In response to public concerns about poor water quality, the City of Lake Stevens and Snohomish County sponsored a limnological study of the lake. This study (Kramer, Chin & Mayo 1987) found that the lake was in the midst of rapid cultural eutrophication as a result of nutrient loading from both the watershed and the lake bottom sediments.

---

[1]Senior Planner, Snohomish County Public Works, Surface Water Management, 2930 Wetmore Avenue, Everett, WA 98201

Study Results

The 1987 study revealed that phosphorus is the nutrient in shortest supply (the limiting nutrient), so that even small additions of phosphorus result in nuisance algae growth in the lake. The annual phosphorus loading to the lake is 2000 kg of which 82% comes from internal sources--the lake sediments and waterfowl. Past watershed activities, such as logging, farming, and widespread on-site sewage systems, have deposited large amounts of phosphorus in the lake. When dissolved oxygen is depleted in the hypolimnion, which occurs every year during the summer stratified period, phosphorus is released from the lake sediments and recycled in the lake.

The limnological study also examined existing sources of nonpoint pollution in the watershed. The investigators identified widespread sources of phosphorus contributions from the suburban land uses around the lake. "Every parcel of land investigated was found to be contributing to the decline in the lake's water quality due to nonpoint pollution." (Kramer, Chin & Mayo 1987)

A lake restoration plan was developed to address both the internal cycling of phosphorus from the lake sediments as well as the watershed sources of phosphorus. Both approaches are necessary to improve the water quality of Lake Stevens.

In-lake Treatment

The lake restoration plan identified two projects to address the internal cycling of phosphorus from the sediments. First, a hypolimnetic aeration system, consisting of four partial-lift (submerged) aerators, has been designed for the lake. The purpose of the aeration is to maintain an oxygenated environment in the deepest bowl of the lake, thus restricting the anoxic release of phosphorus into the water column. Installation is planned for summer 1993.

In addition, application of aluminum sulfate (alum) to the hypolimnion is planned for 1994. The alum will remove phosphorus from the water column by flocculation and will settle to form a physical cap on bottom sediments. Both in-lake projects are being funded by 75% grants from the State of Washington. The cost of in-lake restoration, not including design, is about $1 million.

Watershed Water Quality Plan

Another key element of the lake restoration effort is a plan identifying measures to control watershed sources of phosphorus. Again with the help of state funding, the City of Lake Stevens and Snohomish County developed the Lake Stevens Watershed Water Quality Management Plan (Kramer, Chin & Mayo 1989).

The watershed plan divided pollutant sources into ten categories: ditch erosion/management, increased runoff volumes, land use washoff, construction activity, septic systems, illegal storm drain connections, spills of hazardous substances, livestock access to streams, timber harvesting, and miscellaneous sources.

For each of these categories, the plan identified best management practices and other measures to control phosphorus pollution. A total of 87 measures were recommended. The plan also listed costs, implementation steps, and the responsibilities of each jurisdiction.

The watershed management measures include six small capital projects--stream restorations and biofiltration swales. Other measures address the protection of three major wetlands near the lake and installation of fencing along stream corridors. Special studies of ditches, storm drains, and septic systems; citizen water quality monitoring; best management practices for homes; BMP's for road and drainage maintenance; public education; and water quality revisions to local codes are also included.

Because of the complex nature of the watershed actions, the plan recommended establishing a watershed steward to orchestrate implementation. The steward would coordinate the activities called for in the plan while, at the same time, serving as a sort of forest ranger for the watershed. The steward would have daily contact with citizens and respond to water quality problems.

Watershed Plan Implementation

Several key measures recommended in the watershed plan have been successfully implemented since 1989. Citizen volunteers are monitoring the lake and major streams for nutrients and other water quality parameters. One linear wetland/biofiltration swale has been constructed. Most of the remaining capital projects have been designed and await construction. Public education efforts, such as information booths, lake restoration Jeopardy games, a series of water quality tip sheets, and a video, have been successful.

However, the agencies have found it extremely difficult to implement the remainder of the measures. There appear to be four general reasons for this:

COMPLEXITY. Some measures, such as a septic survey, require coordinated efforts from several agencies. Other actions, such as code revisions, involve multiple departments and a complex political process. For example, Snohomish County's proposed drainage and erosion control standards have been in limbo for four years because of their linkage to controversial stream and wetland buffer requirements being juggled by two departments, citizen committees, and elected officials.

COMMONPLACE NATURE OF POLLUTANT SOURCES. Most of the activities that contribute phosphorus to the lake are not illegal. They are pervasive in everyday life. Restricting or eliminating the use of lawn fertilizers, or cleaning up after the family dog, require changes in life-long habits. Such changes are not made easily. Continuing public education and a community stewardship ethic are necessary to motivate changed behaviors.

LIMITED FUNDING. Implementing all non-structural measures in the watershed plan would cost over $100,000 per year. Since 1989 nearly all available funds from local jurisdictions have been devoted to designing and constructing the in-lake measures and watershed capital projects. Funds to implement the watershed measures are simply not available.

LACK OF FOCUS. Specific agency staff have not been assigned to implement watershed measures. The citizen's group at Lake Stevens was successful in getting the local agencies committed to lake restoration, but since then, citizens have not stepped in to take responsibility for implementing the plan. This points to the need for a watershed steward as the focal point for implementation.

Conclusion

Restoring Lake Stevens requires actions in the lake to reduce nutrient cycling and in the watershed to control nonpoint pollution. Watershed measures have been difficult to implement because of institutional roadblocks, limited funding, and slow-to-change habits. Even at the most important lake in the county, it is still an open question if citizens and agencies are willing to take the steps necessary to protect water quality.

Appendix. References

Kramer, Chin & Mayo. (1987). *Lake Stevens Restoration Phase IIA*.
Kramer, Chin & Mayo. (1989). *Watershed Water Quality Management Plan*. Prepared for the City of Lake Stevens and Snohomish County.

DESIGN OF AQUATIC TREATMENT SYSTEMS
Robert B. Aldrich [1]

INTRODUCTION AND ABSTRACT

This paper describes the design parameters and operating procedures used for a nonpoint pollution/stormwater treatment facility known as an Aquatic Treatment System (ATS), to be located in the City of Lacey, Washington. The facility is essentially a flow-through biofilter, utilizing a detention area and constructed wetland for polishing treated stormwater prior to discharge.

The design parameters were compiled from existing literature sources (Gearhardt et al, Hammer et al, Horner et al, and Schueler et al) and were based on pilot plant studies conducted in 1985 (DOE, 1985). The facility can be adapted to existing wet detention ponds or proposed detention ponds. Common operational parameters include the need for consistent water supplies, adequate land area for wetland treatment, and proper hydraulic head conditions.

DISCUSSION

Factors involved in design, including siting and major components, are discussed below.

Site Considerations

Proper siting is the most important factor in the success of an ATS. Siting considerations include:

- *Presence of base-flow or groundwater for at least 10 months per year.* Adequate water supplies for plant growth are necessary throughout the year, although many wetland species can withstand periodic dry soil conditions Conversely, the design of the facility must allow complete drainage for maintenance purposes.
- *Space for facility sizing.* Adequate area must be provided for sufficient detention storage, maintenance access, and inlet/outlet facilities. In addition, facilities in residential areas may require land area for visual buffers.
- *Sufficient elevation difference between inlet and outlet to attain hydraulic head to drive the system.* At least one to two feet of elevation difference must be available between the inlet and outlet to derive hydraulic head sufficient to drive the system. As it is a gravity flow facility, the more elevation difference between components, the more easily water will flow through the facility.
- *Presence of a impermeable layer beneath components of the facility.* Contact with groundwater sources may be necessary to ensure survival of wetland plant

---

[1] Robert B. Aldrich is a watershed analyst with the firm of Kramer, Chin, and Mayo, Inc, in Seattle, Washington.

species within the facility. However, contamination of groundwater must be prevented by installing an impermeable layer beneath components of the facility receiving first-flush plug flows.
- *Position within the subbasin.* Maximum benefits from an ATS can be achieved by placing the facility at a strategic location in the basin, such as downstream from known loading sources, or near the inlet of a tributary to a receiving water.
- *Whether the facility is to be new or a retrofit of an existing facility.* Although not of prime importance in siting, facility size can be influenced to a great degree by whether the facility is to be new construction or a retrofit of an existing system in terms of cost, degree of difficulty in design, flexibility and applicability.

DESIGN CONSIDERATIONS

Design considerations for the City of Lacey ATS system are listed and described in detail below.

Design Storm

The design event for water quality treatment, as required by the stormwater manual released by the Washington State Department of Ecology (Ecology), is two-thirds of the 2-year, 24-hour event for present and future conditions. The facility was designed to detain and treat this event and bypass larger events. All overflow structures were designed to convey the 100 year, 24-hour event.

Major System Components

Although seemingly complex, the Lacey ATS can be broken down into component parts to be analyzed individually. The major components of the facility are listed and discussed in more detail below.

- Flow Distribution Manhole/Energy Dissipator
- Primary Treatment Cell
- Flow Control/Containment Berms
- Constructed Wetland
- Outfall Structure.

Flow Distribution Manhole/Energy Dissipator

The flow distribution manhole/energy dissipator will be located at the uppermost portion of the facility. The dissipator is designed to convey flows from the outfall to the primary treatment cell. The flow distribution system will consist of an oversized manhole with risers to achieve the requisite elevation for flow conveyance. Energy dissipation will be provided by the size of the manhole and the outfall pad located outside the manhole.

An overflow orifice will be set on the downstream face of the structure to prevent surcharging of the drop manhole. Trash racks will be placed to prevent vandalism, plugging of the facility, and entry by unauthorized personnel.

Primary Treatment Cell

The purpose of the primary treatment cell is to receive and treat the first-flush and subsequent plug flows of concentrated pollutants from the stormwater conveyance system.

The primary treatment cell is an intensively planted constructed wetland. A packed macrophyte bed that lines the bottom of the cell bed will be planted with wetland vegetation known to attenuate relatively high concentrations of oil and grease in stormwater discharges (Horner et al. 1991). A 12-inch-diameter perforated pipe will be placed in the existing channel that conveys flows through the proposed site, and covered with quarry spalls. This pipe will serve to completely drain the structure for periodic maintenance and will be controlled with a valve at the downstream end. A gravel substrate approximately 6 inches deep will be placed in the bottom of the cell. Approximately 6 inches of sand will be placed atop the gravel to serve as a planting medium for emergent vegetation. Species such as *Typha latifolia* (common cattail) and *Scirpus acutus* (hardstem bulrush) will be planted in the treatment bed.

Flow Control/Containment Berms

The flow control/containment berms form the primary and secondary treatment cells. The flow control/containment berms include a berm spanning the downstream portion of the proposed detention site approximately in line with the existing topographic features. This berm will have a 3:1 side slope rising from ground level to a minimum of four feet above ground level at the initial overflow. The emergency overflow will be set at one foot above this elevation.

A second flow control berm will be located upstream from the first downstream berm, at an area where the topography distinctly changes from flat floodplain area to steep, incised ravine area. The second berm will have a 3:1 side slope rising from ground level to a minimum of six feet above ground level at the initial overflow elevation. The emergency overflow will be set at one foot above this elevation.

A flow-directing berm will be placed within the cell to increase contact time. The flow-directing berm will have a 2:1 side slope rising from ground level to a minimum of 3 feet above ground level.

The berms will be constructed with compacted soil. Vegetation such as *Salix spp.* (willow) will be planted on the faces of the berms. The vegetation will help stabilize the berms, provide embankment protection during peak events, and provide habitat amenities for wildlife. Both the upstream and downstream sides of the embankment will be armored to prevent erosion caused by overtopping during extreme events. An armored spillway will be placed in the containment berms to control overtopping and anchor the level control structures. The outfall weir will be designed to accommodate extreme events to reduce the likelihood of overtopping at berms.

Constructed Wetland

The constructed wetland consists of two treatment cells that will operate as polishing ponds for treated effluent from the primary treatment cell. The cells are intended to remove pollutants such as nitrogen, BOD, colloidal suspensions, sediments, fecal coliform bacteria, and metals.

Each cell will be planted with appropriate wetland vegetation. The open water areas will be planted with aquatic macrophytes such as *Lemna minor* (common duckweed) and *Nuphar Polysepalum* (yellow pond lily). Emergent vegetation will be planted on the fringes of the open water area, and scrub/shrub communities will be planted in the transition zone between the wetland and the upland.

Outfall Control Structures

The outfall structure will consist of a broad-crested weir, armored with rip-rap on the upstream and downstream faces. A concrete pad will form the spillway/overflow portion of the structure, with stoplog supports (e.g., removable bollards) formed into the pad. Bollards in the weir structures will be removable to facilitate maintenance. The outfall structure on the downstream end of the facility is identical to the outfall structure that connects the primary treatment cell to the constructed wetland.

EXPECTED LEVELS OF WATER QUALITY IMPROVEMENT

The facility is expected to be consistently efficient at removing nitrogen, organics, sediments, and oil and grease. Removal percentages could be as high as 90 percent for nitrogen (seasonally) and sediments; 70 to 80 percent for fecal coliforms; and 80 to 95 percent for metals and hydrocarbons (Schueler 1990; Hammer et al. 1990). Theoretical pollutant removal expectations for the individual cells and the facility as a whole are shown in Table 1.

TABLE 1. POLLUTANT REMOVAL EXPECTATIONS FOR AQUATIC TREATMENT SYSTEM

Cell	Pollutant	Expected Percent Removal
Primary	Hydrocarbons	70 to 80
	Nitrogen	70 to 85 (seasonal)
	Bacteria	50 to 70
	Sediments	70 to 85
Constructed Wetland	Hydrocarbons	70 to 80
	Nitrogen	60 to 80 (seasonal)
	Bacteria	50 to 70 for fecals
	Sediments	60 to 80
Combined	Hydrocarbons	80 to 95
	Nitrogen	60 to 90 (seasonal)
	Bacteria	70 to 80
	Sediments	70 to 90

Source: Hammer et al. 1990

OPERATION AND MAINTENANCE CONSIDERATIONS

To track operating efficiency, flow and water quality monitoring should be conducted in all cells of the facility. In addition, periodic maintenance should be performed.

Maintenance of the facility should consist of the following tasks:

- Remove invader/weed species
- Periodically thin vegetation in the packed macrophyte bed
- Lubricate slide gates
- Clean downdrain systems
- Remove sediments
- Harvest vegetation in the packed macrophyte bed and wetland areas.

- Replace landscaping
- Thin or replace vegetation on berms
- Replace and interplant wetland vegetation
- Clean and inspect diversion structure
- Irrigate during low rainfall months.

SELECTION OF OPTIMAL BEST MANAGEMENT PRACTICES (BMP's)
Thomas R. Sear[1], M. ASCE and Ronald L. Wycoff[2], M. ASCE

Abstract

The National Pollutant Discharge Elimination System (NPDES) urban storm water program requires municipalities to remove non-point source (NPS) pollution to the "maximum extent practicable", which implies that the selection of best management practices (BMP's) should include a recognition of fiscal constraints. A design technique is presented that allows facility designers to select the most appropriate combination of BMP's; and, minimize cost for fixed pollutant removal goals, or maximize pollutant removal for a fixed monetary budget. This optimization technique is based on the application of production theory, which requires the development of production functions and cost functions for each candidate BMP.

Introduction

The NPDES storm water program requires that municipal separate storm sewer systems (MS4) develop a storm water management program that reduces the impact of non-point source pollution on receiving water bodies. The NPDES program states that the level of pollutant removal should be provided to the "maximum extent practicable" (MEP), which implies that the method used to select and implement BMP's should include an economic analysis and a recognition of fiscal constraints. Given the wide range of structural and non-structural BMP's available, design techniques are required that will allow storm water facility designers to select the most appropriate combination of treatment alternatives. Ideally, the final BMP combination will minimize cost, given fixed pollutant removal goals; or, maximize the pollutants removed, given a fixed monetary budget.

---

[1]Water Resources Engineer, CH2M HILL, 225 E. Robinson Street, Suite 405, Orlando, Florida 32801, (407/423-0030), (Fax: 407/839-5901)

[2]Senior Water Resources Engineer, CH2M HILL 7201 NW 11th Place, Gainesville, Florida 32605, (904/331-2442), (Fax: 904/331-5320)

This technical paper presents a design technique that allows the facility planner to identify optimal BMP combinations; and, presents the results of a case study using this approach, which demonstrates the economic benefit that can result from such an analysis. The optimization technique is based on the application of production theory to storm water pollution control, which requires the development of production and cost functions for each candidate BMP. Production functions relate the level of effort expended to the level of pollution removal achieved; and, are developed in part through the use of continuous simulation runoff files. The Storm Water Management Model (SWMM4) is used to simulate land surface runoff, given the hydrologic characteristics of the drainage basin and an extended period of meteorological data. SWMM runoff files and specific pollutant removal techniques are then used to develop site specific production functions.

Analysis

BMP production functions are in general nonlinear and governed by the "law of diminishing returns"; whereby increasing levels of effort will result in marginally smaller pollutant reductions. A BMP "level of effort" may be the size of a storm water treatment basin, the frequency of street sweeping, or some other characteristic that relates both to the pollution control achieved and to annual cost.

A case study was prepared to demonstrate the production function design technique. The SWMM4 model was used to generate a ten year record of simulated land surface runoff using hourly rainfall data obtained from Jacksonville, Florida. Model parameters were assigned for an 100 acre watershed, with type "A" hydrologic soils and a directly connected impervious area of 25 percent. Production functions were developed for three different BMP's, including dry detention basins, infiltration basins, and street sweeping. Example production and annual cost functions for infiltration basins and dry detention systems are presented on Figures 1 and 2 respectively. The shape of the street sweeping production function is similar to the shape of the functions illustrated in Figure 1. However, for street sweeping, the level of effort is expressed in terms of the sweeping frequency. Annual costs are computed based on the annual number of curb miles swept and the unit cost of street sweeping ($/curb mile).

In this case study the annual cost of infiltration basins are the same as the annual cost of dry detention basins of equal volume (Figure 2). However, the simulated performance of infiltration basins is significantly better than the simulated performance of dry detention basins (Figure 1). Therefore, infiltration systems, where technically and environmentally feasible, are more cost effective than dry detention basins.

Even though infiltration basins are obviously more cost effective than dry detention basins, several questions remain. First, where infiltration basins are infeasible, what is the least cost combinations of dry detention and street sweeping ? Second,

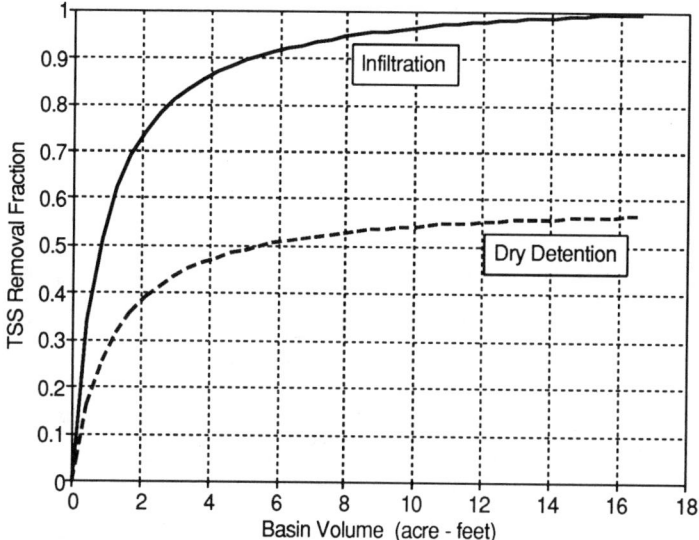

Figure 1. Production Functions for Infiltration and Dry Detention Basins

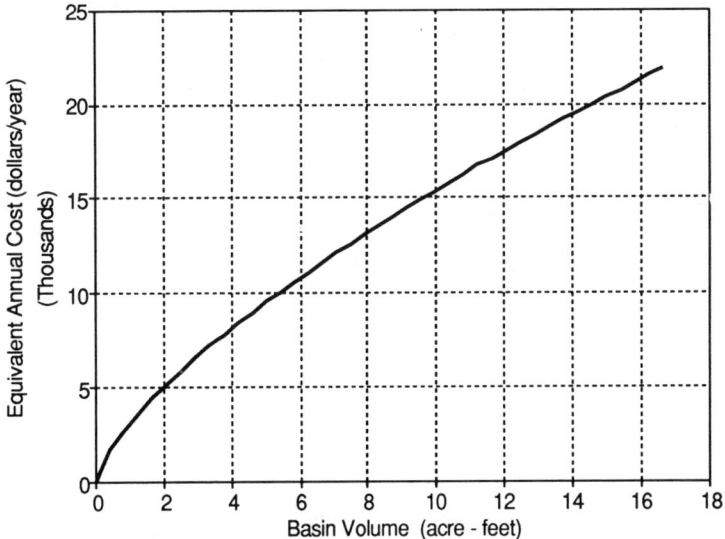

Figure 2. Annual Cost Function for Infiltration or Dry Detention Basins

should street sweeping be used in conjunction with infiltration basins to minimize the overall pollution control program cost ? Given cost and production functions for each candidate technology, these questions can be answered by economic optimization as summarized in the following table.

Table 1
Results of BMP Economic Optimization Analysis

TSS Removal Percent	Street Sweeping & Dry Detention			Street Sweeping & Infiltration		
	Sweeping Interval	Basin Volume ac.-ft.	Minimum Annual Cost $/yr.	Sweeping Interval	Basin Volume ac.-ft.	Minimum Annual Cost $/yr.
50	none	5.4	$10,100	none	0.8	$2,700
60	11 days	7.9	$22,500	none	1.1	$3,500
70	3 days	10.4	$45,400	none	1.7	$4,600
80	1.2 days	16.0	$102,600	none	2.8	$6,300
90	Infeasible			none	5.2	$9,800

When considering the street sweeping and dry detention combination, detention basins alone will provide the least cost solution for areawide TSS removals of 50 % or less. For TSS removals between 60 and 80 percent the least cost solution includes both street sweeping and detention basins. Also, overall removals of 90 % or greater are not feasible. Considering street sweeping and infiltration basins, infiltration basins alone provides the least cost solution for all levels of control.

Production theory analysis also allows the decision maker to evaluate economic trade offs at any given level of control. For example, a 70 % reduction in TSS discharge can be accomplished by providing an intensive street sweeping program (i.e. sweeping all streets every 3 days) and a large (10.4 acre feet) dry detention basin, at a cost of $45,400 per year. Alternatively, if infiltration basins are feasible and acceptable, the same level of control could be provided by a small (1.7 acre feet) infiltration basin at a cost of only $4,600 per year. These trade offs should be fully evaluated and known before finial storm water control decisions are made.

Conclusions

The NPDES storm water program will require a significant expenditure of public funds in order to apply a wide range of storm water treatment BMP's. This technical paper presents an approach to planning level decisions that allow storm water designers to select an optimal combination of BMP's, given the water quality goals, and regulatory and fiscal constraints. Such an approach will allow local governments to obtain greater water quality benefits for the funds expended and help assure the long term success of their storm water management program.

URBAN RUNOFF CONSIDERATIONS IN THE DESIGN OF WASTEWATER MANAGEMENT PROGRAMS FOR COASTAL AREAS

Larry A. Roesner[1], Fellow ASCE

Abstract

The continuing increase in the population of the nation's coastal areas is placing additional stress on coastal aquatic systems faster than it can be reduced through improved wastewater management practices. One of the pollution sources is urban runoff. The cost of removing pollutants from runoff in existing urbanized areas is extremely expensive; possible cost savings can occur if future wastewater management planning will jointly consider stormwater and wastewater in developing plans and programs for protection of coastal receiving water environments.

Introduction

At the present time, more than 35 percent (90 million persons) of the US population live in coastal areas (NOAA, 1990). And the growth is expected to continue to increase well into the 21st century. Moreover, these areas are the most densely populated areas in the US, as illustrated in Figure 1.

According to the USEPA, pollution from diffuse sources has become the largest single factor preventing attainment of water quality standards nationwide (USEPA, 1991). As a result we may expect to see increased emphasis by federal and state environmental regulation agencies on the mitigation of pollution from stormwater runoff. The current stormwater NPDES program for urban areas, and the recent tightening of USEPA's national strategy for control of combined sewer overflows is direct evidence of this emphasis.

---

[1]Senior Vice President, Camp Dresser & McKee Inc., 1950 Summit Park Drive, Suite 300, Orlando, Florida 32810-5934

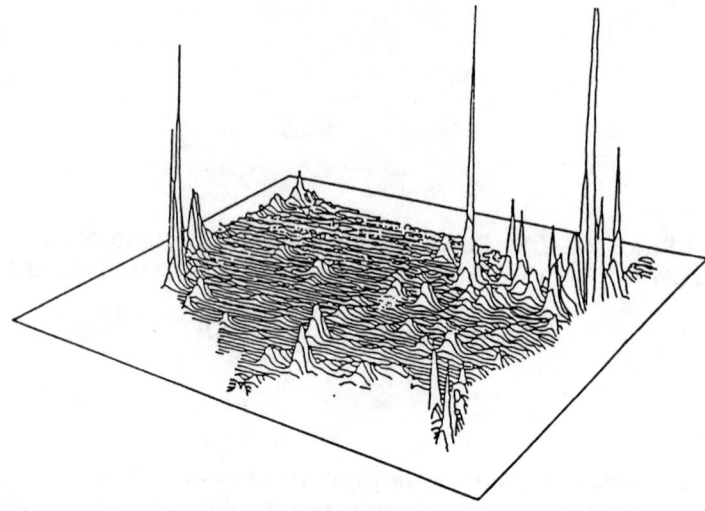

Figure 1
Distribution of Population in the United States by Region (NOAA; 1990)

Runoff quality from new development can probably be adequately controlled through a combination of source controls and structural controls that can be build into new development as it occurs. But in existing urban areas, the mitigation of pollution in urban runoff has potential to be a difficult and extremely costly undertaking. This paper presents some ideas on how joint planning for water quality management of both stormwater and wastewater in existing urbanized watersheds may result in more cost-effective reduction of wet weather pollution loads to receiving waters. They are however, not consistent with current regulatory thinking or with traditional wastewater management agency policies; therefore implementation will require a lot of convincing and possibly some modification of current rules and regulations governing construction and operation of municipal wastewater systems. The discussion distinguishes between areas with separated systems and those with combined sewer systems.

Use of POTWs for Stormwater Treatment in Separated Areas

Urban runoff is an intermittent and highly variable phenomenon; the constituent concentrations within the runoff vary, sometimes by orders of magnitude, both within an event and between events (see Figure 2 for example). In most areas of the country, precipitation varies significantly with the seasons. This is in significant contrast to wastewater systems that have relatively constant flows and quality throughout the year. Thus, conventional wastewater treatment processes do not work well for stormwater treatment, unless the stormwater is captured and slowly bled into the wastewater stream. This concept is used in several municipalities with combined sewer systems to mitigate combined sewer overflows (CSOs). But it has technical merit for use on a limited basis for reducing pollution loads from existing urban development into sensitive receiving waters, such as bathing beaches or ecologically sensitive embayments where circulation and mixing processes are weak.

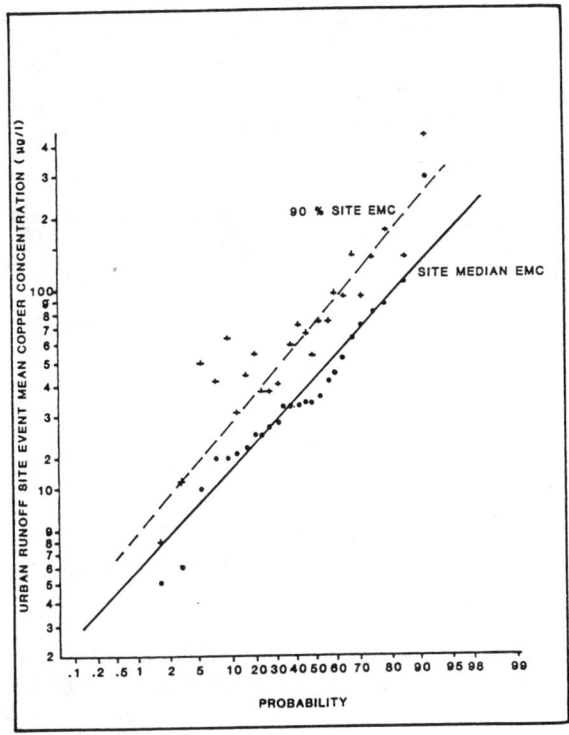

Figure 2
Site Probability Based Upon EMC (Mancini and Plummer 1986)

To illustrate, Figure 3 shows the percent capture of urban runoff, on an annual basis, for six cities as a function of basin storage volume. The curves were developed from 10 to 30-year time series of hourly rainfall data, using the hydrologic model STORM (HEC,1977), and assuming a drawdown rate of one basin volume per 24 hours. The figure shows that in most localities, capturing the first 1/4 to 1/2 inch of runoff (the equivalent of providing storage for 0.02 -

0.04 acre-ft/acre or 6500 - 13000 gal/acre) will result in capture of 80 - 90 percent of the annual runoff. This level of control will result in significant reduction of the pollution load to the receiving waters, yet the overflow frequency will be 3 - 10 times per year, so that larger storms still discharge to the receiving waters and produce the positive benefits that the flushing action of larger storms produces in the receiving waters.

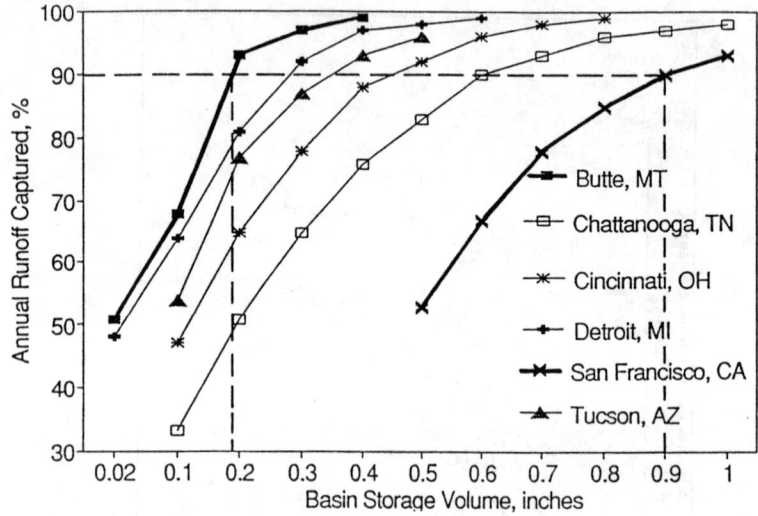

Figure 3
Runoff Capture Efficiency Versus Unit Storage Volume (Roesner et al. 1991)

At a cost of $0.50 - $0.75 per gallon of storage, the cost of such facilities will run about $3250 - $9750 per acre to capture the runoff. Added costs are the increased O&M at the municipal wastewater treatment plant. Thus, while the cost is not cheap, it should be considered and compared as an alternative to other methods for removing pollutants from runoff.

The problem with this approach is institutional and attitudinal. USEPA and state regulations generally do not allow any direct connection of a stormwater system to a sanitary system, and wastewater treatment plant operators traditionally throttle back the inflow to their treatment plants when flows increase beyond a certain level. But these regulations and operating procedures were developed before anyone ever seriously considered that urban runoff might someday need to be treated. It isn't that the sewer systems or wastewater treatment plants can not handle additional flows. Table 1 shows excess capacity, above sanitary flow, for sewer systems in 5 cities. Albeit they are combined systems the capacities shown for the interceptors and the secondary treatment processes are typical of wastewater plants for separate systems. The table does not show primary treatment capacity, but it is generally 2 - 3 times the secondary capacity. Enhanced primary treatment could be given to storm flows during wet weather periods without upsetting the biological secondary units at most plants. The cost effectiveness of this method for stormwater quality management should at least be evaluated as a part of overall wastewater management planning.

Table 1
Typical Combined Sewer System Capacities[1]

City	Combined Sewers	Interceptors	Secondary Treatment
Chattanooga, TN	1.42	0.063	0.020
Cincinnati, OH	1.14	0.015	0.003
Detroit, MI	1.02	0.024	0.006
London, England	0.71	0.043	0.006
New Bedford, MA[2]	0.39	0.026	0.000

[1]Capacities are expressed in cfs/ac or in/hr

[2]Secondary Facilities at New Bedford are not on-line

Treating Stormwater in Combined Sewer Areas

It is interesting that while combined sewer trunk systems were designed to provide both wastewater transport and stormwater drainage, the system interceptors and treatment plants were designed to handle little or no runoff. Most cities that have unimproved combined sewer systems can treat 20 - 30 percent of the runoff generated within them. Optimizing the system with low cost improvement to the transport and treatment system can increase efficiency to 50 - 60 percent capture and treatment, if primary treatment facilities at the plant are used to their capacity during wet weather periods and flows in excess of secondary capacity are bypassed after primary treatment. While some regulatory agencies are beginning to allow this, many still insist that any flow that enters the headworks of the POTW must receive secondary treatment. Thus, in most combined sewer systems, treatment plant and pump station operators throttle back inflows during wet weather to keep inflow to the POTW below the maximum secondary capacity of the plant; as a result, millions of gallons of combined untreated runoff and wastewater overflow untreated into the receiving waters, rather than receive primary treatment at the POTW. This thinking must change.

Another attitude that must change is that no additional stormwater can be introduced to a combined sewer system, even though it is generated in the area served by the combined system. Giving this runoff some sort of treatment by some "Stormwater BMP" is not nearly as effective in removing pollutants as would be enhanced primary treatment at the POTW.

Finally, when master planning for CSO mitigation, the planning process should look at all wet weather sources of pollution to the affected receiving waters; and, a plan that cost-effectively reduces these wet weather loads to a level that provides the desired beneficial uses in the receiving waters should be developed, irrespective of jurisdictional boundaries and type of system. A study of CSOs in Boston showed that even if overflow events were reduced to one per year, Boston Harbor would still experience periods of high coliform levels during wet weather periods (CH_2M Hill, 1992). This is due to the presence of coliforms in the discharges of separate systems in the area that drain to the same receiving waters. But the control or mitigation of coliform pollution from these sources was never considered in the master plan. It is likely that these systems could have been integrated into the overall master plan at significant cost savings to what will be required to correct them individually. This was the case in Milwaukee, Wisconsin where the Metropolitan Sewer District was able to use a single integrated facility plan to solve both CSO problems and sanitary sewer overflow problems at considerable cost savings in comparison to independent solutions (MMSD, 1980).

REFERENCES

CH$_2$M-Hill. Final Combined Sewer Overflow Facilities Plan and Final Environmental Impact Report. Volume II, Recommended Plan. Prepared for the Massachusetts Water Resources Authority. Boston, MA. September 20, 1992.

Hydrologic Engineering Center (HEC). Storage, Treatment, Overflow, Runoff Model "STORM", Users Manual, HEC, Davis, CA, 1977.

Mancini, John L. and Alan H. Plummer. "Urban Runoff and Water Quality Criteria", Urban Runoff Quality Impact and Quality Enhancement Technology. ASCE, New York, NY, 1986 pg. 133-149.

Milwaukee Metropolitan Sewerage District (MMSD). 1980. MMSD Wastewater System Plan Planning Report. Milwaukee, WI: Program Office, Milwaukee Water Pollution Abatement Program. June 1980.

National Oceanic and Atmospheric Administration (NOAA). 50 Years of Population Change Along the Nation's Coasts, 1960-2010. NOAA, Rockville, MD, 1990.

Roesner, L. A., E. H. Burgess, and J. A. Aldrich. "The Hydrology of Urban Runoff Quality Management," in Proceedings of the 18th National Conference on Water Resources Planning and Management/Symposium on Water Resources, May 20 - 22, New Orleans, LA, ASCE, New York, NY.

U.S. Environmental Protection Agency (USEPA), Proposed Development and Approved Guidance -- State Coastal Nonpoint Pollution Control, USEPA, Washington, D.C., 1991.

Urban Stormwater Runoff Control System
Village of Skokie, Illinois

Eddy H. Nakai, P.E.[1]
Robert W. Carr, P.E.[2]

Abstract

Combined sewer systems in many communities lack the capacity to convey stormwater runoff from even small to moderate storms. This inadequate capacity, in turn, causes basement flooding and combined sewer overflows. Traditional solutions to these problems are often the separation of the system into storm and sanitary components or the construction of large diameter relief sewers or tunnels. Unfortunately, the high costs of these solutions have often deterred the resolution of the combined sewer problems.

A runoff control system (RCS) provides a different approach to correcting combined sewer flooding problems by adding stormwater detention to the existing combined sewer system. A carefully engineered system of runoff control facilities will limit the rate of stormwater runoff allowed into a combined system to the maximum hydraulic capacity of the combined sewers.

The completely urbanized Village of Skokie, Illinois is implementing a RCS program. The system is an optimum combination of on-street storage, detention facilities and relief sewers. Nearly half of the required detention storage was accomplished by on-street ponding. Preliminary engineering studies recommended a runoff control system designed for a 10 year recurrence interval storm. The program consists of six components: flow regulators, street berms, storm collector sewers, subsurface stormwater detention tanks, surface detention basins and combined relief sewers. The cost of the recommended system is approximately one third the cost of a conventional relief sewer system providing similar protection.

---

[1] Municipal Engineer, Village of Skokie, 5127 Oakton Street, Skokie, IL 60072, Member ASCE.

[2] Water Resources Project Manager, SEC Donohue, 1020 North Broadway, Suite 400, Milwaukee, WI 53202.

URBAN RUNOFF SYSTEM

Project Description

The Village of Skokie (population 60,000) is located adjacent to and directly north of the City of Chicago. Virtually all of the Village's sewers are combined sewers. Combined sewer surcharging during wet weather, caused by undersized trunk and branch sewers, results in backups into basements throughout the Village. Traditional solutions were evaluated, but were cost prohibitive.

The 5,510 acre Village is divided into three sewer districts: the 1,255 acre Howard Street Sewer District (HSSD) in the southern part of the Village; the 2,300 acre Main Street Sewer District (MSSD) in the central part of the Village; and the 1,955 acre Emerson and Lake Street Sewer District (ELSSD) in the northern part. Land use in the Village is 80 percent residential, 10 percent industrial and 10 percent commercial. The land in the Village generally slopes eastward toward the North Shore Channel. Slopes vary from 0.1 to 1 percent and the overall slope in many areas of the Village is a flat 0.2 percent.

Precipitation is distributed throughout the year with an average annual total of 33.3 inches. For a one hour storm, the 1-, 10- and 100-year recurrence interval rainfall amounts are 1.19, 1.94 and 2.80 inches, respectively. For a 24 hour storm, the 1-, 10- and 100-year amounts are 2.21, 3.86 and 6.70 inches, respectively.

Combined sewage is carried from the Village through three 84-inch trunk sewers to the interceptor sewer owned and maintained by the Metropolitan Water Reclamation District of Greater Chicago (MWRDGC). When the interceptor capacity is exceeded, each trunk sewer overflows first to the MWRDGC TARP (Tunnel and Reservoir Plan) system and then to the North Shore Channel.

Analysis

A four-phase approach was used to identify the individual portions of the system and their relationship to the flooding problems.

The Static Condition Analysis (Phase 1) determined whether flood levels in the North Shore Channel would cause basement flooding which could not be resolved by the RCS. The results of this analysis showed that there were no significant areas in which basement flooding would result solely from high stages in the North Shore Channel.

The Sewer Capacity Analysis (Phase 2) determined the capacity available for carrying stormwater runoff. The maximum allowable regulated runoff rates ranged from 0.08 to 1.0 cfs per acre. Results of this analysis determined the capacity of the existing sewer system and formed the basis for the design of the runoff control system.

The Street Ponding Analysis (Phase 3) determined the location and extent of intentional street ponding which could be achieved through flow regulators and minor street grade modifications (berms). Higher release rates were allowed on streets with little storage available. Conversely, on streets which could pond large stormwater volumes, the release rates were regulated to use the available storage capacity. The product of this analysis included delineation of street ponding elevations and identification of the volume of additional runoff which must be detained in off street locations.

The Storage Alternative Analysis (Phase 4) was conducted to determine the most cost effective method of detention. Off-street detention was used to store runoff in excess of street ponding capacity and where street ponding was not feasible. In most on-street ponding areas, excess detention storage was accomplished with subsurface detention facilities in the street right-of-way. Where possible, excess runoff from ponding areas was conveyed to more cost effective detention facilities. In many locations, oversized storm sewers with restricted outlets were found to be more cost effective than collector storm sewers to a subsurface detention facility.

Recommended Plan

Design criteria developed for the recommended RCS plan included: 1) Design for the 10-year recurrence interval storm; 2) Reduce sewer surcharging to prevent sewer backup into the basements; 3) Use available street ponding capacity without causing flood damage to adjacent private developments; 4) Minimize street flooding on state and county highways and Village arterial streets; 5) Use gravity drained system, wherever possible; 6) Store excess runoff first on Village streets, second in Village off-street areas and last in underground storage facilities; and 7) Disconnect downspouts from the sewer system and discharge to land surface or storm sewers.

The recommended runoff control system for the Village of Skokie consists of 2,009 flow regulators, 871 street berms, 57 subsurface tanks, 21 surface basins, and 28 in-pipe storage facilities for a total of 106 storage facilities. In addition to the storage facilities, 64,230 feet of storm collector sewers and 28,950 feet of combined relief sewers will be constructed. A total storage volume of 6,897,900 cubic feet is required; 3,302,600 cubic feet (48 percent) in on-street storage and 3,595,300 cubic feet (52 percent) in storage facilities. The estimated cost of these facilities is $66 million in 1992 prices.

Implementation

Construction in the HSSD began in late 1983 and was completed in 1986. Construction in the MSSD began in 1988 and is scheduled for completion in 1997. Construction in the ELSSD began in 1989 and is scheduled for completion in 1997. The planning, design and construction engineering continues to be a joint effort of both the Village and SEC Donohue. The Village has designed roadway berms, fabricated and installed the flow regulators and managed the construction of berms, subsurface detention facilities and combined relief sewer. SEC Donohue has designed flow regulators, roadway berms, storm sewers, detention facilities and combined relief sewers.

Summary

The Village of Skokie is not the only community with problems resulting from an outdated combined sewer system. The Skokie RCS is tailored to the combination of topographic, hydrologic, and hydraulic characteristics in Skokie, Illinois. The study techniques, modeling criteria and storage alternatives can be applied to other urban areas experiencing similar problems with surcharging of combined sewers, storm sewers or sanitary sewers. A well engineered on-street runoff control system eliminates adverse impacts and is virtually unnoticeable to residents on both wet and dry days.

References

Donohue & Associates, Inc. Preliminary Engineering Runoff Control Program, Howard Street Sewer District. Prepared for the Village of Skokie, Illinois, July, 1982.

Donohue & Associates, Inc. Preliminary Engineering Runoff Control Program, Howard Street Sewer District - Including Addendum. Prepared for the Village of Skokie, Illinois, March, 1984.

Donohue & Associates, Inc. Preliminary Engineering Runoff Control Program, Emerson and Lake Street Sewer District. Prepared for the Village of Skokie, Illinois, May, 1987.

Donohue & Associates, Inc. Preliminary Engineering Runoff Control Program, Main Street Sewer District. Prepared for the Village of Skokie, Illinois, May, 1987.

STORAGE/TREATMENT ISOQUANTS FOR CSO CONTROL PLANNING
Ronald L. Wycoff, M. ASCE [1] and Lester E. Lee, M. ASCE [2]

Abstract

A quantitative procedure which identifies the technically efficient combinations of storage volume and treatment rate necessary to achieve various levels of combined sewer overflow (CSO) pollution control is presented. This procedure, along with site specific cost functions and economic optimization analysis, can be used to determine the least cost combination of CSO storage and treatment facility requirements.

Introduction

Storage and treatment can operate independently or in combination to reduce CSO. Storage volume can be added to an existing combined sewer system to capture CSO for subsequent treatment at existing facilities. Alternatively, CSO treatment facilities can be constructed to increase wet-weather treatment capacity. Application of either technology will reduce untreated overflow. However, in many cases the optimum or least cost CSO control system uses a combination of both additional storage volume and additional treatment capacity. In order to analyze, plan, and design the most cost-effective storage/treatment systems for CSO control, it is necessary to determine how storage and treatment can work in combination to reduce wet-weather pollution.

In production theory, isoquants are lines of equal quantities (outputs) resulting from the application of two inputs. Therefore, in CSO control planning, a storage/treatment isoquant is defined as the relationship between storage volume and treatment rate which will achieve an equal level of overflow reduction or pollution control. This paper present a new and general algorithm which may be used to estimate the storage/treatment isoquant given the pollution control performance of each individual technology (storage and treatment) acting alone. Derivation of the isoquant algorithm is based on the application of production theory to wet-weather pollution control planning.

----

[1] Senior Water Resources Engineer, CH2M Hill Inc., P.O. Box 147009, Gainesville, Florida 32614.

[2] Senior Environmental Engineer and Project Manager, City of Portland, Bureau of Environmental Services, 1120 S.W. 5th Ave., Room 400, Portland Oregon 97204.

Storage and Treatment Production Functions

Pollution control performance relationships, or "production functions", for storage and treatment are required to apply the isoquant algorithm. A production function relates the level of effort applied to the pollution reduction achieved, for a given CSO control technology. Many pollution control production functions exhibit the law of diminishing returns, and may be represented by the following general equation.

$$Y_p = X_p/(a + b*X_p) \qquad (1)$$

Where; Y_p equals the process output or pollution control achieved, X_p equals the level of effort or input applied, and a and b are best fit parameters derived by linear regression analysis.

Site specific CSO treatment and storage production functions may be derived from analysis of long term continuous hydrologic simulation applied to the combined sewer planning area. Figures 1 and 2 illustrate these functions for Portland Oregon. The level of effort for the treatment production function (Figure 1) is the daily treatment rate (Qt), divided by the annual CSO volume (Vcso) available to the treatment unit. The resulting output is the overflow reduction factor for treatment (ORFt), which is defined as the fraction of the total CSO volume treated.

Figure 2 illustrates the normalized production function for storage. In this case, the level of effort is expressed as the storage volume provided, (Vs), divided by the annual CSO volume (Vcso); and the output is the fraction of CSO volume captured by the storage unit (ORFs). Both production functions define the performance of each technology operating independently.

Storage/Treatment Isoquants

Storage and treatment may be used in combination to reduce CSO. Additional treatment capacity is used first to process flow as it arrives. Once the available treatment capacity is used to its maximum, then wet-weather storage is used to capture additional flow for processing by the treatment unit when capacity once again becomes available. The combined performance of the storage/treatment system can be estimated by application of the following equation.

$$ORFs/t = ORFt + (1 - ORFt)*ORFs' \qquad (2)$$

Where; ORFs/t equals the fraction of the total annual CSO volume processed by the storage/treatment system, ORFt equals the fraction of the original CSO volume processed directly by the treatment unit; and ORFs' is the fraction of the CSO volume not process directly by the treatment unit, which is captured by the storage unit for subsequent treatment. The individual storage and treatment production function equations are used to estimate ORFt and ORFs'. Estimates of OFRt are

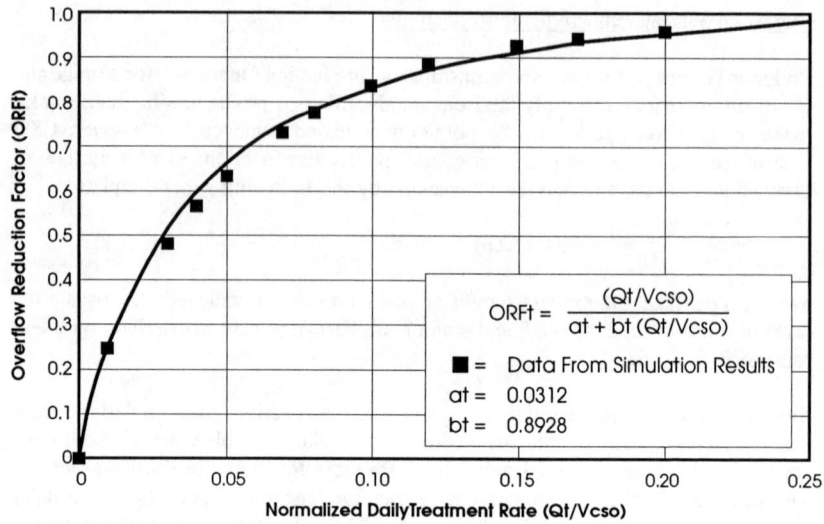

Figure 1. Treatment Production Function for Portland, Oregon.

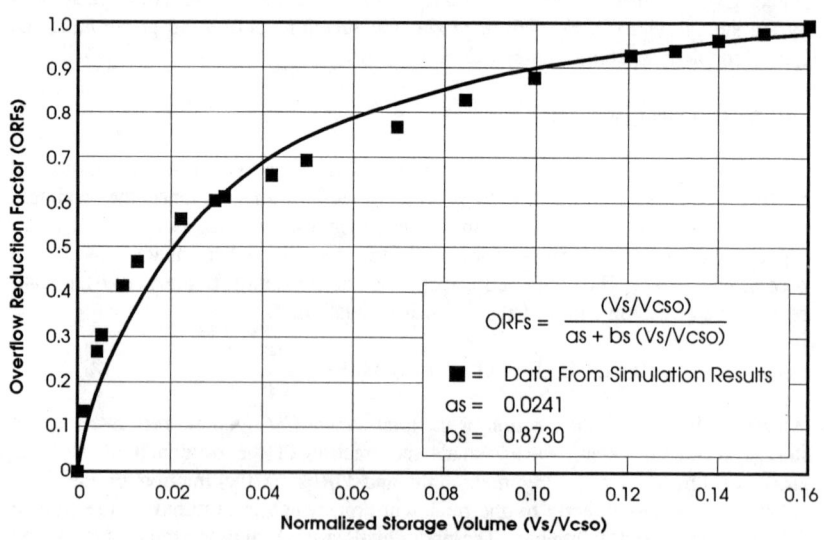

Figure 2. Storage Production Function for Portland, Oregon.

computed directly based on the total annual CSO volume (Vcso). Estimates of ORFs' are computed based on the annual CSO volume available to the storage unit (Vcso'). This volume is a function of the amount treated and is computed as follows.

$$Vcso' = (1 - ORFt)*Vcso \qquad (3)$$

Equations 2 and 3 can then be used to construct storage/treatment isoquants as illustrated in Figure 3. These isoquants define the technically feasible combinations of treatment rate and storage volume which will reduce the total untreated CSO volume by 80, 90, or 95 %. For example, Figure 3 indicates that a storage/treatment system with a daily treatment rate equal to 3 % of the annual CSO volume along with a storage unit with a total volume equal to 5 % of the total annual CSO volume will capture and treat about 95 % of the total annual CSO volume. Isoquants for any desired CSO reduction can be constructed using this technique.

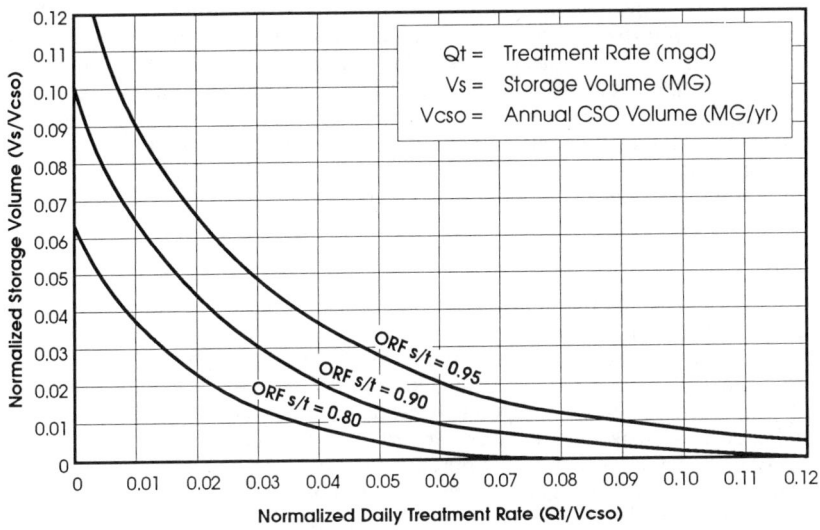

Figure 3. Storage/Treatment Isoquants for Portland, Oregon.

Application to CSO Control Planning

The procedure described above can be used to identify combinations of storage volume and treatment rate which will achieve a given level of CSO control. If annual cost functions for storage and treatment are also known, then the isoquants may be searched for the least cost combinations. This analysis can be used to focus attention on those storage/treatment combinations which will likely achieve the desired CSO control at minimum cost.

Combined Sewer Overflow Abatement Planning Based on SWMM EXTRAN Modeling

David M. Wood,[1] Associate Member, ASCE and
David Crawford,[2] P.E., Member, ASCE

Abstract

This paper discusses use of the Storm Water Management Model (SWMM), with the EXTRAN block as the hydraulic transport mechanism, as a tool for developing a facilities plan for abating the impacts of 42 combined sewer overflows (CSOs) in Portland, Maine. The authors' strategy involved developing and calibrating a combined sewer model for assessing the system, planning and implementing a flow monitoring program, and using the results of the monitoring program to revise and enhance the final model. A description of modeling strategies adopted and insights gained from using SWMM and EXTRAN are presented.

Introduction

The City of Portland, Maine, like many communities in the northeast served by combined sewers, is being required to address the problem of CSOs. Portland is required under an Administrative Consent Agreement with the State of Maine Department of Environmental Protection to develop and implement a prioritized, long-term program to evaluate and abate its CSOs. The city adopted a strategy whereby a detailed system modeling exercise using limited calibration data preceded a comprehensive field program of data collection. The initial modeling characterized the system and provided a basis for developing and fine-tuning the monitoring strategy. Data acquired during the monitoring program were then used to revise, enhance, calibrate, and verify the final model such that it could be confidently used to develop a recommendation for abating CSOs throughout the system.

---

[1]   Water Resources Engineer, 115 Perimeter Center Place, Suite 700, Atlanta, GA 30346-1278.

[2]   Water Resources Engineer, 825 N.E. Multnomah, Suite 1300, Portland, OR 97232-2146.

Portland's Combined Sewer System

The Portland combined sewer system conveys flow from essentially four sources: base sanitary flow (BSF), storm water runoff (runoff), groundwater infiltration (GWI), which leaks into the system when groundwater tables are high, and rainfall-derived infiltration/inflow (RDII), which enters the sewer system at locations other than runoff inflow locations, e.g., roof gutter leaders, basement sump pumps, and leaks and cracks exposed to temporarily elevated groundwater levels following storms. The Portland system has a particularly high RDII component.

Flows are routed to 51 regulators that control activity at 42 CSOs, which discharge to five distinct receiving waters. The regulators have a specific capacity to convey flow to interceptors and in-line pump stations with continued conveyance to the wastewater treatment facility (WWTF). Inflow exceeding regulator capacity is discharged directly to receiving waters through secondary conduits. The regulators typically consist of a weir that diverts incoming flow to an orifice connected to interceptors. When the hydraulic grade at the regulator exceeds the weir elevation, CSOs occur. Other hydraulic components of the Portland system include a siphon and six major pump stations.

Combined Sewer Model

The authors developed a rainfall/runoff/transport model that fully simulates the hydrology and hydraulics of the Portland system. The model uses the RUNOFF block of SWMM to generate surface runoff hydrographs, which are in turn routed through the transport system with the EXTRAN block. The authors chose to include only the regulators and conduits downstream of the regulators in the hydraulic (EXTRAN) portion of the model. All pipes upstream of the regulators were included in the surface runoff portion (RUNOFF).

Sixty-three subcatchments were included in the RUNOFF block. Data used to develop the hydrologic parameters for each subcatchment included topographic, aerial, and land use maps; information from previous studies; and information garnered from city staff. A groundwater subroutine, which is part of the RUNOFF block, was used to generate the GWI and RDII components of inflow to the sewer system. The groundwater routine incorporates subsurface processes into the watershed simulation by routing surface infiltration through the unsaturated and saturated zones. Losses and outflow from the unsaturated zone are via deep percolation, saturated zone evapotranspiration, or groundwater outflow. Groundwater outflow becomes an input to Channel/Gutter flow of the RUNOFF block.

Because of project scope and budget limitations, the authors were unable to quantify the majority of the groundwater input parameters required by the subroutine. However, because an abundance of calibration data generated during the monitoring program were available, the authors were able to adjust controlling parameters of the groundwater subroutine such that the receding limb of predicted hydrographs, caused by GWI and RDII, correlated reasonably well with measured data.

The authors initially attempted to use the SWMM TRANSPORT block for routing flows through the combined sewer system. The TRANSPORT block, which uses a kinematic wave open channel routing approach, quickly proved to be inadequate for modeling Portland's complicated system. The hydraulically complex regulators contribute to widespread surcharging and propagate backwater effects for significant distances upstream, thereby hydraulically linking several regulators. The Portland system also contains branched and looping pipes, siphons, and pump stations. An additional aspect of the Portland system is its many tide-gated outfalls, which are submerged at elevated tide. The system model must therefore be able to account for time-varying outflow boundary conditions.

The Portland system required the use of a fully dynamic routing approach using the full form of the St. Venant gradually varied flow equations. The EXTRAN model was selected. The system can be subdivided into two hydraulically discrete regions, incorporating drainage areas upstream of two major pump stations which discharge to the WWTF. The two regions include a total of 187 conduits, 262 junctions (including storage locations), 39 orifices, 42 weirs, and 6 pump stations. The most recent version of the SWMM supported by the EPA Center for Exposure Assessment Modeling (CEAM), version 4.20, was acquired for use in modeling the Portland system. However, because the data arrays were insufficiently dimensioned to accommodate the Portland application, the source code was modified and recompiled. Modifications and enhancements of the SWMM had been implemented as part of previous CH2M HILL CSO abatement projects.

System Modeling

EXTRAN allows the user a choice between three solution techniques for the gradually varied, one-dimensional flow equations for open channels: the explicit method, the enhanced explicit method, and the iterative method. Explicit methods are susceptible to instability and often require short time steps. The practical difference between the three techniques is that the enhanced explicit and iterative methods allow longer time steps than the explicit solution. EXTRAN documentation recommends that (1) the maximum allowable time step be determined by the shortest, smallest pipe having high inflows and that (2) a time step of 10 seconds be considered small enough to prevent numerical instabilities.

Because the overall goal of the Portland CSO project was to recommend a strategy to reduce the frequency and volume of CSO occurrence, it was necessary to establish an extended precipitation period, sufficiently representative of average or normal conditions, for input to the model. An extended period was necessary because the goal includes reducing the frequency of overflow and not just large volumes associated with design precipitation events. Furthermore, it was found that CSO frequency and volume are strongly dependent on antecedent conditions. Wet antecedent conditions increase the amount of RDII to the system, decrease the time to peak, and exhaust in-system storage capacity. These conditions tend to cause CSO events more often and with larger volumes than would have occurred under identical precipitation events with dry antecedent conditions.

A one-year period was selected as the continuous precipitation record. This precipitation was input to the model for analysis of existing and improved conditions. The period was selected by analyzing storm sequences with respect to volume, intensity, and duration within a particular year and comparing to the period of record. The period most nearly representing the historical average was determined to be 1966. The raw precipitation record for 1966 was correlated with hourly temperature records for the year and adjusted to account for snowfall and snowmelt. During precipitation periods where the temperature fell below zero degrees Celsius, a simple snowmass buildup accounting routine was activated. During subsequent periods of above-freezing temperatures, the snowmass was liquefied, until depletion, according to typical melt rates.

Because the system simulations were to be run for the one-year period, the authors attempted to use the iterative numerical solution technique. However, consistent instabilities were apparent in the form of flow oscillations, unrealistically excessive velocities, and large system continuity errors. Attempts to reduce these instabilities by (1) increasing the length of small pipes, (2) increasing the maximum number of iterations, and (3) tightening surcharge and flow convergence criterion were largely ineffective. In addition, the pipe-lengthening remedy introduces greater volumes to the system, which could potentially distort the prediction of CSO events.

The authors eventually selected the explicit solution technique and accepted the longer run times. The full simulation took 60 and 90 hours on a 486 machine running at 33 Mhz for the two system regions, but the solutions were stable. System continuity error using the explicit method for the Portland system typically ranged from 0 to +- 5 percent. Other stability checks included inspecting the maximum velocities in and total volumes flowing through each of the conduits during the simulation. The latter provided a check for unstable flow oscillations which, over an extended period, can sum to unrealistic volumes.

Conclusion

A sophisticated model has been developed and has been used as a tool for predicting the frequency, volume, and duration of CSOs occurring during a 1-year precipitation series. The model successfully uses the EXTRAN block of SWMM to simulate the hydraulics of the Portland combined sewer system and offers a method for investigating various facility plan alternatives. The model was used to develop and recommend alternatives to satisfy regulatory CSO abatement mandates.

Receiving Water Quality Bases For Evaluating CSO Control Alternatives

Ken C. Hall[1] and William A. Kreutzberger[2]

Introduction

Waters receiving combined sewer overflow (CSO) loadings in the vicinity of Bangor, Maine include the Kenduskeag Stream and the Penobscot River. The Bangor combined sewer system serves a total area of 1,584 ha (3,916 ac) and includes the central business district and residential, commercial, and light industrial areas. The City's CSO facilities planning effort was performed as four major tasks: field monitoring, CSO modeling, estimating impacts on receiving waters, and evaluating alternative scenarios for addressing the CSOs. This paper summarizes the third task. The primary objective of the study was to develop a water quality based approach for evaluating CSO control alternatives.

Approach

The initial assessment of CSO impacts on receiving waters was developed without detailed water quality modeling in the tidally influenced river and stream. The Penobscot River is also impacted by the City of Brewer's CSOs (across the river), upstream municipal and industrial discharges, and nonpoint source runoff. An extensive wet- and dry-weather monitoring program provided information to characterize the quality of the CSO and receiving waters. A model of the wet-weather service area and interceptor system, developed in a previous task, provided annual and summer season estimates of average CSO frequency, volume, and duration. This information, in conjunction with existing water quality and flow data, and current water use classifications and state water quality standards, provided the bases for the evaluation.

---

[1] Member; Water Resources Engineer, CH2M HILL, 115 Perimeter Center Place, N.E., Suite 700, Atlanta, GA, 30346-1278 (404)604-9182 ext 461 FAX(404)604-9183

[2] Water Resources Department Manager, CH2M HILL, Atlanta (404)604-9182 ext 415

The assessment of impacts on receiving waters included eight major steps: (1) event mean concentrations (EMCs) and loads to each receiving water for each pollutant were computed; (2) plots were developed to indicate the presence of a first flush effect; (3) the statistical correlation between TSS and metals in CSO was analyzed; (4) toxicity bioassays were performed for combined sewer samples collected during storm events; (5) receiving water data were statistically analyzed to determine conditions upstream and downstream of CSOs during both dry and wet weather; (6) dye and drogue studies were performed to assess the mixing characteristics of the receiving waters; (7) CSO EMCs were compared to EPA's final acute values (FAVs) for selected parameters; and (8) the probability of CSO events occurring during extreme low-flow events, and the corresponding volume of CSO and receiving waters during these events, were compared.

Results

The analysis of CSO quality data indicated no statistically significant difference between median levels of parameters recorded at the seven sample locations, with drainage areas varying in both size and land use. Service area-wide EMCs were developed for the parameters, as shown in Table 1. Results were generally of the same magnitude as those observed in other areas with CSOs. Loads were then developed for the parameters for each receiving water.

Table 1
SUMMARY OF COMBINED SEWER MEAN CONCENTRATIONS

Parameter	Units	EMC[a] Bangor	Portland ME[b]	MWRA Boston[c]	Portland OR[d]	EPA[e]
BOD$_5$	mg/L	24.5	34	90	30	115
Cadmium	µg/L	0.61	0.6	-	-	-
Chromium	µg/L	12	—	-	-	-
Copper	µg/L	30	38	85	-	-
E. coli[f]	CFU/100 ml	156,000	430,000	-	23,000	-
Fecal Coliform[f]	CFU/100 ml	181,000	—	680,000	163,000	670,000
Nickel	µg/L	11	—	-	-	-
Oil & Grease	mg/L	11.2	—	-	-	-
Lead	µg/L	57	44	110	-	370
Settleable Solids	ml/L	2.1	—	-	-	-
Silver	µg/L	1.8	—	-	-	-
TPH	mg/L	3.45	—	-	-	-
TSS	mg/L	250	217	188	148	370
Zinc	µg/L	114	95	110	110	-

[a] Event Mean Concentration--flow weighted
[b] Portland, ME, CSO Facilities Plan, CH2M HILL, 1992
[c] MWRA CSO Facilities Plan, CH2M HILL, 1989
[d] Portland, OR, CSO Facilities Plan, CH2M HILL, 1990
[e] Nonweighted average of data collected in Des Moines, Milwaukee, New York City, Racine, Rochester. Summary appeared in EPA document EPA-600/8-77-0 14, September, 1977.
[f] Geometric Mean computed for bacteria samples

A positive correlation of metals with TSS concentrations was confirmed in the regression analysis at the 99.9 percent confidence level. The presence of a first flush effect was also indicated for TSS and metals, but not for E. coli. Figure 1 compares the percent total TSS load with the percent total flow volume at one CSO sampling station for all four storm events. Curves lying above the diagonal line indicate the presence of a first flush effect. Acute biotoxicity tests (Ceriodaphnia and fathead minnows) were performed on selected samples. In the biotoxicity tests for CSO (Ceriodaphnia and fathead minnows acute test), only one of eight 96-hour tests in undiluted CSO showed greater than 50 percent mortality of fathead minnows. The higher toxicity of this one test was attributed to uncharacteristically high levels of ammonia in the sample.

The receiving water sampling and analysis effort revealed that median E. coli levels in both waters are higher downstream of Bangor CSO discharges during wet-weather events. Several individual samples exceeded state water quality criteria for E. coli. However, most data on toxicants in the receiving waters, during both dry and wet weather, were below detectable levels.

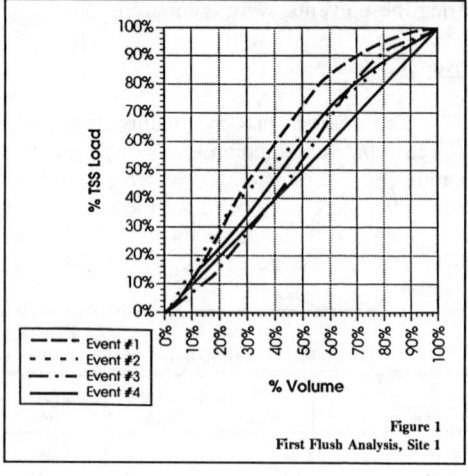

Figure 1
First Flush Analysis, Site 1

The EMCs for the five metals were lower than EPA's FAVs, indicating little potential for toxicity even if dilution in the receiving water is neglected. Table 2 shows CSO dilution in the receiving water volume (under several low flow conditions) during storm events. The far right column gives an indication of how frequently the low flow conditions in the receiving waters coincide with CSO events. The probability of a significant rainfall event occurring during a low flow period, when flow is less than the August median, is very low. For the Penobscot River, rainfall exceeded 0.13 cm (0.05 inch) on only 4.7 percent of days when the river flow was less than the August median. This percentage drops to 0.03 percent if the flow was less than the 7Q10. The corresponding values for the Kenduskeag Stream are even lower, with 2.4 percent when less than the August median flow and 0.05 percent when less than the 7Q10 flow.

Table 2
COMPARISON OF CSO AND RECEIVING WATER VOLUMES DURING STORM EVENTS

Flow Scenario	RW Flow[a] (cfs)	CSO Volume[b] (MG/event)	Duration[b] (hr/event)	RW Volume (MG/event)	CSO Contribution (% Total)	Data Record[c] (days)	Days Q <RW Flow & Precip. >0.05 in.[d] (days)	(%)
Penobscot River						1979-88 3389		
25%[e] of 7Q10	825	13.4	9	200.0	6.3	-	1	0.03
7Q10	3300	13.4	9	799.8	1.6	-	1	0.03
25% of August Median[f]	1625	13.4	9	393.8	3.3	-	158	4.7
August Median	6500	13.4	9	1575.4	0.8	-	158	4.7
Annual Median	9100	10.2	10	2450.6	0.4	-	362	10.7
Kenduskeag Stream						1970-80 4018		
7Q10	2.5	6.8	8	0.5	93	-	2	0.05
August Median	18	6.8	8	3.9	64	-	97	2.4
Annual Median	130	5.2	8	28.0	16	-	488	12.1

[a]Raw receiving water flow data obtained from USGS.
[b]CSO volume and duration data based on CH2M HILL model of Bangor system.
[c]Data period selected as most recent, relatively complete data set. Size of record was limited to approximately 10 years because of computer's internal memory limitations when manipulating large spreadsheets.
[d]Precipitation data extracted from National Climatic Data Center Database.
[e]CH2M HILL estimates (based on dye studies) that initial short-term mixing of CSO in Penobscot occurs in 25% of flow volume.
[f]Medians from TM 7, Bangor, ME CSO Facilities Plan.

Conclusions

Dilution calculations, CSO EMC data, and biotoxicity results indicate little potential for acute toxicity concerns in the receiving waters near the CSOs for the Kenduskeag, and even less potential on the Penobscot. The probability of low-flow and rainfall events occurring simultaneously was shown to be very low. A greater likelihood exists of E. coli levels exceeding state standards. Therefore, water quality based controls should focus on reduction of E. coli levels. It is likely that acceptable, cost-effective controls for E. coli (e.g., optimized combinations of storage and treatment) will result in additional control of toxicants as well. While a first flush effect was observed for TSS and metals, the results for E. coli did not show a similar pattern. Therefore in the subsequent alternative analysis controls designed to capture, divert, or minimize the first flush will have far less impact on E. coli loads than TSS and metals loads. The correlation between TSS and metals indicates that alternative performance based on TSS control will also be indicative of toxicant control.

Acknowledgements

The authors gratefully acknowledge the support of the City of Bangor, Maine, for guiding and funding this study.

APPLICATION OF THE STORM WATER MANAGEMENT MODEL IN EVALUATING COMBINED SEWER OVERFLOWS

by

Rajat Roy Chaudhury[1], Raymond M. Wright[2], Igor Runge[3] and Daniel W. Urish[4]

ABSTRACT

This paper summarizes the application of the Stormwater Management Model (SWMM) to isolate the contributions of Combined Sewer Overflows (CSOs) to a receiving water.

INTRODUCTION

The Providence sewer system consists of combined and separate sewer systems, approximately 45 miles of sewer interceptors, 65 CSO outfalls, pumping stations, inverted siphons and a treatment facility. During wet weather conditions, the CSO outfalls discharge raw sewage into the Moshassuck, Providence, Seekonk, West and Woonasquatucket Rivers. These rivers flow into the Upper Narragansett Bay, a major resource of Rhode Island. The CSO discharges contribute towards the closing of the Upper Narragansett Bay to shellfishing for a period of 5 days if the total rain in a day exceeds 0.5 inches.

The City of Providence has been divided by the Narragansett Bay Commission, the sewer management authority, into six management areas. As part of the CSO abatement strategy, the NBC initiated studies that resulted in calibrated and validated models for the six areas using the Stormwater Management Model (SWMM) (Huber and Dickenson 1988), subsequently, leading to a system wide model (Wright et al. 1992).

The system wide model was applied to characterize the contribution of the CSOs in Providence to the Providence River and the Upper Narragansett Bay. This was performed by determining the predicted overflow volumes for a wet weather study (Wright et al. 1991) sponsored by the

---

1 Graduate Research Assistant, University of Rhode Island, Kingston, RI 02881
2 Associate Professor, University of Rhode Island, Kingston, RI 02881
3 Assistant Professor, University of North Carolina, Charlotte, NC 28223
4 Professor, University of Rhode Island, Kingston, RI 02881

STORMWATER MANAGEMENT MODEL 775

Narragansett Bay Project (NBP) and the U. S. Environmental Protection Agency (USEPA). The loadings from the CSOs were compared to the other major sources.

SYSTEM WIDE MODEL APPLICATION

A major task of the Narragansett Bay Project (NBP), a federally funded program, was to develop a water quality management plan for Narragansett Bay. The NBP research studies included a wet weather study (Wright et al. 1991). The NBP wet weather study was designed to investigate all significant wet weather discharges into

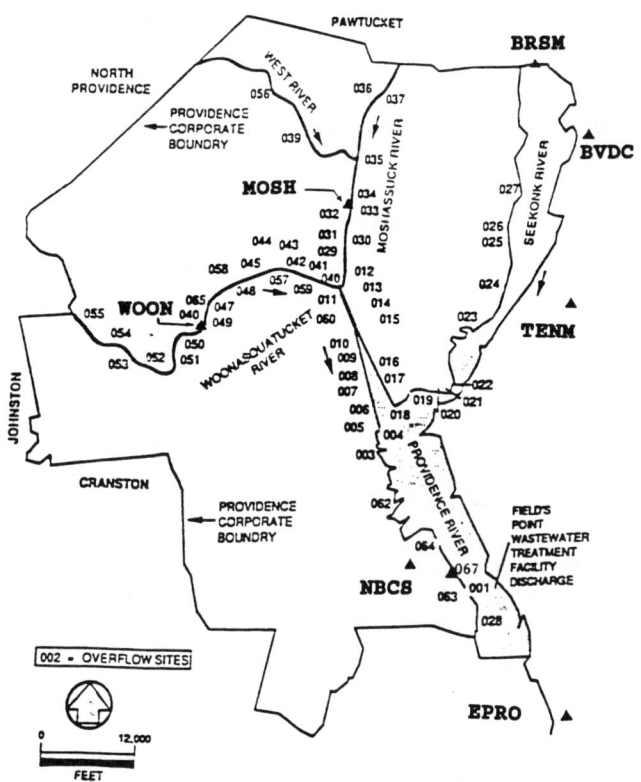

FIGURE 1. Providence CSOs and NBP Sampling Stations

the Providence River and Upper Narragansett Bay (Figure 1). This included five tributaries (Blackstone - BRSM, Moshassuck - MOSH, Pawtuxet - PAWT, Ten Mile - TENM and Woonasquatucket - WOON Rivers), three point sources (NBC facilities at Bucklin Point - BVDC and Fields Point - NBCS, and East Providence Sewage Treatment Plant - EPRO), the two bypasses at the NBC facilities and one CSO (OF-067). These discharges contribute directly to the Providence River. Three storm events were captured for this study, for which mass loadings from the major sources were calculated. Wet and dry weather loadings were evaluated for each station, resulting in a ranking of the major pollutant sources to the Bay. This study, however, could not isolate the contributions of the Providence CSOs.

The application of the system wide model to these three storms provides additional information that will allow the specific estimate of the Providence CSOs contribution to the Bay. The procedure used for this application is detailed below:

1) The first step was to apply the three NBP rainfall hyetographs to the system wide model. The overflow volumes for each of these events were recorded.

2) The pollutant loadings from the CSOs situated upstream of the Woonasquatucket and Moshassuck River stations were established using the average concentrations of water quality samples for CSO's during the 1990 study (Wright et al. 1992). These CSO pollutant loadings were deducted from the wet loadings at WOON and MOSH as estimated by Wright et al (1991). This established the pollutant loadings from the tributaries at the Providence Corporate Boundary.

3) The total CSO loadings were then evaluated and ranked against the other sources contributing to the Upper Narragansett Bay. Rankings of the discharges to the Providence River for TSS, three trace metals (Cu, Pb and Ni) and two nutrients (nitrate and phosphate) are summarized in Table 1.

Fecal coliform loadings from the CSOs alone account for 92.1% of the loadings, while fecal coliform loadings from sources under the control of NBC account for 98.3%. While TSS, Cu, Pb, and Ni ranked 3, 2, 2, and 4, respectively, nutrient (nitrate and phosphate) loadings from CSOs were relatively insignificant.

CONCLUSIONS

Combined sewer overflow contributions to a receiving water under wet weather conditions were made using calibrated and validated SWMM models. Over 90% of the fecal coliform loadings were estimated to be a result of CSO loadings.

Table 1. Ranking and Percent Contributions of Pollutant Discharges to the Providence River and Upper Narragansett Bay

Site	F.C.	TSS	NO3	PÓ4	Cu	Pb	Ni
Blackstone R.	4	1	1	1	1	1	1
Moshassuck R.	6	5	5	8	8	5	8
Pawtuxet R.	5	2	2	2	3	3	3
Ten Mile R.	7	8	3	4	4	6	2
Woonasquatucket R.	8	4	6	7	7	4	6
BVDC Bypass	3	7	7	5	6	8	7
CSOs	1	3	4	6	2	2	4
NBC Bypass	2	6	5	3	5	7	5
Blackstone R.	1	38	66	57	43	48	33
Moshassuck R.	0	3	3	1	4	4	3
Pawtuxet R.	1	35	17	21	14	14	15
Ten Mile R.	0	2	10	5	9	3	23
Woonasquatucket R.	0	6	4	3	4	7	5
BVDC Bypass	2	2	0	4	5	2	3
CSOs	93	11	0	3	16	19	13
NBC Bypass	3	3	0	7	7	3	5

F.C.=Fecal Coliforms; TSS=total suspended solids; NO3=dissolved nitrate+nitrite; PO4=dissolved orthophosphate; BOD5=5 day biochemical oxygen demand; Cu=total copper; Pb=total lead; Ni=total nickel.

ACKNOWLEDGEMENTS

The system wide modeling for the Providence sewer system was sponsored by the Narragansett Bay Commission, Providence, RI.

REFERENCES

1. Huber, W.C. and Dickenson, R.E. (1988). Stormwater Management Model, Version 4: User's Manual. Environmental Research Laboratory, Office of Research and Development, U.S. Environmental Protection Agency, Athens, Georgia 30613.

2. Wright et al. (1991). Problem Assessment and Source Identification and Ranking of Wet Weather Discharges Entering the Providence and Seekonk Rivers, Narragansett Bay Project, Providence, RI and U.S. EPA, Boston, MA

3. Wright et al. (1992). System Wide Modeling for the Providence Sewer System, Narragnsett Bay Commission, Providence, RI.

THE KETTERING COMMUNITY DEMONSTRATION PROJECT NON-POINT POLLUTION CONTROL AND ENVIRONMENTAL ENHANCEMENT PROGRAM

David B. Ennis, P.E.[1], Michael L. Clar, P.E.[2], Larry S. Coffman[3]

ABSTRACT

With the advent of the U.S. EPA's municipal/industrial NPDES and Costal Zone Management programs the development of effective non-point pollution control programs has become the most complex and perplexing challenge facing local and state stormwater authorities. Prince George's County Maryland's Department of Environmental Resources under the auspices of the U.S. EPA's Chesapeake Bay Nonpoint Source Program and Maryland's Department of the Environment is conducting a comprehensive non-point pollution reduction pilot project. The project's scope is to develop, implement and measure the effectiveness of a multifaceted community based non-point pollution control program within a developed urban watershed. A variety of traditional and innovative structural stormwater treatment practices and nonstructural source reduction strategies have been integrated into the project.

INTRODUCTION

The project, located in the Kettering subdivision at

---

[1] Project Manager, Engineering Technologies Associates, Inc. 3458 Ellicott Center Drive, Ellicott City, Md. 21043

[2] Principal, Engineering Technologies Associates, Inc. 3458 Ellicott Center Drive, Ellicott City, Md. 21043

[3] Assistant Branch Manager, Prince George's County Government, Watershed Protection Branch, 9400 Peppercorn Place, Suite 600, Landover, Md. 20785

the intersection of Md. Rt. 214 (Central Avenue) and Maryland Rt. 193 (Watkins Park Drive) in Prince George's County, Maryland consists of approximately 500 acres of mixed use land. The present land use consists of approximately 55% single family residential property with 37% woodlands and less than 8% townhouse and commercial lands. The ultimate development would result in the woodlands being developed as two acre residential lots. Currently, this drainage basin has no storm water management controls. The residential and commercial areas are drained by a network of reinforced concrete pipes which outfalls into a 1800' long concrete swale.

PROJECT COMPONENTS AND DESCRIPTION

Because of the multifarious nature of non-point pollution sources, a complex and comprehensive combination of treatment and source control approaches must be used to reduce pollutant loads to the maximum extent possible. Some of these management techniques would include: retrofit of the existing stormwater control systems; implementation of new and innovative BMP's; public education; public participation; public/private partnerships; preventive maintenance plans; multi-agency participation and cooperation and cost sharing arrangements.

The Kettering non-point pollution control project was conceived as a multi-faceted holistic management program tailored to the watershed's unique urban/suburban land use characteristics, specific runoff problems and socioeconomic make-up. The watershed provides excellent opportunities to implement and monitor a wide range of traditional and innovative non-point pollution control devices and strategies. The project components include:

1. **Wetland Creation** - Two emergent/forested wetland systems will be constructed with a combined area of approximately 3 acres. The primary purpose of these systems is to provide end-of-pipe treatment of the pollutant laden first flush of storm water runoff. Additionally, these systems will provide enhanced aquatic and wildlife habitat and increased passive recreational opportunities for the local residents.

2. **Retrofit Existing Concrete Channel** - The lower end of the watershed is drained by a concrete lined channel. The bottom of the channel will be modified by construction of a series of weirs, to encourage the deposition of sediment, establishment of a natural-like stream bed and development of emergent wetland vegetation for additional water quality benefits.

3. **Retrofit Existing Storm Drain Inlets** - The storm drainage inlets within the watershed will be modified by constructing simple weirs and screens to provide catchment areas for sediment, trash and debris.

4. **Reforestation** - Trees and shrubs will be planted throughout the watershed in such areas as: road right-of-ways; residential open space; adjacent to the concrete channel; within residential yards and as part of the wetland creation projects. These trees will be beneficial in nutrient uptake, thermal pollution reduction and habitat enhancement.

5. **Observation Trail** - The project area lends itself to the creation of a trail system to facilitate observation of the development, operation and monitoring of the wetland areas. The trail system will also be used by the area residents for walking, jogging, bird watching and education/interpretation purposes.

6. **Retrofit Commercial Development** - The Watkins Park Plaza shopping center was developed without on-site water quality controls. Three measures are proposed to reduce non-point pollution loads from this area. First, existing storm drain inlets will be modified to enhance sediment catchment. Second, existing green space areas will be converted to stormwater control devices by regrading and landscaping to provide retention and treatment of the first flush of surface runoff. Third, a grounds maintenance plan will be developed to minimize the use of fertilizers and chemicals and increase the frequency of sweeping to remove trash and debris.

7. **Public Education** - Perhaps one of the most effective methods to reduce non-point pollution is to change human behavior through education programs on the proper use, handling and disposal of potential pollutants. A series of education programs will be conducted in the community to reduce pollution associated with lawn care (fertilizer and pesticides), car care (oil, antifreeze and washing), household hazardous waste, illegal dumping and yard waste.

8. **Public Participation** - Community involvement is essential to the success of this project. To encourage and nurture this involvement a citizens advisory committee has been developed to organize environmental enhancement projects such as tree plantings, stream monitoring, and clean up programs. The County will provide technical assistance when necessary and encourage the continued existence of the

environmental advisory committee upon completion of the demonstration project.

9. **Monitoring** - Several types of long term monitoring programs are proposed to measure the overall effectiveness of the pollution control measures, they include:

    o   Chemical monitoring for nutrients, pesticides, petroleum products and heavy metals.

    o   Biological monitoring of macroinvertebrates and fish to assess water quality, biological diversity and the overall health of the wetlands and receiving streams.

    o   Surveys of the area residents will be used to assess educational needs, estimate pollutant loads and determine the effectiveness of the public educational programs.

    o   Photographic records of all wetland and stream restoration work will document the development of the wetland and stream restoration projects.

10. **Modeling** - The water quality data and survey results will be used to develop and calibrate two computer models. An HSPF water quality model will be used to analyze the water quality changes resulting from all the program components on a watershed wide basis. The second model will be used to assist in the design of wetland systems to maximize water quality benefits.

CONCLUSIONS

The type, range and magnitude of water quality control measures to be implemented as part of this project represents the sum total of our County's knowledge and experience in the art of urban water quality control. By focusing all of what we know on one relatively small watershed, it is anticipated that this project will clearly demonstrate the effectiveness of our current structural and nonstructural stormwater management technology in reducing pollutants from existing urban land uses.

STORM WATER MANAGEMENT IN THE GREATER NEW ORLEANS AREA

Kent B. Dussom[1], Gordon C. Austin[2], Marnie Winter[3]

ABSTRACT

Two parishes, Orleans and Jefferson, located within the Greater New Orleans area have been identified by EPA as urbanized areas requiring storm water permits. Storm water management has played an important role in the development of both these urban areas due to the native elevation, topography and rainfall that both areas receive. Although they share similar challenges, each has a different history, and therefore, they have developed distinctively different and unique operations. Despite these differences, the two parishes are working together to provide comprehensive solutions to storm water quality management within the metropolitan area.

INTRODUCTION

The Metropolitan New Orleans area is principally composed of Orleans and the urbanized area of Jefferson Parish. This area experiences some of the highest rainfall in the United States, averaging over sixty inches per year.

The majority of the land mass is a result of years of soil deposition by the Mississippi River overflowing its natural banks and changing its course. The land is therefore very flat, and many of the soils are alluvial in nature. The first development in the area began along a ridge on the north bank of the Mississippi River which is today the French Quarter area.

The development of the area has long been dependent on the drainage of the area. Development followed the natural topography of the area, beginning along the ridges near the Mississippi River and extending away from the River, north toward Lake Pontchartrain, and south toward Barataria.

---

[1] Senior Engineer, James M. Montgomery, Consulting Engineers, Inc., 3501 North Causeway Boulevard, Metairie, Louisiana 70002

[2] Chief, Environmental Affairs Division, Sewerage and Water Board of New Orleans

[3] Director, Environmental and Development Control Department, Jefferson Parish, Louisiana

PHYSICAL DIFFERENCES

The physical differences between the drainage systems within New Orleans and Jefferson Parish are a result of the era of development. Both areas utilize a canal and pumped system within a completely encircling levee system. The levee system is necessary to protect the area, the majority of which is below sea level, from rising surrounding water that results from heavy storms or hurricanes.

Plans for the drainage system in New Orleans began before 1900. One major purpose was to improve health conditions by removing the sewage that collected in the ditches and canals. Until that time, the City had no separate sewage collection system. With the establishment of the Sewerage and Water Board, near the turn of the century, plans were developed for constructing separate sewerage collection and drainage systems. By 1920, artificial levees were constructed and with the aid of the horizontal screw pump developed by Mr. Baldwin Wood, new areas of the City were drained and developed. The system has continued to develop, and today serves over 55,000 acres of industrial, commercial and residential areas through hundreds of miles of canals and 21 major pumping stations. Most of the major canals that traverse the City are enclosed underground conduits that transport storm water to the discharge pumping stations.

Jefferson Parish also experienced improved drainage in the early 1900's, which included the Harvey Canal which connected the Mississippi River with southerly natural streams. Most of these improvements were made to increase barge transportation to the plantations that existed in Jefferson Parish. It was not until the 1950's that Jefferson Parish experienced significant urbanization. The drainage system at the time was a result of "cuts" made to improve drainage for agriculture. As urbanization occurred, these ditches and canals were expanded to handle the increased storm water flows. Most of the canals within Jefferson are open. The drainage system consists of 280 miles of canals with 39 major pumping stations.

There are significant differences in the operations of the drainage system as well. Most of the Jefferson Parish pump stations are located near the exterior of the drainage area, that is either along Lake Pontchartrain or near the discharge points in the Barataria Basin. In New Orleans, however, most of the pump stations are located within the center of the City. They have long discharge canals that reach out to the receiving waters. In addition, the Sewerage and Water Board system has the capability of transporting water to different stations, depending on the rainfall event and the current condition of each station.

COMMON GOALS

The historical primary focus of the drainage systems for both Orleans and Jefferson have been toward flood protection. The systems have been designed to remove the large volume of water that accumulates within the protection levee system as a result of both rain water and groundwater. There have been secondary water quality benefits as a result of the drainage system operation. For instance, in order to protect the pump impellers from damage, nearly all pump stations contain bar screens for large solids removal. In addition, in order to improve the hydraulic capacity of the conveyance system, pipelines and ditches are regularly maintained and cleaned of debris. While purely a function of providing increased flood protection, the secondary benefit of these practices is improved water quality.

Another indirect benefit to storm water quality is provided through improvements to the sewerage collection system. The local soils have a high potential for subsidence, which causes sewerage connection problems. In addition, the large rainfalls also place a burden on the sewerage collection system because of increased flows due to infiltration and inflow. Again, because of the lack of topographical relief, the sewerage collection system contains an abundance of lift stations and pump stations. When these become overloaded, overflows and bypasses occur. Both parishes have on-going programs to reduce infiltration and inflow and to correct damage or misaligned sewerage pipes.

Each parish has completed the Part 1 Permit Application for Storm Water Discharges. These documents focused on a review of existing data and a cursory investigation into illegal and illicit discharges. The production of these documents revealed several interesting aspects. First, the problem of identifying the proper authorities for responsibility of storm water quality is not trivial. In New Orleans, the responsibility is primarily the City and the Sewerage and Water Board, however several other local, state and federal authorities exist in the City, each having a potential to affect storm water quality. Jefferson Parish has several enclave municipalities, some incorporated and others not. Secondly, the process of detecting illegal and illicit connections through investigating dry weather flows is not practical in this area. This is partly due to the heavy rainfall that occurs in the area as well as the high groundwater table that produces a nearly constant flow in most pipes and canals.

Jefferson Parish has recently begun the process of completing the Part 2 permit application. The Sewerage and Water Board of New Orleans has already submitted the Part 2 permit application for the City of New Orleans to EPA. The initial plan for development of the Part 2 application relied on the collection and analysis of storm event samples in order to identify primary constituents of concern. However, due to inconsistent rainfall patterns and stringent sampling requirements, a limited set of data was available by the mandated deadline. The development plan was revised to include new evaluation criteria. Based on research of management plans from a variety of other areas, the best management plans (BMPs) to be included in the overall storm water management plan for New Orleans focused on three criteria: 1) keep the existing BMPs in place; 2) utilize low cost, source control type BMPs that have effects on a wide-area basis and are suitable for implementation by the Sewerage and Water Board or other co-permittees; and 3) develop programs to encourage land owners, businesses, and industries to implement small-area BMPs on a site specific basis.

Because of the need for aggressive storm water management to maintain adequate flood control, the existing programs go a long way toward addressing water quality concerns. Hence, there is a heavy reliance on the existing programs in the overall storm water management plans for the area. The biggest new program that both parishes will jointly participate in will be the public education and awareness programs. In addition, other local non-profit organizations may aide in the overall public education process.

Utilization of Roadway Crossings as BMP's in Urban Areas

G.V. Loganathan [1], Member, ASCE, E.W. Watkins [2],
A.B. Small [3], and D.F. Kibler [4], Member, ASCE

Abstract

Roadway crossings of natural waterways in the urban setting can be used as Best Management Practices (BMP's) by retrofitting existing outlet structures. Pollutant removal is contingent on inundation of a portion of the flood plain to provide extended detention. The design parameters, namely the detention basin storage and withdrawal rate, should be chosen to maximize the detention time within practical limits. An analytical procedure is formulated for preliminary planning estimates of the design parameters.

Introduction

Simultaneous growth in urban areas and increasing concern about the impacts of stormwater pollution have reduced the options available to municipalities to find suitable locations for BMP detention facilities. An interesting option being considered is the utilization of roadway crossings of natural waterways in the urban setting. Extended detention can be provided by retrofitting existing culverts to inundate a portion of the flood plain. This extended detention provides an environment for pollutants found in stormwater to settle in a process analogous to sediment basins [Randall et al. (1987)]. This option is very attractive to the municipalities charged with reducing nonpoint-source pollution impacts because of the minimal cost. In this paper a novel approach for estimating the average detention time resulting from a series of random runoff events detained by a control device is presented. Also, a formulation based upon a weighted pollutant concentration is presented for selecting the desired pond volume and withdrawal rate.

---

[1]Assoc. Prof., [2,3]Graduate Students, [4]Professor and Head, Dept. of Civil Engrg., Virginia Polytech. Inst. and State Univ., Blacksburg, VA 24061.

Determining Detention Time

The sequence of runoff events that arrive at a detention facility may be idealized as a cyclic process. Each cycle consists of a runoff volume, X_1, a duration of runoff, X_2, and an interevent time, X_3. If X_1 is larger than the available storage, the entire runoff event will not be contained within the BMP pool and overflow, Y, will occur [Loganathan et al. (1985); Watkins (1993)]. In Figure 1 the top of the smaller riser is the top of the BMP pool.

The time a volume of water resides in an extended detention pond is a function of how much water is in the pond and how fast this volume is released. Also, if the next event arrives too soon, the entire volume of the previously captured event does not set the potential detention time. If we define the volume of water in the pond at the end of runoff event n, V_n, the detention time for that event can be defined as:

$$D_n = \min\left(\frac{V_n}{a}, X_3\right) \quad (1)$$

where a is the withdrawal rate from the pond.

The detention time is defined in terms of constant parameters that describe the hydraulics of the pond and random variables with hypothesized exponential distributions (X_1, X_2, and X_3) that describe the hydrologic conditions. The probability distribution and the expected value of the detention time can therefore be determined. The expected detention time, E(D), is subsequently used to determine the pollutant removal efficiency of the extended detention pond. E(D) will be a function of the volume of storage in the BMP pool, b, the withdrawal rate from the pond, a, the mean volume of runoff, $\overline{X}_1$, the mean duration of runoff, $\overline{X}_2$, and the mean interevent time $\overline{X}_3$ [Watkins (1993)].

A two year continuous simulation of a partially developed urban watershed in northern Virginia was performed using the EPA Stormwater Management Model (SWMM). The two year continuous hydrograph was routed through a proposed retrofitted culvert (See Figure 1) using the Storage/Treatment Block in SWMM. Stochastic model results were compared to results obtained from SWMM output. (See Table 1). Because detention time is not a direct output from SWMM, the detention

Table 1. SWMM Results vs. Stochastic Model Results

Avg. Det.Time(SWMM)	=13.37 hrs	Bypass Fraction (SWMM)	= 0.628
E(D) (Model)	=14.97 hrs	P(Y>0) (Model)	= 0.632
Fraction D=b/a (SWMM)	= 0.413	Fraction D=0 (SWMM)	= 0.160
P(D=b/a) (Model)	= 0.509	P(D=0) (Model)	= 0.133

time for each event was calculated according to equation (1). V_n was determined for each event by correlating storage volume with peak discharge values obtained from SWMM output.

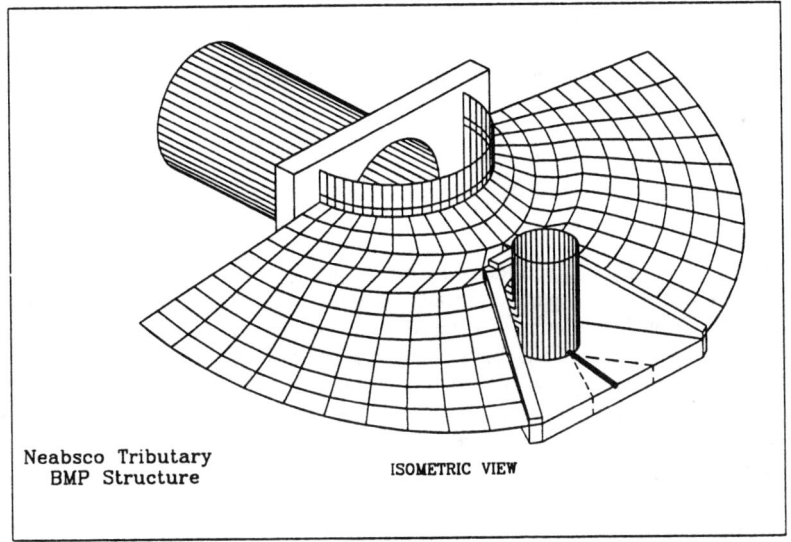

Neabsco Tributary BMP Structure   ISOMETRIC VIEW

Figure 1. BMP Retrofit

Estimating Pollutant Removal

As an aid for selecting the parameters, namely the BMP volume, b and the withdrawal rate, a, the following formulation is presented. If the concentration of the pollutant entering the pond is considered to be constant, the weighted concentration of the pollutant leaving the pond can be expressed as follows:

$$C_w = \frac{E(X_1 - Y)C_{out} + E(Y)C_{ini}}{E(X_1)} \qquad (2)$$

which is analogous to [Loganathan et al. (1985)]

$$C_w = \left\{P(Y=0)\right\}C_{out} + \left\{1 - P(Y=0)\right\}C_{ini} \qquad (3)$$

where: C_w = the weighted concentration of pollutant x leaving the pond; $P(Y=0)$ = the probability of no overflow (water bypassing the BMP pool), which is equivalent to the fraction of water that goes through the BMP orifice and receives "treatment"; C_{out} = the concentration of the pollutant leaving the BMP pool; and C_{ini} = the concentration of the pollutant entering the pond. C_{out} is defined as a fraction of C_{ini} such that $C_{out} = C_{ini} (1 - \eta_{BMP})$, where η_{BMP} is the pollutant removal efficiency in the BMP pool and is a function of the expected detention time, E(D). Substituting, and rearranging equation (3) becomes

$$\eta = \left\{P(Y=0)\right\} \eta_{BMP} \qquad (4)$$

in which, $\eta = (1-C_w/C_{ini})$ and the first term in the right hand side of equation (4) is equivalent to the flow capture efficiency of the extended detention pond. Since the concentration of the pollutant is assumed to be constant, $P(Y=0)$ is also equivalent to the pollutant load that is subject to treatment. The second term in equation (4) represents the fraction of the pollutant load entering the BMP pool that is removed. The relationship between η_{BMP} and E(D) can be determined by using appropriate relationships between detention time and removal efficiency [Randall et al. (1982); Schuler (1987)].

Conclusions

An analytical model has been presented to determine detention time and estimate pollutant removal. The novel aspect of the model is that it accounts for the random storage fluctuations due to the random runoff process. This methodology can be utilized in stormwater management planning to evaluate proposed retrofit sites. Quick estimates of required BMP storage and withdrawal rate for a specified level of pollutant removal can be made and proposed sites may be accepted or rejected based on the estimated performance.

References

Loganathan, G.V., Delluer, J.W., and Segarra, R.I. (1985). "Planning Detention Storage for Stormwater Management." *J. Water Resour. Plng. Mgmt.*, ASCE, 111(4), 382-398.

Randall, C.W., Ellis, K., Grizzard, T.J., and Knocke, W.R. (1982). "Urban Runoff Pollutant Removal by Sedimentation." *Proceedings of the Conference on Stormwater Detention Facilities.* ASCE, pp. 205-219.

Schuler, T.R. (1987). *Controling Urban Runoff: A Practical Manual for Planning and Designing Urban BMP's.* p. 3.12, Metropolitan Washington Council of Governments.

Watkins, E.W. (1993). "A Probabilistic Approach for Estimating Detention Time in Extended Detention Stormwater Ponds.", M.S. Thesis, Dept. of Civil Engrg., Virginia Polytech. Inst. & State Univ., Blacksburg, VA 24061.

Characterizing Urban Runoff Quantity and Quality

G. Padmanabhan and Louis P. Erdrich[1]

Abstract

A method using simulated flows and measured pollutant concentrations was successfully applied to estimate pollutant loadings resulting from runoff during storm events from an urban area. The method was applied to a storm sewer outfall draining an area with a predominantly commercial land use into a natural water course. The method involves collection of runoff samples during the event, observation of rainfall amount and the use of a rainfall-runoff computer model. Runoff samples were analyzed for conventional as well as priority toxic pollutants. Delivered load and mean concentrations of pollutants were estimated for several rain events. The study indicated that the mean concentrations for oxygen-consuming and nutrient type of pollutants were higher than that of comparable sites studied in the National Urban Runoff Program (NURP). The quality of the runoff from the selected basin routinely exceeded point source discharge and surface water quality standards. Also discussed are some issues pertaining to comprehensive assessment of the pollution potential of storm water discharging from several outfalls into receiving waters.

Introduction

The design of urban storm water drainage systems has traditionally focused on the efficient removal of storm runoff without regard to the quality of the runoff discharged into the receiving waters. Typically studies to determine the effects of drainage

---

[1]Respectively, Associate Professor and former graduate student, Department of Civil Engineering and Construction, North Dakota State University, Fargo, ND 58105.

systems and to assess the quality of runoff are undertaken after construction of the system. Many existing drainage networks were installed well before there was any significant concern for the quality of receiving waters. However, recently storm water runoff has been recognized as a significant source of nonpoint source pollution of receiving waters (EPA, 1988; Novotny, 1985). In order to assess the impact of runoff on receiving waters, the water quality and quantity of the runoff must be characterized first. Characterizing the quality of runoff requires adequate sampling and analysis of the runoff flow.

This paper discusses the strategy used in the initial efforts of an on-going study to characterize the storm runoff from the City of Fargo, North Dakota discharging into the receiving waters of the Red River of the North. The approach used is to employ a well-established computer model to predict discharges and manual sampling of runoff to measure pollutant concentrations at the outfalls. The information generated from computer modeling and manual sampling can be used to estimate mass flows, mean concentrations and event loads from the drainage area contributory to the outfall.

Sampling and Modeling

The entire Fargo-Moorhead area is served by a system of storm water drainage networks that eventually empty into the Red River. There are more than 70 outfalls which empty directly into the Red River including county ditches that drain the west of Fargo and the east of Moorhead. The drainage ditches have some 90 outfalls. It would be a formidable undertaking to sample runoff from even a fraction of the total number of outfalls. Characterizing runoff requires numerous samples from each outfall. The focus of this paper is on a drainage area in downtown Fargo which contributes to a major outfall into the Red River. There are approximately 400 inlets and over 25,000 feet of collector pipes or laterals which serve to connect the inlet structures and conduct runoff toward the outfall. Most of the laterals are built of reinforced concrete pipe ranging in diameter from 12 to 66 inches. The system drains an area of approximately 54 acres. The land use is predominantly commercial. The impervious area is approximately 85%.

The computer model ILLUDRAIN, a rainfall-runoff event simulastor was used for predicting outfall discharge from the drainage area under study. Three rainfall events were examined. Rainfall during the events was monitored from the start of the rainfall event and samples of runoff were taken during the ensuing runoff. The samples were analyzed for the conventional pollutants. The simulated hydrographs and the measured pollutographs were coupled together to estimate event Estimated Mean Concentrations (EMC) and pollutant loads. In addition to the grab samples used above, one-hour composite samples were collected for one storm event and two snowmelt events. These samples were analyzed not only for conventional pollutants but also for heavy metals.

Results and Conclusions

Table 1. Range of Values for Discharge Waters

Constituent	Treated Sewage (mg/l)	Mean EMC's NURP Sites (mg/l)	Mean EMC's StudyOutfall* (mg/l)
Total Solids	640 - 1167	----	1049
Dissolved Solids	625 - 1116	----	890
Suspended Solid	15 - 51	22 - 41	140
BOD	2 - 70	8 - 19	112
COD	31 - 155	40 - 18	672
TOC	13 - 20	----	175
Nitrate	.25 - 38	.356 - 1.	8.6
Phosphate	6.2 - 9.6	.105 - .7	3.4
	(µg/l)	(µg/l)	(µg/l)
Mercury	---	----	.46 - 20.1
Cadmium	<20	----	1.06 - 1.37
Copper	31 - 38	11 - 104	14 - 17
Zinc	85 - 190	37 - 1416	184 - 249
Nickel	<10	----	10 - 29.8
Lead	<20	46 - 409	78.1 - 125
Chromium	36 - 70	----	5.44 - 19.6

* Average EMC's based on total loads/total volumes

The contribution of TSS and COD are rather high from this outfall when compared to discharge standards. The event of 7/17/89 delivered an estimated 254 lbs of TSS over 8.5 hours. The quality of the runoff from the selected basin is routinely impaired beyond point source discharge standards and surface water quality standards. The basin also exhibited mean concentrations that were somewhat higher than comparable sites studied in the NURP for oxygen consuming and nutrient type of pollutants. The suspended solids were in the range of those sites studied by the NURP, Heavy metals were detected in the runoff from this basin and at times exceeded surface water quality standards. the quality of runoff water from this basin exhibits pollutant concentrations greater than typical wastewater treatment plant discharges.

Characterization of the outfall discharges may be accomplished by the method used in this study where a rainfall-runoff event simulator is calibrated, verified and then coupled with the measured pollutant concentrations. This may be done on a prioritized basis where the major outfalls are studied first. They may be selected according to land use and other features that might facilitate transferability of data to other areas that are yet to be studied. The pollutants monitored should include a range of conventional pollutants and heavy metals. Additional pollutants can be decided on a site specific basis. Pollutographs need not be run on all pollutants, rather flow weighted samples may be analyzed for all pollutants with pollutographs constructed for selected pollutants requiring analytically less rigorous effort for detection.

APPENDIX - References

1. Abel, P.D. <u>Water Pollution Biology</u>, Ellis Horwood Limited and Halstead Press of John Wiley and Sons, 1989.

2. Environmental Protection Agency, "National Pollutant Discharge Elimination System Permit Application Regulations for Storm Water Discharges." <u>Federal Register</u>/Vol. 53, No. 235, December 7, 1988.

3. Novotny, Vladimir, (Ed.) Proceedings of the Symposium "Nonpoint Pollution: 1988 - Policy, Economy Management and Appropriate Technology", American Water Resources Associations, Milwaukee, Wisconsin, 1988.

4. U.S. Environmental Protection Agency, Water Planning Division, <u>Results of the Nationwide Urban Runoff Program,</u> Volume I, II & III, 1983.

STORAGE OF COMBINED SEWER OVERFLOW: HOW EFFECTIVE IS IT ANYWAY?

Khamis A. Al-Omari,[1] A.M., ASCE

Abstract

The purpose of this paper is to illustrate the importance of modeling long term rainfall records to evaluate the effectiveness of combined sewer overflow storage. This paper will present a case study on the Grand Rapids, Michigan Combined Sewer System to show the application of the USEPA's Stormwater Management Model (SWMM) in a continuous mode to predict the volume and number of overflows over an extended period.

Introduction

In April 1989, the City retained Black & Veatch of Detroit, Michigan, with Fishbeck, Thompson, Carr & Huber, Inc., of Ada, Michigan, to conduct a comprehensive study of the municipal wastewater collection system within the combined sewer area. The purpose of the study was to develop a Long Term Combined Sewer Overflow (CSO) Control Program (Black & Veatch et al., 1990).

The City of Grand Rapids, Michigan, owns and operates a wastewater treatment plant (WWTP) which serves the City and eleven suburban communities. The 1989 service area is about 87,936 acres, with 83,115 acres served by separate sanitary sewers and storm sewers. The remaining 4,821 acres within the City, or about 5.5 percent of the total service area, is served by combined sewers. The combined sewer area is naturally divided by the Grand River which discharges into Lake Michigan about 40 miles west of the City. The west side combined sewer area accounts for 1,949 acres; the east side accounts for 2,872 acres.

---

[1]Project Engineer, Black & Veatch, 211 West Fort Street, Suite 2200, Detroit, Michigan 48226

The peak hydraulic capacities at the WWTP for primary and secondary treatment are 150 and 90 mgd, respectively. The excess primary effluent, up to 60 mgd, is routed to a 11 MG retention basin at the WWTP. The effective treatment rate at the WWTP, based on recent records, is 138 mgd. Allowing for 20 mgd from the Southeast Interceptor, the effective wet weather treatment rate at the main pumping station upstream of the WWTP (Market Avenue Pumping Station or MAPS) is 118 mgd.

Under the current state policy, City of Grand Rapids is required to provide transport and secondary treatment for flows generated during storms up to and including the one-year, one-hour storm event (Total rainfall = 1.1 inches). It also requires the equivalent of primary treatment by providing 30 minute detention for all flows greater than the one-year, one-hour storm event, up to and including the ten-year, one-hour storm event (Total rainfall = 1.8 inches). This policy requires providing a 44 MG storage capacity for the combined sewer area. Recently, the City has completed the construction of a 30.4 MG retention basin on the east side of the Grand River and has implemented a complete sewer separation program on the west side of the river.

Modeling Approach

The objective of the continuous SWMM simulation was to predict the performance of the combined sewer system (CSS) based on historical rainfall records. Performance would be expressed in terms of average annual number and volume of CSO for various levels of CSO control. Continuous SWMM simulates long term precipitation records to generate runoff rates based on hourly data. Hourly precipitation records for the period 1963-88 were obtained from the National Weather Service (NWS). The nearest station is at the Kent County International Airport (NWS Station #203333). The detailed analysis included the periods 1978-80 and 1986-88 with average annual rainfall of 35.3 and 38.4 inches, respectively. The two periods were selected as representative of average and extreme annual rainfall. The annual rainfall ranged from 32.6 inches in 1979 to 46.6 inches in 1988.

For purposes of predicting annual bypass events, an effective treatment rate of 138 mgd was established at the WWTP. This figure was based on actual treatment rates during bypass events for the period May - September 1989. Base flow was estimated to be 57 mgd. Thus, the effective treatment rate of runoff flow from the CSS area was estimated at 81 mgd. The simplification assumes that all hydrographs from the CSS tributaries were aggregated at a single manhole (effectively the MAPS). The calculated bypass rates were thus the total inflow rate from the three tributary areas minus the 118 mgd treatment rate. Based on the above assumptions, continuous SWMM was calibrated using single event hourly precipitation data for 1989 and the NWS data for 1988.

Summary and Conclusions

Continuous SWMM was used to predict the volume and number of overflows over an extended period. Hourly precipitation records from 1978-80 and 1986-88 were utilized, as representative of average and extreme rainfall years, respectively.

A logical flow diagram was established for filling, dewatering and treatment of CSO for each precipitation event during the two 3-year periods of record. Priority was given to dewatering the 11 MG retention basin at the WWTP, followed by dewatering the captured CSO at other modeled retention basins. Storage volumes ranged from 0 to 50 MG. Annual number and volumes of CSO were calculated for the simulated precipitation records.

Figures I and II show the results of the analyses of the 1978-80 and 1986-88 records. Figure I shows the predicted long-term number of events per year for the two rainfall records, while Figure II shows the long-term volume of CSO for the same periods. If no storage is provided, the estimated ranges of annual number and volume of CSO range from 41 to 36 events and from 260 to 400 MG, respectively. The estimated annual number of CSO events is further reduced to 1 to 5 events per year when 30 MG of storage (currently on-line) is analyzed. The estimated annual volume of overflow ranges from 30 to 125 MG. This represents a reduction of 86 percent in annual CSO events and a 69 percent reduction in annual CSO volume from existing conditions for the extreme rainfall record of 1986-88.

Finally, if 44 MG of storage is considered, as would be required under the current state policy, the estimated annual number of CSO events would range from 1 to 3, and the estimated annual CSO volume would range from 10 to 90 MG. This would represent a 92 percent reduction in annual CSO events and a 78 percent reduction in annual CSO volume for the 1986-88 rainfall record.

The conclusion reached based on this analysis is the diminishing effectiveness of additional control (storage) beyond about 25 MG. However, much of the required 44 MG storage volume under the current state policy would be used very little, and the marginal impact on the annual number and volume of CSO becomes minimal.

Appendix I - Conversion Factors to SI Units

1 ft = 0.3048 m; 1 in = 0.0254 m; 1 mi = 1609.34 m; 1 acre = 4046.86 m^2; 1 mgd = 0.04381 m^3/s; 1 MG = 3785.41 m^3.

Appendix II - References

Black & Veatch and Fishbeck, Thompson, Carr & Huber, Inc., *Report on Combined Sewer System Study*, prepared for the City of Grand Rapids, Michigan, October 1990.

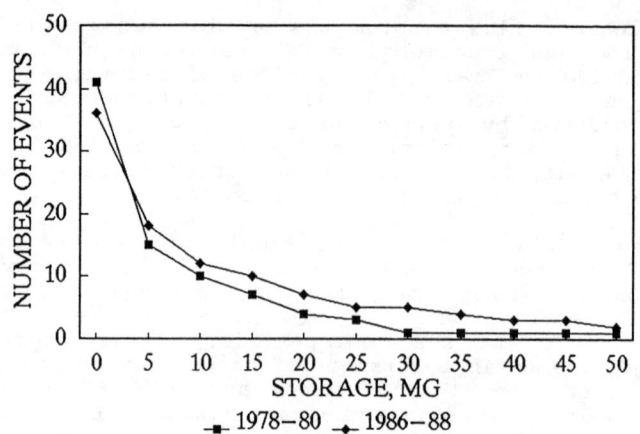

Figure I - Annual Number of Combined Sewer Overflow Events

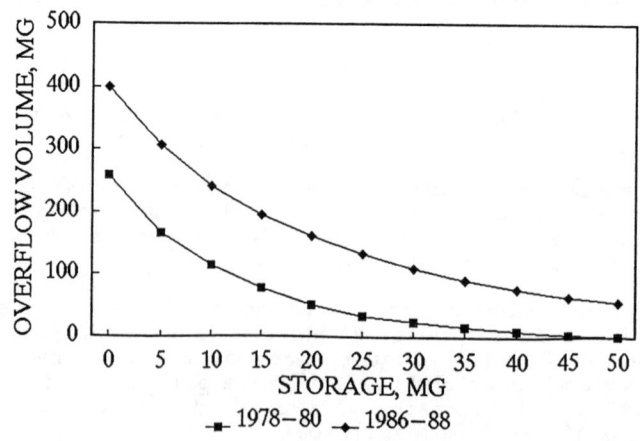

Figure II - Annual Volume of Combined Sewer Overflow

The Fairfield-Suisun Urban Runoff Management Program:
The Approach of Small Communities to NPDES Permitting

Jill C. Bicknell (M. ASCE)[1] and Michael J. Barnes[2]

Abstract

The Fairfield-Suisun Sewer District is responsible for NPDES permit acquisition and maintenance of the storm drainage facilities of the Cities of Fairfield and Suisun City, California. The combined population of the two cities is only 100,000; however, because the drainage system discharges to sensitive receiving waters in San Francisco Bay, the local Regional Water Quality Control Board has required the District to prepare a comprehensive storm water management program. The District and its consultants developed a unique approach which minimizes District costs and meets Regional Board deadlines. The approach consists of the following key elements:

- Begin developing a storm water management plan early;
- Draw upon existing resources developed by other agencies;
- Delay intensive storm water monitoring and field screening until specific needs have been better defined;
- Identify and build on existing District pollution control programs;
- Make maximum use of District staff and laboratory facilities;
- Interface with the District's GIS for efficient data management.

Introduction

The degradation of water quality due to the transport of pollutants in storm water runoff to receiving waters, particularly in urbanized areas, has recently received increased nationwide attention as a major pollution problem. In November 1990, the U.S. Environmental Protection Agency (EPA) published its final rule for storm water discharge regulation under the National Pollutant

---

[1]Water Resources Engineer, Kennedy/Jenks Consultants, 303 Second Street, San Francisco, CA 94107

[2]Deputy District Engineer, Fairfield-Suisun Sewer District, P.O. Box 507, Suisun City, CA 94585

Discharge Elimination System (NPDES). The regulations require municipalities with populations greater than 100,000 to obtain NPDES permits for discharges from municipal separate storm sewer systems (MS4s).

In California, the San Francisco Bay Regional Water Quality Control Board (Regional Board), which has authority to issue NPDES permits, began an aggressive urban runoff management program in advance of the publication of the EPA regulations because of the potential impacts of pollution from urban runoff on the sensitive receiving waters of San Francisco Bay. The Regional Board's urban runoff control strategy, as outlined in the 1991 amendments to its Basin Plan for San Francisco Bay, calls for the development and implementation of "baseline control programs" in all cities and counties in the region and "comprehensive control programs" in selected cities and counties. Baseline control programs focus on source control, i.e., preventing the increase of pollutants in discharges from MS4s. The baseline program must include: operation and maintenance programs for storm drain systems; ordinances for construction site runoff control; public education programs to control improper use and disposal of hazardous materials; and policies to limit pollutant discharges from new developments. Comprehensive programs focus on the prevention of water quality problems and remediation of existing problems. In addition to baseline control program elements, comprehensive program elements must include characterization of urban runoff discharges and implementation of urban runoff management plans, which is consistent with the Federal NPDES regulations for municipal storm water discharges.

The Fairfield-Suisun Program

The Cities of Fairfield and Suisun City (Cities) in Solano County, California have been identified by the Regional Board in its 1991 Basin Plan as areas which need to develop and implement comprehensive control programs for urban runoff. Although the combined population of the two cities is only 100,000, they are rapidly growing areas of Solano County. In addition, the Cities' drainage systems discharge to Suisun Marsh, a sensitive marsh in the upper reaches of San Francisco Bay which contains 10 percent of California's wetlands. As part of their comprehensive control program, the Cities are required to submit Parts 1 and 2 of the NPDES permit application, equivalent to the requirements of the EPA storm water discharge permitting regulations, to the Regional Board by March' 31, 1993 and March 31, 1994, respectively. Under an agreement with both Cities, the Fairfield-Suisun Sewer District (District) is responsible for maintaining the storm drainage systems for the Cities as well as obtaining and implementing an NPDES permit for the region.

The District is developing an Urban Runoff Management Program (URMP) which includes the following tasks during FY 1992/93:

- Development of an accurate and updated system map;
- Development and implementation of a limited monitoring program for discharge characterization and comparison with other Bay area urban runoff data;

- Development and implementation of a limited field screening program for identifying non-storm water or illicit discharges;
- Development of a preliminary Storm Water Management Plan to describe activities that the District and the Cities are currently conducting to control storm water pollution and future pollution prevention activities.

Unique Small Agency Approach

The District and its consultants, Kennedy/Jenks Consultants and EOA, Inc., developed a unique approach to the URMP in order to minimize District costs and meet Regional Board deadlines. Some of the important elements of this approach are described below.

The District assumed that the quality of its urban runoff is most likely similar to that of urban runoff from other Bay area communities, which have performed a substantial amount of sampling in the past few years. Therefore, the District's initial sampling efforts were limited to two sites, which were used to verify this assumption and generally characterize the urban runoff in the Fairfield-Suisun area. Instead of expending much of its limited funds to characterize urban runoff, the District is focusing most of its initial efforts on development of a master plan to reduce the pollutant load in urban runoff.

Field work such as dry weather outfall screening and storm water sampling was initially done by the District's technical staff that are involved with development of the master plan. This "hands-on" experience will facilitate development of a plan that accounts for actual field conditions. In addition, the technical staff was able to refine the field procedures as they gathered information from the field. This approach is less costly than the typical approach in which technical staff train and supervise field staff, especially since the field procedures are still evolving. Once field procedures are better defined, future field work will likely be performed by District field staff at lower cost.

Another cost saving measure was the use of a civil engineering student to assist with the field screening work conducted in the summer of 1992. This was a "win-win" situation since the student gained valuable work experience and the District gained relatively inexpensive assistance from a student with the proper technical background to quickly learn the field procedures.

Because the District is also responsible for wastewater collection and treatment in the region, urban runoff pollution control will be integrated with the District's present pollution control activities to manage water pollution in the region on a watershed basis. In particular, work will be closely coordinated with the District's Source Control Program to determine whether discharges should go to the sewer or storm drain.

Since many entities in the Bay area are faced with similar problems related to urban runoff control, an organization of storm water dischargers was formed in the Bay area to exchange information and technical data related to urban runoff issues. The District is participating in this organization, which will allow the sharing of public education materials, water quality data bases, procedures manuals and other information. This will reduce the cost of producing specific materials for the District's program.

In the long term, the District anticipates managing its urban runoff data in a regional geographic information system (GIS). This will be an efficient approach to data management, as urban runoff data are critically linked to geographic location. Currently, digital orthophotos are being developed for use as a base map and the storm drainage system is being digitized for use as an overlay within the GIS.

Conclusion

The specific process for obtaining an NPDES storm water discharge permit and implementing a storm water management plan will depend on the requirements of local regulatory agencies. However, elements of the Fairfield-Suisun Sewer District's approach may be useful to other small agencies who are trying to comply with storm water regulations. Recommendations are as follows:

1. Begin developing a storm water management plan early in the process.
2. Draw upon existing resources developed by other agencies in the area.
3. Delay intensive storm water monitoring and field screening until specific needs have been better defined.
4. Identify and build on existing pollution control and maintenance programs.
5. Make maximum use of in-house staff and facilities.
6. Get technical staff who are developing the storm water management plan involved in field work in order for them to gain knowledge of the drainage system and make refinements to the plan.
7. Use available GIS mapping and data bases for efficient data management.

References

EOA, Inc, Scope of Work for Cities in San Mateo County (California) to Obtain and Prepare for a Municipal Storm Water NPDES Permit, May 1992.

Fairfield-Suisun Urban Runoff Management Program - Progress Report, Letter from the Fairfield-Suisun Sewer District to the California Regional Water Quality Control Board, San Francisco Bay Region, October 28, 1992.

State Water Resources Control Board and Regional Water Quality Control Board, Water Quality Control Plan Report-San Francisco Bay Basin, 1975; 1986 Amendments; Proposed 1991 Amendments.

U.S. Environmental Protection Agency, National Pollutant Discharge Elimination System Permit Application Regulations for Storm Water Discharges, 40 CFR Parts 122, 123, and 124, 16 November 1990.

Seattle's Storm Water Application

Neil F. Thibert[1], P.E., Member, A.S.C.E.

Abstract

Seattle is considered a large city under the National Pollutant Discharge Elimination System (NPDES) storm water permit regulations and was required to submit a storm water permit application on November 16, 1992. Assembling Seattle's application took months of work by several people, but this effort was much easier than it could have been. City staff had followed the development of the regulations closely and began preparing in advance. An existing drainage control ordinance was revised to provide adequate legal authority. Extensive mapping of the City's storm drains already existed. Sampling for storm water characterization was started in 1986. Public education and source control programs were under way. In putting this application together, Seattle outlined a series of programs that accomplish the intent of the regulations. The application has to provide adequate information to the Washington Department of Ecology to help them write a permit that addresses the specific water quality problems facing Seattle. The needs in a highly urbanized area are quite different from those in a rural area that is experiencing rapid growth, and the application has to provide enough information to allow the permitting agency to tailor the permit to those needs.

Introduction

With a population of 516,259, Seattle is considered a large city under the National Pollutant Discharge Elimination System (NPDES) storm water permit regulations. The City was required to submit a storm water permit application, in two parts, to the Washington Department of Ecology by November 16, 1992. Assembling

---

[1]Senior Civil Engineer, Planning Section, Seattle Engineering Department, Drainage and Wastewater Utility, Room 660 Dexter Horton building, 710 Second Avenue, Seattle, WA 98104

Seattle's application took months of work by several people, but this effort was easier than it could have been. City staff began following the development of the regulations in 1985 and prepared for the permit in advance. In 1988, the Seattle Drainage and Wastewater Utility and a team of consultants completed a Comprehensive Drainage Plan. The Plan establishes the Utility's goals and policies and lays out a program for managing storm water runoff. The City established a drainage utility in 1989 to provide a dedicated funding mechanism.

Legal Authority

The NPDES Permit Application Regulations require cities and counties over 100,000 population to demonstrate that they have the legal authority to control discharges to the municipal separate storm sewer system. Seattle has used a number of ordinances to control pollutants and has required detention since 1979. Additional authority was obtained by adopting a revised Stormwater, Grading and Drainage Control Code. Seattle consolidated the storm water requirements of the NPDES regulations and the Puget Sound Water Quality Management Plan into this one Code.

Mapping

Seattle has an extensive mapping system based on 200-scale maps (one inch = 200 feet), although ditches and culverts are still not adequately covered. Each map covers one section (one square mile), with topography and buildings from 1956 aerial photos. They show street rights-of-way, road surfaces, and creeks. There are several versions, each with a different part of the City's infrastructure, including one set for sewers and one for storm drains. These maps show each length of pipe, the pipe diameter, manhole locations, and invert elevations at the manholes. For greater detail, Seattle maintains side sewer cards for each block in the city. The Seattle Engineering Department is developing a Geographic Information System (GIS) to provide even better information in the future. The City also produced a series of maps specifically for this permit application.

This mapping system is a major tool in Seattle's ability to trace contaminant sources and control spills, as well as maintain the drainage network. The application includes a map of all storm drain outfalls. DWU traced the outline of each drainage system on topographic maps in order to calculate the tributary area and annual discharge. The Utility determined the land use in each drainage basin and used this information to calculate annual pollutant loads. For the purpose of this application, ditch and creek drainage systems are counted as outfalls at the point where they discharge to major receiving waters. The area of Seattle tributary to each of the city's creeks is included in the calculation of mean annual storm water runoff.

Proposed Management Program

Seattle's application outlines a series of programs that accomplish the intent of the regulations - to prohibit non-stormwater discharges into storm drains and reduce the discharge of pollutants to the maximum extent practicable. It has to provide adequate information to help Ecology tailor the permit conditions to the specific water quality problems facing Seattle. Because Seattle is heavily developed already, storm water controls on new development will have little effect on water quality and stream flows. In contrast to rapidly developing areas, erosion from construction sites is not a concern in Seattle except in a few cases.

Reducing the input of contaminants to the municipal storm drain system from industrial sites is a high priority because it can have a large effect with minimal cost. Seattle uses educational programs and inspection of drainage control facilities city-wide to encourage proper storm water management practices. Source control programs provide door to door inspections in the industrial areas. The Stormwater, Grading and Drainage Control Code provides enhanced enforcement capability when contaminant sources are identified.

Most contaminants from commercial land uses, as well as a good share of those from industrial and residential land uses, comes from street and parking lot runoff. Although municipalities have little direct control over motor vehicle contamination, Seattle is taking steps to deal with street and parking lot runoff. Forty percent of Seattle's drainage revenue is used for maintenance practices to reduce contamination from street runoff. DWU crews maintain and clean storm drains and outfalls. The Utility contracts with street maintenance crews to perform activities related to water quality such as cleaning of ditches, culverts, and catch basins. The City has constructed a few grassed swales and regional detention facilities to capture pollutants in road runoff. The Code requires developers to install best management practices (BMPs) to collect sediments and floatables in runoff from parking lots.

Watershed planning and flood control projects are done in the ditch and creek drainage basins. Water quality monitoring is done in both creeks and storm drain systems to characterize runoff from various land uses and to assist with source control programs and capital projects.

Seattle began storm water sampling and flow monitoring in 1986. DWU was able to use several previous samples for the characterization data because they met most of the application requirements. The Part 1 application proposed sampling just two additional sites for three storms each to complete the sampling requirements for Part 2. Seattle was still unable to complete all of the sampling in time for the Part 2 application because of weather constraints.

Seattle has 130,000 single family residences. A few bad practices multiplied by that number of households can add up to a lot of pollution. Since it's impractical to visit every home, DWU will rely on public education and programs such as storm drain stenciling, motor oil recycling, and household hazardous waste collection facilities to reduce contamination from residential areas. Seattle's water quality education program reaches 20,000 school children each year through a teacher's guide, video, field trips, and a salmon rearing project in every grade school.

Assessment of Controls

There is no way to adequately quantify the reduction in pollutant load from these water quality programs at this time. DWU will reduce the obvious sources of pollution through inspections and implementation of best management practices. Some background level of contamination will continue from sources such as natural soil constituents, rooftops, galvanized buildings, and air pollution. The number of structural best management practices employed will depend largely on the amount of new development and redevelopment that occurs. Contamination from motor vehicles has been reduced to some extent already, but is not expected to change dramatically in the next five years. The largest part of the area covered by this permit has residential land uses. However, any reduction in pollution here may be small compared to the amount of contamination washing off of streets and parking lots. Industrial land uses with high levels of contamination hold the greatest potential for improvement, but it's too early to tell what impact these programs will have. DWU will monitor the runoff from industrial areas during the term of the permit to help with source tracing and to quantify any reductions in contamination resulting from those efforts.

Fiscal Resources

The City receives $9.4 million per year from a drainage fee of $36.57 for single family residences. All other properties are charged based on parcel size and percent of impervious surface. Seattle spent about $350,000 in sampling and staff time to develop this permit application. This does not include a cost of about $1 million to establish the drainage utility and develop a comprehensive drainage plan. The cost of the field screening at 100 major outfalls was $45,000. The cost of sampling for characterization data was about $2,500 per sample at ten sites for three storms each or $75,000, including labor and lab analysis. With equipment and a van, it comes to about $100,000. Seattle now has six full time water quality staff to take samples, interpret the data, and conduct source control investigations. As many as 18 people were involved in assembling the various pieces of information needed for this application, including mapping and calculating drainage areas, annual discharges, and pollutant loads. Another $100,000 in staff time was expended to write the new Stormwater, Grading, and Drainage Control Code.

The time frames and expected costs in the Federal Register are unrealistic. Seattle was able to complete this application on time with in-house staff by getting a head start and working closely with the permitting agency. The application describes an achievable five-year strategy for dealing with the most pressing water quality problems and progressing toward the goal of clean water. The emphasis here is on source control and education rather than new development and infrastructure.

NPDES Municipal Storm Water Permit
A Utility Approach

Jeff Niermeyer, P.E.[1] Member, ASCE

Abstract

Salt Lake City Department of Public Utilities (SLCPU) is the agency responsible for the operation of the drainage facilities in Salt Lake City. The drainage function is established as an enterprize fund of Salt Lake City. The procedure for meeting the requirements of the Municipal Separate Storm Sewer NPDES permit application from a utility's perspective are discussed.

Introduction

The true impact of the NPDES permit process on municipal manpower allocations and budgets is being felt by communities targeted under the Phase I implementation of the 1987 amendment to the Clean Water Act. Phase I includes all municipalities with populations greater than 100,000 as determined by the 1990 census. Salt Lake City with a 1990 census population of 159,936 is required to complete the Part 1 and Part 2 permit application under the time schedule established for a medium sized system. Part 1 of the application was submitted on May 18, 1992 and Part 2 must be submitted by May 17, 1993. Salt Lake City recognized early on that the requirements of the NPDES storm water permit would place a tremendous impact on municipal budgets and personnel. A consultant was retained to study the various options of financing the EPA mandated storm water quality program and much needed drainage improvements. A 20 member citizen committee was established to evaluated the alternatives recommended by the consultant team. The results of this study indicated that a storm water utility concept would provide the dedicated funding necessary to develop a comprehensive storm water quality and capital improvements program. A storm water utility was established in July of 1991

---

[1] Drainage Engineer, Salt Lake City Department of Public Utilities
1530 South West Temple Salt Lake City, Utah 84124

and all the drainage functions for Salt Lake City were transferred to the Salt Lake City Department of Public Utilities (SLCPU). The SLCPU is also responsible for the sanitary sewer and culinary water system for Salt Lake City.

The implementation of the NPDES permit requirements under a utility concept offers several advantages:

- Dedicated flexible funding to implement storm water quality programs.

- Utilizing the existing sanitary sewer industrial pretreatment program to extend pollution prevent to storm water runoff from industrial sites.

- Analytical laboratory services can be done at the SLCPU in-house laboratory.

- Environmental expertise is available within the department.

- The GIS system established for the utility can be used to track water quality data.

Funding Flexibility

Part 2 of the NPDES storm water permit process requires that Salt Lake City provide a fiscal analysis of the necessary capital, operational and maintenance expenditures necessary to accomplish the activities of the storm water management program. The commitment of resources is for the five year duration of the permit. With the ever increasing demand on municipal general fund budgets it is difficult to fund a major new program. A storm water utility enterprise fund concept provides the dedicated long term funding that is critical to the success of the storm water management program. Other financing techniques such as impact fees, grants /loans and plan review fees can be used to offset new facilities or direct service costs. However, they do not provided the revenue stream required to support a comprehensive storm water management plan. The utility concept where the user fee is based on a measurable indicator of runoff such as impervious area is a fairer method of allocating storm water cost than an ad valorem tax. The cost of providing storm water service is placed on those that use the system regardless of their tax status.

A dedicated revenue stream allows Salt Lake City to approach storm water management from a long term prospective with the ultimate goal of significantly reducing the pollution of surface waters by storm water runoff.

Existing Pollution Prevention Programs

SLCPU has taken a very progressive approach to pollution prevention. An extensive pretreatment program has been developed for the sanitary sewer utility. The direction of the storm water management program adopted by Salt Lake City will focus on the prevention of storm water pollution at the source. This focus will greatly reduce the need for costly structural pollution controls. The existing pretreatment program personnel actively work with industrial and commercial accounts to reduce the pollutant loads to the sanitary sewer. It is a natural extension to incorporate industrial storm water pollution prevention into this program. The benefits include reduced training time, administrative costs and regulatory impact on system users. The pretreatment personnel are familiar with the area industries and potential storm water pollution sources. A small increase in staffing of the pretreatment program can provide large dividends in reduced storm water pollution and resulting clean-up or treatment costs.

The department is sponsoring a series of pollution prevention workshops aimed at minimizing pollution production at the source. This program directly benefits the storm water management program. The spinoff impact of all the water quality programs implemented within the department will have similar effects.

Analytical Laboratory

The federal storm water regulation requires that wet weather monitoring be conducted on a minimum of 5 sites for 3 storms during the preparation of Part 2 of the permit application. Wet weather storm water monitoring and analytical testing for pollutants is very expensive. A typical suite of analytical tests costs $3,000 per wet weather station per storm. Also, on-going monitoring and testing will be a major requirement of the storm water permit. The incorporation of the storm water utility within the Department of Public Utilities provides access to the existing analytical laboratory operated by the Department. The use of an in-house labaratory reduces the cost of the testing and provides better turn around times on sampling results.

Environmental Expertise

The implementation of a comprehensive storm water program requires knowledgeable people with a strong background in environmental issues. The existing expertise developed within the department to meet environmental issues faced by the water and sanitary sewer utilities provides a major benefit to the storm water utility in developing a storm water management program. The department has NPDES permitting experience for discharges from the sewage and water treatment plants. Working relationships with the regulatory agencies are established. These relationships provide an increased level of communication and understanding during the permit application period. The department also has broad experience in watershed management, pollution prevention, environmental testing, and storm water quality monitoring associated with watershed protection. This expertise resource allows Salt lake City to approach storm water quality management much further along the learning curve.

GIS Mapping

SLCPU is implementing a Geographic information system (GIS) to provide an integral management tool for handling the vast amount of information that is available for each of the utilities. Storm water quality management is a perfect application for GIS. The system allows large amounts of attribute information to be attached to individual parcels within the City. The storm water quality management program use of the GIS will include tracking impervious area, standard industrial code (SIC), storm drain connections, illicit connection inspection and enforcement, maintenance activities and other information that is needed to support the program. The information will be readily available and can be used to document compliance with the terms of the storm water NPDES permit.

Implementing GIS is very expensive and labor intensive. Cost sharing by the three utilities within the Department of Public Utilities reduces the overall cost of the storm water management program yet allows for a state of the art information retrieval system.

Summary

Implementing the storm water management program within the frame work of the Department of Public Utilities offers significant advantages over a stand alone program. Existing expertise, programs and facilities greatly reduce the overall startup cost of a comprehensive storm water quality program. The existing knowledge of EPA regulations and working relationship with regulators greatly benefits the storm water program during the permit application period.

Both Salt Lake City and the businesses within the City benefit by having one agency responsible for the coordination of water quality programs including watershed management, irrigation, storm water, treated sewage effluent and drinking water.

Development and Implementation of Stormwater Utilities in Texas Cities

C. Diane Palmer, P.E.[1]

ABSTRACT

As more Cities begin to understand and feel the financial impact of EPA's Non-Point Discharge Elimination System (NPDES) stormwater regulations on their budgets, many are turning to stormwater utilities to generate new funding for stormwater quality projects and to replace dwindling general fund revenues for flood control projects. In Texas, specific enabling legislation allowing cities to set up stormwater utilities can result in a utility structure that is somewhat different from those in other states.

STORMWATER UTILITIES IN TEXAS

The changes to the Texas Local Government Code (Code) requires a minimum of three (3) public notices and two (2) public hearings. Once the utility is established, it can not be abandoned for a minimum of five (5) years. Several unique features of the Code can foster structures different from other utilities in other states.

Benefitted Property

The Code requires that a property receive stormwater service provided by the municipality levying the fee and that the property be "benefitted" by receiving water, wastewater or electric service provided by the municipality. As a result, most Texas Cities forming stormwater utilities under the Code are not able to charge customers outside their municipal boundaries; eventhough, areas upstream may contribute substantially to the municipality's stormwater quality and flood control problems.

---

[1] Manager, KPMG Peat Marwick, National Utilities Consulting Practice. 200 Crescent Court, Suite 300, Dallas, Texas 75201. (214) 754-2536.

Exemptions

The Code stipulates that "vacant" property shall be exempt from stormwater charges. Additionally, the utility may exempt properties of the state, county, municipality and school districts. Most Texas cities are not choosing to exempt these public properties.

Drainage Related Parameters

The stormwater fee must be based on drainage related parameters. These parameters may vary for each city, since the criteria by which flood control structures and stormwater quality improvement facilities are designed varies within each region and municipality. Of the various criteria that may be used, substantial research data exists to relate the impervious area on a property to both stormwater quantity and stormwater quality. As a result, many cities have chosen to use the measurement of improved area on the property as the primary means by which the costs of stormwater service can be equitably distributed among customers.

The Texas Code specifically allows a municipality to utilize the records of the tax appraisal district to determine the improved area on a parcel, as well as, meet the Code stipulation which states that tracts and parcels within the service area must be inventoried. Using the tax appraisal district records to design stormwater rates can present several problems depending on the sophistication of the database within each district. The land improvement records in many counties do not include information on tax exempt properties and often do not include complete information on residential properties. Consequently, many cities have opted to utilize geographic information system (GIS) technologies to determine improved area, such as satellite or low-level, multi-spectral imagery which can be used to identify impervious area for classes of land use and/or for each individual parcel. Although these GIS methodologies can provide more precise data for designing rates, these tools can be significantly more costly and require substantial lead time to develop data prior to being able to bill for stormwater charges.

Billing of Stormwater Fees

The Code requires that the stormwater fee be identified separately on the bill and stipulates that no deposits can be required for stormwater service. These requirements, as well as, the need to revamp existing billing systems to incorporate and track new customer billing data, such as the area of improvement on each property, frequently necessitates substantial changes to many existing billing systems. The time and expense required to reprogram the existing billing and accounting system for receivables and for development of the customer billing database are often the most complex and time consuming processes in implementing a new stormwater utility.

The data conversion effort involves identifying those properties which can be billed, determining the most feasible system for billing accounts, and establishing what drainage related parameters are available or can be developed for calculation of the fee for each parcel of land. Typically the effort required to convert the data for stormwater billing requires both automated and manual activities. The billing system software modification effort involves designing, developing, programming and testing the functional requirements for stormwater billing. Often the capacity of the existing billing system with regard to its ability to accept software changes and increased data storage demands may be the determining factors in the implementation cost and schedule for initiation of new stormwater charges.

Stormwater Only Accounts

In most cities, stormwater billing is added to the water/wastewater utility billing system and can be assessed against the tenant or owner of the property. However, some municipalities also own and operate the electric utility, which allows those cities to expand their customer base to include properties with wells or septic tanks and large parking lots that may have electric security lights but no water or wastewater service. These "stormwater only accounts" require special bill handling and payment compliance procedures.

Accounting and Customer Service

Under the Code, income must be segregated and identifiable in municipal accounts but may be transferred to the general fund for stormwater services budgeted within the general fund. Revenues charged for future costs of service may not be transferred to the general fund.

Customer service policies for stormwater must be established and customer service representatives be trained to respond to questions about the new utility. The Code allows water, wastewater or electric service to be cut-off for non-payment of stormwater billings, which facilitates payment compliance but may increase customer service interfaces and billing/payment documentation.

Cost-of-service and Rate Design

The cost-of-service on which the rates must be based can include future, as well as, existing expenditures for stormwater service, unlike standard regulatory practices for water and wastewater utilities. The cost-of-service may include costs of real property and real property rights; construction, repair, and maintenance of system and equipment; professional services; machinery, equipment, furniture, and facilities; funding and financing cost, interest, and start-up cost of constructing facilities; funding of future drainage system construction; debt service and reserve requirements; and administrative costs.

The rates must be "nondiscriminatory, equitable, and reasonable" and must be directly related to drainage parameters. The service units must be based on an inventory of the lots and tracts within the service area and may be derived from tax plats and assessment rolls.

Stormwater Organization

One of the more difficult aspects of establishing a stormwater utility is the management of the stormwater organization, particularly when staffs responsible for providing stormwater services may be housed in multiple departments within the municipality. Quite often, all stormwater activities are not carved from other departments into one organization, such as the spill response activities of fire departments. Specific attention must be paid to assuring that the funds collected for providing stormwater services are actually expended on those services. In cities where a specific stormwater organization is being developed, managers must be aware of the need to manage the change process and address the fears of personnel who may be asked to operate under a new framework of responsibility and authority.

Community Awareness

The community awareness program ushering in the stormwater utility may be vital to garnering support from the City's management and its customers. Council members in some Texas cities have been ejected from office for supporting a "rain tax", while other communities have rallied around their new utilities as a way to address citizens' awakening concern over environmental issues. Often the difference in these successes and failures has been a good community awareness campaign which involves direct communication with focus groups who may oppose the utility.

SUMMARY

Several large and medium cities in Texas have instituted stormwater utilities due to significant problems with general fund budget deficits. Many of these cities are phasing in the cost-of-service of stormwater so as to avoid rate shock; but, these cities have accomplished the significant step of instituting a financial framework which will result in a stable and long-term funding source for stormwater quality programs and flood control activities.

WHEN THE WELL RUNS DRY
PAYING FOR STORM WATER

Shaun Pigott[1]

Abstract

Much will be said during this conference about the need for various structural and nonstructural approaches toward mitigating problems related to storm water quality and quantity management. The not so shocking conclusion of this paper is that both approaches entail a significant commitment of resources in terms of staff time, equipment, and capital. The shock comes when the engineering analyses are undertaken and the corrective actions planned without a parallel evaluation of how these programs will be funded. The fact remains that many jurisdictions do not have the financial resources necessary to undertake a comprehensive storm water management program.

Overview

One approach toward storm water funding which has been successfully implemented in an estimated 125 jurisdictions throughout the Northwest and the nation is the "public utility" methodology. This approach implies a "fee for service" structure by which storm water ratepayers as opposed to taxpayers support the operations of the storm water program. The basis for establishing most storm water service charges and rates is the degree to which a customer uses or contributes runoff to the storm water system. The degree of use is typically related to the property's contribution of storm water to the system owned, operated and maintained by the jurisdiction. The need for comprehensive approaches toward storm water management stems from the alteration

---

[1] Principal, Shaun Pigott Associates, 1045 N.W. Bond, Bend, OR. 97701 (503) 383 1960

of natural storm water conditions and from the development of "impervious areas." Thus, the degree or amount of use, and therefore the size or amount of the charge, is in proportion to the magnitude of the property owner's contribution to the total volume of storm water runoff.

What services does a storm water customer buy? The services include: the reduction of hazards to property and life resulting from uncontrolled storm water runoff, such as erosion; improvements in the general health and welfare through reduction of standing water, and other undesirable storm water conditions; improvement of the water quality within the system and its receiving waters; enhancement of water related habitats; and the elimination of potentially harmful land alteration or development activities which may negatively impact the storm water system or its receiving waters. Consequently, there are general services from a storm water program which accrue to all property owners and customers in the service area as well as specific services which are in approximate proportion to the amount of runoff from a given developed property. The amount of runoff generated from a specific site depends upon the specific conditions at that site which may include gross area, developed area, land use, slope, soil type or intensity of development.

Local examples of jurisdictions successfully implementing storm water utilities in the Puget Sound area include King County, Seattle, Tacoma, Snohomish County, Auburn, Kent, Tukwila, Lynnwood, Gig Harbor, Puyallup, and Bainbridge Island. Oregon also has numerous storm water utilities including Portland, Washington County, Eugene, Salem, Ashland, Lake Oswego, West Linn, and Clackamas County.

Application

There are several alternative methods used by local governments throughout the country for funding storm water management. The methods used are typically selected based on a number of factors including:

1. the scope of program to be funded;

2. the authority available though state or local statutes to impose a funding method;

3. existing local funding policies and practices;

4. the general financial health of the jurisdiction;

5. the local political atmosphere in which elected officials must make funding decisions; and,

6. special causes or results of existing storm water conditions.

More and more the selected method is some type of service charge, thus establishing a "utility" or special district/authority with the power to impose a specific fee for service. The primary reasons supporting the decision to implement a service charge structure are as follows:

o **Fair**..properties are measured and evaluated based on contribution to the storm water system..the greater the impervious area development on a parcel, the greater the service provided to that customer ..courts have held this action to be rational and not arbitrary

o **Dependable**..revenues provide a reliable source of support for storm water operations which is predictable and uniform..the consistency and reliability of this revenue flow enables the utility to develop sources of revenue beyond service charges ..consistency allows accurate forecasting of revenues which enables better planning for physical improvements to the storm water control system

o **Dedicated**..under most laws, revenues raised through the surface water utility must be expended on maintenance, operation and/or capital improvements to the storm water system only

o **Legally Defensible**..state supreme courts have held that surface water utilities do have the authority to establish rates based on a contribution to run-off methodology AND rates based on impervious surface measurement are not considered arbitrary

o **Understandable**..the rate structure is usually very straightforward as it is based on the amount of impervious surface or runoff contribution

There are four basic methods currently in use for establishing storm water utility service charges:

Extent of Impervious Surface - The rates under this approach are in direct proportion to the measured, estimated, or assumed extent of impervious area for each parcel of land. Impervious surface is that land occupied by roofed buildings, pavement or similar surfaces.

Rational Method - Under this approach the rate is determined by a runoff factor or coefficient which is deemed to be appropriate for the type of land and the nature of the improvements on each parcel.

Flat Fee - This mechanism utilizes a constant or uniform fee applied against each customer on a community-wide basis. In most cases, the flat fee is used mainly because of its administrative simplicity.

Combination - Some communities have chosen to utilize a flat fee for parts of the service charge calculation such as administration, accounting, planning and use other approaches for maintenance and capital costs.

In the case of the impervious surface or the rational method, land use may be taken into consideration. The rate is graduated to correspond with the increasing intensity of actual land use. The categories of intensity may reflect zoning ordinance classifications, actual use or be based on some other classification specifically adopted in the rate-setting ordinance. Actual calculations or ranges of density can also be applied.

The issue of incorporating water quality variables into the storm water rate structure is currently in the developmental stage. Pending further investigation into the possible use of standard industrial classification (SIC) codes as an element of the water quality rate formula, the use of density of impervious coverage factors may be the most equitable/defensible method for establishing water quality charges.

Summary

Methods for funding storm water management must keep pace with the increasing demands being placed on local jurisdictions to better control the quantity and quality of runoff. The funding structure best able to equitably allocate cost back to those receiving service is the utility approach. Now a "generally accepted" form of public finance, establishing a fee for storm water can be accomplished through a carefully planned process that informs the public, involves elected officials and clearly quantifies the need.

STORM WATER UTILITY EXPERIENCE
IN BELLEVUE, WASHINGTON

Damon Diessner [1]

Abstract

Surface water management is seen as a flood protector by most, a savior of the environment by some and a subversion of property rights by others. As recently as 20 years ago, there were no cities in the nation with departments solely devoted to storm water runoff. Surface water management was usually part of a city's overall public works program and primarily an adjunct to road maintenance. The typical urban storm drain system kept drainage out of sight and out of mind unless there was a flooding emergency. In 1974, the Northwest's first Storm and Surface Water Utility was formed in response to citizen concerns that urbanization was destroying the city's streams and threatening stream property. Other cities followed suit, and today there are hundreds of storm water utilities nationwide.

Introduction

The City of Bellevue is located in the Puget Sound region of Washington state. Incorporated in 1953, it has grown dramatically from a population of 6,000 and a land area of five square miles to over 89,000 residents and 30 square miles today. Shaped by glaciation, elevations range from near sea level to approximately 1,200 ft. Precipitation averages 35-40 inches per year.

In the 1960's, the area began experiencing rapid growth and development. Residents who had originally located here for the natural, environmentally-oriented flavor began to view certain changes in their community as undesirable. As rural and suburban areas developed, the

---
[1] Director, Storm and Surface Water Utility, City of Bellevue, P.O. Box 90012, Bellevue, WA 98009-9012

cherished open space and stream systems deteriorated. Accelerated surface water runoff from more impervious surfaces caused increased flooding, property damage, streambank erosion and diminishing salmon runs.

Formation of the Utility.

Several things occurred which helped pave the way for the formation of a storm water utility. The City Council appointed a citizen committee to recommend standards and procedures for preserving Bellevue's streams. In 1965, state law was changed to allow the establishment of utilities as a funding mechanism for storm water control. Later, a bond issue was passed which included funds for a new storm drainage study. The study examined and later recommended establishing a utility to oversee all local government activities in surface water management.

In 1974, the City Council passed an ordinance establishing the utility. The mission given to the new Storm and Surface Water Utility was to "...manage the storm and surface water system in Bellevue, to maintain a hydrologic balance, to prevent property damage, and to protect water quality for the safety and enjoyment of citizens and the preservation and enhancement of wildlife habitat."

Staff began preparing a Drainage Master Plan to address the most immediate issues of flooding and in-stream scour. The plan examined a range of alternative solutions from construction of large storm sewers to the use of open streams and on-site flood control. An approach was finally selected using an integrated network of open stream channels and pipes for conveyance, with lakes, wetlands, ponds and regional detention basins for peak storage and water quality control. The "open stream concept" for various elements of the city-wide drainage system ranged from four to ten times less costly than traditional storm sewer improvements and has proven to be more environmentally sensitive to the stream ecosystem.

A final hurdle in the establishment of the Utility was setting a rate structure to fund programs. After significant citizen input, the City Council decided to base drainage rates on the estimated amount of runoff individual properties contribute to the total drainage system. Each property is classified according to its "intensity of development" (impervious surface). The classification combined with total property area determines the service charge which is billed every two months. As of January 1, 1993, a typical single family household is billed $18.10 every two months for 10,000-12,000 square feet of property. In recognizing the flood control and water quality benefits of wetlands, changes were later added to the rate structure which benefit properties encompassing wetlands.

Storm and Surface Water Utility Programs

The Utility's programs have changed and expanded in response to customer demand over the years. With successful flood control systems in place, more effort is now focused on water quality controls. Currently, there are six major programs:

Capital Improvement Program. In addition to constructing storm sewers and bridges, the Utility improves stream channels for carrying capacity, stability, wildlife habitat, and migratory fish passage. A series of eleven flood control dams have been constructed within the Bellevue stream system to provide protection for the 24-hour one in one hundred year storm event.

Operations and Maintenance. Flood control gates are operated remotely by a central computer, thus freeing maintenance crews to respond to individual citizen needs during severe storms. The Utility operates a 24-hour emergency telephone line to respond to flooding, pollution events, or other surface water related emergencies. Maintenance staff serve as consultants to private property owners regarding private drainage systems and also perform routine maintenance of public systems such as cleaning, repairing and replacing system components.

Water Quality Control. Study has shown that water quality benefits can be achieved by cleaning inlet sumps typical of the Bellevue system when 60% of capacity is reached. The City has adopted these maintenance frequency standards for public and private drainage systems. The Utility also maintains and operates a number of oil/water separators to reduce introduction of floatable material to the drainage system. Within its service area, the Utility has implemented a variety of surface water treatment for lake restoration and protection programs. Some of these measures include aeration, alum treatment, biofiltration, soil filtration and wetland plant harvesting. Regionally, Bellevue coordinates research and enforcement efforts with state and other local water quality authorities.

Development Services. The Utility enforces a number of codes related to land use and construction through a permitting process. On-site storm water controls are required to provide protection for the 24-hour, one in one hundred storm event. Temporary erosion and sedimentation controls are required on all construction sites. Floodplains, wetlands, stream corridors, and steep slopes are protected from development. Regulations are enforced through a Civil Infraction Code process.

Administration. The Administration program involves financial management, rate administration, comprehensive drainage planning, general administration and support for the City Council and an Environmental Services Commission.

Education. The most visible educational activities are the "Stream Team" and "Business Partners for Clean Water" programs which provide workshops and volunteer monitoring activities for citizens. Business Partners involves five categories of local businesses in developing water quality action programs for the work site. Other public involvement activities involving water quality and fish habitat protection include storm drain stenciling, salmon rearing and release projects, integrated pest management seminars, and stream rehabilitation.

Challenges of a Storm Water Utility.

The public may resent paying for a so-called "rain tax." Since successful flood control is the absence of something, residents do not always feel that they are receiving services for their money. Furthermore, initial start-up costs of a storm water utility can be significant and results may not be immediately forthcoming. Similarly, water quality benefits are usually not seen right away. Even where obvious water quality indicators exist, such as with fish populations, it may be awhile until positive effects are realized. Involving the public during the planning stages of forming a utility and a strong public education program will help customers recognize the many benefits of a storm water utility.

Strengths of a Storm Water Utility.

The road to creating a storm and surface water utility can be long and sometimes bumpy, but the rewards are significant. Funding is secure since state law prohibits utility revenues from being used for purposes other than surface water management. Competition for general tax revenues with other public programs such as public safety, parks and public works is eliminated. Utilities in Washington state are able to issue revenue bonds without a public vote. Tax exempt properties are not exempt from utility charges (Washington State highways and Bellevue streets are two of the city's significant ratepayers.) And, finally, customers have a central contact and someone to hold accountable when reporting drainage emergencies or seeking advice.

Cross Section 90's - A Profile of User Fee Funded Stormwater Utility Practices in the U.S.

Robert B. Benson

The task of securing adequate ad valorem tax funding to support stormwater drainage system operating and capital needs has always been difficult at best. In many communities, ad valorem taxes provide an inadequate and often erratic source of funding at a time when public and governmental safety and environmental expectations are rising and federal grant funding is little more than a memory.

In response to this dilemma, an innovative and practical solution has evolved, which is quietly but rapidly revolutionizing the stormwater management industry across the nation. That solution is the establishment of user fee funded stormwater management utilities. Relatively unheard of only a decade ago, such self supporting enterprise operations are quickly gaining widespread acceptance and popularity. To date several dozen communities have supplemented or replaced tax funding of stormwater system operations with more adequate, equitable and dependable user fee funding. Others are supplementing sanitary sewer fees with stormwater user charges to provide more equitable financing of CSO related improvement needs. Many more are currently exploring or are already in the process of making such transitions. Among all of these groups, there is great interest in learning more about the operating and financial practices of their "sister" user fee funded stormwater utilities across the Country. However, until recently very little information had been gathered and disseminated regarding these relatively new utility enterprises.

In late 1991 the presenter's employer as a public service conducted a comprehensive survey of user fee funded utility operations across the U.S. Fifty-four selected utilities of all sizes from nineteen states are represented. The summarized results of the survey were presented in a 1992 report which was widely distributed without cost to public representatives of the stormwater management industry. Interest and response to the survey and report was so great that an expanded nationwide survey is scheduled to be conducted early in 1993. It is anticipated that the findings of this resulting updated survey will be available for introduction by the presenter at the conference. The 1993 survey is expected to address current stormwater management utility practices in the respective areas of:

- ✓ Organization/Administration
- ✓ Operations
- ✓ Finance
- ✓ User Fees
- ✓ Best Management Practices

The greatest concerns of stormwater utility administrators will also be examined. Survey results will be presented with accompanying colorful illustrations and visual graphics in a fast paced format. The presentation will be designed to provide the listener with a reasonably complete and clear profile of current practices and trends in this rapidly growing segment of the stormwater industry.

Continuous Simulation Modeling for Sewer Systems

Eric C. M. Bergstrom, P.E.[1]

Abstract

The peaking-factor method of sanitary sewer design was developed to simplify the calculations in determining design flow rates. This method of analysis is conservative to the point that a large margin of safety is often provided. For new sewers, the peaking factor method can allow for the uncertainties of eventual development. However, in evaluating constructed systems or designing improvements to serve older development, the peaking-factor method of analysis may overestimate the actual flows and subsequently lead the engineer to overlook capacity available in existing sewers. This can result in the construction of unnecessary facilities.

As an alternative, wastewater flows in a sewer collection system may be modeled using continuous simulation methods. This type of analysis routes diurnal curves or hydrographs through a sewer collection system. Continuous simulation computer software models available today can be used to reflect significant features of a system, such as long, minimally sloped interceptors. This type of modeling is more precise than the peaking-factor method in the depiction of flow rates. For an existing system, when a continuous simulation model is used in conjunction with flow monitoring, hourly flow rates into a treatment facility can also be predicted fairly accurately. In the design of new systems, an appropriate safety factor may be applied to the peak estimated flow rate determined from the model.

The computer software models and techniques available today make it unnecessary to simplify sewer design calculations. This paper addresses the application of continuous simulation modeling for sewer analysis.

Sanitary Wastewater Diurnal Curves

Flow patterns or diurnal curves for wastewater facilities are frequently very consistent. In addition, the flow patterns for weekdays can usually be distinguished from weekends and holidays. These curves are also dependent upon the land use of the area served.

Figure 1 presents the diurnal curves for an interceptor serving a primarily residential neighborhood. The weekday curve is characterized by a high early-morning peak, while the weekend curve has a smaller peak later in the morning.

---

[1] Project Manager, R. W. Beck and Associates, 2101 Fourth Avenue, Suite 600, Seattle, Washington, 98121.

The diurnal curve for January 1, 1992, has also been presented in Figure 1 to illustrate fluctuations observed on holidays. The later peak of this curve reflects the late night celebration characteristic of New Year's Eve.

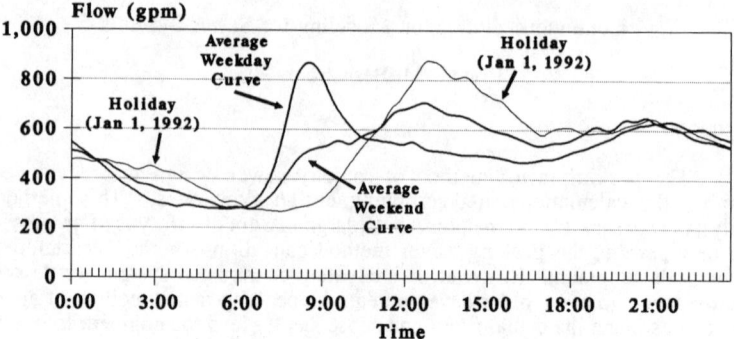

Figure 1 - Residential Diurnal Curves

Figure 2 presents the curves for flows originating from a commercial area. The high midday flows during the week, as opposed to the weekend, indicate the business activities of the area served. The sources tributary to the system have a consistently high flow of cooling water. This is reflected in the steady base flow, even during the weekend.

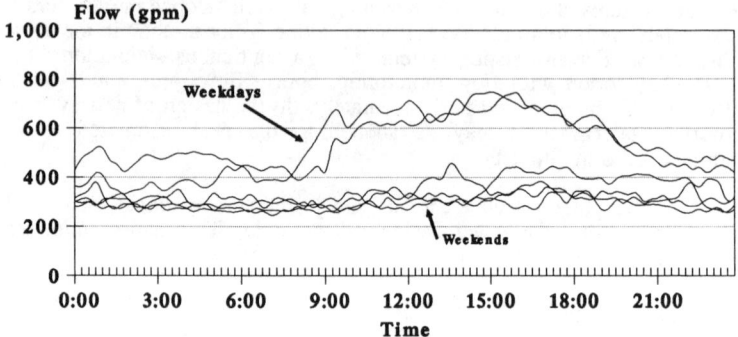

Firure 2 - Commercial Diurnal Curves

Sewers are designed to provide capacity for the peak flows. For the residential neighborhood of Figure 1 this would be 900 gpm and for the commercial area of Figure 2 the peak flow would be 825 gpm. It important to observe that the peak weekday residential flow occurs around 8:00 a.m. whereas the peak commercial flow may occur anytime between 9:00 a.m. and 6:00 p.m.

Sewer Analysis Using Peaking Factors

The traditional method of analysis for sewer systems is the peaking-factor method. In this type of analysis, a peaking factor is selected based on the estimated average daily sanitary flow tributary to a given point in the wastewater collection system. There are several published peaking factor curves, such as those found in the *Water Pollution Control Federation Manual of Practice No. 9* (WPCF MOP No.9). The average daily sanitary flow is multiplied by the peaking factor to estimate a peak sanitary flow rate. Estimates of infiltration and inflow are then added to the estimated peak sanitary flow to provide a peak flow rate.

There are several drawbacks to using a peaking factor method of analysis. Because peaking-factor curves do not typically distinguish between commercial, industrial, and residential flows, they often do not accurately reflect the trend in the system. Commercial and industrial areas have significantly different peaking factors than residential areas. Furthermore, the peaking factor for a residential area with a retirement community is significantly different than residential areas with two-working-person families. Most of the peaking factor curves from the WPCF MOP No. 9 were developed from flow data collected prior to 1960 and do not reflect changing social patterns and commercial activity that have modified water usage. Another problem is that the peaking-factor analysis does not account for time considerations such as the layout or features of a system, which can affect transport time and flow rates. Finally, this analysis does not consider the time of day at which the peak occurs.

Sewer Analysis Using Continuous Simulation Modeling

Within the last 10 years, computers have evolved to become significant tools for engineers. Before this, traditional methods of analysis were used because it was not feasible to analyze a system in a way that actually reflected wastewater flows. When computers first became available for analysis, many software programs adopted the traditional methods of analysis. For example, many of the first water system network analysis programs were based on the Hardy-Cross method. Eventually, software based on solving simultaneous equations was developed and applied to water system network analysis. Early sewer system analysis programs adapted the peaking-factor method of analysis.

The alternative, continuous simulation modeling, more accurately reflects the actual flows and configuration of a system. This method uses representative diurnal curves based on the land use for each tributary area. The diurnal curves are routed through the modeled collection system to account for transit time. As flows are moved through the system, the diurnal curves are merged. In this way, hydrographic analysis provides flow estimates and diurnal curves that are much more representative of both the actual system and future flows. If flow data, such as the information collected during flow monitoring, are available, the model may be calibrated. Because of the complexity of the required calculations, it was not feasible to consider this method of analysis prior to the development of the computer. Figure 3 illustrates how accurately a continuous simulation model can predict the actual flows for a wastewater treatment plant.

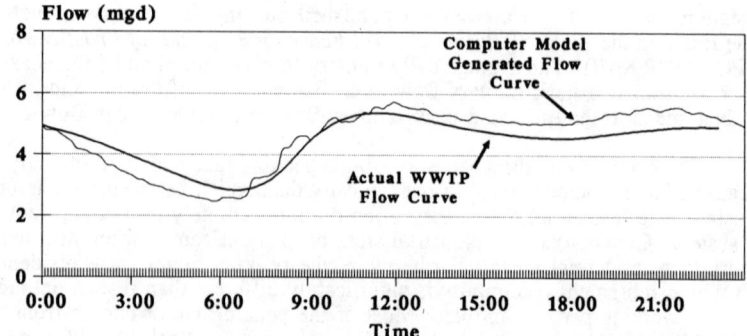

Figure 3 - Diurnal Curves (Wastewater Treatment Plant)

For a combined sewer system, stormwater hydrographs may be merged with the normal dry weather curves. The stormwater hydrograph may be determined from hydrologic models such as the event based Santa Barbara Urban Hydrograph (SBUH) method or the continuous simulation Hydrologic Simulation Program Fortran (HSPF) method. For many combined systems the normal sanitary flow is insignificant compared to the stormwater runoff. In these cases, use of the model basically becomes an exercise in routing storm flows.

Conclusions

For new sewers, the peaking-factor method yields a conservative estimate of design flows which provides an element of protection against the uncertainties of future development or activities. However, this method provides little information on the operation of the sewers.

As an alternative, a continuous simulation model could used to develop a more accurate projection of flow. An appropriate safety factor would then be applied to the projected flows for design purposes. It is recommended that conservative safety factors be applied since the designed pipe would typically result in an increase of only one pipe size, if at all. The risks associated with undersizing a sewer would not justify the material costs saved during construction. Providing an interceptor one size less than that recommended would also restrict future flexibility to accommodate changes in land use.

The most valuable application of continuous simulation analysis is for existing systems, where the calibrated model is effective in establishing the actual loadings are on the sewer system. This can provide a basis for scheduling construction of new interceptors. The author has found that this type of analysis, supported by flow monitoring, allows sewer utilities to make optimum use of their existing facilities and construction of new sewers can often be postponed several years.

Several modeling software programs are available for sewer system analysis. One such program is HYDRA, which allows a system to be analyzed by either continuous simulation methods or through the use of a peaking-factor. These programs enhance the ability of the engineer in analyzing sewer systems and provide better information from which decisions may be made.

Case Study: Storm Water Analysis of Manatee Pocket in Martin County, Florida

By William C.H. Wang, M.ASCE, Robert A. Laura, M.ASCE, E. Scott Webber, P.E.[1]

ABSTRACT: Storm water pollutant loading rates from a 6,000 acre urban watershed were analyzed for source determination of contamination to the receiving water. The receiving water body, Manatee Pocket, is a tidal estuary with minimal flushing. Over the years, the estuary exhibited an elevated degree of pollution in the form of heavy metals, nutrients, sediments and bacteriological contamination. Sampling of actual storm events at seven (7) stations in the watershed and one (1) station at the receiving water were performed over two and a half (2½) year period for pollutant source determination. Storm events during the dry season and the wet season were sampled. The quality of the stormwater runoff was measured. Nineteen water quality parameters were analyzed at each sampling station. These parameters were chosen based on existing problems identified in the estuary. Water quality sampling was conducted at the onset of the storm water runoff ("first flush"), at the peak of the runoff, and near the end of the hydrograph. An analytical methodology was developed to estimate the pollutant loadings from each of the seven sub-basin within the watershed. Physical calculations of the loading rate was done thru the use of a spreadsheet model. Presently all results are preliminary, however, this methodology shows promise to be an effective tool in setting basin pollutant removal priorities.

INTRODUCTION: Manatee Pocket is part of the St. Lucie Estuary located in Martin County, along the coast of east central Florida. It's location at the western corner of the confluence of St. Lucie and Indian River sets the Pocket up to be an important resource of natural habitat for aquatic organisms as well as recreational and commercial activities.

Concerns about the Pocket's water quality become an important aspect of the Indian River Lagoon Surface Water Improvement and Management (SWIM) Plan, which was commissioned by both the St. Johns River Water Management District (SJRWMD) and the South Florida Water Management District (SWFWMD) in 1989. The SWIM plan along with studies of the general area indicate three major cause of surface water deterioration in the Manatee Pocket: 1) Storm water runoff into the pocket from upland, resulting in introduction of pollutants into the bay; 2) Poor flushing action in the estuary causing long water residence time inducing oxygen depletion and particulate settlement; and 3) Spills and other direct discharge from commercial and marine related industries as well as the general public from shorelines adjacent to the Pocket.

---

[1]William C.H. Wang, Senior Engineer II, Law Environmental, Inc., Fort Lauderdale, Florida (305) 771-2147. Robert A. Laura, Senior Civil Engineer, South Florida Water Management District, West Palm Beach, Florida (407) 687-6238. E. Scott Webber, Martin County Drainage Engineer, Martin County, Stuart, Florida (407) 288-5927.

Goal: As this paper is part of a larger water quality study commissioned by Martin County in cooperation with SFWMD, it is the intention of the authors to limit the scope of quantifying storm water related runoff pollution entering into the Pocket.

Project Description: A general location map of the Pocket is shown in Figure 1. The contributing areas are subdivided into ten (10) basins. A summary of land use of each basin is in Table 1. Each basin is numbered correspondingly with its sampling station. The areas shown in Table 1 are preliminary and may be subject to change.

It is noted that only eight of the ten basins are sampled for drainage discharging into the Manatee Pocket. No storm water samples were taken from the other two basin location since no discernable discharge point can be located. Storm drainage in these area are through infiltration and overland flow.

Storms taken over a two and half year period consisted of four baseline and storm event samples; two in the dry and two in wet season. The four storms date and rainfall totals used to calculate the pollutant loadings are: 1) January 15-16, 1991 (4.04"); 2) April 25, 1991, (1.29"); 3) June 24, 1992, (0.8"); and 4) August 24-25, 1992 (2.2").

Sampling stations consisted of a staff gage and where available, stage level recorders. Stations 2 and 4 were chosen for the placement of continuous stage level recorders to determine storm hydrographs. Although more recorder locations are desired, it is thought that the two are representative for other basins hydrograph determination through a comparison system which account for different land uses and basin geometries.

Sampling were done by manual grabs with sequence of samples taken in the order of stations 8, 5, 6, 3, 4, 2 and 1. This sequence was set based on individual basin geometry, size and shape. Three representative samples were gathered for each storm, representing the rising limb, the peak and the recession limb of the hydrograph.

The water quality parameters were selected from past studies done in the general area and based on pollutants likely to be generated from local industries. They are: **Metals:** Cadmium, Cooper, Mercury, and Zinc; **Nutrients:** Ammonia-N, Nitrite, Nitrate and Ortho-Phosphate; **Physical:** TSS, ISS, Conductivity, Turbidity, Temperature and Ph; **Oxygen demanding:** DO and COD; **Toxics/Organics:** Oil & Grease and Chlordane; **Biological:** Total Coliform.

Analytical Methodology: In order to achieve the specified goals, it was necessary to collect storm event runoffs for different seasons, perform the laboratory analysis and determine the loadings. The sequence of events in the project plan was simply defined as 1) through gage information, compute stage discharges for each station 2) determine each of the basin characteristics and relate it to a Soil Conservation Services' (SCS) CN numbers. 3) through the use of CN, determine runoff for each storm in each of the basins and 4) combine the storm volume calculated with laboratory analysis to determine a loading rate for each storm event.

In performing the analysis, several issues had to be resolved. They were 1) separation of specific land use from mixed land use basins 2) estimation of storm runoff volume at stations where there were no stage level recorders, and 3) application of individual laboratory analytical results to the storm volumes.

In the case of separation of land use, it was determined that Station 8 is hydrograph would be easily delineated from the hydrograph of Station 4 since it was sampled individually. Station 8

STORMWATER ANALYSIS

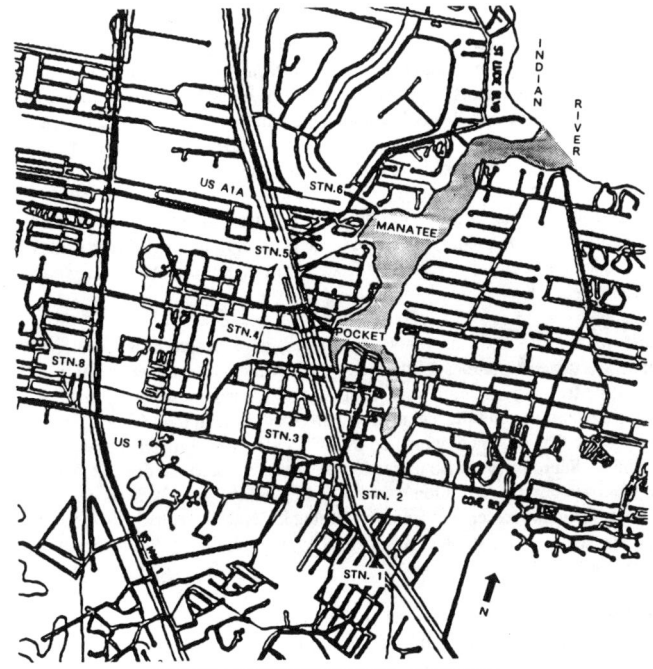

FIGURE 1 MANATEE POCKET LOCATION MAP

BASIN No	BASIN NAME	GAGED DRAINAGE AREA	TOTAL DRAINAGE AREA	LDR	MDR	HDR	MH	COM	IND	REC	VAC	AGR	WET	TOTAL %
1	EAST FORK MANATEE CK.	1347	1565	15	40	5					30		10	100
2	SOUTH FORK MANATEE CK.	687	723	10	30	10	5	10			30		5	100
3	SOUTHWEST PRONG	167	364		35	20		15	20		10			100
4	SALERNO CREEK	1149	1173		25		35	30	5			5		100
5	WEST PRONG	383	383	2	5		5	5	60	3	20			100
6	CROOKED CREEK	106	106	30	10			10	50					100
7	MANATEE POCKET	N/A	*	55	30		5	10						100
8	HIBISCUS PARK	814	1173		90			10						100
9	W.MANATEE POCKET	N/A	579											
10	ROCKY POINT	N/A	579											

LEGEND

LDR LOW DENSITY RESIDENTIAL
MDR MEDIUM DENSITY RESIDENTIAL
HDR HIGH DENSITY RESIDENTIAL
MH   MOBIL HOME
COM COMMERCIAL
IND INDUSTRIAL
REC RECREATIONAL
VAC VACANT
AGR AGRICULTURAL
WET WETLAND

N/A = NOT APPLICABLE

* = THE 179 ACRES OF WATER BODY DOES NOT COUNT AS PART OF THE DRAINAGE BASIN THAT ENTERS INTO THE POCKET

TABLE 1 DRAINAGE BASIN CHARACTERISTICS

a predominantly residential basin, discharge completely into the Station 4 basin. It's separation is desirable for residential pollutants loading rate determination. The separation of hydrographs for Station 8 from Station 4 was based on combined factors of land size and CN number for each of the two basins. These factors represented a runoff potential between Stations 4 and 8. In order to have an appropriate Station 8 runoff volume, the duration of storm runoff from Station 8 have to be estimated. The split was not based on observed flow during the storm primarily because flows were not taken simultaneously at both stations.

It was difficult to separate land use in other basins because they were too diverse. Review of individual basins shows no predominate land use type exist in basins 1 thru 6 except that 60% was combined residential land use in basin 1 and 65% was combined commercial/industrial land use in basin 5.

In response to the problems of runoff estimation, a procedure was developed utilizing the hydrographs available at Stations 2 and 4 and the CN numbers utilized by SCS. The procedure involved first performing a CN analysis within basin 2, 4 and 8 only by estimation of a basin CN from the basin's land use and the CN numbers obtained through the SCS TR-55 publication. The resultant textbook CN was then compared to a CN calculated through the use of hydrographs at Station 2 and 4. Station 8 was also utilized in this case as a separate distinct land use since it was separable from the combined Station 4 and 8 hydrograph. Thus, for station 2, 4, and 8 one can calculate an adjustment percentage from the textbook CN to the calculated CN. This adjustment percentage was then applied to other basin's CN adjustment through a combined weighting factor described below.

The weighting factor was based on the relationship between stations with no hydrograph and stations 2, 4 and 8. It was assumed that this factor relates the basin characteristics to the actual CN. This factor takes into account: 1) basin geometry, 2) basin area; 3) basin land use, and 4) degree of basin's channelization, with each of these factor accounting for 25% of the total. A basin's CN adjustment was then estimated based on likeness of these factors to that of basin 2, 4, and 8. Based on the degree of likeness, the adjustment percentages in basins 2, 4, and 8 was then applied to adjust the textbook CN value obtained from the particular basin. Once the CN was adjusted, it was then used for runoff volume determination.

In the process of determination of the storm event's loading rate, it was found that the proper application of the three laboratory analysis results must be made in relationship to the runoff volume that the sample represented. Since each sample represented a zone of discharge on the hydrograph, the pollutant concentrations of the sample was uniformly applied to the volume generated from this zone. Thus for a storm event, four different loading rates were calculated; the rising limb, the peak and the recession limb of a storm hydrograph and the flow-weighted average. It is anticipated that the design of treatment facilities for these discharge will be based on the peak loading rate.

The determination of the above four different loadings was possible only for those stations that have hydrograph recorded. This is because the sample could be associated with a range of discharge levels only when one has a hydrograph. It is not clear whether one can establish these same type of loading rates except for the flow weighted average for stations other than 2, 4, and 8.

Summary: This paper presented a methodology by which stormwater pollutant loading rate can be calculated. Although the use of this methodology was specifically for the Manatee Pocket, it shows promise to be applicable for other similar studies. The analytical results obtained thus far are quite extensive but are in preliminary form. Final results and the water quality study report are expected to be available in early 1993.

MANAGING STORMWATER WITH A MICROCONTROLLER OPERATED SYSTEM

Edward McCarthy, Ph.D., P.E.[1]

ABSTRACT

Concepts of managing urban stormwater runoff with a microcontroller operated flow control system are presented. The microcontroller system allows flexible and precise flow control from stormwater detention reservoirs and is compared to a conventional flow control structure. Performance characteristics including peak runoff rate control, water quality improvement through increased detention time, minimization of detention reservoir overflow, onsite design refinement and flood routing are discussed. The microcontroller system has greater potential than conventional flow control structures in minimizing detention reservoir overflow and can be used to make onsite system adjustments based on observed performance and to optimize routing of flood water where downstream conveyance capacity is a constraint.

INTRODUCTION

Uncontrolled stormwater runoff from developed areas can cause flooding, channel erosion, sedimentation as well as degradation of wildlife habitat and water quality. Development can dramatically change the hydrology of a site. Roads, driveways, sidewalks, roofs and lawns cause greater volumes of stormwater runoff and at rates higher than under natural conditions. To control peak runoff rates from developed areas, stormwater is typically collected, routed to a detention reservoir and released to the downstream system at a designed rate (Fig. 1). The design release rate is often determined by the capacity of the downstream system and is frequently limited to a proportion of the predeveloped runoff rate.

This paper discusses management of stormwater with a microcontroller operated flow control system (McCarthy, 1992). The microcontroller system includes an adjustable flow restrictor, one or more sensors, a microcontroller, microprocessor algorithms and a motor (Fig. 2). The adjustable flow restrictor precisely controls stormwater flow. The sensors monitor environmental parameters and send signals to the microcontroller. The microcontroller interprets the signals and applies microprocessor algorithms to control flow by adjusting the restrictor using the motor.

ACHIEVING STORMWATER MANAGEMENT OBJECTIVES

The release rate from a detention reservoir is conventionally controlled by a fixed placement flow restrictor consisting of an orifice plate, weir or gate. Fixed placement flow restrictors are configured to discharge stormwater at the design release rate when the detention reservoir is full. As the storage level in the detention reservoir decreases, the hydraulic head on the flow restrictor also decreases, resulting in a decrease in the release rate from the detention reservoir. More storage volume is required to detain runoff from a storm

---

[1] Manager, Water Resources Department, GEODIMENSIONS, INC., 10230 NE Points Drive, Suite 220, Kirkland, WA 98033

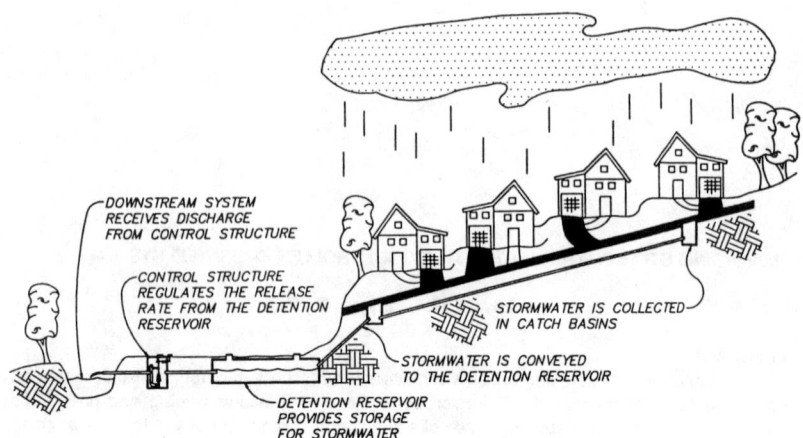

Fig. 1. Stormwater collected in developed areas is conveyed to a detention reservoir before being released at a controlled rate to the downstream system.

than would be required if discharge from the detention reservoir was maintained at the design release rate for all storage levels. Ineffective use of available stormwater reservoir volume increases the costs of construction projects. The value of the real estate occupied by the stormwater facility is an additional cost which may be more significant than stormwater facility construction costs. The microcontroller system allows the design release rate to be maintained for all storage levels in the reservoir. The system accomplishes this by adjusting the restrictor flow area to compensate for changes in hydraulic head. Reservoir storage volume efficiency can be increased by over 40% with precise flow control. The microcontroller system provides the opportunity to economically upgrade existing stormwater management facilities which are not operating at current standards. Because the microcontroller system makes better use of detention reservoir storage volume, existing undersized detention reservoirs can be retrofitted to perform more effectively.

Stormwater detention reservoirs are typically designed to detain stormwater runoff from a storm of a specified return interval. Storms of 10-, 50- or 100-year return intervals are commonly used for sizing detention reservoirs (King County Department of Public Works, 1990). Stormwater inflow, in excess of the design capacity of the detention reservoir, bypasses the flow restrictor through an overflow outlet. Overflow from detention reservoirs is not uncommon and results in significantly higher flow rates than those regulated by the flow restrictor. The microcontroller system, on the other hand, can be programmed and installed to minimize the occurrence of reservoir overflow. This can be accomplished by implementing a sensor to sense conditions indicative of the rate of change in the reservoir storage level. If the reservoir is near capacity and the rate of inflowing water is apt to create overflow conditions, the microcontroller instructs the motor to adjust the flow restrictor to increase the rate of flow. The incremental flow released to prevent overflow is at a rate less than that likely to occur in the event of overflow.

Conventional flow restrictors can be blocked or clogged by debris carried in the stormwater, causing the detention reservoir to fill to capacity and then overflow. The microcontroller system allows the flow restrictor to clear itself of debris. When conditions indicative of debris blockage are monitored, the microcontroller instructs the motor to oscillate the flow restrictor between the open and closed positions to dislodge debris.

The design of the detention reservoir and the flow control structure is typically based on analytical methods using hydrologic models. The accuracy of the models and procedures used in the design process varies. Once in place, there is typically no convenient method for adjusting the design of fixed placement flow restrictors to improve the operation of the system. The microcontroller system, on the other hand, allows adjustment of the flow restrictor to refine initial designs based on actual performance. The microcontroller can be programmed to collect various types of data to evaluate the actual performance of the control system. Inadequacies in flow restrictor design or control logic can be identified and corrected with automated or interactive programming of the microcontroller. Adjustment of the microcontroller system allows the system to be used for a variety of sites and conditions.

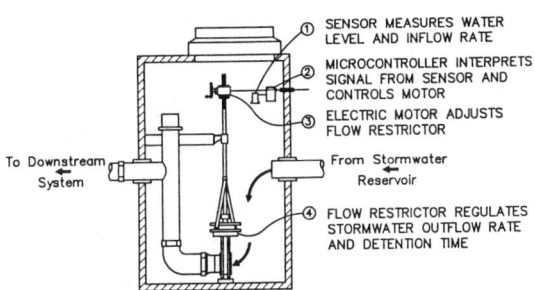

Fig. 2. The microcontroller operated system allows precise and flexible control of stormwater flow.

Fixed placement flow restrictors remain in an open position. No means exist to provide increased detention time under conditions which so permit. As a consequence, fixed placement flow restrictors do not allow substantial improvements in water quality. Substantial gains in water quality can be achieved by detaining low volume storm events and first-flush runoff for longer periods of time (Washington Department of Ecology, 1991). Longer detention time allows pollutant carrying suspended solids in the stormwater to settle to the bottom of the detention reservoir. Longer detention of first-flush runoff is critical because this stormwater carries relatively high concentrations of pollutants which have accumulated on the watershed surface since the previous storm event. The microcontroller system can be programmed to identify low volume storm events and first-flush runoff and provide longer detention time by decreasing or completely closing the restrictor flow area. Detention time can be increased from a matter of hours, as with fixed placement systems, to several days. Knowledge of seasonal rainfall patterns can be used in the control logic to further improve peak runoff rate control and water quality.

Downstream flooding, erosion and sedimentation problems can often be avoided with controlled stormwater routing. By managing stormwater from individual watersheds in coordination with one another, available downstream conveyance capacity can be used most efficiently. A network of microcontroller systems can be linked to a central controller which monitors downstream conveyance capacity and available stormwater storage capacity (Fig. 3). The central controller schedules stormwater release and operates under the constraints of minimizing reservoir overflow and not exceeding the capacity of the downstream system.

SUMMARY

In many developed areas, the cumulative impacts of inadequate stormwater management practices are becoming more evident. Expanding urbanization places increased

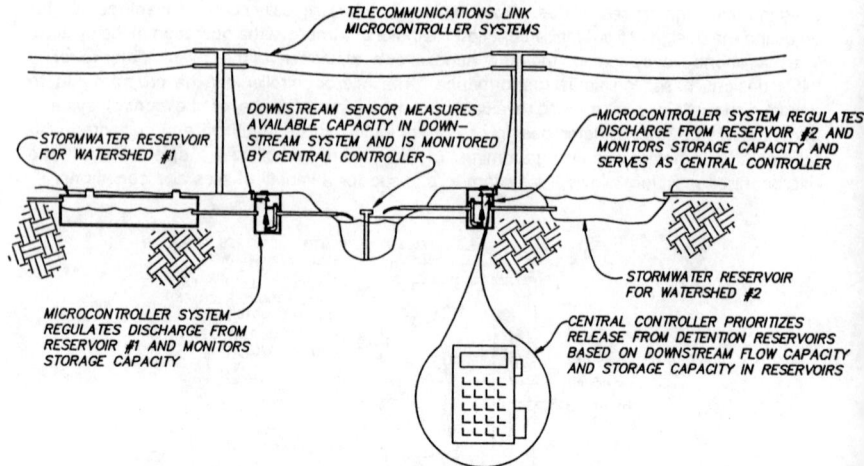

Fig. 3. Stormwater routing optimizes the storage capacity of detention reservoirs and the conveyance capacity of downstream systems to minimize flooding and erosion.

pressure on surrounding water resources and stormwater management requirements are becoming more stringent to minimize downstream impacts. Conventional fixed placement flow restrictors have limited effectiveness in managing stormwater and achieving other water management objectives. A system is needed which has the versatility to meet a variety of stormwater management objectives including more effective use of available storage volume, improving water quality through increased detention time, allowing site specific refinement of operation and routing stormwater.

REFERENCES

King County Department of Public Works. 1990. Surface Water Design Manual. King County, Washington.

McCarthy, E.J., 1992. Stormwater Control System. U.S. Patent Application. Patent status: pending.

Washington Department of Ecology. 1991. Stormwater Management Manual for the Puget Sound Basin. Olympia, Washington.

Trout Spawning Habitat Mitigation:
A Constructed Example

Gregory Koonce[1]

ABSTRACT

A constructed channel for trout spawning habitat is described. Conceptual feasibility, design considerations, construction techniques, and use by spawning trout are highlighted for a channel in Southwestern Montana's Madison River Drainage.

INTRODUCTION

Wade Lake is a natural 240 acre lake located in the Beaverhead National Forest of Southwestern Montana. A trophy trout fishery (source of the current state record brown trout) has been maintained by the Montana Department of Fish, Wildlife and Parks through periodic stocking of hatchery fish. Stocking of hatchery fish was considered necessary because little natural reproduction occurred in the tributaries to the lake. In an effort to reduce stocking costs and comply with Montana's policy of managing for naturally reproducing trout populations, the U.S. Forest Service considered creating a spawning channel.

In October 1989, a feasibility study for the construction of a spawning channel was completed. Design and construction of a 600 foot channel was completed in the fall of 1991. Post construction monitoring of habitat utilization and spawning activities indicate that reproduction success to date has been favorable.

---

[1] Fisheries Biologist, Inter-Fluve, Inc., 1020 Wasco St., Suite I, Hood River, OR 97031

CONCEPTUAL FEASIBILITY

An area of springs and seeps at the east end of the lake was identified as a usable water source for a constructed channel. Given the steep and soggy working conditions in the spring and seep area, a spawning channel within the immediate area was not considered feasible. To the west of the spring area, mature fir trees, steep topography and poor access also precluded a channel. Land to the east of the spring area was steep and lacked a flat shoreline margin, but offered the only practical location for a spawning channel.

To develop a flat area of sufficient size to construct a channel, earth from the east hillslope of Wade Lake would have to be moved down slope into the lake. Water from the spring area could be diverted onto the constructed flat and into a meandering channel to provide spawning habitat.

CHANNEL DESIGN

A topographic survey was conducted and included areas under the lake surface. Design of the final contours produced a gently sloping flat adjacent to the existing shoreline. A total of 16,000 cubic yards of material would need to be bulldozed from the surrounding hill side into the lake to form a 0.8 acre flat. Up to 12 feet of fill would be required within the lake.

A boulder weir, a headgate with an adjustable gate valve and a pipeline were designed to divert water from the spring area into the spawning channel.

Channel configuration and design was based on an assumed diversion from the spring area of 12 cubic feet per second (cfs). Using known trout preferences for spawning habitat as a guide, channel hydraulic characteristics were derived from reiterations of the Manning equation. Appropriate hydraulic parameters and characteristics are listed below.

Discharge	12 cfs;
Slope	1%
Manning roughness	0.08
Depth (over riffles)	0.6 feet
Velocity (over riffles)	1.3 feet per second
Channel Width	16 feet
Substrate	gravel/cobble
Channel banks	Erosion cloth/vegetation

Complete design configurations for the 0.8 acre lake fill resulted in a channel 600 feet in length and 16 feet in width.

Channel substrate was designed to accommodate spawning, egg incubation and juvenile trout rearing. Round, screened, unwashed, non-crushed gravel and cobble varying in diameter from 1/4 to 3 inches was specified. Size composition of spawning substrates was based upon published reports of sediment size distributions in excavated nests. A moderate percentage of fine-sized material in the gravel mixture was also considered beneficial. The finer materials (those less than 1/2 inch), would prevent too much water from flowing through the gravel relative to that flowing over.

In an attempt to anticipate seepage losses through the lake fill material, the hillside soils were analyzed. The hillside soil was determined to have a sufficient silt/clay fraction to preclude the need for an impermeable layer (synthetic or clay liner) under the immediate channel. Compaction along the channel alignment was specified to minimize seepage losses.

Stream channel margins were designed to be resistant to erosion, foster growth of riparian vegetation, provide trout cover and habitat for terrestrial insects (a trout food source). Organic erosion control fabric was specified for applications above water level on all banks. Vegetative planting and seeding rates were specified under the fabric. Gravel and boulders were specified below water level to further protect channel margins.

Erosion control during construction required placement of a turbidity filter/barrier in the lake. This barrier was designed to encompass the construction area and minimize siltation and turbidity in the lake. Design specifications included a floating cableway with a non-woven geotextile fabric curtain attached to the cableway and anchored to the bottom with weights.

Erosion control following construction was specified to prevent rill erosion and wave action on the new shoreline. For the excavated hillslope, the final contours were smoothed and cleated with cat tracks perpendicular to slope. Hydroseeding and a mulching layer of wood pulp followed. The new shoreline was treated with a combination of hydroseeding the upper slopes, organic geotextile and willow plantings from lower slopes to lake water surface and gravel/cobble backfill from lake water surface and below.

CONSTRUCTION TECHNIQUES

Once the turbidity curtain was installed around the construction site, topsoil from the hillside was stripped and stockpiled. Lake fill was then placed around the margin of the fill area. When this perimeter fill was complete, the remainder of the fill area (center) was pumped dry. Compaction of the fill was facilitated through pumping; had the fill simply proceeded into the lake, compaction would have been difficult. Furthermore, encompassing the fill boundary resulted in less turbidity within the fabric curtain enclosure.

Controlling turbidity within the fabric curtain boundary was important, as water from the spring had to pass through the fabric and into the lake. If the fabric clogged, excessive head pressure would cause the fabric curtain to fail.

Channel construction proceeded after the fill was completely in place. The channel was over-excavated and compacted to allow for the placement of the gravel/cobble stream bed substrate. Additional gravel was placed along the lake margin for shoreline erosion protection. Stockpiled soil was then spread on the lake fill. Following completion of the channel, the stream and lake margins were seeded and covered with organic erosion cloth. The turbidity curtain was removed and later the entire site was hydroseeded with native plants and grasses, including alpine wildflowers.

TROUT UTILIZATION

The constructed spawning channel provided about 1,900 square feet of adult and juvenile habitat and a maximum of 2,900 square feet of spawning area. Based on 2,900 square feet of spawning area, 12 square feet of room per adult pair with 2,000 eggs per female, and a 10% survival rate to out-migration, this channel could theoretically produce about 48,000 fry.

Personnel from the Montana Department of Fish Wildlife and Parks observed several pairs of brown trout spawning in the stream during the fall after construction (November 1991). Utilization of the channel in the spring of 1992 by rainbow trout was heavy, with approximately two hundred spawning sites recorded. Observations on brown trout utilization this fall is incomplete; however, heavy utilization is anticipated.

Assessment and Mitigation for Endangered Vernal Pool Invertebrates

E.J. Koford[1]

ABSTRACT: The USFWS is considering declaring five species of vernal pool invertebrates endangered. Among these are the Vernal Pool Fairy Shrimp (Branchinecta lynchi), the California Linderiella (Linderiella occidentalis), and the tadpole shrimp (Lepidurus packardi). The native habitats for these species are pristine vernal pools in the Central Valley of California. Unfortunately for developers they also occur in a variety of non-natural habitats, including roadside ditches, bulldozer scrapes, and in seasonal depressions along railroad lines. The protection afforded by the Federal Endangered Species Act does not discriminate among natural and culturally altered habitats, therefore avoiding impacts to these species in altered habitats will become increasingly difficult. Developers will be required to survey for them, adjust construction areas around locations where they are present, and include mitigation measures such as avoidance, replacement of their habitat, and monitoring. Mitigation can consist of avoiding habitats, or transplanting these species by moving either soil or adults to new locations temporarily as a "seed bank," then replacing them and the ephemeral pool structure after construction or other disturbance. In all cases monitoring for several years should be done to ensure that these species persist after project completion.

Introduction

The United States Fish and Wildlife Service (USFWS) is considering declaring five species of vernal pool invertebrates endangered. Among these are the Vernal Pool Fairy Shrimp (Branchinecta lynchi), the California Linderiella (Linderiella occidentalis), and the tadpole shrimp (Lepidurus packardi). The native habitats for these species are vernal pools, particularly in the Central Valley of California. None are known to occur in running waters. The most significant threat to these species is loss of habitat in rapidly expanding agricultural and urban developments in California.

---

[1] Biologist, Ebasco Environmental, 2525 Natomas Park Drive, Suite 250, Sacramento, CA 95833-2933

Fairy shrimp, tadpole shrimp, and linderiella (collectively called fairy shrimp) are small (11-27 mm) invertebrates uniquely adapted to survive yearly flooding and desiccation. They develop and mature quickly, breed, and produce resting eggs which are resistant to cold, heat, and desiccation. As resting eggs they survive the arid summer until the following winter rains, when they hatch and initiate a new life cycle.

The USFWS proposed to determine endangered status pursuant to the Endangered Species Act of 1973 (Act), based largely on the rapid loss of vernal pool habitats in California. The proposed determination resulted in Ebasco Environmental gathering additional information on the distribution and status of these species. The results demonstrated that these vernal pool invertebrates occur in a variety of altered habitats, including roadside ditches, bulldozer scrapes, and in numerous seasonal depressions along railroad right-of-ways. The Act, however, does not discriminate among natural and culturally altered habitats, and avoiding impacts to these species in altered habitats is becoming increasingly difficult. Whether or not these species are awarded endangered status, they will continue to be high profile species in California's Central Valley, and project developers will need to be prepared to survey for them, avoid construction where they are present, and include mitigation measures such as avoidance, replacement of their habitat, and monitoring.

SURVEYING FOR FAIRY SHRIMP

These species are relatively easy to detect if surveyed during the March-May period when ephemeral pools are present. They can be easily netted with a short-handled dip net, with mesh size of up to 2 mm. Generic identification of the species of concern can be conducted in the field, as the group of species are not difficult to recognize. This is all most assessments will require. This can be accomplished by a competent biologist employing a standard dissecting scope. Identification to species level requires a specialist as these forms have an uncertain and controversial taxonomy. However, under permit, specimens can be sent to a competent specialist for identification. A collection permit from the California Department of Fish and Game (CDFG) is necessary to collect specimens, and if the species is listed (a determination is expected in November 1993) a Memorandum of Understanding from CDFG is also required.

Surveys for the resting eggs of tadpole shrimp during the dry season are possible, but are much more difficult and subject to question. The technique involves sifting soil from suitable habitat and looking for resting eggs. Identification to species using the resting eggs is not generally possible.

AVOIDANCE OF FAIRY SHRIMP HABITAT

Regardless of the fairy shrimp species ultimately listed, avoidance is the best option for preventing "take" of the species. If the project schedule does not allow for field surveys to substantiate the presence or absence of the species, avoidance of all low lying areas, swales, ditches or other ephemeral pools will avoid impacts to these species. However, because fairy shrimp have been found in railroad side depressions, tire tracks in mud, and other exceptionally small puddles, avoidance of these habitats may not be practical.

Temporal avoidance will in all likelihood become a standard part of construction permits within the expected habitat of fairy shrimp. This would limit construction to the dry season, when the shrimp exist as resting eggs. Once the ephemeral pools are dry, vehicles can traverse the area without harm to the encysted eggs. Construction practices such as trenching, which disturb the surface soil would not be acceptable without further mitigation.

MITIGATION

Mitigation for fairy shrimp is a new practice, with only a few cases of successful implementation. If avoiding fairy shrimp habitat is not possible, there has been some success at relocating these species by moving either soil or adults. In principle, the objective is to remove and stockpile the soil containing the encysted eggs. This could consist of scraping away approximately 5 cm of soil in the ephemeral pool. In the case of temporary disturbance this material can be stored onsite, then redistributed on the reconstructed depression surface. If disturbance is permanent, a new pool can be constructed and inoculated with soil containing resting eggs. This latter technique of constructing artificial ephemeral pools has been accomplished only occasionally, and can only be recommended in principle. Vernal pools and seasonal depressions that support fairy shrimp contain a complex of animal and plant species and hydrologic interactions that are poorly understood and difficult to successfully duplicate.

A third technique, that of transplanting either soil or adults to pools devoid of fairy shrimp should generally be discouraged. In such cases, there are reasons (such as water quality, soil chemistry or hydrology) why fairy shrimp are absent from such pools. Also, the population genetics of these species are poorly known, and transplantation may confound the scientific record.

In all cases of mitigation, several years of annual monitoring would be required in order to determine whether these species persist after project completion. Monitoring would consist of one or more dip net surveys in the following spring to evaluate both presence and composition of the invertebrate community.

INTERACTIVE DECISION SUPPORT FOR HYDROLOGIC, HYDRAULIC, & INSTREAM FLOW CRITERIA

By Marshall Flug[1], M.ASCE; & Darrell G. Fontane[2], M.ASCE

Abstract

Most often, impacts on natural resources from alternative river discharges are not quantified. The personal computer (PC) can provide access to useful techniques for analyzing multiobjective systems. This includes algorithms to quantify relations between flow, seasonality, and ecologic attributes for instream flow analysis. A spreadsheet based package is used to analyze and integrate hydrologic, hydraulic, and natural resource attribute information. Interconnected and linked modules operate as a decision support system (DSS). The computer package provides for a rational and consistent evaluation of flow alternatives (i.e., hydrographs) based upon a quantitative measure of each objective (i.e., attribute benefit function) for the hydraulic characteristics (i.e., cross-section) of the river system. A final component of the DSS is a multicriteria decision analysis matrix for analyzing tradeoffs among resources.

Introduction

Management of regulated river systems is particularly complex when natural and instream flow issues are encountered. Typically, water resource professionals are engineers or hydrologists associated with water management agencies concerned with out-of-stream water uses for irrigation, domestic and industrial supply, or instream for power generation, flood control, and water quality. These traditional water uses are generally analayzed by experienced professionals utilizing well established hydrologic, hydraulic, network modeling, and other systems analysis methodologies. The benefits associated with each of these water uses are quantifiable in terms of economic benefits. Quite often water allocation decisions for multiobjective projects are based on maximizing the overall economic return subject to the legal, institutional, and physical constraints imposed by the system. A review of how social and environmental objectives fit into water resources planning in the United States is given by Hobbs et al. (1989). For instream, riparian, and natural resource water uses, economic valuation is difficult and often not the proper basis to determine water

---

[1]Research Hydrologist, Water Resources Division, National Park Service, 1201 Oakridge Dr., Suite 250, Ft. Collins, CO 80525.
[2]Associate Professor, Civil Engineering Department, Colorado State University, Ft. Collins, CO 80523.

allocations. Decision analysis methods can be used to assess tradeoffs and to establish acceptable levels for a variety of instream flow criteria, given hydraulic characteristics and different hydrololgic flow patterns in a river.

Natural resource managers and agencies charged with protecting riparian, aquatic, recreational, and environmental resources are generally not well versed as water resources professionals, and do not have access to staff who routinely analyze hydrologic data. Therefore the capability to perform hydrologic analyses, for example computation of the mean or median flow, or determination of the frequency of high or low flow events, is generally beyond the reach of natural resource land management agencies. In a similar regard, these agencies are without direct access to hydraulic modeling techniques to analyze stream or river cross-sections that relate water depth and flow velocity to river discharge. The natural resource manager's expertise mainly resides in determining relationships that exist between alternative flows and the utility value derived by each instream attribute. Remember that this is typically not a value computed in terms of dollars; rather, activities such as fishing and white water boating may more appropriately be defined as the number of user days per year. These attribute relationships are often computed using visitor survey data. Methods to create dimensionless benefit utility curves as a function of water depth, flow velocity, or river discharge can prove helpful in managing rivers for a variety of conflicting water resource needs. Without basic water resources tools in hydrology and hydraulics, the natural resource manager is left at a distinct disadvantage for determining resource attribute response to various flow alternatives that correspond to different water regulation schemes.

Decision Support System (DSS) Framework

Resource management decisions in river systems often include a determination or selection of a best flow release scheme (i.e., hydrograph) with due consideration given to numerous aquatic and riparian resources, recreational uses, endangered species protection, and other legislated uses of the water. Resource managers, most probably trained in biology, forestry, or other professional fields, are often at a loss to analyze hydrologic and hydraulic data important to this decision making process. To overcome obstacles due to this lack of access to modeling and analysis tools, a personal computer package that operates in a spreadsheet environment was developed as a **Decision Support System (DSS)**. **One component** of the system helps to quantify the relationship between a dependent resource attribute and river discharge with parameters of flow velocity and water depth. This component thereby helps establish utility value functions for the resources. A **second component** performs a hydrologic analysis on historic records of river discharge data, and provides for the creation of alternative flow regimes based on the user's input. A **third component** computes hydraulic characteristics at a given river cross-section, with a user option to partition the discharge into subareas, for computing depth and velocity. By automatically passing data between modules (i.e., linking the modules together) within the spreadsheet package, the interconnected model operates as a DSS. The computer package provides for a rational and consistent evaluation of flow alternatives (i.e., hydrographs) based upon a quantitative measure of each objective (i.e., attribute benefit function) for the hydraulic characteristics (i.e., cross-section) of the river system. A **fourth component** of the DSS is a multicriteria decision analysis matrix for analyzing tradeoffs among the multiple resources.

Each of the four system modules performs calculations in separate disciplines regularly encountered within river systems. The wide array of social, cultural, recreational, natural, and environmental objectives in regulated river systems, presents a difficult tradeoff analysis problem. It is hoped that this DSS will find acceptance and provide insight to future resource decision making. The user friendly environment designed for the DSS should help an inexperienced user to evaluate data, display information, and provide documentation to impacts of hydrology and hydraulics on the resource management decision making process.

Individual DSS Components

Resource Benefit Functions. The first step in resolving resource management conflicts is to identify the specific resources of concern. To help quantify the resource benefits as a function of water discharge variables, a model component called CONBEN was developed. This component helps to define tabular valued utility functions and also displays a graph for a resource attribute (e.g., kayaking) as a function of both depth and velocity on a given river section. For example, kayaking benefits versus water depth would typically start at zero until the water depth approached one foot (i.e., enough to float the kayak), with the benefit then increasing to a maximum value of one at about three feet (i.e., at which the depth is adequate for rolling the kayak over) and staying at a benefit of one, unless deeper water was considered a safety risk. The user friendly module CONBEN contains a few functions that can be drawn upon as a template for modifying and defining other new resource benefit functions. A discussion on how to develop standardized benefit or utility functions is given by Flug and Montgomery (1988). In simplest terms the maximum benefit value is set equal to one, while the minimum benefit value is zero. The standardizing (i.e., dimensionless units) of the benefit values is most useful when comparing or performing tradeoffs among multiple resource attributes. The mere defining of the utility functions allows for a rational and consistent procedure for evaluating resource impacts under alternative river discharges.

Hydrologic Analysis. An engineer, hydrologist, or water resources professional is trained to compute mean annual flow, evaluate flow frequency, and generate alternative flow scenarios for a given river system. Unfortunately, this is not true for most natural resource managers. A second DSS model component called CONHYDRO was developed to read files of historic flow data and perform some routine flow analyses while providing output in tabular and graphic forms. The mean, standard deviation, maximum, minimum, and frequency of the historic flow record, or a user defined partial period of record, are computed and displayed. In addition, CONHYDRO is designed to provide five flow alternative scenarios for use in the overall water resources management decision process. Three flow alternatives presently built into CONHYDRO include the mean, and the mean plus or minus one standard deviation. These alternatives provide limited statistical measures of the hydrologic data. The remaining two alternatives are left to the user to define, either by entering specific data or by using menu options that make it easy to select specific years of data from the historic flow record. Graphic displays for comparing all five flow alternatives are provided. The user can interact with the model components and change data values to help analyze different scenarios easily and quickly.

Hydraulic Calculations. Since most aquatic, riparian and natural resource attributes are impacted by variables such as water depth, velocity and substrate, the model component CONHYDRA was developed to compute cross-section hydraulics. Presently CONHYDRA operates on one cross-section at a time, but work is underway to expand this component to include channel routing between several cross-sections. CONHYDRA, with a user friendly menu format, solicits input of the cross-section geometry and allows the user to identify up to five subareas that the flow can be divided into, thereby allowing for critical influences of sand or gravel bars, for example. This component provides tabular and graphic display of the cross-section geometry as well as the stage-discharge relationship. By selecting one of the five flow alternatives from the component CONHYDRO, the water depth and flow velocity are computed for each river discharge value. These computed values can be displayed, but more importantly, are stored for use in the overall water resources decision making process by coupling with data in CONBEN.

Multicriteria Decision Analysis Matrix. A multicriteria decision analysis module was developed to provide a framework for resolving resource management conflicts associated with a myriad of environmental type objectives. Most of the needed resource and water dependent flow data are available by combining hydrologic flows from CONHYDRO with hydraulic parameters from CONHYDRA and determining the utility (i.e., benefit) from resource functions within CONBEN. The decision analysis component, called CONMCW, uses a value based weighted average method described by Flug and Ahmed (1990) for a river system application. The weighted average portion of the multicriteria decision analysis is input to CONMCW by letting the user identify the relative importance of each of the multiple resources under consideration. By combining weighted scores for each resource benfit value subject to each flow alternative, a Payoff or Impact Matrix can be created. Fontane and Flug (1991) describe the steps for developing the payoff matrix. The organization of information in the Impact Matrix provides a format for analyzing resource impacts, visualizing differences from one alternative to another, and hopefully, facilitating concensus building.

References

Flug, M. and Montgomery, R. H. 1988. "Modeling Instream Recreational Benefits." *Water Resources Bulletin*, AWRA, 24(5), 1073-1081.

Flug, M. and Ahmed, J. 1990. "Prioritizing Flow Alternatives For Social Objectives." *J. Water Resources Planning & Management*, ASCE, 116(5), 610-624.

Fontane, D. G. and Flug, M. 1991. "Introduction to Multi Criterion Methods and Selected Software." In: *Water Resources Planning and Management and Urban Water Resources*, J. Anderson (Editor), ASCE, NY, 449-453.

Hobbs, B. J., Stakhiv, E. Z., and Grayman, W. M. 1989. "Impact Evaluation Procedures: Theory, Practice, and Needs." *J. Water Resources Planning & Management*, ASCE, 115(1), 2-21.

RESERVOIR MANAGEMENT AND THERMAL POWER GENERATION

By Yulianti,[1], and Barbara J. Lence,[2], Associate Member, ASCE

Abstract

Thermoelectric power stations on regulated rivers may cause violations in stream temperature standards if the dilution water from the upstream reservoir is not adequate. However, it is possible that such violations would occur over a short period or that there exists an acceptable temperature violation range. A minmax model is developed for determining the minimum value of the maximum violation of the temperature standard, the reservoir storage, and release levels for reservoir-river systems. The model is applied to the Shellmouth Reservoir in Manitoba, Canada.

Introduction

Optimization approaches have been introduced for managing multipurpose reservoirs (see, e.g., Yeh, 1985). More recently, Lence et al.(1991) describe a Goal Programming (GP) model that minimizes deviations from targets for reservoir storage and of thermal power generation, subject to water quality constraints for temperature. The water quality constraints limit the amount of power that may be generated by a thermal power station on a regulated river in order to maintain the stream temperature standard under a given reservoir release level. This paper presents an approach for determining the storage level and releases from a reservoir that minimize the maximum violation of the temperature standard due to the thermal plant effluent, and the maximum deviations from targets for reservoir release and storage levels. The model also fulfills the water demands downstream, the users upstream, and the flood protection goals at all times.

Minmax Model

The model developed in this work is formulated as:

---

[1] Graduate Research Assistant, Department of Civ. Engrg., University of Manitoba, Winnipeg, Manitoba R3T 2N2, phone (204) 474 6837
[2] Asst. Prof., Dept of Civ. Engrg., University of Manitoba, Winnipeg, Manitoba R3T 2N2, phone (204) 474 8990.

$$\text{minimize } Z_1 + Z_2 + Z_3 + Z_4 \tag{1}$$

Subject to:
continuity

$$S_i^j - S_{i-1}^j - I_i^j + R_i^j + E_i^j = 0 \quad \forall\; i = 1,\ldots, n,\; j=1,\ldots, m \tag{2}$$

storage at the beginning of each year equal to ending storage in the previous year

$$S_o^j - S_n^{j-1} = 0 \quad \forall\; j = 2,\ldots, m \tag{3}$$

initial storage in the first year equal to ending storage in the final year

$$S_o^1 - S_n^m = 0 \tag{4}$$

definition of maximum storage deviations

$$S_i^j - S_{max} - Z_1 \leq 0 \quad \forall\; i=1,\ldots,n,\; j=1,\ldots,m \tag{5}$$

$$S_{min} - S_i^j - Z_2 \leq 0 \quad \forall\; i=1,\ldots,n,\; j=1,\ldots,m \tag{6}$$

definition of water required in downstream river reaches
for individual reaches of the river

$$F_i^{jk} + \sum_{a=1}^{k-1} TI_i^{ja} - \sum_{a=1}^{k} D_i^{ja} = 0 \quad \forall\; i=1,\ldots,n,\; j=1,\ldots,m,\; k=2,\ldots,K \tag{7}$$

for the first reach of the river

$$F_i^{j1} - D_i^{j1} = 0 \quad \forall\; i = 1,\ldots,n,\; j=1,\ldots,m \tag{8}$$

limits on release required to satisfy downstream water needs

$$R_i^j - \sum_{k=1}^{K} F_i^{jk} \geq 0 \quad \forall\; i=1,\ldots,n,\; j=1,\ldots,m \tag{9}$$

definition of maximum release deviation

$$R_i^j - R_{max} - Z_3 \leq 0 \quad \forall\; i=1,\ldots,n,\; j=1,\ldots,m \tag{10}$$

magnitude of stream discharge at the site of the thermoelectric power station

$$R_i^j + \sum_{a=1}^{t-1} TI_i^{ja} - \sum_{a=1}^{t} D_i^{ja} - Q_i^j = 0 \quad \forall\; i=1,\ldots,n,\; j=1,\ldots,m \tag{11}$$

Sum of the stream discharge deficit in each discharge

deficit segment equal to the total discharge deficit

$$\sum_{p=1}^{P} VD_{pi}^{j} - L_i + Q_i^j \leq 0 \quad \forall \ i=1,\ldots,n, \ j=1,\ldots,m \quad (12)$$

definition of temperature violation

$$\sum_{p=1}^{P} \alpha_{pi}^{j} VD_{pi}^{j} - Z_4 \leq 0 \quad \forall \ i=1,\ldots,n, \ j=1,\ldots,m \quad (13)$$

where: Z_1=maximum deviation above the upper storage target, Z_2=maximum deviation below the lower storage target, Z_3=maximum deviation of release above the downstream channel capacity, Z_4=maximum deviation above the water temperature standard, i=month of the year, j=year, m=number of years in a management scenario, and n=number of months in the year over which the reservoir is managed. S_i^j=ending storage, I_i^j=inflow into the reservoir, R_i^j=release from the reservoir, E_i^j=water evaporation from the reservoir, and Q_i^j=discharge available at the thermoelectric station in month i and year j. S_{min}=lower bound and S_{max}=the upper bound of reservoir storage, respectively. K=number of reaches in the river, k, a, and t=reach number, F_i^{jk}=amount of water required to meet the demand, TI_i^{jk}=tributary inflow, and D_i^{jk}=demand for water in river reach k in month i and year j. P=number of segments in, and p=segment in the piecewise linearized discharge deficit versus temperature violation curve, VD_{pi}=discharge deficit in segment p of the discharge versus temperature violation curve in month i, L_i=maximum possible discharge at the site of the thermoelectric power station in month i, and α_{pi}^j=slope of the discharge deficit versus temperature violation curve, in segment p in month i and year j.

Constraint Equations 2 - 6 limit the storage level values and define maximum storage deviations from upper and lower storage bounds. Constraint Equations 7 - 10 define the amount of release required to meet downstream demands and the maximum release deviation from the channel capacity requirement. Equation 11 defines the magnitude of discharge at the thermoelectric power station. Equations 12 and 13 are based on the assumption that a piecewise linear function exists which describes the temperature violation that would result under a given flow condition assuming that the thermoelectric power station was meeting its monthly demand for power. Since such a function would be concave it is necessary to transform it by defining the temperature violation that would result for a discharge deficit below some maximum flow level, L_i, in month i. In this case, $L_i - Q_i^j$ is the discharge deficit at the site of the thermoelectric power station and $\Sigma \alpha_{pi}^j VD_{pi}^j$ is the temperature violation that occurs in month i and year j if the power station meets all of its demands.

Application

The minmax model is applied to a case study based on

the Shellmouth Reservoir and Dam which regulate flows in the Assiniboine River in Southwestern Manitoba. The reservoir is used for flood control, municipal and industrial water supplies for the Cities of Brandon and Portage La Prairie, Manitoba, irrigation and farm water supplies on the Assiniboine River, dilution of municipal and industrial waste from various industries and the Cities of Brandon, Portage La Prairie, and Winnipeg, recreation during summer and dilution of waste heat from a Manitoba Hydro thermoelectric power station in the City of Brandon. The physical description of the reservoir, the hydrologic conditions for the system, the reservoir storage targets and the monthly water supply land power demands are given in Yang et al.(1991) and Lence et al.(1991).

The piecewise linear relationship between the temperature standard violation and the streamflow deficit at the thermoelectric power station is based on the stream temperature simulation model described by Lence et al. (1991). Nine years of inflow data were used for the management period of the model. These data were selected from the 31 years of historical record from 1957 to 1987. Three type of hydrologic year were chosen, dry, average, and wet years. In this study, the optimization model was solved for the case were one dry year is followed by one average year and one wet year and this sequence is repeated three times.The beginning of the year is considered to be the month of April since the reservoir receives the largest portion of its inflow in the early spring.

The minmax model gives a maximum temperature violation of 2.94 $^{\circ}$C which occurs in the month of May in every dry year. If the thermal generation occurs at the desired level, the temperature that would result in May is 14.94 $^{\circ}$C. Since the mean background temperature in May is 13.27 $^{\circ}$C, the average stream temperatures increases only by 1.67 $^{\circ}$C. Depending on the type of fish species present, this deviation in temperature may or may not change the water quality significantly. The temperature deviation in the months of June and September during one of the dry years is 0.34 and 0.44 $^{\circ}$C, respectively. Since the temperature standard in June and September is high (i.e., 23 $^{\circ}$C and 25 $^{\circ}$C, respectively), this violation may also be acceptable.

References

Lence, B. J., Latheef, M. I., Burn, D. H., "Reservoir management and thermal power generation," *Journal of Water Resources Planning and Management, 118*(4), 388-405, 1992.

Yang, Y., Burn, D. H. and Lence, B. J. (1991), "Development of a framework for the selection of a reservoir operating policy," accepted for publication in *Canadian Journal of Civil Engineering*, 1991.

Yeh, W. W.G., "Reservoir management and operation models: A state of the art review," *Water Resources Research*, 21(12), 1797-1818, 1985.

A Raft System for Large River Hydraulic Measurements

Scott D. Wilcox[1] and Ted M. Frink[2]

Abstract: Collection of hydraulic data in large rivers is a difficult, time consuming, and occasionally perilous task. High water velocities, considerable depths, and wide channels present obstacles which preclude use of many traditional methods for flow measurement. A pontoon raft system we developed for use on large, high gradient rivers incorporates a number of features that allow hydraulic measurements to be made safely and efficiently under rugged and remote field conditions. The system consists of two 16-foot (4.88 m) pontoons connected by a lightweight raft frame, a boom assembly for raising and lowering hydraulic equipment, and a cable guide or tethering arrangement for stationing the raft at intervals across a transect. The system is stable and maneuverable in high velocity and turbulent flow conditions, and can be readily transported to remote sites using a variety of methods.

Collection of hydraulic data for physical habitat simulation of large rivers is a difficult, time consuming, and occasionally perilous task. High water velocities and/or excessive depths preclude wading across the transects, and greater channel widths make data collection and field crew coordination more difficult. Remote study sites and safety concerns add to the challenges facing a data collector.

A traditional approach to collecting hydraulic data on large rivers is the rigid boat system described in Instream Flow Paper No. 5 (Bovee and Milhous 1978). This system employees an aluminum boat attached through oarlocks or a USGS suspension system to a cable strung across the river. The system provides a safe and relatively stable working platform for collecting data on many large rivers. However, the rigid boat system has numerous drawbacks in certain situations. Narrow canyons and high gradient rivers with poor road access may make launching of a rigid boat impossible.

---

[1]Principal Scientist, Ebasco Environmental, 2525 Natomas Park Drive, Suite 250, Sacramento, CA 95833

[2]Fisheries Scientist, Ebasco Environmental, 2525 Natomas Park Drive, Suite 250, Sacramento, CA 95833

High gradient, boulder strewn rivers or shallow riffles are not navigable in a rigid boat, and turbulent flow conditions at a study site may make use of a rigid boat impractical, unsafe, or both. In such situations, a raft system can provide a practical, safe alternative working platform for making hydraulic measurements.

A "spider boat" raft system we designed for use on large, high gradient rivers incorporates a number of features we have found effective while collecting hydraulic data at over a dozen sites throughout the western United States. The primary components of the system include two sixteen foot (4.88 m), 30 inch (0.76 m) diameter raft tubes joined together by a frame of rigid, lightweight steel tubing (Figure 1). The tubes are secured to the frame by a combination of wrap around straps and/or D-rings on the raft tubes. The floor is constructed of solid or plywood planks with a non-skid coating. An inclined plank at the back serves as a seat for the oarsman or note taker.

We use two different systems for securing the raft on the river and keeping it on station. The first method is a standard USGS suspension system, with the cross piece which holds the steel suspension cable bolted to either side of the raft frame, and the boom extending forward between the raft tubes (Figure 1). The raft can be ferried along the suspension cable by hand, or by appropriate placement of the oars to serve as a rudder. The suspension cable can be secured to the shore using a variety of anchors: trees, boulders, fence posts, pitons, etc.

A second suspension system consists of a "high line" or "static line" (we have used a PMI brand, 16 mm diameter, kern mantle weave nylon rope) secured well above the water surface and upstream of the cross section(s) to be measured. The raft is tethered to a high line pulley with rope and carabiners, and is maneuvered with assistance from personnel on either bank via ropes attached to the high line pulley (Figure 2). Several transects in close proximity to each other can be measured from one high line by adjusting the length of the tether to the boat. The attachment of the tether line to the boat can be made at various locations on the raft tubes or frame, depending on the configuration of the raft and demands of the river. A quick release mechanism for the raft is strongly recommended, using some type of clamp, jumar or Gibbs Ascender, or (as a last resort) a sharp knife.

This raft system is highly maneuverable and stable even in turbulent conditions and high water velocities (we have used it successfully at velocities of up to 4.3 mps). The parallel raft tubes can straddle large standing waves or areas of turbulent flow, and the extremely low displacement keeps the drag on the suspension system to a minimum (which helps minimize the "yo-yo" tendency of the boat when stationed in the middle of a wide transect). A final (but critical) attribute of the system is that the raft is impossible to swamp, and very difficult to turn over.

The components of the raft system can be transported to the study site using a variety of methods. The raft frame breaks down into smaller components which can be easily transported in the back of a small pickup truck. The raft tubes and frame will also fit in a cargo net which can be lifted into a site by helicopter. For the truly desperate, modifications to the raft frame could be made to enable transporting the system on pack animals.

Hydraulic data collection on large rivers is frequently a challenging task. Use of a versatile, stable system such as this can facilitate getting the task accomplished with a minimum amount of difficulty and danger.

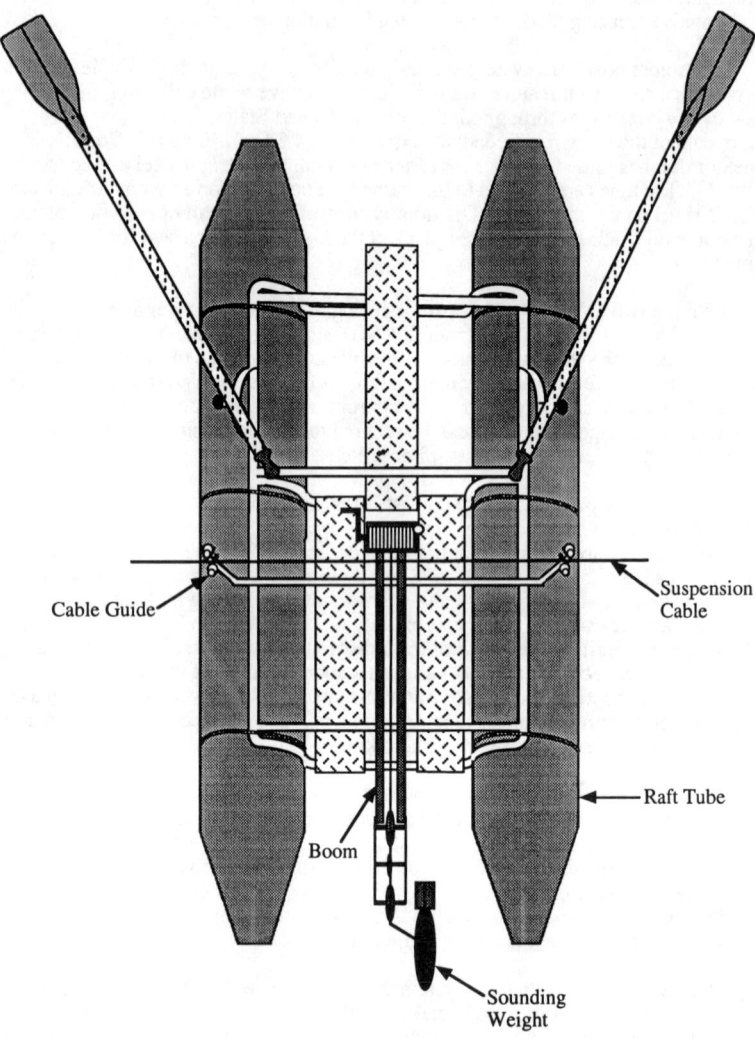

Figure 1. Plan view of "spider boat" raft system with suspension cable and sounding weight boom.

RAFT SYSTEM

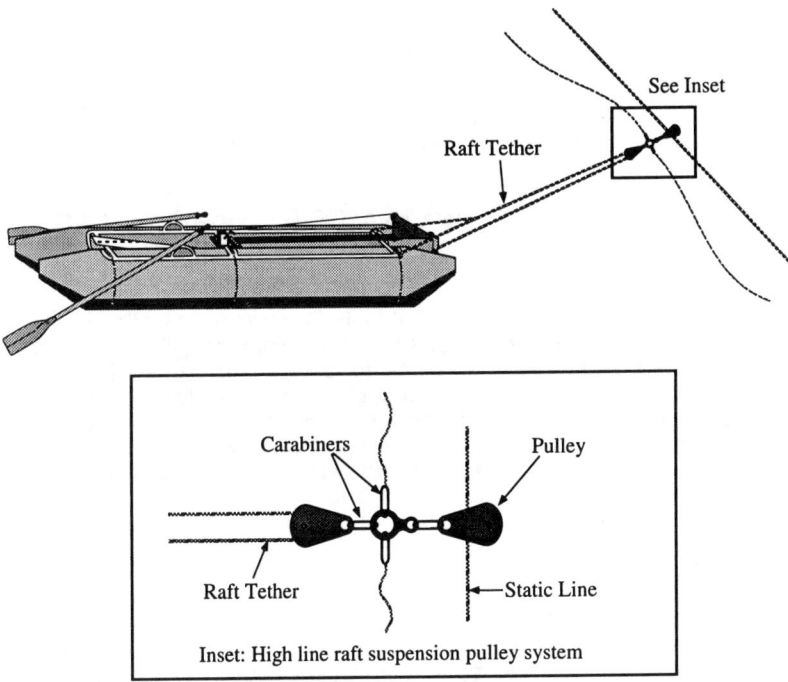

Figure 2. High line rope raft suspension system and detail of tether and pulley configuration for boat positioning.

Appendix References

Bovee, K.D. and R. Milhous. 1978. Hydraulic Simulation in Instream Flow Studies: Theory and Techniques. Instream Flow Information Paper No. 5. Fort Collins, Colorado, Cooperative Instream Flow Service Group, FWS/OBS-78/33.

Alternatives Analysis Using Two-Dimensional Modeling for the Owensboro Bridge and Approaches

M.A. Ports[1], F. ASCE, T.G. Turner[2], M. ASCE, D.C. Froehlich[3], M. ASCE

Introduction

The Kentucky Transportation Cabinet, Department of Highways in cooperation with the Indiana Department of Transportation and the Federal Highway Administration plans to design and construct a new bridge over the Ohio River near Owensboro, Kentucky. Due to concerns about scour, redistribution of flow, and the complex flow conditions at the proposed four mile crossing and recognizing that addressing these issues is important for longer term operation and maintenance, a hydraulic model with two-dimensional capabilities was utilized. The crossing will consist of a main span over the main channel and one or more relief structures in the Kentucky overbank.

This hydraulic analysis was accomplished in three phases using FESWMS, the Finite Element Surface Water Modelling System: Two-Dimensional Flow in a Horizontal Plane, (Froehlich, 1990). FESWMS applies the finite element method and solves the system of equations that govern two-dimensional flow in a horizontal plane. The first two phases consisted of applying the FESWMS model to the proposed site including model calibration and validation, determining baseline (existing) conditions, and assessment of the original proposed configuration of the main span and flood plain crossing. These topics are presented in a paper published in the proceedings of the ASCE Water Forum (Ports, 1992).

This paper presents the formulation, simulation, and comparison of alternative bridge configurations with emphasis on the practical use of the results from the two-dimensional hydraulic analysis.

---

[1] Principal Water Resources Engineer, Parsons Brinckerhoff, 301 North Charles Street, Baltimore, Maryland 21201

[2] Civil Engineer, Parsons Brinckerhoff, 301 North Charles Street, Baltimore, Maryland 21201

[3] Assistant Professor, Department of Civil Engineering, University of Kentucky, Lexington, Kentucky 40506

Criteria

Three basic hydraulic criteria were developed for use in formulating and evaluating the various configurations of bridges. The criteria evolved while several of the early bridge configurations in the study were developed. In order of importance, the criteria are:

1. Limit the maximum velocity upstream or downstream from a relief bridge to 3.5 fps preferred and 4.0 fps maximum.

2. Minimize the increase in water surface elevation (backwater) created by the proposed roadway - preferably no increase in the water surface four miles upstream.

3. Reduce redistribution of flow - preferably 10% or less of overbank flow being redistributed into the main channel per AASHTO guidelines.

The velocity criteria was adopted based on concerns about scour depth at the structures, erosion and resulting damage to downstream agricultural fields, observations by adjacent property owners of unusual flow and sedimentation phenomenon, and the presence of fine silty soils. The backwater criteria is based on a compromise between the State of Kentucky criteria which allows a one foot increase and the State of Indiana which permits zero increase. This criteria was applied for steady state simulations of the 100 year flood event (885,000 cfs).

Optimization

The process of developing each alternative started by defining a goal or general objective and a layout of bridges, (e.g., provide seven relatively short relief structures or provide three relatively long relief structures). Obviously, the local topography and existing drainage constrained the location and size of the alternative relief structure. An iterative process of running a configuration and making adjustments based on those results to improve performance for the next run was used to then optimize each alternative. Generally, six to twelve configurations were run for each alternative.

The results from the hydraulic analysis were then plotted as velocity or unit flow vectors, water surface contours, and lines of flux. The velocity vectors showing magnitude and direction of flow were very useful for evaluating bridge performance and identifying problems areas.

The summary table summarizes several of the more than fourteen alternatives which were evaluated. The original configuration of bridges which consisted of a main span of 4,505 feet and five relief structures totaling 2,000 feet did not satisfy any of the criteria. The original configuration was derived prior to adopting this criteria. Initial attempts to improve performance (see Alternative 3) involved opening the main span. This resulted in only marginal improvements in the performance of the relief structures.

HYDRAULIC ANALYSIS FOR OWENSBORO BRIDGE
COMPARISON OF ALTERNATIVE BRIDGE CONFIGURATIONS

LENGTH OF BRIDGE (FEET) / MAXIMUM VELOCITY (FPS) OUTSIDE THE RIGHT-OF-WAY

	ADDITIONAL MAIN SPAN	RD	BR1	RD	BR1A	RD	BR2	RD	BR3	RD	BR4	RD	BR4A	RD	BR5	TOTAL-LENGTH	BACK-WATER
ORIGINAL DESIGN	0 2.95	2,813	400 5.04			842	500 5.69	2,586	300 5.73	1,689	500 6.00			2,676	300 6.46	2,000	0.251
ALT. 3	1,150 3.51	1,663	400 4.90			842	500 5.61	2,586	300 6.49	1,689	500 5.69			2,676	300 7.14	3,150	0.234
ALT. 4	470 2.92	2,340	1,490 3.97					2,850	1,760 3.79					1,825	1,870 4.96	5,590	0.144
ALT. 5	470 2.97	2,195	730 3.97			1,155	670 3.73	1,725	710 4.12	1,240	940 3.61			1,415	1,600 4.79	5,120	0.143
ALT. 8	470 2.86	2,195	730 4.19			1,360	800 3.86	1,405	940 4.17	1,725	1,370 4.10					4,310	0.176
ALT. 9M	0 2.49	2,665	730 4.19			1,150	670 3.91	1,740	950 4.06	2,160	670 3.95			1,370	950 4.17	3,970	0.153
ALT. 10	470 2.92	2,005	730 3.99			1,555	620 3.97	1,355	710 4.19	1,685	500 3.94	885	730 3.64	965	860 4.02	4,620	0.145
ALT. 12	470 2.91	2,195	550 4.02	840	250 3.01	645	420 4.07	1,785	750 4.12	1,225	500 4.05	1,355	510 3.75	720	850 4.12	4,300	0.150
ALT. 13	5,550 3.41													3,350	4,150 3.41	9,700	0.114
ALT. 14M	0 2.55	2,660	730 3.99			1,160	830 3.46	1,570	710 3.99	1,470	940 3.81			2,030	950 4.13	4,160	0.147

BACKWATER (FT) IS AT A LOCATION APPROXIMATELY FOUR MILES UPSTREAM.
RD LENGTHS REPRESENT ROADWAY ON FILL BETWEEN BRIDGES (FT). FOR EXAMPLE: ALTERNATIVE 4, BRIDGES 1 AND 3 ARE 2850 FT. APART.

Conclusions

After evaluating numerous configurations several general observations were made. First, it was very difficult to satisfy the velocity criteria with less than 4,000 feet in total length of relief structures. Second, all the relief structures are interdependent. Reducing the length of one structure increases the flow and velocity at adjacent structures. Third, spacing the relief structures evenly across the floodplain while conforming to the local topography worked best toward satisfying the criteria.

Through a selection process which considered other factors such as cost, scour, maintenance, right-of-way required, and impact to agriculture Alternatives 9M and 13 were selected as the finalists. Alternative 9M consists of five evenly spaced relief structures with a total length of 3,970 feet. Alternative 13 consists of two long relief structures totaling 9,700 feet. Alternative 13 was developed based on a lower unit cost which can be realized for a trestle-type structure of this length. Currently FHWA is reviewing the results of the alternatives analysis.

References Cited

Froehlich, D.C., 1990. *Finite Element Surface Water Modeling System: Two-Dimensional Flow in a Horizontal Plane.* User's Manual, Department of Civil Engineering, University of Kentucky, Lexington, Kentucky.

Lee, J.K., D.C. Froehlich, J.J. Gilbert, and G.J. Gregg, 1982. *Two-Dimensional Analysis of Bridge Backwater.* Proceedings of the Conference Applying Research to Hydraulic Practice, American Society of Civil Engineers, New York, New York.

Lee, J. K., D. C. Froehlich, J.J. Gilbert and G.J. Wiche, 1983. *A Two-Dimensional Finite Element Model Study of Backwater and Flow Distribution at the I-10 Crossing of the Pearl River Near Slidell, Louisiana.* Water Resources Investigations Report 82-4119, U.S. Geological Survey, NSTL Station, Mississippi.

Ports, M.A., T. G. Turner, D.C. Froehlich, 1992. *Two-Dimensional Hydraulic Analysis of the Owensboro Bridge and Approaches.* Proceedings of the Water Resources Section of the 1992 Water Forum, American Society of Civil Engineers, New York, New York.

EFFECTS OF RESERVOIR MANAGEMENT ON ESTABLISHMENT OF RIPARIAN AND WETLAND VEGETATION IN NORTHEASTERN WASHINGTON. Clayton J. Antieau, non-member of ASCE[1].

The effects of various reservoir pool management scenarios on establishment of riparian and wetland vegetation in arid regions of the Pacific Northwest are poorly known. These effects are becoming an important issue as greater attention focuses on past losses of habitat for sensitive fish and wildlife species as caused by dam construction. Additional attention is drawn to reservoir management because of its role in creating and maintaining new habitat for these and other species, and for its role in managing infestations of noxious aquatic and riparian weeds.

This study quantified characteristics of riparian and wetland vegetation associated with two reservoirs on the Spokane River near Spokane, Washington. The study attempts to correlate differences between the two plant community assemblages to differences in the operational management histories of the two reservoirs.

The Nine Mile and Long Lake Hydroelectric Development projects and their respective reservoirs are adjacent facilities owned by Washington Water Power and operated under licenses issued by the Federal Energy Regulatory Commission. Both projects were built at approximately the same time (1908 and 1914, respectively), and have been operated since that time to generate electricity. Nine Mile reservoir extends about 5.5 miles (9.2 km) and has a capacity of approximately 5,800 acre-feet (7,154,300 m^3). The Long Lake reservoir extends about 21.0 miles (33.9 km) and has a maximum capacity of approximately 253,000 acre-feet (312,075,500 m^3).

The Nine Mile project is operated on a run-of-the-river basis, which results in seasonal changes in the Nine Mile reservoir elevation. The reservoir's base operation elevation is approximately 1606.5 feet (490 m) above sea level. However, the pool drops during the summer and rises above the base level during high flows of winter and early spring. In contrast, Long Lake

---

[1]Clayton J. Antieau, Senior Wetland Ecologist, Ebasco Environmental, 10900 Northeast 8th Street, Bellevue, Washington 98004-4405. (206) 451-4174.

reservoir is operated at a more constant elevation throughout the year, but experiences at least a 10-foot (3 m) winter drawdown approximately once every 5-7 years.

This study quantified the areal extent and composition of existing wetland and riparian vegetation in both reservoirs. These data were analyzed in terms of species diversity, nativity, habitat diversity, and spatial considerations. Results of these analyses were then compared to the maximum and minimum monthly pool elevations for the respective reservoirs.

Nine Mile reservoir primarily supports a well-developed non-wetland riparian vegetation along its existing shoreline. These primarily forested habitats are dominated by a small suite of dominant non-native species, including crack willow (*Salix fragilis*), American elm (*Ulmus americana*), box elder (*Acer negundo*), reed canarygrass (*Phalaris arundinacea*), common tansy (*Tanacetum vulgare*), and climbing nightshade (*Solanum dulcamara*). Presumably, these weed-dominated riparian habitats exist as a result of the environmental perturbation caused by the operation regime of the Nine Mile reservoir, characterized by pool elevation fluctuations during parts of the growing season.

Less than 5.0 acres (2.0 hectares) of emergent wetland habitat and less than 0.5 acre (0.2 hectare) of aquatic bed vegetation exist in Nine Mile reservoir. In contrast, Long Lake reservoir supports comparatively large acreages of well-developed emergent and aquatic bed wetland vegetation dominated by a high diversity of native plant species. Aquatic bed wetlands typically require surface water for optimum growth and reproduction, and are usually best developed in relatively permanent water or under conditions of repeated flooding (Cowardin *et al.* 1978). The current operation regime of Nine Mile reservoir involves a substantial drawdown during the peak of the growing season, which creates large areas of emersed or shallow submersed unconsolidated substrate.

This study suggests development/growth of wetland vegetation are favored by relatively constant levels of inundation. This finding is supported in work by Hiltibran (1982), Gosselink and Turner (1984), Greening and Gerritsen (1987), and Godshalk and Barko (1988). This study also suggests that, in general, drawdowns during the growing season may displace or eliminate aquatic macrophytes. This situation also has been described by Harris and Marshall (1963), Kadlec and Wentz (1974), and Rørslett (1989). Although numerous variables determine the establishment of wetland habitats, these initial results indicate that development of aquatic bed and other wetland habitats in Nine Mile reservoir could be encouraged by stabilizing the pool elevation, particularly during the growing season.

Study results also indicate reservoirs with fluctuating water levels during the growing season may promote establishment of non-native weedy species. This situation would contribute to the spread of troublesome and noxious weeds into native upland, wetland, and riparian plant communities, potentially displacing or preempting more desirable native vegetation.

LITERATURE CITED

Cowardin, L.M., V. Carter, F.C. Golet, and E.T. LaRoe. 1979. Classification of wetlands and deepwater habitats of the United States. U.S. Fish and Wildlife Service publication FWS/OBS-79-31.

Godshalk, G.L. and J.W. Barko. 1988. Effects of winter drawdown on submersed aquatic plants in Eau Galle Reservoir, Wisconsin. Proceedings, 22nd Annual Meeting of the Aquatic Plant Control Research Program. Miscellaneous Paper A-88-5.

Gosselink, J.G. and R.E. Turner. 1984. The role of hydrology in freshwater wetland ecosystems. *In*: R.E. Good, D.F. Whigham, and R.L. Simpson, eds. Freshwater wetlands: ecological processes and management potential. Academic Press, New York, New York.

Greening, H.S. and J. Gerritsen. 1987. Changes in macrophyte community structure following drought in the Okefenokee Swamp, Georgia, U.S.A. Aquatic Botany 28:113-128.

Harris, S.W. and W.H. Marshall. 1963. Ecology of water-level manipulations on a northern marsh. Ecology 44(2):331-343.

Hiltibran, R.C. 1982. The effect of drawdown on aquatic macrophytes in Allerton Lake. Proceedings of the North Central Weed Control Conference 37:122 (abstract).

Kadlec, J.A. and A. Wentz. 1974. State-of-the-art survey and evaluation of marsh plant establishment techniques: induced and natural. Volume I: report of research. U.S. Army Corps of Engineers Waterways Experiment Station Contract Report D-74-9.

Rørslett, B. 1989. An integrated approach to hydropower impact assessment. Hydrobiologia 175:65-82.

Urban Runoff and The Environment
(Symposium Editor: J.J. Warwick)

Introduction

One of the most perplexing and complex water quality problems facing us today is the management of urban runoff. As point sources are brought under control with enhanced treatment and reuse technologies, uncontrolled non-point source flows comprise an increasingly larger proportion of total pollutant loading. This 3-day symposium is comprised of nine oral sessions plus a poster session. The symposium is jointly sponsored by the Urban Water Resources Committee and the Urban Water Resources Research Council.

The first session presents various mathematical modeling approaches used for assessing runoff quantity and quality. The second session addresses environmental impacts, including the effects of urbanization on freshwater wetlands in the Puget Sound region. In the third session, selection and implementation of best management practices (BMPs) on the East and West Coast will be discussed. The fourth session addresses combined sewer overflow, a major urban runoff concern for older cities. The fifth, sixth, and seventh sessions present case studies from the eastern, central and western portions of the United States, respectively. In session eight, stormwater utility experiences in Texas, Washington, Oregon, Utah, and Missouri will be discussed. The ninth and final oral session will be a panel discussion of regulatory issues with six representatives from three EPA regions: VI (South-central U.S.), IX (Western U.S.), and X (Northwestern U.S.). Each region will be represented by a pair of panelists who will discuss both stormwater permitting and total maximum daily load (TMDL) determination. A variety of related topics are also presented in a in a separate poster session which completes the symposium.

Impact of Spatial and Temporal Data Limitations on the Modeling of Runoff Quantity and Quality

J.J. Warwick[1], M. ASCE, and J. Litchfield[2]

Abstract

A small subbasin within an urbanized watershed (Reno, Nevada) was intensively monitored for runoff quantity and quality. The EPA Stormwater Management Model (SWMM) was applied to simulate the observed, in-pipe, runoff hydrographs and associated pollutographs. At this time only the runoff quantity portion of the project has been completed. SWMM performance was evaluated for a variety of spatial aggregation scales from a fine level (1/4 acre with all inlets and pipes) to a rather coarse scale which treated the entire 86 acre subbasin as a single hydrologic unit. ARC/INFO Geographic Information System (GIS) software was used to determine pertinent hydrology modeling parameters (e.g., total percent impervious, slope, etc.). Results demonstrated that the coarsest level of spatial aggregation was unsatisfactory, while the medium and finest levels of detail both performed adequately.

Introduction

The Water Quality Act of 1987 requires the U.S. Environmental Protection Agency (EPA) to establish regulations regarding storm water discharges. U.S. EPA regulations encourage system-wide cooperation in assessing nonpoint source

---

[1] Associate Director, Graduate Program in Hydrology, University of Nevada-Reno, 1000 Valley Road, Reno, Nevada 89512.

[2] Graduate Student, Graduate Program in Hydrology, University of Nevada-Reno, 1000 Valley Road, Reno, Nevada 89512.

problem areas and in evaluating watershed management programs. To this end, stormwater models will be increasingly applied to estimate urban nonpoint source pollutant contributions and associated impacts. In this paper, the Stormwater Management Model (SWMM, Huber, et al., 1983) is applied to an urbanized residential area located in Reno, Nevada. ARC/INFO GIS (ESRI, 1992) coverages of the study area were provided by the Washoe County Department of Transportation. Modeling results to date are limited to the quantification of SWMM hydrograph simulation.

Methodology & Results

The SWMM model was calibrated and verified for three different spatial aggregation levels. The coarsest level of spatial definition (Scenarios 1A and 1B) treated the entire subbasin as one hydrologic unit. The finest level of spatial resolution utilized 30 subbasins draining into 23 inlets, with 36 pipes and 14 connecting manholes. The intermediate level of spatial aggregation utilized 4 subbasins draining into 4 inlets, with 11 pipes and 8 connecting manholes.

Calibration was achieved by minimizing a total error statistic (ET) that incorporates a root mean square error (RMS) expression to quantify overall prediction errors plus an additional component associated with peak flow computation. The relative magnitude of each component is controlled by assignment of a simple weighting term

$$ET = \left\{ (1.0 - W_t) * \left[\sum_{i=1}^{n_m} (Q_{m_i} - Q_{c_i})^2 / n_m \right]^{1/2} \right\} + \left\{ W_t * |Q_{mp} - Q_{cp}| \right\} \quad \text{(Eq. 1)}$$

where: ET = total error statistic (ft^3/sec); W_t = weighting factor; n_m = number of actual flow measurements; Q_m = measured flow (ft^3/sec); Q_c = computed flow (ft^3/sec); Q_{mp} = measured peak flow (ft^3/sec); and Q_{cp} = computed peak flow (ft^3/sec). A subjective balance between fitting the peak and the overall hydrograph shape was achieved by assigning W_t a value equal to 0.2. Figure 1 summarizes the calibration of SWMM where only the ratio of total to effective percent imperviousness

(RTEFF) was adjusted, except for Scenario 1B which included the additional adjustment of subbasin width to match the observed and predicted peak runoff time. Obviously Scenario 1B failed miserably and will not be continued hereafter.

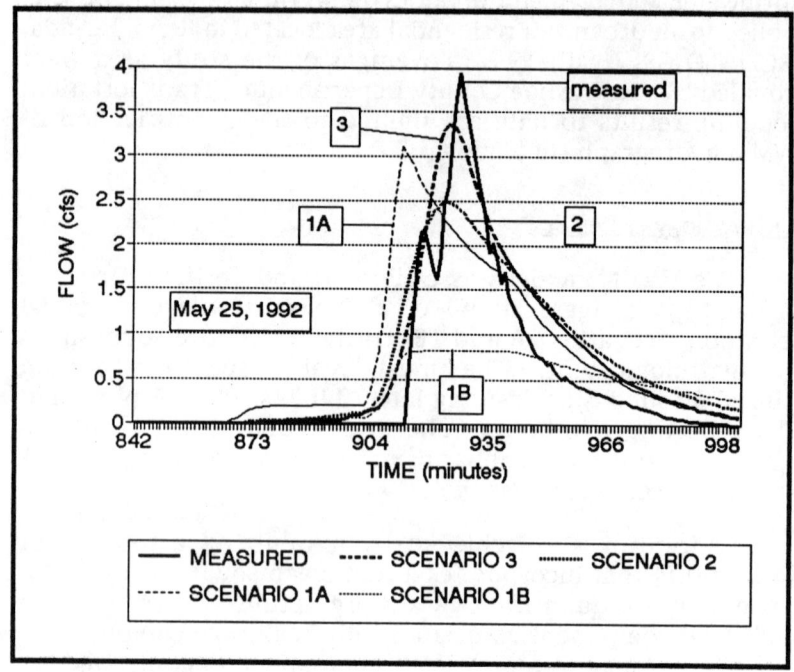

Figure 1: Summary of SWMM Calibration

A forecasting efficiency expression, developed by Loague and Freeze (1985), was used to quantify water quantity predictive performance

$$E_f = \frac{\left\{\left[\sum_{j=2}^{m}\left(Q_{m_i} - Q_{avg}\right)^2\right] - \left[\sum_{j=2}^{m}\left(Q_{m_i} - Q_{c_i}\right)^2\right]\right\}}{\sum_{j=2}^{m}\left(Q_{m_i} - Q_{avg}\right)^2} \quad \text{(Eq. 2)}$$

where: E_f = forecasting efficiency (j=2 since one event was used for calibration); m = total number of events (m=5); and Q_{avg} = average flow value over all verification events (ft^3/sec). Loague and Freeze applied unit hydrograph procedures to three watersheds where they only obtained peak flow forecast efficiencies (EF-Peak) of 0.07, -0.04, -0.28 and total volume forecasts (EF-Volume) of 0.21, 0.01, -0.34. Table 1 summarizes the forecasting efficiencies for the three level of spatial aggregation.

Table 1: Calibration and verification results.

Scenario	RTEFF	ET (cfs)	EF-Peak	EF-Volume
1A	100.	0.712	0.277	0.472
2	83.	0.644	0.900	0.856
3	79.	0.395	0.771	0.894

Conclusions

The coarsest level of spatial aggregation (Scenario 1A) did not perform adequately, while the medium and fine levels (Scenarios 2 and 3, respectively) both performed reasonably well. Additional storm events will need to be simulated before any significant distinction can be established between the two finer levels of spatial representation.

References

1) Environmental Systems Research Institute (1992). *ARC/INFO Users Guide (Version 6.1)*, ESRI, Redlands, California.
2) Huber, W.C., Heaney, J.P., Nix, S.J., Dickinson, R.E. and Polman, D.J., (1983). "Stormwater Management Model User's Manual, Version III." U.S. EPA, Cincinnati, Ohio, 359 pp.
3) Loague, K.M. and Freeze, A.R., (1985). "A Comparison of Rainfall-Runoff Modeling Techniques on Small Upland Catchments." Water Resources Research, Vol. 21, No. 2, pp 229-248.

Projecting Customers Using A GIS Land Use Forecasting Model

Jacqueline Borrego [1]

Abstract

The Salt River Project originally developed the Land Use Forecasting Model to understand the pattern of urbanization of agricultural and desert land in metropolitan Phoenix. The model allocates future land uses to forty-acre parcels based on factors influencing development such as access to freeways, proximity to railroads, and adjacency to intersections. Projecting the distribution of future growth by land use type allows us to study the impact of future water and electrical customers on our system.

Introduction

The Salt River Project (SRP), a water and power utility, developed the Land Use Model (model) to more accurately forecast customers since SRP's service area does not conform to the municipal boundaries defined in the county population and employment forecasts. The model, though, permits a wide range of applications involving water resource management, environmental analysis, transportation development, demand-side implementation and facilities planning. The most extensive application of the model forecasts is SRP's Water Demand Forecasts.[2]

Geographical Information System

Integrating the model with a geographical information system (GIS)[3], ARC/INFO, produces an interactive environment which accelerates the analyst's understanding of the forces of urban change and the relative importance of various development-influencing factors. The model allows the analyst to view and modify land use data, parcel

---

[1] Urban Economist, Strategic Planning, Salt River Project, P.O. Box 52025, Phoenix, Arizona 85072-2025

[2] See (Nichols, 1992) for a more complete discussion of SRP's Water Demand Forecast

[3] "A GIS is not simply a computer system for making maps..... A GIS is an analysis tool. The major advantage of a GIS is that it allows you to identify the relationships between map features." (Understanding GIS, 1991)

attributes, preference surfaces and development patterns at every step of the forecasting process.

Methodology

The forecasting process is divided into five basic steps (figure 1):
- Digitizing land use data for the base year
- Estimating the magnitude of growth using population and employment forecasts.
- Defining the factors influencing development for each of the ten land uses
- Weighting the factors according to how they influence the location of a land use
- Running the forecast routine to produce forecasts of ten land uses for 20 years in five-year increments.

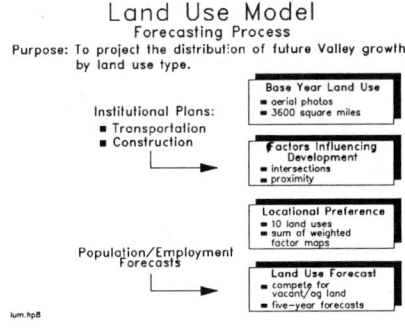

Figure 1

Base Year Land Use

The model defines the metro area in terms of physical characteristics such as low-density commercial, mobile homes, general industrial. Homogeneous land use classes have similar locational characteristics; general industrial uses usually prefer to locate near railways and freeways, medium-density residential uses prefer not to locate at intersections.

Data collection of base year data is achieved by entering a picture of the geographical landscape into the computer from aerial photographs and other sources. A brief description of the land use classes is listed in Table 1.

Regional Control Totals

The model predicts the distribution of growth, not the magnitude of growth. Base year acreages for each of the ten land use classes are extracted using the GIS, and are increased for each forecast period

Table 1: Land Use Classes

Residential
Low-density............ Houses on larger than half acre
Medium-density........ Typical subdivision homes
High-density.......... Apartments, townhomes, condos
Mobile homes

Commercial
Low-density........... Neighborhood retail, services
Medium-density........ Strip commercial development
High-density.......... Mall, hospitals, high-rise office

Industrial
Light.................Industrial and business parks
General...............Conventional industrial

Vacant.................Vacant land available for development

Agriculture............ Crops, dairies, stockyards

according to population and employment growth rates.

Development Factor Maps

The type of development which occurs within a vacant parcel of land depends on the attributes of the parcel, the characteristics of adjacent and surrounding parcels, and the accessibility of the parcel to other land uses within the region. Some factors influencing development include proximity to transportation facilities and adjacency to highways, freeways and intersections.

Preference Maps

Once the factor maps are defined they are linked to the ten land use classes according to their functional relationships. A preference map, reflecting observable locational behavior, is generated for each of the ten land use classes, defining where the land use prefers to locate based on positively and negatively weighted development factors. For example, builders of subdivision homes (medium-density residential) prefer to develop land near, but not adjacent to freeways, accessible to employment opportunities but not adjacent to major streets, intersections and railroads. The medium-residential preference map is partially composed of positive weights for three factor maps: employment accessibility, household proximity and recent residential growth areas; and negative weights for five factor maps: highway adjacency, railroad proximity, intersections, industrial proximity, and proximity to high-density commercial.

Forecast Allocation

Regional control totals are allocated by the forecast routine using a choice modelling technique in which the ten land uses compete for vacant, unrestricted land. Although some vacant parcels are useful for only one land use, the majority of developable parcels have the potential for a mixture of residential, commercial and industrial development. The model allows simultaneous competition among parcels to receive an allocation, and among land uses to occupy a parcel.

The competition of land uses is based on a comparison of preference maps for each land use type, but it is constrained by two primary considerations. First, the total acreage of all land uses (including vacant land) allocated to any forty-acre parcel must equal the amount of developable land within the forty-acre parcel. Second, the sum of the allocations of each land use type must equal the regional control total for that type.

Output from the model consists of ten delta land use maps of allocated growth for a five-year forecast period. For the second forecast period, the base year, factor and preference maps are recalculated based on the forecasted growth of the previous period so the allocation in one period will influence the allocation in a subsequent period.

Water Customers

The Land Use Model forecasts are not used to project precise customer numbers for future years. Rather, the forecasts represent urbanization trends and potential hot areas of growth.

Water customers are indirectly projected. Rather than convert land use acreages to customers, the land use class forecasts are analyzed to understand future water needs for different customer classes. For example, the residential, commercial and industrial acreage forecasts represent Municipal and Industrial customers.

Agricultural customers are tracked by studying the agricultural land use forecasts. This land use class is not one of the ten land uses forecasted by the model. Instead, agricultural forecasts are developed by subtracting the residential, commercial and industrial forecasts from the inventory of agricultural land in the base year. Examination of the urbanization of agricultural land aids water resource planners in studies of canal capacity, well-utilization, and water resource plans.

Summary

Disaggregate land use data is projected for ten land use types according to functional relationships between development factors. The most comprehensive use of the forecasts has been the Water Demand Model which allows water demand forecasts to be examined by lateral, canal or dam.

The versatility of the Land Use Model lies in its ability to manipulate spatially disaggregated data; the accuracy and reliability of the model are attributable to the simulataneous competition of land uses for vacant land; and the success of the model resides in the diverse applications by its users.

References

Nichols R. S., "Small Area Water Demand Forecasting at the Salt River Project", Proceedings of a Conference on Water Management in the 90's: A Time for Innovation, American Society of Civil Engineers, May 3-5, 1993.

Environmental Systems Research Institute, 1991, *Understanding GIS*. Redlands: ESRI.

SUBJECT INDEX
Page number refers to first page of paper

Africa, 518
Agricultural engineering, 514
Agriculture, 332, 698, 702, 705, 709
Alabama, 316
Algorithms, 392
Aquatic habitats, 75, 79, 117, 531, 563, 835, 839
Aqueducts, 396
Aquifers, 8, 324, 581, 585, 601, 605, 725
Arizona, 220, 571, 644, 694, 702, 721
Asia, 514

Bank erosion, 96
Basins, 8, 83, 92, 142, 244, 248, 260, 264, 288, 316, 412, 471, 483
Bayesian analysis, 276
Bays, 177, 181
Boston, 502
Bridge design, 854

Calibration, 296
California, 12, 30, 43, 122, 205, 224, 288, 296, 320, 332, 336, 396, 416, 563, 609, 657, 670, 678, 686, 698, 797, 839
Canada, 531, 846
Canals, 514
Cartography, 220
Catchments, 384
Channel design, 835
Channel erosion, 92
Channel improvements, 92
Channel morphology, 96
Chesapeake Bay, 778
Chicago, 189, 193, 364
Civil engineering, 304
Civil engineers, 308, 312
Climatic changes, 416, 421, 425
Coastal environment, 416
Coastal management, 751, 778
Codes, 344, 352
Colorado, 67, 284, 340
Colorado River, 63, 71, 83, 201, 248, 702, 705
Columbia River, 1, 4, 55, 142, 147, 151, 164, 168, 169, 173, 356, 709
Combined sewers, 758, 762, 766, 770, 774, 793

Communication skills, 38
Community planning, 151, 495, 498, 502, 530, 619, 623, 627
Computer analysis, 842
Computer applications, 216
Computer models, 24, 147, 156, 185, 224, 224, 264, 280, 284, 288, 296, 376, 463, 789, 823
Computer programs, 709
Computerized control systems, 831
Computerized simulation, 24
Computers, 304
Conservation, 713, 717, 721, 839
Construction management, 635
Contaminants, 256
Contamination, 12, 220, 292, 433, 531, 567, 571, 575, 581, 585, 589, 609, 827
Contracts, 34
Cost analysis, 368
Cost effectiveness, 67

Damage prevention, 380
Dams, 506, 522
Databases, 455
Decision making, 51, 151, 252, 268, 276
Decision support systems, 252, 264, 280, 288, 842
Demand, 228, 495, 644, 649, 686
Design, 88, 130, 388, 392, 534, 631
Design standards, 328
Detention basins, 92, 134, 138, 185, 742, 785
Detention reservoirs, 831
Developing countries, 510
Dominican Republic, 522
Drainage systems, 328, 805, 817, 821
Drawdown, 169
Dredge spoil, 109
Droughts, 122, 268, 336, 445, 451, 455, 463, 467, 471, 475, 479, 483, 487, 491, 495, 498, 502, 674, 686, 694

Ecology, 88, 88, 563
Economic analysis, 51, 59, 63, 67, 169, 173, 526, 713, 717, 721
Economic impact, 51, 59, 63, 67, 320
Ecosystems, 292, 531
Education, 308, 312

Endangered species, 1, 59, 156, 160, 164, 168, 173, 320
Energy conservation, 709
Engineering, 88, 304
Engineering education, 308, 312
Environmental engineering, 4, 109, 113, 156
Environmental impacts, 47, 63, 63, 71, 75, 79, 101, 117, 160, 164, 169, 332, 601, 858, 861
Environmental issues, 320, 510, 613, 619
Environmental planning, 47, 88, 101, 252, 778
Erosion control, 396
Estuaries, 181, 827
Ethiopia, 518
Eutrophication, 292, 738
Evaluation, 47, 168, 734

Feasibility studies, 674
Federal government, 408
Finite element method, 725
Fish habitats, 92, 117, 122, 173
Fish management, 1, 51, 59, 67, 75, 142, 147, 156, 160, 164, 168, 169, 173, 835
Fish protection, 1, 122, 160, 164, 168, 169, 173
Fish reproduction, 75, 835
Fisheries, 67
Flood control, 4, 47, 101, 372, 376, 831
Flood damage, 380, 396
Flood frequency, 384
Flooding, 380
Floodwalls, 380
Florida, 20, 39, 105, 113, 138, 316, 324, 412, 601, 827
Flow characteristics, 666, 850
Flow control, 831
Flow patterns, 224, 256, 725, 729, 766, 823
Flushing, 670
Forecasting, 228, 235, 240, 280, 284, 416, 644, 649, 666, 734, 866
Frequency distribution, 455
Fresh water, 181, 690
Funding allocations, 813
Fuzzy sets, 268

Geographic information systems, 216, 220, 224, 866

Georgia, 316
Geothermal energy, 526
Global warming, 416, 425
Government, 408, 498
Great Lakes, 530, 531
Ground water, 216, 601
Ground-water flow, 224, 725
Ground-water management, 8, 12, 433, 514, 563, 605, 639, 702, 705
Ground-water pollution, 12, 220, 567, 575, 609
Ground-water quality, 585, 589, 597

Hazardous waste sites, 220, 575
Heavy metals, 292
History, 304, 308
Hydraulic models, 842, 854
Hydraulics, 850
Hydroelectric power, 55, 526
Hydroelectric powerplants, 24, 63, 197, 201, 372, 534, 538, 543
Hydroelectric resources, 142, 156, 168, 526
Hydrologic aspects, 522
Hydrologic data, 380, 455
Hydrologic models, 105, 147, 185, 235, 280, 296, 384, 416, 421, 425, 518, 729, 842
Hypolimnion, 738

Iceland, 526
Illinois, 189, 193, 364, 758
Industrial wastes, 682
Infiltration rate, 593
Inflow, 181, 231, 256
Infrared analysis, 404
Infrastructure, 404, 510
Inspection, 364
Instream flow, 71, 75, 79, 352, 842, 850
International waters, 506
Irrigation, 514, 682, 698, 702, 705, 709
irrigation water, 674
Irrigation water, 678

Kansas, 471
Kentucky, 649, 854
Kinematics, 593

Lakes, 4, 20, 185, 506, 738
Land fill, 729
Land usage, 866
Leachates, 729

SUBJECT INDEX

Legislation, 440
Levees, 380
Licensing, 30, 538, 543
Linear programming, 197, 201, 205
Litigation, 316
Local governments, 408, 433
Louisiana, 782

Maine, 766, 770
Maintenance, 364, 368
Management methods, 34
Management planning, 47, 433
Management training, 38
Marketing, 55, 356
Markov chains, 272
Marshes, 105
Maryland, 96, 778
Massachusetts, 491, 502
Master plans, 284
Mathematical models, 256, 400, 487, 575, 593, 597, 601, 734
Mercury, 292
Michigan, 793
Microcomputers, 831
Migration, 147
Mining, 396
Minority groups, 312
Mississippi River, 782
Missouri, 471
Modeling, 24, 142, 147, 156, 181, 185, 189, 193, 197, 205, 224, 235, 240, 244, 248, 252, 256, 260, 272, 288, 296, 400, 451, 463, 479, 487, 491, 495, 518, 571, 575, 593, 597, 649, 690, 734, 766, 774, 823, 846, 854, 862
Models, 264, 729
Montana, 835
Municipal government, 408
Municipal wastes, 657, 670
Municipal water, 678

National Weather Service, 280
Network analysis, 571
Network design, 388, 392
Neural networks, 649
Nevada, 16
New Jersey, 653
New York City, 729
New York, State of, 653, 717
Nile River, 518
Noise control, 631

Nonlinear programming, 593
Nonpoint pollution, 738, 742, 747, 778, 785, 809, 827, 861

Object-oriented languages, 252, 260, 451, 479, 483, 487, 491
Oceans, 690
Odor control, 631
Optimization, 388, 392, 567, 747, 846
Optimization models, 197, 201, 205, 240, 360, 400, 585, 593
Oregon, 272, 589
Overflow, 762, 766, 770, 774, 793

Partnering, 39
Peaking capacities, 823
Permits, 30
Pesticides, 12
Phreatic surface, 563
Pipe networks, 388, 392
Pipelines, 396, 404, 690
Piping systems, 400
Planning, 16, 20, 30, 43, 101, 205, 228, 231, 284, 300, 324, 328, 336, 360, 429, 433, 440, 445, 451, 463, 467, 471, 475, 479, 491, 495, 498, 502, 510, 530, 538, 543, 548, 551, 619, 623, 644, 649, 666, 674, 686, 713, 770, 774, 785
Plumbing fixtures, 657, 670, 694
Policies, 244, 248, 260
Political factors, 498
Pollution, 571
Pollution abatement, 12, 126, 130, 134, 530, 531, 581, 585, 609, 827
Pollution control, 138, 738, 742, 747, 751, 758, 778, 785, 789, 797, 801, 805, 813
Polygons, 216
Ponds, 101, 368
Pontoons, 850
Pools, 839
Potable water, 440, 571
Potomac River, 463
Power, 55
Power supplies, 526
Powerplants, 63, 197, 522, 534, 538, 543
Precipitation, 455
Pricing, 713, 717
Probabilistic models, 276
Project control, 34

Project evaluation, 613
Project management, 34, 619, 623, 627, 631, 635
Project planning, 613, 619, 623, 627
Public opinion, 538
Public policy, 1, 16, 30
Public works, 34, 613
Pumping, 400, 567

Rain water, 694
Rainfall-runoff relationships, 235, 240
Rates, 653, 713, 717, 721, 809
Receiving waters, 770, 774
Recharge wells, 593, 597
Reclaimed water, 674, 678, 694
Recovery planning, 59, 151, 156, 160, 164, 168, 169, 173
Recreational facilities, 96, 538, 627
Regional planning, 429, 543, 548, 551, 555, 559
Regression models, 181
Rehabilitation, 360, 364, 400, 581
Reliability analysis, 212
Remote sensing, 404
Reporting, 38
Research needs, 300
Reservoir management, 71, 79, 240, 846, 858
Reservoir operation, 1, 20, 47, 169, 205, 212, 228, 231, 268
Reservoir performance, 117
Reservoir sites, 117
Reservoir storage, 79, 372
Reservoir system regulation, 134, 197, 205, 212, 506
Reservoir systems, 24, 122, 522
Reservoirs, 20, 455
Resource management, 51, 59, 67, 83, 88, 101, 109, 142, 160, 164, 300, 304, 304, 320, 344, 352, 835, 842
Restoration, 96, 531, 585
Retention, 126, 130
Retention basins, 368
Retrofitting, 372, 670, 785
Rhode Island, 774
Rice, 705
Riparian land, 858
Riparian rights, 348
Risk, 280
River basin development, 83
River basins, 142, 244, 248, 260, 264, 272, 288, 376, 412, 471, 475, 483, 518
River crossings, 854
River regulation, 4, 55, 71, 122, 147, 151, 160, 164, 244, 260, 421, 487, 506, 842, 846
River systems, 244, 248, 260, 264
Rivers, 850
Roads, 785
Runoff, 130, 729, 751, 758, 762, 789
Runoff forecasting, 235, 240, 425
Rural areas, 514

Sanitary engineering, 823
Scheduling, 197
Seasonal variations, 717, 721
Secondary systems, 682
Sediment control, 75
Sewer design, 823
Sewers, 360, 364
Simulation, 380, 567
Simulation models, 8, 24, 147, 156, 189, 193, 201, 220, 244, 248, 252, 260, 264, 376, 451, 479, 483, 823
Soil water, 425, 593
Solutes, 256
Spatial data, 862
Speeches, 38
State government, 348, 408
State laws, 344, 348, 352, 356
State planning, 475
Statistical analysis, 212, 228, 713
Statistical models, 181, 189, 193, 201, 666
Stochastic models, 228, 240, 425
Stochastic processes, 268, 725
Storm drainage, 797
Storm drains, 801
Storm runoff, 126, 130, 134, 380, 384
Storm sewers, 747, 751, 758, 762, 766, 789, 793, 797, 805
Stormwater, 734, 742, 827
Stormwater management, 105, 113, 126, 130, 134, 138, 328, 368, 747, 751, 758, 762, 766, 770, 774, 778, 782, 785, 789, 793, 797, 801, 805, 809, 813, 817, 821, 831, 861, 862
Stream channels, 96
Stream improvement, 96
Streamflow, 75, 455, 518
Streamflow forecasting, 231, 240
Structural analysis, 522
Subsurface investigations, 585

SUBJECT INDEX

Surface irrigation, 514
Surface waters, 8, 113, 185, 328, 412, 597, 639, 817
Surveys, 538
System analysis, 24, 300

Temperature, 117, 122, 846
Texas, 71, 201, 376, 425, 639, 809
Thermoelectric power generation, 197, 846
Thermography, 404
Tidal waters, 177
Trichloroethylene, 571, 581
Two-dimensional models, 854

Uncertainty analysis, 142, 231, 268, 272, 276
Unsteady flow, 729
Urban development, 510
Urban planning, 16, 43, 177, 332, 336, 429, 433
Urban roads, 785
Urban runoff, 105, 138, 177, 185, 747, 751, 758, 762, 766, 778, 782, 789, 797, 801, 805, 809, 813, 817, 821, 827, 831, 861, 862
Urban studies, 109, 113, 605
Urbanization, 92, 510
urbanization, 866
U.S. Army Corps of Engineers, 39, 101, 109, 316, 479
User fees, 809, 813, 821
Utah, 805
Utilities, 805, 809, 813, 817, 821

Vegetation, 126, 563, 858
Virginia, 475
Volatile organic chemicals, 609
Volcanoes, 4

Washington, 92, 177, 185, 212, 328, 356, 372, 421, 429, 467, 483, 534, 548, 551, 555, 559, 575, 581, 605, 619, 623, 627, 631, 635, 674, 682, 738, 742, 801, 817, 858
Waste treatment, 272
Wastewater, 360, 364
Wastewater disposal, 682
Wastewater management, 138, 657, 666, 670, 751, 758, 762, 766, 770, 774, 793, 823
Wastewater treatment, 613, 619, 623, 627, 631, 635, 674, 678, 682
Water allocation policy, 8, 340, 344, 348, 356
Water balance, 518
Water conduits, 396
Water conservation, 639, 653, 657, 670, 694, 698, 702, 705, 709, 713, 717, 721
Water consumption, 653
Water costs, 702, 709, 713, 717, 721
Water demand, 644, 649, 686, 866
Water distribution, 205, 388, 392, 400, 404, 514
Water flow, 725
Water law, 16, 248, 344, 352, 356, 440
Water management, 8, 12, 20, 30, 39, 43, 71, 75, 79, 83, 147, 228, 235, 272, 304, 308, 312, 316, 320, 332, 336, 348, 352, 445, 451, 455, 463, 467, 471, 475, 479, 487, 491, 495, 498, 502, 510, 548, 551, 555, 559, 605, 639, 649, 698, 747, 842
Water meters, 653
Water pipelines, 396
Water plans, 686
Water policy, 16, 506, 538, 543
Water pollution, 220, 571, 575, 589, 862
Water pollution control, 12, 126, 130, 134, 138, 189, 193, 252, 292, 433, 531, 581, 609, 738, 742, 747, 751, 758, 762, 778, 789, 797, 801, 805, 809, 813, 827
Water quality, 126, 151, 177, 189, 193, 252, 296, 412, 530, 601, 734, 770, 789, 861, 862
Water quality control, 272, 288, 548, 559, 585, 589, 782, 813
Water quality standards, 328, 433
Water resources, 308, 312
Water resources development, 39, 613
Water resources management, 16, 20, 30, 43, 55, 109, 151, 181, 216, 231, 268, 276, 284, 292, 300, 304, 324, 408, 416, 421, 429, 445, 451, 463, 467, 471, 475, 479, 483, 487, 495, 498, 502, 506, 510, 530, 534, 601, 605, 609, 644, 866
Water reuse, 674, 678, 682, 682, 694
Water rights, 216, 316, 344, 348, 356
Water storage, 690
Water supply, 284, 320, 324, 332, 336, 340, 440, 498, 502, 601

Water supply forecasting, 8, 201, 228, 429, 445, 451, 463, 467, 483, 491, 644, 686
Water supply systems, 388, 392, 400, 404, 548, 551, 555, 559, 605, 639, 690
Water table, 563
Water temperature, 117, 122
Water transfer, 332, 336, 340, 344, 356
Water treatment, 742
Water use, 83, 332, 348, 694, 698
Watershed management, 408, 412, 738
Watersheds, 96
Waterways, 189, 193
Weather, 416, 421, 425
Wells, 567, 581
West Virginia, 487, 495
Wetlands, 88, 105, 109, 113, 613, 742, 858
Wildlife conservation, 101, 839
Wildlife habitats, 109
Wildlife management, 51, 164
Wisconsin, 530
Women, 312

AUTHOR INDEX
Page number fefers to first page of paper

Abbott, Mark J., 601
Adams, Robert, 639
Ahmed, Shabbir, 729
Akan, A. Osman, 134, 597
Aldrich, Robert B., 742
Al-Omari, Khamis A., 793
Al-Sharif, Munjed, 8
Alverson, James E., 30
Anderson, Robert, 581
Antieau, Clayton J., 858
Anton, Walt, 613
Arends, Michael T., 88
Austin, Gordon C., 782
Au-Yeung, Yin, 510

Babcock, Steven D., 467
Bakall, Ergun, 396
Bale, A. E., 292
Banton, David, 581, 589
Barnes, Michael J., 797
Barritt-Flatt, P. E., 197
Barry, Chips, 613
Bart, Michael J., 471
Bathala, Chenchayya T., 396
Baumli, George R., 320
Beard, Leo R., 304
Behrens, Jon, 248
Behrens, Jon S., 244, 260
Ben Jemaa, Fethi, 585
Benedict, Jim, 635
Benson, Robert B., 821
Bergstrom, Eric C. M., 823
Bicknell, Jill C., 797
Bitter, Susan, 126, 130
Bjork, Jack C., 92
Black, Peter E., 408
Block, Tim, 613
Block, Timothy J., 623
Blomquist, William, 498
Blundon, Edward G., 721
Borrego, Jacqueline, 866
Bowling, Chet, 122
Boyer, Jean M., 252
Bradley, Jeffrey B., 75
Brazil, L. E., 235
Breithaupt, Stephen A., 288, 296
Brendecke, Charles M., 147
Brittain, Richard G., 694

Bruns, B. W., 639
Buchleiter, Gerald W., 709
Budai, Chris, 4
Burmeister, William E., 12
Burnaroos, Pat, 548
Burnham, Michael W., 380

Calmer, John, 666
Campbell, Donald E., 514
Carley, Robert L., 332
Carpenter, Jeffrey, 284
Carr, David, 105
Carr, Robert W., 758
Carr, Thomas, 702
Caselton, W. F., 276
Chapra, Steven C., 252
Charles, Thomas, 284
Chaudhury, Rajat Roy, 774
Chieh, Shih-Huang, 224, 601
Clar, Michael, 126, 130
Clar, Michael L., 778
Coffman, Larry, 126, 130
Coffman, Larry S., 778
Corapcioglu, M. Yavuz, 575
Cormie, A. D., 197
Cox, William E., 348
Crawford, David, 766
Crockett, Timothy A., 88
Cromer, Marc V., 224, 601
Cunnane, Mark, 589
Cunningham, David, 177
Curtis, P. Douglas, 518

Dandy, Graeme C., 392
Dardeau, E. A., Jr., 101
Davidoff, Baryohay, 698
Davis, Ray Jay, 344
Dawson, Alexandra D., 502
Day, G. N., 235
Day, Harold J., 530
DeGeorge, John F., 288, 296
DeWitt, Jeffrey S., 721
Diba, Ali, 205
Diessner, Damon, 817
Dotson, Harry W., 380
Dragoon, Ken, 156
Duckstein, L., 268
Dussom, Kent B., 782

Dzurik, Andrew, 316

Eaton, Brad, 284
Eimstad, Robert, 734
Elliott, William G., 502
Ellis, Bob, 613
Emelity, Mark A., 694
Ennis, David B., 778
Erdrich, Louis P., 789
Erickson, Christopher R., 471
Evancho, Jane, 429
Every, David, 613

Fahmy, Hussam, 244, 248
Farris, Paula, 43
Fayegh, D. A., 24
Fazio, John, 164
Feagin, Nancy, 559
Feher, George G., 113
Fischenich, J. Craig, 101
Fisher, Selene, 451
Flug, Marshall, 83, 506, 842
Fontane, Darrell G., 842
Ford, David T., 376
Fordham, John W., 16
Frink, Ted M., 850
Froehlich, D. C., 854

Garvey, Jeffrey, 416
Geiselman, Jim, 168
German, Murray, 531
Goetz, James G., 631
Goetz, Jim, 613
Gracie, James W., 96
Gradilone, Frank, III., 717
Green, David B., 682
Green, Raymond, 126, 130
Green, Raymond I., 96
Greene, J. J., 388
Grigg, Neil S., 300

Hall, Ken C., 770
Hall, Wayne, 316
Hamilton, Joel R., 59
Hampton, Timothey, 609
Hansen, William, 47
Hantush, Mohamed M., 725
Harper, Martin, 682
Harper, Martin E., 177
Harpman, David A., 63
Harris, Jonathan, 609
Hasson, David S., 713

Hayes, Donald F., 88
Haynes, Michael J., 534
Heath, Erick, 666
Hightower, Jackie, 551
Hilmoe, Dave, 613
Hobbs, Noel, 284
Howard, Charles D. D., 228
Howe, Deborah, 538
Hubley, Robie O., 502
Huppert, Daniel D., 173

Israel, Morris, 336

Jackson, William L., 83
Jacobi, Lee A., 332
Jain, Ashu, 649
James, L. Douglas, 308
Jangaard, Loren, 212
Johnson, Carl, 189
Johnson, Donn Michael, 67
Johnson, Lynn E., 280
Johnson, Peggy A., 518
Johnson, Perry, 122
Jones, Clain, 177, 682

Kabir, Jobaid, 705
Kallemeyn, Larry W., 506
Karamouz, Mohammed, 240
Karpiscak, Martin M., 694
Keller, Kent, 575
Kenner, Scott J., 412
Keyes, Allison M., 451, 479
Khanbilvardi, Reza M., 729
Kibler, D. F., 785
Killen, J. Russell, 376
Kim, Joong Hoon, 400
Klamt, Robert R., 288
Klingeman, Peter C., 160
Koford, E. J., 839
Konen, Thomas P., 657
Koonce, Gregory, 835
Kreutzberger, William A., 770

Lamb, Berton L., 352
Lampe, John B., 686
Landin, M. C., 109
Lange, Kelly, 429
Langlow, Kristie, 613
Lansey, Kevin E., 593
Laughlin, Jay, 613
Laura, Robert A., 827
Laurine, D. P., 235

Lee, Lester E., 762
Leitman, Steve, 316
Lence, Barbara J., 272, 846
Lesniak, John, 619
Liou, Jim C. P., 220, 571
Litchfield, J., 862
Litchfield, James, 160
Lochen, Thomas J., 475
Loganathan, G. V., 388, 785
Loucks, Daniel P., 244, 260
Louie, Peter, 205
Lukas, Andrew B., 185
Lund, Guy S., 522
Lund, Jay R., 336
Luo, W., 276
Lynch, Christopher J., 372, 483

Macaitis, Bill, 189, 193, 364
Macdonald, Neil, 543
Maddaus, William, 666
Mahjoub, Manouchehr, 205
Malone, Kevin, 169
Mao, Ning, 181
Mariño, Miguel A., 585, 725
Maristany, Agustin E., 138
Marr, John K., 734
Martin, Quentin W., 71, 201
Martin, R. Bruce, 657
Mays, Larry W., 181, 400, 593
McCarthy, Edward, 831
McClung, E. Morris, 396
McConnaha, Willis E., 142
McCrodden, Brian, 240
McDowell, Bruce D., 498
McLeod, John, 639
McMahon, George, 240
Menard, Rosemary, 613
Mercer, Gary, 193
Miller, C. Lynn, 113
Millet, Jacqueline A., 686
Molzahn, Robert E., 368
Moncrief, W. Jeffery, 396
Muniz, Albert, 324
Munves, Susan, 670
Murphy, Laurie J., 392
Murphy, Thomas D., 421
Mushtaq, Hasan, 593

Nakai, Eddy H., 758
Nichols, Robert S., 644
Niermeyer, Jeff, 805
North, Gerald R., 425

Nvule, Daniel N., 491

Olsen, Darryll, 51
Orlob, G. T., 292
Orlob, Gerald T., 296
Ormsbee, Lindell, 649
Ormsbee, Lindell E., 231

Padmanabhan, G., 789
Paintal, Amreek, 364
Palmer, C. Diane, 809
Palmer, Richard, 240
Palmer, Richard N., 451, 479
Palmer, S. Clayton, 63
Parent, E., 268
Parkinson, David B., 674
Parrish, Kenneth D., 101
Patin, T. R., 109
Pehrson, Ralph W., 38
Perkins, Craig, 670
Peterson, Dave, 12
Phelps, Donald, 356
Pigott, Shaun, 813
Pizzimenti, John J., 169
Pongavanam, Srinivasan, 657
Ports, M. A., 854
Punnett, Richard E., 487, 495
Pyne, R. David G., 324

Railsback, S. F., 117
Randle, Timothy J., 63
Reierson, Tim, 216
Reitsma, Rene, 264
Restrepo, Pedro J., 256
Reyna, Santiago M., 360
Riddle, Hank, 216
Roe, Cary M., 185
Roesner, Larry A., 751
Rosa, Duane J., 526
Rothenberg, Mark D., 717
Ruff, James, 164
Runge, Igor, 774
Runkel, Robert L., 252, 256
Rushton, Betty, 105
Russell, S. O., 24

Sadden, Brian, 169
Sager, John, 4
Salem, Adil J., 39
Sandoval, V. Bruce, 421
Saur, Bill, 575
Sautins, Andy, 264

Schaefer, Richard L., 328
Schiewe, Roger, 168
Schuler, Roderick E., 30
Scott, John F., 340
Sear, Thomas R., 747
Seoane, Rafael S., 384, 425
Shafer, John M., 567
Sheikh, Bahman, 678
Sherk, George William, 352
Shrestha, P. L., 292
Simonson, Eileen R., 502
Simpson, Angus R., 392
Small, A. B., 785
Smayda, Thomas, 177
Smith, David B., 55
Smith, Donald J., 151
Snell, Jonathan, 589
Spahr, Mark D., 555
Stedinger, Jery R., 240
Steiner, Roland C., 463
Stern, Jeffrey H., 328
Stiles, James M., 487, 495
Stimac, Mike, 538, 543
Strzepek, Kenneth, 248
Sullivan, Linda, 627
Suryanarayana, Seshadri, 597
Swanson, William R., 224
Szabad, Candace, 96

Takyi, Andrews K., 272
Tanovan, Bolyvong S., 151
Tao, Pen C., 653
Tappel, Paul, 169
Thibert, Neil F., 801
Thormodsgard, Paul E., 530
Thornhill, Paul D., 639
Toms, Ed A., 522
Tracy, John C., 8
Trost, Sharon M., 324
Trusler, Scott, 440
Turner, T. G., 854

Urish, Daniel W., 774

Valdes, J. B., 235
Valdés, Juan, 240
Valdés, Juan B., 384, 425
Vanegas, Jorge A., 360
Variakojis, John, 193
Varljen, Mark D., 567
Vearil, James W., 20
Velehradsky, John E., 1
Vomero, Lisa T. M., 34
von Braun, Margrit, 220, 571
von Lindern, Ian, 220
von Lindern, Ian H., 571

Waddle, Terry, 79
Walesh, Stuart G., 312
Wallis, James R., 455
Wang, William C. H., 827
Warfel, Michael R., 605
Warwick, J. J., 861, 862
Watkins, E. W., 785
Way, A. William, 92
Webber, E. Scott, 827
Weber, Jack, 666
Wehrend, Steve, 264
Weil, Gary J., 404
Werick, William J., 445
Wilcox, Scott D., 850
Willey, R. G., 151
Williams, Gene N., 738
Winter, E. Robert, 690
Winter, Marnie, 782
Wistrom, Theresa, 288
Witten, Jon D., 433
Wodrich, James V., 674
Wood, David M., 766
Woodruff, Edwin J., 173
Wright, Raymond M., 774
Wycoff, Ronald L., 747, 762

Yates, Eugene B. (Gus), 563
Yeh, William W-G., 205
Yulianti, 846

A

361685